Foundation Design

Foundation Design

Principles and Practices

Second Edition

Donald P. Coduto

Professor of Civil Engineering
California State Polytechnic University, Pomona

Prentice Hall
Upper Saddle River, New Jersey 07458
http://www.prenhall.com

Library of Congress Cataloging-in-Publication Data
Coduto, Donald P.
    Foundation design : principles and practices / Donald P. Coduto.—2nd ed.
      p. cm.
    ISBN 0-13-589706-8
    1. Foundations.   I. Title.

TA775.C63 2001
624.1'5—dc21

                          00-038553

Vice president and editorial director: *Marcia Horton*
Acquisitions editor: *Laura Curless*
Editorial assistant: *Dolores Mars*
Senior marketing manager: *Danny Hoyt*
Production editor: *Pine Tree Composition*
Executive managing editor: *Vince O'Brien*
Managing editor: *David A. George*
Art director: *Jayne Conte*
Art editor: *Adam Velthaus*
Cover designer: *Bruce Kenselaar*
Manufacturing manager: *Trudy Pisciotti*
Manufacturing buyer: *Dawn Murrin*
Vice president and director of production and manufacturing: *David W. Riccardi*

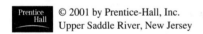 © 2001 by Prentice-Hall, Inc.
Upper Saddle River, New Jersey

Reprinted with corrections January, 2001

The author and publisher of this book have used their best efforts in preparing
this book. These efforts include the development, research, and testing of the
theories to determine their effectiveness.

Printed in the United States of America
10

ISBN 0-13-589706-8

Prentice-Hall International (UK) Limited, *London*
Prentice Hall of Australia Pty. Limited, *Sydney*
Prentice-Hall Canada Inc., *Toronto*
Prentice-Hall Hispanoamericana, S.A., *Mexico*
Prentice-Hall of India Private Limited, *New Delhi*
Prentice-Hall of Japan, Inc., *Tokyo*
Pearson Education Asia Pte. Ltd., *Singapore*
Editora Prentice-Hall do Brasil, Ltda., *Rio de Janeiro*

Contents

Preface

Foundation Design: Principles and Practices is primarily intended for use as a textbook in undergraduate and graduate-level foundation engineering courses. It also serves well as a reference book for practicing engineers. As the title infers, this book covers both "principles" (the fundamentals of foundation engineering) and "practices" (the application of these principles to practical engineering problems). Readers should have already completed at least one university-level course in soil mechanics, and should have had at least an introduction to structural engineering.

This second edition contains many improvements and enhancements. These have been the result of comments and suggestions from those who used the first edition, my own experience using it at Cal Poly Pomona, and recent advances in the state-of-the-art. The chapters on deep foundations have been completely reorganized and rewritten, and new chapters on reliability-based design and sheet pile walls have been added. Extraneous material has been eliminated, and certain analysis methods have been clarified and simplified. The manuscript was extensively tested in the classroom before going to press. This classroom testing allowed me to evaluate and refine the text, the example problems, the homework problems, and the software.

Key features of this book include:

- Integration with *Geotechnical Engineering: Principles and Practices* (Coduto, 1999), including consistent notation, terminology, analysis methods, and coordinated development of topics. However, readers who were introduced to geotechnical engineering using another text can easily transition to this book by reviewing the material in Chapters 3 and 4.

- Consideration of the geotechnical, structural, and construction engineering aspects of the design process, including emphasis on the roles of each discipline and the interrelationships between them.

- Frequent discussions of the sources and approximate magnitudes of uncertainties, along with comparisons of predicted and actual behavior.

- Use of both English and SI units, because engineers in North America and many other parts of the world need to be conversant in both systems.

- Integration of newly-developed Excel spreadsheets for foundation analysis and design. These spreadsheet files may be downloaded from the Prentice Hall website (www.prenhall.com/coduto). They are introduced only after the reader learns how to perform the analyses by manual computations.
- Extensive use of example problems, many of which are new to this edition.
- Inclusion of carefully developed homework problems distributed throughout the chapters, with comprehensive problems at the end of each chapter. Many of these problems are new or revised.
- Discussions of recent advances in foundation engineering, including Statnamic testing, load and resistance factor design (LRFD), and applications of the cone penetration test (CPT).
- Inclusion of extensive bibliographic references for those wishing to study certain topics in more detail.
- An instructor's manual is available to faculty. It may be obtained from your Prentice Hall campus representative.

ACKNOWLEDGMENTS

Many friends, colleagues, and other professionals contributed to this work. Much of the book is the product of their stimulating discussions, constructive reviews, and support. Professor Joseph Caliendo of Utah State University was especially helpful. Stanley Vitton (Michigan Technological University), Samuel Clemence (Syracuse University), Richard Handy (Iowa State University), Raymond Moore (University of Nebraska), John Horvath (Manhattan College), José Pires (University of California, Irvine), Paul Chan (New Jersey Institute of Technology), William and Sandra Houston (Arizona State University), and others reviewed part or all of the manuscripts for the first or second editions and provided many useful suggestions and comments. Iraj Noorany (San Diego State University), Michael O'Neill (University of Houston), Major William Kitch (U.S. Air Force Academy), James Olson (University of Vermont), Samuel Paikowsky (University of Massachusetts, Lowell), Richard W. Stephenson (University of Missouri, Rolla), William Kovacs (University of Rhode Island), Dan Burgess, Rick Drake (Fluor Daniel), Bengt Fellenius (Urkkada Technology), Mohamad Hussein (Goble, Rausche, Likins), and others also provided useful suggestions and advice.

A special note of thanks goes to the foundation engineering students at Cal Poly University who used various draft manuscripts of this book as a makeshift text. Their constructive comments and suggestions have made this book much more useful, and their proofreading has helped eliminate mistakes.

I welcome any constructive comments and suggestions from those who use this book. Please mail them to me at the Civil Engineering Department, Cal Poly University, Pomona, CA 91768.

Donald P. Coduto

Notation and
Units of Measurement

There is no universally accepted notation in foundation engineering. However, the notation used in this book, as described in the following table, is generally consistent with popular usage.

| Symbol | Description | Typical Units | | Defined |
		English	SI	on Page
A	Cross-sectional area	ft^2	m^2	438
A	Base area of foundation	ft^2	m^2	155
A_0	Initial cross-sectional area	in^2	mm^2	95
A_1	Cross-sectional area of column	in^2	mm^2	337
A_2	Base area of frustum	in^2	mm^2	337
A_b	Area of bottom of enlarged base	ft^2	m^2	528
A_f	Cross-sectional area at failure	in^2	mm^2	95
A_s	Steel area	in^2	mm^2	323
a_θ	Factor in N_q equation	Unitless	Unitless	178
B	Width of foundation	ft-in	mm	146
B'	Effective foundation width	ft-in	m	275
B_b	Diameter at base of foundation	ft	m	548
B_s	Diameter of shaft	ft	m	547
b	Unit length	ft	m	156
b_c, b_q, b_y	Base inclination factors	Unitless	Unitless	186
b_0	Length of critical shear surface	in	mm	310
C_1	Depth factor	Unitless	Unitless	235
C_2	Secondary creep factor	Unitless	Unitless	235
C_3	Shape factor	Unitless	Unitless	235
C_A	Aging factor	Unitless	Unitless	122
C_B	SPT borehole diameter correction	Unitless	Unitless	119
C_C	Compression index	Unitless	Unitless	66

C_{OCR}	Overconsolidation correction factor	Unitless	Unitless	122
C_P	Grain size correction factor	Unitless	Unitless	122
$C_{p\phi}$	Passive pressure factor	Unitless	Unitless	602
C_R	SPT rod length correction	Unitless	Unitless	119
C_r	Recompression index	Unitless	Unitless	67
C_s	SPT sampler correction	Unitless	Unitless	119
C_s	Side friction coefficient	Unitless	Unitless	535
C_t	Toe coefficient	Unitless	Unitless	534
C_w	Hydroconsolidation coefficient	Unitless	Unitless	709
c	Wave velocity in pile	ft/s	m/s	571
c	Column or wall width	in	mm	302
c'	Effective cohesion	lb/ft^2	kPa	84
c'_{adj}	Adjusted effective cohesion	lb/ft^2	kPa	198
c_T	Total cohesion	lb/ft^2	kPa	85
D	Depth of foundation	ft-in	mm or m	146
D_{50}	Grain size at which 50% is finer	Unitless	mm	122
D_{min}	Minimun required embedment depth	ft	m	593
D_r	Relative density	percent	percent	51
D_w	Depth from ground surface to groundwater table	ft	m	188
d	Effective depth	in	mm	306
d	Bolt diameter	in	mm	344
d	Vane diameter	in	mm	131
d_b	Reinforcing bar diameter	in	mm	306
d_c, d_q, d_γ	Depth factors	Unitless	Unitless	184
E	Portion of steel in center section	Unitless	Unitless	333
E	Modulus of elasticity	lb/in^2	MPa	231
E_m	SPT hammer efficiency	Unitless	Unitless	119
E_s	Equivalent modulus of elasticity	lb/ft^2	kPa	231
E_u	Undrained modulus of elasticity	lb/ft^2	kPa	226
EI	Expansion index	Unitless	Unitless	673
e	Eccentricity	ft	m	159
e	Void ratio	Unitless	Unitless	49
e	Base of natural logarithms	2.7183	2.7183	XXX
e_0	Initial void ratio	Unitless	Unitless	66
e_B	Eccentricity in the B direction	ft	m	165
e_L	Eccentricity in the L direction	ft	m	165
e_{max}	Maximum void ratio	Unitless	Unitless	51
e_{min}	Minimum void ratio	Unitless	Unitless	51
F	Factor of Safety	Unitless	Unitless	190
F_a	Allowable axial stress	lb/in^2	MPa	439

F_b	Allowable flexural stress	lb/in^2	MPa	439
F_v	Allowable shear stress	lb/in^2	MPa	439
f_a	Average normal stress due to axial load	lb/in^2	MPa	438
f_b	Normal stress in extreme fiber due to flexural load	lb/in^2	MPa	438
f_c'	28-day compressive strength of concrete	lb/in^2	MPa	303
f_{pc}	Effective prestress on gross section	lb/in^2	MPa	448
f_s	Unit side friction resistance	lb/ft^2	kPa	513
$(f_s)_m$	Mobilized unit side-friction resistance	lb/ft^2	kPa	544
f_{sc}	CPT cone side friction	T/ft^2	MPa or kg/cm^2	124
f_v	Shear stress	lb/in^2	MPa	439
f_y	Yield strength of steel	lb/in^2	MPa	303
G_h	Horizontal equivalent fluid density	lb/ft^3	kN/m^3	770
G_s	Specific gravity of solids	Unitless	Unitless	49
G_v	Vertical equivalent fluid density	lb/ft^3	kN/m^3	771
g_c, g_q, g_y	Ground inclination factors	Unitless	Unitless	186
H	Thickness of soil stratum	ft	m	60
H	Wall height	ft	m	759
H_c	Critical height	ft	m	767
H_{fill}	Thickness of fill	ft	m	64
I_1, I_2	Influence factors	Unitless	Unitless	226
I_ϵ	Strain influence factor	Unitless	Unitless	234
I_p	Plasticity index	Unitless	Unitless	56
I_r	Rigidity index	Unitless	Unitless	501
i_c, i_g, i_γ	Load inclination factors	Unitless	Unitless	185
I_σ	Stress influence factor	Unitless	Unitless	210
K	Coefficient of lateral earth pressure	Unitless	Unitless	61
K_a	Coefficient of active earth pressure	Unitless	Unitless	760
K_p	Coefficient of passive earth pressure	Unitless	Unitless	762
k	Factor in computing depth factors	Unitless	Unitless	184
k_s	Coefficient of subgrade reaction	lb/in^3	kN/m^3	356
L	Length of foundation	ft-in	mm	146
L'	Effective foundation length	ft-in	m	275
LL	Liquid limit (see w_L)	Unitless	Unitless	54
l	Cantilever distance	in	mm	322
l_d	Development length	in	mm	318
l_{dh}	Development length for hook	in	mm	337
M	Moment load	ft-k	kN-m	15
M_c	Characteristic moment load	ft-lb	kN-m	601
M_D	Driving moment	ft-lb	kN-m	796

M_g	Applied moment to pile group	ft-lb	kN-m	616
M_{max}	Maximum moment	ft-lb	kN-m	603
M_n	Nominal moment load capactity	ft-k	kN-m	21
M_R	Resisting moment	ft-lb	kN-m	796
N	Number of piles in a group	Unitless	Unitless	538
N	SPT blow count recorded in field	Blows/ft	Blows/300 mm	116
$(N_l)_{60}$	SPT blow count corrected for field procedures and overburden stress	Blows/ft	Blows/300 mm	120
N_σ	Bearing capacity factor	Unitless	Unitless	502
N_{60}	SPT blow count corrected for field procedures	Blows/ft	Blows/300 mm	119
N_c, N_q, N_γ	Bearing capacity factors	Unitless	Unitless	178
N_c^*, N_q^*, N_γ^*	Bearing capacity factors	Unitless	Unitless	501
N_u	Uplift bearing capacity factor	Unitless	Unitless	527
OCR	Overconsolidation ratio	Unitless	Unitless	69
P	Normal load	k	kN	15
P_a	Allowable downward load capacity	k	kN	467
P_a	Normal force acting on a wall under active conditions	lb	kN	759
P_{ag}	Allowable load capacity of pile group	k	kN	538
P_{upward}	Upward load	k	kN	470
$(P_a)_{upward}$	Allowable upward load capacity	k	kN	470
P_D	Driving force	lb	kN	791
P_f	Axial load at failure	lb	N	95
PI	Plasticity index (see I_P)	Unitless	Unitless	54
P_0	Normal force acting on a wall under at-rest conditions	lb	kN	751
PL	Plastic limit (see w_P)	Unitless	Unitless	54
P_n	Nominal normal load capacity	k	kN	21
P_{nb}	Nominal bearing capacity	k	kN	336
P_p	Normal force acting on a wall under passive conditions	lb	kN	759
P_R	Resisting force	lb	kN	791
P_s	Side-friction resistance	k	kN	466
P_t	Toe-bearing resistance	k	kN	466
P_t'	Net toe-bearing resistance	k	kN	407
P_u	Factored normal load	k	kN	21
P_{ult}	Ultimate downward load capacity	k	kN	481
p	Lateral soil resistance per unit length of foundation	lb	kN	587
Q_c	Compressibility factor	Unitless	Unitless	128
q	Bearing pressure	lb/ft^2	kPa	154

q'	Net bearing pressure	lb/ft^2	kPa	158
q	Quake	in.	mm	566
q_a	Allowable bearing capacity	lb/ft^2	kPa	190
q_A	Allowable bearing pressure	lb/ft^2	kPa	262
q_c	CPT cone resistance	T/ft^2	MPa or kg/cm^2	124
q_E	Effective cone resistance	T/ft^2 or MPa	kg/cm^2	533
q_{EG}	Factor in Eslami and Fellenius method	T/ft^2	MPa	534
q_{equiv}	Equivalent bearing pressure	lb/ft^2	kPa	275
q_{max}	Maximum bearing pressure	lb/ft^2	kPa	162
q_{min}	Minimum bearing pressure	lb/ft^2	kPa	162
q_t'	Net unit toe-bearing resistance	lb/ft^2	kPa	500
$(q_t')_m$	Mobilized unit toe-bearing resistance	lb/ft^2	kPa	544
q_{tr}'	Reduced net unit toe-bearing resistance	lb/ft^2	kPa	506
q_u	Unconfined compressive strength	lb/ft^2	kPa	511
q_{ult}	Ultimate bearing capacity	lb/ft^2	kPa	176
r	Distance from centerline of cap	in	mm	615
r	Rigidity factor	Unitless	Unitless	219
R_f	Friction ratio	Unitless	Unitless	124
R_I	Moment of inertia ratio	Unitless	Unitless	601
RQD	Rock quality designation	Unitless	Unitless	511
R_u	Ultimate resistance	k	kN	566
S	Slope of foundation	radians	radians	587
S	Elastic section modulus	in^3	mm^3	438
S	Number of stories	Unitless	Unitless	109
S	Degree of saturation	percent	percent	49
S	Column spacing	ft	m	33
S_0	Degree of saturation before wetting	percent	percent	678
S_1, S_3	Allowable lateral soil pressure	lb/ft^2	kPa	593
s	Shear strength	lb/ft^2	kPa	84
s	Center-to-center spacing of piles	in	mm	540
s	Pile set	in	mm	560
s_c, s_q, s_γ	Shape factors	Unitless	Unitless	184
s_u	Undrained shear strength	lb/ft^2	kPa	89
T	Torsion load	k-ft	m-kN	15
T	Thickness of foundation	ft-in	mm	146
T_f	Torque at failure	in-lb	N-m	131
TMI	Thornthwaite moisture index	Unitless	Unitless	666
t	Time	yr	yr	235
t	Age of soil (since time of deposition)	yr	yr	122

u	Displacement of pile	in	mm	564
u	Pore water pressure	lb/ft^2	kPa	59
u_2	Pore water pressure behind cone point	lb/ft^2	kPa	533
u_D	Pore water pressure at bottom of foundation	lb/ft^2	kPa	155
u_e	Excess pore water pressure	lb/ft^2	kPa	59
u_h	Hydrostatic pore water pressure	lb/ft^2	kPa	58
V	Shear load	k	kN	15
V_a	Shear force under active condition	k	kN	759
V_c	Nominal shear capacity of concrete	lb	kN	309
V_c	Characteristic shear load	lb	kN	601
V_{fa}	Allowable footing shear load capacity	k	kN	276
V_n	Nominal shear load capacity	k	kN	21
V_{nc}	Nominal shear capacity on critical surface	lb	kN	309
V_p	Shear force under passive condition	k	kN	159
V_s	Nominal shear capacity of reinforcing steel	lb	kN	309
V_u	Factored shear load	k	kN	21
V_{uc}	Factored shear force on critical surface	lb	kN	309
W_f	Weight of foundation	lb	kN	154
W_r	Hammer ram weight	lb	kN	560
w	Moisture content	percent	percent	49
w_L	Liquid limit	Unitless	Unitless	54
w_P	Plastic limit	Unitless	Unitless	54
w_s	Shrinkage limit	Unitless	Unitless	54
y	Lateral deflection	in	mm	598
z	Depth below ground surface	ft	m	564
z_c	Depth to centroid of soil resistance	ft	m	544
z_f	Depth below to bottom of foundation	ft	m	210
z_I	Depth to imaginary footing	ft	m	552
z_w	Depth below to groundwater table	ft	m	58
α	Wetting coefficient	Unitless	Unitless	678
α	Adhesion factor	Unitless	Unitless	522
α	Slope of footing bottom	deg	deg	183
α	Inclination of wall from vertical	deg	deg	764
β	Side friction factor in β method	Unitless	Unitless	516
β	Slope of ground surface	deg	deg	759
β	Reliability index	Unitless	Unitless	724
β_0, β_1	Correlation factors	Unitless	Unitless	233
γ	Ratio of steel cage diameter to drilled shaft diameter	Unitless	Unitless	455
γ	Unit weight	lb/ft^3	kN/m^3	49
γ	Load Factor	Unitless	Unitless	21

γ_b	Buoyant unit weight	lb/ft^3	kN/m^3	49
γ_d	Dry unit weight	lb/ft^3	kN/m^3	49
γ_{fill}	Unit weight of fill	lb/ft^3	kN/m^3	64
γ_w	Unit weight of water	lb/ft^3	kN/m^3	49
γ'	Effective unit weight	lb/ft^3	kN/m^3	188
$\Delta\sigma_z$	Change in vertical stress	lb/ft^2	kPa	64
δ	Total settlement	in	mm	29
δ_a	Allowable total settlement	in	mm	29
δ_c	Consolidation settlement	in	mm	72
δ_D	Differential settlement	in	mm	31
δ_{Da}	Allowable differential settlement	in	mm	31
δ_d	Distortion settlement	in	mm	224
δ_e	Settlement due to elastic compression	in	mm	544
δ_u	Settlement required to mobilize ultimate resistance	in	mm	544
δ_w	Heave or settlement due to wetting	in	mm	680
ϵ_{50}	Axial strain at which 50 percent of the soil strength is mobilized	Unitless	Unitless	603
ϵ_f	Strain at failure	Unitless	Unitless	95
ϵ_w	Strain due to wetting	Unitless	Unitless	676
η	Factor in Shields' chart	Unitless	Unitless	286
η	Group efficiency factor	Unitless	Unitless	538
θ	Factor in Converse-Labarre formula	Unitless	Unitless	539
θ_a	Allowable angular distortion	radians	radians	33
λ	Lightweight concrete factor	Unitless	Unitless	319
λ	Factor in Shields' chart	Unitless	Unitless	286
λ	Vane shear correction factor	Unitless	Unitless	131
λ	Equivalent passive fluid density	lb/ft^3	kN/m^3	276
λ	Factor in Evans and Duncan's charts	Unitless	Unitless	602
λ_a	Allowable equivalent passive fluid density	lb/ft^3	kN/m^3	276
μ	Coefficient of friction	Unitless	Unitless	276
μ_a	Allowable coefficient of friction	Unitless	Unitless	276
μ_C	Mean ultimate capacity	k	kN	723
μ_L	Mean load	k	kN	723
υ	Poisson's ratio	Unitless	Unitless	502
ρ	Mass density	lb$_m$/ft^3	kg/m^3	564
ρ	Steel ratio	Unitless	Unitless	317
ρ_s	Ratio of volume of spiral reinforcement to total volume of core	Unitless	Unitless	459
σ	Total stress	lb/ft^2	kPa	60
σ	Normal pressure imparted on wall from soil	lb/ft^2	kPa	760

σ'	Effective stress	lb/ft^2	kPa	60
σ_C	Standard deviation of ultimate capacity	k	kN	724
σ_c'	Preconsolidation stress	lb/ft^2	kPa	67
σ_{hs}	Horizontal swelling pressure	lb/ft^2	kPa	691
σ_L	Standard deviation of load	k	kN	727
σ_m'	Preconsolidation margin	lb/ft^2	kPa	69
σ_p	Representative passive pressure	lb/ft^2	kPa	602
σ_t	Threshold collapse stress	lb/ft^2	kPa	708
σ_x	Horizontal total stress	lb/ft^2	kPa	61
σ_x'	Horizontal effective stress	lb/ft^2	kPa	61
σ_z	Vertical total stress	lb/ft^2	kPa	60
σ_z'	Vertical effective stress	lb/ft^2	kPa	60
σ_{z0}'	Initial vertical effective stress	lb/ft^2	kPa	64
σ_{zD}'	Effective stress at depth D below the ground surface	lb/ft^2	kPa	178
σ_{zD}	Total stress at depth D below the ground surface	lb/ft^2	kPa	175
σ_{zf}'	Final effective stress	lb/ft^2	kPa	64
σ_{zp}'	Initial vertical effective stress at depth of peak strain influence factor	lb/ft^2	kPa	234
τ	Shear stress imparted on wall from soil	lb/ft^2	kPa	760
ϕ	Resistance factor	Unitless	Unitless	21
ϕ'	Effective friction angle	deg	deg	82
ϕ'_{adj}	Adjusted effective friction angle	deg	deg	198
ϕ_T	Total friction angle	deg	deg	85
ϕ_w	Wall-soil interface friction angle	deg	deg	763
Ψ	Three dimensional adjustment factor	Unitless	Unitless	225
Ψ	Factor in Shields' chart	Unitless	Unitless	286

Foundation Design

Part A

General Principles

1

Foundations in Civil Engineering

THE PARABLE OF THE WISE AND FOOLISH BUILDERS

I will show you what he is like who comes to me and hears my words and puts them into practice. He is like a man building a house, who dug down deep and laid the foundation on rock. When a flood came, the torrent struck that house but could not shake it, because it was well built. But the one who hears my words and does not put them into practice is like a man who built a house on the ground without a foundation. The moment the torrent struck that house, it collapsed and its destruction was complete.

Luke 6:47–49 NIV (circa AD 60)

A wise engineer once said "A structure is no stronger than its connections." Although this statement usually invokes images of connections between individual structural members, it also applies to those between a structure and the ground that supports it. These connections are known as its *foundations*. Even the ancient builders knew that the most carefully designed structures can fail if they are not supported by suitable foundations. The Tower of Pisa in Italy (perhaps the world's most successful foundation "failure") reminds us of this truth.

Although builders have recognized the importance of firm foundations for countless generations, and the history of foundation construction extends for thousands of years, the discipline of *foundation engineering* as we know it today did not begin to develop until the late nineteenth century.

1.1   THE EMERGENCE OF MODERN FOUNDATION ENGINEERING

Early foundation designs were based solely on precedent, intuition, and common sense. Through trial-and-error, builders developed rules for sizing and constructing foundations. For example, load-bearing masonry walls built on compact gravel in New York City during the nineteenth century were supported on spread footings that had a width 1.5 times that of the wall. Those built on sand or stiff clay were three times the width of the wall (Powell, 1884).

These empirical rules usually produced acceptable results as long as they were applied to structures and soil conditions similar to those encountered in the past. However, the results were often disastrous when builders extrapolated the rules to new conditions. This problem became especially troublesome when new methods of building construction began to appear during the late nineteenth century. The introduction of steel and reinforced concrete led to a transition away from rigid masonry structures to more flexible frame structures. These new materials also permitted buildings to be taller and heavier than before. In addition, as good sites became occupied, builders were forced to consider sites with poorer soil conditions, and these sites made foundation design and construction much more difficult. Thus, the old rules for foundation design no longer applied.

The Eiffel Tower

The Eiffel Tower in Paris is an excellent example of a new type of structure in which the old rules for foundations no longer applied. It was originally built for the Paris Universal Exposition of 1889 and was the tallest structure in the world. Alexandre Gustave Eiffel, the designer and builder, was very conscious of the need for adequate foundations, and clearly did not want to create another Leaning Tower of Pisa (Kerisel, 1987).

The Eiffel Tower is adjacent to the Seine River, and is underlain by difficult soil conditions, including uncompacted fill and soft alluvial soils. Piers for the nearby Alma bridge, which were founded in this alluvium, had already settled nearly 1 m. The tower could not tolerate such settlements.

Eiffel began exploring the subsurface conditions using the crude drilling equipment of the time, but was not satisfied with the results. He wrote: "What conclusions could one reasonably base on the examination of a few cubic decimeters of excavated soil, more often than not diluted by water, and brought to the surface by the scoop?" (Kerisel, 1987). Therefore, he devised a new means of exploring the soils, which consisted of driving a 200-mm diameter pipe filled with compressed air. The air kept groundwater from entering the tube, and thus permitted recovery of higher quality samples.

Eiffel's studies revealed that the two legs of the tower closest to the Seine were underlain by deeper and softer alluvium, and were immediately adjacent to an old river channel that had filled with soft silt. The foundation design had to accommodate these soil conditions, or else the two legs on the softer soils would settle more than the other two, causing the tower to tilt toward the river.

Figure 1.1 Two legs of the Eiffel Tower in Paris are underlain by softer soils, and thus could have settled more than the other two. Fortunately, Eiffel carefully explored the soil conditions, recognized this problem, and designed the foundations to accommodate these soil conditions. His foresight and diligence resulted in a well-designed foundation system that has not settled excessively.

Based on his study of the soil conditions, Eiffel placed the foundations for the two legs furthest from the river on the shallow but firm alluvial soils. The bottom of these foundations were above the groundwater table, so their construction proceeded easily. However, he made the foundations for the other two legs much deeper so they too were founded on firm soils. This required excavating about 12 m below the ground surface (6 m below the groundwater table). As a result of Eiffel's diligence, the foundations have safely supported the tower for over one hundred years, and have not experienced excessive differential settlements.

Further Developments

As structures continued to become larger and heavier, engineers continued to learn more about foundation design and construction. Instead of simply developing new empirical rules, they began to investigate the behavior of foundations and develop more rational methods of design, thus establishing the discipline of foundation engineering.

A significant advance came in 1873 when Frederick Baumann, a Chicago architect, published the pamphlet *The Art of Preparing Foundations, with Particular Illustration of* the *"Method of Isolated Piers" as Followed in Chicago* (Baumann, 1873). He appears to be the first to recommend that the base area of a foundation should be proportional to the applied load, and that the loads should act concentrically upon the foundation. He also gave allowable bearing pressures for Chicago soils and specified tolerable limits for total and differential settlements.

The growth of geotechnical engineering, which began in earnest during the 1920s, provided a better theoretical base for foundation engineering. It also provided improved methods of exploring and testing soil and rock. These developments continued throughout the twentieth century. Many new methods of foundation construction also have been developed, making it possible to build foundations at sites where construction had previously been impossible or impractical.

Today, our knowledge of foundation design and construction is much better than it was one hundred years ago. It is now possible to build reliable, cost-effective, high-capacity foundations for all types of modern structures.

1.2   THE FOUNDATION ENGINEER

Foundation engineering does not fit completely within any of the traditional civil engineering subdisciplines. Instead, the foundation engineer must be multidisciplinary and possess a working knowledge in each of the following areas:

- **Structural engineering**—A foundation is a structural member that must be capable of transmitting the applied loads, so we must also understand the principles and practices of structural engineering. In addition, the foundation supports a structure, so we must understand the sources and nature of structural loads and the structure's tolerance of foundation movements.
- **Geotechnical engineering**—All foundations interact with the ground, so the design must reflect the engineering properties and behavior of the adjacent soil and rock. Thus, the foundation engineer must understand geotechnical engineering. Most foundation engineers also consider themselves to be geotechnical engineers.
- **Construction engineering**—Finally, foundations must be built. Although the actual construction is performed by contractors and construction engineers, it is very important for the design engineer to have a thorough understanding of construction methods and equipment to develop a design that can be economically built. A knowledge of construction engineering is also necessary when dealing with problems that develop during construction.

This book focuses primarily on the design of foundations, and thus emphasizes the geotechnical and structural engineering aspects. Discussions of construction methods and equipment are generally limited to those aspects that are most important to design engineers. Other aspects which are primarily of interest to contractors, such as scheduling, detailed equipment selection procedures, construction safety, and cost estimating, are beyond the scope of this book.

1.3   UNCERTAINTIES

An unknown structural engineer suggested the following definition for structural engineering:

> Structural engineering is the art and science of molding materials we do not fully understand into shapes we cannot precisely analyze to resist forces we cannot accurately predict, all in such a way that the society at large is given no reason to suspect the extent of our ignorance.

We could apply the same definition, even more emphatically, to foundation engineering. In spite of the many advances in foundation engineering theory, there are still many gaps in our understanding. In general, the greatest uncertainties are the result of our limited knowledge of the soil conditions. Although foundation engineers use various investigation and testing techniques in an attempt to define the soil conditions beneath the site of a proposed foundation, even the most thorough investigation program encounters only a small portion of the soils and relies heavily on interpolation and extrapolation.

Limitations in our understanding of the interaction between a foundation and the soil also introduce uncertainties. For example, how does side friction resistance develop along the surface of a pile? How does the installation of a pile affect the engineering properties of the adjacent soils? These and other questions are the subjects of continued research.

It also is difficult to predict the actual service loads that will act on a foundation, especially live loads. Design values, such as those that appear in building codes, are usually conservative.

Because of these and other uncertainties, the wise engineer does not blindly follow the results of tests or analyses. These tests and analyses must be tempered with precedent, common sense, and engineering judgment. Foundation engineering is still both an art and a science. It is dangerous to view foundation engineering, or any other type of engineering, as simply a collection of formulas and charts to be followed using some "recipe" for design. This is why it is essential to understand the *behavior* of foundations and the basis and limitations of the analysis methods.

Rationalism and Empiricism

Since we do not fully understand the behavior of foundations, most of our analysis and design methods include a mixture of rational and empirical techniques. Rational techniques are those developed from the principles of physics and engineering science, and are useful ways to describe mechanisms we understand and are able to quantify. Conversely, empirical techniques are based primarily on experimental data and thus are especially helpful when we have a limited understanding of the physical mechanisms.

Methods of analyzing foundation problems often begin as simple rational models with little or no experimental data to validate them, or highly empirical techniques that reflect only the most basic insight into the mechanisms that control the observed behavior. Then, as engineers use these methods we search for experimental data to calibrate the rational methods and discernment to understand the empirical data. These efforts are intended to improve the accuracy of the predictions.

One of the keys to successful foundation engineering is to understand this mix of rationalism and empiricism, the strengths and limitations of each, and how to apply them to practical design problems.

Factors of Safety

In spite of the many uncertainties in foundation analysis and design, the public expects engineers to develop reliable and economical designs in a timely and efficient manner. Therefore, we compensate for these uncertainties by using factors of safety in our designs.

Although it's tempting to think of designs that have a factor of safety greater than some standard value as being "safe" and those with a lower factor of safety as "unsafe," it's better to view these two conditions as having different degrees of reliability or different probabilities of failure. All foundations can fail, but some are more likely to fail than others. The design factor of safety defines the engineer's estimate of the best compromise between cost and reliability. It is based on many factors, including the following:

- Required reliability (i.e., the acceptable probability of failure).
- Consequences of a failure.
- Uncertainties in soil properties and applied loads.
- Construction tolerances (i.e., the potential differences between design and as-built dimensions).
- Ignorance of the true behavior of foundations.
- Cost-benefit ratio of additional conservatism in the design.

Factors of safety in foundations are typically greater than those in the superstructure because of the following:

- Extra weight (another consequence of conservatism) in the superstructure increases the loads on members below, thus compounding the cost increases. However, the foundation is the lowest member in a structure, so additional weight does not affect other members. In fact, additional weight may be a benefit because it increases the foundation's uplift capacity.
- Construction tolerances in foundations are wider than those in the superstructure, so the as-built dimensions are often significantly different than the design dimensions.
- Uncertainties in soil properties introduce significantly more risk.
- Foundation failures can be more costly than failure in the superstructure.

Typical design values are presented throughout this book. These values are based primarily on precedent and professional judgement, and have generally provided suitable foundation designs.

As further research helps us better understand the behavior of foundations and the reliability of our analysis and design methods, design values of the factor of safety will

◀ Plate A
The John Hancock Building in Boston

▼ Plate B
A Cable-stayed bridge

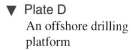

▲ Plate C
Petroleum storage tanks

▼ Plate D
An offshore drilling platform

◀ **Plate E**
Cracks in the interior walls of a house that experienced excessive differential settlements

▼ **Plate F**
A bridge that failed because of scour beneath the foundations

▼ **Plate H**
Breakage of a pile foundation during construction (*Goble Rausche Likins and Associates, Inc.*)

▲ **Plate G**
Failure of a building due to liquifaction of the underlying soils (*Earthquake Engineering Research Center Library, University of California Berkeley, Steinbrugee Collection*)

◀ Plate I
Drilling an exploratory boring

▼ Plate J
A golf course community in Palm Springs, California

▼ Plate K
Building a drilled shaft foundation (*ADSC: The International Association of Foundation Drilling*)

Plate L ▶
Using a vibrofloat to install a stone column

◀ Plate M
A soldier pile wall and a foundation system under construction

▼ Plate N
A retaining wall made from a series of closely-spaced drilled shafts

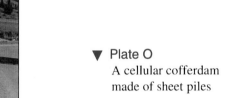

▼ Plate O
A cellular cofferdam made of sheet piles

▲ Plate P
A gabion wall

probably be modified accordingly. The introduction of load and resistance factor design (LRFD) into foundation design, which are discussed in Chapters 2 and 21, may provide a vehicle for these reliability assessments.

Accuracy of Computations

Those who are new to the field of foundation engineering often make the mistake of expressing the results of computations using too many significant figures. For example, to claim that the predicted settlement of a spread footing is 12.214 mm suggests an accuracy that is well beyond that which is possible using normal exploration and testing methods. Such practices give a false sense of security.

As a general rule, perform most foundation engineering calculations to three significant figures and express the final results and designs to two significant figures. For example, the settlement figure just quoted would best be stated as 12 mm, keeping in mind that the true precision may be on the order of ±50%.

1.4 BUILDING CODES

Building codes govern the design and construction of nearly all foundations. Although foundation engineering is not as codified as some other areas of civil engineering, we must be familiar with the applicable regulations for a particular project.

Most foundation construction in the United States, other than highway and railroad bridges, has been governed by one of the three "model" codes:

- The *Uniform Building Code* (ICBO, 1997), which is used primarily in states west of the Mississippi River
- The *National Building Code* (BOCA, 1996), which is used primarily in the midwestern and northeastern states
- The *Standard Building Code* (SBCCI, 1997), which is used primarily in the southeastern states

Some states and cities have adopted their own codes, or use modifications of these model codes to accommodate their own needs and circumstances. For example, the city of New Orleans generally follows the Standard Building Code, but has its own foundation design requirements that reflect the exceptionally poor soil conditions there.

These three model codes have been merged into a single code, the *International Building Code* or *IBC* (ICC, 2000), which is the first true "national" building code in the United States. The ICBO, BOCA, and SBCCI codes will no longer be updated, and will eventually become obsolete as local jurisdictions adopt the IBC.

The model codes and the IBC focus primarily on buildings, towers, tanks, and other similar structures. However, these codes do not address the design of highway bridges because they have substantially different requirements. Most highway bridges in the United States and Canada are designed under the provisions of the *Standard Specifications for*

Highway Bridges published by the American Association of State Highway and Transportation Officials (AASHTO, 1996).

Other codes also govern the design of foundations for certain projects in certain localities. These include:

- *The National Building Code of Canada* (CCC, 1995), which governs the design of buildings and other structures in Canada.
- The *Ontario Highway Bridge Design Code* (MTO, 1991), which is particularly noteworthy in its use of LRFD design for foundations.
- *Planning, Designing and Constructing Fixed Offshore Platforms* (API, 1996, 1997) governs the design of offshore drilling platforms.

Virtually all of these codes rely on other specialty codes. The two most important ones for foundation engineering are:

- *Building Code Requirements for Structural Concrete* (ACI 318-99 and 318M-99), published by the American Concrete Institute (ACI, 1999).
- *Manual of Steel Construction*, published by the American Institute of Steel Construction (AISC, 1989, 1995).

The various codes are sometimes conflicting and contradictory, and one could write an entire book devoted exclusively to code provisions and their interpretation. Although this book does refer to various code requirements as appropriate, it is not a comprehensive commentary and is not a substitute for the code books. These code references are identified in brackets. For example, [IBC 1801.1] refers to section 1801.1 of the International Building Code.

Building codes represent *minimum* design requirements. Simply meeting code requirements does not necessarily produce a satisfactory design, especially in foundation engineering. Often, these requirements must be exceeded and, on occasion, it is appropriate to seek exceptions from certain requirements. In addition, many important aspects of foundation engineering are not even addressed in the codes. Therefore, think of codes as guides, not dictators, and certainly not as a substitute for engineering knowledge, judgment, or common sense.

1.5   CLASSIFICATION OF FOUNDATIONS

As indicated on the first page of this chapter, we will be using the term *foundation* to describe the structural elements that connect a structure to the ground. These elements are made of concrete, steel, wood, or perhaps other materials. We will divide foundations into two broad categories: *shallow foundations* and *deep foundations*, as shown in Figure 1.2. Shallow foundations transmit the structural loads to the near-surface soils; deep foundations transmit some or all of the loads to deeper soils. These two categories are discussed in Chapters 5 to 10 and 11 to 17, respectively.

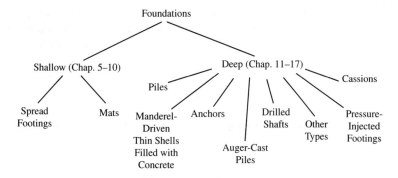

Figure 1.2 Classification of foundations.

Some engineers, especially those involved in the design and construction of dams, use the term *foundation* to describe the underlying soil or rock. However, we will not use this definition.

KEY TO COLOR PHOTOGRAPHS

The following pages contain color photographs showing foundations and earth-retaining structures.

First Page: Foundations in Civil Engineering

All civil engineering structures require foundations. Here are some examples:

A. Buildings impart their weight and other loads, such as wind or seismic forces, onto their foundations. These loads must be safely and economically transmitted into the ground. Large buildings, such as this one, require extensive foundation systems that represent a significant part of the total construction cost.

B. Bridges also generate large structural loads, but unlike buildings they focus these loads on very few supports. For example, this cable-stayed bridge concentrates nearly all of its structural loads on the tower foundations. In addition, many bridge foundations must be built in difficult conditions, such as in the middle of a river, and thus require special design and construction methods.

C. These petroleum storage tanks are part of a refinery complex, and are typical heavy industrial structures. When filled with oil, they are much heavier than a building of comparable height and thus require a high-capacity foundation system.

D. Offshore drilling platforms are subjected to structural loads similar to those in buildings and bridges, as well as significant environmental loads. For example, those in the Gulf of Mexico must survive hurricane-level ocean currents and winds. Many of these loads are especially difficult for the foundation engineer because they act horizontally. The construction of such foundations also is difficult because the water is often very deep and because these platforms are located in the open sea.

Second Page: Foundation Failures

Foundation-related problems can be very costly. Here are some examples of foundation failures and their consequences:

E. The foundations for this single family residence experienced excessive differential settlement, which produced 15-mm wide cracks in the interior drywall. Such problems can be very expensive to repair, but can normally be avoided by careful design and construction practices.

F. This highway bridge crosses a river that normally carries very little water, and the center of the bridge was supported on foundations built in the middle of the river. Unfortunately, during a period of heavy rain, the flow rate in the river increased dramatically, which washed away some of the soils in the river bed (a process called scour) and undermined the foundations. This caused large settlements in the center of the bridge, as shown in the photograph. This problem could have been avoided by building the foundation at a depth below the potential scour zone.

G. The soils beneath this building in Niigata, Japan, liquefied during a magnitude 7.5 earthquake. Since the building was supported on shallow foundations, it sank into the liquefied ground. This failure could have been avoided by first recognizing the presence of liquefiable soil, and then improving the soil or using a different type of foundation.

H. Foundation problems also can develop during construction. This pile was seriously damaged while it was being driven, and is an example of a foundation design that did not properly consider constructibility issues. Such problems can usually be avoided by conducting pile driveability analyses before construction, and using the results of such analyses to develop designs that satisfy both the structural and constructibility requirements.

Third Page: Design and Construction Methods

Engineers and contractors have developed methods of designing and building foundations and earth-retaining structures that are both safe and economical. These photographs illustrate some of these methods.

I. Foundation engineers use drill rigs such as this to explore the soil and rock conditions beneath a construction site and to obtain samples that are later brought to a laboratory for evaluation and testing. Information gained from these efforts is then used to design the foundations.

J. Sometimes the process of developing a site introduces important changes in the underlying soils. This golf course community in Palm Springs, California, is an example because it includes extensive turf areas that must be irrigated, which is a dramatic change from the natural arid conditions shown in the upper portion of the photograph. Some of this irrigation water soaks into the underlying soil and may cause it to compress, possibly producing large settlements. This potential problem must be considered during the design and construction of foundations at such sites.

K. Drilled shafts are one type of deep foundation. In this photograph, a drill rig is creating a cylindrical hole in the ground. The contractor will then insert a reinforcing steel cage and fill the hole with concrete.

L. Structures built on sites underlain by poor soils are usually supported on foundations that accommodate the existing conditions. However, sometimes it is more cost-effective to first improve the soils, then support the structure on a more modest foundation system. Many soil improvement methods are available, each of which is best suited to particular soil conditions and projects. In this case, a series of stone columns are being installed using a vibroflot. These cylindrical columns of gravel improve the load-bearing ability of the soil beneath the site of a new water tank.

Fourth Page: Earth-Retaining Structures

Many civil engineering projects require earth-retaining structures to maintain a difference in elevation between adjacent ground surfaces. The kind of earth-retaining structure to use in a particular circumstance depends on the required height, the soil conditions, and many other factors. Here are some examples:

M. This soldier pile wall is being used to provide temporary support for a building construction site. It consists of horizontal timber lagging that spans across a series of vertical steel beams embedded into the ground, and permits construction of the building foundations in their proper location. These foundations will support the building and the basement wall, which will then act as the permanent earth-retaining structure.

N. The design for this construction site used a series of closely-spaced drilled shaft foundations to form an earth-retaining structure. Once the shafts had been constructed, the soil inside the construction site was excavated to the basement level and building construction began. In this case, the shafts also will serve as the permanent earth-retaining structure.

O. This sheet pile cofferdam has been constructed in the middle of a river to enable construction of a bridge foundation. It restrains the lateral force from 30 feet of water. The foundation construction is visible in the bottom of the photo.

P. Gabions consist of wire baskets filled with gravel and assembled to form an earth-retaining structure such as the one shown in this photograph. This method requires no specialized equipment, and can be very cost-effective in certain situations.

2

Performance Requirements

If a builder builds a house for a man and does not make its construction
firm, and the house which he has built collapses and causes the death of
the owner of the house, that builder shall be put to death.

From *The Code of Hammurabi,* Babylon, circa 2000 B.C.

One of the first steps in any design process is to define the performance requirements. What functions do we expect the final product to accomplish? What are the appropriate design criteria? What constitutes acceptable performance, and what would be unacceptable?

A common misconception, even among some engineers, is that foundations are either perfectly rigid and unyielding, or they are completely incapable of supporting the necessary loads and fail catastrophically. This "it's either black or white" perspective is easy to comprehend, but it is not correct. All engineering products, including foundations, have varying *degrees* of performance that we might think of as various shades of gray. The engineer must determine which shades are acceptable and which are not. Leonards (1982) defined *failure* as "an unacceptable difference between expected and observed performance." For example, consider an engineer who designs a foundation such that it is not expected to settle more than 1 inch when loaded. If it actually settles 1.1 inch, the engineer will probably not consider it to have failed because the difference between the expected and observed performance is small and well within the design factor of safety. However, a settlement of 10 inches would probably be unacceptable and therefore classified as a failure.

Foundation performance standards are not the same for all structures or in all locations. For example, foundation settlements that produce 3-mm wide cracks in walls would probably be unacceptable in an expensive house, and may generate a lawsuit, whereas the same settlements and cracks in a heavy industrial building would probably not even be noticed.

This chapter examines performance requirements for structural foundations. It begins by discussing design loads, then progresses to discussions of performance requirements in each of the following categories:

- Strength requirements
- Serviceability requirements
- Constructibility requirements
- Economic requirements

2.1   DESIGN LOADS

The foundation design process cannot begin until the *design loads* have been defined. These are the loads imparted from the superstructure to the foundation.

Types and Sources

There are four different types of design loads:

- *Normal loads*, designated by the variable P
- *Shear loads*, designated by the variable V
- *Moment loads*, designated by the variable M
- *Torsion loads*, designated by the variable T

Each of these is shown in Figure 2.1.

Normal loads are those that act parallel to the foundation axis. Usually this axis is vertical, so the normal load becomes the vertical component of the applied load. It may act either downward (compression) or upward (tension).

Shear loads act perpendicular to the foundation axis. They may be expressed as two perpendicular components, V_x and V_y. Moment loads also may be expressed using two perpendicular components, M_x and M_y. Sometimes torsion loads, T, also are important, such as with cantilever highway signs. However, in most designs the torsion loads are small and may be ignored.

Most foundations, especially those that support buildings or bridges, are designed primarily to support downward normal loads, so this type of loading receives the most attention in this book. However, other types of loads also can be important, and in some cases can control the design. For example, the design of foundations for electrical transmission towers is often controlled by upward normal loads induced by overturning moments on the tower.

Design loads also are classified according to their source:

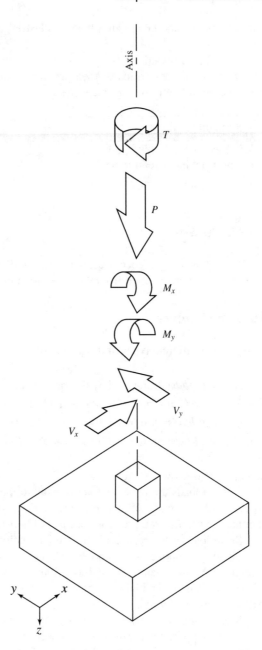

Figure 2.1 Types of structural loads acting
on a foundation.

- *Dead loads* (*D*) are those caused by the weight of the structure, including permanently installed equipment.
- *Live loads* (*L*) are those caused by the intended use and occupancy. These include loads from people, furniture, inventory, maintenance activities, moveable partitions, moveable equipment, vehicles, and other similar sources.
- *Snow loads* (*S*) and *rain loads* (*R*) are a special type of live load caused by the accumulation of snow or rain. Sometimes rain loads caused by ponding (the static accumulation of water on the roof) are considered separately.
- *Earth pressure loads* (*H*) are caused by the weight and lateral pressures from soil or rock, such as those acting on a retaining wall.
- *Fluid loads* (*F*) are those caused by fluids with well-defined pressures and maximum heights, such as water in a storage tank.
- *Earthquake loads* (*E*) are the result of accelerations from earthquakes.
- *Wind loads* (*W*) are imparted by wind onto the structure.
- *Self-straining loads* (*T*) are those caused by temperature changes, shrinkage, moisture changes, creep, differential settlement, and other similar processes.
- *Impact loads* (*I*) are the result of vibratory, dynamic, and impact effects. Impact loads from vessels are especially important in some bridge and port facilities.
- *Stream flow loads* (*SF*) and *ice loads* (*ICE*) are caused by the action of water and ice in bodies of water, and are especially important in bridges, offshore drilling platforms, and port facilities.
- *Centrifugal* (*CF*) and *braking loads* (*BF*) are caused by the motion of vehicles moving on the structure. Centrifugal forces occur when the vehicle is turning, such as on a curved bridge, while braking forces are those transmitted to the structure when a vehicle brakes.

We will identify each load source using subscripts. For example, P_D is a dead normal load, M_E is a moment load caused by an earthquake, etc.

Structural engineers compute the dead loads by simply summing the weights of the structural members. These weights can be accurately predicted and remain essentially constant throughout the life of the structure, so the design dead load values should be very close to the actual dead loads. Incorrect design dead loads are usually due to miscommunication between the structural engineer and others. For example, remodeling may result in new walls, or HVAC (heating, ventilation, and air conditioning) equipment may be heavier than anticipated. Dead loads also can differ if the as-built dimensions are significantly different from those shown on the design drawings.

Some design loads must be computed from the principles of mechanics. For example, fluid loads in a tank are based on fluid statics. The remaining design loads are usually dictated by codes. For example, the BOCA National Building Code (BOCA, 1996) specifies a floor live load of 40 lb/ft^2 for design of classrooms (which is the same as the live load used for prisons!). Most of these code-based design loads are conservative, which is appropriate. This means the real service loads acting on a foundation are probably less than the design loads.

Methods of Expression

There are two methods of expressing and working with design loads: The *allowable stress design* (ASD) method, and the *load and resistance factor design* (LRFD) method. We will be using both of them in this book.

Allowable Stress Design (ASD)

When using the *allowable stress design (ASD) method* (also known as the *working stress design method*), the design loads reflect conservative estimates of the actual service loads. All of the geotechnical analyses in this book, except for those in Chapter 21, use the ASD method. Some of the structural analyses also use ASD.

Evaluating the Design Load

The design load is the most critical combination of the various load sources, as defined by codes. For example, the ANSI/ASCE *Minimum Design Loads for Buildings and Other Structures* (ASCE, 1996a), defines the ASD design load as the greatest of the following four load combinations [ANSI/ASCE 2.4.1]:

$$\boxed{D} \tag{2.1}$$

$$\boxed{D + L + F + H + T + (L_r \text{ or } S \text{ or } R)} \tag{2.2}$$

$$\boxed{D + L + (L_r \text{ or } S \text{ or } R) + (W \text{ or } E)} \tag{2.3}$$

$$\boxed{D + (W \text{ or } E)} \tag{2.4}$$

Other codes use different equations for computing the ASD design load, so it is important to check the applicable code for each project.

Equations 2.1 to 2.4 apply to all types of loads (normal, shear, and moment). For example, when evaluating the normal load, Equation 2.4 produces $P = P_D + P_W$ or $P = P_D + P_E$, whichever is greater. When adding these loads, be sure to consider their direction. For example, when evaluating the normal load, some components may induce compression, while others induce tension.

Equation 2.1 governs only when some of the loads act in opposite directions. For example, if a certain column is subjected to a 500 kN compressive dead load and a 100 kN tensile live load, the design load would be 500 kN per Equation 2.1.

Alternative Method of Evaluating Wind and Seismic Loads

Some codes (e.g., IBC 1605.3.2 and Table 1804.2; UBC 1612.3.2 and Table 18-I-A; BOCA 1805.2) permit greater allowable load capacities in structural materials and soil when considering load combinations that include wind or seismic components. Usually this increased capacity is one-third greater than the static load capacity. For example, if a particular foundation has an allowable load capacity of 600 kN when subjected to static loads (Equations 2.1 and 2.2), then the allowable load capacity under wind or seismic loads (Equations 2.3 or 2.4) would be $600 \times 1.33 = 800$ kN. Even if this increase is not specifically authorized by the prevailing code, foundation engineers normally have the authority to use it in foundation design. Section 8.4 discusses this topic in more detail.

One way to implement this criterion is to size the foundation twice: first using the design load from Equations 2.1 and 2.2 with the static allowable load capacity, and second using the design load from Equations 2.3 and 2.4 with the wind/seismic load capacity, then using the larger of the two designs. However, this method is tedious and time-consuming. An alternative approach is to divide the loads computed in Equations 2.3 and 2.4 by 1.33 (or multiply them by $1/1.33 = 0.75$) instead of increasing the capacity by a factor of 1.33. Then the loads from all four equations can be compared to the same allowable capacity and the foundation needs to be sized only once. Thus, for ASD design of foundations, we rewrite Equations 2.3 and 2.4 as follows:

$$\boxed{0.75[D + L + (L_r \text{ or } S \text{ or } R) + (W \text{ or } E)]} \qquad (2.3a)$$

$$\boxed{0.75[D + (W \text{ or } E)]} \qquad (2.4a)$$

Therefore, when designing foundations using ASD, we will normally compute the design load as the largest of Equations 2.1, 2.2, 2.3a, or 2.4a, then size the foundation using an allowable bearing pressure that does not include the 33 percent increase.

Example 2.1

A column carries the following vertical compressive loads: $P_D = 2100$ kN downward, $P_L = 1400$ kN downward, and $P_W = 600$ kN upward. Using the ASD load combinations, compute the design normal load for use in foundation design.

Solution

Using Equations 2.1, 2.2, 2.3a, and 2.4a:

$$P = P_D = 2100 \text{ kN}$$

$$P = P_D + P_L + P_F + P_H + P_T + (P_{Lr} \text{ or } P_S \text{ or } P_R)$$

$$= 2100 \text{ kN} + 1400 \text{ kN} + 0 + 0 + 0 + 0$$

$$= 3500 \text{ kN} \qquad \Leftarrow Governs$$

$$P = 0.75[P_D + P_L + (P_{Lr} \text{ or } P_S \text{ or } P_R) + (P_W \text{ or } P_E)]$$

$$= 0.75[2100 \text{ kN} + 1400 \text{ kN} + 0 - 600 \text{ kN}]$$

$$= 2175 \text{ kN}$$

$$P = 0.75[P_D + (P_W \text{ or } P_E)] = 0.75[2100 \text{ kN} - 600 \text{ kN}] = 1125 \text{ kN}$$

Therefore, the design normal load is **3500 kN downward** $\Leftarrow Answer$

Commentary

Notice how this solution considers downward (compressive) loads to be positive and upward (tensile) loads negative. This is the sign convention customarily used by geotechnical engineers. However, structural engineers normally use the opposite sign convention (tension is positive, compression is negative). This difference can be a source of confusion, so it is important to always be conscious of which sign convention is being used.

The ASD design process then compares the design load with the allowable load, which is the ultimate capacity divided by a factor of safety. For example, if the foundation that supports the column described in Example 2.1 has an ultimate downward capacity of 9000 kN, and is being designed for a factor of safety of 2, the allowable downward capacity is 9000/2 = 4500 kN. Since this is greater than the design load of 3500 kN, the design is satisfactory, or perhaps somewhat overdesigned.

Example 2.2

The column described in Example 2.1 will be supported by a group of four steel H-pile foundations. These H-piles are similar to wide flange beams, and are driven vertically into the ground. The piles will be made of A36 steel ($F_y = 248$ MPa) and the allowable compressive stress, F_a, is 0.50 F_y. Considering only the stresses in the steel, determine the required cross-sectional area of each pile.

Solution

$$P/\text{pile} = \frac{3500 \text{ kN}}{4} = 875 \text{ kN}$$

$$F_a = 0.50 \, F_y = (0.50)(248 \text{ MPa}) = 124 \text{ MPa} = 124,000 \text{ kPa}$$

$$A = \frac{P}{F_a} = \left(\frac{875 \text{ kN}}{124,000 \text{ kPa}}\right)\left(\frac{1000 \text{ mm}}{\text{m}}\right)^2 = \textbf{7056 mm}^2 \qquad \Leftarrow Answer$$

Commentary

The next step in the design process would be to select a standard H-pile section that has a cross-sectional area of at least 7056 mm^2. We will discuss this step in Chapter 17.

This analysis considers only the stress in the steel. A complete design would also need to consider the load transfer between the pile and the ground, as discussed later in this book.

Load and Resistance Factor Design (LRFD)

The *load and resistance factor design* (*LRFD*) method (also known as the *ultimate strength design* method) uses a different approach. It applies *load factors*, γ, most of which are greater than one, to the nominal loads to obtain the *factored load*, *U*. In the case of normal loads, the factored load P_u is:

$$P_u = \gamma_1 P_D + \gamma_2 P_L + \ldots \qquad (2.5)$$

Design codes present a series of equations in the form of Equation 2.5, each with a different load combination, and define the factored load as the largest load computed from these equations. The LRFD method also applies a *resistance factor*, ϕ (also known as a *strength reduction factor*) to the ultimate capacity from a strength limit analysis. Nearly all resistance factors are less than one. Finally, the design must satisfy the following criteria:

$$P_u \leq \phi P_n \qquad (2.6)$$

Where:

P_u = factored normal load

γ = load factor

P_D = normal dead load

P_L = normal live load

ϕ = resistance factor

P_n = nominal normal load capacity

Similar equations also are used for shear and moment loads. Chapter 21 discusses LRFD in more detail.

American Concrete Institute (ACI) Code

The first widely accepted LRFD code in North America was developed by the American Concrete Institute (ACI) for the design of reinforced concrete. In its current form (ACI, 1999), this code defines the factored load as the largest of those computed from the following equations [ACI 9.2]:

$$U = 1.4D + 1.7L \qquad (2.7)$$

$$U = 0.75(1.4D + 1.4T + 1.7L) \tag{2.8}$$

$$U = 0.9D + 1.4F \tag{2.9}$$

$$U = 1.4D + 1.7L + 1.4F \tag{2.10}$$

$$U = 1.4D + 1.7L + 1.7H \tag{2.11}$$

$$U = 0.9D + 1.3W \tag{2.12}$$

$$U = 0.9D + 1.43E \tag{2.13}$$

$$U = 0.75(1.4D + 1.7L + 1.7W) \tag{2.14}$$

$$U = 0.75(1.4D + 1.7L + 1.87E) \tag{2.15}$$

$$U = 0.9D + 1.7H \tag{2.16}$$

$$U = 1.4(D + T) \tag{2.17}$$

Once again, these load combinations apply to all types of loads, and thus are expressed as P_U, M_U, and so forth. Note how the 0.75 reduction factor, as discussed under ASD, is already included in Equations 2.14 and 2.15.

ANSI/ASCE and AISC Codes

The American National Standards Institute (ANSI) and the American Society of Civil Engineers (ASCE) have developed a reliability-based LRFD standard for computing the factored load. It has been published as ANSI/ASCE 7-95 *Minimum Design Loads for Buildings and Other Structures* (ASCE, 1996a). This standard forms the basis for LRFD design of steel structures (AISC, 1995), and is accepted as an alternative method for concrete structures (ACI, 1999). It will probably become the universal standard in North America for structures other than bridges.

According to the ANSI/ASCE standard, the factored load is the largest load computed from the following formulas [ANSI/ASCE 2.3.2]:

$$U = 1.4D \tag{2.18}$$

$$U = 1.2(D + F + T) + 1.6(L + H) + 0.5(L_r \text{ or } S \text{ or } R) \tag{2.19}$$

$$U = 1.2D + 1.6(L_r \text{ or } S \text{ or } R) + (0.5L \text{ or } 0.8W) \qquad (2.20)$$

$$U = 1.2D + 1.3W + 0.5L + 0.5(L_r \text{ or } S \text{ or } R) \qquad (2.21)$$

$$U = 1.2D + 1.0E + 0.5L + 0.2S \qquad (2.22)$$

$$U = 0.9D + (1.3W \text{ or } 1.0E) \qquad (2.23)$$

In certain special cases, these formulas are modified slightly (see ASCE, 1996). The International Building Code (ICC, 2000) uses a modified version of these formulas [IBC 1605].

AASHTO Code

The American Association of State Highway and Transportation Officials (AASHTO) have developed a separate standard for computing the factored load (Nowak, 1995; AASHTO, 1996). This standard is based on loadings for bridges, which are significantly different than those on buildings and other structures. In addition, the probabilistic evaluations of these loads are different from those for buildings. Therefore, the AASHTO standard produces factored loads that are different from both the ACI and ANSI/ASCE standards.

Example 2.3

Solve Examples 2.1 and 2.2 using LRFD with the ANSI/ASCE load factors and a resistance factor of 0.70.

Solution

The factored load is governed by Equation 2.19:

$$P_U = 1.2(D + F + T) + 1.6(L + H) + 0.5(L_r \text{ or } S \text{ or } R)$$

$$= 1.2(2100) + 1.6(1400) + 0 + 0$$

$$= 4760 \text{ kN}$$

$$P_u/\text{pile} = \frac{4760 \text{ kN}}{4} = 1190 \text{ kN}$$

$$P_n = F_a A = (248{,}000 \text{ kPa}) A$$

$$P_u \leq \phi P_n$$

$$1190 \text{ kN} \leq (0.70)(248{,}000 \text{ kPa}) \, A$$

$$A \geq \textbf{6855 mm}^2 \qquad \Leftarrow Answer$$

Commentary

In this case, the computed area is slightly less than that from the ASD analysis.

2.2 STRENGTH REQUIREMENTS

Once the design loads have been defined, we need to develop foundation designs that satisfy several performance requirements. The first category is *strength requirements*, which are intended to avoid catastrophic failures. There are two types: geotechnical strength requirements and structural strength requirements.

Geotechnical Strength Requirements

Geotechnical strength requirements are those that address the ability of the soil or rock to accept the loads imparted by the foundation without failing. The strength of soil is governed by its capacity to sustain shear stresses, so we satisfy geotechnical strength requirements by comparing shear stresses with shear strengths and designing accordingly.

In the case of spread footing foundations, geotechnical strength is expressed as the *bearing capacity* of the soil. If the load-bearing capacity of the soil is exceeded, the resulting shear failure is called a *bearing capacity failure*, as shown in Figure 2.2. Later in this book we will examine such failures in detail, and learn how to design foundations that have a sufficient factor of safety against such failures.

Geotechnical strength analysis are almost always performed using allowable stress design (ASD) methods. However, LRFD-based design is beginning to appear, as discussed in Chapter 21.

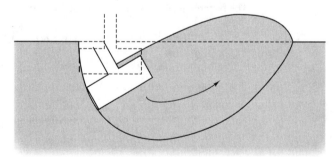

Figure 2.2 A bearing capacity failure beneath a spread footing foundation. The soil has failed in shear, causing the foundation to collapse.

Structural Strength Requirements

Structural strength requirements address the foundation's structural integrity and its ability to safely carry the applied loads. For example, pile foundations made from A36 steel are normally designed for a maximum allowable compressive stress of 12,600 lb/in². Thus, the thickness of the steel must be chosen such that the stresses induced by the design loads do not exceed this allowable value. Foundations that are loaded beyond their structural capacity will, in principle, fail catastrophically.

Structural strength analyses are conducted using either ASD or LRFD methods, depending on the type of foundation, the structural materials, and the governing code.

QUESTIONS AND PRACTICE PROBLEMS

Unless otherwise stated all ASD design loads should be computed using Equations 2.1, 2.2, 2.3a, and 2.4a.

2.1 A proposed column has the following design loads:

Axial load: $P_D = 200$ k, $P_L = 170$ k, $P_E = 50$ k, $P_W = 60$ k (all compression)

Shear load: $V_D = 0$, $V_L = 0$, $V_E = 40$ k, $V_W = 48$ k

Compute the design axial and shear loads for foundation design using ASD.

2.2 Repeat Problem 2.1 using LRFD with the ACI load factors.

2.3 A certain foundation will experience a bearing capacity failure when it is subjected to a downward load of 2200 kN. Using ASD with a factor of safety of 3, determine the maximum allowable load that will satisfy geotechnical strength requirements.

2.4 A steel pile foundation with a cross-sectional area of 15.5 in² and $F_y = 50$ k/in² is to carry axial compressive dead and live loads, of 300 and 200 k, respectively. Using LRFD with the ANSI/ASCE load factors and a resistance factor of 0.75, determine whether this pile satisfies structural strength requirements for axial compression.

2.3   SERVICEABILITY REQUIREMENTS

Foundations that satisfy strength requirements will not collapse, but they still may not have adequate performance. For example, they may experience excessive settlement. Therefore, we have the second category of performance requirements, which are known as *serviceability requirements*. These are intended to produce foundations that perform well when subjected to the service loads. These requirements include:

- **Settlement**—Most foundations experience some downward movement as a result of the applied loads. This movement is called *settlement*. Keeping settlements within tolerable limits is usually the most important foundation serviceability requirement.
- **Heave**—Sometimes foundations move upward instead of downward. We call this upward movement *heave*. The most common source of heave is the swelling of expansive soils.
- **Tilt**—When settlement or heave occurs only on one side of the structure, it may begin to *tilt*. The Leaning Tower of Pisa is an extreme example of tilt.
- **Lateral movement**—Foundations subjected to lateral loads (shear or moment) deform horizontally. This *lateral movement* also must remain within acceptable limits to avoid structural distress.
- **Vibration**—Some foundations, such as those supporting certain kinds of heavy machinery, are subjected to strong vibrations. Such foundations need to accommodate these vibrations without experiencing resonance or other problems.
- **Durability**—Foundations must be resistant to the various physical, chemical, and biological processes that cause deterioration. This is especially important in waterfront structures, such as docks and piers.

Failure to satisfy these requirements generally results in increased maintenance costs, aesthetic problems, diminished usefulness of the structure, and other similar effects.

Settlement

The vertical downward load is usually the greatest load acting on foundations, and the resulting vertical downward movement is usually the largest and most important movement. We call this vertical downward movement *settlement*. Sometimes settlement also occurs as a result of other causes unrelated to the presence of the foundation, such as consolidation due to the placement of a fill.

Although foundations with zero settlement would be ideal, this is not an attainable goal. Stress and strain always go together, so the imposition of loads from the foundation always cause some settlement in the underlying soils. Therefore, the question that faces the foundation engineer is not *if* the foundation will settle, but rather defining the *amount* of settlement that would be tolerable and designing the foundation to accommodate this requirement. This design process is analogous to that for beams where the deflection must not exceed some maximum tolerable value.

Structural Response to Settlement

Structures can settle in many different ways, as shown in Figure 2.3a. Sometimes the settlement is uniform, so the entire structure moves down as a unit. In this case, there is no damage to the structure itself, but there may be problems with its interface with the adjacent ground or with other structures. Another possibility is settlement that varies linearly

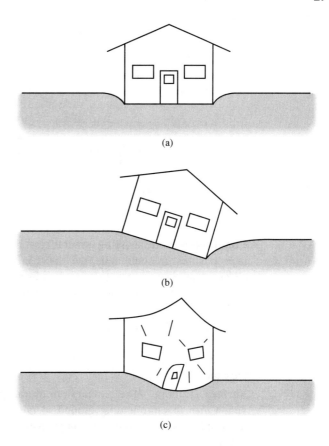

(a)

(b)

Figure 2.3 Modes of settlement; (a) uniform; (b) tilting with no distortion; (c) distortion.

(c)

across the structure as shown in Figure 2.3b. This causes the structure to tilt. Finally, Figure 2.3c shows a structure with irregular settlements. This mode distorts the structure and typically is the greatest source of problems.

The response of structures to foundation settlement is very complex, and a complete analysis would require consideration of many factors. Such analyses would be very time-consuming, are thus not practical for the vast majority of structures. Therefore, we simplify the problem by describing settlement using only two parameters: total settlement and differential settlement (Skempton and MacDonald, 1956; Polshin and Tokar, 1957; Burland and Wroth, 1974; Grant, et al., 1974; Wahls, 1981; Wahls, 1994; Frank, 1994).

Total Settlement

The *total settlement*, δ, is the change in foundation elevation from the original unloaded position to the final loaded position, as shown in Figure 2.4.

Structures that experience excessive total settlements might have some of the following problems:

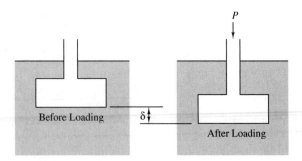

Figure 2.4 Total settlement in a spread footing foundation.

- **Connections with existing structures**—Sometimes buildings must join existing structures. In such cases, the floors in the new building must be at the same elevation as those in the existing building. However, if the new building settles excessively, the floors will no longer match, causing serious serviceability problems.
- **Utility lines**—Buildings, tanks, and many other kinds of structures are connected to various utilities, many of which are located underground. If the structure settles excessively, the utility connections can be sheared or distorted. This is especially troublesome with gravity flow lines, such as sewers.
- **Surface drainage**—The ground floor of buildings must be at a slightly higher elevation than the surrounding ground so rainwater does not enter. However, settlement might destroy these drainage patterns and cause rainwater to enter the structure.
- **Access**—Vehicles and pedestrians may need to access the structure, and excessive settlement might impede them.
- **Aesthetics**—Excessive settlement may cause aesthetic problems long before there is any threat to structural integrity or serviceability.

Some structures have sustained amazingly large total settlements, yet remain in service. For example, many buildings have had little or no ill effects even after settling as much as 250 mm (10 in). Others have experienced some distress, but continue to be used following even greater settlements. Some of the most dramatic examples are located in Mexico City, where buildings have settled more than 2 m (7 ft) and are still in use. Some bridges, tanks, and other structures also might tolerate very large settlements. However, these are extreme examples. Normally engineers have much stricter performance requirements.

Table 2.1 presents typical design values for the allowable total settlement, δ_a. These values already include a factor of safety, and thus may be compared directly to the pre-

TABLE 2.1   TYPICAL ALLOWABLE TOTAL SETTLEMENTS
FOR FOUNDATION DESIGN

Type of Structure	Typical Allowable Total Settlement, δ_a	
	(in)	(mm)
Office buildings	0.5–2.0 (1.0 is the most common value)	12–50 (25 is the most common value)
Heavy industrial buildings	1.0–3.0	25–75
Bridges	2.0	50

dicted settlement. The design meets total settlement requirements if the following condition is met:

$$\boxed{\delta \leq \delta_a} \tag{2.24}$$

Where:

δ = total settlement of foundation

δ_a = allowable total settlement

Methods of computing δ are covered later in this book.

When using Table 2.1, keep the following caveats in mind:

1. Customary engineering practice in this regard varies significantly between regions, which is part of the reason for the wide ranges of values in this table. For example, an office building in one state may need to be designed for a total settlement of 12 mm (0.5 in), while the same building in another state might be designed using 25 mm (1.0 in). It is important to be aware of local practices, since engineers are normally expected to conform to them. However, it also is important to recognize that local practices in some areas are overly conservative and produce foundations that are more expensive than necessary.

2. Foundations with large total settlements also tend to have large differential settlements. Therefore, limits on allowable differential settlement often indirectly place limits on total settlement that are stricter than those listed in Table 2.1. Chapter 7 discusses this aspect in more detail.

If the predicted settlement, δ, is greater than δ_a, we could consider any or all of the following measures:

- **Adjust the foundation design** — Adjustments in the foundation design often will solve problems with excessive settlement. For example, the settlement of spread footing foundations can be reduced by increasing their width.
- **Use a more elaborate foundation**—For example, we might use piles instead of spread footings, thus reducing the settlement.
- **Improve the properties of the soil**—Many techniques are available to do this; some of them are discussed in Chapter 18.
- **Redesign the structure so it is more tolerant of settlements**—For example, flexible joints could be installed on pipes, as shown in Figure 2.5.

Differential Settlement

Engineers normally design the foundations for a structure such that all of them have the same computed total settlement. Thus, in theory, the structure will settle uniformly. Unfortunately, the actual performance of the foundations will usually not be exactly as predicted, with some of them settling more than expected and other less. This discrepancy between predicted behavior and actual behavior has many causes, including the following:

- **The soil profile may not be uniform across the site**—This is nearly always true, no matter how uniform it might appear to be.
- **The ratio between the actual load and the design load may be different for each column**—Thus, the column with the lower ratio will settle less than that with the higher ratio.

Figure 2.5 This flexible-extendible pipe coupler has ball joints at each end and a telescoping section in the middle. These couplers can be installed where utility lines enter structures, thus accommodating differential settlement, lateral movement, extension, and compression. For example, an 8-in (203 mm) diameter connector can accommodate differential settlements of up to 15-in (380 mm). (Photo courtesy of EBAA Iron Sales, Inc.).

- **The ratio of dead load to live load may be different for each column**— Settlement computations are usually based on dead-plus-live load, and the foundations are sized accordingly. However, in many structures much of the live load will rarely, if ever, occur, so foundations that have a large ratio of design live load to design dead load will probably settle less than those carrying predominantly dead loads.

- **The as-built foundation dimensions may differ from the plan dimensions**—This will cause the actual settlements to be correspondingly different.

The *differential settlement*, δ_D, is the difference in total settlement between two foundations or between two points on a single foundation. Differential settlements are generally more troublesome than total settlements because they distort the structure, as shown in Figure 2.3c. This causes cracking in walls and other members, jamming in doors and windows, poor aesthetics, and other problems. If allowed to progress to an extreme, differential settlements could threaten the integrity of the structure. Figures 2.6 and 2.7 show examples of structures that have suffered excessive differential settlement.

Therefore, we define a maximum *allowable differential settlement*, δ_{Da}, and design the foundations so that:

$$\boxed{\delta_D \le \delta_{Da}} \tag{2.25}$$

Figure 2.6 This wood-frame house was built on an improperly compacted fill, and thus experienced excessive differential settlements. The resulting distortions produced drywall cracks up to 15-mm wide, as shown in this photograph, along with additional cracks in the exterior walls and the floor slab.

Figure 2.7 This old brick building has experienced excessive differential settlements, as evidenced by the diagonal cracks in the exterior wall.

In buildings, δ_{Da} depends on the potential for jamming doors and windows, excessive cracking in walls and other structural elements, aesthetic concerns, and other similar issues. The physical processes that cause these serviceability problems are very complex, and depend on many factors, including the type and size of the structure, the properties of the building materials and the subsurface soils, and the rate and uniformity of the settlement (Wahls, 1994). These processes are much too complex to model using rational structural analyses, so engineers depend on empirical methods. These methods are based on measurements of the actual differential settlements in real buildings and assessments of their performance.

Comprehensive studies of differential settlements in buildings include Skempton and MacDonald (1956), Polshin and Tokar (1957), and Grant et al. (1974). Skempton and MacDonald's work is based on the observed performance of ninety eight buildings of various types, forty of which had evidence of damage due to excessive settlements. Polshin and Tokar reported the results of 25 years of observing the performance of structures in the Soviet Union and reflected Soviet building codes. The study by Grant et al. encompassed data from ninety five buildings, fifty six of which had damage.

Table 2.2 presents a synthesis of these studies, expressed in terms of the allowable angular distortion, θ_a. These values already include a factor of safety of at least 1.5, which is why they are called "allowable." We use them to compute δ_{Da} as follows:

$$\boxed{\delta_{Da} = \theta_a S} \qquad (2.26)$$

Where:

δ_{Da} = allowable differential settlement

θ_a = allowable angular distortion (from Table 2.2)

S = column spacing (horizontal distance between columns)

Empirical data also suggests that typical buildings have architectural damage at $\theta \approx 1/300$ and structural damage at $\theta \approx 1/150$ (Stephenson, 1995). However, both these values and those in Table 2.2 especially depend on the type of exterior cladding, because cracks in the cladding can allow water to enter the structure and damage the interior. For example, corrugated metal siding tolerates much more differential settlement than stucco. In addition, it may be wise to use lower δ_a and δ_{Da} values for warehouses, tanks, and other structures in which the live load represents a large portion of the total load and does not occur until after the structure is complete.

Be sure to consider local practice and precedent when developing design values of δ_{Da}. Engineers in some areas routinely design structures to accommodate relatively large settlements, and may be willing to accept some long-term maintenance costs (i.e., repairing minor cracking, rebuilding entranceways, etc.) in exchange for reduced construction costs. However, in other areas, even small settlements induce lawsuits, so foundations are designed to meet stricter standards.

TABLE 2.2   ALLOWABLE ANGULAR DISTORTION, θ_a (COMPILED FROM WAHLS, 1994; AASHTO, 1996; AND OTHER SOURCES)

Type of Structure	θ_a
Steel tanks	1/25
Bridges with simply-supported spans	1/125
Bridges with continuous spans	1/250
Buildings that are very tolerant of differential settlements, such as industrial buildings with corrugated steel siding and no sensitive interior finishes.	1/250
Typical commercial and residential buildings.	1/500
Overhead traveling crane rails.	1/500
Buildings that are especially intolerant of differential settlement, such as those with sensitive wall or floor finishes.	1/1000
Machinery[a]	1/1500
Buildings with unreinforced masonry load-bearing walls Length/height ≤ 3 Length/height ≥ 5	1/2500 1/1250

[a] Large machines, such as turbines or large punch presses, often have their own foundation, separate from that of the building that houses them. It often is appropriate to discuss allowable differential settlement issues with the machine manufacturer.

Methods for computing δ_D are discussed later in this book within the chapters related to the type of foundation being considered.

Design Load

Serviceability requirements are dictated by the performance of the structure under the actual service loads. This is quite different from strength requirements, which are concerned with avoiding failure under extreme loading events. Therefore, settlement analyses are based on the unfactored static working loads (i.e., the larger of Equations 2.1 to 2.4). This is true regardless of whether ASD or LRFD methods are used to satisfy strength requirements.

Sometimes engineers perform settlement analyses using only the static loads (i.e., the larger of Equations 2.1 and 2.2). This is because wind or seismic loads represent extreme events. In addition, these loads generally have a very short duration (i.e., the peak wind gusts and the peak earthquake accelerations are very short), so the soil may not have enough time to respond in the same way it does for long-term loads. Even if there were no additional settlements, some nonstructural cracking will probably occur during anyway as a result of swaying in the superstructure. Very few building owners would object to patching a few cracks after a design-level earthquake or hurricane.

Example 2.4

A steel-frame office building has a column spacing of 20 ft. It is to be supported on spread footings founded on a clayey soil. What are the allowable total and differential settlements?

Solution

Per Table 2.1, use $\delta_a = 1.0$ in
Per Table 2.2, $\theta_a = 1/500$ $\Leftarrow$ *Answer*

$$\delta_{Da} = \theta_a S$$
$$= (1/500)(20)$$
$$= 0.04 \text{ ft} = 0.5 \text{ in} \quad \Leftarrow Answer$$

Rate of Settlement

It also is important to consider the rate of settlement and how it compares with the rate of construction. Foundations in sands settle about as rapidly as the loads are applied, whereas those in saturated clays move much more slowly.

In some structures, much of the load is applied to the foundation before the settlement-sensitive elements are in place. For example, settlements of bridge piers that occur before the deck is placed are far less important than those that occur after. Buildings may not be sensitive to differential settlements until after sensitive finishes, doors, and other architectural items are in place, yet if the foundation is in sand, most of the settlement may have already occurred by then.

However, other structures generate a large portion of their loads after they are completed. For example, the greatest load on a water storage tank is its contents.

Heave

Sometimes foundations move upward instead of downward. This kind of movement is called *heave*. It may be due to applied upward loads, but more often it is the result of external forces, especially those from expansive soils. The design criteria for heave are the same as those for settlement. However, if some foundations are heaving while others are settling, then the differential is the sum of the two.

Tilt

Excessive tilt is often a concern in tall, rigid structures, such as chimneys, silos, and water towers. To preserve aesthetics, the tilt, ω, from the vertical should be no more than 1/500 (7 min of arc). Greater tilts would be noticeable, especially in taller structures and those that are near other structures. In some cases, stricter limits on tilt are appropriate, especially for exceptionally tall structures. For comparison, the Leaning Tower of Pisa has a tilt of about 1/10.

Lateral Movement

Foundations subjected to lateral loads have corresponding lateral movements. These movement also have tolerable limits. For bridge foundations, Bozozuk (1978) recommended maximum lateral movements of 25 mm (1 in).

Vibration

Foundations that support large machinery are sometimes subjected to substantial vibratory loads. Such foundations must be designed to accommodate these vibratory loads without introducing problems, such as resonance.

Durability

Soil can be a very hostile environment to place engineering materials. Whether they are made of concrete, steel, or wood, structural foundations may be susceptible to chemical and/or biological attack that can adversely affect their integrity.

Corrosion of Steel

Under certain conditions, steel can be the object of extensive corrosion. This can be easily monitored when the steel is above ground, and routine maintenance, such as painting, will usually keep corrosion under control. However, it is impossible to inspect underground steel visually, so it is appropriate to be concerned about its potential for corrosion and long-term integrity.

Owners of underground steel pipelines are especially conscious of corrosion problems. They often engage in extensive corrosion surveys and include appropriate preventive measures in their designs. These procedures are well established and effective, but should they also be used for steel foundations such as H-piles or steel pipe piles?

For corrosion assessment, steel foundations can be divided into two categories: those in marine environments and those in land environments. Both are shown in Figure 2.8.

Steel foundations in marine environments have a significant potential for corrosion, especially those exposed to salt water. Studies of waterfront structures have found that steel is lost at a rate of 0.075 to 0.175 mm/yr (Whitaker, 1976). This corrosion occurs most rapidly in the tidal and splash zones (Dismuke et. al., 1981) and can also be very extensive immediately above the sea floor; then it becomes almost negligible at depths more than about 0.5 m (2 ft) below the sea floor. Such structures may also be prone to abrasion from moving sand, ships, floating debris, and other sources. It is common to protect such foundations with coatings or jackets, at least through the water and splash zones.

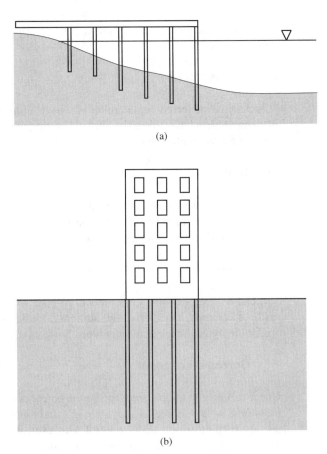

(a)

(b)

Figure 2.8 (a) Marine environments include piers, docks, drilling platforms, and other similar structures where a portion of the foundation is exposed to open water. (b) Land environments include buildings and other structures that are built directly on the ground and the entire foundation is buried.

However, the situation in land environments is quite different. Based on extensive studies, Romanoff (1962, 1970) observed that no structural failures have been attributed to the corrosion of steel piles in land environments. One likely reason for this excellent performance record is that piles, unlike pipelines, can tolerate extensive corrosion, even to the point of occasionally penetrating through the pile, and remain serviceable.

Romanoff also observed that piles founded in natural soils (as opposed to fills) experienced little or no corrosion, even when the soil could be identified as potentially corrosive. The explanation for this behavior seems to be that natural soils contain very little free oxygen, an essential ingredient for the corrosion process.

However, fills do contain sufficient free oxygen and, under certain circumstances, can be very corrosive environments. Therefore, concern over corrosion of steel piles in land environments can normally be confined to sites where the pile penetrates through fill. Some fills have very little potential for corrosion, whereas others could corrode steel at rates of up to 0.08 mm/yr (Tomlinson, 1987), which means that a typical H-pile section could lose half of its thickness in about 50 years.

Schiff (1982) indicated that corrosion would be most likely in the following soil conditions:

- High moisture content
- Poorly aerated
- Fine grained
- Black or gray color
- Low electrical resistivity
- Low or negative redox potential
- Organic material present
- High chemical content
- Highly acidic
- Sulfides present
- Anaerobic microorganisms present

Areas where the elevation of the groundwater table fluctuates, such as tidal zones, are especially difficult because this scenario continually introduces both water and oxygen to the pile. Contaminated soils, such as sanitary landfills and shorelines near old sewer outfalls, are also more likely to have problems.

One of the most likely places for corrosion on land piles is immediately below a concrete pile cap. Local electrical currents can develop because of the change in materials, with the concrete acting as a cathode and the soil as an anode. Unfortunately, this is also the most critical part of the pile because the stresses are greatest there.

If the foundation engineer suspects that corrosion may be a problem, it is appropriate to retain the services of a corrosion engineer. Detailed assessments of corrosion and the development of preventive designs are beyond the expertise of most foundation engineers.

The corrosion engineer will typically conduct various tests to quantify the corrosion potential of the soil and consider the design life of the foundation to determine whether any preventive measures are necessary. Such measures could include the following:

- Use a different construction material (i.e., concrete, wood).
- Increase the thickness of steel sections by an amount equal to the anticipated deterioration.
- Cover the steel with a protective coating (such as coal tar epoxy) to protect it from the soil. This method is commonly used with underground tanks and pipes, and has also been successfully used with pile foundations. However, consider the possibility that some of the coating may be removed by abrasion when the pile is driven into the ground, especially when sands or gravels are present. Coatings can also be an effective means of combatting corrosion near pile caps, as discussed earlier. In this case, the coating is applied to the portion of the steel that will be encased in the concrete, thus providing the electrical insulation needed to stop or significantly slow the corrosion process.
- Provide a *cathodic protection system*. Such systems consist of applying a DC electrical potential between the foundation (the cathode) and a buried sacrificial metal (the anode). This system causes the corrosion to be concentrated at the anode and protects the cathode (see Figure 2.9). Rectifiers connected to a continuous power source provide the electricity. These systems consume only nominal amounts of electricity. In some cases, it is possible to install a self-energizing system that generates its own current.

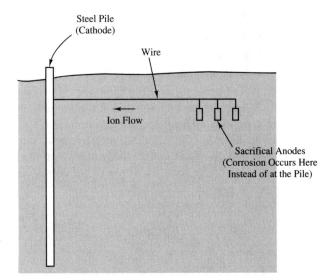

Figure 2.9 Use of a cathodic protection system to protect steel foundations from corrosion.

Sulfate Attack on Concrete

Buried concrete is usually very resistant to corrosion and will remain intact for many years. However, serious degradation can occur in concrete subjected to soils or groundwater that contain high concentrations of sulfates (SO_4). These sulfates can react with the cement to form calcium sulfoaluminate (ettringite) crystals. As these crystals grow and expand, the concrete cracks and disintegrates. In some cases, serious degradation has occurred within 5 to 30 years of construction. Although we do not yet fully understand this process (Mehta, 1983), engineers have developed methods of avoiding these problems.

We can evaluate a soil's potential for sulfate attack by measuring the concentration of sulfates in the soil and/or in the groundwater and comparing them with soils that have had problems with sulfate attack. Soils with some or all of the following properties are most likely to have high sulfate contents:

- Wet
- Fine-grained
- Black or gray color
- High organic content
- Highly acidic or highly alkaline

Some fertilizers contain a high concentration of sulfates that may cause problems when building in areas that were formerly used for agricultural purposes. The same is true for some industrial wastes. It is often wise to consult with corrosion experts in such cases. Seawater also has a high concentration: about 2300 ppm.

If the laboratory tests indicate that the soil or groundwater has a high sulfate content, design the buried concrete to resist attack by using one or more of the following methods:

- **Reduce the water:cement ratio**—This reduces the permeability of the concrete, thus retarding the chemical reactions. This is one of the most effective methods of resisting sulfate attack. Suggested maximum ratios are presented in Table 2.3.
- **Increase the cement content**—This also reduces the permeability. Therefore, concrete that will be exposed to problematic soils should have a cement content of at least 6 sacks/yd^3 (564 lb/yd^3 or 335 kg/m^3).
- **Use sulfate-resisting cement**—Type II low-alkali and type V portland cements are specially formulated for use in moderate and severe sulfate conditions, respectively. Pozzolan additives to a type V cement also help. Type II is easily obtained, but type V may not be readily available in some areas. Table 2.3 gives specific guidelines.
- **Coat the concrete with an asphalt emulsion**—This is an attractive alternative for retaining walls or buried concrete pipes, but not for foundations.

TABLE 2.3   USE OF SULFATE-RESISTING CEMENTS AND LOW WATER:CEMENT RATIOS TO AVOID SULFATE ATTACK OF CONCRETE (Adapted from Kosmatka and Panarese, 1988, and PCA, 1991). Used with permission of the Portland Cement Association.

Water-Soluble Sulfates in Soil (% by weight)	Sulfates in Water (ppm)	Sulfate Attack Hazard	Cement Type	Maximum Water:Cement Ratio
0.00–0.10	0–150	Negligible	—	—
0.10–0.20	150–1500	Moderate	II	0.50
0.20–2.00	1500–10,000	Severe	V	0.45
>2.00	>10,000	Very severe	V plus pozzolan	0.45

Decay of Timber

The most common use of wood in foundations is timber piles. The lifespan of these piles varies depending on their environment. Even untreated timber piles can have a very long life if they are continually submerged below the groundwater table. This was illustrated when a campanile in Venice fell in 1902. The submerged timber piles, which had been driven in A.D. 900, were found to be in good condition and were used to support the replacement structure (Chellis, 1962). However, when located above the groundwater table, timber can be subject to deterioration from several sources (Chellis, 1961), including:

- **Decay** caused by the growth of fungi. This process requires moisture, oxygen, and favorable temperatures. These conditions are often most prevalent in the uppermost 2 m (6 ft) of the soil. If the wood is continually very dry, then decay will be limited due to the lack of moisture.
- **Insect attack**, including termites, beetles, and marine borers.
- **Fire**, especially in marine structures.

The worst scenario is one in which the piles are subjected to repeated cycles of wetting and drying. Such conditions are likely to be found near the groundwater table because it usually rises and sinks with the seasons and near the water surface in marine applications where splashing and tides will cause cyclic wetting and drying.

To reduce problems of decay, insect attack, and fungi growth, timber piles are usually treated before they are installed. The most common treatment consists of placing them in a pressurized tank filled with creosote or some other preserving chemical. This *pressure treatment* forces some of the creosote into the wood and forms a thick coating on the outside, leaving a product that is almost identical to many telephone poles. When the piles are fully embedded into soil, creosote-treated piles normally have a life at least as long as the design life of the structure.

Timber piles also will lose part of their strength if they are subjected to prolonged high temperatures. Therefore, they should not be used under hot structures such as blast furnaces.

QUESTIONS AND PRACTICE PROBLEMS

2.5 A seven-story steel-frame office building will have columns spaced 7 m on center and will have typical interior and exterior finishes. Compute the allowable total and differential settlements for this building.

2.6 A two-story reinforced concrete art museum is to be built using an unusual architectural design. It will include many tile murals and other sensitive wall finishes. The column spacing will vary between 5 and 8 m. Compute the allowable total and differential settlements for this buiding.

2.7 A 40 ft × 60 ft one-story agricultural storage building will have corrugated steel siding and no interior finish or interior columns. However, it will have two roll-up doors. Compute the allowable total and differential settlement for this building.

2.8 A sandy soil has 0.03 percent sulfates. Evaluate the potential for sulfate attack of concrete exposed to this soil and recommend preventive design measures, if needed.

2.4   CONSTRUCTIBILITY REQUIREMENTS

The third category of performance requirements is *constructibility*. The foundation must be designed such that a contractor can build it without having to use extraordinary methods or equipment. There are many potential designs that might be quite satisfactory from a design perspective, but difficult or impossible to build.

For example, Chapter 11 of this book discusses different types of deep foundations. One of these, a *pile foundation*, consists of a prefabricated pole that is driven into the ground using a modified crane called a pile driver. The pile driver lifts the pile into the air, then drives it into the ground, as shown in Figure 2.10. Therefore, piles can be installed only at locations that have sufficient headroom. Fortunately, the vast majority of construction sites have plenty of headroom.

Nevertheless, a design engineer who is not familiar with this construction procedure might ask for piles to be installed at a location with minimal headroom. For example, the pile design shown in Figure 2.11 is unbuildable because it is impossible to fit the required installation equipment into such a small space.

Karl Terzaghi expressed this concept very succinctly when he said:

Do not design on paper what you have to wish into the ground.

Figure 2.10 Pile foundations are installed using a pile driver, such as this one. The pile is lifted into the vertical section, which is called the leads, then driven into the ground with the pile hammer. Thus, the pile driver must be slightly taller than the pile to be installed.

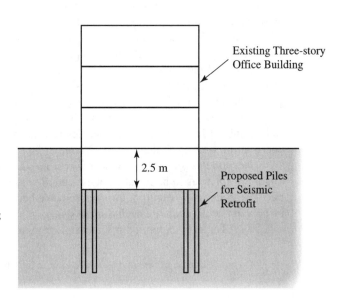

Figure 2.11 As part of a seismic retrofit project, a design engineer has called for installing 450-mm diameter, 9-m long prestressed concrete pile foundations to be installed beneath the basement of an existing building. This pile foundation design is unbuildable because the required pile-driving equipment would not fit in the basement, and because there is not enough room to set the pile upright.

This is why it is important for design engineers to have at least a rudimentary understanding of construction.

2.5 ECONOMIC REQUIREMENTS

Foundation designs are usually more conservative than those in the superstructure. This approach is justified for the following reasons:

- Foundation designs rely on our assessments of the soil and rock conditions. These assessments always include considerable uncertainty.
- Foundations are not built with the same degree of precision as the superstructure. For example, spread footings are typically excavated with a backhoe and the sides of the excavation becomes the "formwork" for the concrete, compared to concrete members in the superstructure that are carefully formed with plywood or other materials.
- The structural materials may be damaged when they are installed. For example, cracks and splits may develop in a timber pile during hard driving.
- There is some uncertainty in the nature and distribution of the load transfer between foundations and the ground, so the stresses at any point in a foundation are not always known with as much certainty as might be the case in much of the superstructure.
- The consequences of a catastrophic failure are much greater.
- The additional weight brought on by the conservative design is of no consequence, because the foundation is the lowest structural member and therefore does not affect the dead load on any other member. Additional weight in the foundation is actually beneficial in that it increases its uplift resistance.

However, gross overconservatism is not warranted. An overly conservative design can be very expensive to build, especially with large structures where the foundation is a greater portion of the total project cost. This also is a type of "failure": the failure to produce an economical design.

The nineteenth-century engineer Arthur Wellington once said that an engineer's job is that of "doing well with one dollar which any bungler can do with two." We must strive to produce designs that are both safe and cost-effective. Achieving the optimum balance between reliability (safety) and cost is part of good engineering.

Designs that minimize the required quantity of construction materials do not necessarily minimize the cost. In some cases, designs that use more materials may be easier to build, and thus have a lower overall cost. For example, spread footing foundations are usually made of low-strength concrete, even though it makes them thicker. In this case, the savings in materials and inspection costs are greater than the cost of buying more concrete.

SUMMARY

Major Points

1. The foundation engineer must determine the necessary performance requirements before designing a foundation.

2. Foundations must support various types of structural loads. These can include normal, shear, moment, and/or torsion loads. The magnitude and direction of these loads may vary during the life of the structure.

3. Loads also are classified according to their source. These include dead loads, live loads, wind loads, earthquake loads, and several others.

4. Design loads may be expressed using either the allowable stress design (ASD) or the load and resistance factor design (LRFD) method. It is important to know which method is being used, because the design computations must be performed accordingly.

5. Strength requirements are those that are intended to avoid catastrophic failure. There are two kinds: geotechnical strength requirements and structural strength requirements.

6. Serviceability requirements are those intended to produce foundations that perform well when subjected to the service loads. These requirements include settlement, heave, tile, lateral movement, vibration, and durability.

7. Settlement is often the most important serviceability requirement. The response of structures to settlements is complex, so we simplify the problem by considering two types of settlement: total settlement and differential settlement. We assign maximum allowable values for each, then design the foundations to satisfy these requirements.

8. Durability is another important serviceability requirement. Foundations must be able to resist the various corrosive and deteriorating agents in soil and water.

9. Foundations must be buildable, so design engineers need to have at least a rudimentary understanding of construction methods and equipment.

10. Foundation designs must be economical. Although conservatism is appropriate, excessively conservative designs can be too needlessly expensive to build.

Vocabulary

Allowable differential settlement	Earthquake load	Self-straining load
Allowable total settlement	Economic requirement	Serviceability requirement
Allowable angular distortion	Failure	Settlement
	Fluid load	Shear load
Allowable stress design	Geotechnical strength requirement	Snow load
Braking load		Stream flow loads
Cathodic protection	Heave	Strength requirement
Centrifugal load	Impact load	Structural strength requirement
Column spacing	Lateral movement	
Constructibility requirement	Live load	Sulfate attack
	Load factor	Tilt
Dead load	Load and resistance factor design	Torsion load
Design load		Total settlement
Differential settlement	Moment load	Vibration
Durability	Normal load	Wind load
Earth pressure load	Performance requirement	
	Resistance factor	

COMPREHENSIVE QUESTIONS AND PRACTICE PROBLEMS

2.9 A certain clayey soil contains 0.30 percent sulfates. Would you anticipate a problem with concrete foundations in this soil? Are any preventive measures necessary? Explain.

2.10 A series of 50-ft long steel piles are to be driven into a natural sandy soil. The groundwater table is at a depth of 35 ft below the ground surface. Would you anticipate a problem with corrosion? What additional data could you gather to make a more informed decision?

2.11 A one-story steel warehouse building is to be built of structural steel. The roof is to be supported by steel trusses that will span the entire 70 ft width of the building and supported on columns adjacent to the exterior walls. These trusses will be placed 24 ft on center. No interior columns will be present. The walls will be made of corrugated steel. There will not be any roll-up doors. Compute the allowable total and differential settlements.

2.12 The grandstands for a minor league baseball stadium are to be built of structural steel. The structural engineer plans to use a very wide column spacing (25 m) to provide the best spectator visibility. Compute the allowable total and differential settlements.

2.13 The owner of a 100-story building purchased a plumb bob with a very long string. He selected a day with no wind, and then gently lowered the plumb bob from his penthouse office win-

Figure 2.12 Proposed department store for Problem 2.14.

dow. When it reached the sidewalk, it was 1.0 m from the side of the building. Is this building tilting excessively? Explain.

2.14 A two-story department store identical to the one in Figure 2.12 is to be built. This structure will have reinforced masonry exterior walls. The ground floor will be slab-on-grade. The reinforced concrete upper floor and roof will be supported on a steel-frame with columns 50 ft on-center. Compute the allowable total and differential settlements for this structure.

3

Soil Mechanics

Measure it with a micrometer,
mark it with chalk,
cut it with an axe.

An admonition to maintain a consistent degree of precision throughout
the analysis, design, and construction phases of a project.

Engineers classify earth materials into two broad categories: *rock* and *soil*. Although both materials play an important role in foundation engineering, most foundations are supported by soil. In addition, foundations on rock are often designed much more conservatively because of the rock's greater strength, whereas economics prevents overconservatism when building foundations on soil. Therefore, it is especially important for the foundation engineer to be familiar with soil mechanics.

Users of this book should already have acquired at least a fundamental understanding of the principles of soil mechanics. This chapter reviews these principles, emphasizing those that are most important in foundation analyses and design. Relevant principles of rock mechanics are included in later chapters within the context of specific applications. *Geotechnical Engineering: Principles and Practices* (Coduto, 1999), the companion volume to this book, explores all of these topics in much more detail.

3.1 SOIL COMPOSITION

One of the fundamental differences between soil and most other engineering materials is that it is a *particulate material*. This means that it is an assemblage of individual particles rather than being a *continuum* (a continuous solid mass). The engineering properties of soil, such as strength and compressibility, are dictated primarily by the arrangement of these particles and the interactions between them, rather than by their internal properties.

Another important characteristic that differentiates soil from most other materials is that it can contain all three phases of matter (solid, liquid, and gas) simultaneously. The solid portion (the particles) includes one or more of the following materials:

- *Rock fragments* such as granite, limestone, and basalt.
- *Rock minerals* such as quartz, feldspar, mica, and gypsum.
- *Clay minerals* such as kaolinite, smectite, and illite.
- *Organic matter* such as decomposed plant materials.
- *Cementing agents* such as calcium carbonate.
- *Miscellaneous materials* such as man-made debris.

Liquids and/or gasses fill the voids between the solid particles. The liquid component is usually water, but it also could contain various chemicals in solution. The latter could come from natural sources, such as calcite leached from limestone, or artificial sources such as gasoline from leaking tanks or pipes. Likewise, the gas component is usually air, but also could consist of other materials, such as methane. For simplicity, we will refer to these components as "water" and "air."

A special exception to this three-phase structure is the case of a saturated soil at a temperature below the freezing point of water. These frozen soils are essentially a completely solid material and require special analysis and design techniques.

Weight-Volume Relationships

A knowledge of the relative proportions of solids, water, and air can give important insights into the engineering behavior of a particular soil. A *phase diagram*, as shown in Figure 3.1, describes these proportions.

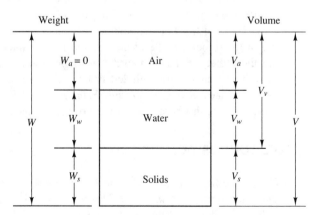

Figure 3.1 A phase diagram describes the relative proportions of solids, water, and air in a soil. The weight of the air, W_a, is negligible.

Geotechnical engineers have developed several standard parameters to define these proportions, and these parameters form much of the basic vocabulary of soil mechanics. Table 3.1 contains the definitions of weight-volume parameters commonly used in foundation engineering, along with typical numerical values. These values are not absolute limits and unusual soils may have properties outside these ranges.

These weight-volume parameters are related to each other, and many formulas are available to express these relationships. Some of the more useful formulas include the following:

$$e = \frac{w\, G_s}{S} \tag{3.1}$$

$$e = \frac{G_s\, \gamma_w}{\gamma_d} - 1 \tag{3.2}$$

TABLE 3.1 DEFINITIONS AND TYPICAL VALUES OF COMMON SOIL WEIGHT-VOLUME PARAMETERS

Parameter	Symbol	Definition	Typical Range	
			English	SI
Unit weight	γ	$\dfrac{W}{V}$	90–130 lb/ft^3	14–20 kN/m^3
Dry unit weight	γ_d	$\dfrac{W_s}{V}$	60–125 lb/ft^3	9–19 kN/m^3
Unit weight of water	γ_w	$\dfrac{W_w}{V_w}$	62.4 lb/ft^3	9.8 kN/m^3
Buoyant unit weight	γ_b	$\gamma_{sat} - \gamma_w$	28–68 lb/ft^3	4–10 kN/m^3
Degree of saturation	S	$\dfrac{V_w}{V_v} \times 100\%$	2–100%	2–100%
Moisture content	w	$\dfrac{W_w}{W_s} \times 100\%$	3–70%	3–70%
Void ratio	e	$\dfrac{V_v}{V_s}$	0.1–1.5	0.1–1.5
Porosity	n	$\dfrac{V_v}{V} \times 100\%$	9–60 %	9–60 %
Specific gravity of solids	G_s	$\dfrac{W_s}{V_s \gamma_w}$	2.6–2.8	2.6–2.8

γ_{sat} is the unit weight, γ, when $S = 100\%$.

$$\boxed{\gamma_d = \frac{\gamma}{1 + w}} \qquad\qquad (3.3)$$

$$\boxed{w = S\left(\frac{\gamma_w}{\gamma_d} - \frac{1}{G_s}\right)} \qquad\qquad (3.4)$$

Convert any parameters expressed as a percentage into decimal form before using them in these formulas.

Geotechnical engineers often use the term "density" instead of "unit weight" for the variable γ. Although density is technically mass/volume (not weight/volume), consider these two terms to be synonymous when used in the context of geotechnical engineering. However, this book uses only the more correct term "unit weight." Table 3.2 presents typical unit weights for various soils.

Most weight-volume parameters must be determined directly or indirectly from laboratory tests on soil samples. However, it is usually not necessary to measure the specific gravity of solids, G_s. For most projects, we can assume $G_s = 2.65$ for clays or 2.70 for sands.

TABLE 3.2   TYPICAL UNIT WEIGHTS

Soil Type and Unified Soil Classification (See Figure 3.3)	Typical Unit Weight, γ			
	Above Groundwater Table		Below Groundwater Table	
	(lb/ft³)	(kN/m³)	(lb/ft³)	(kN/m³)
GP—Poorly-graded gravel	110–130	17.5–20.5	125–140	19.5–22.0
GW—Well-graded gravel	110–140	17.5–22.0	125–150	19.5–23.5
GM—Silty gravel	100–130	16.0–20.5	125–140	19.5–22.0
GC—Clayey gravel	100–130	16.0–20.5	125–140	19.5–22.0
SP—Poorly-graded sand	95–125	15.0–19.5	120–135	19.0–21.0
SW—Well-graded sand	95–135	15.0–21.0	120–145	19.0–23.0
SM—Silty sand	80–135	12.5–21.0	110–140	17.5–22.0
SC—Clayey sand	85–130	13.5–20.5	110–135	17.5–21.0
ML—Low plasticity silt	75–110	11.5–17.5	80–130	12.5–20.5
MH—High plasticity silt	75–110	11.5–17.5	75–130	11.5–20.5
CL—Low plasticity clay	80–110	12.5–17.5	75–130	11.5–20.5
CH—High plasticity clay	80–110	12.5–17.5	70–125	11.0–19.5

Example 3.1

A 0.320 ft^3 sample of a certain soil has a weight of 38.9 lb, a moisture content of 19.2%, and a specific gravity of solids of 2.67. Find its void ratio and degree of saturation.

Solution

$$\gamma = \frac{W}{V} = \frac{38.9 \text{ lb}}{0.320 \text{ ft}^3} = 121.6 \text{ lb/ft}^3$$

$$\gamma_d = \frac{\gamma}{1+w} = \frac{121.6 \text{ lb/ft}^3}{1+0.192} = 102.0 \text{ lb/ft}^3$$

$$e = \frac{G_s \gamma_w}{\gamma_d} - 1 = \frac{(2.67)(62.4 \text{ lb/ft}^3)}{102.0 \text{ lb/ft}^3} - 1 = \mathbf{0.633} \qquad \Leftarrow Answer$$

$$e = \frac{w G_s}{S} \rightarrow S = \frac{w G_s}{e} = \frac{(0.192)(2.67)}{0.633} = \mathbf{81.0\%} \qquad \Leftarrow Answer$$

Relative Density

The *relative density*, D_r, is a convenient way to express the void ratio of sands and gravels. It is based on the void ratio of the soil, e, the *minimum index void ratio*, e_{min}, and the *maximum index void ratio*, e_{max}:

$$\boxed{D_r = \frac{e_{max} - e}{e_{max} - e_{min}} \times 100\%} \tag{3.5}$$

The values of e_{min} and e_{max} are determined by conducting standard laboratory tests [ASTM D4254]. The in-situ e could be computed from the unit weight of the soil using Equation 3.2, but accurate measurements of the unit weight of clean sands and gravels are difficult or impossible to obtain. Therefore, engineers often obtain D_r from correlations based on in-situ tests, as described in Chapter 4.

If a soil has a relative density of 0%, it is supposedly in its loosest possible condition, while at 100 % it is supposedly in its densest possible condition. Although it is possible for natural soils to have relative densities outside this range, such conditions are very unusual. A relationship between consistency of sands and gravels and relative density is shown in Table 3.3.

Do not confuse relative density with relative compaction. The latter is based on the Proctor compaction test [ASTM D1557] and is typically used to evaluate compacted fills. Although these two parameters measure similar soil properties, and both are expressed as a percentage, they are not numerically equal.

The relative density applies only to sands and gravels with less than 15 percent fines. It can be an excellent indicator of the engineering properties of such soils, and it is therefore an important part of many analysis methods. However, other considerations,

TABLE 3.3   CONSISTENCY OF COARSE-GRAINED
SOILS VARIOUS RELATIVE DENSITIES (Adapted
from Lambe and Whitman, 1969)

Relative Density, D_r (%)	Classification
0–15	Very loose
15–35	Loose
35–65	Medium dense[a]
65–85	Dense
85–100	Very dense

[a]Lambe and Whitman used the term "medium," but "medium dense"
is better because "medium" usually refers to the grain size distribution.

such as stress history, mineralogical content, grain-size distribution, and fabric (the configuration of the particles), also affect the engineering properties.

Particle Size

Because soil is a particulate material, it is natural to consider the size of these particles
and their effect on the behavior of the soil. Several different classification schemes are
available, but the one published by ASTM (American Society for Testing and Materials)
is the most common system used by geotechnical engineers. This system classifies soil
particles, as shown in Table 3.4.

TABLE 3.4   ASTM PARTICLE SIZE CLASSIFICATION (Per ASTM D2487)

Sieve Size		Particle Diameter		Soil Classification	
Passes	Retained on	(in)	(mm)		
	12 in	> 12	> 350	Boulder	Rock
12 in	3 in	3–12	75.0–350	Cobble	Fragments
3 in	3/4 in	0.75–3	19.0–75.0	Coarse gravel	
3/4 in	#4	0.19–0.75	4.75–19.0	Fine gravel	
#4	#10	0.079–0.19	2.00–4.75	Coarse sand	Soil
#10	#40	0.016–0.079	0.425–2.00	Medium sand	
#40	#200	0.0029–0.016	0.075–0.425	Fine sand	
#200		< 0.0029	< 0.075	Fines (silt + clay)	

Most natural soils contain a wide variety of particle sizes and thus do not fall completely within any of the categories listed in Table 3.4. Thus, the distribution of particle sizes in a particular soil is most easily expressed in the form of a *grain-size distribution curve*, which is a plot of the percentage of the dry soil by weight that is smaller than a certain particle diameter vs. the particle diameter.

Clays

Soils that consist of silt, sand, or gravel are primarily the result of physical and mild chemical weathering processes, and retain much of the chemical structure of their parent rocks. However, this is not the case with clay soils because they have experienced extensive chemical weathering and have been changed into a new material quite different from the parent rocks.

One of the important consequences of this extensive weathering is that individual clay particles are extremely small (less than 2×10^{-3} mm in diameter). They cannot be seen with optical microscopes, and require an electron microscope for scientific study. This small size is significant because the weights of individual particles are small compared to the ionic forces between particles, so their engineering behavior (shear strength, compressibility, etc.) depends more on these interparticle forces. This is quite different from coarser soils, whose behavior depends primarily on gravitational forces. Another important difference between clays and other soils is that they offer much more resistance to the flow of water. This low *hydraulic conductivity* impacts many of its engineering properties.

Because of these many differences, we often have different analysis methods for clays. Many of the discussions in this book treat clays and clayey silts separately from sands, nonplastic silts, and gravels.

Plasticity and the Atterberg Limits

The moisture content, w, is a basic and useful indicator of soil properties, especially in cohesive soils. For example, clays with a low moisture content are stronger and less compressible than those with a high moisture content.

In 1911, the Swedish soil scientist Albert Atterberg (1846–1916) developed a series of tests to evaluate the relationship between moisture content and soil consistency. In the 1930s, Karl Terzaghi and Arthur Casagrande adapted these tests for civil engineering purposes and they soon became a routine part of geotechnical engineering. This series includes three separate tests: the *liquid limit test*, the *plastic limit test*, and the *shrinkage limit test* (ASTM D427 and D4318). Together they are known as the *Atterberg limits tests*. Figure 3.2 shows qualitative descriptions of the changes in consistency of a fine-grained soil that occur as its moisture content changes. Dry cohesive soils are hard and brittle, whereas wet cohesive soils are soft and pliable. Although these changes in consistency are gradual, the Atterberg limits tests define the boundaries between the various

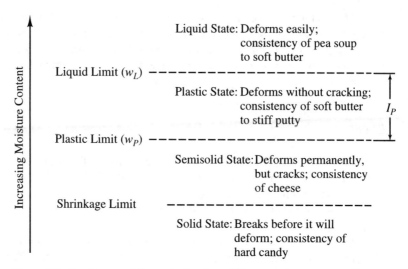

Figure 3.2   Consistency of fine-grained soils at different mositure contents (Sowers, 1979).

states in a somewhat arbitrary but standardized way. The test results are expressed in terms of the moisture content with the percent sign dropped:

w_s = shrinkage limit (or SL)

w_p = plastic limit (or PL)

w_L = liquid limit (or LL)

By comparing the moisture content of a soil with its Atterberg limits, an engineer could gain a qualitative sense of its consistency. For example, if a certain soil has a liquid limit of 55, a plastic limit of 20, and a moisture content of 25%, then it would have a consistency comparable to that of a stiff putty (i.e., it is slightly wetter than the plastic limit).

The liquid limit and plastic limit tests are part of many laboratory test programs. They are inexpensive and the results can be quite useful. In contrast, the shrinkage limit has little practical significance for engineers and is rarely measured.

When the soil is at a moisture content between the liquid limit and the plastic limit, it is said to be in a *plastic state*. It can be easily molded without cracking or breaking. The children's toy play-dough has a similar consistency. This property relates to the amount and type of clay in the soil. The *plasticity index,* PI or I_p, is a measure of the range of moisture contents that encompass the plastic state:

$$\boxed{I_p = w_L - w_p}$$ (3.6)

Soils with a large clay content retain this plastic state over a wide range of moisture contents, and thus have a high plasticity index. The opposite is true of silty soils. Clean sands and gravels are considered to be nonplastic (NP).

The Atterberg limits give the engineer a feel for the soil's behavior, and they form the basis for the Unified Soil Classification System (described in the next section). Engineers also have developed empirical correlations between the Atterberg limits and soil properties such as compressibility and shear strength.

3.2 SOIL CLASSIFICATION

Standardized systems of classifying soil are very important, and many such systems have been developed. A proper classification reveals much useful information to a foundation engineer.

Unified Soil Classification System

The most common soil classification system for foundation engineering problems is the *Unified Soil Classification System* (USCS) [ASTM D2487]. This system assigns a two or four-letter *group symbol* to the soil, along with standardized descriptions called the *group name*. Several potential group names are associated with each group symbol. For example, a certain soil might be classified as "SC–Clayey sand," where SC is the group symbol and clayey sand is the group name.

The first letter of the group symbol tells the general type of soil:

G = gravel
S = sand
M = silt
C = clay
O = organic

The second letter is a supplementary description:

W = well-graded
P = poorly-graded
M = silty
C = clayey
L = low plasticity
H = high plasticity

A special group symbol, Pt, is assigned to peat, which is a highly organic soil that is generally unsuitable for supporting structural foundations.

Figure 3.3 presents an abbreviated outline of the USCS group symbols. See *Geotechnical Engineering: Principles and Practices* or ASTM D2487 for a complete explanation of group symbols and group names.

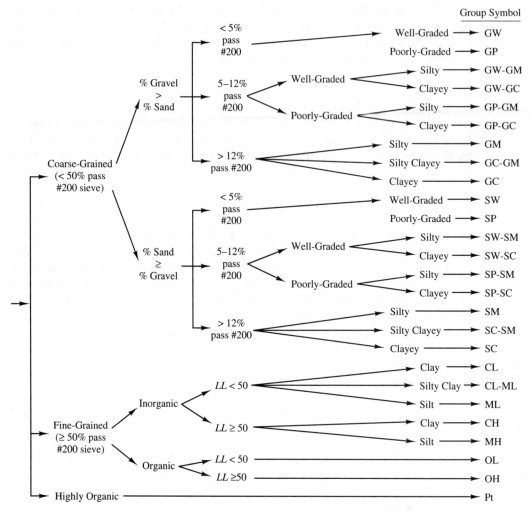

Figure 3.3 Abbreviated outline of the Unified Soil Classification System (Adapted from ASTM D2487. Used with permission of ASTM).

QUESTIONS AND PRACTICE PROBLEMS

3.1 Explain the difference between moisture content and degree of saturation.

3.2 A certain saturated sand ($S = 100\%$) has a moisture content of 25.1% and a specific gravity of solids of 2.68. It also has a maximum index void ratio of 0.84 and a minimum index void ratio of 0.33. Compute its relative density and classify its consistency.

3.3 Consider a soil that is being placed as a fill and compacted using a sheepsfoot roller (a piece of construction equipment). Will the action of the roller change the void ratio of the soil? Explain.

3.4 A sample of soil has a volume of 0.45 ft^3 and a weight of 53.3 lb. After being dried in an oven, it has a weight of 45.1 lb. It has a specific gravity of solids of 2.70. Compute its moisture content and degree of saturation before it was placed in the oven.

3.3   GROUNDWATER

The presence of water in soil has a dramatic impact on its behavior. Karl Terzaghi (1939) once said:

> . . . in engineering practice, difficulties with soils are almost exclusively due not to the soils themselves but to the water contained in their voids. On a planet without any water there would be no need for soil mechanics.

The adverse effects of water in foundation engineering problems include:

- Softening of clay bonds with resulting decrease in strength and increase in compressibility
- Reduction of effective stress, with corresponding decrease in shear strength
- Expansion (in the case of certain clays) or collapse (in the case of certain loose, dry soils) when dry soils become wetted
- Hydrostatic uplift pressures

Therefore, the assessment of groundwater conditions is an important part of foundation engineering.

Groundwater Table

Groundwater conditions are often very complex, and can be the subject of extensive investigation and monitoring. These complexities can include *artesian* conditions, *perched groundwater*, and complex interbedded *aquicludes* and *aquifers*. However, in this book we will consider only simple groundwater conditions defined solely by the position of a horizontal *groundwater table* (also called the *phreatic surface*). It represents the level to which groundwater would rise in an observation well, as shown in Figure 3.4. In this simple groundwater profile, all of the soil voids below the groundwater table are completely filled with water ($S = 100\%$).

Pore Water Pressure

The *pore water pressure*, u, is the pressure in the water within the soil "pores" (voids). There are two sources of pore water pressure: hydrostatic pore water pressure and excess pore water pressure.

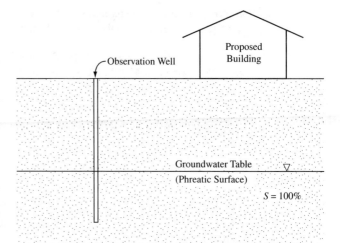

Figure 3.4 Simple groundwater conditions defined solely by the position of the groundwater table. An observation well is a perforated pipe placed in a boring for the purpose of locating and monitoring the groundwater table.

Hydrostatic Pore Water Pressure

The *hydrostatic pore water pressure*, u_h, is that solely caused by the force of gravity acting on the pore water. In the simple groundwater conditions just defined, the hydrostatic pore water pressure is:

$$u_h = \gamma_w z_w \qquad (3.7)$$

Where:

u_h = hydrostatic pore water pressure at a point in the ground

γ_w = unit weight of water = 62.4 lb/ft³ = 9.8 kN/m³

z_w = vertical distance from groundwater table to the point

For example, if the groundwater table is at elevation 215 ft and the point is at elevation 190 ft, then $u_h = \gamma_w z_w = (62.4 \text{ lb/ft}^3)(215 - 190 \text{ ft}) = 1600 \text{ lb/ft}^2$.

Excess Pore Water Pressure

If the soil is in the process of settling, heaving, or shearing, the resulting squeezing or expansion of the soil voids can produce an additional pressure called the *excess pore water pressure*, u_e. The magnitude of u_e depends on many factors, including:

- The rate of application of the normal and/or shear loads that are causing the void size to change

- The change in the void size
- The rate of drainage into or out of the voids

The excess pore water pressure is positive when the voids are in the process of becoming smaller, as when the soil is settling, and negative when the voids are becoming larger, as when the soil is heaving. Shear strains can produce either positive or negative excess pore water pressures.

The pore water pressure, u, is the sum of the hydrostatic and excess pore water pressures:

$$u = u_h + u_e \tag{3.8}$$

Since excess pore water pressures are the consequence of settling, heaving, or shearing, they dissipate when these processes are completed. Therefore, excess pore water pressures are always temporary phenomena.

Although it is sometimes possible to explicitly consider excess pore water pressures in geotechnical analyses (such as when performing consolidation rate analyses), we will not be doing so in this book. Our discussions of excess pore water pressures will be limited to helping us understand the physical processes that control soil behavior. All of the pore water pressure computations in this book will consider only the hydrostatic pressure. Thus, for our purposes:

$$\boxed{u = \gamma_w z_w} \tag{3.9}$$

The impact of excess pore water pressures on soil properties will be considered implicitly, as described later in this chapter.

3.4 STRESS

Many foundation engineering analyses require a knowledge of the stresses in the soil. These stresses have two kinds of sources:

- *Geostatic stresses* are the result of the force of gravity acting directly on the soil mass.
- *Induced stresses* are caused by external loads, such as foundations.

Both normal and shear stresses may be present in a soil. These are represented by the variables σ and τ, respectively. We will identify the direction of normal stresses using subscripts σ_x and σ_y for the two perpendicular horizontal stresses (which, for foundation design purposes are usually assumed to be equal in magnitude) and σ_z is the vertical stress.

Normal stresses in the ground are nearly always compressive, so geotechnical engineers use a sign convention opposite that used by structural engineers: compressive stresses are positive, while tensile stresses are negative. This way we virtually always work with positive numbers.

When using English units, express stresses in units of lb/ft^2 in the soil and lb/in^2 in structural members. With SI units, use kPa in the soil and MPa in structural members. Engineers in some non-SI metric countries use units of kg$_f$/cm^2 or bars (1 bar = 100 kPa).

Geostatic Stresses

The most important geostatic stress is the vertical compressive stress because it is directly caused by gravity. Geotechnical engineers compute this stress more often than any other. The geostatic vertical total stress, σ_z, is:

$$\sigma_z = \Sigma \, \gamma H$$

(3.10)

Where:

σ_z = geostatic vertical total stress

γ = unit weight of soil stratum

H = thickness of soil stratum

We carry this summation from the ground surface down to the point at which the stress is to be computed. This computation is similar to the procedure for computing pressure in a body of water, except for the additional consideration that unit weight may vary with depth.

Part of the vertical total stress is carried by the solid particles, and the rest is carried by the pore water. We are especially interested in the part carried by the solids, and have given it a special name: the *effective stress*, σ'. It is computed as follows:

$$\sigma' = \sigma - u$$

(3.11)

In the case of vertical stress:

$$\sigma'_z = \sigma_z - u$$

(3.12)

Where:

σ_z' = vertical effective stress

σ_z = vertical total stress

u = pore water pressure

Combining equations 3.10 and 3.12 gives:

$$\sigma'_z = \Sigma \, \gamma H - u$$

(3.13)

Note how the distribution of force between the solids and water is *not* proportional to their respective cross-sectional areas.

Horizontal Stress

The geostatic horizontal stress also is important for many engineering analyses. For example, the design of retaining walls depends on the horizontal stresses in the soil being retained. It may be expressed either as total horizontal stress, σ_x, or effective horizontal stress, σ_x'. The ratio of the horizontal to vertical effective stresses is defined as the *coefficient of lateral earth pressure, K*:

$$K = \frac{\sigma_x'}{\sigma_z'}$$
(3.14)

The value of K in undisturbed ground is known as the *coefficient of lateral earth pressure at rest*, K_0, and can vary between about 0.2 and 6. Typical values are between 0.35 and 0.7 for normally consolidated soils and between 0.5 and 3 for overconsolidated soils.[1] The most accurate way to determine K_0 is by measuring σ_x in-situ using methods such as the pressuremeter, dilatometer, or stepped blade, and combining it with computed values of σ_z and u. However, these methods are not commonly used in North America, and are probably justified only for critical projects. A less satisfactory method is to measure K_0 in the laboratory using a triaxial compression machine or other equipment. Unfortunately, this method suffers from problems because of sample disturbance, stress history, and other factors.

The most common method of assessing K_0 is by using empirical correlations with other soil properties. Several such correlations have been developed, including the following by Mayne and Kulhawy (1982), which is based on laboratory tests on 170 soils that ranged from clay to gravel. This formula is applicable only when the ground surface is level:

$$K_0 = (1 - \sin\phi')\text{OCR}^{\sin\phi'}$$
(3.15)

Where:

K_0 = coefficient of lateral earth pressure at rest

ϕ' = effective friction angle (see definition later in this chapter)

OCR = overconsolidation ratio (see definition later in this chapter)

Chapter 23 discusses lateral earth pressures in more detail.

---

[1]The terms "normally consolidated soils" and "overconsolidated soils" are defined later in this chapter.

Induced Stresses

Induced stresses are those caused by external loads, such as structural foundations. These stresses are a very important part of foundation engineering, and most of the related geotechnical analyses focus on the soil's response to these stresses. We will discuss them in the context of the various types of foundations.

Example 3.2

The soil profile beneath a certain site consists of 5.0 m of silty sand underlain by 13.0 m of clay. The groundwater table is at a depth of 2.8 m below the ground surface. The sand has a unit weight of 19.0 kN/m^3 above the groundwater table and 20.0 kN/m^3 below. The clay has a unit weight of 15.7 kN/m^3, an effective friction angle of 35°, and an overconsolidation ratio of 2. Compute σ_x, σ'_x, σ_z, and σ_z. at a depth of 11.0 m.

Solution

$$\sigma_z = \Sigma \gamma H$$

$$= (19.0 \text{ kN/m}^3)(2.8 \text{ m}) + (20.0 \text{ kN/m}^3)(2.2 \text{ m}) + (15.7 \text{ kN/m}^3)(6.0 \text{ m})$$

$$= \textbf{191 kPa} \quad \Leftarrow Answer$$

$$u = \gamma_w z_w = (9.8 \text{ kN/m}^3)(8.2 \text{ m}) = 80 \text{ kPa}$$

$$\sigma'_z = \sigma_z - u = 191 \text{ kPa} - 80 \text{ kPa} = \textbf{111 kPa} \quad \Leftarrow Answer$$

$$K_0 = (1 - \sin \phi')\text{OCR}^{\sin\phi'} = (1 - \sin 35°)2^{\sin 35°} = 0.635$$

$$\sigma'_x = K\sigma'_z = (0.635)(111 \text{ kPa}) = \textbf{70 kPa} \quad \Leftarrow Answer$$

$$\sigma_x = \sigma'_x + u = 70 \text{ kPa} + 80 \text{ kPa} = \textbf{150 kPa} \quad \Leftarrow Answer$$

QUESTIONS AND PRACTICE PROBLEMS

3.5 A site is underlain by a soil that has a unit weight of 18.7 kN/m^3 above the groundwater table and 19.9 kN/m^3 below. The groundwater table is located at a depth of 3.5 m below the ground surface. Compute the total vertical stress, pore water pressure, and effective vertical stress at the following depths below the ground surface:

   a. 2.2 m

   b. 4.0 m

   c. 6.0 m

3.6 The subsurface profile at a certain site is shown in Figure 3.5. Compute u, σ_x, σ_z, σ_x', and σ_z' at Point A.

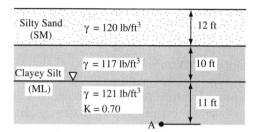

Figure 3.5 Soil profile for Problem 3.6.

3.5   COMPRESSIBILITY AND SETTLEMENT

Settlement requirements often control the design of foundations, as discussed in Chapter 2, so we need to have methods of computing the settlement in soils. In this section we will review the physical processes that control settlement and the methods of computing settlement due to the weight of extensive fills. Chapters 7 and 14 use these concepts to compute settlements due to the loads applied to foundations.

Physical Processes

Settlement is the result of various physical processes. These include the following:

- **Consolidation settlement**—When a fill is placed on the ground, the vertical stress in the underlying soil increases. This increase causes the solid particles to pack together more tightly, which in turn causes the soil to settle. We call this process *consolidation*, and the resulting settlement is *consolidation settlement*.

- **Secondary compression settlement**—The process called *secondary compression* is the result of creep, viscous behavior of the clay-water system, the compression and decomposition of organic matter, and other physical and chemical processes. *Secondary compression settlement* is negligible in sands and in overconsolidated clays. It can be important in some normally consolidated clays, and is usually very important in organic soils.

- **Distortion settlement**—When heavy loads are applied over a small area, the soil can deform laterally. Similar lateral deformations also can occur near the perimeter of larger loaded areas. These deformations produce *distortion settlement*, which can be important in foundations. Chapter 7 discusses this topic in more detail.

- **Settlement caused by compression or collapse of underground mines, sinkholes, or tunnels**—These can be important at some sites, but are beyond the scope of this book.

- **Settlement caused by hydrocollapse**—This occurs in certain soils when they become wetted. We will discuss this phenomena in Chapter 20.

- **Settlement or heave caused by wetting or drying of expansive soils**—This occurs in certain clayey soils, and is discussed in Chapter 19.

Consolidation

Consolidation is usually the most important source of settlement, and it is the only one we will discuss in this chapter. The consolidation process begins with the application of a load, which produces an increase in the vertical stress, $\Delta\sigma_z$, in the underlying ground. If the load is caused by the weight of an extensive fill, then:

$$\boxed{\Delta\sigma_z = \gamma_{fill}H_{fill}}$$

        (3.16)

Where:

    $\Delta\sigma_z$ = change in vertical stress in the ground beneath the fill

    γ_{fill} = unit weight of the fill

    H_{fill} = thickness of the fill

If the length and width of the fill are large (i.e., an *extensive* fill), then $\Delta\sigma_z$ is essentially constant with depth.

      The introduction of $\Delta\sigma_z$ causes an immediate increase in the vertical total stress, σ_z, in the underlying soils. In addition, if these soils are saturated, $\Delta\sigma_z$ also causes an equal amount of excess pore water pressure, u_e. In other words, immediately after the fill is placed, its weight is carried entirely by the pore water. This means the vertical effective stress immediately after loading is unchanged from its pre-construction value.

      The presence of these excess pore water pressures produces a hydraulic gradient, which forces some of the pore water to flow out of the soil. As the water flows out, the soil consolidates and $\Delta\sigma_z$ is slowly transferred from the pore water to the solids. Eventually, the excess pore water pressure becomes zero and the vertical effective stress increases to its final value, σ_{zf}':

$$\boxed{\sigma_{zf}' = \sigma_{z0}' + \Delta\sigma_z}$$

        (3.17)

Where:

    σ_{z0}' = initial vertical effective stress at a point in the soil beneath the fill

    σ_{zf}' = final vertical effective stress at a point in the soil beneath the fill

    $\Delta\sigma_z$ = change in vertical stress

The consolidation settlement, δ_c, is the result of this increase in vertical effective stress.

Consolidation (Oedometer) Tests

To predict the magnitude of δ_c in a soil, we need to know its stress–strain properties. This normally required obtaining soil samples in the field, bringing them to the laboratory, subjecting them to a series of loads, and measuring the corresponding settlements. We do

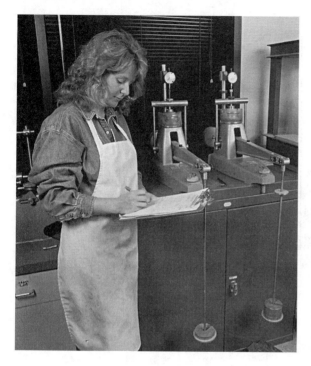

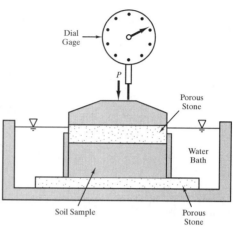

Figure 3.6 (a) Performing consolidation tests in the laboratory. The two consolidometers use the weights in the foreground to load the samples; (b) cross section of a consolidometer.

this by conducting a *consolidation test* (also known as an *oedometer test*), which is performed in a *consolidometer* (or *oedometer*) as shown in Figure 3.6.

Because we are mostly interested in the engineering properties of natural soils as they exist in the field, consolidation tests are usually performed on high quality "undisturbed" samples. It is fairly simple to obtain these samples in soft to medium clays, but quite difficult in clean sands. It also is important for samples that were saturated in the field to remain so during storage and testing, because irreversible changes can occur if they are allowed to dry.

Sometimes engineers need to evaluate the consolidation characteristics of proposed compacted fills, and do so by performing consolidation tests on samples that have been remolded and compacted in the laboratory. These tests are usually less critical because well-compacted fills generally have a low compressibility.

Test Procedure and Results

The test procedure consists of applying a series of normal loads to the sample, allowing it to consolidate under each load, and measuring the corresponding settlements. The results are presented in a semilogarithmic plot as shown in Figure 3.7. This plot represents consolidation both in terms of vertical strain, ϵ_z, and change in void ratio, e. The first part of the test results, indicated by Curve AB, is called the *recompression curve*. The middle part, indicated by Curve BC, is called the *virgin curve*. Upon reaching Point C, the sample is progressively unloaded, thus producing the *rebound curve*, CD.

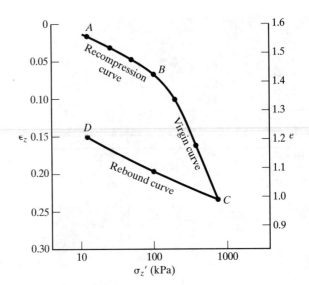

Figure 3.7 Results of a laboratory consolidation test. The initial void ratio, e_0, is 1.60.

The slope of the virgin curve is called the *compression index*, C_c:

$$C_c = -\frac{d\,e}{d\log\sigma_z'}$$

(3.18)

Since the virgin curve is a straight line (on a semilogarithmic plot), we can obtain a numerical value for C_c by selecting any two points, a and b, on this line, as shown in Figure 3.8, and rewriting Equation 3.18 as:

$$C_c = -\frac{e_a - e_b}{(\log\sigma_z')_b - (\log\sigma_z')_a}$$

(3.19)

Equation 3.19 is used when the test results have been plotted in terms of void ratio $(e - \sigma_z')$. Alternatively, if the data is plotted only in vertical strain form $(\epsilon_z - \sigma_z')$ form then:

$$\frac{C_c}{1 + e_0} = \frac{(\epsilon_z)_b - (\epsilon_z)_a}{(\log\sigma_z')_b - (\log\sigma_z')_a}$$

(3.20)

Where:

e_0 = initial void ratio (i.e., at beginning of test)

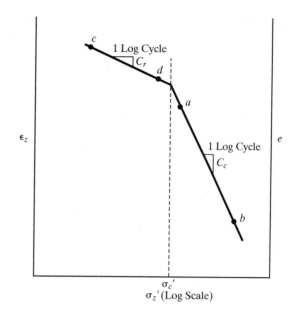

Figure 3.8 The slopes of consolidation curves on a semilogarithmic e vs. σ_z' plot are C_c and C_r. The break in slope occurs at the preconsolidation stress, σ_c'.

When using Equations 3.19 and 3.20, it is convenient to select points a and b such that $\log (\sigma_z')_b = 10 \log (\sigma_z')_a$. This makes the denominator of both equations equal to 1, which simplifies the computation. This choice also demonstrates that C_c could be defined as the reduction in void ratio per tenfold increase (one log-cycle) in effective stress.

In theory, the recompression and rebound curves have nearly equal slopes, but the rebound curve is more reliable because it is less sensitive to sample disturbance effects. This slope, which we call the *recompression index*, C_r, is defined in the same way as C_c and can be found using Equations 3.21 or 3.22 with points c and d on the decompression curve.

Using void ratio data:

$$C_r = \frac{e_c - e_d}{(\log \sigma_z')_d - (\log \sigma_z')_c} \qquad (3.21)$$

Using strain data:

$$\frac{C_r}{1 + e_0} = \frac{(\epsilon_z)_d - (\epsilon_z)_c}{(\log \sigma_z')_d - (\log \sigma_z')_c} \qquad (3.22)$$

Another important parameter from the consolidation test results is the stress that corresponds to the break-in-slope in Figure 3.9. This is called the *preconsolidation stress*, σ_c', and represents the greatest vertical effective stress the soil has ever experienced. The value of σ_c' obtained from the consolidation test represents only the conditions at the point where the sample was obtained. If the sample had been taken at a different eleva-tion, the preconsolidation stress would change accordingly.

Consolidation Status in the Field

Normally Consolidated, Overconsolidated, and Underconsolidated Soils

When performing consolidation analyses, we need to compare the preconsolidation stress, σ_c', with the initial vertical effective stress, σ_{z0}'. The former is determined from laboratory test data, as described earlier, while the later is determined using Equation 3.13 with the original field conditions (i.e., without the new load) and the original hydrostatic pore water pressures (i.e., Equation 3.9). Both values must be determined at the same depth, which normally is the depth of the sample on which the consolidation test was performed. Once these values have been determined, we need to assess which of the following three conditions exist in the field:

1. If $\sigma_{z0}' \approx \sigma_c'$, then the vertical effective stress in the field has never been higher than its current magnitude. This condition is known as being *normally consolidated* (NC). For example, this might be the case at the bottom of a lake, where sediments brought in by a river have slowly accumulated. In theory these two values should be exactly equal, but both are subject to error because of sample disturbance and other factors. Therefore, we will consider the soil to be normally consolidated if they are equal within about ± 20%.

2. If $\sigma_{z0}' < \sigma_c'$, then the vertical effective stress in the field was once higher than its current magnitude. This condition is known as being *overconsolidated* (OC) or *pre-consolidated*. There are many processes that can cause a soil to become overconsol-idated, including (Brumund, et al., 1976):
   - Extensive erosion or excavation such that the ground surface was once much higher than its current elevation.
   - Surcharge loading from a glacier, which has since melted.
   - Surcharge loading from a structure, such as a storage tank, which has since been removed.
   - Increases in the pore water pressure, such as from a rising groundwater table.
   - Dessiccation (drying) and the resulting surface tension forces in the remaining pore water.
   - Chemical changes in the soil.

   The term *overconsolidated* can be misleading since it implies there has been exces-sive consolidation. Although there are a few situations, such as cut slopes, where heavily overconsolidated soils are less desirable, overconsolidation is almost always a good thing.

3. If $\sigma_{z0}' > \sigma_c'$, the soil is said to be *underconsolidated*, which means the soil is still in the process of consolidating under a previously applied load. We will not be dealing with this case.

TABLE 3.5   CLASSIFICATION OF SOIL COMPRESSIBILITY

$\dfrac{C_c}{1+e_0}$ or $\dfrac{C_r}{1+e_0}$	Classification
0–0.05	Very slightly compressible
0.05–0.10	Slightly compressible
0.10–0.20	Moderately compressible
0.20–0.35	Highly compressible
> 0.35	Very highly compressible

Table 3.5 gives a classification of soil compressibility based on C_c / $(1+e_0)$ for normally consolidated soils or C_r / $(1+e_0)$ for overconsolidated soils.

Overconsolidation Margin and Overconsolidation Ratio

The σ_c' values from the laboratory only represent the preconsolidation stress at the sample depth. To estimate σ_c' at other depths in the same strata (i.e., in a soil strata with the same geologic origin), compute the *overconsolidation margin*, σ_m', at the sample depth using:

$$\boxed{\sigma_m' = \sigma_c' - \sigma_{z0}'}\tag{3.23}$$

The overconsolidation margin should be approximately constant in a strata with common geologic origins, so we can estimate the preconsolidation stress at other depths in that strata by using Equation 3.23 with σ_{z0}' at the desired depth. In normally consolidated soils, $\sigma_m' = 0$. Table 3.6 presents typical ranges of σ_m'.

Another useful parameter is the *overconsolidation ratio* or OCR:

$$\boxed{\text{OCR} = \frac{\sigma_c'}{\sigma_{z0}'}}\tag{3.24}$$

TABLE 3.6   TYPICAL RANGES OF OVERCONSOLIDATION MARGINS

Overconsolidation Margin, σ_m'		Classification
(kPa)	(lb/ft^2)	
0	0	Normally consolidated
0–100	0–2000	Slightly overconsolidated
100–400	2000–8000	Moderately overconsolidated
> 400	> 8000	Heavily overconsolidated

Unlike the overconsolidation margin, the OCR is not constant with depth in a given strata. For normally consolidated soils, OCR = 1.

Example 3.3

A proposed fill is to be placed on the soil profile shown in Figure 3.9. A consolidation test has been performed on a sample obtained from Point A, and this test indicates the preconsolidation pressure at that point is 200 kPa. Determine whether the silty clay stratum is normally consolidated or overconsolidated, and compute the overconsolidation margin and overconsolidation ratio. The proposed fill has not yet been placed.

Solution

At Point A:

$$\sigma'_{z0} = \Sigma \gamma H - u$$
$$= (19.1 \text{ kN/m}^3)(1.2 \text{ m}) + (20.2 \text{ kN/m}^3)(1.0 \text{ m})$$
$$+ (17.8 \text{ kN/m}^3)(5.0 \text{ m}) - (9.8 \text{ kN/m}^3)(6.0 \text{ m})$$
$$= 73 \text{ kPa}$$

$\sigma'_{z0} < \sigma'_c$ by more than 20%, so **soil is overconsolidated**   ⇐ *Answer*

$$\sigma'_m = \sigma'_c - \sigma'_{z0} = 200 \text{ kPa} - 73 \text{ kPa} = \textbf{127 kPa} \qquad ⇐ Answer$$

$$\text{OCR} = \frac{\sigma'_c}{\sigma'_{z0}} = \frac{200 \text{ kPa}}{73 \text{ kPa}} = \textbf{2.7} \qquad ⇐ Answer$$

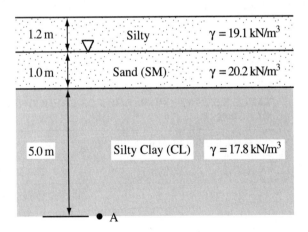

Figure 3.9 Soil profile for Example 3.3.

Compressibility of Sands and Gravels

The principles of consolidation apply to all soils, but the consolidation test is primarily applicable to clays and silts. It is very difficult to perform reliable consolidation tests on most sands because they are more prone to sample disturbance, and this disturbance has a significant effect on the test results. Clean sands are especially troublesome. Gravels have similar sample disturbance problems, plus their large grain size requires very large samples and a very large consolidometer, neither of which is practical.

Fortunately, sands and gravels subjected to static loads are much less compressible than silts and clays, so it often is sufficient to use estimated values of C_c or C_r in lieu of laboratory tests. For sands, these estimates can be based on the data gathered by Burmister (1962) as interpreted in Table 3.7. He performed a series of consolidation tests on samples reconstituted to various relative densities. Engineers can estimate the in-situ relative density using the methods described in Chapter 4, then select an appropriate $C_c/(1 + e_0)$ from this table. Note that all of these values are "very slightly compressible" as defined in Table 3.5.

For saturated overconsolidated sands, $C_r / (1+e_0)$ is typically about one-third of the values listed in Table 3.7, which makes such soils nearly incompressible. Compacted

TABLE 3.7 TYPICAL CONSOLIDATION PROPERTIES OF SATURATED NORMALLY CONSOLIDATED SANDY SOILS AT VARIOUS RELATIVE DENSITIES (Adapted from Burmister, 1962)

Soil Type	$C_c / (1+e_0)$					
	$D_r = 0\%$	$D_r = 20\%$	$D_r = 40\%$	$D_r = 60\%$	$D_r = 80\%$	$D_r = 100\%$
Medium to coarse sand, some fine gravel (SW)	—	—	0.005	—	—	—
Medium to coarse sand (SW/SP)	0.010	0.008	0.006	0.005	0.003	0.002
Fine to coarse sand (SW)	0.011	0.009	0.007	0.005	0.003	0.002
Fine to medium sand (SW/SP)	0.013	0.010	0.008	0.006	0.004	0.003
Fine sand (SP)	0.015	0.013	0.010	0.008	0.005	0.003
Fine sand with trace fine to coarse silt (SP-SM)	—	—	0.011	—	—	—
Find sand with little fine to coarse silt (SM)	0.017	0.014	0.012	0.009	0.006	0.003
Fine sand with some fine to coarse silt (SM)	—	—	0.014	—	—	—

fills can be considered to be overconsolidated, as can soils that have clear geologic evidence of preloading, such as glacial tills. Therefore, many settlement analyses simply consider the compressibility of such soils to be zero. If it is unclear whether a soil is normally consolidated or overconsolidated, it is conservative to assume it is normally consolidated.

Very few consolidation tests have been performed on gravelly soils, but the compressibility of these soils is probably equal to or less than those for sand, as listed in Table 3.7.

Another characteristic of sands and gravels is their high hydraulic conductivity, which means any excess pore water drains very quickly. Thus, the rate of consolidation is very fast, and typically occurs nearly as fast as the load is applied. Therefore, if the load is caused by a fill, the consolidation of these soils may have little practical significance.

Another way to assess the compressibility of sands is to use in-situ tests. We will discuss these test methods in Chapter 4, and will apply them to sand compressibility in Chapter 7. This method is especially useful for settlements due to loads on foundations.

Consolidation Settlement Predictions

The purpose of performing consolidation tests is to define the stress–strain properties of the soil and thus allow us to predict consolidation settlements in the field. We perform this computation by projecting the laboratory test results (as contained in the parameters C_c, C_r, e_0, and σ_c') back to the field conditions.

For simplicity, the discussions of consolidation settlement predictions in this chapter consider only the case of one-dimensional consolidation, and we will be computing only the ultimate consolidation settlement. *One-dimensional consolidation* means only vertical strains occur in the soil (i.e., $\epsilon_x = \epsilon_y = 0$). This is a reasonable assumption when computing settlements due to the weight of fills, but is not quite true for settlements due to loads on foundations. We will return to this issue in Chapter 7. The *ultimate consolidation settlement* is the value of δ_c at the end of the consolidation process.

Normally Consolidated Soils ($\sigma_{z0}' \approx \sigma_c'$)

If $\sigma_{z0}' \approx \sigma_c'$, the soil is, by definition, normally consolidated. Thus, the initial and final conditions are as shown in Figure 3.10, and the compressibility is defined by C_c, the slope of the virgin curve.

To compute the ultimate consolidation settlement, we divide the soil into layers, compute the settlement of each layer, and sum as follows:

$$\delta_c = \Sigma \frac{C_c}{1 + e_0} H \log \left(\frac{\sigma_{zf}'}{\sigma_{z0}'} \right) \qquad (3.25)$$

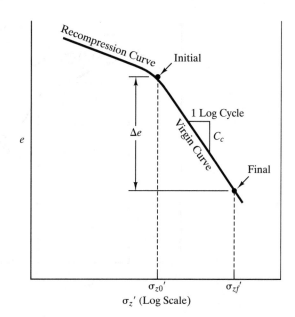

Figure 3.10 Consolidations of normally consolidated soils.

Where:

δ_c = ultimate consolidation settlement at the ground surface

C_c = compression index

e_0 = initial void ratio

H = thickness of soil layer

σ_{z0}' = initial vertical effective stress

σ_{zf}' = final vertical effective stress

When using Equation 3.25, compute σ_{z0}' and σ_{zf}' at the midpoints of each layer.

Overconsolidated Soils — Case I ($\sigma_{z0}' < \sigma_{zf}' \le \sigma_c'$)

If both σ_{z0}' and σ_{zf}' do not exceed σ_c', the entire consolidation process occurs on the recompression curve as shown in Figure 3.11. The analysis is thus identical to that for normally consolidated soils except we use the recompression index, C_r instead of the compression index, C_c:

$$\delta_c = \Sigma \frac{C_r}{1 + e_0} H \log \left(\frac{\sigma_{zf}'}{\sigma_{z0}'} \right)$$ (3.26)

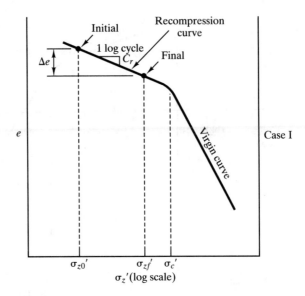

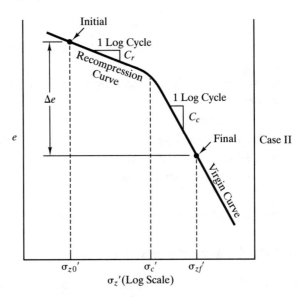

Figure 3.11 Consolidation of overconsolidated soils.

Overconsolidated Soils — Case II ($\sigma_{z0}' < \sigma_c' < \sigma_{zf}'$)

If the consolidation process begins on the recompression curve and ends on the virgin curve, as shown in Figure 3.11, then the analysis must consider both C_c and C_r:

$$\delta_c = \Sigma \left[\frac{C_r}{1 + e_0} H \log \left(\frac{\sigma_c'}{\sigma_{z0}'} \right) + \frac{C_c}{1 + e_0} H \log \left(\frac{\sigma_{zf}'}{\sigma_c'} \right) \right] \qquad (3.27)$$

This condition is quite common, since many soils that might appear to be normally consolidated from a geologic analysis actually have a small amount of overconsolidation (Mesri, Lo, and Feng, 1994).

Ultimate Consolidation Settlement Analysis Procedure

Use the following procedure to compute δ_c caused by the weight of extensive fills:

1. Beginning at the original ground surface, divide the soil profile into strata, where each stratum consists of a single soil type with a common geologic origin. For example, one stratum may consist of a dense sand, while another might be a soft-to-medium clay. Continue downward with this process until you have passed through all of the compressible strata (i.e., until you reach bedrock or some very hard soil). For each stratum, identify the unit weight, γ. Note: Boring logs usually report the dry unit weight, γ_d, and moisture content, w, but we can compute γ from this data using Equation 3.3. Also define the location of the groundwater table.

2. Each clay or silt stratum must have results from at least one consolidation test (or at least estimates of these results). Using the techniques described earlier, determine if each stratum is normally consolidated or overconsolidated, then assign values for $C_c/(1+e_0)$ and/or $C_r/(1+e_0)$. For each overconsolidated stratum, compute σ_m' using Equation 3.23 and assume it is constant throughout that stratum. For normally consolidated soils, set $\sigma_m' = 0$.

3. For each sand or gravel stratum, assign a value for $C_c/(1+e_0)$ or $C_r/(1+e_0)$ based on the information in Table 3.7.

4. For any very hard stratum, such as bedrock or glacial till, that is virtually incompressible compared to the other strata, assign $C_c = C_r = 0$.

5. Working downward from the original ground surface (i.e., do not consider any proposed fills), divide the soil profile into horizontal layers. Begin a new layer whenever a new stratum is encountered, and divide any thick strata into multiple layers. When performing computations by hand, each strata should have layers no more than 2 to 5 m (5 to 15 ft) thick. Thinner layers are especially appropriate near the ground surface, because the strain is generally larger there. Computer-based computations can use much thinner layers throughout the entire depth, and achieve slightly more precise results.

6. Tabulate the following parameters at the midpoint of each layer:
For normally consolidated strata:

$$\sigma'_{z0}$$

$$\sigma'_{zf}$$

$$C_c/(1 + e_0)$$

$$H$$

For overconsolidated strata:

$$\sigma'_{z0}$$

$$\sigma'_{zf}$$

$$\sigma'_c = \sigma'_{z0} + \sigma'_m$$

$$C_c/(1 + e_0)$$

$$C_r/(1 + e_0)$$

$$H$$

It is not necessary to record these parameters in incompressible strata.

Normally we compute σ'_{z0} and σ'_{zf} using the hydrostatic pore water pressures (Equation 3.9) with no significant seepage force, and this is the only case we will consider in this book. However, if preexisting excess pore water pressures or significant seepage forces are present, they should be evaluated. Sometimes this may require the installation of piezometers to obtain accurate information on the in-situ pore water pressures.

7. Using Equation 3.25, 3.26, or 3.27, compute the consolidation settlement for each layer, then sum to find δ_c. Note that each layer will not necessarily use the same equation. If σ'_c is only slightly greater than σ'_{z0} (perhaps less than 20 percent greater), then it may not be clear if the soil is truly overconsolidated, or if the difference is only an apparent overconsolidation caused by uncertainties in assessing these two values. In such cases, it is acceptable to use either Equation 3.25 (normally consolidated) or 3.27 (overconsolidated case II).

Example 3.4

A 3.0 m deep compacted fill is to be placed over the soil profile shown in Figure 3.12. A consolidation test on a sample from point A produced the following results:

$$C_c = 0.40$$

$$C_r = 0.08$$

$$e_0 = 1.10$$

$$\sigma'_c = 70.0 \text{ kPa}$$

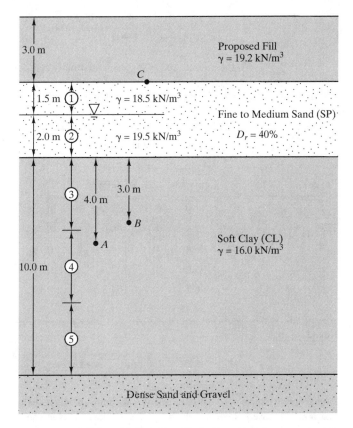

Figure 3.12 Soil profile for Example 3.4.

Compute the ultimate consolidation settlement due to the weight of this fill.

Solution

$$\sigma'_{zf} = \sigma'_{z0} + \Delta\sigma_z$$
$$= \sigma'_{z0} + \gamma_{fill}H_{fill}$$
$$= \sigma'_{z0} + (19.2 \text{ kN/m}^3)(3.0 \text{ m})$$
$$= \sigma'_{z0} + 57.6 \text{ kPa}$$

Compute the initial vertical stress at the sample location:

$$\sigma'_{z0} = \Sigma \gamma H - u$$
$$= (18.5 \text{ kN/m}^3)(1.5 \text{ m}) + (19.5 \text{ kN/m}^3)(2.0 \text{ m})$$
$$+ (16.0 \text{ kN/m}^3)(4.0 \text{ m}) - (9.8 \text{ kN/m}^3)(6.0 \text{ m})$$
$$= 72.0 \text{ kPa}$$

$$\frac{C_c}{1 + e_0} = \frac{0.40}{1 + 1.10} = 0.190$$

At the sample $\sigma'_c \approx \sigma'_{z0} \rightarrow \therefore$ normally consolidated

For the sand strata, use $C_c/(1+e_0) = 0.008$, per Table 3.7

		At midpoint of layer				
Layer	H (m)	σ'_{z0} (kPa) Eqn 3.13	σ'_{zf} (kPa)	$\dfrac{C_c}{1 + e_0}$	Eqn.	$(\delta_c)_{ult}$ (mm)
1	1.5	13.9	71.5	0.008	3.25	8
2	2.0	37.4	95.0	0.008	3.25	6
3	3.0	56.4	114.0	0.19	3.25	174
4	3.0	75.0	132.6	0.19	3.25	141
5	4.0	96.7	154.3	0.19	3.25	154
					$\delta_c =$	483

Round off to:

$$\delta_c = \textbf{480 mm} \qquad \Leftarrow Answer$$

Notice how we have used the same analysis for soils above and below the groundwater table, and both are based on saturated $C_c / (1 + e_0)$ values. This is conservative (although in this case, very slightly so) because the soils above the groundwater table are probably less compressible.

Example 3.5

An 8.5-m deep compacted fill is to be placed over the soil profile shown in Figure 3.13. Consolidation tests on samples from points A and B produced the following results:

	Sample A	Sample B
C_c	0.25	0.20
C_r	0.08	0.06
e_0	0.66	0.45
$\sigma_c{}'$	101 kPa	510 kPa

Compute the ultimate consolidation settlement caused by the weight of this fill.

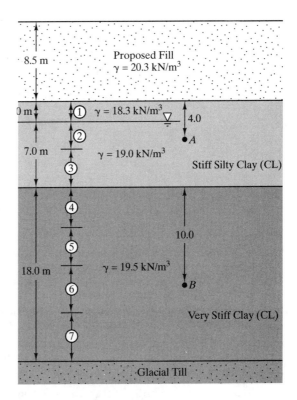

Figure 3.13 Soil profile for Example 3.5.

Solution

$$\sigma'_{zf} = \sigma'_{z0} + \Delta\sigma_z$$

$$= \sigma'_{z0} + \gamma_{fill}H_{fill}$$

$$= \sigma'_{z0} + (20.3 \text{ kN/m}^3)(8.5 \text{ m})$$

$$= \sigma'_{z0} + 172.6 \text{ kPa}$$

At sample A:

$$\sigma'_{z0} = \Sigma \gamma H - u$$

$$= (18.3 \text{ kN/m}^3)(2.0 \text{ m}) + (19.0 \text{ kN/m}^3)(2.0 \text{ m}) - (9.8 \text{ kN/m}^3)(2.0 \text{ m})$$

$$= 55.0 \text{ kPa}$$

$$\sigma'_{zf} = \sigma'_{z0} + \gamma_f H_f$$

$$= 55.0 \text{ kPa} + 172.6 \text{ kPa}$$

$$= 227.6 \text{ kPa}$$

$$\sigma'_{z0} < \sigma'_c \text{ and } \sigma'_{zf} > \sigma'_c \qquad \therefore \text{ overconsolidated case II}$$

At sample B:

$$\sigma'_{z0} = \Sigma \, \gamma H - u$$

$$= (18.3 \text{ kN/m}^3)(2.0 \text{ m}) + (19.0 \text{ kN/m}^3)(7.0 \text{ m})$$

$$+ (19.5 \text{ kN/m}^3)(10.0 \text{ m}) - (9.8 \text{ kN/m}^3)(17.0 \text{ m})$$

$$= 198.0 \text{ kPa}$$

$$\sigma'_{zf} = \sigma'_{z0} + \gamma_{fill} H_{fill}$$

$$= 198.0 \text{ kPa} + 172.6 \text{ kPa}$$

$$= 370.6 \text{ kPa}$$

$$\sigma'_{z0} < \sigma'_c \text{ and } \sigma'_{zf} \leq \sigma'_c \qquad \therefore \text{ overconsolidated case I}$$

Layer	H (m)	At midpoint of layer			$\dfrac{C_r}{1 + e_0}$	$\dfrac{C_c}{1 + e_0}$	Eqn.	δ_c (mm)
		σ_{z0}' (kPa) Eqn. 3.13	σ_c' (kPa) Eqn. 3.21	σ_{zf}' (kPa)				
1	2.0	18.3	64.3	190.9	0.05	0.15	3.27	196
2	3.0	50.4	96.4	223.0	0.05	0.15	3.27	206
3	4.0	82.6	128.6	255.2	0.05	0.15	3.27	217
4	4.0	120.4	—	293.0	0.04	0.14	3.26	62
5	4.0	159.2	—	331.8	0.04	0.14	3.26	51
6	5.0	202.8	—	375.4	0.04	0.14	3.26	53
7	5.0	251.4	—	424.0	0.04	0.14	3.26	45
							$\delta_c =$	830

$$\delta_c = \textbf{830 mm} \qquad \Leftarrow Answer$$

Notice how most of the compression occurs in the upper strata, which is overconsolidated case II (i.e., some of the compression occurs along the virgin curve). The lower strata has much less compression even though it is twice as thick because it is overconsolidated case I and all of the compression occurs on the recompression curve, and because the ratio of σ_{zf}/σ_{z0} is smaller.

QUESTIONS AND PRACTICE PROBLEMS

3.7 A 2-m thick fill is to be placed on the soil shown in Figure 3.14. Once it is compacted, this fill will have a unit weight of 19.5 kN/m^3. Compute the ultimate consolidation settlement.

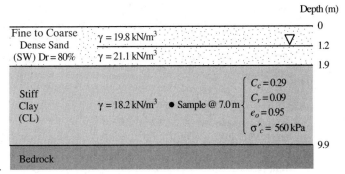

Figure 3.14 Soil profile for Problem 3.7.

3.6 STRENGTH

Structural foundations also induce significant shear stresses into the ground. If these shear stresses exceed the shear strength of the soil or rock, failure occurs. Therefore, we must be able to assess these shear stresses and strengths and design foundations such that the shear stresses are sufficiently smaller than the shear strength.

Sources of Shear Strength in Soil

The shear strength of common engineering materials, such as steel, is controlled by their molecular structure. Failure generally requires breaking the molecular bonds that hold the material together, and thus depends on the strength of these bonds. For example, steel has very strong molecular bonds, and thus has a high shear strength, while plastic has much weaker bonds and a correspondingly lower shear strength.

However, the physical mechanisms that control shear strength in soil are much different. Soil is a particulate material, so shear failure occurs when the stresses between the particles are such that they slide or roll past each other as shown in Figure 3.15. It is not

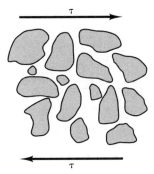

Figure 3.15 Shear failure in soil occurs when the shear stresses are large enough to make the particles roll and slide past each other.

necessary to fracture individual particles. Therefore, the shear strength depends on interactions between the particles, not on their internal strength.

When viewed on a microscopic scale, these interactions are very complex. For simplicity, we will divide them into two broad categories: frictional strength and cohesive strength.

Frictional Strength

Frictional strength is similar to classic sliding friction problems from basic physics. The force that resists sliding is equal to the normal force multiplied by the coefficient of friction, μ, as shown in Figure 3.16.

However, instead of using the coefficient of friction, μ, geotechnical engineers prefer to describe frictional strength using the *effective friction angle* (or *effective angle of internal friction*), ϕ', where:

$$\phi' = \tan^{-1}\mu \tag{3.28}$$

The value of ϕ' depends on both the frictional properties of the individual particles and the interlocking between particles. These are affected by many factors, including:

- **Mineralogy**—The effective friction angle in sands made of pure quartz is typically 30 to 36°. However, the presence of other minerals can change ϕ'. For example, sands contain significant quantities of mica, which is much smoother than quartz, have a smaller ϕ'. These are called *micaceous sands*. Clay minerals are typically even weaker (ϕ' values as low as 4° have been measured in pure montmorillonite).
- **Shape**—The friction angle of angular particles is much higher than that of rounded ones.
- **Gradation**—Well graded soils typically have more interlocking between the particles, and thus a higher friction angle than those that are poorly graded. For example, GW soils typically have ϕ' values about 2° higher than comparable GP soils.

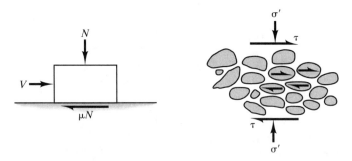

Figure 3.16 Comparison between friction on a sliding block and frictional strength in a soil.

- **Void ratio**—Decreasing the void ratio, such as by compacting a soil with a sheeps-foot roller, also increases interlocking, which results in a higher ϕ'.
- **Organic material**—Organics introduce many problems, including a decrease in the friction angle.

The impact of water on frictional strength is especially important, and many shear failures are induced by changes in the groundwater conditions. However, many people mistakenly believe water-induced changes in shear strength are primarily caused by lubrication effects. Although the process of wetting some dry soils can induce lubrication, the resulting decrease in ϕ' is very small. Sometimes the introduction of water has an antilubricating effect (Mitchell, 1993) and causes a small increase in ϕ'. Nevertheless, focusing on these small effects tends to obscure another far more important process: the impact of groundwater on the effective stress, σ'.

As the groundwater table rises, the pore water pressure, u, increases and the effective stress, σ', decreases. This causes a reduction in frictional strength, just as a reduction in the normal force, N, reduces the frictional resistance μN acting on the bottom of the sliding block in Figure 3.16. This is the primary way groundwater influences frictional strength in soils.

Cohesive Strength

Some soils have shear strength even when the effective stress, σ', is zero, or at least *appears* to be zero. This strength is called the *cohesive strength*, and we describe it using the variable c', the *effective cohesion*. There are two types of cohesive strength: true cohesion and apparent cohesion (Mitchell, 1993).

True cohesion is shear strength that is truly the result of bonding between the soil particles. These bonds include the following:

- *Cementation* is chemical bonding caused by the presence of cementing agents, such as calcium carbonate ($CaCO_3$) or iron oxide (Fe_2O_3) (Clough, et al., 1981). Even small quantities of these agents can provide significant cohesive strengths. *Caliche* is an example of a heavily cemented soil that has a large cohesive strength. Cementation also can be introduced artificially using Portland cement or special chemicals.
- *Electrostatic and electromagnetic attractions* hold particles together. However, these forces are very small and probably do not produce significant shear strength in soils.
- *Primary valence bonding* (*adhesion*) is a type of cold welding that occurs in clays when they become overconsolidated.

Apparent cohesion is shear strength that appears to be caused by bonding between the soil particles, but it is really frictional strength in disguise. Sources of apparent cohesion include the following:

- *Negative pore water pressures* that have not been considered in the stress analysis. These negative pore water pressures are present in soils above the groundwater table.
- *Negative excess pore water pressures caused by dilation*. Some soils tend to *dilate* or expand when they are sheared. In saturated soils, this dilation draws water into the voids. However, sometimes the rate of shearing is more rapid than the rate at which water can flow (i.e., the voids are trying to expand more rapidly than they can draw in the extra water). This is especially likely in saturated clays, because their hydraulic conductivity is so low. When this occurs, large negative excess pore water pressures can develop in the soil.
- *Apparent mechanical forces* are those caused by particle interlocking, and can develop in soils where this interlocking is very difficult to overcome.

Geotechnical engineers often use the term "cohesive soil" to describe clays. Although this term is convenient, it also is very misleading (Santamarina, 1997). Most of the so-called cohesive strength in clays is really apparent cohesion caused by pore water pressures that are negative, or at least less than the hydrostatic pore water pressure. In such soils it is better to think of "cohesive strength" as a mathematical idealization rather than a physical reality.

In cemented soils, cohesive strength really does reflect bonding between the soil particles. In some cases, we may rely on this strength in our designs. However, in other cases, it is wise to ignore this source of strength. For example, if the cementing agent is water-soluble, it may disappear if the soil becomes wetted during the life of the project.

Methods of Shear Strength Analysis

There are two principal methods of conducting shear strength analyses: An effective stress analysis and a total stress analysis. Both methods are used in foundation design.

Effective Stress Analyses

The shear strength in a soil is developed only by the solid particles, because the water and air phases have no shear strength. Therefore, it seems reasonable to evaluate strength problems using the effective stress, σ', because it is the portion of the total stress, σ, carried by the solid particles. When using an effective stress analysis, we describe shear strength using the *Mohr-Coulomb failure criterion*:

$$\boxed{s = c' + \sigma' \tan \phi'}$$
                                                                                    (3.29)

Where:

   s = shear strength

   c' = effective cohesion

   σ' = effective stress acting on the shear surface

   ϕ' = effective friction angle

The values of c' and ϕ' for a particular soil are usually obtained from special laboratory tests, as discussed later in this chapter.

Total Stress Analyses

Analyses based on effective stresses are possible only if we can predict the effective stresses in the field. This is a simple matter when only hydrostatic pore water pressures are present, but this can become very complex when there are excess pore water pressures. For example, when a foundation is placed over a saturated clay, the applied load produces excess pore water pressures as discussed earlier. In addition, some soils also develop additional excess pore water pressures as they are sheared. Often these excess pore water pressures are difficult or impossible to predict. Unfortunately, if we cannot predict the pore water pressure, then we do not know the magnitude of the effective stress and cannot solve Equation 3.29.

Because of these difficulties, geotechnical engineers are sometimes forced to evaluate problems based on total stresses instead of effective stresses. This approach involves reducing the lab data in terms of total stress and expressing it using the parameters c_T and ϕ_T. Equation 3.29 then needs to be rewritten as:

$$\boxed{s = c_T + \sigma \tan \phi_T}$$

(3.30)

Where:

s = shear strength
c_T = total cohesion
σ = total stress
ϕ_T = total friction angle

This method assumes the pore water pressures developed in the lab are the same as those in the field, and thus are implicitly incorporated into c_T and ϕ_T. This assumption introduces some error in the analysis, but it becomes an unfortunate necessity when we cannot predict the magnitudes of excess pore water pressures in the field.

Volume Changes and Excess Pore Water Pressures

When loads such as those from structural foundations are placed on the ground, the resulting normal and shear stresses cause the soil voids to expand or contract. If the soil is saturated, this expansion or contraction produces excess pore water pressures, as discussed earlier. However, this build-up of excess pore water pressures induces a hydraulic gradient that forces some of the water out of the voids and causes these pressures to dissipate. Thus, there are two processes acting simultaneously: the build-up of excess pore water pressure, which depends primarily on the rate of loading, and the dissipation of these pressures, which depends on the rate of drainage.

Geotechnical engineers often evaluate these two competing processes by considering two drainage conditions: the drained condition and the undrained condition.

If the rate of drainage is faster than the rate of loading, then any excess pore water pressures will be small and short-lived. We call this the *drained condition*. It exists when the rate of loading is very slow, the rate of drainage is very high, or both. Figure 3.17a shows the progression of various parameters with time when drained conditions exist.

If drained conditions exist, we assume the pore water pressure always equals the hydrostatic pore water pressure, and thus may be computed using Equation 3.29. This is very convenient and simplifies the computations.

Conversely, if the rate of drainage is slower than the rate of loading, significant excess pore water pressures may develop in the soil. We call this the *undrained condition*, and it is described in Figure 3.17b. It occurs when the load is applied rapidly, or the soil drains slowly.

The undrained condition is more difficult to analyze because we need to account for the excess pore water pressures. This may be done either explicitly or implicitly, as discussed later.

Shear Strength of Saturated Sands and Gravels

The rate of drainage in sands and gravels is very rapid because of their high hydraulic conductivity. In other words, any excess pore water pressures that may develop in these soils dissipate very rapidly because water flows through them very quickly and easily. In addition, most of the load acting on foundations usually consists of dead and live loads, which are applied over a period of days to weeks. This is much slower than the rate of drainage in sands and gravels. Therefore, when designing foundations on sands and gravels, we can nearly always assume drained conditions are present. Therefore, the pore water pressure equals the hydrostatic pore water pressure, as defined in Equation 3.7, and can compute the vertical effective stress using Equation 3.13.

The effective cohesion, c', and effective friction angle, ϕ', are obtained from laboratory tests (as discussed later in this chapter) or from in-situ tests (as discussed in Chapter 4). For clean or silty sands and gravels (USCS group symbols SM, SP, SW, GM, GP, and GW) it is normally best to use $c' = 0$. Some cohesion may be present in clayey sands and gravels (group symbols SC and GC), but it should be used only with great caution, because it may not be present in the field. Figure 3.18 presents typical values of ϕ', and may be used to check the lab or field test results.

Finally, knowing the values of σ', c', and ϕ', we can compute the shear strength using Equation 3.29.

Shear Strength of Saturated Clays and Silts

The hydraulic conductivity of clays is about one million times smaller than that of sands, so the rate of drainage in these soils is very slow—much slower than the rate of loading. Therefore, undrained conditions are typically present in such soils. This means significant excess pore water pressures may be present during and immediately after loading. Eventually these excess pore water pressures dissipate, as shown in Figure 3.17b.

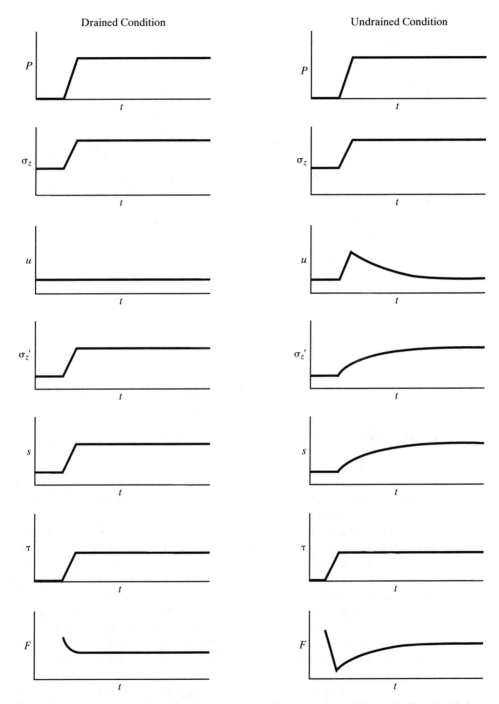

Figure 3.17 Changes in normal and shear stresses, ϕ and τ; shear strength, s; and factor of safety, F, with time at a point in a saturated soil subjected to a load P under drained vs. undrained conditions.

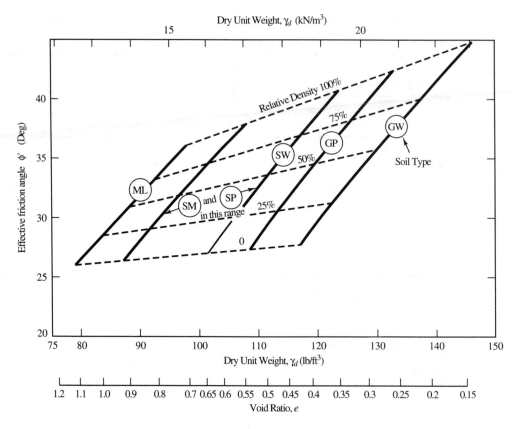

Figure 3.18 Typical ϕ' values for sands, gravels, and silts without plastic fines (Adapted from U.S. Navy, 1982a).

The hydraulic conductivity of silts is greater than that of clays, but it still is much smaller than that in sands. Once again, undrained conditions will prevail, although less time is required for the excess pore water pressures to dissipate.

To understand the impact of these excess pore water pressures, compare the two sets of plots in Figure 3.17, along with Equations 3.13 and 3.29. In both cases, the vertical total stress, σ_z, and the shear stress, τ, increase as the load P is applied. Under drained conditions, the vertical effective stress, σ_z', and the shear strength, s, both rise concurrent with σ_z, so both τ and s reach their peak value at the end of loading.

However, under undrained conditions, the temporary presence of excess pore water pressure (the spike in the u plot) causes a lag in the increase in shear strength. Although τ reaches its peak at the end of loading, s is still low, thus producing a temporary drop in the factor of safety, F. Then, as the excess pore water pressures dissipate, both s and F slowly climb.

The most likely time for failure is immediately after construction (i.e., when F is at a minimum). Therefore, foundations are normally designed to have a certain minimum factor of safety at this critical moment. To accomplish this goal, we need to consider the excess pore water pressure, either explicitly or implicitly.

In principle, we should be able to determine the magnitude of the excess pore water pressure, u_e, and use Equations 3.7, 3.8, 3.13, and 3.29 to compute the shear strength immediately after construction. This method is called an effective stress analysis. We then could use this shear strength as the basis for a foundation design that provides the minimum acceptable factor of safety. Unfortunately, the magnitude of u_e is difficult to determine, especially in overconsolidated soils, so this method is not economically viable for most foundation projects.

Because of this problem, engineers normally resort to performing a total stress analysis and compute the shear strength using Equation 3.30. This equation is based on c_T and ϕ_T, which are determined by evaluating the laboratory test data in terms of total stress instead of effective stress. Presumably, the consequences of the excess pore water pressures are implicit within these values. We then assume the excess pore water pressures in the field are the same as those in the lab, and apply Equation 3.30 to the field conditions without needing to know the magnitude of the excess pore water pressures.

If the soil is truly saturated and truly undrained, then $\phi_T = 0$ (even though $\phi' > 0$) because newly applied loads are carried entirely by the pore water and do not change σ' or s. This is very convenient because the second term in Equation 3.30 drops out and we no longer need to compute σ. We call this a "$\phi = 0$ analysis," and the shear strength is called the *undrained shear strength*, s_u, where $s_u = c_T$. Table 3.8 gives typical values of s_u, which may be used for preliminary analyses or to check laboratory test results.

Usually we assign an appropriate s_u value for each saturated undrained strata based on laboratory or field test results. Many geotechnical analysis methods use this s_u

TABLE 3.8 RELATIONSHIP BETWEEN CONSISTENCY OF COHESIVE SOILS AND UNDRAINED SHEAR STRENGTH (Adapted from Terzaghi & Peck, 1967 and ASTM D2488-90; used with permission).

Consistency	Undrained Shear Strength, s_u (lb/ft^2)	(kPa)	Visual Identification
Very soft	<250	<12	Thumb can penetrate more than 1 in (25 mm)
Soft	250–500	12–25	Thumb can penetrate about 1 in (25 mm)
Medium	500–1000	25–50	Penetrated with thumb with moderate effort
Stiff	1000–2000	50–100	Thumb will indent soil about 1/4 in (8 mm)
Very stiff	2000–4000	100–200	Thumb will not indent, but readily indented with thumbnail
Hard	>4000	>200	Indented by thumbnail with difficulty or cannot indent with thumbnail

value directly. Other analysis methods require the shear strength to be defined using c and ϕ. When using the later type with saturated undrained analyses, we set $c = s_u$ and $\phi = 0$.

In reality, s_u is probably not constant throughout a particular soil strata, even if it appears to be homogeneous. In general, s_u increases with depth because the lower portions of the strata have been consolidated to correspondingly greater loads, and thus have a higher shear strength. The shallow portions also have higher strengths if they had once dried out (desiccated) and formed a crust. Finally, the natural non-uniformities in a soil strata produce variations in s_u. We can accommodate these variations by simply taking an average value, or by dividing the strata into smaller layers.

Shear Strength of Saturated Intermediate Soils

Thus far we have divided soils into two distinct categories. Sands and gravels do not develop excess pore water pressures during static loading, and thus may be evaluated using effective stress analyses and hydrostatic pore water pressures. Conversely, silts and clays do develop excess pore water pressures, and thus require more careful analysis. They also may have problems with sensitivity and creep. Although many "real-world" soils neatly fit into one of these two categories, others behave in ways that are intermediate between these two extremes. Their field behavior is typically somewhere between being drained and undrained (i.e., they develop some excess pore water pressures, but not as much as would occur in a clay).

Although there are no clear-cut boundaries, these intermediate soils typically include those with unified classification SC, GC, SC-SM, or GC-GM, as well as some SM, GM, and ML soils. Proper shear strength evaluations for engineering analyses require much more engineering judgment, which is guided by a thorough understanding of soil strength principles. When in doubt, it is usually conservative to evaluate these soils using the techniques described for silts and clays.

Shear Strength of Unsaturated Soils

Thus far we have only considered soils that are saturated ($S = 100\%$). The strength of unsaturated soils ($S < 100\%$) is generally greater, but more difficult to evaluate. Nevertheless, many engineering projects encounter these soils, so geotechnical engineers need to have methods of evaluating them. This has been a topic of ongoing research (Fredlund and Rahardjo, 1993), and standards of practice are not yet as well established as those for saturated soils.

Some of the additional strength in unsaturated soils is caused by negative pore water pressures. These negative pore water pressures increase the effective stress, and thus increase the shear strength. However, this additional strength is very tenuous and is easily lost if the soil becomes wetted.

Geotechnical engineers usually base designs on the assumption that unsaturated soils could become wetted in the future. This wetting could come from a rising groundwater table, irrigation, poor surface drainage, broken pipelines, or other causes. Therefore, we usually saturate (or at least "soak") soil samples in the laboratory before performing strength tests. This is intended to remove the apparent cohesion and thus simulate the "worst case" field conditions. We then determine the highest likely elevation for the groundwater table, which may be significantly higher than its present location, and compute positive pore water pressures accordingly. Finally, we assume $u = 0$ in soils above the groundwater table.

Laboratory Shear Strength Tests

The shear strength parameters, c' and ϕ' (or c_T and ϕ_T) may be determined by performing laboratory or in-situ tests. This section discusses some of the more common laboratory tests, and Chapter 4 discusses in-situ tests.

Several different laboratory tests are commonly used to measure the shear strength of soils. Each has its advantages and disadvantages, and no one test is suitable for all circumstances. When selecting a test method, we must consider many factors, including the following:

- Soil type
- Initial moisture content and need, if any, to saturate the sample
- Required drainage conditions (drained or undrained)

Direct Shear Test

The French engineer Alexandre Collin may have been the first to measure the shear strength of a soil (Head, 1982). His tests, conducted in 1846, were similar to the modern direct shear test. The test as we now know it [ASTM D3080] was perfected by several individuals during the first half of the twentieth century.

The apparatus, shown in Figure 3.19, typically accepts a 60 to 75 mm (2.5–3.0 in) diameter cylindrical sample and subjects it to a certain effective stress. The shear stress is then slowly increased until the soil fails along the surface shown in the figure. The test is normally repeated on new samples of the same soil until three sets of effective stress and shear strength measurements are obtained. A plot of this data produces values of the cohesion, c, and friction angle, ϕ.

The direct shear test has the advantage of being simple and inexpensive and it is an appropriate method when we need the drained strength of sands. It also can be used to obtain the drained strength of clays, but produces less reliable results because we have no way of controlling the drainage conditions other than varying the speed of the test. The direct shear test also has the disadvantages of forcing the shear to occur along a specific plane instead of allowing the soil to fail along the weakest zone, and it produces nonuniform strains in the sample, which can produce erroneous results in strain-softening soils.

(a)

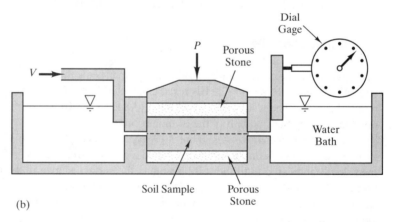

(b)

Figure 3.19 (a) A direct shear machine. The sample is inside the sample holder, directly below the upper dial gauge, (b) Cross section through the sample holder showing the soil sample and shearing action.

Example 3.6

A series of three direct shear tests has been conducted on a certain saturated soil. Each test was performed on a 2.375-inch diameter, 1.00-inch tall sample. The test has been performed slowly enough to produce drained conditions. The results of these tests are as follows:

Test Number	Normal Load (lb)	Shear Load at Failure (lb)
1	75	51
2	150	110
3	225	141

Determine c' and ϕ'.

Solution

$$A = \frac{\pi D^2}{4} = \frac{\pi (2.375 \text{ in})^2}{4} \left(\frac{1 \text{ ft}^2}{144 \text{ in}^2} \right) = 0.0308 \text{ ft}^2$$

Based on this area and the measured forces:

Test Number	σ' (lb/ft^2)	τ (lb/ft^2)
1	2435	1656
2	4870	3571
3	7305	4578

This data is plotted in Figure 3.20. It does not form a perfect line. This is because of experimental error, slight differences in the three samples, true nonlinearity, and other factors. We have drawn a best-fit line through these three points to obtain $c' = 400$ lb/ft^2 and $\phi' = 31°$. Note: The shear area changes as the direct shear test progresses, and a more thorough analysis would account for this change. However, most engineers neglect this change because it has very little effect on the final test results and because it is slightly conservative to do so.

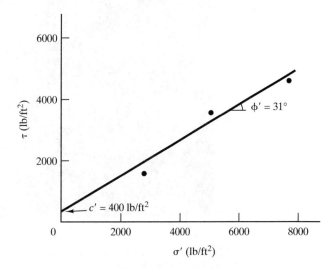

Figure 3.20 Direct shear test results.

Unconfined Compression Test

The unconfined compression test [ASTM D2166], shown in Figure 3.21a, uses a tall, cylindrical sample of cohesive soil subjected to an axial load. This load is applied quickly (i.e., only a couple of minutes to failure) to maintain undrained conditions.

At the beginning of the test, this load and the stresses in the soil are both equal to zero. As the load increases, the stresses in the soil increase, as shown by the Mohr's cir-

(a)

Figure 3.21   (a) Unconfined compression test machine; (b) Force analysis of an unconfined compression test.

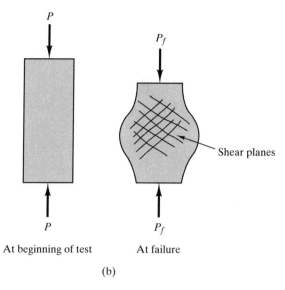

At beginning of test   At failure

(b)

Figure 3.21 *Continued*

cles in Figure 3.22, until the soil fails. The soil appears to fail in compression, and the test results are often expressed in terms of the compressive strength[2]; it actually fails in shear on diagonal planes, as shown in Figure 3.21b. The cross-sectional area of the sample increases as the test progresses, and the area at failure, A_f, is:

$$A_f = \frac{A_0}{1 - \epsilon_f}$$ (3.31)

The undrained shear strength, s_u, is:

$$s_u = \frac{P_f}{2 A_f}$$ (3.32)

Where:

A_f = cross-sectional area at failure

A_0 = initial cross-sectional area

ϵ_f = axial strain at failure

s_u = undrained shear strength

P_f = axial load at failure

---

[2]When reviewing unconfined compression test results, be sure to note whether they are expressed in terms of "compressive" strength or shear strength. In this test, the shear strength is equal to half of the compressive strength.

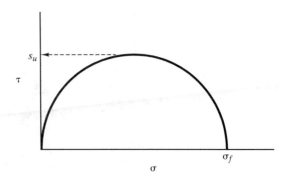

Figure 3.22 Mohr's circle at failure in an
unconfined compression test.

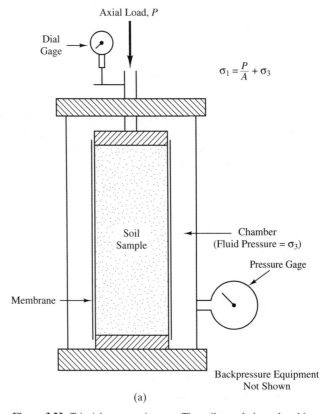

(a)

Figure 3.23 Triaxial compression test. The soil sample is enclosed in a
flexible membrane and subjected to a horizontal chamber pressure, σ_3.
This pressure remains constant during the test. The axial load, P, is then
increased until the soil fails. The dial gauge is used to measure the axial
strain during the test. The chamber in the photograph is covered with a
protective wire mesh.

Figure 3.23 *Continued*

    This test is inexpensive and in common use. It does not force failure to occur along
a predetermined surface, and thus it reflects the presence of weak zones or planes. It usually provides conservative (low) results because the horizontal stress is zero rather than
what is was in the field and because of sample disturbance. Tests of fissured clays are
often misleading unless a large sample is used because the fissures in small samples are
rarely representative of those in the field.

Triaxial Compression Test

The triaxial compression test [ASTM D2850] may be thought of as an extension of the
unconfined compression test. The cylindrical soil sample is now located inside a pressurized chamber that supplies the desired lateral stress, as shown in Figure 3.23. Although
the apparatus required to perform this test is much more complex, it also allows more
flexibility and greater control over the test. It can measure either the drained or undrained
strength of nearly any type of soil. In addition, unsaturated soils can be effectively saturated before testing.

    The three most common types of triaxial compression tests are as follows:

- The ***unconsolidated-undrained (UU) test*** (also known as a *quick* or *Q* test): Horizontal and vertical stresses, usually equal to the vertical stress that was present in
  the field, are applied to the sample. No consolidation is permitted, and the soil is
  sheared under undrained conditions. The result is expressed as an s_u value.

- The *consolidated-drained (CD) test* (also known as a *slow* or *S* test): Horizontal and vertical stresses, usually equal to or greater than the vertical stress that was present in the field, are applied and the soil is allowed to consolidate. Then, it is sheared under drained conditions. Typically, three of these tests are performed at different confining stresses to find the drained values of *c* and ϕ.
- The *consolidated-undrained test* (also known as a *rapid* or *R* test): The initial stresses are applied as with the CD test and the soil is allowed to consolidate. However, the shearing occurs under undrained conditions. The results could be expressed as a value of s_u, but it is also possible to obtain the drained *c* and ϕ by measuring pore water pressures during the test and computing the effective stresses.

QUESTIONS AND PRACTICE PROBLEMS

3.8 Estimate the effective friction angle of the following soils:
  a. Silty sand with a dry unit weight of 110 lb/ft^3.
  b. Poorly-graded gravel with a relative density of 70%.
  c. Very dense well-graded sand.

3.9 Explain the difference between the drained condition and the undrained condition.

3.10 A soil has $c' = 5$ kPa and $\phi' = 32°$. The effective stress at a point in the soil is 125 kPa. Compute the shear strength at this point.

SUMMARY

Major Points

1. Soil is a particulate material, which means it is an assemblage of individual particles. Its engineering properties are largely dependant on the interactions between these particles.
2. Soil can potentially include all three phases of matter (solid, liquid, and gas) simultaneously. It is helpful to measure the relative proportions of these three phases, and we express these proportions using standard weight-volume parameters.
3. The relative density is a special weight-volume parameter often used to describe the void ratio of sands and gravels.
4. The particle sizes in soil vary over several orders of magnitude, from sub-microscopic clay particles to gravel, cobbles, and boulders.
5. Clays are a special kind of soil because of their extremely small particle sizes and because of the special interactions between these particles and between the solids and the pore water.
6. Plasticity describes the relationship between moisture content and consistency in clays and silts. These relationships are quantified using the Atterberg limits.

7. Various soil classification systems are used in civil engineering. The most common system is the Unified Soil Classification System, which uses a standard system of group symbols and group names.

8. Groundwater has a profound impact on soil properties. Groundwater conditions can be very complex, but in this book, we will consider only the simple case of a horizontal groundwater table.

9. The pore water pressure is the pressure in the water within the soil voids. There are two kinds: hydrostatic pore water pressure is due solely to gravity acting on the pore water, while excess pore water pressure is due to the squeezing or expanding of the soil voids. Excess pore water pressures are always temporary.

10. Soils have both normal and shear stresses. They have two sources: Geostatic stresses are those due to the force of gravity acting on the soil mass, while induced stresses are due to applied external loads, such as foundations.

11. Settlement in soils can be the result of various processes. The most important of these is consolidation, which is the rearranging of particles into a tighter packing. Consolidation settlement analyses are based on data from a consolidation test, which is performed in the laboratory.

12. Soils in the field can be either normally consolidated or overconsolidated, depending on the difference between the current vertical effective stress and the preconsolidation stress, which is the greatest past value of this stress.

13. The degree of overconsolidation may be expressed using the overconsolidation margin or the overconsolidation ratio.

14. Shear strength in soil has two sources: frictional strength and cohesive strength.

15. Shear strength analyses may be based on effective stresses or on total stresses. Effective stress analyses are more accurate models of soil behavior, but are difficult to perform when excess pore water pressures are present.

16. The drained conditions are present when the rate of loading is slow compared to the rate of drainage. This is the case in sands and gravels. The undrained condition occurs when the reverse is true, which occurs in silts and clays.

17. Various tests are available to measure shear strength in the laboratory.

Vocabulary

Apparent cohesion	Cohesion	Drained condition
Atterberg limits	Cohesive strength	Dry unit weight
Boulder	Compression index	Effective stress
Clay	Consolidation	Effective stress analysis
Cobble	Consolidation test	Excess pore water pressure
Coefficient of lateral earth pressure	Degree of saturation	Fines
	Direct shear test	Friction angle

Frictional strength

Geostatic stress

Gravel

Groundwater

Groundwater table

Group symbol

Horizontal stress

Hydraulic conductivity

Hydrostatic pore water
pressure

Induced stress

Liquid limit

Moisture content

Normally consolidated

Overconsolidated

Overconsolidation margin

Overconsolidation ratio

Particulate material

Phase diagram

Plastic limit

Plasticity

Pore water pressure

Porosity

Rebound curve

Recompression index

Recompression curve

Relative density

Sand

Settlement

Silt

Specific gravity

Stress

Total stress

Total stress analysis

Triaxial compression test

True cohesion

Unconfined compression
test

Undrained condition

Undrained shear strength

Unified soil classification
system

Unit weight

Virgin curve

Void ratio

Weight-volume
relationship

COMPREHENSIVE QUESTIONS AND PRACTICE PROBLEMS

3.11 Which laboratory tests would be appropriate for finding s_u of a clay?

3.12 Which laboratory tests would be appropriate for finding ϕ' of a sand?

3.13 The following data were obtained from direct shear tests on a series of 60-mm diameter samples of a certain soil:

Test No.	Effective Stress at Failure (kPa)	Shear Strength (kPa)
1	75.0	51.2
2	150.0	82.7
3	225.0	110.1

Find the values of c' and ϕ'.

3.14 The soil profile at a certain site is as follows:

Depth (ft)	γ (lb/ft^3)	c' (lb/ft^2)	ϕ' (deg)	s_u (lb/ft^2)
0–12	119			1000
12–20	126	200	20	
20–32	129	0	32	

The groundwater table is at a depth of 15 ft.

Develop plots of pore water pressure, total stress, effective stress, and shear strength vs. depth. All four of these plots should be superimposed on the same diagram with the parameters on the horizontal axis (increasing to the right) and depth on the vertical axis (increasing downward).

Hint: Because the cohesion and friction angle suddenly change at the strata boundaries, the shear strength also may change suddenly at these depths.

3.15 Repeat Problem 3.14 using the following data:

Depth (m)	γ (kN/m^3)	c' (kPa)	ϕ' (deg)	s_u (kPa)
0–5	18.5			50.0
5–12	20.0	8.4	21	
12–20	20.5	0.0	35	

The groundwater table is at a depth of 7 m.

3.16 A 9 ft thick fill is to be placed on the soil shown in Figure 3.24. Once it is compacted, this fill will have a unit weight of 122 lb/ft^3. Compute the ultimate settlement caused by consolidation of the underlying clay.

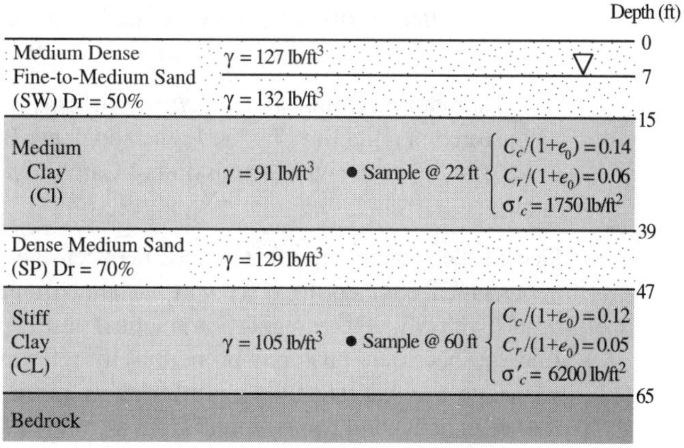

Figure 3.24 Soil profile for Problem 3.16.

4

Site Exploration and Characterization

One of the fundamental differences between the practices of structural engineering and
geotechnical engineering is the way each determines the engineering properties of the materials with which they work. For practical design problems, structural engineers normally
find the necessary material properties by referring to handbooks. For example, if one
wishes to use A36 steel, its engineering properties (strength, modulus of elasticity, etc.)
are well known and can be found from a variety of sources. It is not necessary to measure
the strength of A36 steel each time we use it in a design (although routine strength tests
may be performed later as a quality control function). Conversely, the geotechnical engineer works with soil and rock, both of which are natural materials with unknown engineering properties. Therefore, we must identify and test the materials at each new site before conducting any analyses.

Modern soil investigation and testing techniques have progressed far beyond Bullet's method of beating the earth with wooden rafters. A variety of techniques are available, as discussed in this chapter, yet this continues to be the single largest source of

uncertainties in foundation engineering. Our ability to perform analyses far exceeds our ability to determine the appropriate soil properties to input into these analyses. Therefore, it is very important for the foundation engineer to be familiar with the available techniques, know when to use them, and understand the degree of precision (or lack of precision!) associated with them.

For purposes of this discussion, we will divide these techniques into three categories:

- **Site investigation** includes methods of defining the soil profile and other relevant data and recovering soil samples.
- **Laboratory testing** includes testing the soil samples in order to determine relevant engineering properties.
- **In-situ testing** includes methods of testing the soils in-place, thus avoiding the difficulties associated with recovering samples.

4.1 SITE EXPLORATION

The objectives of the site investigation phase include:

- Determining the locations and thicknesses of the soil strata.
- Determining the location of the groundwater table as well as any other groundwater-related characteristics.
- Recovering soil samples.
- Defining special problems and concerns.

Typically, we accomplish these goals using a combination of literature searches and on-site exploration techniques.

Background Literature Search

Before conducting any new exploration at a project site, gather whatever information is already available, both for the proposed structure and the subsurface conditions at the site. Important information about the structure would include:

- Its location and dimensions.
- The type of construction, column loads, column spacing, and allowable settlements.
- Its intended use.
- The finish floor elevation.
- The number and depth of any basements.
- The depth and extent of any proposed grading.
- Local building code requirements.

The literature search also should include an effort to obtain at least a preliminary idea of the subsurface conditions. It would be very difficult to plan an exploration program with no such knowledge. Fortunately, many methods and resources are often available to gain a preliminary understanding of the local soil conditions. These may include one or more of the following:

- Determining the geologic history of the site, including assessments of anticipated rock and soil types, the proximity of faults, and other geologic features.
- Gathering copies of boring logs and laboratory test results from previous investigations on this or other nearby sites.
- Reviewing soil maps developed for agricultural purposes.
- Reviewing old and new aerial photographs and topographic maps (may reveal previous development or grading at the site).
- Reviewing water well logs (helps establish historic groundwater levels).
- Locating underground improvements, such as utility lines, both onsite and immediately offsite.
- Locating foundations of adjacent structures, especially those that might be impacted by the proposed construction.

At some sites, this type of information may be plentiful, whereas at others it may be scarce or nonexistent.

Field Reconnaissance

Along with the background literature search, the foundation engineer should visit the site and perform a field reconnaissance. Often such visits will reveal obvious concerns that may not be evident from the literature search or the logs of the exploratory borings.

The field reconnaissance would include obtaining answers to such questions as the following:

- Is there any evidence of previous development on the site?
- Is there any evidence of previous grading on the site?
- Is there any evidence of landslides or other stability problems?
- Are nearby structures performing satisfactorily?
- What are the surface drainage conditions?
- What types of soil and/or rock are exposed at the ground surface?
- Will access problems limit the types of subsurface exploration techniques that can be used?
- Might the proposed construction affect existing improvements? (For example, a fragile old building adjacent to the site might be damaged by vibrations from pile driving.)
- Do any offsite conditions affect the proposed development? (For example, potential flooding, mudflows, rockfalls, etc.)

Subsurface Exploration and Sampling

The heart of the site investigation phase consists of exploring the subsurface conditions and sampling the soils. These efforts provide most of the basis for developing a design soil profile. A variety of techniques are available to accomplish these goals.

Exploratory Borings

The most common method of exploring the subsurface conditions is to drill a series of vertical holes in the ground. These are known as *borings* or *exploratory borings* and are typically 75 to 600 mm (3–24 in) in diameter and 3 to 30 m (10–100 ft) deep. They can be drilled with hand augers or with portable power equipment, but they are most commonly drilled using a truck-mounted rig, as shown in Figure 4.1.

A wide variety of drilling equipment and techniques are available to accommodate the various subsurface conditions that might be encountered. Sometimes it is possible to drill an open hole using a *flight auger* or a *bucket auger*, as shown in Figure 4.2. However, if the soil is subject to *caving* (i.e., the sides of the boring fall in) or *squeezing* (the soil moves inward, reducing the diameter of the boring), then it will be necessary to provide some type of lateral support during drilling. Caving is likely to be encountered in clean sands, especially below the groundwater table, while squeezing is likely in soft saturated clays.

One method of dealing with caving or squeezing soils is to use *casing*, as shown in Figure 4.3a. This method involves temporarily lining some or all of the boring with a steel

Figure 4.1 A truck-mounted drill rig.

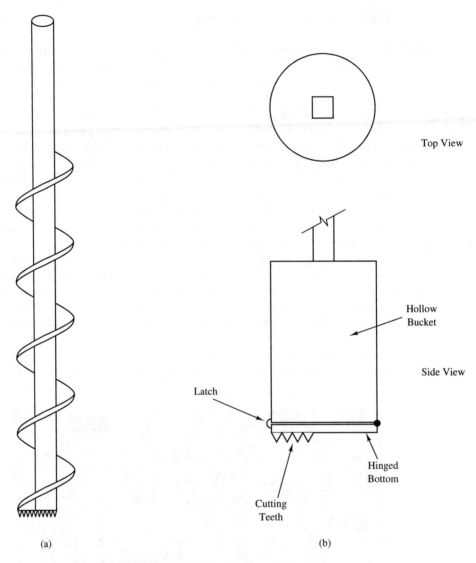

Top View

Hollow
Bucket

Side View

Latch

Hinged
Bottom

Cutting
Teeth

(a)   (b)

Figure 4.2 (a) Flight auger; and (b) Bucket auger.

pipe. Alternatively, we could use a *hollow-stem auger*, as shown in Figure 4.3b. The driller screws each of these augers into the ground and obtains soil samples by lowering sampling tools through a hollow core. When the boring is completed, the augers are re-moved. Finally, we could use a *rotary wash boring*, as shown in Figure 4.3c. These bor-ings are filled with a bentonite slurry (a combination of bentonite clay and water) to provide hydrostatic pressure on the sides of the boring and thus prevent caving.

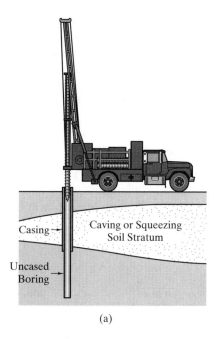

Casing

Uncased
Boring

Caving or Squeezing
Soil Stratum

(a)

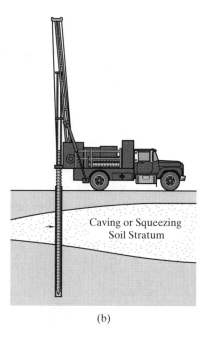

Caving or Squeezing
Soil Stratum

(b)

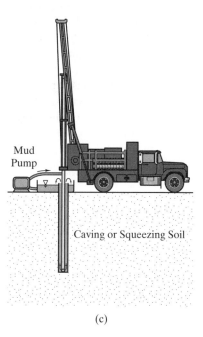

Mud
Pump

Caving or Squeezing Soil

(c)

Figure 4.3 Methods of dealing with caving or squeezing soils: (a) casing; (b) hollow-stem auger; and (c) rotary wash boring.

Drilling through rock, especially hard rock, requires different methods and equipment. Engineers often use *coring*, which recovers intact cylindrical specimens of the rock.

There are no absolute rules to determine the required number, spacing, and depth of exploratory borings. Such decisions are based on the findings from the field reconnaissance, along with engineering judgement and a knowledge of customary standards of practice. This is a subjective process that involves many factors, including:

- How large is the site?
- What kinds of soil and rock conditions are expected?
- Is the soil profile erratic, or is it consistent across the site?
- What is to be built on the site (small building, large building, bridge, etc.)?
- How critical is the proposed project (i.e., what would be the consequences of a failure?)?
- How large and heavy are the proposed structures?
- Are all areas of the site accessible to drill rigs?

Although we will not know the final answers to some of these questions until the site characterization program is completed, we should have at least a preliminary idea based on the literature search and field reconnaissance.

For buildings taller than three stories or 12.2 m (40 ft), the *BOCA National Building Code* (BOCA, 1996) requires at least one boring for every 230 m^2 (2500 ft^2) of built-over area. Table 4.1 presents rough guidelines for determining the normal spacing of exploratory borings at sites not governed by the BOCA code. However, it is important to recognize that there is no single "correct" solution for the required number and depth of borings, and these guidelines must be tempered with appropriate engineering judgement.

Borings generally should extend at least to a depth such that the change in vertical effective stress due to the new construction is less than 10 percent of the initial vertical effective stress. For buildings on spread footing or mat foundations, this criteria is met by following the guidelines in Table 4.2. If fill is present, the borings must extend through it and into the natural ground below, and if soft soils are present, the borings should extend through them and into firmer soils below. For heavy structures, at least some of the bor-

TABLE 4.1   ROUGH GUIDELINES FOR SPACING EXPLORATORY BORINGS
FOR PROPOSED MEDIUM TO HEAVY WEIGHT BUILDINGS, TANKS,
AND OTHER SIMILAR STRUCTURES

| | Structure Footprint Area for Each Exploratory Boring | |
Subsurface Conditions	(m^2)	(ft^2)
Poor quality and/or erratic	100–300	1,000–3,000
Average	200–400	2,000–4,000
High quality and uniform	300–1,000	3,000–10,000

TABLE 4.2   GUIDELINES FOR DEPTHS OF
EXPLORATORY BORINGS FOR BUILDINGS ON SHALLOW
FOUNDATIONS (Adapted from Sowers, 1979)

Subsurface Conditions	Minimum Depth of Borings (S = number of stories; D = anticipated depth of foundation)	
	(m)	(ft)
Poor	$6\,S^{0.7} + D$	$20\,S^{0.7} + D$
Average	$5\,S^{0.7} + D$	$15\,S^{0.7} + D$
Good	$3\,S^{0.7} + D$	$10\,S^{0.7} + D$

ings should be carried down to bedrock, if possible, but certainly well below the depth of any proposed pile foundations.

On large projects, the drilling program might be divided into two phases: a preliminary phase to determine the general soil profile, and a final phase based on the results of the preliminary borings.

The conditions encountered in an exploratory boring are normally presented in the form of a *boring log*, as shown in Figure 4.4. These logs also indicate the sample locations and might include some of the laboratory test results.

Soil Sampling

One of the primary purposes of drilling the exploratory borings is to obtain representative soil samples. We use these samples to determine the soil profile and to perform laboratory tests.

There are two categories of samples: disturbed and undisturbed. A *disturbed sample* (sometimes called a *bulk* sample) is one in which there is no attempt to retain the in-place structure of the soil. The driller might obtain such a sample by removing the cuttings off the bottom of a flight auger and placing them in a bag. Disturbed samples are suitable for many purposes, such as classification and Proctor compaction tests.

A truly *undisturbed sample* is one in which the soil is recovered completely intact and its in-place structure and stresses are not modified in any way. Such samples are desirable for laboratory tests that depend on the structure of the soil, such as consolidation tests and shear strength tests. Unfortunately, the following problems make it almost impossible to obtain a truly undisturbed soil sample:

- Shearing and compressing the soil during the process of inserting the sampling tool.
- Relieving the sample of its in-situ stresses.
- Possible drying and desiccation.
- Vibrating the sample during recovery and transport.

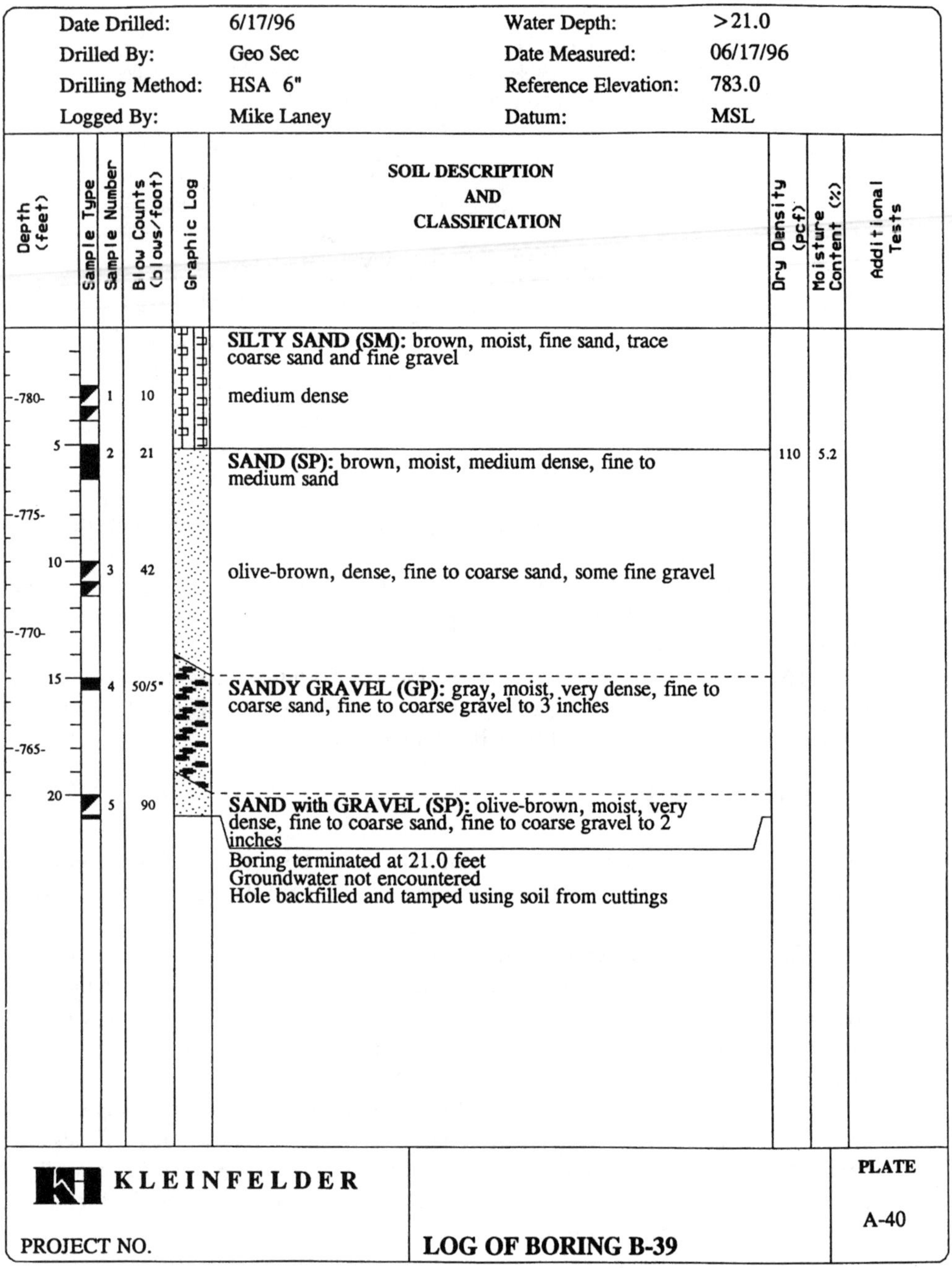

Date Drilled:	6/17/96	Water Depth:	>21.0
Drilled By:	Geo Sec	Date Measured:	06/17/96
Drilling Method:	HSA 6"	Reference Elevation:	783.0
Logged By:	Mike Laney	Datum:	MSL

Depth (feet)	Sample Type	Sample Number	Blow Counts (blows/foot)	Graphic Log	SOIL DESCRIPTION AND CLASSIFICATION	Dry Density (pcf)	Moisture Content (%)	Additional Tests
					SILTY SAND (SM): brown, moist, fine sand, trace coarse sand and fine gravel			
-780-		1	10		medium dense			
5		2	21		**SAND (SP):** brown, moist, medium dense, fine to medium sand	110	5.2	
-775-								
10		3	42		olive-brown, dense, fine to coarse sand, some fine gravel			
-770-								
15		4	50/5"		**SANDY GRAVEL (GP):** gray, moist, very dense, fine to coarse sand, fine to coarse gravel to 3 inches			
-765-								
20		5	90		**SAND with GRAVEL (SP):** olive-brown, moist, very dense, fine to coarse sand, fine to coarse gravel to 2 inches			
					Boring terminated at 21.0 feet Groundwater not encountered Hole backfilled and tamped using soil from cuttings			

KLEINFELDER

PROJECT NO.

LOG OF BORING B-39

PLATE

A-40

Figure 4.4 A boring log. Samples 2 and 4 were obtained using a heavy-wall sampler, and the corresponding blow counts are the number of hammer blows required to drive the sampler. Samples 1, 3, and 5 are standard penetration tests, and the corresponding blow counts are the N_{60} values, as discussed later in this chapter. If the groundwater table had been encountered, it would have been shown on the log (Kleinfelder, Inc.).

   Some soils are more subject to disturbance than others. For example, techniques are available to obtain good quality samples of medium clays, while clean sands are almost impossible to sample without extensive disturbance. However, even the best techniques produce samples that are best described as "relatively undisturbed."

   A variety of sampling tools is available. Some of these tools are shown in Figure 4.5. Those with thin walls produce the least disturbance, but they may not have the integrity needed to penetrate hard soils.

Groundwater Monitoring

The position and movements of the groundwater table are very important factors in foundation design. Therefore, subsurface investigations must include an assessment of groundwater conditions. This is often done by installing an *observation well* in the com-

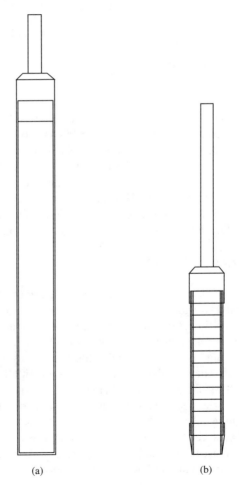

Figure 4.5 Common soil sampling tools: (a) Shelby tube samples have thin walls to reduce sample disturbance; and (b) Ring-lined barrel samplers have thicker walls to withstand harder driving. Both are typically 60–100 mm (2.5–4 in) in diameter.

(a)   (b)

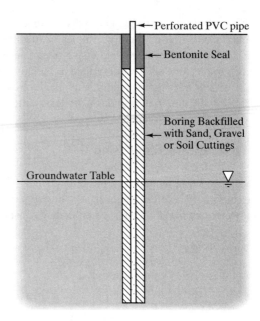

Figure 4.6 A typical observation well.

pleted boring to monitor groundwater conditions. Such wells typically consist of slotted or perforated PVC pipe, as shown in Figure 4.6. Once the groundwater level has stabilized, we can locate it by lowering a probe into the observation well.

We usually compare observation well data with historic groundwater records, or at least consider the season of the year and the recent precipitation patterns to determine the design groundwater level. This design level should represent the worst (i.e., shallowest) groundwater level that is likely to occur during the design life of the project. This level is often shallower than that observed in the observation wells.

Exploratory Trenches

Sometimes it is only necessary to explore the upper 3 m (10 ft) of soil. This might be the case for lightweight structures on sites where the soil conditions are good, or on sites with old shallow fills of questionable quality. Additional shallow investigations might also be necessary to supplement a program of exploratory borings. In such cases, it can be very helpful to dig *exploratory trenches* (also known as *test pits*) using a backhoe as shown in Figure 4.7. They provide more information than a boring of comparable depth (because more of the soil is exposed) and are often less expensive. Disturbed samples can easily be recovered with a shovel, and undisturbed samples can be obtained using hand-held sampling equipment.

Two special precautions are in order when using exploratory trenches: First, these trenches must be adequately shored before anyone enters them (see OSHA Excavation Regulations). Many individuals, including one of the author's former colleagues, have

Figure 4.7 This exploratory trench was dug by the backhoe in the background, and has been stabilized using aluminum-hydraulic shoring. An engineering geologist is logging the soil conditions in one wall of the trench.

been killed by neglecting to enforce this basic safety measure. Second, these trenches must be properly backfilled to avoid creating an artificial soft zone that might affect future construction.

4.2   LABORATORY TESTING

Soil samples obtained from the field are normally brought to a soil mechanics laboratory for further classification and testing. This method is sometimes called *ex-situ testing*. The purpose of the testing program is to determine the appropriate engineering properties of the soil, as follows:

- **Classification, weight-volume, and index tests**—Several routine tests are usually performed on many of the samples to ascertain the general characteristics of the soil profile. These include:
  - Moisture content
  - Unit weight (density)
  - Atterberg limits (plastic limit, liquid limit)
  - Grain-size distribution (sieve and hydrometer analyses)

These tests are inexpensive and can provide a large quantity of valuable information.

- **Shear strength tests**—Virtually all foundation designs require an assessment of shear strength. Common shear strength tests include the direct shear test, the unconfined compression test, and the triaxial compression test, as discussed in Chapter 3. Shear strength also may be determined from in-situ tests, as discussed later in this chapter.

- **Consolidation tests**—Foundation designs also require an assessment of soil compressibility, which provides the necessary data for settlement analyses. In clays and silts, this data is most often obtained by conducting laboratory consolidation tests, as discussed in Chapter 3.

- **Compaction tests**—Sometimes it is necessary to place compacted fills at a site, and place the foundations on these fills. In such cases, we perform Proctor compaction tests to assess the compaction characteristics of the soil.

- **Corrosivity tests**—When corrosion or sulfate attack is a concern, as discussed in Chapter 2, we need to perform special tests such as resistivity tests and sulfate content tests, and then use the results to design appropriate protective measures.

QUESTIONS AND PRACTICE PROBLEMS

4.1 Describe a scenario that would require a very extensive site investigation and laboratory testing program (i.e., one in which a large number of borings and many laboratory and/or in-situ tests would be necessary).

4.2 How would you go about determining the location of the groundwater table in the design soil profile. Recall that this is not necessarily the same as the groundwater table that was present when the borings were made.

4.3 A five-story office building is to be built on a site underlain by moderately uniform soils. Bedrock is at a depth of over 200 m. This building will be 50 m wide and 85 m long, and the foundations will be founded at a depth of 1 m below the ground surface. Determine the required number and depth of the exploratory borings.

4.4 A two-story reinforced concrete building is to be built on a vacant parcel of land. This building will be 100 ft wide and 200 ft long. Based on information from other borings on adjacent properties, you are reasonably certain that the soils below a depth of 5 to 8 feet (1.5 to 2.5 m) are strong and relatively incompressible. However, the upper soils are questionable because several uncompacted fills have been found in the neighborhood. Not only are these uncompacted fills loose, they have often contained various debris such as wood, rocks, and miscellaneous trash. However, none of these deleterious materials is present at the ground surface at this site.

Plan a site investigation program for this project and present your plan in the form of written instructions to your field crew. This plan should include specific instructions regarding what to do, where to do it, and any special instructions. You should presume that the field crew is experienced in soil investigation work, but is completely unfamiliar with this site.

4.3   IN-SITU TESTING

Laboratory tests on "undisturbed" soil samples are no better than the quality of the sample. Depending on the type of test, the effects of sample disturbance can be significant, especially in sands. Fortunately, we can often circumvent these problems by using *in-situ* (in-place) testing methods. These entail bringing the test equipment to the field and testing the soils in-place.

In addition to bypassing sample disturbance problems, in-situ tests have the following advantages:

- They are usually less expensive, so a greater number of tests can be performed, thus characterizing the soil in more detail.
- The test results are available immediately.

However, they also have disadvantages, including:

- Often no sample is obtained, thus making soil classification more difficult.
- The engineer has less control over confining stresses and drainage.

In most cases, we must use empirical correlations and calibrations to convert in-situ test results to appropriate engineering properties for design. Many such methods have been published, and some of them are included in this book. Most of these correlations were developed for clays of low to moderate plasticity or for quartz sands, and thus may not be appropriate for special soils such as very soft clays, organic soils, sensitive clays, fissured clays, cemented soils, calcareous sands, micaceous sands, collapsible soils, and frozen soils.

Some in-situ test methods have been in common use for several decades, while others are relative newcomers. Many of these tests will probably continue to become more common in engineering practice.

Standard Penetration Test (SPT)

One of the most common in-situ tests is the standard penetration test, or SPT. This test was originally developed in the late 1920s and has been used most extensively in North and South America, the United Kingdom, and Japan. Because of this long record of experience, the test is well established in engineering practice. Unfortunately, it is also plagued by many problems that affect its accuracy and reproducibility and is slowly being replaced by other test methods, especially on larger and more critical projects.

Test Procedure

The test procedure was not standardized until 1958 when ASTM standard D1586 first appeared. It is essentially as follows[1]:

1. Drill a 60 to 200 mm (2.5–8 in) diameter exploratory boring to the depth of the first test.

2. Insert the SPT sampler (also known as a *split-spoon sampler*) into the boring. The shape and dimensions of this sampler are shown in Figure 4.8. It is connected via steel rods to a 63.5 kg (140 lb) hammer, as shown in Figure 4.9.

3. Using either a rope and cathead arrangement or an automatic tripping mechanism, raise the hammer a distance of 760 mm (30 in) and allow it to fall. This energy drives the sampler into the bottom of the boring. Repeat this process until the sampler has penetrated a distance of 450 mm (18 in), recording the number of hammer blows required for each 150 mm (6 in) interval. Stop the test if more than fifty blows are required for any of the intervals, or if more than one hundred total blows are required. Either of these events is known as *refusal* and is so noted on the boring log.

4. Compute the N value by summing the blow counts for the last 300 mm (12 in) of penetration. The blow count for the first 150 mm (6 in) is retained for reference purposes, but not used to compute N because the bottom of the boring is likely to be disturbed by the drilling process and may be covered with loose soil that fell from the sides of the boring. Note that the N value is the same regardless of whether the engineer is using English or SI units.

5. Remove the SPT sampler; remove and save the soil sample.

6. Drill the boring to the depth of the next test and repeat steps 2 through 6 as required.

Thus, N values may be obtained at intervals no closer than 450 mm (18 in).

Unfortunately, the procedure used in the field varies, partially due to changes in the standard, but primarily as a result of variations in the test procedure and poor workmanship. The test results are sensitive to these variations, so the N value is not as repeatable as we would like. The principal variants are as follows:

- Method of drilling
- How well the bottom of the hole is cleaned before the test
- Presence or lack of drilling mud
- Diameter of the drill hole
- Location of the hammer (surface type or down-hole type)
- Type of hammer, especially whether it has a manual or automatic tripping mechanism

---

[1]See the ASTM D1586 standard for the complete procedure.

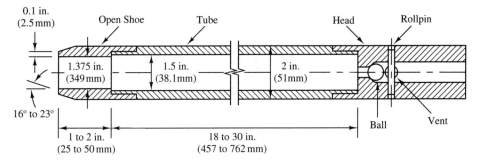

Figure 4.8 The SPT sampler (Adapted from ASTM D1586; Copyright ASTM, used with permission).

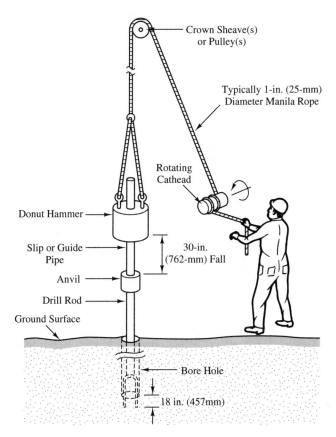

Figure 4.9 The SPT sampler in place in the boring with hammer, rope, and cathead (Adapted from Kovacs, et al., 1981).

- Number of turns of the rope around the cathead
- Actual hammer drop height (manual types are often as much as 25 percent in error)
- Mass of the anvil that the hammer strikes
- Friction in rope guides and pulleys
- Wear in the sampler drive shoe
- Straightness of the drill rods
- Presence or absence of liners inside the sampler (this seemingly small detail can alter the test results by 10 to 30 percent)
- Rate at which the blows are applied

As a result of these variables, the test results will vary depending on the crew and equipment. These variations, as well as other aspects of the test, were the subject of increased scrutiny during the 1970s and 1980s along with efforts to further standardize the "standard" penetration test (DeMello, 1971; Nixon, 1982). Based on these studies, Seed et al. (1985) recommended the following additional criteria be met when conducting standard penetration tests:

- Use the rotary wash method to create a boring that has a diameter between 200 and 250 mm (4–5 in). The drill bit should provide an upward deflection of the drilling mud (tricone or baffled drag bit).
- If the sampler is made to accommodate liners, then these liners should be used so the inside diameter is 35 mm (1.38 in).
- Use A or AW size drill rods for depths less than 15 m (50 ft) and N or NW size for greater depths.
- Use a hammer that has an efficiency of 60 percent.
- Apply the hammer blows at a rate of 30 to 40 per minute.

Fortunately, automatic hammers are becoming more popular. They are much more consistent than hand-operated hammers, and thus improve the reliability of the test.

Although much has been said about the disadvantages of the SPT, it does have at least three important advantages over other in-situ test methods: First, it obtains a sample of the soil being tested. This permits direct soil classification. Most of the other methods do not include sample recovery, so soil classification must be based on conventional sampling from nearby borings and on correlations between the test results and soil type. Second, it is very fast and inexpensive because it is performed in borings that would have been drilled anyway. Finally, nearly all drill rigs used for soil exploration are equipped to perform this test, whereas other in-situ tests require specialized equipment that may not be readily available.

Corrections to the Test Data

We can improve the raw SPT data by applying certain correction factors. The variations in testing procedures may be at least partially compensated by converting the measured N to N_{60} as follows (Skempton, 1986):

TABLE 4.3 SPT HAMMER EFFICIENCIES (Adapted from Clayton, 1990).

Country	Hammer Type (per Figure 4.10)	Hammer Release Mechanism	Hammer Efficiency E_m
Argentina	Donut	Cathead	0.45
Brazil	Pin weight	Hand dropped	0.72
China	Automatic	Trip	0.60
	Donut	Hand dropped	0.55
	Donut	Cathead	0.50
Colombia	Donut	Cathead	0.50
Japan	Donut	Tombi trigger	0.78–0.85
	Donut	Cathead 2 turns + special release	0.65–0.67
UK	Automatic	Trip	0.73
US	Safety	2 turns on cathead	0.55–0.60
	Donut	2 turns on cathead	0.45
Venezuela	Donut	Cathead	0.43

$$N_{60} = \frac{E_m C_B C_S C_R N}{0.60}$$

(4.1)

Where:

N_{60} = SPT N value corrected for field procedures

E_m = hammer efficiency (from Table 4.3)

C_B = borehole diameter correction (from Table 4.4)

C_S = sampler correction (from Table 4.4)

C_R = rod length correction (from Table 4.4)

N = measured SPT N value

Many different hammer designs are in common use, none of which is 100 percent efficient. Some common hammer designs are shown in Figure 4.10, and typical hammer efficiencies are listed in Table 4.3. Many of the SPT-based design correlations were developed using hammers that had an efficiency of about 60 percent, so Equation 4.2 corrects the results from other hammers to that which would have been obtained if a 60 percent efficient hammer was used.

The SPT data also may be adjusted using an *overburden correction* that compensates for the effects of effective stress. Deep tests in a uniform soil deposit will have higher N values than shallow tests in the same soil, so the overburden correction adjusts the measured N values to what they would have been if the vertical effective stress, σ_z', was 100 kPa (2000 lb/ft^2). The corrected value, $(N_1)_{60}$, is (Liao and Whitman, 1985):

TABLE 4.4   BOREHOLE, SAMPLER, AND ROD CORRECTION FACTORS (Adapted from Skempton, 1986).

Factor	Equipment Variables	Value
Borehole diameter factor, C_B	65–115 mm (2.5–4.5 in)	1.00
	150 mm (6 in)	1.05
	200 mm (8 in)	1.15
Sampling method factor, C_S	Standard sampler	1.00
	Sampler without liner (not recommended)	1.20
Rod length factor, C_R	3–4 m (10–13 ft)	0.75
	4–6 m (13–20 ft)	0.85
	6–10 m (20–30 ft)	0.95
	>10 m (>30 ft)	1.00

$$(N_1)_{60} = N_{60} \sqrt{\frac{2000 \text{ lb/ft}^2}{\sigma_z'}}$$

(4.2 English)

$$(N_1)_{60} = N_{60} \sqrt{\frac{100 \text{ kPa}}{\sigma_z'}}$$

(4.2 SI)

Where:

$(N_1)_{60}$ = SPT N value corrected for field procedures and overburden stress

σ_z' = vertical effective stress at the test location

N_{60} = SPT N value corrected for field procedures

Although Liao and Whitman did not place any limits on this correction, it is probably best to keep $(N_1)_{60} \leq 2 N_{60}$. This limit avoids excessively high $(N_1)_{60}$ values at shallow depths.

The use of correction factors is often a confusing issue. Corrections for field procedures are always appropriate, but the overburden correction may or may not be appropriate depending on the procedures used by those who developed the analysis method under consideration. In this book, the overburden correction should be applied only when the analysis procedure calls for an $(N_1)_{60}$ value.

Uses of SPT Data

The SPT N value, as well as many other test results, is only an *index* of soil behavior. It does not directly measure any of the conventional engineering properties of soil and is

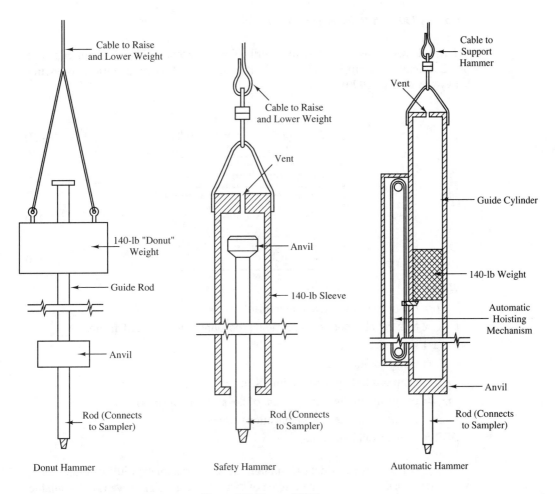

Figure 4.10 Types of SPT hammers.

useful only when appropriate correlations are available. Many such correlations exist, all of which were obtained empirically.

Unfortunately, most of these correlations are very approximate, especially those based on fairly old data that were obtained when test procedures and equipment were different from those now used. In addition, because of the many uncertainties in the SPT results, all of these correlations have a wide margin of error.

Be especially cautious when using correlations between SPT results and engineering properties of clays because these functions are especially crude. In general, the SPT should be used only in sandy soils.

Correlation with Relative Density

The relative density is a useful way to describe the consistency of sands, as discussed in Chapter 3. It may be determined from SPT data using the following empirical correlation (Kulhawy and Mayne, 1990):

$$D_r = \sqrt{\frac{(N_1)_{60}}{C_P C_A C_{OCR}}} \times 100\% \tag{4.3}$$

$$C_P = 60 + 25 \log D_{50} \tag{4.4}$$

$$C_A = 1.2 + 0.05 \log\left(\frac{t}{100}\right) \tag{4.5}$$

$$C_{OCR} = OCR^{0.18} \tag{4.6}$$

Where:

D_r = relative density (in decimal form)

$(N_1)_{60}$ = SPT N value corrected for field procedures and overburden stress

C_P = grain-size correction factor

C_A = aging correction factor

C_{OCR} = overconsolidation correction factor

D_{50} = grain size at which 50 percent of the soil is finer (mm)

t = age of soil (time since deposition) (years)

OCR = overconsolidation ratio

Rarely will we have test data to support all of the parameters in Equations 4.4 to 4.6, so it often is necessary to estimate some of them. If a grain size curve is not available, D_{50} may be estimated from a visual examination of the soil with reference to Table 3.4. Geologists can sometimes estimate the age of sand deposits, but Equation 4.5 is not very sensitive to the chosen value, so a "wild guess" is probably sufficient. A value of $t = 1000$ years is probably sufficient for most analyses. The overconsolidation ratio is rarely known in sands, but values of about 1 in loose sands ($N_{60} < 10$) to about 4 in dense sands ($N_{60} > 50$) should be sufficient for Equation 4.6.

Correlation with Shear Strength

DeMello (1971) suggested a correlation between SPT results and the effective friction angle of uncemented sands, ϕ', as shown in Figure 4.11. This correlation should be used only at depths greater than about 2 m (7 ft).

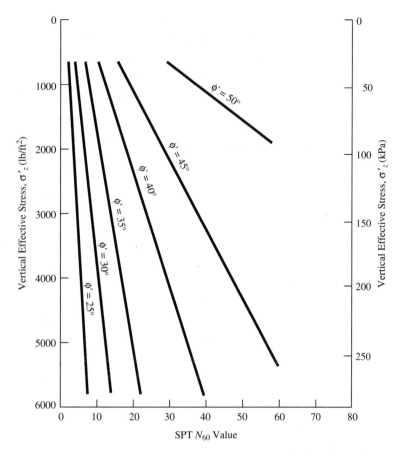

Figure 4.11 Empirical correlation between N_{60} and ϕ' for uncemented sands (Adapted from DeMello, 1971).

Example 4.1

A 6-inch diameter exploratory boring has been drilled through a fine sand to a depth of 19 ft. An SPT N-value of 23 was obtained at this depth using a USA safety hammer with a standard sampler. The boring then continued to greater depths, eventually encountering the ground-water table at a depth of 35 ft. The unit weight of the sand is unknown. Compute $(N_1)_{60}$, ϕ', and D_r at the test location, and use this data to classify the consistency of the sand.

Solution

Per Table 3.2, SP soils above the groundwater table typically have a unit weight of 95 to 125 lb/ft³. The measured N-value of 23 suggests a moderately dense sand, so use $\gamma = 115$ lb/ft³. Therefore, at the sample depth:

$$\sigma_z' = \Sigma \, \gamma H - u = (115 \text{ lb/ft}^3)(19 \text{ ft}) - 0 = 2185 \text{ lb/ft}^2$$

$$N_{60} = \frac{E_m C_B C_S C_R N}{0.60} = \frac{(0.57)(1.05)(1.00)(0.85)(23)}{0.60} = 19.5$$

$$(N_1)_{60} = N_{60} \sqrt{\frac{2000 \text{ lb/ft}^2}{\sigma_z'}} = 19.5 \sqrt{\frac{2000 \text{ lb/ft}^2}{2185 \text{ lb/ft}^2}} = \mathbf{18.7} \qquad \Leftarrow Answer$$

Per Figure 4.11: $\mathbf{\phi' = 40°}$ $\quad \Leftarrow Answer$

D_{50} was not given. However, the soil is classified as a fine sand, and Table 3.4 indicates such soils have grain sizes of 0.075 to 0.425 mm. Therefore, use $D_{50} = 0.2$ mm.

$$C_P = 60 + 25 \log D_{50} = 60 + 25 \log(0.2) = 42.5$$

t was not given, so use $t = 1000$ years.

$$C_A = 1.2 + 0.05 \log \left(\frac{t}{100}\right) = 1.2 + 0.05 \log \left(\frac{1000}{100}\right) = 1.25$$

OCR was not given, so use OCR = 2.5.

$$C_{OCR} = OCR^{0.18} = 2.5^{0.18} = 1.18$$

$$D_r = \sqrt{\frac{(N_1)_{60}}{C_P C_A C_{OCR}}} \times 100\% = \sqrt{\frac{18.7}{(42.5)(1.25)(1.18)}} \times 100\% = \mathbf{55\%} \qquad \Leftarrow Answer$$

Per Table 3.3, the soil is **Medium Dense** $\Leftarrow Answer$

Cone Penetration Test (CPT)

The *cone penetration test* [ASTM D3441], or CPT, is another common in-situ test (Schmertmann, 1978; De Ruiter, 1981; Meigh, 1987; Robertson and Campanella, 1989; Briaud and Miran, 1991). Most of the early development of this test occurred in western Europe in the 1930s and again in the 1950s. Although many different styles and configurations have been used, the current standard grew out of work performed in the Netherlands, so it is sometimes called the *Dutch Cone*. The CPT has been used extensively in Europe for many years and is becoming increasingly popular in North America and elsewhere.

Two types of cones are commonly used: the *mechanical cone* and the *electric cone*, as shown in Figure 4.12. Both have two parts, a 35.7-mm diameter cone-shaped tip with a 60° apex angle and a 35.7 mm diameter $\times$ 133.7 mm long cylindrical sleeve. A hydraulic ram pushes this assembly into the ground and instruments measure the resistance to penetration. The *cone resistance*, q_c, is the total force acting on the cone divided by the projected area of the cone (10 cm^2); and the *cone side friction*, f_{sc}, is the total frictional force acting on the friction sleeve divided by its surface area (150 cm^2). It is common to express the side friction in terms of the *friction ratio*, R_f, which is equal to $f_{sc}/q_c \times 100\%$.

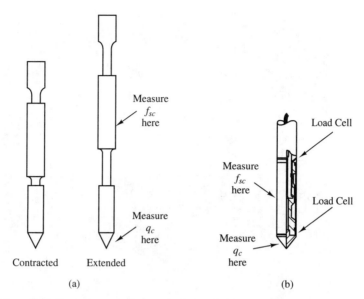

Figure 4.12 Types of cones: (a) A mechanical cone (also known as a Begemann Cone); and (b) An electric cone (also known as a Fugro Cone).

The operation of the two types of cones differs in that the mechanical cone is advanced in stages and measures q_c and f_{sc} at intervals of about 20 cm, whereas the electric cone includes built-in strain gauges and is able to measure q_c and f_{sc} continuously with depth. In either case, the CPT defines the soil profile with much greater resolution than does the SPT.

CPT rigs are often mounted in large three-axle trucks such as the one in Figure 4.13. These are typically capable of producing maximum thrusts of 10 to 20 tons (100–200 kN). Smaller, trailer-mounted or truck-mounted rigs also are available.

The CPT has been the object of extensive research and development (Robertson and Campanella, 1983) and thus is becoming increasingly useful to the practicing engineer. Some of this research effort is now being conducted using cones equipped with pore pressure transducers in order to measure the excess pore water pressures that develop while conducting the test. These are known as *piezocones*, and the enhanced procedure is known as a CPTU test. These devices promise to be especially useful in saturated clays.

A typical plot of CPT results is shown in Figure 4.14.

Although the CPT has many advantages over the SPT, there are at least three important disadvantages:

- No soil sample is recovered, so there is no opportunity to inspect the soils.
- The test is unreliable or unusable in soils with significant gravel content.
- Although the cost per foot of penetration is less than that for borings, it is necessary to mobilize a special rig to perform the CPT.

Figure 4.13 A truck-mounted CPT rig. A hydraulic ram, located inside the truck, pushes the cone into the ground using the weight of the truck as a reaction.

Uses of CPT Data

The CPT is an especially useful and inexpensive way to evaluate soil profiles. Since it retrieves data continuously with depth (with electric cones) or at very close intervals (with mechanical cones), the CPT is able to detect fine changes in the stratigraphy. Therefore, engineers often use the CPT in the first phase of subsurface investigation, saving boring and sampling for the second phase.

The CPT also can be used to assess the engineering properties of the soil through the use of empirical correlations. Those correlations intended for use in cohesionless soils are generally more accurate and most commonly used. They are often less accurate in cohesive soils because of the presence of excess pore water pressures and other factors. However, the piezocone may overcome this problem.

Correlations are also available to directly relate CPT results with foundation behavior. These are especially useful in the design of deep foundations, as discussed in Chapter 14.

Correlation with Soil Classification

Because the CPT does not recover any soil samples, it is not a substitute for conventional exploratory borings. However, it is possible to obtain an approximate soil classification using the correlation shown in Figure 4.15.

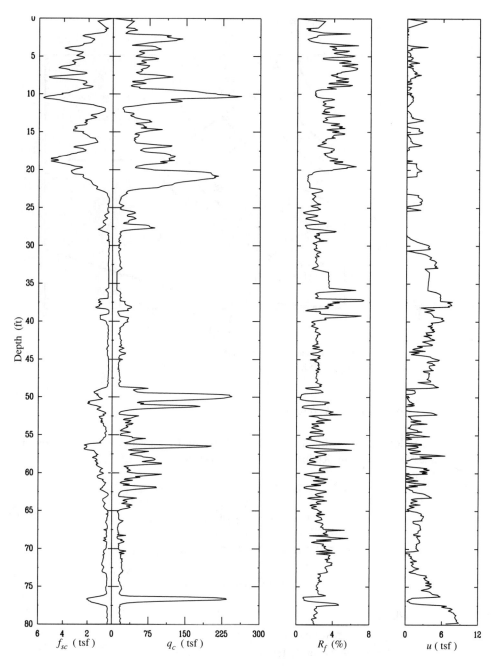

Figure 4.14 Sample CPT test results. These results were obtained from a piezocone, and thus also include a plot of pore water pressure, u, vs. depth. All stresses and pressures are expressed in tons per square foot (tsf). For practical purposes, 1 tsf = 1 kg/cm^2 (Alta Geo Cone Penetrometer Testing Services, Sandy, Utah).

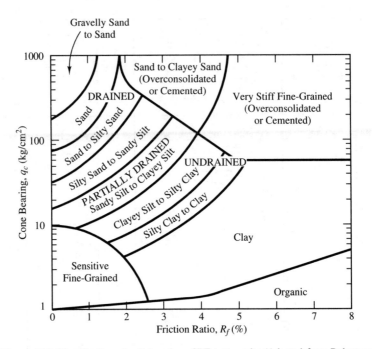

Figure 4.15 Classification of soil based on CPT test results (Adapted from Robertson and Campanella, 1983).

Correlation with Relative Density

The following approximate relationship between CPT results and the relative density of sands (Adapted from Kulhawy and Mayne, 1990):

$$D_r = \sqrt{\left(\frac{q_c}{315 \, Q_c \, \text{OCR}^{0.18}}\right) \sqrt{\frac{2000 \, \text{lb/ft}^2}{\sigma'_z}}} \times 100\% \qquad \text{(4.7 English)}$$

$$D_r = \sqrt{\left(\frac{q_c}{315 \, Q_c \, \text{OCR}^{0.18}}\right) \sqrt{\frac{100 \, \text{kPa}}{\sigma'_z}}} \times 100\% \qquad \text{(4.7 SI)}$$

Where:

q_c = cone resistance (ton/ft^2 or kg/cm^2)

Q_c = compressibility factor

= 0.91 for highly compressible sands

= 1.00 for moderately compressible sands

= 1.09 for slightly compressible sands

For purposes of solving this formula, a sand with a high fines content or a high mica content is "highly compressible," whereas a pure quartz sand is "slightly compressible."

OCR = overconsolidation ratio

σ_z' = vertical effective stress (lb/ft² or kPa)

Correlation with Shear Strength

The CPT results also have been correlated with shear strength parameters, especially in sands. Figure 4.16 presents Robertson and Campanella's 1983 correlation for uncemented, normally consolidated quartz sands. For overconsolidated sands, subtract 1° to 2° from the effective friction angle obtained from this figure.

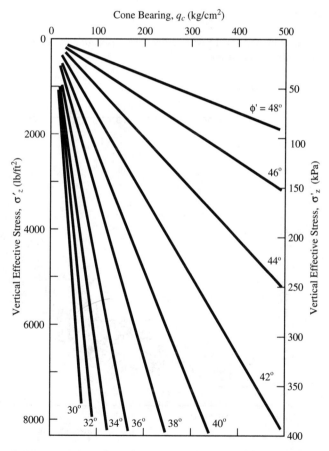

Figure 4.16 Relationship between CPT results, overburden stress and effective friction angle for uncemented, normally consolidated quartz sands (Adapted from Robertson and Campanella, 1983).

Correlation with SPT N-Value

Since the SPT and CPT are the two most common in-situ tests, it often is useful to convert results from one to the other. The ratio q_c/N_{60} as a function of the mean grain size, D_{50}, is shown in Figure 4.17. Note that N_{60} does not include an overburden correction.

Be cautious about converting CPT data to equivalent N values, and then using SPT-based analysis methods. This technique compounds the uncertainties because it uses two correlations—one to convert to N, and then another to compute the desired quantity.

Example 4.2

Using the CPT log in Figure 4.14, classify the soil stratum located between depths of 28 and 48 ft, and compute the equivalent SPT N-value.

Solution

This stratum has $q_c = 15$ tsf and $R_f = 2.0\%$. According to Figure 4.15, this soil is probably a **clayey silt.** $\Leftarrow Answer$

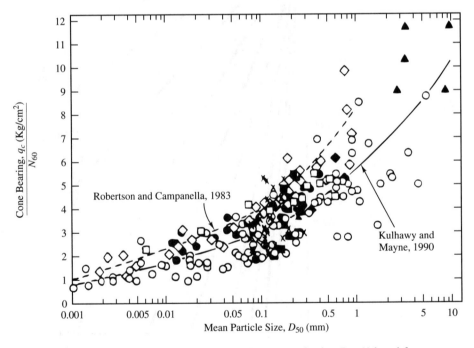

Figure 4.17 Correlation between q_c/N_{60} and the mean grain size, D_{50}. (Adapted from Kulhawy and Mayne, 1990.) Copyright © 1990 Electric Power Research Institute, reprinted with permission.

No grain-size data is available. However, based on the soil classification and Table 3.4, the mean grain size, D_{50}, is probably about 0.02 mm. According to Figure 4.17, $q_c/N_{60} = 1$. Therefore $N_{60} = 15$ $\Leftarrow$ *Answer*

Commentary

Equation 4.8 and Figure 4.16 do not apply to this soil because it is not a sand.

Vane Shear Test (VST)

The Swedish engineer John Olsson developed the vane shear test (VST) in the 1920s to test the sensitive Scandinavian marine clays in-situ. The VST has grown in popularity, especially since the Second World War, and is now used around the world.

   This test [ASTM D2573] consists of inserting a metal vane into the soil, as shown in Figure 4.18, and rotating it until the soil fails in shear. The undrained shear strength may be determined from the torque at failure, the vane dimensions, and other factors. The vane can be advanced to greater depths by simply pushing it deeper (especially in softer soils) or the test can be performed below the bottom of a boring and repeated as the boring is advanced. However, because the vane must be thin to minimize soil disturbance, it is only strong enough to be used in soft to medium cohesive soils. The test is performed rapidly (about 1 minute to failure) and therefore measures only the undrained strength.

   The shear surface has a cylindrical shape, and the data analysis neglects any shear resistance along the top and bottom of this cylinder. Usually the vane height-to-diameter ratio is 2, which, when combined with the applied torque, produces the following theoretical formula:

$$s_u = \frac{6 T_f}{7 \pi d^3} \tag{4.8}$$

Where:

   s_u = undrained shear strength

   T_f = torque at failure

   d = diameter of vane

   However, several researchers have analyzed failures of embankments, footings, and excavations using vane shear tests (knowing that the factor of safety was 1.0) and found that Equation 4.9 often overestimates s_u. Therefore, an empirical correction factor, λ, as shown in Figure 4.19, is applied to the test results:

$$\boxed{s_u = \frac{6 \lambda T_f}{7 \pi d^3}} \tag{4.9}$$

   An additional correction factor of 0.85 should be applied to test results from organic soils other than peat (Terzaghi, Peck, and Mesri, 1996).

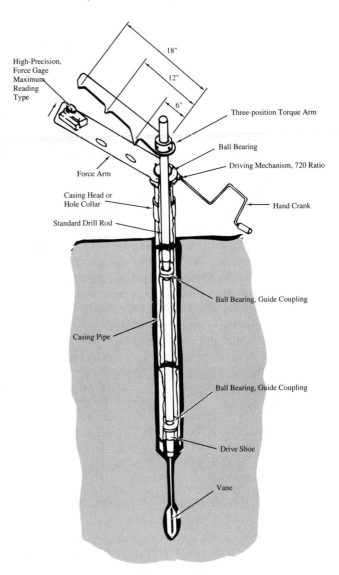

High-Precision,
Force Gage
Maximum
Reading
Type

18"

12"

6"

Three-position Torque Arm

Ball Bearing

Driving Mechanism, 720 Ratio

Force Arm

Casing Head or
Hole Collar

Hand Crank

Standard Drill Rod

Ball Bearing, Guide Coupling

Casing Pipe

Ball Bearing, Guide Coupling

Drive Shoe

Vane

Figure 4.18 The vane shear test (U.S.
Navy, 1982a).

Pressuremeter Test (PMT)

In 1954, a young French engineering student named Louis Ménard began to develop a
new type of in-situ test: the pressuremeter test. Although Kögler had done some limited
work on a similar test some twenty years earlier, it was Ménard who made it a practical
reality.

The pressuremeter is a cylindrical balloon that is inserted into the ground and inflated, as shown in Figures 4.20 and 4.21. Measurements of volume and pressure can be

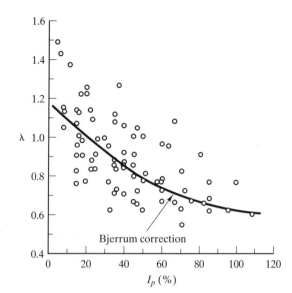

Figure 4.19 Vane shear correction factor, λ (from *Soil Mechanics in Engineering Practice,* 3rd ed. By Terzaghi, Peck, and Mesri; copyright © 1996; used by permission of John Wiley and Sons).

used to evaluate the in-situ stress, compressibility, and strength of the adjacent soil and thus the behavior of a foundation (Baguelin et al., 1978; Briaud, 1992).

The PMT may be performed in a carefully drilled boring or the test equipment can be combined with a small auger to create a self-boring pressuremeter. The latter design provides less soil disturbance and more intimate contact between the pressuremeter and the soil.

The PMT produces much more direct measurements of soil compressibility and lateral stresses than do the SPT and CPT. Thus, in theory, it should form a better basis for settlement analyses, and possibly for pile capacity analyses. However, the PMT is a difficult test to perform and is limited by the availability of the equipment and personnel trained to use it.

Although the PMT is widely used in France and Germany, it is used only occasionally in other parts of the world. However, it may become more popular in the future.

Dilatometer Test (DMT)

The dilatometer (Marchetti, 1980; Schmertmann, 1986b, 1988a, and 1988b), which is one of the newest in-situ test devices, was developed during the late 1970s in Italy by Silvano Marchetti. It is also known as a *flat dilatometer* or a *Marchetti dilatometer* and consists of a 95-mm wide, 15-mm thick metal blade with a thin, flat, circular, expandable steel membrane on one side, as shown in Figure 4.22.

The dilatometer test (DMT) is conducted as follows (Schmertmann, 1986a):

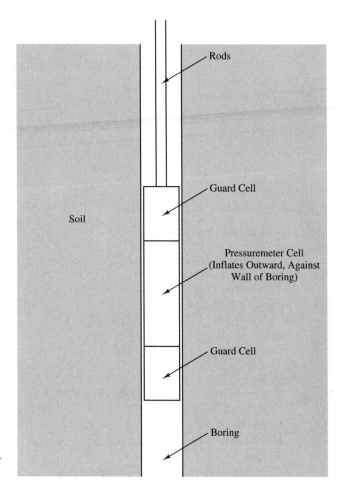

Rods

Guard Cell

Soil

Pressuremeter Cell
(Inflates Outward, Against
Wall of Boring)

Guard Cell

Boring

Figure 4.20 Schematic of the pressuremeter
test.

1. Press the dilatometer into the soil to the desired depth using a CPT rig or some other suitable device.

2. Apply nitrogen gas pressure to the membrane to press it outward. Record the pressure required to move the center of the membrane 0.05 mm into the soil (the *A* pressure) and that required to move its center 1.10 mm into the soil (the *B* pressure).

3. Depressurize the membrane and record the pressure acting on the membrane when it returns to its original position. This is the *C* pressure and is a measure of the pore water pressure in the soil.

4. Advance the dilatometer 150 to 300 mm deeper into the ground and repeat the test. Continue until reaching the desired depth.

Each of these test sequences typically requires 1 to 2 minutes to complete, so a typical *sounding* (a complete series of DMT tests between the ground surface and the desired

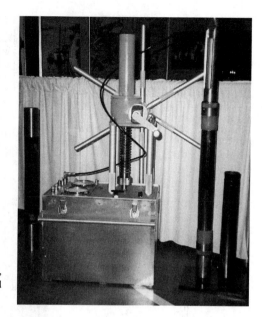

Figure 4.21 A complete pressuremeter set, including three different cell assemblies and the control unit.

depth) may require about 2 hours. In contrast, a comparable CPT sounding might be completed in about 30 minutes.

The primary benefit of the DMT is that it measures the lateral stress condition and compressibility of the soil. These are determined from the A, B, and C pressures and certain equipment calibration factors and expressed as the *DMT indices:*

I_D = material index (a normalized modulus)

K_D = horizontal stress index (a normalized lateral stress)

E_D = dilatometer modulus (theoretical elastic modulus)

Researchers have developed correlations between these indices and certain engineering properties of the soil (Schmertmann, 1988b; Kulhawy and Mayne, 1990), including:

- Classification
- Coefficient of lateral earth pressure, K_0
- Overconsolidation ratio, OCR
- Modulus of elasticity, E, or constrained modulus, M

The CPT and DMT are complementary tests (Schmertmann, 1988b). The cone is a good way to evaluate soil strength, whereas the dilatometer assesses compressibility and in-situ stresses. These three kinds of information form the basis for most foundation engineering analyses. In addition, the dilatometer blade is most easily pressed into the ground

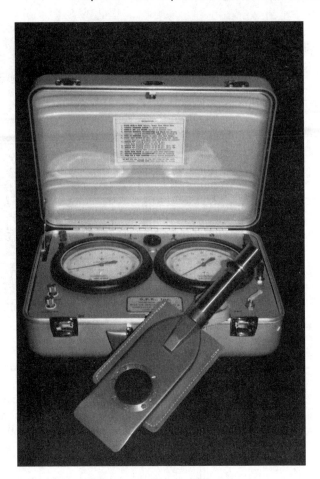

Figure 4.22 The Marchetti dilatometer along with its control unit and nitrogen gas bottle (GPE, Inc., Gainesville, FL).

using a conventional CPT rig, so it is a simple matter to conduct both CPT and DMT tests while mobilizing only a minimum of equipment.

The dilatometer test is a relative newcomer, and thus has not yet become a common engineering tool. Engineers have had only limited experience with it and the analysis and design methods based on DMT results are not yet well developed. However, its relatively low cost, versatility, and compatibility with the CPT suggest that it may enjoy widespread use in the future. It has very good repeatability, and can be used in soft or moderately stiff soils (i.e., those with $N \leq 40$), and provides more direct measurements of stress-strain properties.

Becker Penetration Test

Soils that contain a large percentage of gravel and those that contain cobbles or boulders create problems for most in-situ test methods. Often, the in-situ test device is not able to penetrate through such soils (it meets *refusal*) or the results are not representative because the particles are about the same size as the test device. Frequently, even conventional drilling equipment cannot penetrate through these soils.

One method of penetrating through these very large-grained soils is to use a *Becker hammer drill*. This device, developed in Canada, uses a small diesel pile-driving hammer and percussion action to drive a 135 to 230 mm (5.5–9.0 inch) diameter double-wall steel casing into the ground. The cuttings are sent to the top by blowing air through the casing. This technique has been used successfully on very dense and coarse soils.

The Becker hammer drill also can be used to assess the penetration resistance of these soils using the *Becker penetration test*, which is monitoring the hammer blow-count. The number of blows required to advance the casing 300 mm (1 ft) is the Becker blowcount, N_B. Several correlations are available to convert it to an equivalent SPT N value (Harder and Seed, 1986). One of these correlation methods also considers the bounce chamber pressure in the diesel hammer.

Comparison of In-Situ Test Methods

Each of the in-situ test methods has its strengths and weaknesses. Table 4.5 compares some of the important attributes of the tests described in this chapter.

TABLE 4.5   ASSESSMENT OF IN-SITU TEST METHODS (Adapted from Mitchell, 1978a; used with permission of ASCE)

	Standard Penetration Test	Cone Penetration Test	Pressure-meter Test	Dilatometer Test	Becker Penetration Test
Simplicity and Durability of Apparatus	Simple; rugged	Complex; rugged	Complex; delicate	Complex; moderately rugged	Simple rugged
Ease of Testing	Easy	Easy	Complex	Easy	Easy
Continuous Profile or Point Values	Point	Continuous	Point	Point	Continuous
Basis for Interpretation	Empirical	Empirical; theory	Empirical; theory	Empirical; theory	Empirical
Suitable Soils	All except gravels	All except gravels	All	All except gravels	Sands through boulders
Equipment Availability and Use in Practice	Universally available; used routinely	Generally available; used routinely	Difficult to locate; used on special projects	Difficult to locate; used on special projects	Difficult to locate; used on special projects
Potential for Future Development	Limited	Great	Great	Great	Uncertain

QUESTIONS AND PRACTICE PROBLEMS

4.5 Discuss the advantages of the cone penetration test over the standard penetration test.

4.6 A standard penetration test was performed in a 150-mm diameter boring at a depth of 9.5 m below the ground surface. The driller used a UK-style automatic trip hammer and a standard SPT sampler. The actual blow count, N, was 19. The soil is a normally consolidated fine sand with a unit weight of 18.0 kN/m^3 and $D_{50} = 0.4$ mm. The groundwater table is at a depth of 15 m. Compute the following:

   a. N_{60}

   b. $(N_1)_{60}$

   c. D_r

   d. Consistency (based on Table 3.3)

   e. ϕ'

4.7 Using the cone penetration test data in Figure 4.14, a unit weight of 115 lb/ft^3, and an over-consolidation ratio of 3, compute the following for the soil between depths of 21 and 23 ft. Use a groundwater depth of 15 ft below the ground surface.

   a. Soil classification

   b. D_r (assume the soil has some fines, but no mica)

   c. Consistency (based on Table 3.3)

   d. ϕ'

   e. N_{60} (use an estimated D_{50} of 0.60 mm)

4.4   SYNTHESIS OF FIELD AND LABORATORY DATA

Investigation and testing programs often generate large amounts of information that can be difficult to sort through and synthesize. Real soil profiles are nearly always very complex, so the borings will not correlate and the test results will often vary significantly. Therefore, we must develop a simplified soil profile before proceeding with the analysis. In many cases, this simplified profile is best defined in terms of a one-dimensional function of soil type and engineering properties vs. depth; an idealized boring log. However, when the soil profile varies significantly across the site, one or more vertical cross-sections may be in order.

    The development of these simplified profiles requires a great deal of engineering judgment along with interpolation and extrapolation of the data. It is important to have a feel for the approximate magnitude of the many uncertainties in this process and reflect them in an appropriate degree of conservatism. This judgment comes primarily with experience combined with a thorough understanding of the field and laboratory methodologies.

4.5   ECONOMICS

The site investigation and soil testing phase of foundation engineering is the single largest source of uncertainties. No matter how extensive it is, there is always some doubt whether the borings accurately portray the subsurface conditions, whether the samples are representative, and whether the tests are correctly measuring the soil properties. Engineers attempt to compensate for these uncertainties by applying factors of safety in our analyses. Unfortunately, this solution also increases construction costs.

In an effort to reduce the necessary level of conservatism in the foundation design, the engineer may choose a more extensive investigation and testing program to better define the soils. The additional costs of such efforts will, to a point, result in decreased construction costs, as shown in Figure 4.23. However, at some point, this becomes a matter of diminishing returns, and eventually the incremental cost of additional investigation and testing does not produce an equal or larger reduction in construction costs. The minimum on this curve represents the optimal level of effort.

An example of this type of economic consideration is a decision regarding the value of conducting full-scale pile load tests (see Chapter 13). On large projects, the potential savings in construction costs will often justify one or more tests, whereas on small projects, a conservative design developed without the benefit of load tests may result in a lower total cost.

We also must decide whether to conduct a large number of moderately precise tests (such as the SPT) or a smaller number of more precise but expensive tests (such as the PMT). Handy (1980) suggested the most cost-effective test is the one with a variability

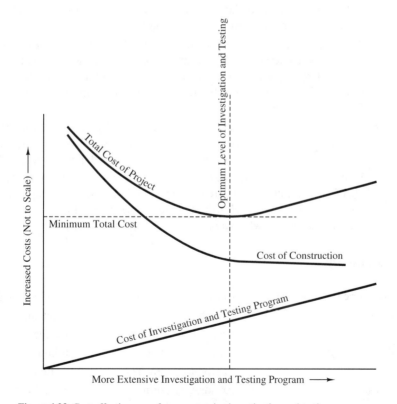

Figure 4.23 Cost effectiveness of more extensive investigation and testing programs.

consistent with the variability of the soil profile. Thus, a few precise tests might be appropriate in a uniform soil deposit, but more data points, even if they are less precise, are more valuable in an erratic deposit.

SUMMARY

Major Points

1. Soil and rock are natural materials. Therefore, their engineering properties vary from site to site and must be determined individually for each project. This process is known as site exploration and characterization.

2. The first step in a site exploration and characterization program typically consists of conducting a background literature search and a field reconnaissance.

3. Site exploration efforts usually include drilling exploratory borings and/or digging exploratory trenches, as well as obtaining soil and rock samples from these borings or trenches.

4. We bring these soil and rock samples to a laboratory to conduct standard laboratory tests. These tests help classify the soil, determine its strength, and assess its compressibility. Other tests might also be conducted.

5. In-situ tests are those conducted in the ground. These techniques are especially useful in soils that are difficult to sample, such as clean sands. In-situ methods include the standard penetration test (SPT), the cone penetration test (CPT), the vane shear test (VST), the pressuremeter test (PMT), the dilatometer test (DMT), and the Becker penetration test.

6. An investigation and testing program will typically generate large amounts of data, even though only a small portion of the soil is actually tested. The engineer must synthesize this data into a simplified form to be used in analyses and design.

7. Investigation and testing always involves uncertainties and risks. These can be reduced, but not eliminated, by drilling more borings, retrieving more samples, and conducting more tests. However, there are economic limits to such endeavors, so we must determine what amount of work is most cost effective.

Vocabulary

Becker penetration test	Cone penetration test	Field reconnaissance
Blow count	Coring	Flight auger
Boring log	Dilatometer test	Hammer
Bucket auger	Disturbed sample	Hollow-stem auger
Casing	Exploratory trench	In-situ test
Caving	Exploratory boring	Observation well

Overburden correction	Shelby tube sampler	Standard penetration test
Pressuremeter test	Site investigation	Undistrubed sample
Rotary wash boring	Squeezing	Vane shear test

COMPREHENSIVE QUESTIONS AND PRACTICE PROBLEMS

4.8 Classify the soil stratum between depths of 66 and 80 ft in Figure 4.14. What is the significance of the spike in the plots at a depth of 77 ft?

4.9 The following standard penetration tests results were obtained in a uniform silty sand:

Depth (m)	N_{60}
1	12
2	13
3	18
5	15

The groundwater table is at a depth of 2.5 m. Assume a reasonable value for γ, then determine ϕ' for each test. Finally, determine a single design ϕ' value for this stratum.

4.10 A series of vane shear tests have been performed on a soft clay stratum. The results of these tests are as follows:

Depth (m)	T_f (N-m)
5.0	9.0
5.5	10.7
7.5	12.0
9.0	14.7

The vane was 60 mm in diameter and 120 mm long. The soil has a liquid limit of 100 and a plastic limit of 30. Compute the undrained shear strength for each test, then develop a plot of undrained shear strength vs. depth. This plot should have depth on the vertical axis, with zero at the top of the plot.

Part B

Shallow Foundation Analysis and Design

5

Shallow Foundations

> *The most important thing is to keep the most*
> *important thing the most important thing.*

Shallow foundations are those that transmit structural loads to the near-surface soils. These include *spread footing foundations* and *mat foundations*. This chapter introduces both types of foundations, then Chapters 6 to 10 discuss the various geotechnical and structural design aspects.

5.1 SPREAD FOOTINGS

A spread footing (also known as a *footer* or simply a *footing*) is an enlargement at the bottom of a column or bearing wall that spreads the applied structural loads over a sufficiently large soil area. Typically, each column and each bearing wall has its own spread footing, so each structure may include dozens of individual footings.

Spread footings are by far the most common type of foundation, primarily because of their low cost and ease of construction. They are most often used in small to medium-size structures on sites with moderate to good soil conditions, and can even be used on some large structures when they are located at sites underlain by exceptionally good soil or shallow bedrock.

Spread footings may be built in different shapes and sizes to accommodate individual needs, as shown in Figure 5.1. These include the following:

- *Square spread footings* (or simply *square footings*) have plan dimensions of $B \times B$. The depth from the ground surface to the bottom of the footing is D and the thickness is T. Square footings usually support a single centrally-located column.

145

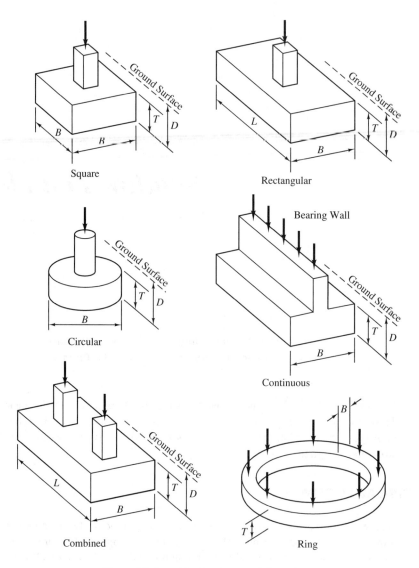

Figure 5.1 Spread footing shapes and dimensions.

- *Rectangular spread footings* have plan dimensions of $B \times L$, where L is the longest dimension. These are useful when obstructions prevent construction of a square footing with a sufficiently large base area and when large moment loads are present.

- *Circular spread footings* are round in plan view. These are most frequently used as foundations for light standards, flagpoles, and power transmission lines. If these foundations extend to a large depth (i.e., D/B greater than about 3), they may behave more like a deep foundation (see Chapter 11).

- *Continuous spread footings* (also known as *wall footings* or *strip footings*) are used to support bearing walls.
- *Combined footings* are those that support more than one column. These are useful when columns are located too close together for each to have its own footing.
- *Ring spread footings* are continuous footings that have been wrapped into a circle. This type of footing is commonly used to support the walls of above-ground circular storage tanks. However, the contents of these tanks are spread evenly across the total base area, and this weight is probably greater than that of the tank itself. Therefore, the geotechnical analyses of tanks usually treat them as circular foundations with diameters equal to the diameter of the tank.

Sometimes it is necessary to build spread footings very close to a property line, another structure, or some other place where no construction may occur beyond one or more of the exterior walls. This circumstance is shown in Figure 5.2. Because such a footing cannot be centered beneath the column, the load is eccentric. This can cause the footing to rotate and thus produce undesirable moments and displacements in the column.

One solution to this problem is to use a *strap footing* (also known as a *cantilever footing*), which consists of an eccentrically loaded footing under the exterior column connected to the first interior column using a *grade beam*. This arrangement, which is similar to a combined footing, provides the necessary moment in the exterior footing to counter the eccentric load. Sometimes we use grade beams to connect all of the spread footings in a structure to provide a more rigid foundation system.

Materials

Before the mid-nineteenth-century, almost all spread footings were made of masonry, as shown in Figure 5.3. *Dimension-stone footings* were built of stones cut and dressed to specific sizes and fit together with minimal gaps, while *rubble-stone footings* were built from random size material joined with mortar (Peck et al., 1974). These footings had very little tensile strength, so builders had to use large height-to-width ratios to keep the flexural stresses tolerably small and thus avoid tensile failures.

Although masonry footings were satisfactory for small structures, they became large and heavy when used in heavier structures, often encroaching into the basement. For example, the masonry footings beneath the nine-story Home Insurance Building in Chicago (built in 1885) had a combined weight equal to that of one of the stories (Peck, 1948). As larger structures became more common, it was necessary to develop footings that were shorter and lighter, yet still had the same base dimensions. This change required structural materials that could sustain flexural stresses.

The *steel grillage footings* used in the ten-story Montauk Block Building in Chicago in 1882, may have been the first spread footings designed to resist flexure. They included several layers of railroad tracks, as shown in Figure 5.4. The flexural strength of the steel permitted construction of a short and lightweight footing. Steel grillage footings, modified to use I-beams instead of railroad tracks, soon became the dominant design. They prevailed until the advent of reinforced concrete in the early twentieth century.

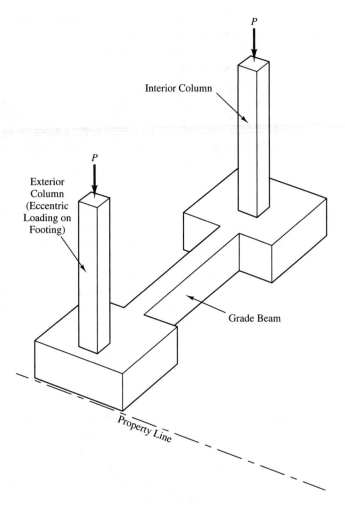

P

Interior Column

P

Exterior
Column
(Eccentric
Loading on
Footing)

Grade Beam

Property Line

Figure 5.2 Use of a strap footing to support
exterior columns when construction cannot
extend beyond the property line.

Figure 5.5 shows a typical reinforced concrete footing. These are very strong, economical, durable, and easy to build. Reinforced concrete footings are much thinner than
the old masonry footings, so they do not require large excavations and do not intrude into
basements. Thus, nearly all spread footings are now made of reinforced concrete.

Construction Methods

Contractors usually use a *backhoe* to excavate spread footings, as shown in Figure 5.6.
Typically, some hand work is also necessary to produce a clean excavation. Once the excavation is open, it is important to check the exposed soils to verify that they are comparable to those used in the design. Inspectors often check the firmness of these soils using a

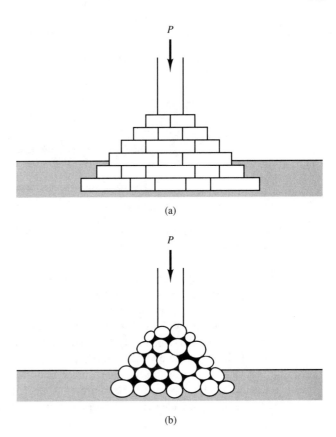

Figure 5.3 (a) Dimension-stone footing, and (b) Rubble-stone footing.

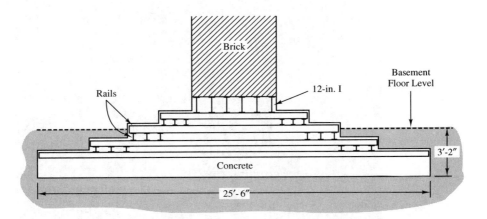

Figure 5.4 Steel grillage footing made from railroad tracks, Montauk Block Building, Chicago, 1882. The concrete that surrounded the steel was for corrosion protection only (Peck, 1948).

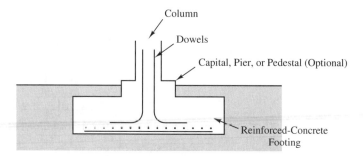

Figure 5.5 Reinforced concrete spread footing.

9-mm (3/8 in) diameter steel probe. If the soil conditions are not as anticipated, especially if they are too soft, it may be necessary to revise the design accordingly.

Most soils have sufficient strength to stand vertically until it is time to pour the concrete. This procedure of pouring the concrete directly against the soil is known as pouring a *neat footing*, as shown in Figure 5.7a. Sometimes shallow wooden forms are placed above the excavation, as shown in Figure 5.7b, so the top of the footing is at the proper elevation. If the soil will not stand vertically, such as with clean sands or gravels, it is necessary to make a larger excavation and build a full-depth wooden form, as shown in Figure 5.7c. This is known as a *formed footing*.

Once the excavation has been made and cleaned, and the forms (if needed) are in place, the contractor places the reinforcing steel. If the footing will support a wood or steel structure, threaded *anchor bolts* and/or steel brackets are embedded into the con-

Figure 5.6 A backhoe making an excavation for a spread footing.

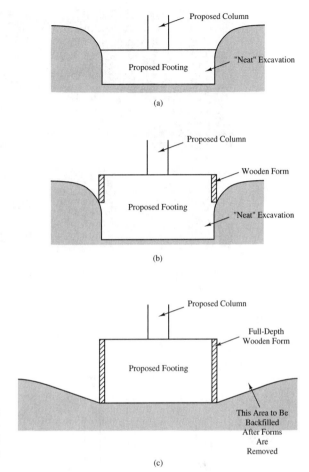

Figure 5.7 Methods of placing concrete in footings: (a) Neat excavation, (b) Neat excavation with wooden forms at the top, (c) Formed footing with full-depth wooden forms.

crete. For concrete or masonry structures, short steel rebars, called *dowels* are placed such that they extend above the completed footing, thus providing for a lap splice with the column or wall steel. Chapter 9 discusses these connections in more detail. Finally, the concrete is placed and, once it has cured, the forms are removed. Figure 5.8 shows a completed spread footing.

5.2   MATS

The second type of shallow foundation is a *mat foundation*, as shown in Figure 5.9. A mat is essentially a very large spread footing that usually encompasses the entire footprint of the structure. They also are known as *raft foundations*. They are always made of reinforced concrete.

Figure 5.8 A completed spread footing. The four bolts extending out of the footing will be connected to the base plate of a steel column.

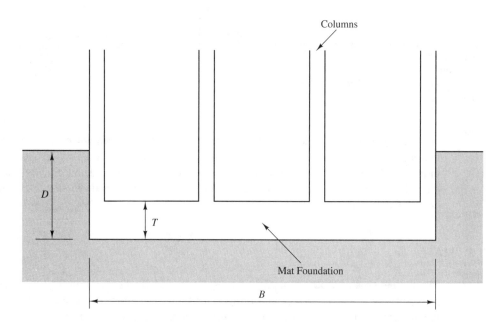

Figure 5.9 A mat foundation.

Foundation engineers often consider mats when dealing with any of the following conditions:

- The structural loads are so high or the soil conditions so poor that spread footings would be exceptionally large. As a general rule of thumb, if spread footings would cover more than 50 percent of the building footprint area, a mat or some type of deep foundation will usually be more economical.
- The soil is very erratic and prone to excessive differential settlements. The structural continuity and flexural strength of a mat will bridge over these irregularities. The same is true of mats on highly expansive soils prone to differential heaves.
- The structural loads are erratic, and thus increase the likelihood of excessive differential settlements. Again, the structural continuity and flexural strength of the mat will absorb these irregularities.
- The lateral loads are not uniformly distributed through the structure and thus may cause differential horizontal movements in spread footings or pile caps. The continuity of a mat will resist such movements.
- The uplift loads are larger than spread footings can accommodate. The greater weight and continuity of a mat may provide sufficient resistance.
- The bottom of the structure is located below the groundwater table, so waterproofing is an important concern. Because mats are monolithic, they are much easier to waterproof. The weight of the mat also helps resist hydrostatic uplift forces from the groundwater.

Many buildings are supported on mat foundations, as are silos, chimneys, and other types of tower structures. Mats are also used to support storage tanks and large machines. The seventy five story Texas Commerce Tower in Houston is one of the largest mat-supported structures in the world. Its mat is 3 m (9 ft 9 in) thick and is bottomed 19.2 m (63 ft) below the street level.

5.3 BEARING PRESSURE

The most fundamental parameter that defines the interface between a shallow foundation and the soil that supports it is the *bearing pressure*. This is the contact force per unit area along the bottom of the foundation. Engineers recognized the importance of bearing pressure during the nineteenth century, thus forming the basis for later developments in bearing capacity and settlement theories.

Distribution of Bearing Pressure

Although the integral of the bearing pressure across the base area of a shallow foundation must be equal to the force acting between the foundation and the soil, this pressure is not necessarily distributed evenly. Analytical studies and field measurements (Schultze,

1961; Dempsey and Li, 1989; and others) indicate that the actual distribution depends on several factors, including the following:

- Eccentricity, if any, of the applied load
- Magnitude of the applied moment, if any
- Structural rigidity of the foundation
- Stress-strain properties of the soil
- Roughness of the bottom of the foundation

Figure 5.10 shows the distribution of bearing pressure along the base of shallow foundations subjected to concentric vertical loads. Perfectly flexible foundations bend as necessary to maintain a uniform bearing pressure, as shown in Figures 5.10a and 5.10b, whereas perfectly rigid foundations settle uniformly but have variations in the bearing pressure, as shown in Figures 5.10c and 5.10d.

Real spread footings are close to being perfectly rigid, so the bearing pressure distribution is not uniform. However, bearing capacity and settlement analyses based on such a distribution would be very complex, so it is customary to assume that the pressure beneath concentric vertical loads is uniform across the base of the footing, as shown in Figure 5.10e. The error introduced by this simplification is not significant.

Mat foundations have a much smaller thickness-to-width ratio, and thus are more flexible than spread footings. In addition, we evaluate the flexural stresses in mats more carefully and develop more detailed reinforcing steel layouts. Therefore, we conduct more detailed analyses to determine the distribution of bearing pressure on mats. Chapter 10 discusses these analyses.

When analyzing shallow foundations, it is customary and reasonable to neglect any sliding friction along the sides of the footing and assume that the entire load is transmitted to the bottom. This is an important analytical difference between shallow and deep foundations, and will be explored in more detail in Chapter 11.

Computation of Bearing Pressure

The *bearing pressure* (or *gross bearing pressure*) along the bottom of a shallow foundation is:

$$q = \frac{P + W_f}{A} - u_D \tag{5.1}$$

Where:

q = bearing pressure

P = vertical column load

W_f = weight of foundation, including the weight of soil above the foundation, if any

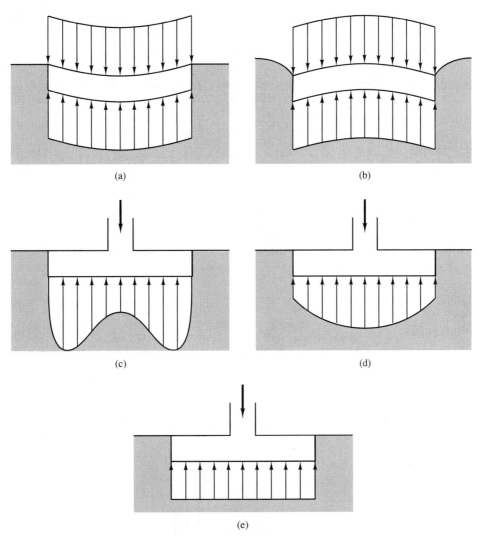

Figure 5.10 Distribution of bearing pressure along the base of shallow foundations subjected to concentric vertical loads: (a) flexible foundation on clay, (b) flexible foundation on sand, (c) rigid foundation on clay, (d) rigid foundation on sand, and (e) simplified distribution (after Taylor, 1948).

A = base area of foundation (B^2 for square foundations or BL for rectangular foundations)

u_D = pore water pressure at bottom of foundation (i.e., at a depth D below the ground surface).

Virtually all shallow foundations are made of reinforced concrete, so W_f is computed using a unit weight for concrete of 150 lb/ft^3 or 23.6 kN/m^3.

The pore water pressure term accounts for uplift pressures (buoyancy forces) that are present if a portion of the foundation is below the groundwater table. If the groundwater table is at a depth greater than D, then set $u_D = 0$.

For continuous footings, we express the applied loads as a force per unit length, such as 2000 kN/m. For ease of computation, we identify this unit length as b, which is usually 1 m or 1 ft as shown in Figure 5.11. Thus, the load is expressed using the variable P/b, and the weight using W_f/b.

The bearing pressures for continuous footings is then:

$$q = \frac{P/b + W_f/b}{B} - u_D \qquad (5.2)$$

Example 5.1

The 5-ft square footing shown in Figure 5.12 supports a column load of 100 k. Compute the bearing pressure.

Solution

Use 150 lb/ft³ for the unit weight of concrete, and compute W_f as if the concrete extends from the ground surface to a depth D (this is conservative when the footing is covered with soil because soil has a lower unit weight, but the error introduced is small).

$$W_f = (5 \text{ ft})(5 \text{ ft})(4 \text{ ft})(150 \text{ lb/ft}^3) = 15{,}000 \text{ lb}$$

$$A = (5 \text{ ft})(5 \text{ ft}) = 25 \text{ ft}^2$$

$$u_D = \gamma_w z_w = (62.4 \text{ lb/ft}^3)(4 \text{ ft} - 3 \text{ ft}) = 62 \text{ lb/ft}^2$$

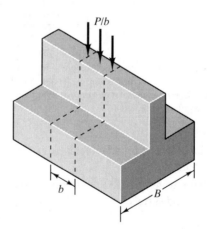

Figure 5.11 Definitions for loads on continuous footings.

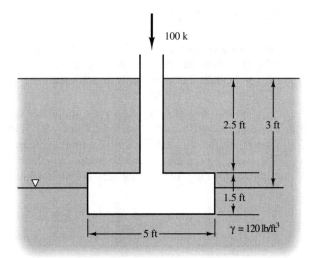

Figure 5.12 Spread footing for Example 5.1.

$$q = \frac{P + W_f}{A} - u_D$$

$$= \frac{100,000 \text{ lb} + 15,000 \text{ lb}}{25 \text{ ft}^2} - 62 \text{ lb/ft}^2$$

$$= \mathbf{4538 \text{ lb/ft}^2} \quad \Leftarrow Answer$$

Example 5.2

A 0.70-m wide continuous footing supports a wall load of 110 kN/m. The bottom of this footing is at a depth of 0.50 m below the adjacent ground surface and the soil has a unit weight of 17.5 kN/m³. The groundwater table is at a depth of 10 m below the ground surface. Compute the bearing pressure.

Solution

Use 23.6 kN/m³ for the unit weight of concrete.

$$W_f/b = (0.70 \text{ m})(0.50 \text{ m})(23.6 \text{ kN/m}^3) = 8 \text{ kN/m}$$

$$u_D = 0$$

$$q = \frac{P/b + W_f/b}{B} - u_D = \frac{110 \text{ kN/m} + 8 \text{ kN/m}}{0.70 \text{ m}} - 0 = \mathbf{169 \text{ kPa}} \quad \Leftarrow Answer$$

Net Bearing Pressure

An alternative way to define bearing pressure is the *net bearing pressure*, q', which is the difference between the gross bearing pressure, q, and the initial vertical effective stress, σ_{z0}', at depth D. In other words, q' is a measure of the increase in vertical effective stress at depth D (Coduto, 1994).

Use of the net bearing pressure simplifies some computations, especially those associated with settlement of spread footings, but it makes others more complex. Some engineers prefer this method, while others prefer to use the gross bearing pressure. Therefore, it is important to understand which definition is being used. In this book we will use only the gross bearing pressure for designing shallow foundations because:

- This is the most commonly used method.
- The presumptive bearing pressures presented in building codes use this method [IBC 1805.4.1.1].
- It is conceptually easier.
- It simplifies the analysis of floating foundations.

Floating Foundations

Mat foundations are often placed in deep excavations, as shown in Figure 5.13. In addition to providing underground space, such as for subterranean parking, this design decreases the bearing pressure because the weight of the foundation is substantially less than the weight of the excavated soil. In other words, the weight of the structure and the foundation is partially offset by the removal of soil from the excavation. This reduction in q significantly reduces the settlement. Such designs are called *floating foundations*, and they are discussed in more detail in Chapter 18.

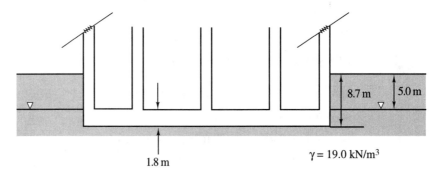

Figure 5.13 A floating foundation.

Example 5.3

The mat foundation in Figure 5.13 is to be 50 m wide, 70 m long, and 1.8 m thick. The sum of the column and wall loads is 805 MN. Compute the average bearing pressure, then compare it with the initial vertical effective stress in the soil immediately below the mat.

Solution

Use $\gamma_{conc} = 23.6$ kN/m^3

$$W_f = (50 \text{ m})(70 \text{ m})(1.8 \text{ m})(23.6 \text{ kN/m}^3) = 149{,}000 \text{ kN}$$

$$u_D = \gamma_w z_{wD} = (9.8 \text{ kN/m}^3)(8.7 \text{ m} - 5.0 \text{ m}) = 36 \text{ kPa}$$

$$A = (50 \text{ m})(70 \text{ m}) = 3500 \text{ m}^2$$

$$q = \frac{P + W_f}{A} - u_D$$

$$= \frac{805{,}000 \text{ kN} + 149{,}000 \text{ kN}}{3500 \text{ m}^2} - 36 \text{ kPa}$$

$$= 237 \text{ kPa} \quad \Leftarrow Answer$$

Note: Mat foundations are neither perfectly flexible nor perfectly rigid. Thus, the actual bearing pressure will vary across the base area. This computation produced the average value.

The vertical effective stress at a depth D before construction was:

$$\sigma'_{zD} = \Sigma \gamma H - u = (19.0 \text{ kN/m}^3)(8.7 \text{ m}) - 36 \text{ kPa} = 129 \text{ kPa} \quad \Leftarrow Answer$$

Thus, the net result of making the excavation and constructing the building will result in a 237 − 129 = 108 kPa increase in the vertical effective stress immediately below the mat. ⇐ *Answer*

Foundations with Eccentric or Moment Loads

Most foundations are built so the vertical load acts through the centroid, thus producing a fairly uniform distribution of bearing pressure. However, sometimes it becomes necessary to accommodate loads that act through other points, as shown in Figure 5.14a. These are called *eccentric loads*, and they produce a non-uniform bearing pressure distribution. The eccentricity, e, of the bearing pressure is equal to:

$$e = \frac{Pe_1}{P + W_f} \tag{5.3}$$

or, for continuous footings:

$$e = \frac{(P/b)\, e_1}{P/b + W_f/b} \tag{5.4}$$

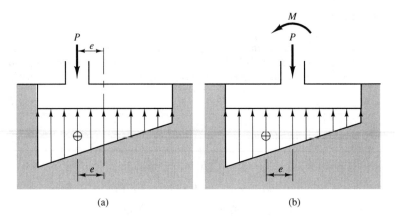

Figure 5.14 (a) Eccentric and (b) moment loads on shallow foundations.

Another similar condition occurs when moment loads are applied to foundations, as shown in Figure 5.14b. These loads also produce non-uniform bearing pressures. In this case, the eccentricity of the bearing pressure is:

$$e = \frac{M}{P + W_f} \tag{5.5}$$

$$e = \frac{M/b}{P/b + W_f/b} \tag{5.6}$$

Where:

e = eccentricity of bearing pressure distribution

P = applied vertical load

P/b = applied vertical load per unit length of foundation

M = applied moment load

M/b = applied moment load per unit length of foundation

e_1 = eccentricity of applied vertical load

W_f = weight of foundation

W_f/b = weight of unit length of foundation

In both cases, we assume the bearing pressure distribution beneath spread footings is linear, as shown in Figure 5.14. This is a simplification of the truth, but sufficiently accurate for practical design purposes. For mat foundations, we perform more detailed analyses as discussed in Chapter 10.

One-Way Loading

If the eccentric or moment loads occur only in the B direction, then the bearing pressure distribution is as shown in Figure 5.15.

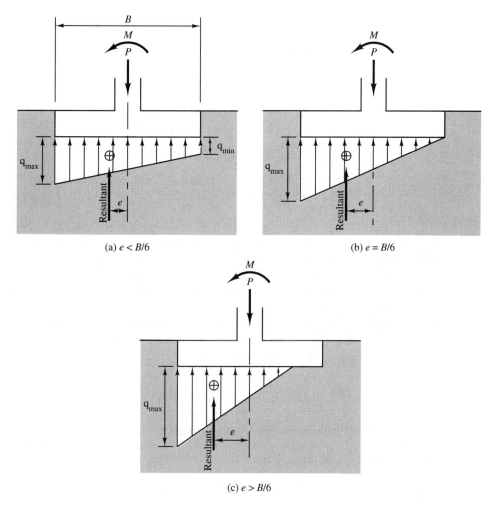

Figure 5.15 Distribution of bearing pressure beneath footings with various eccentricities:
(a) $e < B/6$, (b) $e = B/6$, and (c) $e > B/6$.

If $e \leq B/6$ the bearing pressure distribution is trapezoidal, as shown in Figure 5.15a, and the minimum and maximum bearing pressures are:

$$q_{min} = \left(\frac{P + W_f}{A} - u_D \right) \left(1 - \frac{6e}{B} \right) \tag{5.7}$$

$$q_{max} = \left(\frac{P + W_f}{A} - u_D \right) \left(1 + \frac{6e}{B} \right) \tag{5.8}$$

Where:

q_{min} = minimum bearing pressure

q_{max} = maximum bearing pressure

P = column load

A = base area of foundation

u_D = pore water pressure along base of foundation

e = eccentricity of bearing pressure distribution

B = width of foundation

If the eccentric or moment load is only in the L direction, substitute L for B in Equations 5.7 and 5.8. For continuous footings, substitute P/b and M/b for P and M, respectively, and substitute B for A.

If $e = B/6$ (i.e., the resultant force acts at the third-point of the foundation), then $q_{min} = 0$ and the bearing pressure distribution is triangular as shown in Figure 5.15b. Therefore, so long as $e \leq B/6$, there will be some contact pressure along the entire base area.

However, if $e > B/6$, the resultant of the bearing pressure acts outside the third-point and the pressure distribution is as shown in Figure 5.15c. There can be no tension between the foundation and the soil, so one side of the foundation will lift off the ground. In addition, the high bearing pressure at the opposite side may cause a large settlement there. The net result is an excessive tilting of the foundation, which is not desirable. Therefore, foundations with eccentric or moment loads must satisfy the following condition:

$$\boxed{e \leq \frac{B}{6}} \tag{5.9}$$

This criterion maintains compressive stresses along the entire base area.

For rectangular foundations with the moment or eccentric load in the long direction, substitute L for B in Equation 5.9.

Example 5.4

A 5.0-ft wide continuous footing is subjected to a concentric vertical load of 12.0 k/ft and a moment load of 8.0 ft-k/ft acting laterally across the footing, as shown in Figure 5.16. The groundwater table is at a great depth. Determine whether the resultant force on the base of the footing acts within the middle third and compute the maximum and minimum bearing pressures.

Solution

$$W_f/b = (5.0\text{ ft})(1.5\text{ ft})(150\text{ lb/ft}^3) = 1125\text{ lb/ft}$$

$$e = \frac{M/b}{P/b + W_f/b} = \frac{8,000\text{ ft-lb/ft}}{12,000\text{ lb/ft} + 1125\text{ lb/ft}} = 0.610\text{ ft}$$

$$\frac{B}{6} = \frac{5\text{ ft}}{6} = 0.833\text{ ft}$$

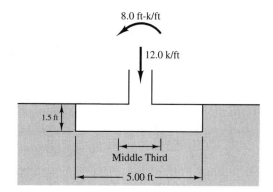

Figure 5.16 Proposed footing for Example 5.4.

$e < B/6$; therefore **the resultant is in the middle third** $\Leftarrow$ *Answer*

$$q_{min} = \left(\frac{P/b + W_f/b}{B} - u_D \right) \left(1 - \frac{6e}{B} \right)$$

$$= \left(\frac{12{,}000 + 1125}{5.0} - 0 \right) \left(1 - \frac{(6)(0.610)}{5.0} \right)$$

$$= 703 \text{ lb/ft}^2 \quad \Leftarrow Answer$$

$$q_{max} = \left(\frac{P/b + W_f/b}{B} - u_D \right) \left(1 + \frac{6e}{B} \right)$$

$$= \left(\frac{12{,}000 + 1125}{5.0} - 0 \right) \left(1 + \frac{(6)(0.610)}{5.0} \right)$$

$$= 4546 \text{ lb/ft}^2 \quad \Leftarrow Answer$$

When designing combined footings, try to arrange the footing dimensions and column locations so the resultant of the applied loads acts through the centroid of the footing. This produces a uniform bearing pressure distribution. Some combined footing designs accomplish this by using a trapezoidal shaped footing (as seen in plan view) with the more lightly loaded column on the narrow side of the trapezoid. When this is not possible, be sure all of the potential loading conditions produce eccentricities no more than $B/6$.

Two-Way Eccentric or Moment Loading

If the resultant load acting on the base is eccentric in both the B and L directions, it must fall within the diamond-shaped kern shown in Figure 5.17 for the contact pressure to be compressive along the entire base of the foundation. It falls within this kern only if the following condition is met:

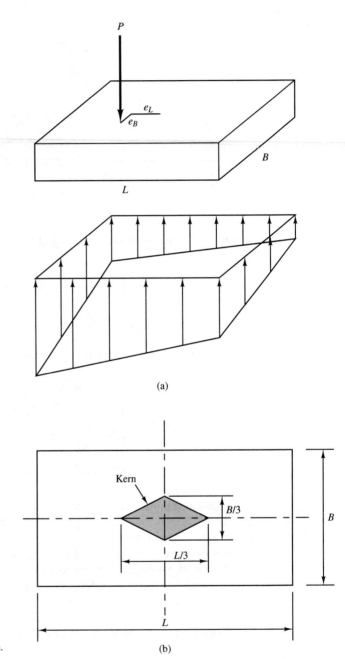

(a)

(b)

Figure 5.17 (a) Pressure distribution beneath spread footing with vertical load that is eccentric in both the B and L directions; (b) To maintain $q \geq 0$ along with the entire base of the footing, the resultant force must be located within this diamond-shaped kern.

$$\boxed{\frac{6\,e_B}{B} + \frac{6\,e_L}{L} \le 1.0}$$

(5.10)

Where:

e_B = eccentricity in the B direction

e_L = eccentricity in the L direction

If Equation 5.10 is satisfied, the magnitudes of q at the four corners of a square or rectangular shallow foundation are:

$$\boxed{q = \left(\frac{P + W_f}{A} - u_D\right)\left(1 \pm \frac{6\,e_B}{B} \pm \frac{6\,e_L}{L}\right)}$$

(5.11)

Example 5.5

The mat foundation shown in Figure 5.18 will support four grain silos. These are cylindrical structures used to store grain. Each of the silos has an empty weight of 29 MN, and can hold up to 110 MN of grain. The mat has a weight of 60 MN. Since each silo is filled independently, the resultant load imposed on the mat does not necessarily act through the centroid. Evaluate the various loading conditions and determine whether eccentric loading requirements will be met. If these requirements are not met, determine the minimum mat width, B, needed to satisfy these requirements.

Solution

1. Check one-way eccentricity

The greatest one-way eccentricity occurs when two adjacent silos are full, and the other two are empty.

$$P = 4(29 \text{ MN}) + 2(110 \text{ MN}) = 336 \text{ MN}$$

$$M = 2(110 \text{ MN})(12 \text{ m}) = 2640 \text{ MN-m}$$

$$e = \frac{M}{P + W_f} = \frac{2640 \text{ MN-m}}{336 \text{ MN} + 60 \text{ MN}} = 6.67 \text{ m}$$

$$e \le \frac{B}{6} \qquad \therefore \text{ OK for one-way eccentricity}$$

2. Check two-way eccentricity

The greatest two-way eccentricity occurs when one silos is full and the other three are empty.

$$\frac{B}{6} = \frac{50 \text{ m}}{6} = 8.33 \text{ m}$$

$$P = 4(29 \text{ MN}) + (110 \text{ MN}) = 226 \text{ MN}$$

$$M_B = M_L = (110 \text{ MN})(12 \text{ m}) = 1320 \text{ MN-m}$$

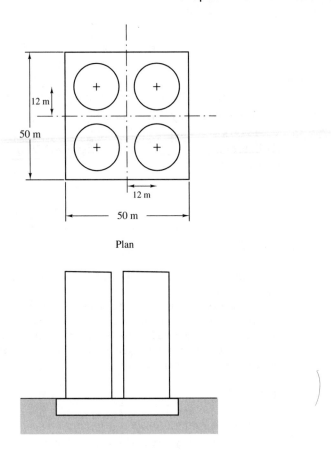

Plan

Elevation

Figure 5.18 Mat foundation and grain silos for Example 5.5.

$$e_B = e_L = \frac{M}{P + W_f} = \frac{1320 \text{ MN-m}}{226 \text{ MN} + 60 \text{ MN}} = 4.62 \text{ m}$$

$$\frac{6e_B}{B} + \frac{6e_L}{L} = \frac{6(4.62 \text{ m})}{50 \text{ m}} + \frac{6(4.62 \text{ m})}{50 \text{ m}} = 1.11 > 1 \qquad \text{Not acceptable}$$

3. Conclusion

Although the foundation is satisfactory for one-way eccentricity, it does not meet the criterion for two-way eccentricity because the resultant is outside the kern. This means the corner of the mat opposite the loaded silo may lift up, causing excessive tilting. Therefore, it is necessary to increase B.   *⇐ Answer*

4. Revised Design

$$\frac{6e_B}{B} + \frac{6e_L}{L} = \frac{6(4.62 \text{ m})}{B} + \frac{6(4.62 \text{ m})}{L} = 1$$

minimum B = minimum L = 55.4 m   ⇐ *Answer*

Therefore, a 55.4 m × 55.4 m mat would be required to keep the resultant within the kern. However, this is only one of many design criteria for mat foundations. Other criteria, as discussed in Chapters 6 to 10, also need to be checked.

SUMMARY

Major Points

1. Shallow foundations are those that transmit structural loads to the near-surface soils. There are two kinds: spread footing foundations and mat foundations.
2. Although other materials have been used in the past, today virtually all shallow foundations are made of reinforced concrete.
3. Spread footings are most often used in small- to medium-size structures on sites with moderate to good soil conditions. Mats are most often used on larger structures, especially those with differential settlement problems and those with foundations below the groundwater table.
4. The bearing pressure is the contact pressure between the bottom of a shallow foundation and the underlying soils.
5. A floating foundation is one where the weight of the foundation is substantially less than the weight of the excavated soils. This occurs in buildings with basements and other similar structures.
6. If the loads applied to a foundation are eccentric, or if moment loads are applied, the resulting bearing pressure distribution also will be eccentric. In such cases, the foundation needs to be designed so the resultant of the bearing pressure is within the middle third of the foundation (for one-way eccentricity) or in a diamond-shaped kern (for two-way eccentricity). This requirement ensures the entire base of the foundation has compressive bearing pressures, and thus avoids problems with uplift.

Vocabulary

Backhoe	Eccentric load	Moment load
Bearing pressure	Floating foundation	Neat footing
Circular spread footing	Formed footing	Net bearing pressure
Combined spread footing	Gross bearing pressure	Raft foundation
Continuous spread footing	Kern	Rectangular spread footing
Dimension-stone footings	Mat	Ring footing

Rubble-stone footings   Spread footing   Steel grillage footing

Shallow foundation   Square spread footing

COMPREHENSIVE QUESTIONS AND PRACTICE PROBLEMS

5.1 What is the difference between a square footing and a continuous footing, and when would each type be used?

5.2 A 400 kN vertical downward column load acts at the centroid of a 1.5-m square footing. The bottom of this footing is 0.4 m below the ground surface and the top is flush with the ground surface. The groundwater table is at a depth of 3 m below the ground surface. Compute the bearing pressure.

5.3 A bearing wall carries a dead load of 5.0 k/ft and a live load of 3.0 k/ft. It is supported on a 3 ft wide, 2 ft deep continuous footing. The top of this footing is flush with the ground surface and the groundwater table is at a depth of 35 ft below the ground surface. Compute the bearing pressure.

5.4 The mat foundation in Figure 5.19 is 45 m wide and 90 m long. It has a weight of 140 MN. The sum of the applied structural loads is 1300 MN. Compute the average bearing pressure with the groundwater table at position A. Then repeat the computation with the groundwater table at position B. Explain why these two values of q are different.

5.5 A 5-ft square, 2-ft deep spread footing is subjected to a concentric vertical load of 60 k and an overturning moment of 30 ft-k. The overturning moment acts parallel to one of the sides of the footing, the top of the footing is flush with the ground surface, and the groundwater table is at a depth of 20 ft below the ground surface. Determine whether the resultant force acts within the middle third of the footing, compute the minimum and maximum bearing pressures, and show the distribution of bearing pressure in a sketch.

5.6 Consider the footing and loads in Problem 5.5, except that the overturning moment now acts at a 45° angle from the side of the footing (i.e., it acts diagonally across the top of the footing). Determine whether the resultant force acts within the kern. If it does, then compute the bear-

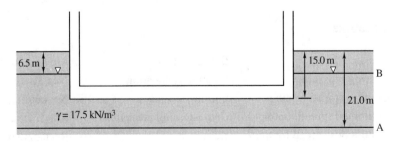

Figure 5.19 Mat foundation for Problem 5.4

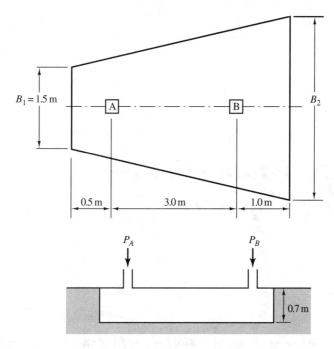

Figure 5.20 Proposed combined footing for Problems 5.7 and 5.8.

ing pressure at each corner of the footing and show the pressure distribution in a sketch similar to Figure 5.17.

5.7 The two columns in Figure 5.20 are to be supported on a combined footing. The vertical dead loads on Columns A and B are 500 and 1400 kN, respectively. Determine the required dimension B_2 so the resultant of the column loads acts through the centroid of the footing and express your answer as a multiple of 100 mm.

5.8 In addition to the dead loads described in Problem 5.7, Columns A and B in Figure 5.20 also can carry vertical live loads of up to 800 and 1200 k, respectively. The live loads vary with time, and thus may be present some days and absent other days. In addition, the live load on each column is independent of that on the other column (i.e., one could be carrying the full live load while the other has zero live load). Using the dimensions obtained in Problem 5.7, and the worst possible combination of live loads, determine if the bearing pressure distribution always meets the eccentricity requirements described in this chapter. The groundwater table is at a depth of 10 m.

5.9 Beginning with $(P + W_f)/A \pm Mc/I$, derive Equations 5.7 and 5.8. Would these equations also apply to circular shallow foundations? Why or why not?

6

Shallow Foundations—
Bearing Capacity

When we are satisfied with the spot fixed on for the site of the city . . .
the foundations should be carried down to a solid bottom, if such can be
found, and should be built thereon of such thickness as may be
necessary for the proper support of that part of the wall which stands
above the natural level of the ground. They should be of the soundest
workmanship and materials, and of greater thickness than the walls
above. If solid ground can be come to, the foundations should go down
to it and into it, according to the magnitude of the work, and the
substruction to be built up as solid as possible. Above the ground of the
foundation, the wall should be one-half thicker than the column it is to
receive so that the lower parts which carry the greatest weight, may be
stronger than the upper part . . . Nor must the mouldings of the bases of
the columns project beyond the solid. Thus, also, should be regulated the
thickness of all walls above ground.

Marcus Vitruvius, Roman Architect and Engineer
1st century B.C.
as translated by Morgan (1914)

Shallow foundations must satisfy various performance requirements, as discussed in
Chapter 2. One of them is called *bearing capacity*, which is a geotechnical strength requirement. This chapter explores this requirement, and shows how to design shallow
foundations so that they do not experience bearing capacity failures.

6.1 BEARING CAPACITY FAILURES

Shallow foundations transmit the applied structural loads to the near-surface soils. In the process of doing so, they induce both compressive and shear stresses in these soils. The magnitudes of these stresses depend largely on the bearing pressure and the size of the footing. If the bearing pressure is large enough, or the footing is small enough, the shear stresses may exceed the shear strength of the soil or rock, resulting in a *bearing capacity failure*. Researchers have identified three types of bearing capacity failures: *general shear failure*, *local shear failure*, and *punching shear failure*, as shown in Figure 6.1. A typical load-displacement curve for each mode of failure is shown in Figure 6.2.

General shear failure is the most common mode. It occurs in soils that are relatively incompressible and reasonably strong, in rock, and in saturated, normally consolidated clays that are loaded rapidly enough that the undrained condition prevails. The failure surface is well defined and failure occurs quite suddenly, as illustrated by the load-displacement curve. A clearly formed bulge appears on the ground surface adjacent to the foundation. Although bulges may appear on both sides of the foundation, ultimate failure occurs on one side only, and it is often accompanied by rotations of the foundation.

The opposite extreme is the punching shear failure. It occurs in very loose sands, in a thin crust of strong soil underlain by a very weak soil, or in weak clays loaded under slow, drained conditions. The high compressibility of such soil profiles causes large settlements and poorly defined vertical shear surfaces. Little or no bulging occurs at the ground surface and failure develops gradually, as illustrated by the ever-increasing load-settlement curve.

Local shear failure is an intermediate case. The shear surfaces are well defined under the foundation, and then become vague near the ground surface. A small bulge may occur, but considerable settlement, perhaps on the order of half the foundation width, is necessary before a clear shear surface forms near the ground. Even then, a sudden failure does not occur, as happens in the general shear case. The foundation just continues to sink ever deeper into the ground.

Vesić (1973) investigated these three modes of failure by conducting load tests on model circular foundations in a sand. These tests included both shallow and deep foundations. The results, shown in Figure 6.3, indicate shallow foundations (D/B less than about 2) can fail in any of the three modes, depending on the relative density. However, deep foundations (D/B greater than about 4) are always governed by punching shear. Although these test results apply only to circular foundations in Vesić's sand and cannot necessarily be generalized to other soils, it does give a general relationship between the mode of failure, relative density, and the D/B ratio.

Complete quantitative criteria have yet to be developed to determine which of these three modes of failure will govern in any given circumstance, but the following guidelines are helpful:

- Shallow foundations in rock and undrained clays are governed by the general shear case.

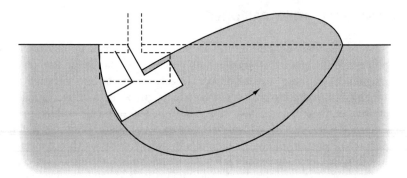

(a) General Shear Failure

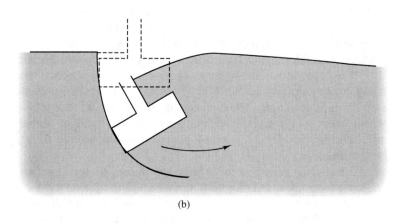

(b)

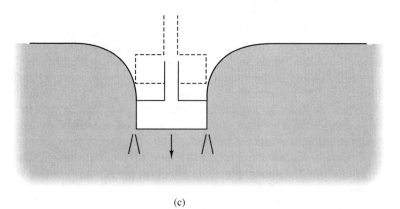

(c)

Figure 6.1 Modes of bearing capacity failure: (a) general shear failure; (b) local shear failure; (c) punching shear failure.

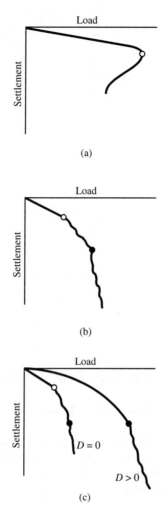

Figure 6.2   Typical load-displacement curves for different modes of bearing capacity failure: (a) general shear failure; (b) local shear failure; (c) punching shear failure. The circles indicate various interpretations of failure. (Adapted from Vesić, 1963).

- Shallow foundations in dense sands are governed by the general shear case. In this context, a dense sand is one with a relative density, D_r, greater than about 67%.
- Shallow foundations on loose to medium dense sands ($30\% < D_r < 67\%$) are probably governed by local shear.
- Shallow foundations on very loose sand ($D_r < 30\%$) are probably governed by punching shear.

For nearly all practical shallow foundation design problems, it is only necessary to check the general shear case, and then conduct settlement analyses to verify that the foundation will not settle excessively. These settlement analyses implicitly protect against local and punching shear failures.

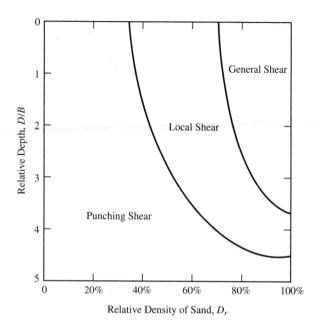

Figure 6.3   Modes of failure of model circular foundations in Chattahoochee Sand (Adapted from Vesić, 1963 and 1973).

6.2   BEARING CAPACITY ANALYSES IN SOIL—GENERAL SHEAR CASE

Methods of Analyzing Bearing Capacity

To analyze spread footings for bearing capacity failures and design them in a way to avoid such failures, we must understand the relationship between bearing capacity, load, footing dimensions, and soil properties. Various researchers have studied these relationships using a variety of techniques, including:

- Assessments of the performance of real foundations, including full-scale load tests.
- Load tests on model footings.
- Limit equilibrium analyses.
- Detailed stress analyses, such as finite element method (FEM) analyses.

Full-scale load tests, which consist of constructing real spread footings and loading them to failure, are the most precise way to evaluate bearing capacity. However, such tests are expensive, and thus are rarely, if ever, performed as a part of routine design. A few such tests have been performed for research purposes.

Model footing tests have been used quite extensively, mostly because the cost of these tests is far below that for full-scale tests. Unfortunately, model tests have their limitations, especially when conducted in sands, because of uncertainties in applying the proper scaling factors. However, the advent of centrifuge model tests has partially overcome this problem.

Limit equilibrium analyses are the dominant way to assess bearing capacity of shallow foundations. These analyses define the shape of the failure surface, as shown in Figure 6.1, then evaluate the stresses and strengths along this surface. These methods of analysis have their roots in Prandtl's studies of the punching resistance of metals (Prandtl, 1920). He considered the ability of very thick masses of metal (i.e., not sheet metal) to resist concentrated loads. Limit equilibrium analyses usually include empirical factors developed from model tests.

Occasionally, geotechnical engineers perform more detailed bearing capacity analyses using numerical methods, such as the finite element method (FEM). These analyses are more complex, and are justified only on very critical and unusual projects.

We will consider only limit equilibrium methods of bearing capacity analyses, because these methods are used on the overwhelming majority of projects.

Simple Bearing Capacity Formula

The limit equilibrium method can be illustrated by considering the continuous footing shown in Figure 6.4. Let us assume this footing experiences a bearing capacity failure, and that this failure occurs along a circular shear surface as shown. We will further assume the soil is an undrained clay with a shear strength s_u. Finally, we will neglect the shear strength between the ground surface and a depth D, which is conservative. Thus, the soil in this zone is considered to be only a surcharge load that produces a vertical total stress of $\sigma_{zD} = \gamma D$ at a depth D.

The objective of this derivation is to obtain a formula for the *ultimate bearing capacity, q_{ult}*, which is the bearing pressure required to cause a bearing capacity failure. By

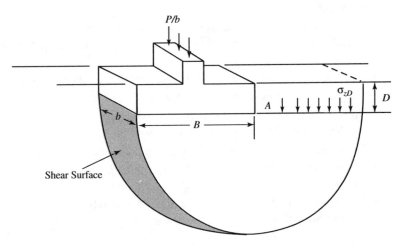

Figure 6.4 Bearing capacity analysis along a circular failure surface.

considering a slice of the foundation of length b and taking moments about Point A, we obtain the following:

$$M_A = (q_{ult}Bb)(B/2) - (s_u\pi Bb)(B) - \sigma_{zD}Bb(B/2) \tag{6.1}$$

$$q_{ult} = 2\pi s_u + \sigma_{zD} \tag{6.2}$$

It is convenient to define a new parameter, called a *bearing capacity factor*, N_c, and rewrite Equation 6.2 as:

$$\boxed{q_{ult} = N_c s_u + \sigma_{zD}} \tag{6.3}$$

Equation 6.3 is known as a *bearing capacity formula*, and could be used to evaluate the bearing capacity of a proposed foundation. According to this derivation, $N_c = 2\pi = 6.28$.

This simplified formula has only limited applicability in practice because it considers only continuous footings and undrained soil conditions ($\phi = 0$), and it assumes the foundation rotates as the bearing capacity failure occurs. However, this simple derivation illustrates the general methodology required to develop more comprehensive bearing capacity formulas.

Terzaghi's Bearing Capacity Formulas

Various limit equilibrium methods of computing bearing capacity of soils were advanced in the first half of the twentieth century, but the first one to achieve widespread acceptance was that of Terzaghi (1943). His method includes the following assumptions:

- The depth of the foundation is less than or equal to its width ($D \leq B$).
- The bottom of the foundation is sufficiently rough that no sliding occurs between the foundation and the soil.
- The soil beneath the foundation is a homogeneous semi-infinite mass (i.e., the soil extends for a great distance below the foundation and the soil properties are uniform throughout).
- The shear strength of the soil is described by the formula $s = c' + \sigma' \tan \phi'$.
- The general shear mode of failure governs.
- No consolidation of the soil occurs (i.e., settlement of the foundation is due only to the shearing and lateral movement of the soil).
- The foundation is very rigid in comparison to the soil.
- The soil between the ground surface and a depth D has no shear strength, and serves only as a surcharge load.
- The applied load is compressive and applied vertically to the centroid of the foundation and no applied moment loads are present.

Terzaghi considered three zones in the soil, as shown in Figure 6.5. Immediately beneath the foundation is a *wedge zone* that remains intact and moves downward with the

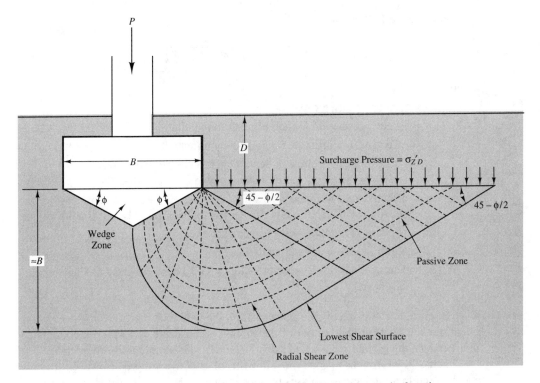

Figure 6.5 Geometry of failure surface for Terzaghi's bearing capacity formulas.

foundation. Next, a *radial shear zone* extends from each side of the wedge, where he took the shape of the shear planes to be logarithmic spirals. Finally, the outer portion is the *linear shear zone* in which the soil shears along planar surfaces.

Since Terzaghi neglected the shear strength of soils between the ground surface and a depth D, the shear surface stops at this depth and the overlying soil has been replaced with the surcharge pressure σ_{zD}'. This approach is conservative, and is part of the reason for limiting the method to relatively shallow foundations ($D \leq B$).

Terzaghi developed his theory for continuous foundations (i.e., those with a very large L/B ratio). This is the simplest case because it is a two-dimensional problem. He then extended it to square and round foundations by adding empirical coefficients obtained from model tests and produced the following bearing capacity formulas:

For square foundations:

$$q_{ult} = 1.3\, c'N_c + \sigma_{zD}'N_q + 0.4\gamma'BN_\gamma \qquad (6.4)$$

For continuous foundations:

$$q_{ult} = c'N_c + \sigma_{zD}'N_q + 0.5\gamma'BN_\gamma \qquad (6.5)$$

For circular foundations:

$$q_{ult} = 1.3\, c'N_c + \sigma'_{zD}N_q + 0.3\gamma'BN_\gamma \qquad (6.6)$$

Where:

q_{ult} = ultimate bearing capacity

c' = effective cohesion for soil beneath foundation

ϕ' = effective friction angle for soil beneath foundation

σ_{zD}' = vertical effective stress at depth D below the ground surface ($\sigma_{zD}' = \gamma D$ if depth to groundwater table is greater than D)

γ' = effective unit weight of the soil ($\gamma = \gamma'$ if groundwater table is very deep; see discussion later in this chapter for shallow groundwater conditions)

D = depth of foundation below ground surface

B = width (or diameter) of foundation

N_c, N_q, N_γ = Terzaghi's bearing capacity factors $= f(\phi')$ (See Table 6.1 or Equations 6.7–6.12.)

Because of the shape of the failure surface, the values of c' and ϕ' only need to represent the soil between the bottom of the footing and a depth B below the bottom. The soils between the ground surface and a depth D are treated simply as overburden.

Terzaghi's formulas are presented in terms of effective stresses. However, they also may be used in a total stress analyses by substituting c_T, ϕ_T, and σ_D for c', ϕ', and σ_D'. If saturated undrained conditions exist, we may conduct a total stress analysis with the shear strength defined as $c_T = s_u$ and $\phi_T = 0$. In this case, $N_c = 5.7$, $N_q = 1.0$, and $N_c = 0.0$.

The Terzaghi bearing capacity factors are:

$$N_q = \frac{a_\theta^2}{2\cos^2(45 + \phi'/2)} \qquad (6.7)$$

$$a_\theta = e^{\pi(0.75 - \phi'/360)\tan\phi'} \qquad (6.8)$$

$$N_c = 5.7 \qquad \text{for } \phi' = 0 \qquad (6.9)$$

$$N_c = \frac{N_q - 1}{\tan\phi'} \qquad \text{for } \phi' > 0 \qquad (6.10)$$

$$N_\gamma = \frac{\tan\phi'}{2}\left(\frac{K_{p\gamma}}{\cos^2\phi'} - 1\right) \qquad (6.11)$$

These bearing capacity factors are also presented in tabular form in Table 6.1. Notice that Terzaghi's N_c of 5.7 is smaller than the value of 6.28 derived from the simple bearing capacity analysis. This difference the result of using a circular failure surface in the simple method and a more complex geometry in Terzaghi's method.

TABLE 6.1 BEARING CAPACITY FACTORS

ϕ' (deg)	Terzaghi (for use in Equations 6.4–6.6)			Vesić (for use in Equation 6.13)		
	N_c	N_q	N_γ	N_c	N_q	N_γ
0	5.7	1.0	0.0	5.1	1.0	0.0
1	6.0	1.1	0.1	5.4	1.1	0.1
2	6.3	1.2	0.1	5.6	1.2	0.2
3	6.6	1.3	0.2	5.9	1.3	0.2
4	7.0	1.5	0.3	6.2	1.4	0.3
5	7.3	1.6	0.4	6.5	1.6	0.4
6	7.7	1.8	0.5	6.8	1.7	0.6
7	8.2	2.0	0.6	7.2	1.9	0.7
8	8.6	2.2	0.7	7.5	2.1	0.9
9	9.1	2.4	0.9	7.9	2.3	1.0
10	9.6	2.7	1.0	8.3	2.5	1.2
11	10.2	3.0	1.2	8.8	2.7	1.4
12	10.8	3.3	1.4	9.3	3.0	1.7
13	11.4	3.6	1.6	9.8	3.3	2.0
14	12.1	4.0	1.9	10.4	3.6	2.3
15	12.9	4.4	2.2	11.0	3.9	2.6
16	13.7	4.9	2.5	11.6	4.3	3.1
17	14.6	5.5	2.9	12.3	4.8	3.5
18	15.5	6.0	3.3	13.1	5.3	4.1
19	16.6	6.7	3.8	13.9	5.8	4.7
20	17.7	7.4	4.4	14.8	6.4	5.4
21	18.9	8.3	5.1	15.8	7.1	6.2
22	20.3	9.2	5.9	16.9	7.8	7.1
23	21.7	10.2	6.8	18.0	8.7	8.2
24	23.4	11.4	7.9	19.3	9.6	9.4
25	25.1	12.7	9.2	20.7	10.7	10.9
26	27.1	14.2	10.7	22.3	11.9	12.5
27	29.2	15.9	12.5	23.9	13.2	14.5
28	31.6	17.8	14.6	25.8	14.7	16.7
29	34.2	20.0	17.1	27.9	16.4	19.3
30	37.2	22.5	20.1	30.1	18.4	22.4
31	40.4	25.3	23.7	32.7	20.6	26.0
32	44.0	28.5	28.0	35.5	23.2	30.2
33	48.1	32.2	33.3	38.6	26.1	35.2
34	52.6	36.5	39.6	42.2	29.4	41.1
35	57.8	41.4	47.3	46.1	33.3	48.0
36	63.5	47.2	56.7	50.6	37.8	56.3
37	70.1	53.8	68.1	55.6	42.9	66.2
38	77.5	61.5	82.3	61.4	48.9	78.0
39	86.0	70.6	99.8	67.9	56.0	92.2
40	95.7	81.3	121.5	75.3	64.2	109.4
41	106.8	93.8	148.5	83.9	73.9	130.2

Terzaghi used a tedious graphical method to obtain values for $K_{p\gamma}$, then used these values to compute N_γ. He also computed values of the other bearing capacity factors and presented the results in plots of N_c, N_γ, and N_q as a function of ϕ'. These plots and tables such as Table 6.1 are still a convenient way to evaluate these parameters. However, the advent of computers and hand-held calculators has generated the need for the following simplified formula for N_γ:

$$N_\gamma \approx \frac{2(N_q + 1)\tan\phi'}{1 + 0.4\sin(4\phi')} \tag{6.12}$$

The author developed Equation 6.12 by fitting a curve to match Terzaghi's. It produces N_γ values within about 10 percent of Terzaghi's values. Alternatively, Kumbhojkar (1993) provides a more precise, but more complex, formula for N_γ.

Example 6.1

A square footing is to be constructed as shown in Figure 6.6. The groundwater table is at a depth of 50 ft below the ground surface. Compute the ultimate bearing capacity and the column load required to produce a bearing capacity failure.

Solution

For purposed of evaluating bearing capacity, ignore the slab-on-grade floor.

For $\phi' = 30°$: $N_c = 37.2$, $N_q = 22.5$, $N_\gamma = 20.1$ (from Table 6.1)

$\sigma'_{zD} = \gamma D - u = (121\ \text{lb/ft}^3)(2\ \text{ft}) - 0 = 242\ \text{lb/ft}^2$

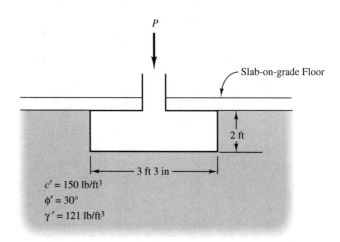

Figure 6.6   Proposed footing for Example 6.1.

$$q_{ult} = 1.3c'N_c + \sigma'_{zD}N_q + 0.4\gamma'BN_\gamma$$

$$= (1.3)(150 \text{ lb/ft}^2)(37.2) + (242 \text{ lb/ft}^2)(22.5) + (0.4)(121 \text{ lb/ft}^3)(3.25 \text{ ft})(20.1)$$

$$= 7254 + 5445 + 3162$$

$$= \mathbf{15{,}900 \text{ lb/ft}^2} \qquad \qquad \Leftarrow Answer$$

$$W_f = (3.25 \text{ ft})(3.25 \text{ ft})(2.0 \text{ ft})(150 \text{ lb/ft}^3) = 3169 \text{ lb}$$

Setting $q = q_{ult}$, using Equation 5.1, and solving for P gives:

$$q = \frac{P + W_f}{A} - u$$

$$15{,}900 \text{ lb/ft}^2 = \frac{P + 3169 \text{ lb}}{(3.25 \text{ ft})^2} - 0$$

$$P = 165{,}000 \text{ lb}$$

$$= \mathbf{165 \text{ k}} \qquad \Leftarrow Answer$$

According to this analysis, a column load of 165 k would cause a bearing capacity failure beneath this footing. Nearly half of this capacity comes from the first term in the bearing capacity formula and is therefore dependent on the cohesion of the soil. Since the cohesive strength is rather tenuous, it is prudent to use conservative values of c in bearing capacity analyses. In contrast, the frictional strength is more reliable and does not need to be interpreted as conservatively.

Example 6.2

The proposed continuous footing shown in Figure 6.7 will support the exterior wall of a new industrial building. The underlying soil is an undrained clay, and the groundwater table is below the bottom of the footing. Compute the ultimate bearing capacity, and compute the wall load required to cause a bearing capacity failure.

Solution

This analysis uses the undrained shear strength, s_u. Therefore, we will use Terzaghi's bearing capacity formula with $c_T = s_u = 120 \text{ kPa}$ and $\phi = 0$.

$$\text{For } \phi = 0: \quad N_c = 5.7, N_q = 1, N_\gamma = 0 \text{ (from Table 6.1)}$$

The depth of embedment, D, is measured from the lowest ground surface, so $D = 0.4 \text{ m}$.

$$\sigma'_{zD} = \gamma D = (18.0)(0.4) = 7.2 \text{ kPa}$$

$$q_{ult} = s_u N_c + \sigma'_{zD}N_q + 0.5\gamma'BN_\gamma$$

$$= (120 \text{ kPa})(5.7) + (7.2)(1) + 0.5\gamma'B(0)$$

$$= \mathbf{691 \text{ kPa}} \qquad \qquad \Leftarrow Answer$$

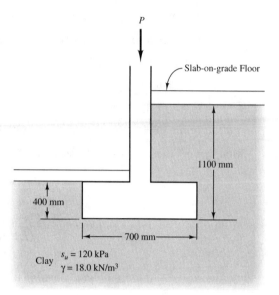

Figure 6.7   Proposed footing for Example
6.2.

The zone above the bottom of the footing is partly concrete and partly soil. The weight of this
zone is small compared to the wall load, so compute it using 21 kN/m³ as the estimated
weighted average for γ:

$$W_f/b = (0.7 \text{ m}) \left(\frac{0.4 \text{ m} + 1.1 \text{ m}}{2} \right) (21 \text{ kN/m}^3) = 11 \text{ kN/m} \quad \Leftarrow Answer$$

Using Equation 5.2:

$$q_{ult} = q = \frac{P/b + W_f/b}{B} - u$$

$$691 \text{ kPa} = \frac{P/b + 11 \text{ kN/m}}{(0.7 \text{ m})} - 0$$

$$\textbf{P = 473 kN/m} \quad \Leftarrow Answer$$

Terzaghi's method is still often used, primarily because it is simple and familiar.
However, it does not consider special cases, such as rectangular footings, inclined loads,
or footings with large depth:width ratios.

Vesić's Bearing Capacity Formulas

The topic of bearing capacity has spawned extensive research and numerous methods of
analysis. Skempton (1951), Meyerhof (1953), Brinch Hansen (1961b), DeBeer and
Ladanyi (1961), Meyerhof (1963), Brinch Hansen (1970), and many others have con-

tributed. The formula developed in Vesić (1973, 1975) is based on theoretical and experimental findings from these and other sources and is an excellent alternative to Terzaghi. It produces more accurate bearing values and it applies to a much broader range of loading and geometry conditions. The primary disadvantage is its added complexity.

Vesić retained Terzaghi's basic format and added the following additional factors:

s_c, s_q, s_γ = shape factors
d_c, d_q, d_γ = depth factors
i_c, i_q, i_γ = load inclination factors
b_c, b_q, b_γ = base inclination factors
g_c, g_q, g_γ = ground inclination factors

He incorporated these factors into the bearing capacity formula as follows:

$$q_{ult} = c'N_c s_c d_c i_c b_c g_c + \sigma'_{zD} N_q s_q d_q i_q b_q g_q + 0.5\gamma' B N_\gamma s_\gamma d_\gamma i_\gamma b_\gamma g_\gamma \qquad (6.13)$$

Once again, this formula is written in terms of the effective stress parameters c' and ϕ', but also may be used in a total stress analysis by substituting c_T and ϕ_T. For undrained total stress analyses, use $c_T = s_u$ and $\phi_T = 0$.

Terzaghi's formulas consider only vertical loads acting on a footing with a horizontal base with a level ground surface, whereas Vesić's factors allow any or all of these to vary. The notation for these factors is shown in Figure 6.8.

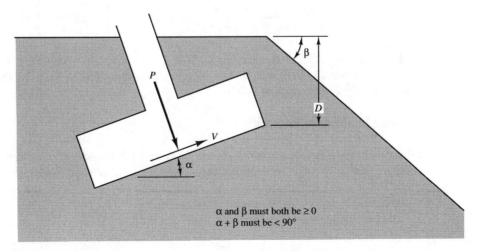

Figure 6.8   Notation for Vesić's load inclination, base inclination, and ground inclination factors. All angles are expressed in degrees.

Shape Factors

Vesić considered a broader range of footing shapes and defined them in his *s* factors:

$$s_c = 1 + \left(\frac{B}{L}\right)\left(\frac{N_q}{N_c}\right) \tag{6.14}$$

$$s_q = 1 + \left(\frac{B}{L}\right)\tan\phi' \tag{6.15}$$

$$s_\gamma = 1 - 0.4\left(\frac{B}{L}\right) \tag{6.16}$$

For continuous footings, $B/L \to 0$, so s_c, s_q, and s_γ become equal to 1. This means the *s* factors may be ignored when analyzing continuous footings.

Depth Factors

Unlike Terzaghi, Vesić has no limitations on the depth of the footing. This method might even be used for deep foundations, although other methods are probably better for reasons discussed in Chapter 14. The depth of the footing is considered in the following depth factors:

$$d_c = 1 + 0.4\,k \tag{6.17}$$

$$d_q = 1 + 2k\tan\phi'\,(1 - \sin\phi')^2 \tag{6.18}$$

$$d_\gamma = 1 \tag{6.19}$$

For relatively shallow foundations ($D/B \leq 1$), use $k = D/B$. For deeper footings ($D/B > 1$), use $k = \tan^{-1}(D/B)$ with the $\tan^{-1}$ term expressed in radians. Note that this produces a discontinuous function at $D/B = 1$.

Load Inclination Factors

The load inclination factors are for loads that do not act perpendicular to the base of the footing, but still act through its centroid (eccentric loads are discussed in Chapter 8). The variable P refers to the component of the load that acts perpendicular to the bottom of the footing, and V refers to the component that acts parallel to the bottom.

The load inclination factors are:

$$i_c = 1 - \frac{mV}{Ac'N_c} \geq 0 \tag{6.20}$$

$$i_q = \left[1 - \frac{V}{P + \dfrac{Ac'}{\tan \phi'}} \right]^m \geq 0 \qquad (6.21)$$

$$i_\gamma = \left[1 - \frac{V}{P + \dfrac{Ac'}{\tan \phi'}} \right]^{m+1} \geq 0 \qquad (6.22)$$

For loads inclined in the *B* direction:

$$m = \frac{2 + B/L}{1 + B/L} \qquad (6.23)$$

For loads inclined in the *L* direction:

$$m = \frac{2 + L/B}{1 + L/B} \qquad (6.24)$$

Where:

V = applied shear load

P = applied normal load

A = base area of footing

c' = effective cohesion (use $c = s_u$ for undrained analyses)

ϕ' = effective friction angle (use $\phi = 0$ for undrained analyses)

B = foundation width

L = foundation length

If the load acts perpendicular to the base of the footing, the *i* factors equal 1 and may be neglected. The *i* factors also equal 1 when $\phi = 0$.

See the discussion in Chapter 8 for additional information on design of spread footings subjected to applied shear loads.

Base Inclination Factors

The vast majority of footings are built with horizontal bases. However, if the applied load is inclined at a large angle from the vertical, it may be better to incline the base of the footing to the same angle so the applied load acts perpendicular to the base. However, keep in mind that such footings may be difficult to construct.

The base inclination factors are:

$$b_c = 1 - \frac{\alpha}{147°}$$ (6.25)

$$b_q = b_\gamma = \left(1 - \frac{\alpha \tan \phi'}{57°}\right)^2$$ (6.26)

If the base of the footing is level, which is the usual case, all of the b factors become equal to 1 and may be ignored.

Ground Inclination Factors

Footings located near the top of a slope have a lower bearing capacity than those on level ground. Vesić's ground inclination factors, presented below, account for this. However, there are also other considerations when placing footings on or near slopes, as discussed in Chapter 8.

$$g_c = 1 - \frac{\beta}{147°}$$ (6.27)

$$g_q = g_\gamma = [1 - \tan \beta]^2$$ (6.28)

If the ground surface is level ($\beta = 0$), the g factors become equal to 1 and may be ignored.

Bearing Capacity Factors

Vesić used the following formulas for computing the bearing capacity factors N_q and N_c:

$$N_q = e^{\pi \tan\phi'} \tan^2(45 + \phi'/2)$$ (6.29)

$$N_c = \frac{N_q - 1}{\tan \phi'}$$ for $\phi' > 0$ (6.30)

$$N_c = 5.14$$ for $\phi = 0$ (6.31)

Most other authorities also accept Equations 6.29 to 6.31, or others that produce very similar results. However, there is much more disagreement regarding the proper value of N_γ. Relatively small changes in the geometry of the failure surface below the footing can create significant differences in N_γ, especially in soils with high friction angles. Vesić recommended the following formula:

$$N_\gamma = 2(N_q + 1) \tan \phi'$$ (6.32)

Vesić's bearing capacity factors also are presented in tabular form in Table 6.1. The application of Vesić's formula is illustrated in Example 6.3 later in this chapter.

QUESTIONS AND PRACTICE PROBLEMS

Note: Unless otherwise stated, all foundations have level bases, are located at sites with level ground surfaces, support vertical loads, and are oriented so the top of the foundation is flush with the ground surface.

6.1 List the three types of bearing capacity failures and explain the differences between them.

6.2 A 1.2-m square, 0.4-m deep spread footing is underlain by a soil with the following properties: $\gamma = 19.2$ kN/m^3, $c' = 5$ kPa, $\phi' = 30°$. The groundwater table is at a great depth.
   a. Compute the ultimate bearing capacity using Terzaghi's method.
   b. Compute the ultimate bearing capacity using Vesić's method.

6.3 A 5 ft wide, 8 ft long, 2 ft deep spread footing is underlain by a soil with the following properties: $\gamma = 120$ lb/ft^3, $c' = 100$ lb/ft^2, $\phi' = 28°$. The groundwater table is at a great depth. Using Vesić's method, compute the column load required to cause a bearing capacity failure.

6.3   GROUNDWATER EFFECTS

The presence of shallow groundwater affects shear strength in two ways: the reduction of apparent cohesion, and the increase in pore water pressure. Both of these affect bearing capacity, and thus need to be considered.

Apparent Cohesion

Sometimes soil samples obtained from the exploratory borings are not saturated, especially if the site is in an arid or semi-arid area. These soils have additional shear strength due to the presence of apparent cohesion, as discussed in Chapter 3. However, this additional strength will disappear if the moisture content increases. Water may come from landscape irrigation, rainwater infiltration, leaking pipes, rising groundwater, or other sources. Therefore, we do not rely on the strength due to apparent cohesion.

In order to remove the apparent cohesion effects and simulate the "worst case" condition, geotechnical engineers usually wet the samples in the lab prior to testing. This may be done by simply soaking the sample, or, in the case of the triaxial test, by backpressure saturation. However, even with these precautions, the cohesion measured in the laboratory test may still include some apparent cohesion. Therefore, we often perform bearing capacity computations using a cohesion value less than that measured in the laboratory.

Pore Water Pressure

If there is enough water in the soil to develop a groundwater table, and this groundwater table is within the potential shear zone, then pore water pressures will be present, the effective stress and shear strength along the failure surface will be smaller, and the ultimate bearing capacity will be reduced (Meyerhof, 1955). We must consider this effect when conducting bearing capacity computations.

When exploring the subsurface conditions, we determine the current location of the groundwater table and worst-case (highest) location that might reasonably be expected during the life of the proposed structure. We then determine which of the following three cases describes the worst-case field conditions:

- Case 1: $D_w \leq D$
- Case 2: $D < D_w < D + B$
- Case 3: $D + B \leq D_w$

All three cases are shown in Figure 6.9.

We account for the decreased effective stresses along the failure surface by adjusting the effective unit weight, γ', in the third term of Equations 6.4 to 6.6 and 6.13 (Vesić, 1973). The effective unit weight is the value that, when multiplied by the appropriate soil thickness, will give the vertical effective stress. It is the weighted average of the buoyant unit weight, γ_b, and the unit weight, γ, and depends on the position of the groundwater table. We compute γ' as follows:

For Case 1 ($D_w \leq D$):

$$\gamma' = \gamma_b = \gamma - \gamma_w \tag{6.33}$$

For Case 2 ($D < D_w < D + B$):

$$\gamma' = \gamma - \gamma_w \left(1 - \left(\frac{D_w - D}{B}\right)\right) \tag{6.34}$$

For Case 3 ($D + B \leq D_w$; no groundwater correction is necessary):

$$\gamma' = \gamma \tag{6.35}$$

In Case 1, the second term in the bearing capacity formulas also is affected, but the appropriate correction is implicit in the computation of σ_D'.

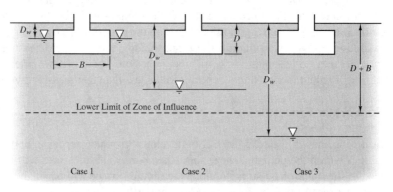

Figure 6.9   Three groundwater cases for bearing capacity analyses.

If a total stress analysis is being performed, do not apply any groundwater correction because the groundwater effects are supposedly implicit within the values of c_T and ϕ_T. In this case, simply use $\gamma' = \gamma$ in the bearing capacity equations, regardless of the groundwater table position.

Example 6.3

A 30-m by 50-m mat foundation is to be built as shown in Figure 6.10. Compute the ultimate bearing capacity.

Solution

Determine groundwater case:

$$D_w = 12 \text{ m}; D = 10 \text{ m}; B = 30 \text{ m} \quad D < D_w < D + B \therefore \text{Case 2}$$

Using Equation 6.34:

$$\gamma' = \gamma - \gamma_w \left(1 - \left(\frac{D_w - D}{B}\right)\right) = 18.5 - 9.8 \left(1 - \left(\frac{12 - 10}{30}\right)\right) = 9.4 \text{ kN/m}^3$$

Use Vesić's method with γ' in the third term. Since $c' = 0$, there is no need to compute any of the other factors in the first term of the bearing capacity equation.

For $\phi' = 30°$: $N_q = 18.4$, $N_\gamma = 22.4$ (from Table 6.1)

$$\sigma'_{zD} = \gamma D - u$$
$$= (18.5 \text{ kN/m}^3)(10 \text{ m}) - 0$$
$$= 185 \text{ kPa}$$

$$s_q = 1 + \left(\frac{B}{L}\right) \tan \phi = 1 + \left(\frac{30}{50}\right) \tan 30° = 1.35$$

$$k = \frac{D}{B} = \frac{10}{30} = 0.33$$

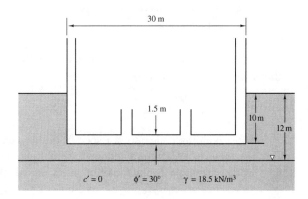

Figure 6.10 Proposed mat foundation for Example 6.3.

$$d_q = 1 + 2k \tan \phi'(1 - \sin \phi')^2$$
$$= 1 + 2(0.33) \tan 30°(1 - \sin 30°)^2$$
$$= 1.10$$

The various i, b, and g factors in Vesić's equation are all equal to 1, and thus may be ignored.

$$q_{ult} = c'N_c s_c i_c b_c g_c + \sigma'_{zD} N_q s_q d_q i_q b_q g_q + 0.5\,\gamma'BN_\gamma s_\gamma d_\gamma i_\gamma b_\gamma g_\gamma$$
$$= 0 + 185\,(18.4)(1.35)(1.10) + 0.5(9.4)(30)(22.4)(0.76)(1)$$
$$= \textbf{7455 kPa} \qquad \Leftarrow Answer$$

Commentary

Because of the large depth and large width, this is a very large ultimate bearing capacity. It is an order of magnitude greater than the bearing pressure produced by the heaviest structures, so there is virtually no risk of a bearing capacity failure. This is always the case with mats on sandy soils. However, mats on saturated clays need to be evaluated using the undrained strength, $c = s_u$, $\phi = 0$, so q_{ult} is much smaller and bearing capacity might be a concern (see the case study of the Fargo Grain Elevator later in this chapter).

6.4 ALLOWABLE BEARING CAPACITY

Nearly all bearing capacity analyses are currently implemented using allowable stress design (ASD) methods. This is true regardless of whether or not load and resistance factor design (LRFD) methods are being used in the structural design. To use ASD, we divide the ultimate bearing capacity by a factor of safety to obtain the *allowable bearing capacity*, q_a:

$$\boxed{q_a = \frac{q_{ult}}{F}} \qquad (6.36)$$

Where:

q_a = allowable bearing capacity

q_{ult} = ultimate bearing capacity

F = factor of safety

We then design the foundation so that the bearing pressure, q, does not exceed the allowable bearing pressure, q_a:

$$\boxed{q \le q_a} \qquad (6.37)$$

Most building codes do not specify design factors of safety. Therefore, engineers must use their own discretion and professional judgment when selecting F. Items to consider when selecting a design factor of safety include the following:

- **Soil type.** Shear strength in clays is less reliable than that in sands, and more failures have occurred in clays than in sands. Therefore, use higher factors of safety in clays.
- **Site characterization data.** Projects with minimal subsurface exploration and laboratory or in-situ tests have more uncertainty in the design soil parameters, and thus require higher factors of safety. However, when extensive site characterization data is available, there is less uncertainty so lower factors of safety may be used.
- **Soil variability.** Projects on sites with erratic soil profiles should use higher factors of safety than those with uniform soil profiles.
- **Importance of the structure and the consequences of a failure.** Important projects, such as hospitals, where foundation failure would be more catastrophic may use higher factors of safety than less important projects, such as agricultural storage buildings, where cost of construction is more important. Likewise, permanent structures justify higher factors of safety than temporary structures, such as construction falsework. Structures with large height-to-width ratios, such as chimneys or towers, could experience more catastrophic failure, and thus should be designed using higher factors of safety.
- **The likelihood of the design load ever actually occurring.** Some structures, such as grain silos, are much more likely to actually experience their design loads, and thus might be designed using a higher factor of safety. Conversely, office buildings are much less likely to experience the design load, and might use a slightly lower factor of safety.

Figure 6.11 shows ranges of these parameters and typical values of the factor of safety. Geotechnical engineers usually use factors of safety between 2.5 and 3.5 for bearing capacity analyses of shallow foundations. Occasionally we might use values as low as 2.0 or as high as 4.0.

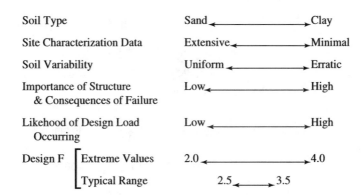

Figure 6.11   Factors affecting the design factor of safety, and typical values of F.

The true factor of safety is probably much greater than the design factor of safety, because of the following:

- The shear strength data are normally interpreted conservatively, so the design values of c and ϕ implicitly contain another factor of safety.
- The service loads are probably less than the design loads.
- Settlement, not bearing capacity, often controls the final design, so the footing will likely be larger than that required to satisfy bearing capacity criteria.
- Spread footings are commonly built somewhat larger than the plan dimensions.

Bearing capacity analyses also can be performed using LRFD, as described in Chapter 21.

Example 6.4

A column has the following design vertical loads: $P_D = 300$ k, $P_L = 140$ k, $P_W = 160$ k will be supported on a spread footing located 3 ft below the ground surface. The underlying soil has an undrained shear strength of 2000 lb/ft² and a unit weight of 109 lb/ft³. The groundwater table is at a depth of 4 ft. Determine the minimum required footing width to maintain a factor of safety of 3 against a bearing capacity failure.

Solution

Determine design working load using Equations 2.1, 2.2, 2.3a, and 2.4a:

$$P_D = 300 \text{ k}$$

$$P_D + P_L = 300 \text{ k} + 140 \text{ k} = 440 \text{ k}$$

$$0.75\,(P_D + P_L + P_W) = 0.75(300 \text{ k} + 140 \text{ k} + 160 \text{ k}) = 450 \text{ k} \leftarrow \text{Controls}$$

$$0.75\,(P_D + P_W) = 0.75\,(300 \text{ k} + 160 \text{ k}) = 345 \text{ k}$$

Using Terzaghi's method:

$$\sigma'_D = \gamma D - u = (109 \text{ lb/ft}^3)(3 \text{ ft}) - 0 = 327 \text{ lb/ft}^2$$

$$q_{ult} = 1.3s_u N_c + \sigma'_D N_q + 0.4\gamma' B N_\gamma$$

$$= 1.3(2000 \text{ lb/ft}^2)(5.7) + (327 \text{ lb/ft}^2)(1) + 0$$

$$= 15{,}147 \text{ lb/ft}^2$$

$$q_a = \frac{q_{ult}}{F} = \frac{15{,}147 \text{ lb/ft}^2}{3} = 5049 \text{ lb/ft}^2$$

$$W_f = 3\,B^2(150 \text{ lb/ft}^3) = 450\,B^2$$

$$q_a = q = \frac{P + W_f}{A} - u \rightarrow 5049 = \frac{450{,}000 + 450\,B^2}{B^2} - 0 \rightarrow B = 9.89 \text{ ft}$$

Round off to the nearest 3 in (with SI units, round off to nearest 100 mm):

$$B = 10 \text{ ft } 0 \text{ in} \qquad \Leftarrow Answer$$

Note: Chapter 8 presents an alternative method of sizing footings subjected to wind or seismic loads.

6.5   SELECTION OF SOIL STRENGTH PARAMETERS

Proper selection of the soil strength parameters, c' and ϕ', can be the most difficult part of performing bearing capacity analyses. Field and laboratory test data is often incomplete and ambiguous, and thus difficult to interpret. In addition, the computed ultimate bearing capacity, q_{ult}, is very sensitive to changes in the shear strength. For example, if a bearing capacity analysis on a sandy soil is based on $\phi' = 40°$, but the true friction angle is only $35°$ (a 13 percent drop), the ultimate bearing capacity will be 50 to 60 percent less than expected. Thus, it is very important not to overestimate the soil strength parameters. This is why most engineers intentionally use a fairly conservative interpretation of field and laboratory test data when assessing soil strength parameters.

Degree of Saturation and Location of Groundwater Table

As discussed in Section 6.3, soils that are presently dry could become wetted sometime during the life of the structure. It is prudent to design for the worst-case conditions, so we nearly always use the saturated strength when performing bearing capacity analyses, even if the soil is not currently saturated in the field. This produces worst-case values of c' and ϕ'. We can do this by saturating, or at least soaking, the samples in the laboratory before testing them.

However, determining the location of the groundwater table is a different matter. We normally attempt to estimate the highest potential location of the groundwater table and design accordingly using the methods described in Section 6.3. The location of the groundwater table influences the bearing capacity because of its effect on the effective stress, σ'.

Drained vs. Undrained Strength

Footings located on saturated clays generate positive excess pore water pressures when they are loaded, so the most likely time for a bearing capacity failure is immediately after the load is applied. Therefore, we conduct bearing capacity analyses on these soils using the undrained shear strength, s_u.

For footings on saturated sands and gravels, any excess pore water pressures are very small and dissipate very rapidly. Therefore, evaluate such footings using the effective cohesion and effective friction angle, c' and ϕ'.

Saturated intermediate soils, such as silts, are likely to be partially drained, and engineers have varying opinions on how to evaluate them. The more conservative approach is to use the undrained strength, but many engineers use design strengths somewhere between the drained and undrained strength.

Unsaturated soils are more complex and thus more difficult to analyze. If the groundwater table will always be well below the ground surface, many engineers use total stress parameters c_T and ϕ_T based on samples that have been "soaked," but not necessarily fully saturated, in the laboratory. Another option is to treat such soils as being fully saturated and analyze them as such.

Collapse of the Fargo Grain Elevator

One of the most dramatic bearing capacity failures was the Fargo Grain Elevator collapse of 1955. This grain elevator, shown in Figure 6.12, was built near Fargo, North Dakota, in 1954. It was a reinforced concrete structure composed of twenty cylindrical bins and other appurtenant structures, all supported on a 52-ft (15.8 m) wide, 218-ft (66.4 m) long, 2-ft 4-in (0.71 m) thick mat foundation.

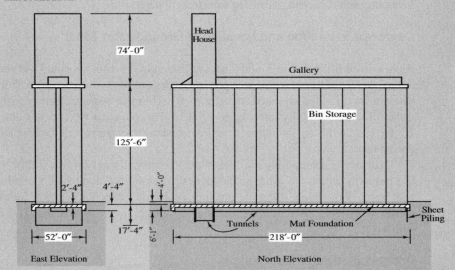

Figure 6.12 Elevation views of the elevator (Nordlund and Deere, 1970; Reprinted by permission of ASCE).

The average net bearing pressure, $q' = q - \sigma_{z0}'$, caused by the weight of the empty structure was 1590 lb/ft² (76.1 kPa). When the bins began to be filled with grain in April 1955, q' began to rise, as shown in Figure 6.13. In this type of structure, the live load (i.e., the grain) is much larger than the dead load; so by mid-June, the average net bearing pressure had tripled and reached 4750 lb/ft² (227 kPa). Unfortunately, as the bearing pressure rose, the elevator began to settle at an accelerating rate, as shown in Figure 6.14.

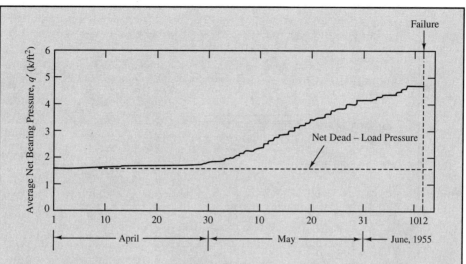

Figure 6.13   Rate of loading (Nordlund and Deere, 1970; Reprinted by permission of ASCE).

Early on the morning of June 12, 1955, the elevator collapsed and was completely destroyed. This failure was accompanied by the formation of a 6 ft (2 m) bulge, as shown in Figure 6.15.

No geotechnical investigation had been performed prior to the construction of the elevator, but Nordlund and Deere (1970) conducted an extensive after-the-fact investigation. They found that the soils were primarily saturated clays with $s_u = 600$–1000 lb/ft^2 (30–50 kPa). Bearing capacity analyses based on this data indicated a net ultimate bearing capacity of 4110 to 6520 lb/ft^2 (197–312 kPa) which compared well with the q' at failure of 4750 lb/ft^2 (average) and 5210 lb/ft^2 (maximum).

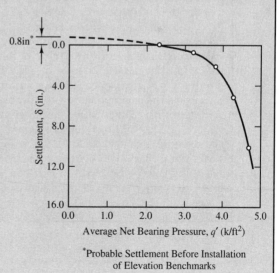

Figure 6.14   Settlement at centroid of mat (Nordlund and Deere, 1970; Reprinted by permission of ASCE).

*Probable Settlement Before Installation of Elevation Benchmarks

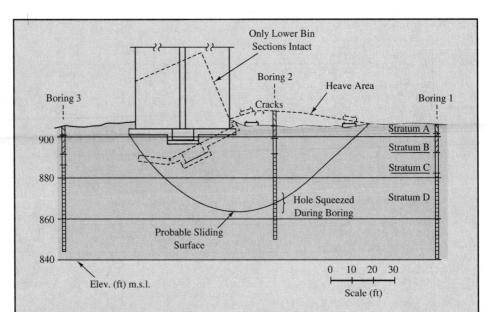

Figure 6.15   Cross section of collapsed elevator (Nordlund and Deere, 1970; Reprinted by permission of ASCE).

The investigation of the Fargo Grain Elevator failure demonstrated the reliability of bearing capacity analyses. Even a modest exploration and testing program would have produced shear strength values that would have predicted this failure. If such an investigation had been performed, and if the design had included an appropriate factor of safety, the failure would not have occurred. However, we should not be too harsh on the designers, since most engineers in the early 1950s were not performing bearing capacity analyses.

Although bearing capacity failures of this size are unusual, this failure was not without precedent. A very similar failure occurred in 1913 at a grain elevator near Winnipeg, Manitoba, approximately 200 miles (320 km) north of Fargo (Peck and Bryant, 1953; White, 1953; Skaftfeld, 1998). This elevator rotated to an inclination of 27° from the vertical when the soil below experienced a bearing capacity failure at an average q' of 4680 lb/ft^2 (224 kPa). The soil profile is very similar to the Fargo site, as is the average q' values at failure.

Geotechnical researchers from the University of Illinois investigated the Winnipeg failure in 1951. They drilled exploratory borings, performed laboratory tests, and computed a net ultimate bearing capacity of 5140 lb/ft^2 (246 kPa). Once again, a bearing capacity analysis would have predicted the failure, and a design with a suitable factor of safety would have prevented it. Curiously, the results of their study were published in 1953, only two years before the Fargo failure. This is a classic example of engineers failing to learn from the mistakes of others.

QUESTIONS AND PRACTICE PROBLEMS

Note: Unless otherwise stated, all foundations have level bases, are located at sites with level ground surfaces, support vertical loads, and are oriented so the top of the foundation is flush with the ground surface.

6.4 A column carrying a vertical downward dead load and live load of 150 k and 120 k, respectively, is to be supported on a 3-ft deep square spread footing. The soil beneath this footing is an undrained clay with $s_u = 3000$ lb/ft^2 and $\gamma = 117$ lb/ft^3. The groundwater table is below the bottom of the footing. Compute the width B required to obtain a factor of safety of 3 against a bearing capacity failure.

6.5 A 120-ft diameter cylindrical tank with an empty weight of 1,900,000 lb (including the weight of the cylindrical mat foundation) is to be built. The bottom of the mat will be at a depth of 2 ft below the ground surface. This tank is to be filled with water. The underlying soil is an undrained clay with $s_u = 1000$ lb/ft^2 and $\gamma = 118$ lb/ft^3, and the groundwater table is at a depth of 5 ft. Using Terzaghi's equations, compute the maximum allowable depth of the water in the tank that will maintain a factor of safety of 3.0 against a bearing capacity failure. Assume the weight of the water and tank is spread uniformly across the bottom of the tank.

6.6 A 1.5-m wide, 2.5-m long, 0.5-m deep spread footing is underlain by a soil with $c' = 10$ kPa, $\phi' = 32°$, $\gamma = 18.8$ kN/m^3. The groundwater table is at a great depth. Compute the maximum load this footing can support while maintaining a factor of safety of 2.5 against a bearing capacity failure.

6.7 A bearing wall carries a dead load of 120 kN/m and a live load of 100 kN/m. It is to be supported on a 400-mm deep continuous footing. The underlying soils are medium sands with $c' = 0$, $\phi' = 37°$, $\gamma = 19.2$ kN/m^3. The groundwater table is at a great depth. Compute the minimum footing width required to maintain a factor of safety of at least 2 against a bearing capacity failure. Express your answer to the nearest 100 mm.

6.8 After the footing in Problem 6.7 was built, the groundwater table rose to a depth of 0.5 m below the ground surface. Compute the new factor of safety against a bearing capacity failure. Compare it with the original design value of 2 and explain why it is different.

6.9 A 5-ft wide, 8-ft long, 3-ft deep footing supports a downward load of 200 k and a horizontal shear load of 25 k. The shear load acts parallel to the 8-ft dimension. The underlying soils have $c_T = 500$ lb/ft^2, $\phi_T = 28°$, $\gamma = 123$ lb/ft^3. Using a total stress analysis, compute the factor of safety against a bearing capacity failure.

6.10 A spread footing supported on a sandy soil has been designed to support a certain column load with a factor of safety of 2.5 against a bearing capacity failure. However, there is some uncertainty in both the column load, P, and the friction angle, ϕ. Which would have the greatest impact on the actual factor of safety: An actual P that is twice the design value, or actual ϕ that is half the design value? Use bearing capacity computations with reasonable assumed values to demonstrate the reason for your response.

6.6   BEARING CAPACITY ANALYSIS IN SOIL—LOCAL AND PUNCHING SHEAR CASES

As discussed earlier, engineers rarely need to compute the local or punching shear bearing capacities because settlement analyses implicitly protect against this type of failure. In addition, a complete bearing capacity analysis would be more complex because of the following:

- These modes of failure do not have well-defined shear surfaces, such as those shown in Figure 6.1, and are therefore more difficult to evaluate.
- The soil can no longer be considered incompressible (Ismael and Vesić, 1981).
- The failure is not catastrophic (refer to Figure 6.2), so the failure load is more difficult to define.
- Scale effects make it difficult to properly interpret model footing tests.

Terzaghi (1943) suggested a simplified way to compute the local shear bearing capacity using the general shear formulas with appropriately reduced values of c' and ϕ':

$$c'_{adj} = 0.67\, c' \tag{6.38}$$

$$\phi'_{adj} = \tan^{-1}(0.67 \tan \phi') \tag{6.39}$$

Vesić (1975) expanded upon this concept and developed the following adjustment formula for sands with a relative density, D_r, less than 67%:

$$\phi'_{adj} = \tan^{-1}\left[(0.67 + D_r - 0.75D_r^2) \tan \phi'\right] \tag{6.40}$$

Where:

c'_{adj} = adjusted effective cohesion

ϕ'_{adj} = adjusted effective friction angle

D_r = relative density of sand, expressed in decimal form ($0 \leq D_r \leq 67\%$)

Although Equation 6.40 was confirmed with a few model footing tests, both methods are flawed because the failure mode is not being modeled correctly. However, local or punching shear will normally only govern the final design with shallow, narrow footings on loose sands, so an approximate analysis is acceptable. The low cost of such footings does not justify a more extensive analysis, especially if it would require additional testing.

An important exception to this conclusion is the case of a footing supported by a thin crust of strong soil underlain by very weak soil. This would likely be governed by punching shear and would justify a custom analysis.

6.7 BEARING CAPACITY ON LAYERED SOILS

Thus far, the analyses in this chapter have considered only the condition where c', ϕ', and γ are constant with depth. However, many soil profiles are not that uniform. Therefore, we need to have a method of computing the bearing capacity of foundations on soils where c, ϕ, and γ vary with depth. There are three primary ways to do this:

1. Evaluate the bearing capacity using the lowest values of c', ϕ', and γ in the zone between the bottom of the foundation and a depth B below the bottom, where B = the width of the foundation. This is the zone in which bearing capacity failures occur (per Figure 6.5), and thus is the only zone in which we need to assess the soil parameters. This method is conservative, since some of the shearing occurs in the other, stronger layers. However, many design problems are controlled by settlement anyway, so a conservative bearing capacity analysis may be the simplest and easiest solution. In other words, if bearing capacity does not control the design even with a conservative analysis, there is no need to conduct a more detailed analysis.

or 2. Use weighted average values of c', ϕ', and γ based on the relative thicknesses of each stratum in the zone between the bottom of the footing and a depth B below the bottom. This method could be conservative or unconservative, but should provide acceptable results so long as the differences in the strength parameters are not too great.

or 3. Consider a series of trial failure surfaces beneath the footing and evaluate the stresses on each surface using methods similar to those employed in slope stability analyses. The surface that produces the lowest value of q_{ult} is the critical failure surface. This method is the most precise of the three, but also requires the most effort to implement. It would be appropriate only for critical projects on complex soil profiles.

Example 6.5

Using the second method described above, compute the factor of safety against a bearing capacity failure in the square footing shown in Figure 6.16.

Solution

Weighting factors
Upper stratum: $1.1/1.8 = 0.611$
Lower stratum: $0.7/1.8 = 0.389$

Weighted values of soil parameters:

$$c' = (0.611)(5 \text{ kPa}) + (0.389)(0) = 3 \text{ kPa}$$

$$\phi' = (0.611)(32°) + (0.389)(38°) = 34°$$

$$\gamma = (0.611)(18.2 \text{ kN/m}^3) + (0.389)(20.1 \text{ kN/m}^3) = 18.9 \text{ kN/m}^3$$

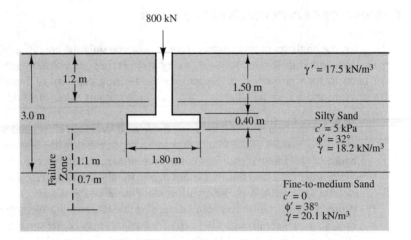

Figure 6.16   Spread footing for Example 6.6.

Groundwater case 1 ($D_w \leq D$)

$$\gamma' = \gamma - \gamma_w = 18.9\ \text{kN/m}^3 - 9.8\ \text{kN/m}^3 = 9.1\ \text{kN/m}^3$$

$$W_f = (1.8\ \text{m})^2(1.5\ \text{m})(17.5\ \text{kN/m}^3) + (1.8\ \text{m})^2(0.4\ \text{m})(23.6\ \text{kN/m}^3) = 116\ \text{kN}$$

$$\sigma_D' = \Sigma\,\gamma H - u$$
$$= (17.5\ \text{kN/m}^3)(1.2\ \text{m}) + (18.2\ \text{kN/m}^3)(0.7\ \text{m}) - (9.8\ \text{kN/m}^3)(0.7\ \text{m})$$
$$= 27\ \text{kPa}$$

$$q = \frac{P + W_f}{A} - u_D = \frac{800\ \text{kN} + 116\ \text{kN}}{(1.8\ \text{m}^2)} - 27\ \text{kPa} = 256\ \text{kPa}$$

Use Terzaghi's formula
For $\phi' = 34°$, $N_c = 52.6$, $N_q = 36.5$, $N_\gamma = 39.6$

$$q_{ult} = 1.3c'N_c + \sigma_D'N_q + 0.4\,\gamma BN_\gamma$$
$$= (1.3)(3)(52.6) + (27)(36.5) + (0.4)(9.1)(1.8)(39.6)$$
$$= 1450\ \text{kPa}$$

$$F = \frac{q_{ult}}{q} = \frac{1450\ \text{kPa}}{256\ \text{kPa}} = \textbf{5.7} \qquad \Leftarrow Answer$$

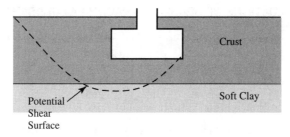

Figure 6.17   Spread footing on a hard crust underlain by softer soils.

> **The computed factor of safety of 5.7 is much greater than the typical minimum values of 2.5 to 3.5. Therefore, the footing is overdesigned as far as bearing capacity is concerned. However, it is necessary to check settlement (as discussed in Chapter 7) before reducing the size of this footing.**   ⇐ *Answer*

Figure 6.17 shows a layered soil condition that deserves special attention: a shallow foundation constructed on a thin crust underlain by softer soils. Such crusts are common in many soft clay deposits, and can be deceiving because they appear to provide good support for foundations. However, the shear surface for a bearing capacity failure would extend into the underlying weak soils. This is especially problematic for wide foundations, such as mats, because they have correspondingly deeper shear surfaces.

This condition should be evaluated using the third method described above. In addition, the potential for a punching shear failure needs to be checked.

6.8   ACCURACY OF BEARING CAPACITY ANALYSES

Engineers have had a few opportunities to evaluate the accuracy of bearing capacity analyses by evaluating full-scale bearing capacity failures in real foundations, and by conducting experimental load tests on full-size foundations.

Bishop and Bjerrum (1960) compiled the results of fourteen case studies of failures or load tests on saturated clays, as shown in Table 6.2, and found the computed factor of safety in each case was within 10 percent of the true value of 1.0. This is excellent agreement, and indicates the bearing capacity analyses are very accurate in this kind of soil. The primary source of error is probably the design value of the undrained shear strength, s_u. In most practical designs, the uncertainty in s_u is probably greater than 10 percent, but certainly well within the typical factor of safety for bearing capacity analyses.

Shallow foundations on sands have a high ultimate bearing capacity, especially when the foundation width, B, is large, because these soils have a high friction angle. Small-model footings, such as those described in Section 6.1, can be made to fail, but it is very difficult to induce failure in large footings on sand. For example, Briaud and

TABLE 6.2   EVALUATIONS OF BEARING CAPACITY FAILURES ON SATURATED CLAYS
(Bishop and Bjerrum, 1960).

Locality	Clay Properties					Computed Factor of Safety F
	Moisture content w	Liquid limit w_L	Plastic limit w_P	Plasticity index I_P	Liquidity index I_L	
Loading test, Marmorerá	10	35	15	20	−0.25	0.92
Kensal Green						1.02
Silo, Transcona	50	110	30	80	0.25	1.09
Kippen	50	70	28	42	0.52	0.95
Screw pile, Lock Ryan						1.05
Screw pile, Newport						1.07
Oil tank, Fredrikstad	45	55	25	30	0.67	1.08
Oil tank A, Shellhaven	70	87	25	62	0.73	1.03
Oil tank B, Shellhaven						1.05
Silo, US	40		20	35	1.37	0.98
Loading test, Moss	9		16	8	1.39	1.10
Loading test, Hagalund	68	55	19	18	1.44	0.93
Loading test, Torp	27	24				0.96
Loading test, Rygge	45	37				0.95

Gibbens (1994) conducted static load tests on five spread footings built on a silty fine sand. The widths of these footings ranged from 1 to 3 m, the computed ultimate bearing capacity ranged from 800 to 1400 kPa, and the load-settlement curves are shown in Figure 6.18. The smaller footings show no indiction of approaching the ultimate bearing capacity, even at bearing pressures of twice q_{ult} and settlements of about 150 mm. The larger footings appear to have an ultimate bearing capacity close to q_{ult}, but a settlement of well over 150 mm would be required to reach it. These curves also indicate the design of the larger footings would be governed by settlement, not bearing capacity, so even a conservative evaluation of bearing capacity does not adversely affect the final design. For smaller footings, the design might be controlled by the computed bearing capacity and might be conservative. However, even then the conservatism in the design should not significantly affect the construction cost.

Therefore, we have good evidence to support the claim that bearing capacity analysis methods as presented in this chapter are suitable for the practical design of shallow foundations. Assuming reliable soil strength data is available, the computed values of q_{ult} are either approximately correct or conservative. The design factors of safety discussed in Section 6.4 appear to adequately cover the uncertainties in the analysis.

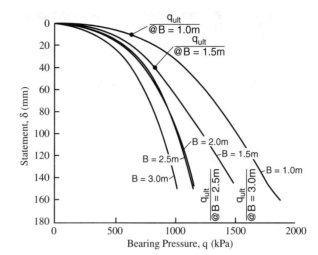

Figure 6.18 Results of static load tests on full-sized spread footings (Adapted from Briaud and Gibbens, 1994).

6.9   BEARING SPREADSHEET

Bearing capacity analyses can easily be performed using a spreadsheet, such as Microsoft Excel. These spreadsheets remove much of the tedium of performing the analyses by hand. For example, to find the required footing width, the engineer can simply input all of the other parameters and, through a rapid process of trial-and-error, find the value of B that produces the required allowable load capacity. Spreadsheets also facilitate "what-if" studies.

A Microsoft Excel spreadsheet called BEARING.XLS has been developed in conjunction with this book. It may be downloaded from the Prentice Hall web site, as described in Appendix B. Figure 6.19 shows a typical screen.

QUESTIONS AND PRACTICE PROBLEMS—SPREADSHEET ANALYSES

6.11 A certain column carries a vertical downward load of 1200 kN. It is to be supported on a 1 m deep, square footing. The soil beneath this footing has the following properties: $\gamma = 20.5$ kN/m^3, $c' = 5$ kPa, $\phi' = 36°$. The groundwater table is at a depth of 1.5 m below the ground surface. Using the BEARING.XLS spreadsheet, compute the footing width required for a factor of safety of 3.5.

6.12 a.   Using the BEARING.XLS spreadsheet, solve Problem 6.7.

  b. This footing has been built to the size determined in Part a of this problem. Sometime after construction, assume the groundwater table rises to a depth of 0.5 m below the ground surface. Use the spreadsheet to determine the new factor of safety.

6.13 A certain column carries a vertical downward load of 424 k. It is to be supported on a 3-ft deep rectangular footing. Because of a nearby property line, this footing may be no more than

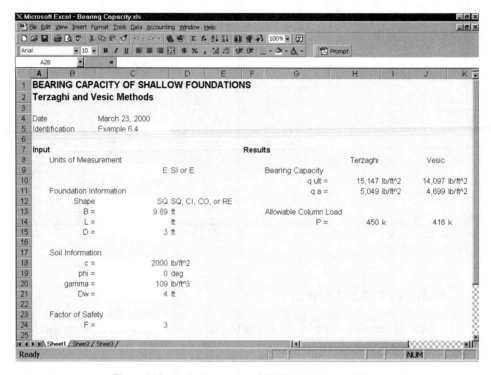

Figure 6.19   Typical screen from BEARING.XLS spreadsheet.

5 ft wide. The soil beneath this footing has the following properties: $\gamma = 124$ lb/ft^3, $c' = 50$ lb/ft^2, $\phi' = 34°$. The groundwater table is at a depth of 6 ft below the ground surface. Using the BEARING.XLS spreadsheet, compute the footing length required for a factor of safety of 3.0.

SUMMARY

Major Points

1. A *bearing capacity failure* occurs when the soil beneath the footing fails in shear. There are three types of bearing capacity failures: general shear, local shear, and punching shear.

2. Most bearing capacity analyses for shallow foundations consider only the general shear case.

3. A variety of formulas have been developed to compute the *ultimate bearing capacity*, q_{ult}. These include Terzaghi's formulas and Vesić's formulas. Terzaghi's are

most appropriate for quick hand calculations, whereas Brinch Hansen's are more useful when greater precision is needed or special loading or geometry conditions must be considered.

4. Shallow groundwater tables reduce the effective stress in the near-surface soils and can therefore adversely affect bearing capacity. Adjustment factors are available to account for this effect.

5. The *allowable bearing capacity*, q_a, is the ultimate bearing capacity divided by a factor of safety. The bearing pressure, q, produced by the unfactored structural load must not exceed q_a.

6. Bearing capacity analyses should be based on the worst-case soil conditions that are likely to occur during the life of the structure. Thus, we typically wet the soil samples in the lab, even if they were not saturated in the field.

7. Bearing capacity analyses on sands and gravels are normally based on the effective stress parameters, c' and ϕ'. However, those on saturated clays are normally based on the undrained strength, s_u.

8. Bearing capacity computations also may be performed for the local and punching shear cases. These analyses use reduced values of c' and ϕ'.

9. Bearing capacity analyses on layered soils are more complex because they need to consider the c' and ϕ' values for each layer.

10. Evaluations of foundation failures and static load tests indicate the bearing capacity analysis methods presented in this chapter are suitable for the practical design of shallow foundations.

Vocabulary

Allowable bearing capacity	Bearing capacity formula	Local shear failure
Apparent cohesion	Bearing capacity failure	Punching shear failure
Bearing capacity factors	General shear failure	Ultimate bearing capacity

COMPREHENSIVE QUESTIONS AND PRACTICE PROBLEMS

6.14 Conduct a bearing capacity analysis on the Fargo Grain Elevator (see sidebar) and back-calculate the average undrained shear strength of the soil. The groundwater table is at a depth of 6 ft below the ground surface. Soil strata A and B have unit weights of 110 lb/ft^3; stratum D has 95 lb/ft^3. The unit weight of stratum C is unknown. Assume that the load on the foundation acted through the centroid of the mat.

6.15 Three columns, A, B, and C, are colinear, 500 mm in diameter, and 2.0 m on-center. They have vertical downward loads of 1000, 550, and 700 kN, respectively, and are to be supported on a single, 1.0 m deep rectangular combined footing. The soil beneath this proposed footing has the following properties: $\gamma = 19.5$ kN/m^3, $c' = 10$ kPa, and $\phi' = 31°$. The groundwater table is at a depth of 25 m below the ground surface.

a. Determine the minimum footing length, L, and the placement of the columns on the footing that will place the resultant load at the centroid of the footing. The footing must extend at least 500 mm beyond the edges of columns A and C.

b. Using the results from part a, determine the minimum footing width, B, that will maintain a factor of safety of 2.5 against a bearing capacity failure. Show the final design in a sketch.

Hint: Assume a value for B, compute the allowable bearing capacity, then solve for B. Repeat this process until the computed B is approximately equal to the assumed B.

6.16 Two columns, A and B, are to be built 6 ft 0 in apart (measured from their centerlines). Column A has a vertical downward dead load and live loads of 90 k and 80 k, respectively. Column B has corresponding loads of 250 k and 175 k. The dead loads are always present, but the live loads may or may not be present at various times during the life of the structure. It is also possible that the live load would be present on one column, but not the other.

These two columns are to be supported on a 4 ft 0 in deep rectangular spread footing founded on a soil with the following parameters: unit weight = 122 lb/ft^3, effective friction angle = 37°, and effective cohesion = 100 lb/ft^2. The groundwater table is at a very great depth.

a. The location of the resultant of the loads from columns A and B depends on the amount of live load acting on each at any particular time. Considering all of the possible loading conditions, how close could it be to column A? To column B?

b. Using the results of part a, determine the minimum footing length, L, and the location of the columns on the footing necessary to keep the resultant force within the middle third of the footing under all possible loading conditions. The footing does not need to be symmetrical. The footing must extend at least 24 in beyond the centerline of each column.

c. Determine the minimum required footing width, B, to maintain a factor of safety of at least 2.5 against a bearing capacity failure under all possible loading conditions.

d. If the B computed in part c is less than the L computed in part b, then use a rectangular footing with dimensions $B \times L$. If not, then redesign using a square footing. Show your final design in a sketch.

6.17 In May 1970, a 70 ft tall, 20 ft diameter concrete grain silo was constructed at a site in Eastern Canada (Bozozuk, 1972b). This cylindrical silo, which had a weight of 183 tons, was supported on a 3 ft wide, 4 ft deep ring foundation. The outside diameter of this foundation was 23.6 ft, and its weight was about 54 tons. There was no structural floor (in other words, the contents of the silo rested directly on the ground).

The silo was then filled with grain. The exact weight of this grain is not known, but was probably about 533 tons. Unfortunately, the silo collapsed on September 30, 1970 as a result of a bearing capacity failure.

The soils beneath the silo are primarily marine silty clays. Using an average undrained shear strength of 500 lb/ft^2, a unit weight of 80 lb/ft^3, and a groundwater table 2 ft below the ground surface, compute the factor of safety against a bearing capacity failure, then comment on the accuracy of the analysis, considering the fact that a failure did occur.

7

Shallow Foundations–Settlement

From decayed fortunes every flatterer shrinks,
Men cease to build where the foundation sinks.

From the seventeenth-century British opera
The Duchess of Malfi by John Webster (1624)

By the 1950s, engineers were performing bearing capacity analyses as a part of many routine design projects. However, during that period many engineers seemed to have the misconception that any footing designed with an adequate factor of safety against a bearing capacity failure would not settle excessively. Although settlement analysis methods were available, Hough (1959) observed that these analyses, if conducted at all, were considered to be secondary. Fortunately, Hough and others emphasized that bearing capacity and settlement do not go hand-in-hand, and that independent settlement analyses also need to be performed. We now know that settlement frequently controls the design of spread footings, especially when B is large, and that the bearing capacity analysis is, in fact, often secondary.

Although this chapter concentrates on settlements caused by the application of structural loads on the footing, other sources of settlement also may be important. These include the following:

- Settlements caused by the weight of a recently placed fill
- Settlements caused by a falling groundwater table
- Settlements caused by underground mining or tunneling
- Settlements caused by the formation of sinkholes
- Settlements caused by secondary compression of the underlying soils
- Lateral movements resulting from nearby excavations that indirectly cause settlements

7.1   DESIGN REQUIREMENTS

The design of most foundations must satisfy certain settlement requirements, as discussed in Chapter 2. These requirements are usually stated in terms of the allowable total settlement, δ_a, and the allowable differential settlement, δ_{Da}, as follows:

$$\boxed{\delta \leq \delta_a} \tag{7.1}$$

$$\boxed{\delta_D \leq \delta_{Da}} \tag{7.2}$$

Where:

> δ = settlement (or total settlement)
>
> δ_a = allowable settlement (or allowable total settlement)
>
> δ_D = differential settlement
>
> δ_{Da} = allowable differential settlement

The design must satisfy both of these requirements.

Note that there is no factor of safety in either equation, because the factor of safety is already included in δ_a and δ_{Da}. The adjective "allowable" always indicates a factor of safety has already been applied. The values of δ_a and δ_{Da} are obtained using the techniques described in Chapter 2. They depend on the type of structure being supported by the foundation, and its tolerance of total and differential settlements. This chapter describes how to compute δ and δ_D for shallow foundations.

Both δ and δ_D must be computed using the unfactored downward load as computed using Equations 2.1 to 2.4. Chapter 8 discusses design loads in more detail.

7.2   OVERVIEW OF SETTLEMENT ANALYSIS METHODS

Analyses Based on Plate Load Tests

Some of the earliest attempts to assess settlement potential in shallow foundations consisted of conducting *plate load tests*. This approach consisted of making an excavation to the depth of the proposed footings, temporarily placing a 1-ft (305 mm) square steel plate on the base of the excavation, and loading it to obtain in-situ load-settlement data. Usually the test continued until a certain settlement was reached, then the foundations were designed using the bearing pressure that corresponded to some specific settlement of the plate, such as 0.5 in or 1.0 cm.

Although plate load tests may seem to be a reasonable approach, experience has proven otherwise. This is primarily because the plate is so much smaller than the foundation, and we cannot always extrapolate the data accurately.

The depth of influence of the plate (about twice the plate width) is much shallower than that of the real footing, as shown in Figure 7.1, so the test reflects only the properties

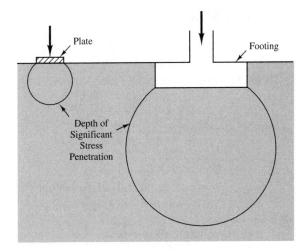

Figure 7.1 The stresses induced by a plate load test do not penetrate very deep into the soil, so its load-settlement behavior is not the same as that of a full-sized footing.

of the near-surface soils. This can introduce large errors, and several complete foundation failures occurred in spite of the use of plate load tests (Terzaghi and Peck, 1967).

In addition, because of the small size of the plate, the test reflects only the properties of the uppermost soils and thus can be very misleading, especially when the soil properties vary with depth. For example, D'Appolonia, et al. (1968) conducted a series of plate load tests in northern Indiana and found that, even after adjusting the test results for scale effects, the plate load tests underestimated the actual settlements by an average of a factor of 2. The test for a certain 12-ft wide footing at the site was in error by a factor of 3.2.

Because of these problems, and because of the development of better methods of testing and analysis, current engineering practice rarely uses plate load tests for foundation design problems. However, these tests are still useful for other design problems, such as those involving wheel loads on pavement subgrades, where the service loads act over smaller areas.

Analyses Based on Laboratory or In-Situ Tests

Today, nearly all settlement analyses are based on the results of laboratory or in-situ tests. The laboratory methods are based on the results of consolidation tests, and thus are primarily applicable to soils that can be sampled and tested without excessive disturbance. This is usually the preferred method for foundations underlain by clayey soils.

In-situ methods are based on standard penetration tests, cone penetration tests, or other in-situ tests. In principle, these methods are applicable to all soil types, but have been most often applied to sandy soils because they are difficult to sample and thus are not well suited to consolidation testing.

This chapter discusses both laboratory and in-situ methods, both of which produce predictions of the total settlement, δ. It also discusses methods of computing the differential settlement, δ_D.

7.3   INDUCED STRESSES BENEATH SHALLOW FOUNDATIONS

The bearing pressure from shallow foundations induces a vertical compressive stress in the underlying soils. We call this stress $\Delta\sigma_z$, because it is the change in stress that is superimposed on the initial vertical stress:

$$\boxed{\Delta\sigma_z = I_\sigma(q - \sigma'_{zD})} \tag{7.3}$$

Where:

$\Delta\sigma_z$ = induced vertical stress due to load from foundation

I_σ = stress influence factor

q = bearing pressure along bottom of foundation

σ_{zD}' = vertical effective stress at a depth D below the ground surface

The q term reflects the increase in vertical stress caused by the applied structural load and the weight of the foundation, while the σ_{zD}' term reflects the reduction in vertical stress caused by excavation of soil to build the foundation. Thus, $\Delta\sigma_z$ reflects the net result of these two effects.

Immediately beneath the foundation, the applied load is distributed across the base area of the foundation, so $I_\sigma = 1$. However, as the load propagates through the ground, it is spread over an increasingly larger area, so $\Delta\sigma_z$ and I_σ decrease with depth, as shown in Figure 7.2.

Boussinesq's Method

Boussinesq (1885) developed the classic solution for induced stresses in an elastic material due to an applied point load. Newmark (1935) then integrated the Boussinesq equation to produce a solution for I_σ at a depth z_f beneath the corner of a rectangular foundation of width B and length L, as shown in Figure 7.3. This solution produces the following two equations:

If $B^2 + L^2 + z_f^2 < B^2 L^2 / z_f^2$:

$$\boxed{\begin{aligned} I_\sigma = \frac{1}{4\pi}&\left[\left(\frac{2BLz_f\sqrt{B^2 + L^2 + z_f^2}}{z_f^2(B^2 + L^2 + z_f^2) + B^2L^2}\right)\left(\frac{B^2 + L^2 + 2z_f^2}{B^2 + L^2 + z_f^2}\right)\right. \\ &\left. + \pi - \sin^{-1}\frac{2BLz_f\sqrt{B^2 + L^2 + z_f^2}}{z_f^2(B^2 + L^2 + z_f^2) + B^2L^2}\right] \end{aligned}} \tag{7.4}$$

Otherwise:

$$\boxed{\begin{aligned} I_\sigma = \frac{1}{4\pi}&\left[\left(\frac{2BLz_f\sqrt{B^2 + L^2 + z_f^2}}{z_f^2(B^2 + L^2 + z_f^2) + B^2L^2}\right)\left(\frac{B^2 + L^2 + 2z_f^2}{B^2 + L^2 + z_f^2}\right)\right. \\ &\left. + \sin^{-1}\frac{2BLz_f\sqrt{B^2 + L^2 + z_f^2}}{z_f^2(B^2 + L^2 + z_f^2) + B^2L^2}\right] \end{aligned}} \tag{7.5}$$

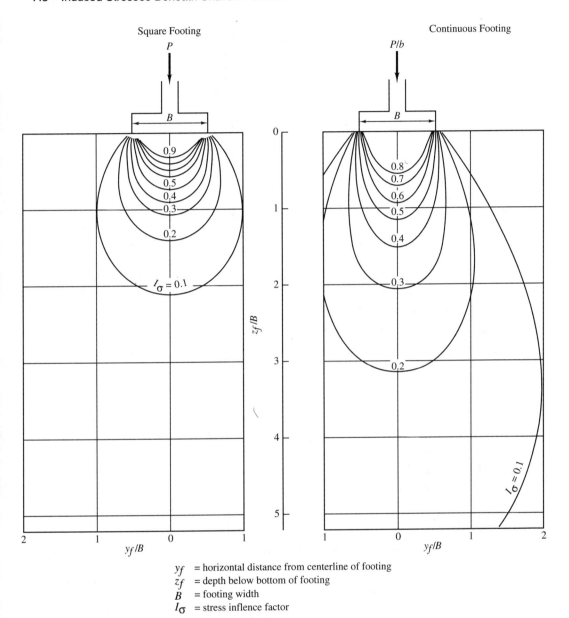

y_f = horizontal distance from centerline of footing
z_f = depth below bottom of footing
B = footing width
I_σ = stress inflence factor

Figure 7.2   Stress bulbs based on Newmark's solution of Boussinesq's equation for square and continuous footings.

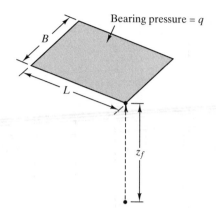

Figure 7.3   Newmark's solution for induced vertical stress beneath the corner of a
rectangular footing.

where:

I_σ = strain influence factor at a point beneath the corner of a rectangular foundation

B = width of the foundation

L = length of the foundation

z_f = vertical distance from the bottom of the foundation to the point (always > 0)

q = bearing pressure

Notes:

1. The $\sin^{-1}$ term must be expressed in radians.
2. Newmark's solution is often presented as a single equation with a $\tan^{-1}$ term, but that equation is incorrect when $B^2 + L^2 + z_f^2 < B^2 L^2 / z_f^2$.
3. It is customary to use B as the shorter dimension and L as the longer dimension, as shown in Figure 7.3.

Example 7.1

A 1.2 m × 1.2 m square footing supports a column load of 250 kN. The bottom of this footing is 0.3 m below the ground surface, the groundwater table is at a great depth, and the unit weight of the soil is 19.0 kN/m³. Compute the induced vertical stress, $\Delta\sigma_z$, at a point 1.5 m below the corner of this footing.

Solution

Unless stated otherwise, we can assume the top of this footing is essentially flush with the ground surface.

$$\sigma'_{zD} = \gamma D - u = (19.0 \text{ kN/m}^3)(0.3 \text{ m}) - 0 = 6 \text{ kPa}$$

$$W_f = (1.2 \text{ m})(1.2 \text{ m})(0.3 \text{ m})(23.6 \text{ kN/m}^3) = 10 \text{ kN}$$

$$q = \frac{P + W_f}{A} - u_D = \frac{250 \text{ kN} + 10 \text{ kN}}{(1.2 \text{ m})^2} - 0 = 181 \text{ kPa}$$

$B^2 + L^2 + z_f^2 = 1.2^2 + 1.2^2 + 1.5^2 = 5.130$
$B^2 L^2 / z_f^2 = (1.2)^2 (1.2)^2 / (1.5)^2 = 0.9216$
$B^2 + L^2 + z_f^2 > B^2 L^2 / z_f^2$. Therefore, use Equation 7.5

$$
\begin{aligned}
I_\sigma &= \frac{1}{4\pi} \left[\left(\frac{2BL z_f \sqrt{B^2 + L^2 + z_f^2}}{z_f^2 (B^2 + L^2 + z_f^2) + B^2 L^2} \right) \left(\frac{B^2 + L^2 + 2z_f^2}{B^2 + L^2 + z_f^2} \right) \right. \\
&\quad \left. + \sin^{-1} \frac{2BL z_f \sqrt{B^2 + L^2 + z_f^2}}{z_f^2 (B^2 + L^2 + z_f^2) + B^2 L^2} \right] \\
&= \frac{1}{4\pi} \left[\left(\frac{2(1.2)(1.2)(1.5) \sqrt{5.130}}{(1.5)^2(5.130) + (1.2)^2(1.2)^2} \right) \left(\frac{(1.2)^2 + (1.2)^2 + 2(1.5)^2}{5.130} \right) \right. \\
&\quad \left. + \sin^{-1} \frac{2(1.2)(1.2)(1.5) \sqrt{5.130}}{(1.5)^2(5.130) + (1.2)^2(1.2)^2} \right] \\
&= 0.146
\end{aligned}
$$

$$
\begin{aligned}
\Delta\sigma_z &= I_\sigma(q - \sigma'_{zD}) \\
&= (0.146)(181 - 6) \\
&= \mathbf{26 \text{ kPa}} \quad \Leftarrow \textit{Answer}
\end{aligned}
$$

Using superposition, Newmark's solution of Boussinesq's method also can be used to compute $\Delta\sigma_z$ at other locations, both beneath and beyond the footing. This technique is shown in Figure 7.4, and illustrated in Example 7.2.

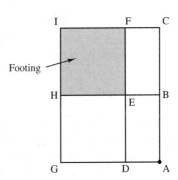

Figure 7.4   Using Newmark's solution and superposition to find the induced vertical stress at any point beneath a rectangular footing.

To Compute Stress at Point A Due to Load from Footing EFHI:
$(\Delta\sigma_v')_A = (\Delta\sigma_v')_{ACGI} - (\Delta\sigma_v')_{ACDF} - (\Delta\sigma_v')_{ABGH} + (\Delta\sigma_v')_{ABDE}$

Example 7.2

Compute the induced vertical stress, $\Delta\sigma_z$, at a point 1.5 m below the center of the footing described in Example 7.1.

Solution

Since Newmark's solution of Boussinesq's equation considers only stresses beneath the corner of a rectangular footing, we must divide the real footing into four equal quadrants. These quadrants meet at the center of the footing, which is where we wish to compute $\Delta\sigma_z$. Since each quadrant imparts one-quarter of the total load on one-quarter of the base area, the bearing pressure is the same as computed in Example 7.1. However, the remaining computations must be redone using $B = L = 1.2$ m/2 $= 0.6$ m.

$B^2 + L^2 + z_f^2 = 0.6^2 + 0.6^2 + 1.5^2 = 2.970$
$B^2L^2/z_f^2 = (0.6)^2(0.6)^2/(1.5)^2 = 0.0576$
$B^2 + L^2 + z_f^2 > B^2L^2/z_f^2$. Therefore, use Equation 7.5

$$
I_\sigma = \frac{1}{4\pi}\left[\left(\frac{2BLz_f\sqrt{B^2 + L^2 + z_f^2}}{z_f^2(B^2 + L^2 + z_f^2) + B^2L^2}\right)\left(\frac{B^2 + L^2 + 2z_f^2}{B^2 + L^2 + z_f^2}\right)\right.
$$

$$
\left. + \sin^{-1}\frac{2BLz_f\sqrt{B^2 + L^2 + z_f^2}}{z_f^2(B^2 + L^2 + z_f^2) + B^2L^2}\right]
$$

$$
= \frac{1}{4\pi}\left[\left(\frac{2(0.6)(0.6)(1.5)\sqrt{2.970}}{(1.5)^2(2.970) + (0.6)^2(0.6)^2}\right)\left(\frac{(0.6)^2 + (0.6)^2 + 2(1.5)^2}{2.970}\right)\right.
$$

$$
\left. + \sin^{-1}\frac{2(0.6)(0.6)(1.5)\sqrt{2.970}}{(1.5)^2(2.970) + (0.6)^2(0.6)^2}\right]
$$

$$
= 0.602
$$

Since there are four identical "sub-footings," we must multiply the computed stress by four.

$$
\Delta\sigma_z = 4 I_\sigma(q - \sigma'_{zD})
$$

$$
= 4(0.602)(181 - 6)
$$

$$
= \mathbf{42\ kPa} \qquad \Leftarrow Answer
$$

Westergaard's Method

Westergaard (1938) solved the same problem Boussinesq addressed, but with slightly different assumptions. Instead of using a perfectly elastic material, he assumed one that contained closely spaced horizontal reinforcement members of infinitesimal thickness, such that the horizontal strain is zero at all points. His equation also can be integrated over an area and thus may be used to compute $\Delta\sigma_z$ beneath shallow foundations (Taylor, 1948).

The Westergaard solution produces $\Delta\sigma_z$ values equal to or less than the Boussinesq values. As Poisson's ratio, v, increases, the computed stress becomes smaller, eventually reaching zero at $v = 0.5$. Although some geotechnical engineers prefer Westergaard, at least for certain soil profiles, Boussinesq is more conservative, and probably more appropriate for most problems.

Simplified Method

The Boussinesq equations are tedious to solve by hand, so it is useful to have simple approximate methods of computing stresses in soil for use when a quick answer is needed, or when a computer is not available.

The following approximate formulas compute the induced vertical stress, σ_z, beneath the center of a shallow foundation.[1] They produce answers that are within 5 percent of the Boussinesq values, which is more than sufficient for virtually all practical problems.

For circular foundations (adapted from Poulos and Davis, 1974):

$$\Delta\sigma_z = \left[1 - \left(\frac{1}{1 + \left(\dfrac{B}{2z_f}\right)^2} \right)^{1.50} \right](q - \sigma'_{zD}) \qquad (7.6)$$

For square foundations:

$$\Delta\sigma_z = \left[1 - \left(\frac{1}{1 + \left(\dfrac{B}{2z_f}\right)^2} \right)^{1.76} \right](q - \sigma'_{zD}) \qquad (7.7)$$

For continuous foundations of width B:

$$\Delta\sigma_z = \left[1 - \left(\frac{1}{1 + \left(\dfrac{B}{2z_f}\right)^2} \right)^{2.60} \right](q - \sigma'_{zD}) \qquad (7.8)$$

For rectangular foundations of width B and length L:

$$\Delta\sigma_z = \left[1 - \left(\frac{1}{1 + \left(\dfrac{B}{2z_f}\right)^{1.38 + 0.62B/L}} \right)^{2.60 - 0.84B/L} \right](q - \sigma'_{zD}) \qquad (7.9)$$

---

[1]Equations 7.4 and 7.5 compute the induced vertical stress beneath the *corner* of the loaded area, while Equations 7.6 to 7.9 compute it beneath the *center* of the loaded area.

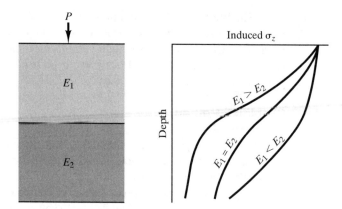

Figure 7.5   Distribution of induced stress, $\Delta\sigma_z$, in layered strata.

Where:

$\Delta\sigma_z$ = induced vertical stress beneath the center of a foundation

B = width or diameter of foundation

L = length of foundation

z_f = depth from bottom of foundation to point

q = bearing pressure

σ_{zD}' = vertical effective stress at a depth D below the ground surface

Stresses in Layered Strata

Thus far, our computations have assumed the soil beneath the foundation is homogeneous, which in this context means the modulus of elasticity, E, shear modulus, G, and Poisson's ratio, ν, are constants. This is an acceptable assumption for many soil profiles, even when there are only slight variations in the soil. However, when the strata beneath the foundation are distinctly stratified, the stress distribution changes.

One common condition consists of a soil layer underlain by a much stiffer bedrock ($E_1 < E_2$), as shown in Figure 7.5. In this case, there is less spreading of the load, so the induced stresses in the soil are greater than those computed by Boussinesq. Conversely, if we have a stiff stratum underlain by one that is softer ($E_1 > E_2$), the load spreading is enhanced and the induced stresses are less than the Boussinesq values.

Usually, engineers do not explicitly consider these effects, but we must be mindful of them to properly interpret settlement analyses. Poulos and Davis (1974) provide methods for computing these stresses in situations where an explicit analysis is warranted. Alternatively, a finite element analysis could be used.

QUESTIONS AND PRACTICE PROBLEMS

7.1 The consolidation settlement computations described in Chapter 3 considered $\Delta\sigma_z$ to be constant with depth. However, in this chapter, $\Delta\sigma_z$ decreases with depth. Why?

7.2 Examine the stress bulbs for square and continuous footings shown in Figure 7.2. Why do those for continuous footings extend deeper than those for square footings?

7.3 A 1500-mm square, 400-mm deep square footing supports a column load of 350 kN. The underlying soil has a unit weight of 18.0 kN/m^3 and the groundwater table is at a depth of 2 m below the ground surface. Compute the change in vertical stress, $\Delta\sigma_z$, beneath the center of this footing at a point 500 mm below the bottom of the footing:

   a. Using the simplified method.

   b. Using Newmark's integration of Boussinesq's method.

7.4 A column that carries a vertical downward load of 120 k is supported on a 5-ft square, 2-ft deep spread footing. The soil below has a unit weight of 124 lb/ft^3 above the groundwater table and 127 lb/ft^3 below. The groundwater table is at a depth of 8 ft below the ground surface.

   a. Develop a plot of the initial vertical stress, σ_{z0}', (i.e., the stress present before the footing was built) vs. depth from the ground surface to a depth of 15 ft below the ground surface.

   b. Using the simplified method, develop a plot of $\Delta\sigma_z$ vs. depth below the center of the bottom of the footing and plot it on the diagram developed in part a.

   c. Using Newmark's integration of Boussinesq's method, compute $\Delta\sigma_z$ at depths of 2 ft and 5 ft below the center of the bottom of the footing and plot them on the diagram.

7.4 SETTLEMENT ANALYSES BASED ON LABORATORY TESTS

Many different physical processes can contribute to settlement of shallow foundations. Some of these, as listed on the first page of this chapter, are beyond the scope of our discussion. However, the most common source of settlement, and usually the only significant source, is *consolidation*. In Chapter 3 we reviewed the process of consolidation, and noted that it is caused by shifting of the solid particles in response to an increase in the vertical effective stress.

To evaluate consolidation settlement, we begin by drilling exploratory borings into the ground and retrieving undisturbed soil samples. We then bring these samples to a soil mechanics laboratory and conduct *consolidation tests*, which measure the stress–strain properties of the soil. The test results are presented in terms of C_c, C_r, e_0, and σ_m', as discussed in Chapter 3. Finally, we perform settlement analyses based on these parameters.

This approach is usable only if we can obtain good-quality samples suitable for consolidation testing. Such samples can easily be obtained in most clayey or silty soils. However, they are very difficult to obtain in clean sands. Therefore, the methods discussed in this section are most applicable to foundations to be supported on clays or silts. Settlement of foundations on sands is most often evaluated using in-situ tests, as discussed in the next section.

We will cover two methods of using consolidation test data to compute total settlement: The *classical method* and the *Skempton and Bjerrum method*.

Classical Method

The *classical method* of computing total settlement of shallow foundations is based on Terzaghi's theory of consolidation. It assumes settlement is a one-dimensional process, in which all of the strains are vertical.

Computation of Effective Stresses

To apply Terzaghi's theory of consolidation, we need to know both the initial vertical effective stress, σ_{z0}', and the final vertical effective stress, σ_{zf}', at various depths beneath the foundation. The values of σ_{z0}' are computed using the techniques described in Section 3.4, and reflect the pre-construction conditions (i.e., without the proposed foundation). We then compute σ_{zf}' using the following equation:

$$\sigma_{zf}' = \sigma_{z0}' + \Delta\sigma_z \qquad (7.10)$$

Foundation Rigidity Effects

According to Figure 7.2, the value of $\Delta\sigma_z$ is greater under the center of a foundation than it is at the same depth under the edge. Therefore, the computed consolidation settlement will be greater at the center.

For example, consider the cylindrical steel water tank in Figure 7.6. The water inside the tank weighs much more than the tank itself, and this weight is supported directly on the plate-steel floor. In addition, the floor is relatively thin, and could be considered to be perfectly flexible. We could compute the settlement beneath both the center and the edge, using the respective values of $\Delta\sigma_z$. The difference between these two is the differential settlement, δ_D, which could then be compared to the allowable differential settlement, δ_{Da}.

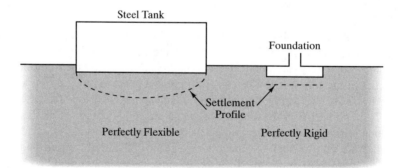

Figure 7.6   Influence of foundation rigidity on settlement. The steel tank on the left is very flexible, so the center settles more than the edge. Conversely, the reinforced concrete footing on the right is very rigid, and thus settles uniformly.

However, such an analysis would not apply to square spread footings, such as the one shown in Figure 7.6, because footings are much more rigid than plate steel tank floors. Although the center of the footing "wants" to settle more than the edge, the rigidity of the footing forces the settlement to be the same everywhere.

A third possibility would be a mat foundation, which is more rigid than the tank, but less rigid than the footing. Thus, there will be some differential settlement between the center and the edge, but not as much as with a comparably-loaded steel tank. Chapter 10 discusses methods of computing differential settlements in mat foundations, and the corresponding flexural stresses in the mat.

When performing settlement analyses on spread footings, we account for this rigidity effect by computing the settlement using $\Delta\sigma_z$ values beneath the center of the footing, then multiplying the result by a *rigidity factor*, r. Table 7.1 presents r-values for various conditions.

Many engineers choose to ignore the rigidity effect (i.e., they use $r = 1$ for all conditions), which is conservative. This practice is acceptable, especially on small or moderate-size structures, and usually has a small impact on construction costs. The use of $r < 1$ is most appropriate when the subsurface conditions have been well defined by extensive subsurface investigation and laboratory testing, which provides the needed data for a more "precise" analysis.

Settlement Computation

We compute the consolidation settlement by dividing the soil beneath the foundation into layers, computing the settlement of each layer, and summing. The top of first layer should be at the bottom of the foundation, and the bottom of the last layer should be at a depth such that $\Delta\sigma_z < 0.10\,\sigma_{z0}'$, as shown in Figure 7.7. Unless the soil is exceptionally soft, the strain below this depth is negligible, and thus may be ignored.

TABLE 7.1   r-VALUES FOR COMPUTATION OF TOTAL SETTLEMENT
AT THE CENTER OF A SHALLOW FOUNDATION, AND METHODOLOGY
FOR COMPUTING DIFFERENTIAL SETTLEMENT

Foundation rigidity	r for computation of δ at center of foundation	Methodology for computing δ_D
Perfectly flexible (i.e., steel tanks)	1.00	Compute $\Delta\sigma_z$ below edge and use $r = 1$.
Intermediate (i.e., mat foundations)	0.85–1.00, typically about 0.90	Use method described in Chapter 10.
Perfectly rigid (i.e., spread footings)	0.85	Entire footing settles uniformly, so long as bearing pressure is uniform. Compute differential settlement between footings or along length of continuous footing using method described in Section 7.7.

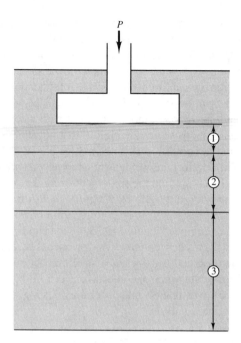

Figure 7.7 The classical method divides the soil beneath the footing into layers. The best precision is obtained when the uppermost layer is thin, and they become progressively thicker with depth.

Since strain varies nonlinearly with depth, analyses that use a large number of thin layers produce a more precise results than those that use a few thick layers. Thus, computer analyses generally use a large number of thin layers. However, this would be too tedious to do by hand, so manual computations normally use fewer layers. For most soils, the guidelines in Table 7.2 should produce reasonable results.

The consolidation settlement equations in Chapter 3 (Equations 3.25–3.27) must be modified by incorporating the r factor, as follows:

For normally consolidated soils ($\sigma_{z0}' \approx \sigma_c'$):

$$\delta_c = r \Sigma \frac{C_c}{1 + e_0} H \log \left(\frac{\sigma_{zf}'}{\sigma_{z0}'} \right) \tag{7.11}$$

For overconsolidated soils—Case I ($\sigma_{zf}' < \sigma_c'$):

$$\delta_c = r \Sigma \frac{C_r}{1 + e_0} H \log \left(\frac{\sigma_{zf}'}{\sigma_{z0}'} \right) \tag{7.12}$$

For overconsolidated soils—Case II ($\sigma_{z0}' < \sigma_c' < \sigma_{zf}'$):

$$\delta_c = r \Sigma \left[\frac{C_r}{1 + e_0} H \log \left(\frac{\sigma_c'}{\sigma_{z0}'} \right) + \frac{C_c}{1 + e_0} H \log \left(\frac{\sigma_{zf}'}{\sigma_c'} \right) \right] \tag{7.13}$$

TABLE 7.2   APPROXIMATE THICKNESSES OF SOIL LAYERS FOR MANUAL
COMPUTATION OF CONSOLIDATION SETTLEMENT OF SHALLOW FOUNDATIONS

	Approximate Layer Thickness	
Layer Number	Square Footing	Continuous Footing
1	$B/2$	B
2	B	$2B$
3	$2B$	$4B$

1. Adjust the number and thickness of the layers to account for changes in soil properties. Locate each layer entirely within one soil stratum.
2. For rectangular footings, use layer thicknesses between those given for square and continuous footings.
3. Use somewhat thicker layers (perhaps up to 1.5 times the thicknesses shown) if the groundwater table is very shallow.
4. For quick, but less precise, analyses, use a single layer with a thickness of about $3B$ (square footings) or $6B$ (continuous footings).

Where:

δ_c = ultimate consolidation settlement

r = rigidity factor (see Table 7.1)

C_c = compression index

C_r = recompression index

e_0 = initial void ratio

H = thickness of soil layer

σ_{z0}' = initial vertical effective stress at midpoint of soil layer

σ_{zf}' = final vertical effective stress at midpoint of soil layer

σ_c' = preconsolidation stress at midpoint of soil layer

As discussed earlier, many engineers choose to ignore the rigidity effect, which means the r factor drops out of these equations.

Example 7.3

The allowable settlement for the proposed square footing in Figure 7.8 is 1.0 in. Using the classical method, compute its settlement and determine if it satisfies this criterion.

Solution

$$W_f = (6 \text{ ft})^2 (2 \text{ ft})(150 \text{ lb/ft}^3) = 10,800 \text{ lb}$$

$$q = \frac{P + W_f}{A} - u_D = \frac{100,000 \text{ lb} + 10,800 \text{ lb}}{(6 \text{ ft})^2} - 0 = 3078 \text{ lb/ft}^2$$

$$\sigma_{zD}' = (115 \text{ lb/ft}^3)(2 \text{ ft}) = 230 \text{ lb/ft}^2$$

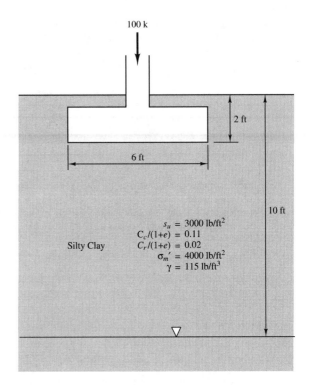

Figure 7.8 Proposed spread footing for Example 7.3.

Using Equations 3.13, 7.7, 7.10, and 7.12 with $r = 0.85$,

Layer No.	H (ft)	z_f (ft)	At midpoint of soil layer σ'_{z0} (lb/ft²)	$\Delta\sigma_z$ (lb/ft²)	σ'_{zf} (lb/ft²)	σ'_c (lb/ft²)	Case	$\dfrac{C_c}{1 + e_0}$	$\dfrac{C_r}{1 + e_0}$	δ_c (in)
1	3.0	1.5	402	2680	3082	4402	OC-I	0.11	0.02	0.54
2	6.0	6.0	920	925	1845	4920	OC-I	0.11	0.02	0.37
3	12.0	15.0	1518	190	1708	5518	OC-I	0.11	0.02	0.13
									$\Sigma =$	1.04

Round off to **$\delta = 1.0$ in** $\Leftarrow$ *Answer*

$\delta \leq \delta_a$, **so the settlement criterion has been satisfied** $\Leftarrow$ *Answer*

Note: In this case, $\sigma_m' > q$, so the soil must be overconsolidated case I. Therefore, there is no need to compute σ_c', or to list the $C_c/(1+e_0)$ values.

Example 7.4

The allowable settlement for the proposed continuous footing in Figure 7.9 is 25 mm. Using the classical method, compute its settlement and determine if it satisfies this criterion.

Solution

$$P = P_D + P_L = 40 \text{ kN/m} + 25 \text{ kN/m} = 65 \text{ kN/m}$$

$$W_f/b = (1.2 \text{ m})(0.5 \text{ m})(23.6 \text{ kN/m}^3) = 14 \text{ kN/m}$$

$$q = \frac{P/b + W_f/b}{B} - u_D = \frac{65 \text{ kN/m} + 14 \text{ kN/m}}{1.2 \text{ m}} - 0 = 66 \text{ kPa}$$

$$\sigma'_{zD} = (18.0 \text{ kN/m}^3)(0.5 \text{ m}) = 9 \text{ kPa}$$

Using Equations 3.13, 3.23, 7.7, 7.10, 7.12, and 7.13 with $r = 0.85$,

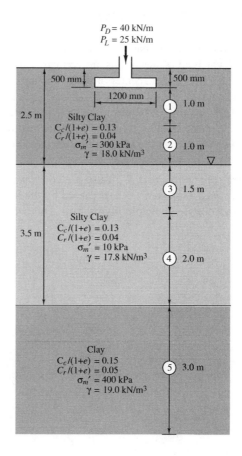

Figure 7.9   Proposed footing for Example 7.4.

Layer No.	H (m)	z_f (m)	At midpoint of soil layer				Case	$\dfrac{C_c}{1 + e_0}$	$\dfrac{C_r}{1 + e_0}$	δ_c (mm)
			σ'_{z0} (kPa)	$\Delta\sigma_z$ (kPa)	σ'_{zf} (kPa)	σ'_c (kPa)				
1	1.0	0.50	18	50	68	318	OC-I	0.13	0.04	20
2	1.0	1.50	36	27	63	336	OC-I	0.13	0.04	8
3	1.5	2.75	51	15	66	61	OC-II	0.13	0.04	10
4	2.0	4.50	65	8	73	75	OC-I	0.13	0.04	3
5	3.0	7.00	87	5	92	487	OC-I	0.16	0.05	3
									$\Sigma =$	44

$\delta = \textbf{44 mm} \Leftarrow Answer$

$\delta > \delta_a$, so **the settlement criterion has not been satisfied** $\quad \Leftarrow Answer$

Skempton and Bjerrum Method

The classical method is based on the assumption that settlement is a one-dimensional process in which all of the strains are vertical. This assumption is accurate when evaluating settlement beneath the center of wide fills, but it is less accurate when applied to shallow foundations, especially spread footings, because their loaded area is much smaller. Therefore, Skempton and Bjerrum (1957) presented another method of computing the total settlement of shallow foundations. This method accounts for three-dimensional effects by dividing the settlement into two components:

- *Distortion settlement*, δ_d, (also called *immediate settlement, initial settlement,* or *undrained settlement*) is that caused by the lateral distortion of the soil beneath the foundation, as shown in Figure 7.10. This settlement is similar to that which occurs

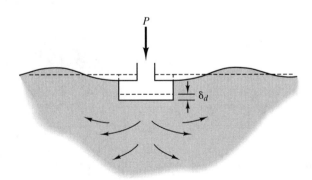

Figure 7.10   Distortion settlement beneath a spread footing.

when a load is placed on a bowl of Jello®, and occurs immediately after application of the load.

- *Consolidation settlement*, δ_c (also known as *primary consolidation* settlement), is that caused by the change in volume of the soil that results from changes in the effective stress.

In addition, Skempton and Bjerrum accounted for differences in the way excess pore water pressures are generated when the soil experiences lateral strain. This is reflected in the parameter ψ, as shown in Figure 7.11.

According to Skempton and Bjerrum's method, the settlement of a shallow foundation is computed as:

$$\delta = \delta_d + \psi\delta_c$$
(7.14)

Where:

δ = settlement

δ_d = distortion settlement (per Equation 7.15)

ψ = three-dimensional adjustment factor (from Figure 7.11)

δ_c = consolidation settlement (per Equations 7.4–7.13)

Based on elastic theory, the distortion settlement is:

$$\delta_d = \frac{(q - \sigma'_{zD})B}{E_u} I_1 I_2$$
(7.15)

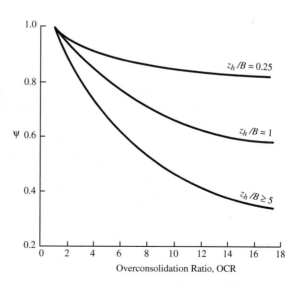

Figure 7.11 ψ factors for Skempton and Bjerrum method (Adapted from Leonards, 1976).

Where:

δ_d = distortion settlement

q = bearing pressure

σ_{zD}' = vertical effective stress at a depth D below the ground surface

B = foundation width

I_1, I_2 = influence factors (per Figure 7.12)

E_u = undrained modulus of elasticity of soil

Janbu, Bjerrum, and Kjaernsli (1956) first proposed this formula. Since then, Christian and Carrier (1978) revised the procedure and Taylor and Matyas (1983) shed additional light on its theoretical basis. The updated influence factors are shown in Figure 7.12.

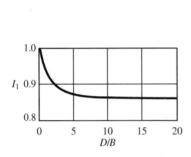
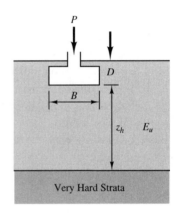

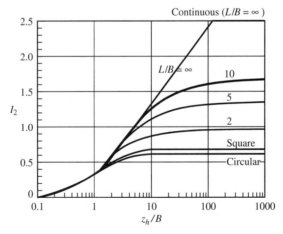

Figure 7.12   Influence factors I_1 and I_2 for use in Equation 7.15 (Adapted from Christian and Carrier, 1978; used with permission). The recommended function for continuous footings is the author's interpretation. z_h is the distance from the bottom of the footing to some very hard strata, such as bedrock. Usually, the z_h/B ratio is very large.

Equation 7.15 implicitly uses a Poisson's ratio of 0.5, which is the usual design value for saturated soils.

The undrained modulus of elasticity, E_u, is the most difficult factor to assess. Soil does not have linear stress-strain properties, so E_u must represent an equivalent linear material. One method of measuring it is to apply incremental loads on an undisturbed sample in a triaxial compression machine and measure the corresponding deformations. Unfortunately, this method tends to underestimate the modulus, sometimes by a large margin (Simons, 1987). It appears that measurements of the modulus are exceptionally sensitive to sample disturbance and the test results can be in error by as much as a factor of 3. Although careful sampling and special laboratory test techniques can reduce this error, direct laboratory testing is generally not a reliable method of measuring the modulus of elasticity.

Normally, geotechnical engineers obtain E_u for this analysis using empirical correlations with the undrained shear strength, s_u. This is convenient because we already have s_u from the bearing capacity analysis, and thus don't need to spend any extra money on additional tests. This correlation is very rough, and the ratio E_u/s_u varies between about 100 and 1500 (Duncan and Buchignani, 1976). However, the following equation should produce δ_d values that are accurate to within ±5 mm, which should be sufficient for nearly all design problems:

$$\boxed{E_u = 300\, s_u}\tag{7.16}$$

If the soil has a large organic content, the modulus may be smaller than suggested by Equation 7.16 and the distortion settlement will be correspondingly higher. For example, Foott and Ladd (1981) reported $E_u \approx 100\, s_u$ for normally consolidated Taylor River Peat at shear stress levels comparable to those that might be found beneath a spread footing.

If the computed distortion settlement is large, it may be necessary to obtain a more precise assessment of E_u, perhaps using pressuremeter or dilatometer tests. In that case, a more sophisticated analysis, such as that proposed by D'Appolonia, Poulos, and Ladd (1971) may be justified.

Example 7.5

Solve Example 7.3 using the Skempton and Bjerrum method.

Solution

The thickness of the silty clay stratum is not given, but it appears to extend to a great depth. Therefore, use a large value of z_h.

Distortion settlement

$$E_u = 300\, s_u = (300)(3000\ \text{lb/ft}^2) = 900{,}000\ \text{lb/ft}^2$$

$$D/B = 2/6 = 0.3 \quad \therefore\ \mathrm{I}_1 = 0.98$$

$$L/B = 1,\ z_h/B = \infty \quad \therefore\ \mathrm{I}_2 = 0.7$$

$$\delta_d = \frac{(q - \sigma'_{zD})B}{E_u} I_1 I_2$$

$$= \frac{(3078 \text{ lb/ft}^2 - 230 \text{ lb/ft}^2)(6 \text{ ft})}{900,000 \text{ lb/ft}^2}(0.98)(0.7)$$

$$= 0.013 \text{ ft}$$

$$= 0.2 \text{ in}$$

Consolidation settlement

Per Figure 7.11, use $\psi = 0.9$

Total settlement

$$\delta = \delta_d + \psi\delta_c = 0.2 \text{ in} + (0.9)(1.0 \text{ in}) = \mathbf{1.1 \text{ in}} \qquad \Leftarrow Answer$$

General Methodology

In summary, the general methodology for computing total settlement of shallow foundations based on laboratory tests is as follows:

1. Drill exploratory borings at the site of the proposed foundations and obtain undisturbed samples of each soil strata. Also use these borings to develop a design soil profile.
2. Perform one or more consolidation tests for each of the soil strata encountered beneath the foundation, and determine the parameters $C_c/(1 + e_0)$, $C_r/(1 + e_0)$, and σ_m' for each strata using the techniques describe in Chapter 3. In many cases, all of the soil may be considered to be a single stratum, so only one set of these parameters is needed. However, if multiple clearly defined strata are present, then each must have its own set of parameters.
3. Divide the soil below the foundation into layers. Usually, about three layers provide sufficient accuracy, but more layers may be necessary if multiple strata are present or if additional precision is required. For square foundations, the bottom of the lowest layer should be about $3B$ to $5B$ below the bottom of the foundation; for continuous foundations it should be about $6B$ to $9B$ below the bottom of the foundation. When using three layers, choose their thicknesses approximately as shown in Table 7.2.
4. Compute σ_{z0}' at the midpoint of each layer.
5. Using any of the methods described in Section 7.3, compute $\Delta\sigma_z$ at the midpoint of each layer. For hand computations, it is usually best to used Equations 7.6 to 7.9.
6. Using Equation 7.10, compute σ_{zf}' at the midpoint of each layer.
7. If the soil might be overconsolidated case II, use Equation 3.23 to compute σ_c' at the midpoint of each layer.

8. Using Equation 7.11, 7.12, or 7.13, compute δ_c for each layer, then sum. Note that the some layers may require the use of one of these equations, while other layers may require another. If using the classical method, this is the computed settlement.

If the analysis is being performed using the Skempton and Bjerrum method, continue with the following steps:

9. Determine the average undrained shear strength, s_u, in the soils between the bottom of the footing and a depth B below the bottom, then use Equation 7.16 to estimate the undrained modulus, E_u.
10. Use Equation 7.15 to compute the distortion settlement, δ_d.
11. Use Figure 7.11 to determine the three-dimensional adjustment coefficient, ψ.
12. Compute the settlement using Equation 7.14.

QUESTIONS AND PRACTICE PROBLEMS

7.5 A proposed office building will include an 8-ft 6-in square, 3-ft deep spread footing that will support a vertical downward load of 160 k. The soil below this footing is an overconsolidated clay (OC case I) with the following engineering properties: $C_c/(1 + e_0) = 0.10$, $C_r/(1 + e_0) = 0.022$, and $\gamma = 113$ lb/ft^3. This soil strata extends to a great depth and the groundwater table is at a depth of 50 ft below the ground surface. Using the classical method with hand computations, determine the total settlement of this footing.

7.6 A 1.0-m square, 0.5-m deep footing carries a downward load of 200 kN. It is underlain by an overconsolidated clay (OC case I) with the following engineering properties: $C_c = 0.20$, $C_r = 0.05$, $e = 0.7$, and $\gamma = 15.0$ kN/m^3 above the groundwater table and 16.0 kN/m^3 below. The groundwater table is at a depth of 1.0 m below the ground surface. The secondary compression settlement is negligible. Using the classical method with hand computations, determine the total settlement of this footing.

7.7 Solve Problem 7.5 using Skempton and Bjerrum's method with $s_u = 3500$ lb/ft^2 and OCR = 3.

7.8 Solve Problem 7.6 using Skempton and Bjerrum's method with $s_u = 200$ kPa and OCR = 2.

7.5 SETTLEMENT SPREADSHEET

Settlement analyses of shallow foundations, as presented in Section 7.4, can be performed using a spreadsheet such as Microsoft® Excel. Such spreadsheet solutions allow the use of much thinner layers, which improves accuracy and flexibility, and allows the analyses to be performed much more quickly. Spreadsheets are especially useful when the engineer wishes to size a footing to satisfy a particular settlement criterion, which is the most common design problem. Such analyses may be performed by quickly trying various values of footing width, B, until the required settlement is obtained.

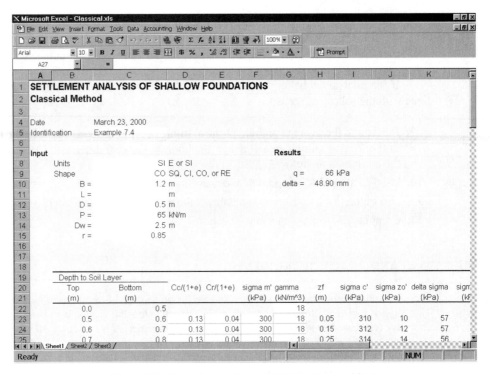

Figure 7.13   Typical screen from SETTLEMENT.XLS spreadsheet.

A Microsoft Excel spreadsheet SETTLEMENT.XLS has been developed in conjunction with this book and may be downloaded from the Prentice Hall website. Downloading instructions are presented in Appendix B. A typical screen is shown in Figure 7.13.

QUESTIONS AND PRACTICE PROBLEMS—SPREADSHEET ANALYSES

7.9 Using the SETTLEMENT.XLS spreadsheet, solve Problem 7.5. Since the soil is stated as being "overconsolidated case I," you should use a large value for σ_m'.

7.10 Using the SETTLEMENT.XLS spreadsheet and the data in Problem 7.5, determine the required footing width to obtain a total settlement of no more than 1.0 in. Select a width that is a multiple of 3 in. Would it be practical to build such a footing?

7.11 Solve Problem 7.6 using the SETTLEMENT.XLS spreadsheet.

7.12 Using the SETTLEMENT.XLS spreadsheet and the data in Problem 7.6, determine the required footing width to obtain a total settlement of no more than 25 mm. Select a width that is a multiple of 100 mm. Would it be practical to build such a footing?

7.13 A proposed building is to be supported on a series of spread footings embedded 36 inches into the ground. The underlying soils consist of silty clays with $C_c/(1 + e_0) = 0.12$, $C_r/(1 + e_0) = 0.030$,

$\sigma_m{}' = 5000$ lb/ft^2, and $\gamma = 118$ lb/ft^3. This soil strata extends to a great depth and the groundwater table is at a depth of 10 ft below the ground surface. The allowable settlement is 1.0 in. Using the SETTLEMENT.XLS spreadsheet, develop a plot of allowable column load vs. footing width.

SETTLEMENT ANALYSES BASED ON IN-SITU TESTS

The second category of settlement analysis techniques consists of those based on in-situ tests. Most of these analyses use results from the standard penetration test (SPT) or the cone penetration test (CPT). However, other in-situ tests, especially the dilatometer test (DMT) and the pressuremeter test (PMT) also may be used.

In principle, settlement analyses based on in-situ tests are suitable for all soil types. However, in practice they are most often used on sandy soils, because they are so difficult to sample. Many different methods have been proposed (for example, Meyerhof, 1965, Burland and Burbidge, 1985), but we will consider only Schmertmann's method.

Schmertmann's Method

Schmertmann's method (Schmertmann, 1970, 1978; and Schmertmann, et al., 1978) was developed primarily as a means of computing the settlement of spread footings on sandy soils. It is most often used with cone penetration test (CPT) results, but can be adapted to other in-situ tests. This method was developed from field and laboratory tests, most of which were conducted by the University of Florida. Unlike many of the other methods, which are purely empirical, the Schmertmann method is based on a physical model of settlement, which has been calibrated using empirical data.

Equivalent Modulus of Elasticity

The classical method of computing foundation settlements described the stress–strain properties using the *compression index*, C_c, for normally consolidated soils, or the *recompression index*, C_r, for overconsolidated soils. Both of these parameters are logarithmic, as discussed in Chapter 3.

Schmertmann's method uses the *equivalent modulus of elasticity*, E_s, which is a linear function and thus simplifies the computations. However, soil is not a linear material (i.e., stress and strain are not proportional), so the value of E_s must reflect that of an equivalent unconfined linear material such that the computed settlement will be the same as in the real soil.

The design value of E_s implicitly reflects the lateral strains in the soil. Thus, it is larger than the *modulus of elasticity*, E (also known as *Young's modulus*), but smaller than the *confined modulus*, M.

E_s From Cone Penetration Test (CPT) Results

Schmertmann developed empirical correlations between E_s and the cone resistance, q_c, from a cone penetration test (CPT). This method is especially useful because the CPT

TABLE 7.3 E_s-VALUES FROM CPT RESULTS [Adapted from Schmertmann, et al. (1978), Robertson and Campanella (1989), and other sources.]

Soil Type	USCS Group Symbol	E_s/q_c
Young, normally consolidated clean silica sands (age < 100 years)	SW or SP	2.5–3.5
Aged, normally consolidated clean silica sands (age > 3000 years)	SW or SP	3.5–6.0
Overconsolidated clean silica sands	SW or SP	6.0–10.0
Normally consolidated silty or clayey sands	SM or SC	1.5
Overconsolidated silty or clayey sands	SM or SC	3

provides a continuous plot of q_c vs. depth, so our analysis can model E_s as a function of depth. Table 7.3 presents a range of recommended design values of E_s/q_c. It is usually best to treat all soils as being young and normally consolidated unless there is compelling evidence to the contrary. Such evidence might include:

- Clear indications that the soil is very old. This might be established by certain geologic evidence.
- Clear indications that the soil is overconsolidated. Such evidence would not be based on consolidation tests on the sand (because of soil sampling problems), but might be based on consolidation tests performed on samples from interbedded clay strata. Alternatively, overconsolidation could be deduced from the origin of the soil deposit. For example, lodgement till and compacted fill are clearly overconsolidated.

When interpreting the CPT data for use in Schmertmann's method, do not apply an overburden correction to q_c.

E_s From Standard Penetration Test (SPT) Results

Schmertmann's method also may be used with E_s values based on the standard penetration test. However, these values are not as precise as those obtained from the cone penetration test because:

- The standard penetration test is more prone to error, and is a less precise measurement, as discussed in Chapter 4.
- The standard penetration test provides only a series of isolated data points, whereas the cone penetration test provides a continuous plot.

Nevertheless, SPT data is adequate for many projects, especially those in which the loads are small and the soil conditions are good.

Several direct correlations between E_s and N_{60} have been developed, often producing widely disparate results (Anagnostopoulos, 1990; Kulhawy and Mayne, 1990). This scatter is probably caused in part by the lack of precision in the SPT, and in part to the influence of other factors beside N_{60}. Nevertheless, the following relationship should produce approximate, if somewhat conservative, values of E_s:

$$\boxed{E_s = \beta_0 \sqrt{OCR} + \beta_1 N_{60}} \qquad (7.17)$$

Where:

   E_s = equivalent modulus of elasticity

   β_0, β_1 = correlation factors from Table 7.4

   OCR = overconsolidation ratio

   N_{60} = SPT N-value corrected for field procedures

Once again, most analyses should use OCR = 1 unless there is clear evidence of overconsolidation.

E_s From Dilatometer Test (DMT) Results

The dilatometer test (DMT) measures the modulus directly, and this data also could be used as the basis for a Schmertmann analysis. Alternatively, Leonards and Frost (1988) proposed a method of combining CPT and DMT data to assess both compressibility and overconsolidation and use the results in a modified version of Schmertmann's method.

E_s From Pressuremeter Test (PMT) Results

The pressuremeter test also measures the modulus, and also could be used with Schmertmann's method. However, special analysis methods intended specifically for pressuremeter data also are available.

Strain Influence Factor

Schmertmann conducted extensive research on the distribution of vertical strain, ϵ_z, below spread footings. He found the greatest strains do not occur immediately below the footing, as one might expect, but at a depth of 0.5 B to B below the bottom of the footing, where B is the footing width. This distribution is described by the *strain influence factor*, I_ϵ, which is a type of weighting factor. The distribution of I_ϵ with depth has been idealized as two straight lines, as shown in Figure 7.14.

TABLE 7.4   FACTORS FOR EQUATION 7.17.

Soil Type	β_0		β_1	
	(lb/ft²)	(kPa)	(lb/ft²)	(kPa)
Clean sands (SW and SP)	100,000	5,000	24,000	1,200
Silty sands and clayey sands (SM and SC)	50,000	2,500	12,000	600

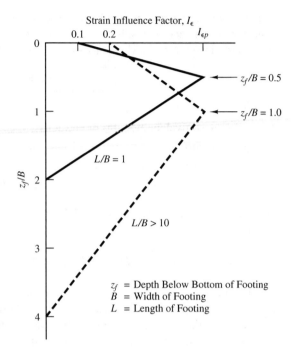

Figure 7.14   Distribution of strain influence factor with depth under square and continuous footings (Adapted from Schmertmann 1978; used with permission of ASCE).

z_f = Depth Below Bottom of Footing
B = Width of Footing
L = Length of Footing

The peak value of the strain influence factor, $I_{\epsilon p}$ is:

$$I_{\epsilon p} = 0.5 + 0.1\sqrt{\frac{q - \sigma'_{zD}}{\sigma'_{zp}}} \qquad (7.18)$$

Where:

$I_{\epsilon p}$ = peak strain influence factor

q = bearing pressure

σ'_{zD} = vertical effective stress at a depth D below the ground surface

σ'_{zp} = initial vertical effective stress at depth of peak strain influence factor (for square and circular foundations ($L/B = 1$), compute σ'_{zp} at a depth of $D+B/2$ below the ground surface; for continuous footings ($L/B \geq 10$), compute it at a depth of $D+B$)

The exact value of I_{ϵ} at any given depth may be computed using the following equations:

Square and circular foundations:

For $z_f = 0$ to $B/2$: $I_{\epsilon} = 0.1 + (z_f/B)(2I_{\epsilon p} - 0.2)$ (7.19)
For $z_f = B/2$ to $2B$: $I_{\epsilon} = 0.667 I_{\epsilon p}(2 - z_f/B)$ (7.20)

Continuous foundations ($L/B \geq 10$):

For $z_f = 0$ to B: $I_\epsilon = 0.2 + (z_f/B)(I_{\epsilon p} - 0.2)$ (7.21)

For $z_f = B$ to $4B$: $I_\epsilon = 0.333 I_{\epsilon p}(4 - z_f/B)$ (7.22)

Rectangular foundations ($1 < L/B < 10$):

$$I_\epsilon = I_{\epsilon s} + 0.111(I_{\epsilon c} - I_{\epsilon s})(L/B - 1)$$ (7.23)

Where:

z_f = depth from bottom of foundation to midpoint of layer

I_ϵ = strain influence factor

$I_{\epsilon c}$ = I_ϵ for a continuous foundation

$I_{\epsilon p}$ = peak I_ϵ from Equation 7.18

$I_{\epsilon s}$ = I_ϵ for a square foundation ≥ 0

The procedure for computing I_ϵ beneath rectangular foundations requires computation of I_ϵ for each layer using the equations for square foundations (based on the $I_{\epsilon p}$ for square foundations) and the I_ϵ for each layer using the equations for continuous foundations (based on the $I_{\epsilon p}$ for continuous foundations), then combining them using Equation 7.23.

Schmertmann's method also includes empirical corrections for the depth of embedment, secondary creep in the soil, and footing shape. These are implemented through the factors C_1, C_2, and C_3:

$$C_1 = 1 - 0.5\left(\frac{\sigma'_{zD}}{q - \sigma'_{zD}}\right)$$ (7.24)

$$C_2 = 1 + 0.2 \log\left(\frac{t}{0.1}\right)$$ (7.25)

$$C_3 = 1.03 - 0.03\, L/B \geq 0.73$$ (7.26)

Where:

δ = settlement of footing

C_1 = depth factor

C_2 = secondary creep factor (see discussion in Section 7.8)

C_3 = shape factor = 1 for square and circular foundations

q = bearing pressure

σ_{zD}' = effective vertical stress at a depth D below the ground surface

I_ϵ = influence factor at midpoint of soil layer

H = thickness of soil layer

E_s = equivalent modulus of elasticity in soil layer

t = time since application of load (yr) ($t \geq 0.1$ yr)

B = foundation width

L = foundation length

These formulas may be used with any consistent set of units, except that t must be expressed in years. If no time is given, use $t = 50$ yr ($C_2 = 1.54$).

Finally, this information is combined using the following formula to compute the settlement, δ:

$$\delta = C_1 \, C_2 \, C_3 \, (q - \sigma'_{zD}) \, \Sigma \, \frac{I_\epsilon H}{E_s} \qquad (7.27)$$

Analysis Procedure

The Schmertmann method uses the following procedure:

1. Perform appropriate in-situ tests to define the subsurface conditions.
2. Consider the soil from the base of the foundation to the depth of influence below the base. This depth ranges from $2B$ for square footings or mats to $4B$ for continuous footings. Divide this zone into layers and assign a representative E_s value to each layer. The required number of layers and the thickness of each layer depend on the variations in the E vs. depth profile. Typically 5 to 10 layers are appropriate.
3. Compute the peak strain influence factor, $I_{\epsilon p}$, using Equation 7.18.
4. Compute the strain influence factor, I_ϵ, at the midpoint of each layer. This factor varies with depth as shown in Figure 7.14, but is most easily computed using Equations 7.19 to 7.23.
5. Compute the correction factors, C_1, C_2, and C_3, using Equations 7.24 to 7.26.
6. Compute the settlement using Equation 7.27.

Example 7.6

The results of a CPT sounding performed at McDonald's Farm near Vancouver, British Columbia, are shown in Figure 7.15. The soils at this site consist of young, normally consolidated sands with some interbedded silts. The groundwater table is at a depth of 2.0 m below the ground surface.

A 375 kN/m load is to be supported on a 2.5 m × 30 m footing to be founded at a depth of 2.0 m in this soil. Use Schmertmann's method to compute the settlement of this footing soon after construction and the settlement 50 years after construction.

Solution

Use $E_s = 2.5 \, q_c$
1 kPa = 0.01020 kg/cm^2
Depth of influence = $D + 4B = 2.0 + 4(2.5) = 12.0$ m

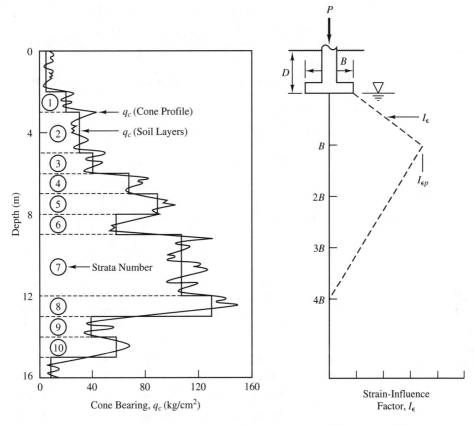

Figure 7.15 CPT results at McDonald's farm (Adapted from Robertson and Campanella, 1988).

Layer No.	Depth (m)	q_c (kg/cm^2)	E_s (kPa)
1	2.0–3.0	20	4,902
2	3.0–5.0	30	7,353
3	5.0–6.0	41	10,049
4	6.0–7.0	68	16,667
5	7.0–8.0	90	22,059
6	8.0–9.0	58	14,216
7	9.0–12.0	108	26,471

$$W_f/b = (2.5 \text{ m})(2.0 \text{ m})(23.6 \text{ kN/m}^3) = 118 \text{ kN/m}$$

$$q = \frac{P/b + W_f/b}{B} - u_D = \frac{375 \text{ kN/m} + 118 \text{ kN/m}}{2.5 \text{ m}} - 0 = 197 \text{ kPa}$$

Use $\gamma = 17 \text{ kN/m}^3$ above groundwater table and 20 kN/m^3 below (from Table 3.2).

$$\sigma'_{zp} (\text{at } z = D + B) = \Sigma\gamma H - u$$

$$= (17 \text{ kN/m}^3)(2 \text{ m}) + (20 \text{ kN/m}^3)(2.5 \text{ cm}) - (9.8 \text{ kN/m}^3)(2.5 \text{ m})$$

$$= 59 \text{ kPa}$$

$$\sigma'_{zD} = \gamma D = (17)(2) = 34 \text{ kPa}$$

$$I_{\epsilon p} = 0.5 + 0.1 \sqrt{\frac{q - \sigma'_{zD}}{\sigma'_{zp}}} = 0.5 + 0.1 \sqrt{\frac{197 \text{ kPa} - 34 \text{ kPa}}{59 \text{ kPa}}} = 0.666$$

Layer No.	E_s (kPa)	z_f (m)	I_ϵ	H (m)	$I_\epsilon H/E_s$
1	4,902	0.5	0.293	1.0	5.98×10^{-5}
2	7,353	2.0	0.573	2.0	15.58×10^{-5}
3	10,049	3.5	0.577	1.0	5.74×10^{-5}
4	16,667	4.5	0.488	1.0	2.93×10^{-5}
5	22,059	5.5	0.399	1.0	1.81×10^{-5}
6	14,216	6.5	0.310	1.0	2.18×10^{-5}
7	26,471	8.5	0.133	3.0	1.51×10^{-5}
				$\Sigma =$	35.73×10^{-5}

$$C_1 = 1 - 0.5 \left(\frac{\sigma'_{zD}}{q - \sigma'_{zD}}\right) = 1 - 0.5 \left(\frac{34 \text{ kPa}}{197 \text{ kPa} - 34 \text{ kPa}}\right) = 0.896$$

$$C_3 = 1.03 - 0.03 \, L/B \geq 0.73$$

$$= 1.03 - 0.03 \, (30/2.5)$$

$$= 0.67$$

Use $C_3 = 0.73$

At $t = 0.1$ yr:

$$C_2 = 1$$

$$\delta = C_1 \, C_2 \, C_3 \, (q - \sigma'_{zD}) \, \Sigma \, \frac{I_\epsilon H}{E_s}$$

$$= (0.896)(1)(0.73)(197 - 34)(35.73 \times 10^{-5})$$

$$= 0.038 \text{ m}$$

$$= \textbf{38 mm} \qquad \Leftarrow Answer$$

At $t = 50$ yr:

$$C_2 = 1 + 0.2 \log \left(\frac{t}{0.1} \right) = 1 + 0.2 \log \left(\frac{50}{0.1} \right) = 1.54$$

$$\delta = 38 \, (1.54) = \textbf{59 mm} \qquad \Leftarrow Answer$$

Simplified Schmertmann Method

If E_s is constant with depth between the bottom of the foundation and the depth of influence ($2 \, z_f /B$ for square and circular foundations to $4 \, z_f /B$ for continuous footings), then Equation 7.24 simplifies to the following:

For square and circular foundations ($L/B = 1$):

$$\delta = \frac{C_1 \, C_2 \, C_3 \, (q - \sigma'_{zD})(I_{\epsilon p} + 0.025)B}{E_s} \qquad (7.28)$$

For continuous footings ($L/B \geq 10$):

$$\delta = \frac{C_1 \, C_2 \, C_3 \, (q - \sigma'_{zD})(2I_{\epsilon p} + 0.1)B}{E_s} \qquad (7.29)$$

Equations 7.28 and 7.29 are especially useful when only minimal subsurface data is available, as is often the case with the SPT, and the soil appears to be fairly homogeneous.

Example 7.7

A 200-k column load is to be supported on a 3-ft deep square footing underlain by a silty sand with an average N_{60} of 28 and $\gamma = 120$ lb/ft^3. The groundwater table is at a depth of 50 ft below the ground surface. The allowable total settlement is 0.75 in. Using the simplified Schmertmann method, determine the required footing width.

Solution

$$E_s = \beta_0 \sqrt{\text{OCR}} + \beta_1 N_{60} = 50,000\sqrt{1} + (12,000)(28) = 386,000 \text{ lb/ft}^2$$

$$\sigma'_{zD} = (120 \text{ lb/ft}^3)(3 \text{ ft}) = 360 \text{ lb/ft}^2$$

Estimate $B = 7$ ft

$$W_f = (7 \text{ ft})^2(3 \text{ ft})(150 \text{ lb/ft}^3) = 22,000 \text{ lb}$$

$$q = \frac{P + W_f}{B^2} - u = \frac{200,000 \text{ lb} + 22,000 \text{ lb}}{(7 \text{ ft})^2} - 0 = 4,530 \text{ lb/ft}^2$$

$$C_1 = 1 - 0.5 \left(\frac{\sigma'_{zD}}{q - \sigma'_{zD}} \right) = 1 - 0.5 \left(\frac{360 \text{ lb/ft}^2}{4530 \text{ lb/ft}^2 - 360 \text{ lb/ft}^2} \right) = 0.957$$

$$C_2 = 1 + 0.2 \log \left(\frac{t}{0.1} \right) = 1 + 0.2 \log \left(\frac{50 \text{ yr}}{0.1} \right) = 1.54$$

$$C_3 = 1$$

$$\sigma'_{zp} @z = 6.5 \text{ ft} = (120 \text{ lb/ft}^3)(6.5 \text{ ft}) = 780 \text{ lb/ft}^2$$

$$I_{\epsilon p} = 0.5 + 0.1 \sqrt{\frac{1 - \sigma'_D}{\sigma'_{zp}}} = 0.5 + 0.1 \sqrt{\frac{4530 \text{ lb/ft}^2 - 360 \text{ lb/ft}^2}{780 \text{ lb/ft}^2}} = 0.731$$

$$\delta = \frac{C_1 C_2 C_3 (q - \sigma'_{zD})(I_{\epsilon p} + 0.025)B}{E_s}$$

$$\frac{0.75 \text{ in}}{12 \text{ in/ft}} = \frac{(0.957)(1.54)(1)\left(\dfrac{200,000 + 450 B^2}{B^2} - 360 \text{ lb/ft}^2 \right)(0.731 + 0.025)B}{386,000 \text{ lb/ft}^2}$$

$$B = 9 \text{ ft } 3 \text{ in}$$

Reevaluate with $B = 8$ ft 9 in

$$W_f = (8.75 \text{ ft})^2(3 \text{ ft})(150 \text{ lb/ft}^3) = 34,500 \text{ lb}$$

$$q = \frac{P + W_f}{B^2} - u = \frac{200,000 \text{ lb} + 34,500 \text{ lb}}{(8.75 \text{ ft})^2} - 0 = 3062 \text{ lb/ft}^2$$

$$C_1 = 1 - 0.5 \left(\frac{\sigma'_{zD}}{q - \sigma'_{zD}} \right) = 1 - 0.5 \left(\frac{360 \text{ lb/ft}^2}{3062 \text{ lb/ft}^2 - 360 \text{ lb/ft}^2} \right) = 0.933$$

$$\sigma'_{zp} @z = 7.37 \text{ ft} = (120 \text{ lb/ft}^3)(7.37 \text{ ft}) = 885 \text{ lb/ft}^2$$

$$I_{\epsilon p} = 0.5 + 0.1 \sqrt{\frac{q - \sigma'_D}{\sigma'_{zp}}} = 0.5 + 0.1 \sqrt{\frac{3062 \text{ lb/ft}^2 - 360 \text{ lb/ft}^2}{885 \text{ lb/ft}^2}} = 0.675$$

$$\delta = \frac{C_1 C_2 C_3 (q - \sigma'_{zD})(I_{\epsilon p} + 0.025)B}{E_s}$$

$$\frac{0.75 \text{ in}}{12 \text{ in/ft}} = \frac{(0.933)(1.54)(1)\left(\dfrac{200,000 + 450\,B^2}{B^2} - 360 \text{ lb/ft}^2\right)(0.675 + 0.025)B}{386,000 \text{ lb/ft}^2}$$

$$B = 8.75 \text{ ft}$$

Use $B = 8$ ft 9 in $\Leftarrow$ *Answer (as a multiple of 3 in)*

Application to Mat Foundations

Schmertmann's method was developed primarily for spread footings, so the various empirical data used to calibrate the method have been developed with this type of foundation in mind. In principle, the method also may be used with mat foundations. However, it tends to overestimate the settlement of mats because their depth of influence is much greater and the equivalent modulus values at these depths is larger than predicted by the methods described earlier in this section.

Therefore, when applying Schmertmann's method to mat foundations, it is best to progressively increase the E_s values with depth, such that E_s at 30 m (100 ft) is about three times that predicted by the methods described earlier in this section.

QUESTIONS AND PRACTICE PROBLEMS

Note: All depths for CPT and SPT data are measured from the ground surface.

7.14 A 250-k column load is to be supported on a 9 ft × 9 ft square footing embedded 2 ft below the ground surface. The underlying soil is a silty sand with an average N_{60} of 32 and a unit weight of 129 lb/ft^3. The groundwater table is at a depth of 35 ft. Using the simplified Schmertmann method, compute the settlement of this footing at $t = 50$ yr.

7.15 A 190-k column load is to be supported on a 10-ft square, 3-ft deep spread footing underlain by young, normally consolidated sandy soils. The results of a representative CPT sounding at this site are as follows:

Depth (ft)	q_c (kg/cm^2)
0.0–6.0	30
6.0–10.0	51
10.0–18.0	65
18.0–21.0	59
21.0–40.0	110

The groundwater table is at a depth of 15 ft; the unit weight of the soil is 124 lb/ft^3 above the groundwater table and 130 lb/ft^3 below. Using Schmertmann's method with hand computations, compute the total settlement of this footing 30 years after construction.

7.16 A 650-kN column load is supported on a 1.5-m wide by 2.0-m long by 0.5 m deep spread footing. The soil below is a well graded, normally consolidated sand with $\gamma = 17.0$ kN/m^3 and the following SPT N_{60} values:

Depth (m)	1.0	2.0	3.0	4.0	5.0
N_{60}	12	13	13	18	22

The groundwater table is at a depth of 25 m. Using Schmertmann's method and hand computations, compute the total settlement at $t = 30$ yr.

7.7   SCHMERTMANN SPREADSHEET

The SETTLEMENT.XLS spreadsheet described earlier in this chapter can be modified to compute settlements using the Schmertmann method. This has been done, and a spreadsheet called SCHMERTMANN.XLS is available from the Prentice Hall website. Figure 7.16 is a typical screen from this spreadsheet.

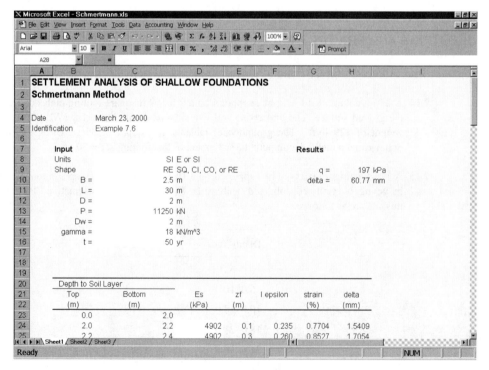

Figure 7.16   Typical screen from SCHMERTMANN.XLS spreadsheet.

QUESTIONS AND PRACTICE PROBLEMS—SPREADSHEET ANALYSES

Note: All depths for CPT and SPT data are measured from the ground surface.

7.17 Solve Problem 7.14 using the SCHMERTMANN.XLS spreadsheet.

7.18 Solve Problem 7.15 using the SCHMERTMANN.XLS spreadsheet.

7.19 Solve Problem 7.16 using the SCHMERTMANN.XLS spreadsheet.

7.20 A 300-k column load is to be supported on a 10-ft square, 4-ft deep spread footing. Cone penetration tests have been conducted at this site, and the results are shown in Figure 7.15. The groundwater table is at a depth of 6 ft, $\gamma = 121$ lb/ft^3, and $\gamma_{sat} = 125$ lb/ft^3.

   a. Compute the settlement of this footing using the SCHMERTMANN.XLS spreadsheet.

   b. The design engineer is considering the use of vibroflotation to densify the soils at this site (see discussion in Chapter 19). This process would increase the q_c values by 70 percent, and make the soil slightly overconsolidated. The unit weights would increase by 5 lb/ft^3. Use the spreadsheet to compute the settlement of a footing built and loaded after densification by vibroflotation.

7.21 A proposed building is to be supported on a series of spread footings embedded 36 inches into the ground. The underlying soils consist of silty sands with $N_{60} = 30$, an estimated overconsolidation ratio of 2, and $\gamma = 118$ lb/ft^3. This soil strata extends to a great depth and the groundwater table is at a depth of 10 ft below the ground surface. The allowable settlement is 1.0 in. Using the SCHMERTMANN.XLS spreadsheet, develop a plot of allowable column load vs. footing width.

7.8 SETTLEMENT OF SHALLOW FOUNDATIONS ON STRATIFIED SOILS

When computing the settlement of shallow foundations on soil profiles that are primarily clays or silts, we normally use the methods based on laboratory tests as discussed in Section 7.4. Conversely, when the soil profile consists primarily of sands, we normally use methods based on in-situ tests, as discussed in Section 7.6. However, when the profile is stratified and includes both types of soil, it can be difficult to determine which method to use.

If the soil profile consists predominantly of clays and silts, then it is probably best to use the methods described in Section 7.4. Determine the values of $C_c/(1 + e_0)$ and $C_r/(1 + e_0)$ for the clayey and silty strata using laboratory consolidation tests, and those for the sandy strata using Table 3.7.

Conversely, if the profile is primarily sandy, then it is probably better to use Schmertmann's method. The equivalent modulus, E_s, for normally consolidated clayey layers may be computed using the following equation:

$$E_s = \frac{2.30\,\sigma_z'}{C_c/(1 + e_0)}$$ (7.30)

For overconsolidated soils, substitute C_r for C_c in Equation 7.30.

Another option is to conduct two parallel analyses, one for the clayey strata using laboratory test data, and another for the sandy strata using in-situ data, and adding the two computed settlements.

7.9   DIFFERENTIAL SETTLEMENT

Differential settlement, δ_D, is the difference in settlement between two foundations, or the difference in settlement between two points on a single foundation. Excessive differential settlement is troublesome because it distorts the structure and thus introduces serviceability problems, as discussed in Chapter 2.

Normally we design the foundations for a structure such that all of them have the same computed total settlement, δ. Therefore, in theory, there should be no differential settlement. However, in reality differential settlements usually occur anyway. There are many potential sources of these differential settlements, including:

- **Variations in the soil profile.** For example, part of structure might be underlain by stiff natural soils, and part by a loose, uncompacted fill. Such a structure may have excessive differential settlement because of the different compressibility of these soil types, and possibly because of settlement due to the weight of the fill. This source of differential settlements is present to some degree on nearly all sites because of the natural variations in all soils, and is usually the most important source of differential settlement.

- **Variations in the structural loads.** The various foundations in a structure are designed to accommodate different loads according to the portion of the structure they support. Normally each would be designed for the same total settlement under its design load, so in theory the differential settlement should be zero. However, the ratio of actual load to design load may not be same for all of the foundations. Thus, those with a high ratio will settle more than those with a low ratio.

- **Design controlled by bearing capacity.** The design of some of the foundations may have been controlled by bearing capacity, not settlement, so even the design settlement may be less than that of other foundations in the same structure.

- **Construction tolerances.** The as-built dimensions of the foundations will differ from the design dimensions, so their settlement behavior will vary accordingly.

The rigidity of the structure also has an important influence on differential settlements. Some structures, such as the steel frame in Figure 7.17, are very flexible. Each foundation acts nearly independent of the others, so the settlement of one foundation has almost no impact on the other foundations. However, other structures are much stiffer, perhaps because of the presence of shear walls or diagonal bracing. The braced steel frame structure in Figure 7.18 is an example of a more rigid structure. In this case, the

Figure 7.17 This steel-frame structure has no diagonal bracing or shear walls, and thus would be classified as "flexible."

structure tends to smooth over differential settlement problems. For example, if one foundation settles more than the others, a rigid structure will redirect some of its load, as shown in Figure 7.19, thus reducing the differential settlement.

Computing Differential Settlement of Spread Footings

There are at least two methods of predicting differential settlements of spread footings. The first method uses a series of total settlement analyses that consider the expected variations in each of the relevant factors. For example, one analysis might consider the best-

Figure 7.18 The diagonal bracing in this steel-frame structure has been installed to resist seismic loads. However, a side benefit is that this bracing provides more rigidity, which helps even out potential differential settlements. Shear walls have a similar effect. The two bays in the center of the photograph have no diagonal bracing, and thus would be more susceptible to differential settlement problems.

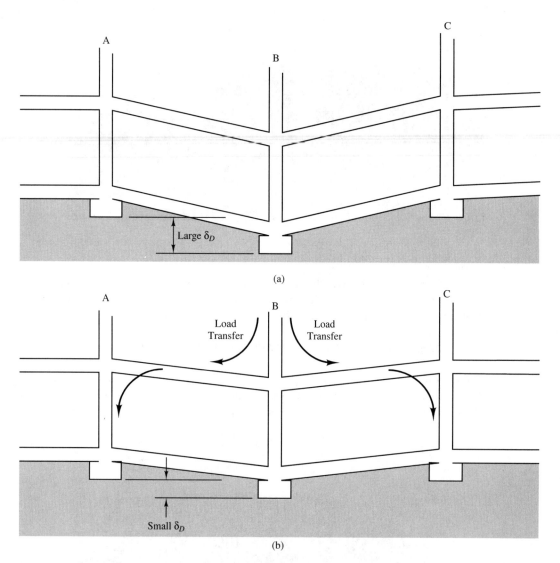

(a)

(b)

Figure 7.19 Influence of structural rigidity on differential settlements: (a) a very flexible structure has little load transfer, and thus could have larger differential settlements; (b) a more rigid structure has greater capacity for load transfer, and thus provides more resistance to excessive differential settlements.

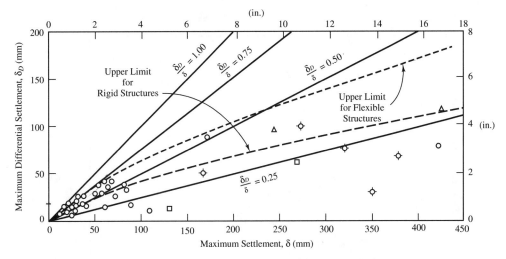

Figure 7.20 Total and differential settlements of spread footings on clays (Adapted from Bjerrum, 1963).

case scenario of soil properties, loading, and so forth, while another would consider the worst-case scenario. The difference between these two total settlements is the differential settlement.

The second method uses δ_D/δ ratios that have been observed in similar structures on similar soil profiles. For example, Bjerrum (1963) compared the total and differential settlements of spread footings on clays and sands, as shown in Figures 7.20 and 7.21. Presumably, this data was obtained primarily from sites in Scandinavia, and thus reflects the very soft soil conditions encountered in that region. This is why much of the data reflects very large settlements.

Sometimes locally-obtained δ_D/δ observations are available. Such data is more useful than generic data, such as Bjerrum's, because it implicitly reflects local soil conditions. This kind of empirical local data is probably the most reliable way to assess δ_D/δ ratios.

In the absence of local data, the generic δ_D/δ ratios in Table 7.5 may be used to predict differential settlements. The values in this table are based on Bjerrum's data and the author's professional judgement, and are probably conservative.

Remedying Differential Settlement Problems

If the computed differential settlements in a structure supported on spread footings are excessive ($\delta_D > \delta_{Da}$), the design must be changed, even if the total settlements are acceptable. Possible remedies include:

- Enlarge all of the footings until the differential settlements are acceptable. This could be done by using the allowable differential settlement, δ_{Da} and the δ_D/δ ratio to compute a new value of δ_a, then sizing the footings accordingly. Example 7.8 illustrates this technique.

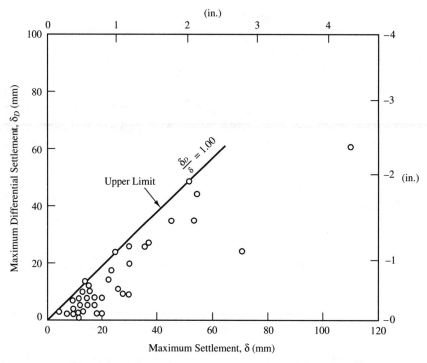

Figure 7.21   Total and differential settlement of spread footings on sands (Adapted from Bjerrum, 1963).

TABLE 7.5   DESIGN VALUES OF δ_D/δ FOR SPREAD FOOTING FOUNDATIONS

	Design Value of δ_D/δ	
Predominant Soil Type Below Footings	Flexible Structures	Rigid Structures
Sandy		
Natural soils	0.9	0.7
Compacted fills of uniform thickness underlain by stiff natural soils	0.5	0.4
Clayey		
Natural soils	0.8	0.5
Compacted fills of uniform thickness underlain by stiff natural soils	0.4	0.3

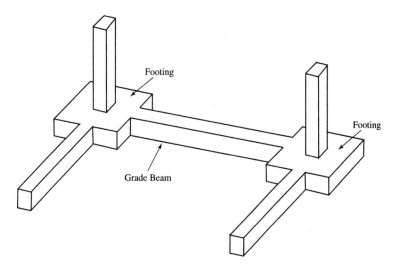

Figure 7.22   Use of grade beams to tie spread footings together.

- Connect the footings with *grade beams*, as shown in Figure 7.22. These beams provide additional rigidity to the foundation system, thus reducing the differential settlements. The effectiveness of this method could be evaluated using a structural analysis.
- Replace the spread footings with a mat foundation. This method provides even more rigidity, and thus further reduces the differential settlements. Chapter 10 discusses the analysis and design of mats.
- Replace the spread footings with a system of deep foundations, as discussed in Chapter 11.
- Redesign the superstructure so that it can accommodate larger differential settlements, so that the structural loads are lower, or both. For example, a masonry structure could be replaced by a wood-frame structure.
- Provide a method of releveling the structure if the differential settlements become excessive. This can be done by temporarily lifting selected columns from the footing and installing shims between the base plate and the footing.
- Accept the large differential settlements and repair any damage as it occurs. For some structures, such as industrial buildings, where minor distress is acceptable, this may be the most cost-effective alternative.

Example 7.8

A "flexible" steel frame building is to be built on a series of spread footing foundations supported on a natural clayey soil. The allowable total and differential settlements are 20 and

12 mm, respectively. The footings have been designed such that their total settlement will not exceed 20 mm, as determined by the analysis techniques described in this chapter. Will the differential settlements be within tolerable limits?

Solution

According to Table 7.5, the δ_D/δ ratio is about 0.8. Therefore, the differential settlements may be as large as (0.8)(20 mm) = 16 mm. This is greater than the allowable value of 12 mm, and thus is unacceptable. Therefore, it is necessary to design the footings such that their total settlement is no greater than (12 mm)/(0.8) = 15 mm. Thus, in this case the allowable total settlement must be reduced to δ_a = 15 mm.

Mats

Because of their structural continuity, mat foundations generally experience less differential settlement, or at least the differential settlement is spread over a longer distance and thus is less troublesome. In addition, differential settlements in mat foundations are much better suited to rational analysis because they are largely controlled by the structural rigidity of the mat. We will cover these methods in Chapter 10.

QUESTIONS AND PRACTICE PROBLEMS

7.22 A steel frame office building with no diagonal bracing will be supported on spread footings founded in a natural clay. The computed total settlement of these footings is 20 mm. Compute the differential settlement.

7.23 A reinforced concrete building with numerous concrete shear walls will be supported on spread footings founded in a compacted sand. The computed total settlement of these footings is 0.6 in. Compute the differential settlement.

7.10   RATE OF SETTLEMENT

Clays

If the clay is saturated, it is safe to assume the distortion settlement occurs as rapidly as the load is applied. The consolidation settlement will occur over some period, depending on the drainage rate.

Terzaghi's theory of consolidation includes a methodology for computing the rate of consolidation settlement in saturated soils. It is controlled by the rate water is able to squeeze out of the pores and drain away. However, because the soil beneath a footing is able to drain in three dimensions, not one as assumed in Terzaghi's theory, the water will drain away more quickly, so consolidation settlement also will occur more quickly. Davis and Poulos (1968) observed this behavior when they reviewed fourteen case histories. In four of these cases, the rate was very much faster than predicted, and in another four cases, the rate was somewhat faster. In the remaining six cases, the rate was very close to or slightly slower than predicted, but this was attributed to the drainage conditions being close to one-dimensional. They also presented a method of accounting for this effect.

Rate estimates become more complex for some partially saturated soils, as discussed in Chapters 19 and 20.

Sands

The rate of settlement in sands depends on the pattern of loading. If the load is applied only once and then remains constant, then the settlement occurs essentially as fast as the load is applied. The placement of a fill is an example of this kind of loading. The dead load acting on a foundation is another example.

However, if the load varies over time, sands exhibit additional settlement that typically occurs over a period of years or decades. The live loads on a foundation are an example, especially with tanks, warehouses, or other structures in which the live load fluctuates widely and is a large portion of the total load.

A series of long-term measurements on structures in Poland (Bolenski, 1973) has verified this behavior. Bolenski found that footings with fairly constant loads, such as those supporting office buildings, exhibit only a small amount of additional settlement after construction. However, those with varying loads, such as storage tanks, have much more long-term settlement. Burland and Burbidge (1985) indicate the settlement of footings on sand 30 years after construction might be 1.5 to 2.5 times as much as the post-construction settlement. This is the reason for the secondary creep factor, C_2, in Schmertmann's equation.

7.11 ACCURACY OF SETTLEMENT PREDICTIONS

After studying many pages of formulas and procedures, the reader may develop the mistaken impression that settlement analyses are an exact science. This is by no means true. It is good to recall a quote from Terzaghi (1936):

> Whoever expects from soil mechanics a set of simple, hard-and-fast rules for settlement computations will be deeply disappointed. He might as well expect a simple rule for constructing a geologic profile from a single test boring record. The nature of the problem strictly precludes such rules.

Although much progress has been made since 1936, the settlement problem is still a difficult one. The methods described in this chapter should be taken as guides, not dictators, and should be used with engineering judgment. A vital ingredient in this judgement is an understanding of the sources of error in the analysis. These include:

- Uncertainties in defining the soil profile. This is the largest single cause. There have been many cases of unexpectedly large settlements due to undetected compressible layers, such as peat lenses.
- Disturbance of soil samples.
- Errors in in-situ tests (especially the SPT).
- Errors in laboratory tests.

- Uncertainties in defining the service loads, especially when the live load is a large portion of the total load.
- Construction tolerances (i.e., footing not built to the design dimensions).
- Errors in determining the degree of overconsolidation.
- Inaccuracies in the analysis methodologies.
- Neglecting soil-structure interaction effects.

We can reduce some of these errors by employing more extensive and meticulous exploration and testing techniques, but there are economic and technological limits to such efforts.

Because of these errors, the actual settlement of a spread footing may be quite different from the computed settlement. Figure 7.23 shows 90 percent confidence intervals for spread footing settlement computations.

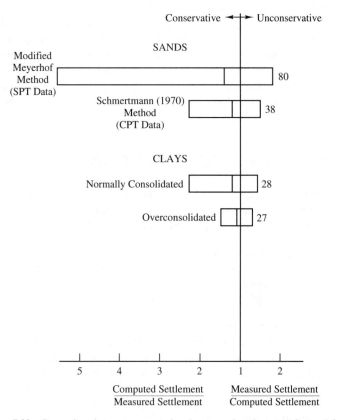

Figure 7.23 Comparison between computed and measured settlements of spread footings. Each bar represents the 90 percent confidence interval (i.e., 90 percent of the settlement predictions will be within this range). The line in the middle of each bar represents the average prediction, and the number to the right indicates the number of data points used to evaluate each method. (Based on data from Burland and Burbridge, 1985; Butler, 1975; Schmertmann, 1970; and Wahls, 1985).

The Leaning Tower of Pisa

During the Middle Ages, Europeans began to build larger and heavier structures, pushing the limits of design well beyond those of the past. In Italy, the various republics erected towers and campaniles to symbolize their power (Kerisel, 1987). Unfortunately, vanity and ignorance often lead to more emphasis on creative architecture than on structural integrity, and many of these structures collapsed. Although some of these failures were caused by poor structural design, many were the result of overloading the soil. Other monuments tilted, but did not collapse. The most famous of these is the campanile in Pisa, more popularly known as the Leaning Tower of Pisa.

Construction of the tower began in the year 1173 under the direction of Bananno Pisano and continued slowly until 1178. This early work included construction of a ring-shaped footing 64.2 ft (19.6 m) in diameter along with three and one-half stories of the tower. By then, the average bearing pressure below the footing was about 6900 lb/ft^2 (330 kPa) and the tower had already begun to tilt. Construction ceased at this level, primarily because of political and economic unrest. We now know that this suspension of work probably saved the tower, because it provided time for the underlying soils to consolidate and gain strength.

Nearly a century later, in the year 1271, construction resumed under the direction of a new architect, Giovanni Di Simone. Although it probably would have been best to tear down the completed portion and start from scratch with a new and larger foundation, Di Simone chose to continue working on the uncompleted tower, attempting to compensate for the tilt by tapering the successive stories and adding extra weight to the high side. He stopped work in 1278 at the seventh cornice. Finally, the tower was completed with the construction of the belfry during a third construction period, sometime between 1360 and 1370. The axis of the belfry is inclined at an angle of 3° from the rest of the tower, which was probably the angle of tilt at that time. Altogether, the project had taken nearly two hundred years to complete.

Both the north and south sides of the tower continued to settle (the tilt has occurred because the south side settled more than the north side), so that by the early nineteenth century, the tower had settled about 2.5 meters into the ground. As a result, the elegant carvings at the base of the columns were no longer visible. To "rectify" this problem, a circular trench was excavated around the perimeter of the tower in 1838 to expose the bottom of the columns. This trench is known as the catino. Unfortunately, construction of the trench disturbed the groundwater table and removed some lateral support from the side of the tower. As a result, the tower suddenly lurched and added about half a meter to the tilt at the top. Amazingly, it did not collapse. Nobody dared do anything else for the next hundred years.

During the 1930s, the Fascist dictator Benito Mussolini decided the leaning tower presented an inappropriate image of the country, and ordered a fix. His workers drilled holes through the floor of the tower and pumped 200 tons of concrete into the underlying soil, but this only aggravated the problem and the tower gained an additional 0.1 degree of tilt.

During most of the twentieth century the tower has been moving at a rate of about 7 seconds of arc per year. By the end of the century the total tilt was about 5.5 degrees to the south which means that the top of the tower structure was 5.2 m (17.0 ft) off of being plumb. The average bearing pressure under the tower is 497 kPa (10,400 lb/ft^2), but the tilting caused its weight to act eccentrically on the foundation, so the bearing pressure is not uniform. By the twentieth century, it ranged from 62 to 930 kPa (1,300–19,600 lb/ft^2), as shown in Figure 7.24. The tower is clearly on the brink of collapse. Even a minor earthquake could cause it to topple, so it became clear that some remedial measure must be taken.

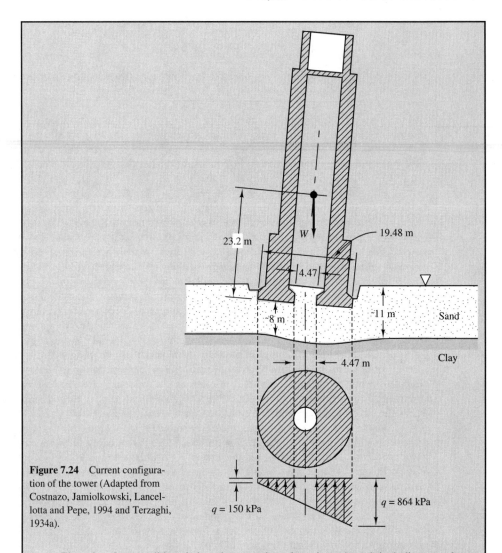

Figure 7.24 Current configuration of the tower (Adapted from Costnazo, Jamiolkowski, Lancellotta and Pepe, 1994 and Terzaghi, 1934a).

The subsurface conditions below the tower have been investigated, including an exhaustive program sponsored by the Italian government that began in 1965. The profile, shown in Figure 7.25, is fairly uniform across the site and consists of sands underlain by fat

This problem has attracted the attention of both amateurs and professionals, and the authorities have received countless "solutions," sometimes at the rate of more than fifty per week. Some are clearly absurd, such as tying helium balloons to the top of the tower, or installing a series of cherub statues with flapping wings. Others, such as large structural supports (perhaps even a large statue leaning against the tower?), may be technically feasible, but aesthetically unacceptable.

In 1990 the interior of the tower was closed to visitors, and in 1993 about 600 tons of lead ingots were placed on the north side of the tower as a temporary stabilization measure.

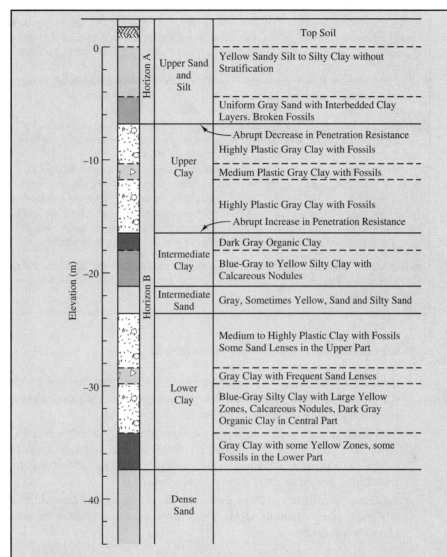

Figure 7.25 Soil profile below tower (Adapted from Mitchell, et al., 1977; Used by permission of ASCE).

Then, in 1995, engineers installed a concrete ring around the foundation and began drilling tiedown anchors through the ring and into the dense sand stratum located at a depth of about 40 ft (see profile in Figure 7.25). The weights caused the underlying soils to compress and slightly reduced the tilt, but construction of the anchors disturbed the soil and produced a sudden increase in the tilt of the tower. In one night the tower moved about 1.5 mm, which is the equivalent to a year's worth of normal movement. As a result, the work was quickly abandoned and more lead ingots were added to the north side.

A period of inactivity followed, but in 1997 an earthquake in nearby Assisi caused a tower in that city to collapse—and that tower was not even leaning! This failure induced a new cycle of activity at Pisa, and the overseeing committee approved a new method of stabilizing the tower: soil extraction.

The method of soil extraction consists of carefully drilling diagonal borings into the ground beneath the north side of the tower and extracting small amounts of soil. The overlying soils then collapse into the newly created void, which should cause the north side of the tower to settle, thus decreasing the tilt. The objective of this effort is to reduce the tilt from 5.5 degrees to 5.0 degrees, which is the equivalent of returning the tower to its position of three hundred years ago. There is no interest in making the tower perfectly plumb.

Soil extraction has been successfully used to stabilize structures in Mexico City, and appears to be the most promising method for Pisa. This process must proceed very slowly, perhaps over a period of months or years, while continuously monitoring the movements of the tower. When this book was published, the soil extraction work had begun and the tilt had been very slightly reduced. If this effort is successful, the temporary lead weights will no longer be necessary, and the life of the tower should be extended for at least three hundred years.

Recommended reference: Costanzo, D.; Jamiolkowski, M., Lancellotta, R., and Pepe, M.C. (1994), *Leaning Tower of Pisa: Description of the Behavior*, Settlement 94 Banquet Lecture, Texas A&M University

We can draw the following conclusions from this data:

- Settlement predictions are conservative more often than they are unconservative (i.e., they tend to overpredict the settlement more often than they underpredict it). However, the range of error is quite wide.

- Settlement predictions made using the Schmertmann method with CPT data are much more precise than those based on the SPT. (Note that these results are based on the 1970 version of Schmertmann's method. Later refinements, as reflected in this chapter, should produce more precise results.)

- Settlement predictions in clays, especially those that are overconsolidated, are usually more precise than those in sands. However, the magnitude of settlement in clays is often greater.

Many of the soil factors that cause the scatter in Figure 7.23 do not change over short distances, so predictions of differential settlements should be more precise than those for total settlements. Therefore, the allowable differential settlement criteria described in Table 2.2 (which include factors of safety of at least 1.5) reflect an appropriate level of conservatism.

SUMMARY

Major Points

1. Foundations must meet two settlement requirements: total settlement and differential settlement.

2. The load on spread footings causes an increase in the vertical stress, $\Delta\sigma_z$, in the soil below. This stress increase causes settlement in the soil beneath the footing.

3. The magnitude of $\Delta\sigma_z$ directly beneath the footing is equal to the bearing pressure, q. It decreases with depth and becomes very small at a depth of about $2B$ below square footings or about $6B$ below continuous footings.

4. The distribution of $\Delta\sigma_z$ below a footing may be calculated using Boussinesq's method, Westergaard's method, or the simplified method.

5. Settlement analyses in clays and silts are usually based on laboratory consolidation tests. The corresponding settlement analysis is an extension of the Terzaghi settlement analyses for fills, as discussed in Chapter 3.

6. Settlement analyses based on laboratory tests may use either the classical method, which assumes one-dimensional consolidation, or the Skempton and Bjerrum method, which accounts for three-dimensional effects.

7. Settlement analyses in sands are usually based on in-situ tests. The Schmertmann method may be used with these test results.

8. Differential settlements may be estimated based on observed ratios of differential to total settlement.

9. Settlement estimates based on laboratory consolidation tests of clays and silts typically range from a 50 percent overestimate (unconservative) to a 100 percent underestimate (conservative).

10. Settlement estimates based on CPT data from sandy soils typically range from a 50 percent overestimate (unconservative) to a 100 percent underestimate (conservative). However, estimates based on the SPT are much less precise.

Vocabulary

Allowable differential settlement	Differential settlement	Schmertmann's method
Allowable settlement	Distortion settlement	Settlement
Boussinesq's method	Induced stress	Skempton and Bjerrum method
Classical method	Plate load test	
	Rigidity	

COMPREHENSIVE QUESTIONS AND PRACTICE PROBLEMS

7.24 A 600-mm wide, 500-mm deep continuous footing carries a vertical downward load of 85 kN/m. The soil has $\gamma = 19\text{kN/m}^3$. Using Boussinesq's method, compute $\Delta\sigma_z$ at a depth of 200 mm below the bottom of the footing at the following locations:

- Beneath the center of the footing
- 150 mm from the center of the footing
- 300 mm from the center of the footing (i.e., beneath the edge)
- 450 mm from the center of the footing

Plot the results in the form of a pressure diagram similar to those in Figure 5.10 in Chapter 5.

Hint: Use the principle of superposition.

7.25 A 3-ft square, 2-ft deep footing carries a column load of 28.2 k. An architect is proposing to build a new 4 ft wide, 2 ft deep continuous footing adjacent to this existing footing. The side of the new footing will be only 6 inches away from the side of the existing footing. The new footing will carry a load of 12.3 k/ft. $\gamma = 119\text{lb/ft}^3$.

Develop a plot of $\Delta\sigma_z$ due to the new footing vs. depth along a vertical line beneath the center of the existing footing. This plot should extend from the bottom of the existing footing to a depth of 35 ft below the bottom of this footing.

7.26 Using the data from Problem 7.25, $C_r/(1 + e_0) = 0.08$ and $\gamma = 119$ lb/ft^3, compute the consolidation settlement of the old footing due to the construction and loading of the new footing. The soil is an overconsolidated (case I) silty clay, and the groundwater table is at a depth of 8 ft below the ground surface.

7.27 Using the SCHMERTMANN.XLS spreadsheet and the subsurface data from Example 7.6, develop a plot of footing width, B, vs. column load, P, for square spread footings embedded 3 ft below the ground surface. Develop a P vs. B curve for each of the following settlements: 0.5 in, 1.0 in, and 1.5 in, and present all three curves on the same diagram.

8

Spread Footings–Geotechnical Design

Your greatest danger is letting the urgent things
crowd out the important.

From *Tyranny of the Urgent* by Charles E. Hummel[1]

This chapter shows how to use the results of bearing capacity and settlement computations, as well as other considerations, to develop spread footing designs that satisfy geotechnical requirements. These are the requirements that relate to the safe transfer of the applied loads from the footing to the ground. Chapter 9 builds on this information, and discusses the structural design aspects, which are those that relate to the structural integrity of the footing and the connection between the footing and the superstructure.

8.1 DESIGN FOR CONCENTRIC DOWNWARD LOADS

The primary load on most spread footings is the downward compressive load, P. This load produces a bearing pressure q along the bottom of the footing, as described in Section 5.3. Usually we design such footings so that the applied load acts through the centroid (i.e., the column is located in the center of the footing). This way the bearing pressure is uniformly distributed along the base of the footing (or at least it can be assumed to be uniformly distributed) and the footing settles evenly.

---

[1]©1967 by InterVarsity Christian Fellowship of the USA. Used by permission of InterVarsity Press-USA.

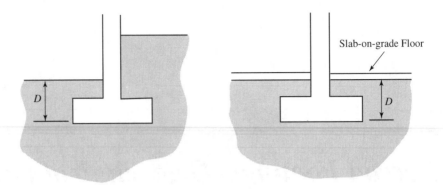

Figure 8.1   Depth of embedment for spread footings.

Footing Depth

The depth of embedment, D, must be at least large enough to accommodate the required footing thickness, T, as shown in Figure 8.1. This depth is measured from the lowest adjacent ground surface to the bottom of the footing. In the case of footings overlain by a slab-on-grade floor, D is measured from the subgrade below the slab.

Tables 8.1 and 8.2 present minimum D values for various applied loads. These are the unfactored loads (i.e., the greatest from Equations 2.1–2.4). These D values are intended to provide enough room for the required footing thickness, T. In some cases, a more detailed analysis may justify shallower depths, but D should never be less than 300 mm (12 in). The required footing thickness, T, is governed by structural concerns, as discussed in Chapter 9.

TABLE 8.1   MINIMUM DEPTH OF EMBEDMENT FOR SQUARE AND RECTANGULAR FOOTINGS

Load P (k)	Minimum D (in)	Load P (kN)	Minimum D (mm)
0–65	12	0–300	300
65–140	18	300–500	400
140–260	24	500–800	500
260–420	30	800–1100	600
420–650	36	1100–1500	700
		1500–2000	800
		2000–2700	900
		2700–3500	1000

TABLE 8.2   MINIMUM DEPTH OF EMBEDMENT
FOR CONTINUOUS FOOTINGS

Load P/b (k/ft)	Minimum D (in)	Load P/b (kN/m)	Minimum D (mm)
0–10	12	0–170	300
10–20	18	170–250	400
20–28	24	250–330	500
28–36	30	330–410	600
36–44	36	410–490	700
		490–570	800
		570–650	900
		650–740	1000

Sometimes it is necessary to use embedment depths greater than those listed in Tables 8.1 and 8.2. This situations include the following:

- The upper soils are loose or weak, or perhaps consist of a fill of unknown quality. In such cases, we usually extend the footing through these soils and into the underlying competent soils.
- The soils are prone to frost heave, as discussed later in this section. The customary procedure in such soils is to extend the footings to a depth that exceeds the depth of frost penetration.
- The soils are expansive. One of the methods of dealing with expansive soils is to extend the footings to a greater depth. This gives them additional flexural strength, and places them below the zone of greatest moisture fluctuation. Chapter 19 discusses this technique in more detail.
- The soils are prone to scour, which is erosion caused by flowing water. Footings in such soils must extend below the potential scour depth. This is discussed in more detail later in this chapter.
- The footing is located near the top of a slope in which there is some, even remote, possibility of a shallow landslide. Such footings should be placed deeper than usual in order to provide additional protection against undermining from any such slides.

Sometimes we also may need to specify a maximum depth. It might be governed by such considerations as:

- Potential undermining of existing foundations, structures, streets, utility lines, etc.
- The presence of soft layers beneath harder and stronger near-surface soils, and the desire to support the footings in the upper stratum.

- A desire to avoid working below the groundwater table, and thus avoid construction dewatering expenses.
- A desire to avoid the expense of excavation shoring, which may be needed for footing excavations that are more than 1.5 m (5 ft) deep.

Footing Width

Sometimes bearing capacity and settlement concerns can be addressed by increasing the footing depth. For example, if the near-surface soils are poor, but those at slightly greater depths are substantially better, bearing capacity and settlement problems might be solved by simply deepening the footing until it reaches the higher quality stratum. However, in more uniform soil profiles, we usually satisfy bearing capacity and settlement requirements by adjusting the footing width, B. Increasing B causes the bearing pressure, q, to decrease, which improves the factor of safety against a bearing capacity failure and decreases the settlement.

Most structures require many spread footings, perhaps dozens of them, so it is inconvenient to perform custom bearing capacity and settlement analyses for each one. Instead, geotechnical engineers develop generic design criteria that are applicable to the entire site, then the structural engineer sizes each footing based on its load and these generic criteria. We will discuss two methods of presenting these design criteria: the *allowable bearing pressure method* and the *design chart method*.

Allowable Bearing Pressure Method

The *allowable bearing pressure, q_A*, is the largest bearing pressure that satisfies both bearing capacity and settlement criteria. In other words, it is equal to the allowable bearing capacity, q_a, or the q that produces the greatest acceptable settlement, whichever is less. Normally we develop a single q_A value that applies to the entire site, or at least to all the footings of a particular shape at that site.

Geotechnical engineers develop q_A using the following procedure:

1. Select a depth of embedment, D, as described earlier in this chapter. If different depths of embedment are required for various footings, perform the following computations using the smallest D.
2. Determine the design groundwater depth, D_w. This should be the shallowest groundwater depth expected to occur during the life of the structure.
3. Determine the required factor of safety against a bearing capacity failure (see Figure 6.11).
4. Using the techniques described in Chapter 6, perform a bearing capacity analysis on the footing with the smallest applied normal load. This analysis is most easily performed using the BEARING.XLS spreadsheet. Alternatively, it may be performed as follows:

     a. Using Equation 5.1 or 5.2, write the bearing pressure, q, as a function of B.

     b. Using Equation 6.4, 6.5, 6.6, or 6.13, along with Equation 6.36, write the allowable bearing capacity, q_a, as a function of B.

     c. Set $q = q_a$ and solve for B.

     d. Using Equation 6.4, 6.5, 6.6, or 6.13, along with Equation 6.36 and the B from Step c, determine the allowable bearing capacity, q_a.

5. Using the techniques described in Chapter 2, determine the allowable total and differential settlements, δ_a and δ_{Da}. Normally the structural engineer performs this step and provides these values to the geotechnical engineer.

6. Using local experience or Table 7.5, select an appropriate δ_D/δ ratio.

7. If $\delta_{Da} \geq \delta_a\,(\delta_D/\delta)$, then designing the footings to satisfy the total settlement requirement ($\delta \leq \delta_a$) will implicitly satisfy the differential settlement requirement as well ($\delta_D \leq \delta_{Da}$). Therefore, continue to Step 8 using δ_a. However, if $\delta_{Da} < \delta_a\,(\delta_D/\delta)$, it is necessary to reduce δ_a to keep differential settlement under control (see Example 7.8). In that case, continue to Step 8 using a revised $\delta_a = \delta_{Da}\,/\,(\delta_D/\delta)$.

8. Using the δ_a value obtained from Step 7, and the techniques described in Chapter 7, perform a settlement analysis on the footing with the largest applied normal load. This analysis is most easily performed using the SETTLEMENT.XLS or SCHMERTMANN .XLS spreadsheets. Determine the maximum bearing pressure, q, that keeps the total settlement within tolerable limits (i.e., $\delta \leq \delta_a$).

9. Set the allowable bearing pressure, q_A equal to the lower of the q_a from Step 4 or q from Step 8. Express it as a multiple of 500 lb/ft^2 or 25 kPa.

If the structure will include both square and continuous footings, we can develop separate q_A values for each.

We use the most lightly loaded footing for the bearing capacity analysis because it is the one that has the smallest B and therefore the lowest ultimate bearing capacity (per Equations 6.4–6.6). Thus, this footing has the lowest q_a of any on the site, and it is conservative to design the other footings using this value.

However, we use the most heavily loaded footing (i.e., the one with the largest B) for the settlement analysis, because it is the one that requires the lowest value of q to satisfy settlement criteria. To understand why this is so, compare the two footing in Figure 8.2. Both of these footings have the same bearing pressure, q. However, since a greater volume of soil is being stressed by the larger footing, it will settle more than the smaller footing. For footings on clays loaded to the same q, the settlement is approximately proportional to B, while in sands it is approximately proportional to $B^{0.5}$. Therefore, the larger footing is the more critical one for settlement analyses.

The geotechnical engineer presents q_A, along with other design criteria, in a written report. The structural engineer receives this report, and uses the recommended q_A to design the spread footings such that $q \leq q_A$. Thus, for square, rectangular, and circular footings:

$$q = \frac{P + W_f}{A} - u_D \leq q_A' \tag{8.1}$$

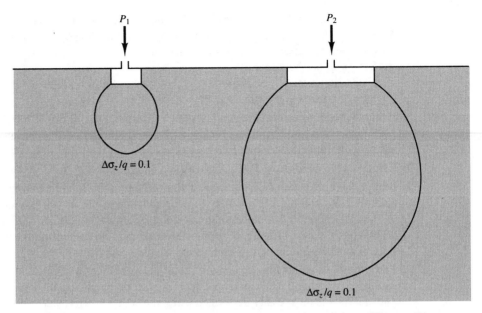

Figure 8.2   These two footings are loaded to the same q, but each has a different width and a correspondingly different P. The larger footing induces stresses to a greater depth in the soil, so it settles more than the smaller footing.

Setting $q = q_A$ and rewriting gives:

$$A = \frac{P + W_f}{q_A + u_D} \qquad (8.2)$$

For continuous footings:

$$B = \frac{P/b + W_f/b}{q_A + u_D} \qquad (8.3)$$

Where:

   A = required base area

         for square footings, $A = B^2$

         for rectangular footings, $A = BL$

         for circular footings, $A = \pi B^2/4$

   B = footing width or diameter

   L = footing length

   P = applied normal load (unfactored)

   P/b = applied normal load per unit length (unfactored)

W_f = weight of foundation

W_f/b = weight of foundation per unit length

b = unit length of foundation (normally 1 m or 1 ft)

q_A = allowable bearing pressure

u_D = pore water pressure along base of footing. $u_D = 0$ if the groundwater table is at a depth greater than D. Otherwise, $u_D = \gamma_w (D - D_w)$.

The footing width must be determined using the unfactored design load (i.e., the largest load computed from Equations 2.1–2.4), even if the superstructure has been designed using the factored load.

Example 8.1

As part of an urban redevelopment project, a new parking garage is to be built at a site formerly occupied by two-story commercial buildings. These old buildings have already been demolished and their former basements have been backfilled with well-graded sand, sandy silt, and silty sand. The lower level of the proposed parking garage will be approximately flush with the existing ground surface, and the design column loads range from 250 to 900 k. The allowable total and differential settlements are 1.0 and 0.6 inches, respectively.

A series of five exploratory borings have been drilled at the site to evaluate the subsurface conditions. The soils consist primarily of the basement backfill, underlain by alluvial sands and silts. The groundwater table is at a depth of about 200 ft. Figure 8.3 shows a design soil profile compiled from these borings, along with all of the standard penetration test N_{60} values.

The basement backfills were not properly compacted and only encompass portions of the site. Therefore, in the interest of providing more uniform support for the proposed spread footing foundations, the upper ten feet of soil across the entire site will be excavated and re-compacted to form a stratum of properly compacted fill. This fill will have an estimated over-consolidation ratio of 3 and an estimated N_{60} of 60. A laboratory direct shear test on a compacted sample of this soil produced $c' = 0$ and $\phi' = 35°$.

Determine the allowable bearing pressures, q_A, for square and continuous footings at this site, then use this q_A to determine the required dimensions for a square footing that will support a 300 k column load.

Solution

Step 1 — Use an estimated D of 3 ft

Step 2 — The groundwater table is very deep, and is not a concern at this site

Step 3 — Use $F = 2.5$

Step 4 — Using the BEARING.XLS spreadsheet with $P = 250$ k, the computed allowable bearing pressure, q_a is 10,500 lb/ft^2

Step 5 — Per the problem statement, $\delta_a = 1.0$ in and $\delta_{Da} = 0.6$ in

Step 6 — Using Table 7.5 and assuming the parking garage is a "flexible" structure, the design value of δ_D/δ is 0.5

Step 7 — $\delta_{Da} > \delta_a (\delta_D/\delta)$, so the total settlement requirement controls the settlement analysis

Step 8 — Using Table 7.3 and Equation 7.17, the equivalent modulus values for each SPT data point are as follows:

Boring No.	Depth (ft)	Soil Type	N_{60}	β_0	β_1	E_s (lb/ft^2)
1	14	SW	20	100,000	24,000	580,000
1	25	SP	104	100,000	24,000	2,596,000
1	35	SP	88	100,000	24,000	2,212,000
1	45	SW	96	100,000	24,000	2,404,000
2	14	SW	44	100,000	24,000	1,156,000
2	25	SP	122	100,000	24,000	3,028,000
3	14	SP & SW	72	100,000	24,000	1,828,000
3	25	SW	90	100,000	24,000	2,260,000
4	15	SW	46	100,000	24,000	1,204,000
4	25	SW	102	100,000	24,000	2,548,000
5	19	SW	92	100,000	24,000	2,308,000
5	29	SP	68	100,000	24,000	1,732,000
5	40	SW	74	100,000	24,000	1,876,000
5	45	SW	60	100,000	24,000	1,540,000
5	49	SW	56	100,000	24,000	1,444,000

The equivalent modulus of the proposed compacted fill is:

$$E_s = B_0 \sqrt{\text{OCR}} + \beta_1 N_{60}$$
$$= 7000 \sqrt{3} + (16,000)(60)$$
$$= 1,100,000 \text{ lb/ft}^2$$

Based on this data, we can perform the settlement analysis using the following equivalent modulus values:

Depth (ft)	E_s (lb/ft^2)
0 – 10	1,100,000
10 – 20	1,000,000
> 20	1,700,000

Using the SCHMERTMANN.XLS spreadsheet with P = 900 k and δ_a = 1.0 in produces q = 6,700 lb/ft^2

Step 9 — 6,700 < 10,500, so settlement controls the design. Rounding to a multiple of 500 lb/ft^2 gives:

$$q_A = 6500 \text{ lb/ft}^2 \quad \Leftarrow Answer$$

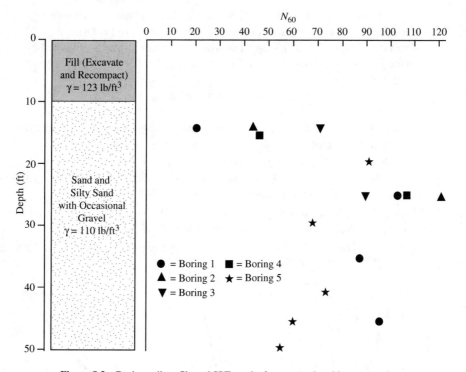

Figure 8.3   Design soil profile and SPT results for proposed parking garage site.

For a 300 k column load, $W_f = (3 \text{ ft})(150 \text{ lb/ft}^3) B^2 = 450 B^2$

$$B = \sqrt{A} = \sqrt{\frac{P + W_f}{q_A + u_D}}$$

$$= \sqrt{\frac{300,000 + 450 B^2}{6500 + 0}}$$

$$= \textbf{7 ft 0 in} \quad \Leftarrow Answer$$

Design Chart Method

The allowable bearing pressure method is sufficient for most small to medium-size structures. However, larger structures, especially those with a wide range of column loads,
warrant a more precise method: the design chart. This added precision helps us reduce
both differential settlements and construction costs.

Instead of using a single allowable bearing pressure for all footings, it is better to use a higher pressure for small ones and a lower pressure for large ones. This method reduces the differential settlements and avoid the material waste generated by the allowable bearing pressure method. This concept is implicit in a design chart such as the one in Figure 8.4.

Use the following procedure to develop design charts:

1. Determine the footing shape (i.e., square, continuous, etc.) for this design chart. If different shapes are to be used, each must have its own design chart.

2. Select the depth of embedment, D, using the guidelines described earlier in this chapter. If different D values are required for different footings, perform these computations using the smallest D.

3. Determine the design groundwater depth, D_w. This should be the shallowest groundwater depth expected to occur during the life of the structure.

4. Select the design factor of safety against a bearing capacity failure (see Figure 6.11).

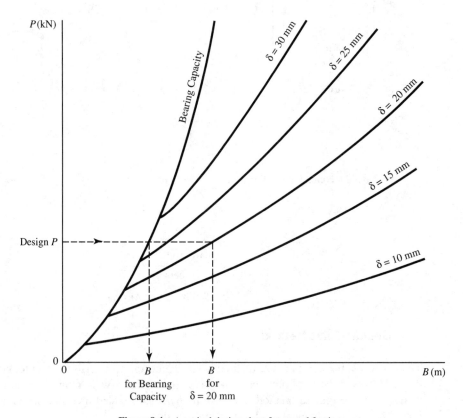

Figure 8.4   A typical design chart for spread footings.

5. Set the footing width B equal to 300 mm or 1 ft, then conduct a bearing capacity analysis and compute the column load that corresponds to the desired factor of safety. Plot this (B, P) data point on the design chart. Then select a series of new B values, compute the corresponding P, and plot the data points. Continue this process until the computed P is slightly larger than the maximum design column load. Finally, connect these data points with a curve labeled "bearing capacity." The spreadsheet developed in Chapter 6 makes this task much easier.

6. Develop the first settlement curve as follows:
    a. Select a settlement value for the first curve (e.g., 0.25 in).
    b. Select a footing width, B, that is within the range of interest and arbitrarily select a corresponding column load, P. Then, compute the settlement of this footing using the spreadsheets developed in Chapters 6 and 7, or some other suitable method.
    c. By trial-and-error, adjust the column load until the computed settlement matches the value assigned in step a. Then, plot the point B, P_a on the design chart.
    d. Repeat steps b and c with new values of B until a satisfactory settlement curve has been produced.

7. Repeat step 6 for other settlement values, thus producing a family of settlement curves on the design chart. These curves should encompass a range of column loads and footing widths appropriate for the proposed structure.

8. Using the factors in Table 7.5, develop a note for the design chart indicating the design differential settlements are ___% of the total settlements.

Once the design chart has been obtained, the geotechnical engineer gives it to the structural engineer who sizes each footing using the following procedure:

1. Compute the design load, P, which is the largest load computed from Equations 2.1, 2.2, 2.3a, or 2.4a. Note that this is the unfactored load, even if the superstructure has been designed using the factored load.

2. Using the bearing capacity curve on the design chart, determine the minimum required footing width, B, to support the load P while satisfying bearing capacity requirements.

3. Using the settlement curve that corresponds to the allowable total settlement, δ_a, determine the footing width, B, that corresponds to the design load, P. This is the minimum width required to satisfy total settlement requirements.

4. Using the δ_D/δ ratio stated on the design chart, compute the differential settlement, δ_D, and compare it to the allowable differential settlement, δ_{Da}.

5. If the differential settlement is excessive ($\delta_D > \delta_{Da}$), then use the following procedure:
    a. Use the allowable differential settlement, δ_{Da}, and the δ_D/δ ratio to compute a new value for allowable total settlement, δ_a. This value implicitly satisfies both total and differential settlement requirements.

    b. Using the settlement curve on the design chart that corresponds to this revised δ_a, determine the required footing width, B. This footing width is smaller than that computed in step 3, and satisfies both total and differential settlement criteria.

6. Select the larger of the B values obtained from the bearing capacity analysis (step 2) and the settlement analysis (step 3 or 5b). This is the design footing width.

7. Repeat steps 1 to 6 for the remaining columns.

These charts clearly demonstrate how the bearing capacity governs the design of narrow footings, whereas settlement governs the design of wide ones.

The advantages of this method over the allowable bearing pressure method include:

- The differential settlements are reduced because the bearing pressure varies with the footing width.
- The selection of design values for total and differential settlement becomes the direct responsibility of the structural engineer, as it should be. (With the allowable bearing pressure method, the structural engineer must give allowable settlement data to the geotechnical engineer who incorporates it into q_A.)
- The plot shows the load-settlement behavior, which we could use in a soil-structure interaction analysis.

Example 8.2

Develop a design chart for the proposed arena described in Example 8.1, then use this chart to determine the required width for a footing that is to support a 300-k column load.

Solution

Bearing capacity analyses (based on BEARING.XLS spreadsheet)

B (ft)	P (k)
2	29
3	74
4	146
5	251
6	395
7	582
8	818
9	1109

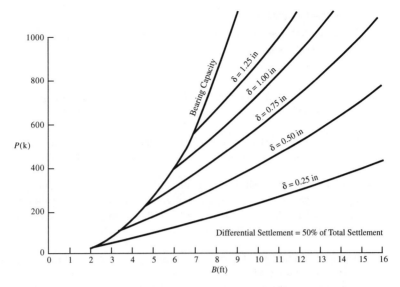

Figure 8.5 Design chart for Example 8.2.

Settlement analyses (based on SCHMERTMANN.XLS spreadsheet)

Column loads to obtain a specified total settlement

	$B = 2$ ft	$B = 7$ ft	$B = 12$ ft	$B = 17$ ft
$\delta = 0.25$ in	32 k	145 k	280 k	450 k
$\delta = 0.50$ in	56 k	260 k	510 k	830 k
$\delta = 0.75$ in	77 k	365 k	720 k	1190 k
$\delta = 1.00$ in	97 k	465 k	925 k	
$\delta = 1.25$ in	115 k	555 k	1120 k	

The result of these analyses are plotted in Figure 8.5.

According to this design chart, a 300-k column load may be supported on a 5 ft, 6 in wide footing. This is much smaller than the 7 ft, 0 in wide footing in Example 8.1.

Use **$B = 5$ ft 6 in**   ⇐ *Answer*

QUESTIONS AND PRACTICE PROBLEMS

8.1 Which method of expressing footing width criteria (allowable bearing pressure or design chart) would be most appropriate for each of the following structures?

   a. A ten-story reinforced concrete building

   b. A one-story wood frame house

   c. A nuclear power plant

   d. A highway bridge

8.2 Explain why an 8-ft wide footing with $q = 3000$ lb/ft^2 will settle more than a 3-ft wide one with the same q.

8.3 Under what circumstances would bearing capacity most likely control the design of spread footings? Under what circumstances would settlement usually control?

8.4 A proposed building will have column loads ranging from 40 to 300 k. All of these columns will be supported on square spread footings. When computing the allowable bearing pressure, q_A, which load should be used to perform the bearing capacity analyses? Which should be used to perform the settlement analyses?

8.5 A proposed building will have column loads ranging from 50 to 250 k. These columns are to be supported on spread footings which will be founded in a silty sand with the following engineering properties: $\gamma = 119$ lb/ft^3 above the groundwater table and 122 lb/ft^3 below, $c' = 0$, $\phi' = 32°$, $N_{60} = 30$. The groundwater table is 15 ft below the ground surface. The required factor of safety against a bearing capacity failure must be at least 2.5 and the allowable settlement, δ_a, is 0.75 in.

   Compute the allowable bearing pressure for square spread footings founded 2 ft below the ground surface at this site. You may use the spreadsheets described in Chapters 6 and 7 to perform the computations, or you may do so by hand. Then, comment on the feasibility of using spread footings at this site.

8.6 A proposed office building will have column loads between 200 and 1000 kN. These columns are to be supported on spread footings which will be founded in a silty clay with the following engineering properties: $\gamma = 15.1$ kN/m^3 above the groundwater table and 16.5 kN/m^3 below, $s_u = 200$ kPa, $C_r/(1+e_0) = 0.020$, $\sigma_m' = 400$ kPa. The groundwater table is 5 m below the ground surface. The required factor of safety against a bearing capacity failure must be at least 3 and the allowable settlement, δ_a, is 20 mm.

   Compute the allowable bearing pressure for square spread footings founded 0.5 m below the ground surface at this site. You may use the spreadsheets described in Chapters 6 and 7 to perform the computations, or you may do so by hand. Then, comment on the feasibility of using spread footings at this site.

8.7 A series of columns carrying vertical loads of 20 to 90 k are to be supported on 3-ft deep square footings. The soil below is a clay with the following engineering properties: $\gamma = 105$ lb/ft^3 above the groundwater table and 110 lb/ft^3 below, $s_u = 3000$ lb/ft^2, $C_r/(1 + e_0) = 0.03$ in the upper 10 ft and 0.05 below. Both soil strata are overconsolidated Case I. The groundwater table is 5 ft below the ground surface. The factor of safety against a bearing capacity failure must be at least 3. Use the spreadsheets described in Chapters 6 and 7 to compute the allowable bearing pressure, q_A. The allowable settlement is 1.4 in.

8.8 Using the information in Problem 8.7, develop a design chart. Consider footing widths of up to 12 ft.

8.9 Several cone penetration tests have been conducted in a young, normally consolidated silica sand. Based on these tests, an engineer has developed the following design soil profile:

Depth (m)	q_c (kg/cm^2)
0–2.0	40
2.0–3.5	78
3.5–4.0	125
4.0–6.5	100

This soil has an average unit weight of 18.1 kN/m^3 above the groundwater table and 20.8 kN/m^3 below. The groundwater table is at a depth of 3.1 m.

Using this data with the spreadsheets described in Chapters 6 and 7, create a design chart for 1.0-m deep square footings. Consider footing widths of up to 4 m and column loads up to 1500 kN, a factor of safety of 2.5, and a design life of 50 years.

Hint: In a homogeneous soil, the critical shear surface for a bearing capacity failure extends to a depth of approximately B below the bottom of the footing. See Chapter 4 for a correlation between q_c in this zone and ϕ'.

8.2   DESIGN FOR ECCENTRIC OR MOMENT LOADS

Sometimes it becomes necessary to build a footing in which the downward load, P, does not act through the centroid, as shown in Figure 8.6a. One example is in an exterior footing in a structure located close to the property line, as shown in Figure 5.2. The bearing pressure beneath such footings is skewed, as discussed in Section 5.3.

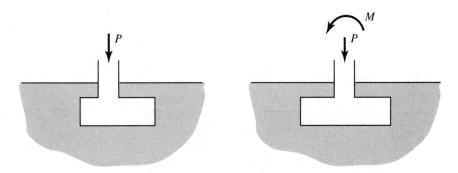

Figure 8.6   (a) Spread footing subjected to an eccentric downward load; (b) Spread footing subjected to a moment load.

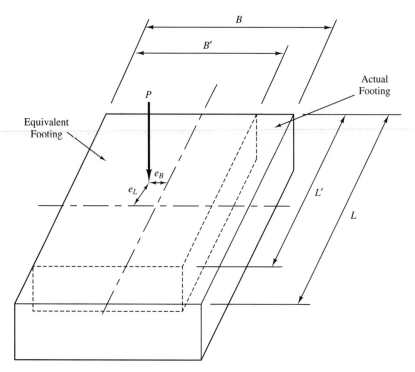

Figure 8.7   Equivalent footing for evaluating the bearing capacity of footings with eccentric or applied moment loads. Note that the equivalent footing has no eccentricity.

Another, more common possibility is a footing that is subjected to an applied moment load, M, as shown in Figure 8.6b. This moment may be permanent, but more often it is a temporary load due to wind or seismic forces acting on the structure. These moment loads also produce a skewed bearing pressure.

Use the following process to design for footings with eccentric or moment loads:

1. Develop preliminary values for the plan dimensions B and L. If the footing is square, then $B = L$. These values might be based on a concentric downward load analysis, as discussed in Section 8.1, or on some other method.

2. Determine if the resultant of the bearing pressure acts within the middle third of the footing (for one-way loading) or within the kern (for two-way loading). The tests for these conditions are described in Equations 5.9 and 5.10, and illustrated in Examples 5.4 and 5.5. If these criteria are not satisfied, then some of the footing will tend to lift off the soil, which is unacceptable. Therefore, any such footings need to be modified by increasing the width or length, as illustrated in Example 5.5.

3. Using the following procedure, determine the effective footing dimensions, B' and L', as shown in Figure 8.7 (Meyerhof, 1963; Brinch Hansen, 1970):
   a. Using Equations 5.3 to 5.6, compute the bearing pressure eccentricity in the B and/or L directions (e_B, e_L).

b. Compute the effective footing dimensions:

$$B' = B - 2e_B \quad\quad\quad (8.4)$$

$$L' = L - 2e_L \quad\quad\quad (8.5)$$

This produces an equivalent footing with an area $A' = B' \times L'$ as shown in Figure 8.7.

4. Compute the equivalent bearing pressure using:

$$q_{equiv} = \frac{P + W_f}{B'L'} - u_D \quad\quad\quad (8.6)$$

5. Compare q_{equiv} with the allowable bearing pressure, q_a. If $q_{equiv} \le q_a$ then the design is satisfactory. If not, then increase the footing size as necessary to satisfy this criterion.

Example 8.3

A 5-ft square, 2-ft deep footing supports a vertical load of 80 k and a moment load of 60 ft-k. The underlying soil has an allowable bearing pressure, q_A, of 3500 lb/ft^2 and the groundwater table is at a great depth. Is this design satisfactory?

Solution

$$W_f = (5 \text{ ft})(5 \text{ ft})(2 \text{ ft})(150 \text{ lb/ft}^3) = 7500 \text{ lb}$$

Using Equation 5.5:

$$e = \frac{M}{P + W_f} = \frac{60 \text{ ft-k}}{80 \text{ k} + 7.5 \text{ k}} = 0.686 \text{ ft}$$

$$\frac{B}{6} = \frac{5 \text{ ft}}{6} = 0.833 \text{ ft}$$

$$e \le \frac{B}{6} \quad \therefore \text{ OK for eccentric loading}$$

$$B' = B - 2e_B = 5 - (2)(0.686) = 3.63 \text{ ft}$$

$$q_{equiv} = \frac{P + W_f}{A} - u_D = \frac{800,000 \text{ lb} + 7,500 \text{ lb}}{(3.63 \text{ ft})(5 \text{ ft})} - 0 = 4821 \text{ lb/ft}^2$$

Since $q_{equiv} > q_A$ **(4821 > 3500), this design is not satisfactory. This is true even though eccentric loading requirement ($e \le B/6$) has been met. Therefore, a larger B is required.** $\Leftarrow$ *Answer*

Further trials will demonstrate that $B = 6$ ft 0 in satisfies all of the design criteria.

8.3   DESIGN FOR SHEAR LOADS

Some footings are also subjected to applied shear loads, as shown in Figure 8.8. These loads may be permanent, as those from retaining walls, or temporary, as with wind or seismic loads on buildings.

Shear loads are resisted by passive pressure acting on the side of the footing, and by sliding friction along the bottom. The allowable shear capacity, V_{fa}, for footings located above the groundwater table at a site with a level ground surface is:

$$V_{fa} = \frac{(P + W_f)\mu + P_p - P_a}{F} \tag{8.7}$$

Passive and active forces are discussed in Chapter 23. However, rather than individually computing them for each footing, it is usually easier to compute λ, which is the net result of the active and passive pressures expressed in terms of an equivalent fluid density. In other words, we evaluate the problem as if the soil along one side of the footing is replaced with a fluid that has a unit weight of λ, then using the principles of fluid statics to compute the equivalent of $P_p - P_a$. Thus, for square footings, Equation 8.7 may be rewritten as:

$$V_{fa} = (P + W_f)\mu_a + 0.5\lambda_a BD^2 \tag{8.8}$$

$$\mu_a = \frac{\mu}{F} \tag{8.9}$$

$$\lambda_a = \frac{\gamma[\tan^2(45° + \phi/2) - \tan^2(45° - \phi/2)]}{F} \tag{8.10}$$

Equation 8.10 considers only the frictional strength of the soil. In some cases, it may be appropriate to also consider the cohesive strength using the techniques described in Chapter 23.

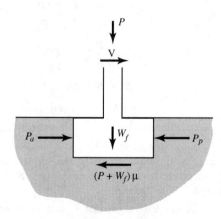

Figure 8.8   Shear load acting on a spread footing.

The footing must then be designed so that:

$$\boxed{V \leq V_a^{'}}$$ (8.11)

Where:

$V =$ applied shear load

$V_{fa} =$ allowable footing shear load capacity

$P =$ downward load acting on the footing

$W_f =$ weight of footing

$B =$ footing width

$D =$ depth of embedment

$\mu =$ coefficient of friction (from Table 8.3 or Equation 8.12)

$\mu_a =$ allowable coefficient of friction

$\lambda =$ equivalent passive fluid density

$\lambda_a =$ allowable equivalent passive fluid density

$\phi =$ friction angle of soil (use ϕ' for drained loading conditions or ϕ_T for undrained loading conditions)

$F =$ factor of safety (typically 1.5 to 2.0 for μ and 2 to 3 for λ)

$$\boxed{\mu = \tan(0.7\phi')}$$ (8.12)

When quoting or using μ and λ, it is important to clearly indicate whether they are ultimate and allowable values. This is often a source of confusion, which can result in compounding factors of safety, or designing without a factor of safety. Normally, geotechnical engineering reports quote allowable values of these parameters.

The engineer also must be careful to use the proper value of P in Equation 8.8. Typically, multiple load combinations must be considered, and the shear capacity must be satisfactory for each combination. Thus, the P for a particular analysis must be the minimum

TABLE 8.3 DESIGN VALUES OF μ FOR CAST-IN-PLACE CONCRETE
(U.S. Navy, 1982b)

Soil or Rock Classification	μ
Clean sound rock	0.70
Clean gravel, gravel-sand mixtures, coarse sand	0.55–0.60
Clean fine-to-medium sand, silty medium-to-coarse sand, silty or clayey gravel	0.45–0.55
Clean fine sand, silty or clayey fine to medium sand	0.35–0.45
Fine sandy silt, nonplastic silt	0.30–0.35
Very stiff and hard residual or overconsolidated clay	0.40–0.50
Medium stiff and stiff clay and silty clay	0.30–0.35

normal load that would be present when the design shear load acts on the footing. For example, if V is due to wind loads on a building, P should be based on dead load only because the live load might not be present when the wind loads occur. If the wind load also causes an upward normal load on the footing, then P would be equal to the dead load minus the upward wind load.

Example 8.4

A 6 ft × 6 ft × 2.5 ft deep footing supports a column with the following design loads: $P = 112$ k, $V = 20$ k. The soil is a silty fine-to-medium sand with $\phi' = 29°$, and the groundwater table is well below the bottom of the footing. Check the shear capacity of this footing and determine if the design will safely withstand the design shear load.

Solution

Per Table 8.3: $\mu = 0.35\text{–}0.45$
Per Equation 8.11: $\mu = \tan[0.7(29)] = 0.37$
∴ Use $\mu = 0.38$

$$\mu_a = \frac{\mu}{F} = \frac{0.38}{1.5} = 0.25$$

$$\lambda_a = \frac{120[\tan^2(45 + 29/2) - \tan^2(45 - 29/2)]}{2} = 152 \text{ lb/ft}^3$$

$$W_f = (6)(6)(2.5)(150) = 13,500 \text{ lb}$$

$$V_{fa} = (112 + 12.5)(0.25) + (0.5)\left(\frac{152}{1000}\right)(6)(2.5^2) = 34 \text{ k} \qquad \Leftarrow Answer$$

$V \le V_{fa}(20 \le 34)$ **so the footing has sufficient lateral load capacity** ⟸ *Answer*

Footings subjected to applied shear loads also have a smaller ultimate bearing capacity, which may be assessed using the i factors in Vesic's method, as described in Chapter 6. This reduction in bearing capacity is often ignored when the shear load is small (i.e., less than about 0.20 P), but it can become significant with larger shear loads.

QUESTIONS AND PRACTICE PROBLEMS

8.10 A square spread footing with $B = 1000$ mm and $D = 500$ mm supports a column with the following design loads: $P = 150$ kN, $M = 22$ kN-m. The underlying soil has an allowable bearing pressure of 200 kPa. Is this design acceptable? If not, compute the minimum required footing width and express it as a multiple of 100 mm.

8.11 A 3 ft × 7 ft rectangular footing is to be embedded 2 ft into the ground and will support a single centrally-located column with the following design loads: $P = 50$ k, $M = 80$ ft-k (acts in long direction only). The underlying soil is a silty sand with $c' = 0$, $\phi' = 31°$, $\gamma = 123$ lb/ft³, and a very deep groundwater table. Using a factor of safety of 2.5, determine if this design is acceptable for bearing capacity.

8.12 A 4-ft square spread footing embedded 1.5 ft into the ground is subjected to the following design loads: $P = 25$ k, $V = 6$ k. The underlying soil is a well-graded sand with $c' = 0$, $\phi' = 36°$, $\gamma = 126$ lb/ft^3, and a very deep groundwater table. Using a factor of safety of 2.5 on bearing capacity, 2 on passive pressure, and 1.5 on sliding friction, determine if this design is acceptable for bearing capacity and for lateral load capacity.

8.4 DESIGN FOR WIND OR SEISMIC LOADS

Many spread footings are subjected to wind or seismic loads in addition to the static loads. For purposes of foundation design, these loads are nearly always expressed in terms of equivalent static loads, as discussed in Section 2.1. In some cases, engineers might use dynamic analyses to evaluate the seismic loads acting on foundations, but such methods are beyond the scope of this book.

Wind and seismic loads are primarily horizontal, so they produce shear loads on the foundations and thus require a shear load capacity evaluation as discussed in Section 8.3. In addition, wind and seismic loads on the superstructure can produce additional normal loads (either downward or upward) on the foundations. These loads are superimposed on the static normal loads. In some cases, such as single pole transmission towers, wind and seismic loads impart moments onto the foundation.

When these loads are expressed as equivalent static loads, the methods of evaluating the load capacity of foundations is essentially the same as for static loads. However, geotechnical engineers usually permit a 33 percent greater load-bearing capacity for load comminations that include wind or seismic components. This increase is based on the following considerations:

- The shear strength of soils subjected to rapid loading, such as from wind or seismic loads, is greater than that during sustained loading, especially in sands. Therefore the ultimate bearing capacity and ultimate lateral capacity is correspondingly larger.
- We are willing to accept a lower factor of safety against a bearing capacity failure or a lateral sliding failure under the transitory loads because these loads are less likely to occur.
- Settlement in soils subjected to transitory loading is generally less than that under an equal sustained load, because the soil has less time to respond.
- We can tolerate larger settlements under transitory loading conditions. In other words, most people would accept some cracking and other minor distress in a structure following a design windstorm or a design earthquake, both of which would be rare events.

For example, if the allowable bearing pressure, q_A, for static loads computed using the technique described in Section 8.1 is 150 kPa, the allowable bearing pressure for load combinations that include wind or seismic components would be $(1.33)(150 \text{ kPa}) = 200 \text{ kPa}$.

Most building codes allow this one-third increase for short term loads [ICBO 1612.3, 1809.2, and Table 18-I-A], [BOCA 1805.2], [ICC 1605.3.2 and Table 1804.2]. In addition, most building codes permit the geotechnical engineer to specify allowable bearing pressures based on a geotechnical investigation, and implicitly allow the flexibility to express separate allowable bearing pressures for short- and long-term loading conditions.

This one-third increase is appropriate for most soil conditions. However, it probably should not be used for foundations supported on soft clayey soils, because they may have lower strength when subjected to strong wind or seismic loading (Krinitzky, et al., 1993). In these soils, the foundations should be sized using a design load equal to the greatest of Equations 2.1 to 2.4 and the q_A value from Chapter 8.

There are two ways to implement this one-third increase in the design process for downward loads:

Method 1:

1. Compute the long duration load as the greater of that produced by Equations 2.1 and 2.2.

2. Size the foundation using the load from Step 1, the q_A from Chapter 8, and Equation 8.2 or 8.3.

3. Compute the short duration load as the greater of that produced by Equations 2.3 and 2.4.

4. Size the foundation using the load from Step 3, 1.33 times the q_A from Chapter 8, and Equation 8.2 or 8.3.

5. Use the larger of the footing sizes from Steps 2 and 4 (i.e., the final design may be controlled by either the long term loading condition or the short term loading condition).

This method is a straightforward application of the principle described above, but can be tedious to implement. The second method is an attempt to simplify the analysis while producing the same design:

Method 2:

1. Compute the design load as the greatest of that produced by Equations 2.1, 2.2, 2.3a, and 2.4a.

2. Size the foundation using the load from Step 1, the q_A from Chapter 8, and Equation 8.2 or 8.3.

Therefore, the author recommends using Method 2.

The design process for shear loads also may use either of these two methods. Once again, it is often easier to use Method 2.

Special Seismic Considerations

Loose sandy soils pose special problems when subjected to seismic loads, especially if these soils are saturated. The most dramatic problem is *soil liquefaction*, which is the sudden loss of shear strength due to the build-up of excess pore water pressures (see Coduto,

Figure 8.9   The soils beneath these apartment buildings in Niigata, Japan liquified during the 1964 earthquake, which produced bearing capacity failures. These failures reportedly occurred very slowly, and the buildings were very strong and rigid, so they remained virtually intact as they tilted. Afterwards, the occupants of the center building were able to evacuate by walking down the exterior wall (Earthquake Engineering Research Center Library, University of California, Berkeley, Steinbrugge Collection).

1999). This loss in strength can produce a bearing capacity failure, as shown in Figure 8.9. Another problem with loose sands, even if they are not saturated and not prone to liquefaction, is earthquake-induced settlement. In some cases, such settlements can be very large.

Earthquakes also can induce landslides, which can undermine foundations built near the top of a slope. This type of failure occurred in Anchorage, Alaska, during the 1964 earthquake, as well as elsewhere. The evaluation of such problems is a slope stability concern, and thus is beyond the scope of this book.

8.5   LIGHTLY-LOADED FOOTINGS

The principles of bearing capacity and settlement apply to all sizes of spread footings. However, the design process can be simplified for lightly-loaded footings. For purposes of geotechnical foundation design, we will define *lightly-loaded footings* as those subjected to vertical loads less than 200 kN (45 k) or 60 kN/m (4 k/ft). These include typical one- and two-story wood-frame buildings, and other similar structures. The foundations for such structures are small, and do not impose large loads onto the ground, so extensive subsurface investigation and soil testing programs are generally not cost-effective. Normally it is less expensive to use conservative designs than it is to conduct extensive investigations and analyses.

Presumptive Allowable Bearing Pressures

Spread footings for lightweight structures are often designed using *presumptive allowable bearing pressures* (also known as *prescriptive bearing pressures*) which are allowable bearing pressures obtained directly from the soil classification. These presumptive bearing pressures appear in building codes, as shown in Table 8.4. They are easy to implement, and do not require borings, laboratory testing, or extensive analyses. The engineer simply obtains the q_A value from the table and uses it with Equation 8.2 or 8.3 to design the footings.

TABLE 8.4   PRESUMPTIVE ALLOWABLE BEARING PRESSURES FROM VARIOUS BUILDING CODES[a,c]

Soil or Rock Classification	Allowable Bearing Pressure, q_A lb/ft^2 (kPa)		
	Uniform Building Code[b] (ICBO, 1997)	National Building Code (BOCA, 1996) and International Building Code (ICC, 2000)	Canadian Code (NRCC, 1990)
Massive crystalline bedrock	4,000–12,000 (200–600)	12,000 (600)	40,000–200,000 (2,000–10,000)
Sedimentary rock	2,000–6,000 (100–300)	6,000 (300)	10,000–80,000 (500–4000)
Sandy gravel or gravel	2,000–6,000 (100–300)	5,000 (250)	4,000–12,000 (200–600)
Sand, silty sand, clayey sand, silty gravel, or clayey gravel	1,500–4,500 (75–225)	3,000 (150)	2,000–8,000 (100–400)
Clay, sandy clay, silty clay, or clayey silt	1,000–3,000 (50–150)	2,000 (100)	1,000–12,000 (50–600)

ICBO values reproduced from the 1997 edition of the *Uniform Building Code*, © 1997, with permission of the publisher, the International Conference of Building Officials. Boca Values copyright 1996, Building Officials and Code Administrators International, Inc., County Club Hills, IL. Reproduced with permission.

[a]The values in this table are for illustrative purposes only and are not a complete description of the code provisions. Portions of the table include the author's interpretations to classify the presumptive bearing values into uniform soil groups. Refer to the individual codes for more details.

[b]The Uniform Building Code values in soil are intended to provide a factor of safety of at least 3 against a bearing capacity failure, and a total settlement of no more than 0.5 in (12 mm) (ICBO, 1997). The lower value for each soil is intended for footings with $B = 12$ in (300 mm) and $D = 12$ in (300 mm) and may be increased by 20 percent for each additional foot of width and depth to the maximum value shown. Exception: No increase for additional width is allowed for clay, sandy clay, silty clay, or clayey silt.

[c]The Standard Building Code (SBCCI, 1997) does not include any presumptive allowable bearing pressures.

Presumptive allowable bearing pressures have been used since the late nineteenth century, and thus predate bearing capacity and settlement analyses. Today they are used primarily for lightweight structures on sites known to be underlain by good soils. Although presumptive bearing pressures are usually conservative (i.e., they produce larger footings), the additional construction costs are small compared to the savings in soil exploration and testing costs.

However, it is inappropriate to use presumptive bearing pressures for larger structures founded on soil because they are not sufficiently reliable. Such structures warrant more extensive engineering and design, including soil exploration and testing. They also should not be used on sites underlain by poor soils.

Minimum Dimensions

If the applied loads are small, such as with most one- or two-story wood-frame structures, bearing capacity and settlement analyses may suggest that extremely small footings would be sufficient. However, from a practical perspective, very small footings are not acceptable for the following reasons:

- Construction of the footing and the portions of the structure that connect to it would be difficult.
- Excavation of soil to build a footing is by no means a precise operation. If the footing dimensions were very small, the ratio of the construction tolerances to the footing dimensions would be large, which would create other construction problems.
- A certain amount of flexural strength is necessary to accommodate nonuniformities in the loads and local inconsistencies in the soil, but an undersized footing would have little flexural strength.

Therefore, all spread footings should be built with certain minimum dimensions. Figure 8.10 shows typical minimums. In addition, building codes sometimes dictate other minimum dimensions. For example, the *Uniform Building Code* and the International Building Code stipulate certain minimum dimensions for footings that support wood-frame structures. The minimum dimensions for continuous footings are presented in Table 8.5, and those for square footings are presented in Note 3 of the table. This code also allows the geotechnical engineer to supercede these minimum dimensions [UBC 1806.1, IBC 1805.21].

Potential Problems

Although the design of spread footings for lightweight structures can be a simple process, as just described, be aware that such structures are not immune to foundation problems. Simply following these presumptive bearing pressures and code minimums does not necessarily produce a good design. Engineers need to know when these simple design guidelines are sufficient, and when additional considerations need to be included.

Most problems with foundations of lightweight structures are caused by the soils below the foundations, rather than high loads from the structure. For example, founda-

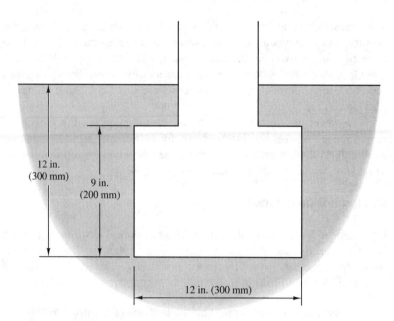

Figure 8.10   Minimum dimensions for spread footings. If the footing is reinforced, the thickness should be at least 12 in (3000 mm).

TABLE 8.5   MINIMUM DIMENSIONS FOR CONTINUOUS FOOTINGS THAT SUPPORT WOOD-FRAME BEARING WALLS PER UBC AND IBC (ICBO, 1997 and ICC, 2000)

Number of floors supported by the foundation	Thickness of Foundation Wall		Footing Width, B		Footing Thickness, T		Footing Depth Below Undisturbed Ground Surface, D	
	(mm)	(in)	(mm)	(in)	(mm)	(in)	(mm)	(in)
1	150	6	300	12	150	6	300	12
2	200	8	375	15	175	7	450	18
3	250	10	450	18	200	8	600	24

1. Where unusual conditions or frost conditions are found, footings and foundations shall be as required by UBC Section 1806.1 or IBC Section 1805.2.1.

2. The ground under the floor may be excavated to the elevation of the top of the footing.

3. Interior stud bearing walls may be supported by isolated footings. The footing width and length shall be twice the width shown in this table and the footings shall be spaced not more than 6 ft (1829 mm) on center.

4. In Seismic Zone 4, continuous footings shall be provided with a minimum of one No. 4 bar top and bottom.

5. Foundations may support a roof in addition to the stipulated number of floors. Foundations supporting roofs only shall be as required for supporting one floor.

tions placed in loose fill may settle because of the weight of the fill or because of infiltration of water into the fill. Expansive soils, collapsible soils, landslides, and other problems also can affect foundations of lightweight structures. These problems may justify more extensive investigation and design effort.

QUESTIONS AND PRACTICE PROBLEMS

8.13 A certain square spread footing for an office building is to support the following downward design loads: dead load = 800 kN, live load = 500 kN, seismic load = 400 kN. The 33 percent increase for seismic load capacity is applicable to this site.

a. Compute the design load.

b. Using the design chart from Example 8.2, determine the required width of this footing such that the total settlement is no more than 20 mm.

8.14 A three-story wood-frame building is to be built on a site underlain by sandy clay. This building will have wall loads of 1900 lb/ft on a certain exterior wall. Using the minimum dimensions presented in Table 8.4 and presumptive bearing pressures from the International Building Code as presented in Table 8.5, compute the required width and depth of this footing. Show your final design in a sketch.

8.6   FOOTINGS ON OR NEAR SLOPES

Vesic's bearing capacity formulas in Chapter 6 are able to consider footings near sloping ground, and we also could compute the settlement of such footings. However, it is best to avoid this condition whenever possible. Special concerns for such situations include:

- The reduction in lateral support makes bearing capacity failures more likely.
- The foundation might be undermined if a shallow (or deep!) landslide were to occur.
- The near-surface soils may be slowly creeping downhill, and this creep may cause the footing to move slowly downslope. This is especially likely in clays.

However, there are circumstances where footings must be built on or near a slope. Examples include abutments of bridges supported on approach embankments, foundations for electrical transmission towers, and some buildings.

Shields, Chandler, and Garnier (1990) produced another solution for the bearing capacity of footings located on sandy slopes. This method, based on centrifuge tests, relates the bearing capacity of footings at various locations with that of a comparable footing with $D = 0$ located on level ground. Figures 8.11 to 8.13 give this ratio for 1.5:1 and 2:1 slopes.

The Uniform Building Code and the International Building Code require setbacks as shown in Figure 8.14. We can meet these criteria either by moving the footing away from the slope or by making it deeper.

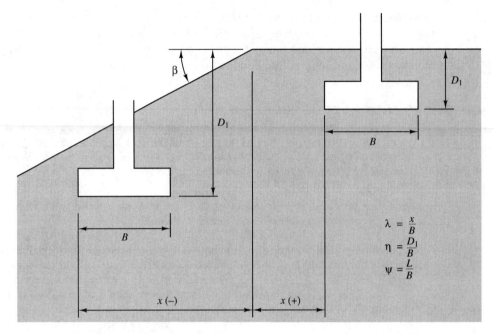

Figure 8.11   Definition of terms for computing bearing capacity of footings near or on sandy slopes (Adapted from Shields, Chandler and Garnier, 1990; Used by permission of ASCE).

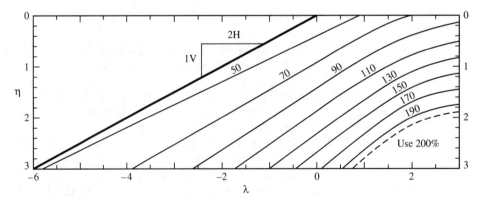

Figure 8.12   Bearing capacity of footings near or on a 2H:1V sandy slopes. The contours are the bearing capacity divided by the bearing capacity of a comparable footing located at the surface of level ground, expressed as a percentage. (Adapted from Shields, Chandler and Garnier, 1990; Used by permission of ASCE).

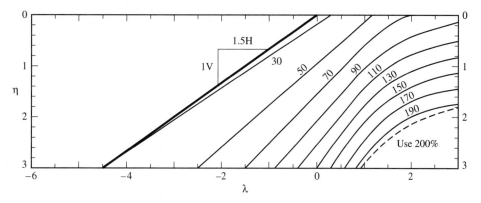

Figure 8.13 Bearing capacity of footings near or on a 1.5H:1V sandy slopes. The contours are the bearing capacity divided by the bearing capacity of a comparable footing located at the surface of level ground, expressed as a percentage. (Adapted from Shields, Chandler and Garnier, 1990; Used by permission of ASCE).

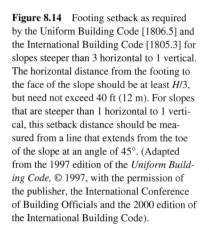

Figure 8.14 Footing setback as required by the Uniform Building Code [1806.5] and the International Building Code [1805.3] for slopes steeper than 3 horizontal to 1 vertical. The horizontal distance from the footing to the face of the slope should be at least $H/3$, but need not exceed 40 ft (12 m). For slopes that are steeper than 1 horizontal to 1 vertical, this setback distance should be measured from a line that extends from the toe of the slope at an angle of 45°. (Adapted from the 1997 edition of the *Uniform Building Code,* © 1997, with the permission of the publisher, the International Conference of Building Officials and the 2000 edition of the International Building Code).

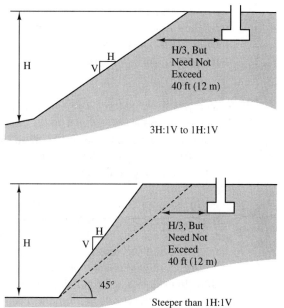

However, this setback criteria does not justify building foundations above unstable slopes. Therefore, we also should perform appropriate slope stability analyses to verify the overall stability.

8.7 FOOTINGS ON FROZEN SOILS

In many parts of the world, the air temperature in the winter often falls below the freezing point of water ($0°$ C) and remains there for extended periods. When this happens, the ground becomes frozen. In the summer, the soils become warmer and return to their unfrozen state. Much of the northern United States, Canada, central Europe, and other places with similar climates experience this annual phenomenon.

The greatest depth to which the ground might become frozen at a given locality is known as the *depth of frost penetration*. This distance is part of an interesting thermodynamics problem and is a function of the air temperature and its variation with time, the initial soil temperature, the thermal properties of the soil, and other factors. The deepest penetrations are obtained when very cold air temperatures are maintained for a long duration. Typical depths of frost penetration in the United States are shown in Figure 8.15.

These annual freeze–thaw cycles create special problems that need to be considered in foundation design.

Frost Heave

The most common foundation problem with frozen soils is *frost heave*, which is a rising of the ground when it freezes.

When water freezes, it expands about 9 percent in volume. If the soil is saturated and has a typical porosity (say, 40 percent), it will expand about $9\% \times 40\% \approx 4\%$ in volume when it freezes. In climates comparable to those in the northern United States, this could correspond to surface heaves of as much as 25 to 50 mm (1–2 in). Although such heaves are significant, they are usually fairly uniform, and thus cause relatively little damage.

However, there is a second, more insidious source of frost heave. If the groundwater table is relatively shallow, capillary action can draw water up to the frozen zone and form ice lenses, as shown in Figure 8.16. In some soils, this mechanism can move large quantities of water, so it is not unusual for these lenses to produce ground surface heaves of 300 mm (1 ft) or more. Such heaves are likely to be very irregular and create a hummocky ground surface that could extensively damage structures, pavements, and other civil engineering works.

In the spring, the warmer weather permits the soil to thaw, beginning at the ground surface. As the ice melts, it leaves a soil with much more water than was originally present. Because the lower soils will still be frozen for a time, this water temporarily cannot drain away, and the result is a supersaturated soil that is very weak. This condition is often the cause of ruts and potholes in highways and can also effect the performance of shallow foundations and floor slabs. Once all the soil has thawed, the excess water drains down and the soil regains its strength. This annual cycle is shown in Figure 8.17.

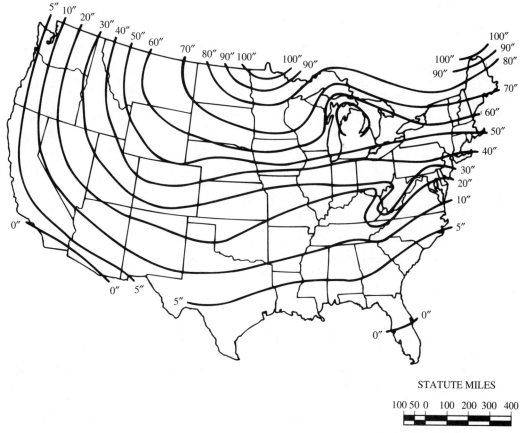

Figure 8.15   Approximate depth of frost penetration in the United States (U.S. Navy, 1982a).

To design foundations in soils that are prone to frost heave, we need to know the depth of frost penetration. This depth could be estimated using Figure 8.15, but as a practical matter it is normally dictated by local building codes. For example, the Chicago Building Code specifies a design frost penetration depth of 1.1 m (42 in). Rarely, if ever, would a rigorous thermodynamic analysis be performed in practice.

Next, the engineer will consider whether ice lenses are likely to form within the frozen zone, thus causing frost heave. This will occur only if both of the following conditions are met:

1. There is a nearby source of water; and
2. The soil is *frost-susceptible*.

To be considered frost-susceptible, a soil must be capable of drawing significant quantities of water up from the groundwater table into the frozen zone. Clean sands and

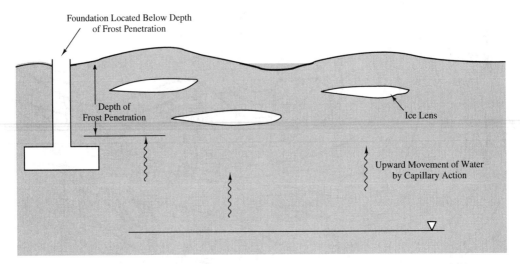

Figure 8.16   Formation of ice lenses. Water is drawn up by capillary action and freezes when it nears the surface. The frozen water forms ice lenses that cause heaving at the ground surface. Foundations placed below the depth of frost penetration are not subject to heaving.

gravels are not frost-susceptible because they are not capable of significant capillary rise. Conversely, clays are capable of raising water through capillary rise, but they have a low permeability and are therefore unable to deliver large quantities of water. Therefore, clays are capable of only limited frost heave. However, intermediate soils, such as silts and fine sands, have both characteristics: They are capable of substantial capillary rise and have a high permeability. Large ice lenses are able to form in these soils, so they are considered to be very frost-susceptible.

The U.S. Army Corps of Engineers has classified frost-susceptible soils into four groups, as shown in Table 8.6. Higher group numbers correspond to greater frost susceptibility and more potential for formation of ice lenses. Clean sands and gravels (i.e., < 3% finer than 0.02 mm) may be considered non frost-susceptible and are not included in this table.

The most common method of protecting foundations from the effects of frost heave is to build them at a depth below the depth of frost penetration. This is usually wise in all soils, whether or not they are frost-susceptible and whether or not the groundwater table is nearby. Even "frost-free" clean sands and gravels often have silt lenses that are prone to heave, and groundwater conditions can change unexpectedly, thus introducing new sources of water. The small cost of building deeper foundations is a wise investment in such cases. However, foundations supported on bedrock or interior foundations in heated buildings normally do not need to be extended below the depth of frost penetration.

Builders in Canada and Scandinavia often protect buildings with slab-on-grade floors using thermal insulation, as shown in Figure 8.18. This method traps heat stored in the ground during the summer and thus protects against frost heave, even though the

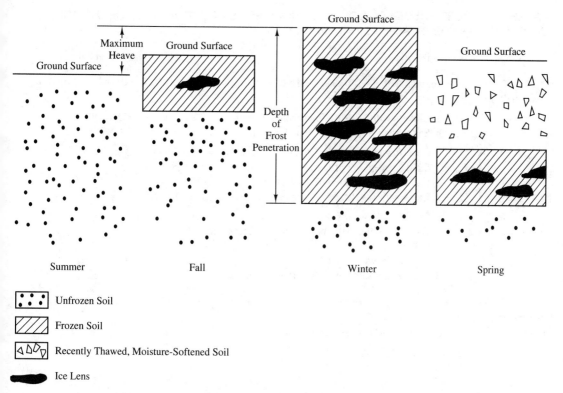

![Unfrozen Soil key]	Unfrozen Soil
![Frozen Soil key]	Frozen Soil
![Recently Thawed key]	Recently Thawed, Moisture-Softened Soil
![Ice Lens key]	Ice Lens

Figure 8.17   Idealized freeze-thaw cycle in temperate climates. During the summer, none of the ground is frozen. During the fall and winter, it progressively freezes from the ground surface down. Then, in the spring, it progressively thaws from the ground surface down.

foundations are shallower than the normal frost depth. Both heated and nonheated buildings can use this technique (NAHB, 1988 and 1990).

Alternatively, the natural soils may be excavated to the frost penetration depth and replaced with soils that are known to be frost-free. This may be an attractive alternative for unheated buildings with slab floors to protect both the floor and the foundation from frost heave.

Although frost heave problems are usually due to freezing temperatures from natural causes, it is also possible to freeze the soil artificially. For example, refrigerated buildings such as cold-storage warehouses or indoor ice skating rinks can freeze the soils below and be damaged by frost heave, even in areas where natural frost heave is not a concern (Thorson and Braun, 1975). Placing insulation or air passages between the building and the soil and/or using non–frost-susceptible soils usually prevents these problems.

A peculiar hazard to keep in mind when foundations or walls extend through frost-susceptible soils is *adfreezing* (CGS, 1992). This is the bonding of soil to a wall or foundation as it freezes. If heaving occurs after the adfreezing, the rising soil will impose a large

TABLE 8.6 FROST SUSCEPTIBILITY OF VARIOUS SOILS ACCORDING TO THE U.S. ARMY CORPS OF ENGINEERS (Adapted from Johnston, 1981).

Group	Soil Types[a]	USCS Group Symbols[b]
F1 (least susceptible)	Gravels with 3–10% finer than 0.02 mm	GW, GP, GW-GM, GP-GM
F2	a. Gravels with 10–20% finer than 0.02 mm b. Sands with 3–15% finer than 0.02 mm	GM, GW-GM, GP-GM SW, SP, SM, SW-SM, SP-SM
F3	a. Gravels with more than 20% finer than 0.02 mm b. Sands, except very fine silty sands, with more than 15% finer than 0.02 mm c. Clays with $I_P > 12$, except varved clays	GM, GC SM, SC CL, CH
F4 (most susceptible)	a. Silts and sandy silts b. Fine silty sands with more than 15% finer than 0.02 mm c. Lean clays with $I_P < 12$ d. Varved clays and other fine-grained, banded sediments	ML, MH SM CL, CL-ML

[a]I_P = Plasticity Index (explained in Chapter 3).
[b]See Chapter 3 for an explanation of USCS group symbols.

upward load on the structure, possibly separating structural members. Placing a 10-mm (0.5 in) thick sheet of rigid polystyrene between the foundation and the frozen soil reduces the adfreezing potential.

Permafrost

In areas where the mean annual temperature is less than 0°C, the penetration of freezing in the winter may exceed the penetration of thawing in the summer and the ground can become frozen to a great depth. This creates a zone of permanently frozen soil known as *permafrost*. In the harshest of cold climates, such as Greenland, the frozen ground is continuous, whereas in slightly "milder" climates, such as central Alaska, central Canada, and much of Siberia, the permafrost is discontinuous. Areas of seasonal and continuous permafrost in Canada are shown in Figure 8.19.

In areas where the summer thaws occur, the upper soils can be very wet and weak and probably not capable of supporting any significant loads, while the deeper soils remain permanently frozen. Foundations must penetrate through this seasonal zone and well into the permanently frozen ground below. It is very important that these foundations be designed so that they do not transmit heat to the permafrost, so buildings are typically built with raised floors and a ducting system to maintain subfreezing air temperatures between the floor and the ground surface.

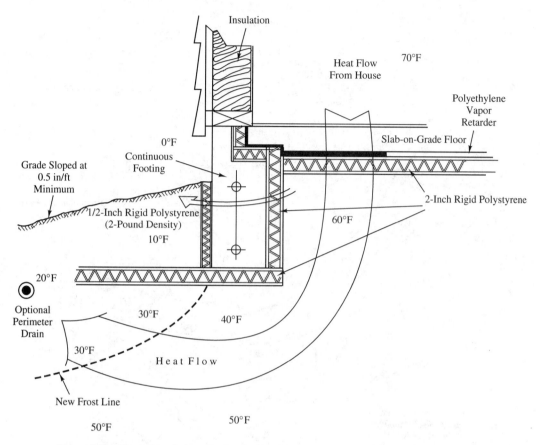

Figure 8.18   Thermal insulation traps heat in the soil, thus protecting a foundation from frost heave (NAHB, 1988, 1990).

The Alaska Pipeline project is an excellent example of a major engineering work partially supported on permafrost (Luscher et. al, 1975).

8.8   FOOTINGS ON SOILS PRONE TO SCOUR

Scour is the loss of soil because of erosion in river bottoms or in waterfront areas. This is an important consideration for design of foundations for bridges, piers, docks, and other structures, because the soils around and beneath the foundation could be washed away.

Scour around the foundations is the most common cause of bridge failure. For example, during spring 1987, there were seventeen bridge failures caused by scour in the northeastern United States alone (Huber, 1991). The most notable of these was the collapse of the Interstate Route 90 bridge over Schoharie Creek in New York (Murillo,

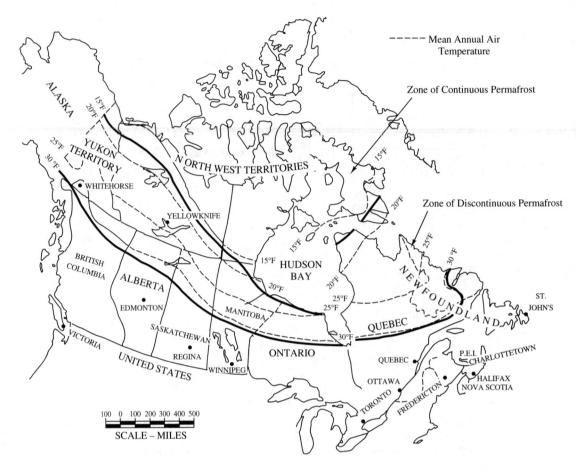

Figure 8.19   Zones of continous and discontinuous permafrost in Canada (Adapted from Crawford and Johnson, 1971).

1987), a failure that killed ten people. Figures 8.20 and 8.21 show another bridge that collapsed as a result of scour.

Scour is part of the natural process that moves river-bottom sediments downstream. It can create large changes in the elevation of the river bottom. For example, Murphy (1908) describes a site on the Colorado River near Yuma, Arizona, where the river bed consists of highly erodible fine silty sands and silts. While passing a flood, the water level at this point rose 4.3 m (14 ft) and the bottom soils scoured to depths of up to 11 m (36 ft)! If a bridge foundation located 10.7 m (35 ft) below the river bottom had been built at this location, it would have been completely undermined by the scour and the bridge would have collapsed.

Figure 8.20   One of the mid-channel piers supporting this bridge sank about 1.5 m when it was undermined by scour in the river channel.

Figure 8.21   Deck view of the bridge shown in Figure 8.20. The lanes on the right side of the fence are supported by a separate pier that was not undermined by the scour.

Scour is often greatest at places where the river is narrowest and constrained by levees or other means. Unfortunately, these are the locations most often selected for bridges. The presence of a bridge pier also creates water flow patterns that intensify the scour. However, methods are available to predict scour depths (Richardson et al., 1991) and engineers can use preventive measures, such as armoring, to prevent scour problems (TRB, 1984).

8.9   FOOTINGS ON ROCK

In comparison to foundations on soil, those on bedrock usually present few difficulties for the designer (Peck, 1976). The greatest problems often involve difficulties in construction, such as excavation problems and proper removal of weathered or disturbed material to provide good contact between the footing and the bedrock.

The allowable bearing pressure on rock may be determined in at least four ways (Kulhawy and Goodman, 1980):

- Presumptive allowable bearing pressures
- Empirical rules
- Rational methods based on bearing capacity and settlement analyses
- Full-scale load tests

When supported on good quality rock, spread footings are normally able to support moderately large loads with very little settlement. Engineers usually design them using presumptive bearing pressures, preferably those developed for the local geologic conditions. Typical values are listed in Table 8.7.

If the rock is very strong, the strength of the concrete may govern the bearing capacity of spread footings. Therefore, do not use an allowable bearing value, q_a, greater than one-third of the compressive strength of the concrete ($0.33 f_c'$).

When working with bedrock, be aware of certain special problems. For example, soluble rocks, including limestone, may have underground cavities that might collapse, causing *sinkholes* to form at the ground surface. These have caused extensive damage to buildings in Florida and elsewhere.

Soft rocks, such as siltstone, claystone, and mudstone, are very similar to hard soil, and often can be sampled, tested, and evaluated using methods developed for soils.

TABLE 8.7 TYPICAL PRESUMPTIVE ALLOWABLE BEARING PRESSURES
FOR FOUNDATIONS ON BEDROCK (Adapted from US Navy, 1982b)

Rock Type	Rock Consistency	Allowable Bearing Pressure, q_A	
		(lb/ft²)	(kPa)
Massive crystalline igneous and metamorphic rock: Granite, diorite, basalt, gneiss, thoroughly cemented conglomerate	Hard and sound (minor cracks OK)	120,000–200,000	6000–10,000
Foliated metamorphic rock: Slate, schist	Medium hard, sound (minor cracks OK)	60,000–80,000	3000–4000
Sedimentary rock: Hard-cemented shales, siltstone, sandstone, limestone without cavities	Medium hard, sound	30,000–50,000	1500–2500
Weathered or broken bedrock of any kind; compaction shale or other argillaceous rock in sound condition	Soft	16,000–24,000	800–1200

QUESTIONS AND PRACTICE PROBLEMS

8.15 A 4 ft square, 2 ft deep spread footing carries a compressive column load of 50 k. The edge of this footing is 1 ft behind the top of a 40 ft tall, 2H:1V descending slope. The soil has the following properties: $c = 200$ lb/ft², $\phi = 31°$, $\gamma = 121$ lb/ft³, and the groundwater table is at a great depth. Compute the factor of safety against a bearing capacity failure and comment on this design.

8.16 Classify the frost susceptibility of the following soils:
  a. Sandy gravel (GW) with 3% finer than 0.02 mm.
  b. Well graded sand (SW) with 4% finer than 0.02 mm.
  c. Silty sand (SM) with 20% finer than 0.02 mm.
  d. Fine silty sand (SM) with 35% finer than 0.02 mm.
  e. Sandy silt (ML) with 70% finer than 0.02 mm.
  f. Clay (CH) with plasticity index = 60.

8.17 A compacted fill is to be placed at a site in North Dakota. The following soils are available for import: Soil 1 - silty sand; Soil 2 - lean clay; Soil 3 - Gravelly coarse sand. Which of these soils would be least likely to have frost heave problems?

8.18 Would it be wise to use slab-on-grade floors for houses built on permafrost? Explain.

8.19 What is the most common cause of failure in bridges?

8.20 A single-story building is to be built on a sandy silt in Detroit. How deep must the exterior footings be below the ground surface to avoid problems with frost heave?

SUMMARY

Major Points

1. The depth of embedment, D, must be great enough to accommodate the required footing thickness, T. In addition, certain geotechnical concerns, such as loose soils or frost heave, may dictate an even greater depth.

2. The required footing width, B, is a geotechnical problem, and is governed by bearing capacity and settlement criteria. It is inconvenient to satisfy these criteria by performing custom bearing capacity and settlement computations for each footing, so we present the results of generic computations in a way that is applicable to the entire site. There are two methods of doing so: the allowable bearing pressure method and the design chart method.

3. Footings subjected to eccentric or moment loads need to be evaluated using the "equivalent footing."

4. Footings can resist applied shear loads through passive pressure and sliding friction.

5. Wind and seismic loads are normally treated as equivalent static loads. For most soils, load combinations that include wind or seismic components may be evaluated using a 33 percent greater allowable bearing pressure.

6. The design of lightly-loaded footings is often governed by minimum practical dimensions.

7. Lightly-loaded footings are often designed using an presumptive allowable bearing pressure, which is typically obtained from the applicable building code.

8. The design of footings on or near slopes needs to consider the sloping ground.

9. Footings on frozen soils need special considerations. The most common problem is frost heave, and the normal solution is to place the footing below the depth of frost penetration.

10. Footings in or near riverbeds are often prone to scour, and must be designed accordingly.

11. Rock usually provides excellent support for spread footings. Such footings are typically designed using a presumptive allowable bearing pressure.

Vocabulary

Allowable bearing pressure	Equivalent footing	Permafrost
Concentric downward load	Frost heave	Presumptive allowable bearing pressure
Depth of frost penetration	Frost-susceptible soil	Scour
Design chart	Lightly-loaded footing	Shear load
Eccentric load	Moment load	

COMPREHENSIVE QUESTIONS AND PRACTICE PROBLEMS

8.21 A 2.0-m square, 0.5-m deep spread footing carries a concentrically applied downward load of 1200 kN and a moment load of 300 m-kN. The underlying soil has an undrained shear strength of 200 kPa. The design must satisfy the eccentric load requirements described in Chapter 5, and it must have a factor of safety of at least 3 against a bearing capacity failure. Determine if these requirements are being met. If not, adjust the footing dimensions until both requirements have been satisfied.

8.22 The soil at a certain site has the following geotechnical design parameters: $q_A = 4000$ lb/ft^2, $\mu_a = 0.28$, and $\lambda_a = 200$ lb/ft^3. The groundwater table is at a depth of 20 ft. A column that is to

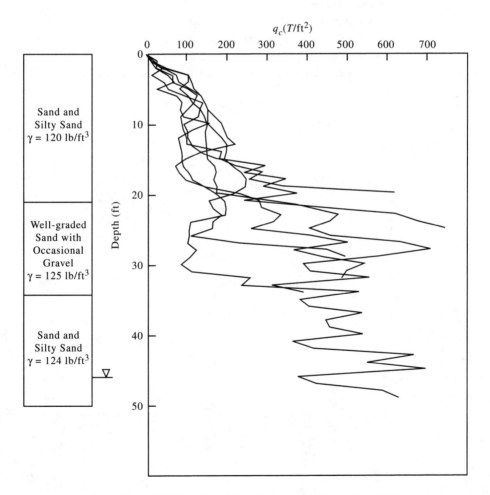

Figure 8.22 CPT data and synthesis of boring for Problems 8.23 and 8.24.

be supported on a square spread footing on this soil will impart the following load combinations onto the footing: $P = 200$ k, $V = 18$ k.

Determine the required footing width and depth of embedment.

8.23 Six cone penetration tests and four exploratory borings have been performed at the site of a proposed warehouse building. The underlying soils are natural sands and silty sands with occasional gravel. The CPT results and a synthesis of the borings are shown in Figure 8.22. The warehouse will be supported on 3 ft deep square footings that will have design downward loads of 100 to 600 k. The allowable total settlement is 1.0 in and the allowable differential settlement is 0.5 in. Using this data with reasonable factors of safety, develop values of q_A, μ_a, and λ_a. Use Figure 4.16 to estimate the friction angle.

8.24 Using the design values in Problem 8.23, determine the required width of a footing that must support the following load combinations:

Load combination 1: $P_D = 200$ k, $P_L = 0$, $V = 0$

Load combination 2: $P_D = 200$ k, $P_L = 0$, $V_E = 21$ k

Load combination 3: $P_D = 200$ k, $P_L = 240$, $V_E = 40$ k

Load combination 4: $P_D = 200$ k, $P_L = 240$, $P_E = -20$ k, $V_E = 40$ k

9

Spread Footings—
Structural Design

Foundations ought to be twice as thick as the wall to be built on them; and regard in this should be had to the quality of the ground, and the largeness of the edifice; making them greater in soft soils, and very solid where they are to sustain a considerable weight.

The bottom of the trench must be level, that the weight may press equally, and not sink more on one side than on the other, by which the walls would open. It was for this reason the ancients paved the said bottom with tivertino, and we usually put beams or planks, and build on them.

The foundations must be made sloping, that is, diminished in proportion as they rise; but in such a manner, that there may be just as much set off one side as on the other, that the middle of the wall above may fall plumb upon the middle of that below: Which also must be observed in the setting off of the wall above ground; because the building is by this method made much stronger than if the diminutions were done any other way.

Sometimes (especially in fenny places, and where the columns intervene) to lessen the expence, the foundations are not made continued, but with arches, over which the building is to be.

It is very commendable in great fabricks, to make some cavities in the thickness of the wall from the foundation to the roof, because they give vent to the winds and vapours, and cause them to do less damage to the building. They save expence, and are of no little use if there are to be circular stairs from the foundation to the top of the edifice.

The First Book of Andrea Palladio's Architecture (1570),
as translated by Isaac Ware (1738)

The plan dimensions and minimum embedment depth of spread footings are primarily geotechnical concerns, as discussed in Chapters 6 to 8. Once these dimensions have been set, the next step is to develop a structural design that gives the foundation enough integrity to safely transmit the design loads from the structure to the ground. The structural design process for reinforced concrete foundations includes:

- Selecting a concrete with an appropriate strength
- Selecting an appropriate grade of reinforcing steel
- Determining the required foundation thickness, T, as shown in Figure 9.1
- Determining the size, number, and spacing of the reinforcing bars
- Designing the connection between the superstructure and the foundation.

The structural design aspects of foundation engineering are far more codified than are the geotechnical aspects. These codes are based on the results of research, the performance of existing structures, and the professional judgment of experts. Engineers in North America use the *Building Code Requirements for Structural Concrete* (ACI 318-99 and ACI 318M-99) for most projects. This code is published by the American Concrete Institute (ACI, 1999). The most notable alternative to ACI is the *Standard Specifications for Highway Bridges* published by the American Association of State Highway and Transportation Officials (AASHTO, 1996). The model building codes (ICBO, 1997; BOCA, 1996; SBCCI, 1997; ICC, 2000) contain additional design requirements.

This chapter covers the major principles of structural design of spread footings, and often refers to specific code requirements, with references shown in brackets [ ]. How-

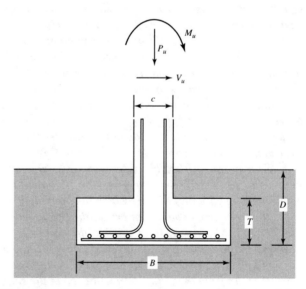

Figure 9.1   Cross section of a spread footing showing applied loads, reinforcing steel, and relevant dimensions.

ever, it is not a comprehensive discussion of every code provision, and thus is not a substitute for the code books.

9.1   SELECTION OF MATERIALS

Unlike geotechnical engineers, who usually have little or no control over the engineering properties of the soil, structural engineers can, within limits, select the engineering properties of the structural materials. In the context of spread footing design, we must select an appropriate concrete strength, f_c', and reinforcing steel strength, f_y.

When designing a concrete superstructure, engineers typically use concrete that has $f_c' = 20\text{--}35$ MPa (3000–5000 lb/in^2). In very tall structures, f_c' might be as large as 70 MPa (10,000 lb/in^2). The primary motive for using high-strength concrete in the superstructure is that it reduces the section sizes, which allows more space for occupancy and reduces the weight of the structure. These reduced member weights also reduce the dead loads on the underlying members.

However, the plan dimensions of spread footings are governed by bearing capacity and settlement concerns and will not be affected by changes in the strength of the concrete; only the thickness, T, will change. Even then, the required excavation depth, D, may or may not change because it might be governed by other factors. In addition, saving weight in a footing is of little concern because it is the lowest structural member and does not affect the dead load on any other members. In fact, additional weight may actually be a benefit in that it increases the uplift capacity.

Because of these considerations, and because of the additional materials and inspection costs of high strength concrete, spread footings are usually designed using an f_c' of only 15–20 MPa (2000–3000 lb/in^2). For footings that carry relatively large loads, perhaps greater than about 2000 kN (500 k), higher strength concrete might be justified to keep the footing thickness within reasonable limits, perhaps using an f_c' as high as 35 MPa (5000 lb/in^2).

Since the flexural stresses in footings are small, grade 40 steel (metric grade 300) is usually adequate. However, this grade is readily available only in sizes up through #6 (metric #22), and grade 60 steel (metric grade 420) may be required on the remainder of the project. Therefore, engineers often use grade 60 (metric grade 420) steel in the footings for reinforced concrete buildings so only one grade of steel is used on the project. This makes it less likely that leftover grade 40 (metric grade 300) bars would accidently be placed in the superstructure.

9.2   BASIS FOR DESIGN METHODS

Before the twentieth century, the design of spread footings was based primarily on precedent. Engineers knew very little about how footings behaved, so they followed designs that had worked in the past.

The first major advance in our understanding of the structural behavior of reinforced concrete footings came as a result of full-scale load tests conducted at the University of Illinois by Talbot (1913). He tested 197 footings in the laboratory and studied the mechanisms of failure. These tests highlighted the importance of shear in footings.

During the next five decades, other individuals in the United States, Germany, and elsewhere conducted additional tests. These tests produced important experimental information on the flexural and shear resistance of spread footings and slabs as well as the response of new and improved materials. Richart's (1948) tests were among the most significant of these. He tested 156 footings of various shapes and construction details by placing them on a bed of automotive coil springs that simulated the support from the soil and loaded them using a large testing machine until they failed. Whitney (1957) and Moe (1961) also made important contributions.

A committee of engineers (ACI-ASCE, 1962) synthesized this data and developed the analysis and design methodology that engineers now use. Because of the experimental nature of its development, this method uses simplified, and sometimes arbitrary, models of the behavior of footings. It also is conservative.

As often happens, theoretical studies have come after the experimental studies and after the establishment of design procedures (Jiang, 1983; Rao and Singh, 1987). Although work of this type has had some impact on engineering practice, it is not likely that the basic approach will change soon. Engineers are satisfied with the current procedures for the following reasons:

- Spread footings are inexpensive, and the additional costs of a conservative design are small.

- The additional weight that results from a conservative design does not increase the dead load on any other member.

- The construction tolerances for spread footings are wider than those for the superstructure, so additional precision in the design probably would be lost during construction.

- Although perhaps crude when compared to some methods available to analyze superstructures, the current methods are probably more precise than the geotechnical analyses of spread footings and therefore are not the weak link in the design process.

- Spread footings have performed well from a structural point-of-view. Failures and other difficulties have usually been due to geotechnical or construction problems, not bad structural design.

- The additional weight of conservatively designed spread footings provides more resistance to uplift loads.

Standard design methods emphasize two modes of failure: shear and flexure. A shear failure, shown in Figure 9.2, occurs when part of the footing comes out of the bottom. This type of failure is actually a combination of tension and shear on inclined failure surfaces. We resist this mode of failure by providing an adequate footing thickness, T.

Figure 9.2 "Shear" failure in a spread footing loaded in a laboratory (Talbot, 1913). Observe how this failure actually is a combination of tension and shear.

A flexural failure is shown in Figure 9.3. We analyze this mode of failure by treating the footing as an inverted cantilever beam and resisting the flexural stresses by placing tensile steel reinforcement near the bottom of the footing.

9.3 DESIGN LOADS

The structural design of spread footings is based on LRFD methods (ACI calls it *ultimate strength design* or *USD*), and thus uses the factored loads as defined in Equations 2.7 to 2.15. Virtually all footings support a compressive load, P_u, and it should be computed

Figure 9.3 Flexural failure in a spread footing loaded in a laboratory (Talbot, 1913).

without including the weight of the footing because this weight is evenly distributed and thus does not produce shear or moment in the footing. Some footings also support shear (V_u) and/or moment (M_u) loads, as shown in Figure 9.1, both of which must be expressed as the factored load. This is often a point of confusion, because the geotechnical design of the same footing is normally based on ASD methods, and thus use the unfactored load, as defined in Equations 2.1 to 2.4. In addition, the geotechnical design must include the weight of the footing.

Therefore, when designing spread footings, be especially careful when computing the load. The footing width, B, is based on geotechnical requirements and is thus based on the unfactored load, as discussed in Chapter 8, whereas the thickness, T, and the reinforcement are structural concerns, and thus are based on the factored load. Examples 9.1 and 9.2 illustrate the application of these principles.

9.4 MINIMUM COVER REQUIREMENTS AND STANDARD DIMENSIONS

The ACI code specifies the minimum amount of concrete cover that must be present around all steel reinforcing bars [7.7]. For concrete in contact with the ground, such as spread footings, at least 70 mm (3 in) of concrete cover is required, as shown in Figure 9.4. This cover distance is measured from the edge of the bars, not the centerlines. It provides proper anchorage of the bars and corrosion protection. It also allows for irregularities in the excavation and accommodates possible contamination of the lower portion of the concrete.

Sometimes it is appropriate to specify additional cover between the rebar and the soil. For example, it is very difficult to maintain smooth footing excavation at sites with loose sands or soft clays, so more cover may be appropriate. Sometimes contractors place a thin layer of lean concrete, called a *mud slab* or a *leveling slab*, in the bottom of the footing excavation at such sites before placing the steel, thus providing a smooth working surface.

For design purposes, we ignore any strength in the concrete below the reinforcing steel. Only the concrete between the top of the footing and the rebars is considered in our analyses. This depth is the *effective depth*, d, as shown in Figure 9.4.

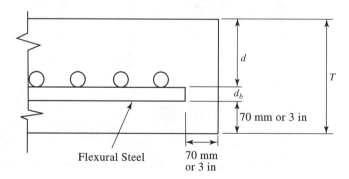

Figure 9.4 In square spread footings, the effective depth is the distance from the top of the concrete to the contact point of the flexural steel.

Footings are typically excavated using backhoes, and thus do not have precise as-built dimensions. Therefore, there is no need to be overly precise when specifying the footing thickness T. Round it to a multiple of 3 in or 100 mm (i.e., 12, 15, 18, 21, etc. inches or 300, 400, 500, etc. mm). The corresponding values of d are:

$$\boxed{d = T - 3 \text{ in} - d_b} \qquad \text{(9.1 English)}$$

$$\boxed{d = T - 70 \text{ mm} - d_b} \qquad \text{(9.1 SI)}$$

Where d_b is the nominal diameter of the steel reinforcing bars (see Table 9.1).

ACI [15.7] requires d be at least 6 in (150 mm), so the minimum acceptable T for reinforced footings is 12 in or 300 mm.

9.5   SQUARE FOOTINGS

This section considers the design of square footings supporting a single centrally-located column. Other types of footings are covered in subsequent sections.

In most reinforced concrete design problems, the flexural analysis is customarily done first. However, with spread footings, it is most expedient to do the shear analysis first. This is because it is not cost-effective to use shear reinforcement (stirrups) in most

TABLE 9.1   DESIGN DATA FOR STEEL REINFORCING BARS

Bar Size Designation		Available Grades		Nominal Dimensions			
				Diameter, d_b		Cross-Sectional Area, A_s	
English	SI	English	SI	(in)	(mm)	(in²)	(mm²)
#3	#10	40, 60	300, 420	0.375	9.5	0.11	71
#4	#13	40, 60	300, 420	0.500	12.7	0.20	129
#5	#16	40, 60	300, 420	0.625	15.9	0.31	199
#6	#19	40, 60	300, 420	0.750	19.1	0.44	284
#7	#22	60	420	0.875	22.2	0.60	387
#8	#25	60	420	1.000	25.4	0.79	510
#9	#29	60	420	1.128	28.7	1.00	645
#10	#32	60	420	1.270	32.3	1.27	819
#11	#36	60	420	1.410	35.8	1.56	1006
#14	#43	60	420	1.693	43.0	2.25	1452
#18	#57	60	420	2.257	57.3	4.00	2581

spread footings, and because we neglect the shear strength of the flexural steel. The only source of shear resistance is the concrete above the flexural reinforcement, so the effective depth, d, as shown in Figure 9.4, must be large enough to provide sufficient shear capacity. We then perform the flexural analysis using this value of d.

Designing for Shear

ACI defines two modes of shear failure, *one-way shear* (also known as *beam shear* or *wide-beam shear*) and *two-way shear* (also known as *diagonal tension shear*). In the context of spread footings, these two modes correspond to the failures shown in Figure 9.5. Although the failure surfaces are actually inclined, as shown in Figure 9.2, engineers use these idealized vertical surfaces to simplify the computations.

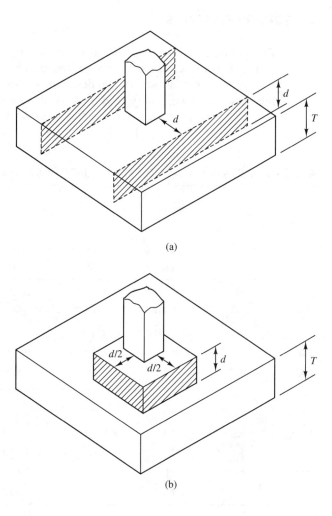

Figure 9.5   The two modes of shear failure: (a) one-way shear, and (b) two-way shear.

Various investigators have suggested different locations for the idealized critical shear surfaces shown in Figure 9.5. The ACI code [11.12.1] specifies that they be located a distance d from the face of the column for one-way shear and a distance $d/2$ for two-way shear.

The footing design is satisfactory for shear when it satisfies the following condition on all critical shear surfaces:

$$\boxed{V_{uc} \leq \phi \, V_{nc}} \tag{9.2}$$

Where:

V_{uc} = factored shear force on critical surface

ϕ = resistance factor for shear = 0.85

V_{nc} = nominal shear capacity on the critical surface

The nominal shear load capacity, V_{nc}, on the critical shear surface is [11.1]:

$$\boxed{V_{nc} = V_c + V_s} \tag{9.3}$$

Where:

V_c = nominal shear load capacity of concrete

V_s = nominal shear load capacity of reinforcing steel

For spread footings, we neglect V_s and rely only on the concrete for shear resistance.

Two-Way Shear

The footing may be subjected to applied normal, moment, and shear loads, P_u, M_u, and V_u, all of which produce shear forces on the critical shear surfaces.

To visualize the shear force on the critical surface, V_{uc}, caused by the applied normal load, P_u, we divide the footing into two blocks, one inside the shear surface and the other outside, as shown in Figure 9.6. The factored normal load, P_u, is applied to the top of the inner block and is transferred to a uniform pressure acting on the base of both blocks. Some of this load is transferred to the soil beneath the inner block, while the remainder must pass through the critical shear surface and enters the soil beneath the lower block. Only the later portion produces a shear force on the critical shear surface. In other words, the percentage of P_u that produces shear along the critical surfaces is the ratio of the base area of the outer block to the total base area.

If an applied moment load, M_u, is present, it produces an additional shear force on two opposing faces of the inner block, as shown in Figure 9.7. The shear force on one of the faces acts in the same direction as the shear force induced by the normal load, while that on the other face acts in the opposite direction. Therefore, the face with both forces

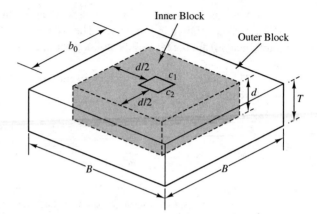

Figure 9.6   The inner block is the portion of the footing inside the critical section for two-way shear. The factored shear force acting along the perimeter of this block, V_{uc}, must not exceed ϕV_n. The factored shear force, V_{uc}, is the portion of the factored column load, P_u, that must pass through the outside surfaces of the inner block before reaching the ground.

acting in the same direction has the greatest shear force, and thus controls the design. The force on this face is:

$$
\boxed{
\begin{aligned}
V_{uc} &= \left(\frac{P_u}{4} + \frac{M_u}{c + d} \right) \left(\frac{\text{base area of outer block}}{\text{total base area}} \right) \\
&= \left(\frac{P_u}{4} + \frac{M_u}{c + d} \right) \left(\frac{B^2 - (c + d)^2}{B^2} \right)
\end{aligned}
}
\tag{9.4}
$$

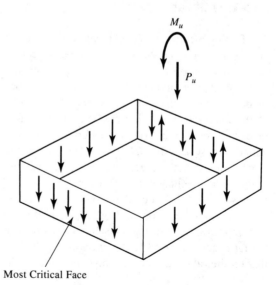

Figure 9.7   Distribution of shear forces on the critical shear surfaces for two-way shear when the footing is subjected to both normal and moment loads.

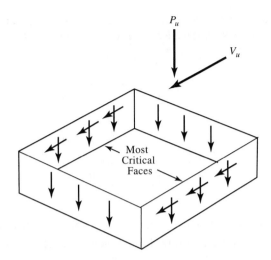

Figure 9.8 Distribution of shear forces on the critical shear surfaces for two-way shear when the footing is subjected to both normal and shear loads.

If an applied shear load, V_u, is present and it acts in the same direction as the moment load (which is the usual case), it produces a shear force on the other two faces, as shown in Figure 9.8. If we assume the applied shear force is evenly divided between these two faces, then the shear force on each face is:

$$
\begin{aligned}
V_{uc} &= \left(\frac{\text{base area of outer block}}{\text{total base area}} \right) \sqrt{ \left(\frac{P_u}{4} \right)^2 + \left(\frac{V_u}{2} \right)^2 } \\
&= \left(\frac{B^2 - (c + d)^2}{B^2} \right) \sqrt{ \left(\frac{P_u}{4} \right)^2 + \left(\frac{V_u}{2} \right)^2 }
\end{aligned}
\tag{9.5}
$$

Where:

V_{uc} = factored shear force on the most critical face

P_u = applied normal load

M_u = applied moment load

V_u = applied shear load

c = column width or diameter (for concrete columns) or base plate width (for steel columns)

d = effective depth

B = footing width

The design should be based on the larger of the V_{uc} values obtained from Equations 9.4 and 9.5, thus accounting for applied normal, shear, and/or moment loads.

For square footings supporting square or circular columns located in the interior of the footing (i.e., not on the edge or corner), the nominal two-way shear capacity is [ACI 11.12.2.1]:

$$V_{nc} = V_c = 4\, b_0\, d \sqrt{f_c'} \qquad \text{(9.6 English)}$$

$$V_{nc} = V_c = \frac{1}{3}\, b_0\, d\, \sqrt{f_c'} \qquad \text{(9.6 SI)}$$

$$b_0 = c + d \qquad \text{(9.7)}$$

Where:

V_{nc} = nominal two-way shear capacity on the critical section (lb, N)

V_c = nominal two-way shear capacity of concrete (lb, N)

b_0 = length of critical shear surface = length of one face of inner block

c = column width (in, mm)

d = effective depth (in, mm)

f_c' = 28-day compressive strength of concrete (lb/in², MPa)

Other criteria apply if the column has another shape, or if it is located along edge or corner of the footing [ACI 11.12.2.1]. Special criteria also apply if the footing is made of prestressed concrete [ACI 11.12.2.2], but spread footings are rarely, if ever, prestressed.

The objective of this analysis is to find the effective depth, d, that satisfies Equation 9.2. Both V_{uc} and V_{nc} depend on the effective depth, d, but there is no direct solution. Therefore, it is necessary to use the following procedure:

1. Assume a trial value for d. Usually a value approximately equal to the column width is a good first trial. When selecting trial values of d, remember T must be a multiple of 3 in or 100 mm, as discussed in Section 9.4, so the corresponding values of d are the only ones worth considering. Assuming $d_b \approx 1$ in (25 mm), the potential values of d are 8, 11, 14, 17, etc. inches or 200, 300, 400, etc. mm.

2. Compute V_{uc} and V_{nc}, and check if Equation 9.2 has been satisfied.

3. Repeat Steps 1 and 2 as necessary until finding the smallest d that satisfies Equation 9.2.

4. Using Equation 9.1 with $d_b = 1$ in or 25 mm, compute the footing thickness, T. Express it as a multiple of 3 in or 100 mm. T must be at least 12 in or 300 mm.

The final value of d_b will be determined as a part of the flexural analysis, and may be different from the 1 in or 25 mm assumed here. However, this difference is small compared to the construction tolerances, so there is no need to repeat the shear analysis with the revised d_b.

One-Way Shear

Two-way shear always governs the design of square footings subjected only to vertical loads. There is no need to check one-way shear in such footings. However, if applied shear and/or moment loads are present, both kinds of shear need to be checked.

To analyze this situation, we will make the following assumptions:

- The shear stress caused by the applied vertical load, P_u, is uniformly distributed across the two critical vertical planes shown in Figure 9.5a.
- The shear stress on the vertical planes caused by the applied moment load, M_u, is expressed by the flexure formula, $\sigma = Mc/I$, and thus is greatest at the left and right edges of these planes.
- The shear stress caused by the applied shear load is uniformly distributed across the planes.
- The applied normal, moment, and shear loads must be multiplied by $(B - c - 2d)/B$ before applying them to the critical vertical planes. This factor is the ratio of the footing base area outside the critical planes to the total area, and thus reflects the percentage of the applied loads that must be transmitted through the critical vertical planes.
- The maximum shear stress on the critical vertical surfaces is the vector sum of those due to the applied normal, moment, and shear loads.
- The factored shear stress on the critical vertical surfaces is the greatest shear stress multiplied by the area of the shear surfaces. This may be greater than the integral of the shear stress across the shear surfaces, but is useful because it produces a design that keeps the maximum shear stress within acceptable limits.

Based on these assumptions, we compute the factored shear force on the critical vertical surfaces, V_{uc}, as follows:

$$V_{uc} = \left(\frac{B - c - 2d}{B} \right) \sqrt{\left(P_u + \frac{6 M_u}{B} \right)^2 + V_u^2} \qquad (9.8)$$

Where:

V_{uc} = shear force on critical shear surfaces
B = footing width
c = column width
d = effective depth
P_u = applied normal load
M_u = applied moment load
V_u = applied shear load

The nominal one-way shear load capacity on the critical section [11.3.1.1] is:

$$\boxed{V_{nc} = V_c = 2\, b_w d\, \sqrt{f'_c}}$$ (9.9 English)

$$\boxed{V_{nc} = V_c = \frac{1}{6}\, b_w d\, \sqrt{f'_c}}$$ (9.9 SI)

Where:

V_{nc} = nominal one-way shear capacity on the critical section (lb, N)

V_c = nominal one-way shear capacity of concrete (lb, N)

b_w = length of critical shear surface = $2B$ (in, mm)

d = effective depth (in, mm)

f'_c = 28-day compressive strength of concrete (lb/in², MPa)

Once again, the design is satisfactory when Equation 9.2 has been satisfied.

Both V_{uc} and V_{nc} depend on the effective depth, d, which must be determined using Equations 9.8 and 9.9 with the procedure described under two-way shear. The final design value of d is the larger of that obtained from the one-way and two-way shear analyses.

The final value of d_b will be determined as a part of the flexural analysis, and may be different from the 1 in or 25 mm assumed here. However, this difference is small compared to the construction tolerances, so there is no need to repeat the shear analysis with the revised d_b.

Example 9.1— Part A

A 21-inch square reinforced concrete column carries a vertical dead load of 380 k and a vertical live load of 270 k. It is to be supported on a square spread footing that will be founded on a soil with an allowable bearing pressure of 6500 lb/ft². The groundwater table is well below the bottom of the footing. Determine the required width, B, thickness, T, and effective depth, d.

Solution

Unfactored load—Equation 2.2 governs

$$P = P_D + P_L + \ldots = 380\,k + 270\,k + 0 = 650\,k$$

Per Table 8.1, use $D = 36$ in

$$W_f = B^2 D \gamma_c = B^2\,(3\text{ ft})(150\text{ lb/ft}^3) = 450\,B^2$$

$$B = \sqrt{\frac{P + W_f}{q_A + u_D}} = \sqrt{\frac{650{,}000 + 450\,B^2}{6500 + 0}}$$

$$B = 10.36\text{ ft} \;\rightarrow\; \text{use } B = 10\text{ ft }6\text{ in }\;(126\text{ in})$$

Factored load (Equation 2.7 governs)

$$P_u = 1.4\,P_D + 1.7\,P_L = (1.4)(380) + (1.7)(270) = 991\text{ k}$$

Because of the large applied load and because this is a large spread footing, we will use $f_c' = 4000$ lb/in^2 and $f_y = 60$ k/in^2.

Since there are no applied moment or shear loads, there is no need to check one-way shear. Determine required thickness based on a two-way shear analysis.

Try $T = 24$ in:

$$d = T - 1 \text{ bar diameter} - 3 \text{ in} = 24 - 1 - 3 = 20 \text{ in}$$

$$V_{uc} = \left(\frac{P_u}{4} + \frac{M_u}{c + d}\right)\left(\frac{B^2 - (c + d^2)}{B^2}\right)$$

$$= \left(\frac{991{,}000 \text{ lb}}{4} + 0\right)\left(\frac{(126 \text{ in})^2 - (21 \text{ in} + 20 \text{ in})^2}{(126 \text{ in})^2}\right)$$

$$= 221{,}500 \text{ lb}$$

$$b_0 = c + d = 21 + 20 = 41 \text{ in}$$

$$V_{nc} = 4\,b_0 d\,\sqrt{f_c'}$$

$$= 4(41 \text{ in})(20 \text{ in})\sqrt{4000 \text{ lb/in}^2}$$

$$= 207{,}400 \text{ lb}$$

$$\phi\,V_{nc} = (0.85)(207{,}400 \text{ lb}) = 176{,}300 \text{ lb}$$

$$V_{uc} > \phi\,V_{nc} \quad \therefore \text{ Not acceptable}$$

Try $T = 27$ in:

$$d = T - 1 \text{ bar diameter} - 3 \text{ in} = 27 - 1 - 3 = 23 \text{ in}$$

$$V_{uc} = \left(\frac{P_u}{4} + \frac{M_u}{c + d}\right)\left(\frac{B^2 - (c + d)^2}{B^2}\right)$$

$$= \left(\frac{991{,}000 \text{ lb}}{4} + 0\right)\left(\frac{(126 \text{ in})^2 - (21 \text{ in} + 23 \text{ in})^2}{(126 \text{ in})^2}\right)$$

$$= 217{,}500 \text{ lb}$$

$$b_0 = c + d = 21 + 23 = 44 \text{ in}$$

$$V_{nc} = 4\,b_0\,d\sqrt{f_c'}$$
$$= 4(41 \text{ in})(23 \text{ in})\sqrt{4000 \text{ lb/in}^2}$$
$$= 256{,}000 \text{ lb}$$

$$\phi\,V_n = (0.85)(256{,}000 \text{ lb}) = 217{,}600 \text{ lb}$$

$$V_{uc} \leq \phi\,V_{nc} \quad \textbf{OK}$$

$$\therefore \text{ Use } \boldsymbol{B = 10 \text{ ft } 6 \text{ in}; \ T = 27 \text{ in}; \ d = 23 \text{ in}} \quad \Leftarrow \textit{Answer}$$

Note 1: In this case, V_{uc} is almost exactly equal to ϕV_{nc}. However, since we are considering only certain values of d, it is unusual for them to match so closely.

Note 2: The depth of embedment of 3 ft, as obtained from Table 8.1, is not needed here (unless frost depth or other concerns dictate it). We could use $D = 30$ in and still have plenty of room for a 27 inch-thick footing.

Designing for Flexure

Once we have completed the shear analysis, the design process can move to the flexural analysis.

ACI Flexural Design Standards

Reinforcing Steel

Concrete is strong in compression, but weak in tension. Therefore, engineers add reinforcing steel, which is strong in tension, to form *reinforced concrete*. This reinforcement is necessary in members subjected to pure tension, and those that must resist *flexure* (bending). Reinforcing steel may consist of either *deformed bars* (more commonly known as *reinforcing bars*, or *rebars*) or *welded wire fabric*. However, wire fabric is rarely used in foundations.

Manufacturers produce reinforcing bars in various standard diameters, typically ranging between 9.5 mm (3/8 in) and 57.3 mm (2 1/4 in). In the United States, the English and metric bars are the same size (i.e., we have used a soft conversion), and are identified by the *bar size* designations in Table 9.1.

Rebars are available in various strengths, depending on the steel alloys used to manufacture them. The two most common bar strengths used in the United States are:

- Grade 40 bars (also known as metric grade 300), which have a yield strength, f_y, of 40 k/in^2 (300 MPa)
- Grade 60 bars (also known as metric grade 420), which have a yield strength, f_y, of 60 k/in^2 (420 MPa)

Flexural Design Principles

The primary design problem for flexural members is as follows: Given a factored moment on the critical surface, M_{uc}, determine the necessary dimensions of the member and the necessary size and location of reinforcing bars. Fortunately, flexural design in foundations is simpler than that for some other structural members because geotechnical concerns dictate some of the dimensions.

The amount of steel required to resist flexure depends on the *effective depth, d*, which is the distance from the extreme compression fiber to the centroid of the tension reinforcement, as shown in Figure 9.9.

The nominal moment capacity of a flexural member made of reinforced concrete with $f_c' \leq 30$ MPa (4000 lb/in^2) as shown in Figure 9.9 is:

$$M_n = A_s f_y \left(d - \frac{a}{2} \right)$$ (9.10)

$$a = \frac{\rho \, d f_y}{0.85 \, f_c'}$$ (9.11)

$$\boxed{\rho = \frac{A_s}{b \, d}}$$ (9.12)

Setting $M_u = \phi \, M_n$, where M_u is the factored moment at the section being analyzed, and solving for A_s gives:

$$\boxed{A_s = \left(\frac{f_c' \, b}{1.176 \, f_y} \right) \left(d - \sqrt{d^2 - \frac{2.353 \, M_{uc}}{\phi f_c' \, b}} \right)}$$ (9.13)

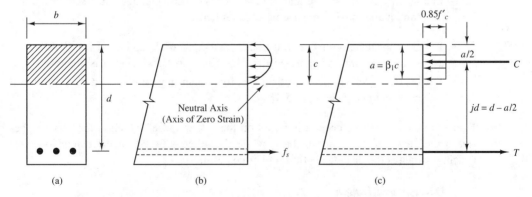

Figure 9.9 The reinforcing bars are placed in the portion of the member that is subjected to tension. (a) Cross section, (b) actual stress distribution, and (c) equivalent rectangular stress distribution. The effective depth, d, is the distance from the extreme compression fiber to the centroid of the tension reinforcement. β_1 is an empirical factor that ranges between 0.65 and 0.85. (Adapted from MacGregor, 1996).

Where:

A_s = cross-sectional area of reinforcing steel

f_c' = 28-day compressive strength of concrete

f_y = yield strength of reinforcing steel

ρ = steel ratio

b = width of flexural member

d = effective depth

ϕ = 0.9 for flexure in reinforced concrete

M_{uc} = factored moment at the section being analyzed

Two additional considerations also enter the design process: minimum steel and maximum steel. The minimum steel in footings is governed by ACI 10.5.4 and 7.12.2, because footings are treated as "structural slabs of uniform thickness" (MacGregor, 1996). These requirements are as follows:

For grade 40 (metric grade 300) steel $A_s \geq 0.0020\, A_g$

For grade 60 (metric grade 420) steel $A_s \geq 0.0018\, A_g$

Where:

A_g = gross cross-sectional area

The ρ_{min} criteria in ACI 10.5.1 do not apply to footings.

The maximum steel requirement [10.3] is intended to maintain sufficient ductility. It never governs the design of simple footings, but it may be of concern in combined footings or mats.

We can supply the required area of steel, computed using Equation 9.13 by any of several combinations of bar size and number of bars. This selection must satisfy the following minimum and maximum spacing requirements:

- The clear space between bars must be at least equal to d_b, 25 mm (1 in), or 4/3 times the nominal maximum aggregate size [3.3.2 and 7.6.1], whichever is greatest.
- The center-to-center spacing of the reinforcement must not exceed $3T$ or 500 mm (18 in), whichever is less [10.5.4].

Notice how one of these criteria is based on the "clear space" which is the distance between the edges of two adjacent bars, while the other is based on the center-to-center spacing, which is the distance between their centerlines.

Development Length

The rebars must extend a sufficient distance into the concrete to develop proper anchorage [ACI 15.6]. This distance is called the *development length*. Provides the clear spacing between the bars is at least $2d_b$, and the concrete cover is at least d_b, the ratio of the minimum required development length, l_d, to the bar diameter, d_b is [ACI 12.2.3]:

$$\boxed{\frac{I_d}{d_b} = \frac{3}{40} \frac{f_y}{\sqrt{f_c'}} \frac{\alpha \beta \gamma \lambda}{\left(\dfrac{c + K_{tr}}{d_b}\right)}}$$ (9.14 English)

$$\boxed{\frac{I_d}{d_b} = \frac{9}{10} \frac{f_y}{\sqrt{f_c'}} \frac{\alpha \beta \gamma \lambda}{\left(\dfrac{c + K_{tr}}{d_b}\right)}}$$ (9.14 SI)

$$K_{tr} = \frac{A_{tr} f_{yt}}{1500\, s\, n}$$ (9.15 English)

$$K_{tr} = \frac{A_{tr} f_{yt}}{10\, s\, n}$$ (9.15 SI)

For spread footings, use $K_{tr} = 0$, which is conservative.

Where:

l_d = minimum required development length (in, mm)

d_b = nominal bar diameter (in, mm)

f_y = yield strength of reinforcing steel (lb/in^2, MPa)

f_{yt} = yield strength of transverse reinforcing steel (lb/in^2, MPa)

f_c' = 28-day compressive strength of concrete (lb/in^2, MPa)

α = reinforcement location factor

   α = 1.3 for horizontal reinforcement with more than 300 mm (12 in) of fresh concrete below the bar

   α = 1.0 for all other cases

β = coating factor

   β = 1.5 for epoxy coated bars or wires with cover less than $3d_b$ or clear spacing less than $6d_b$

   β = 1.2 for other epoxy coated bars or wires

   β = 1.0 for uncoated bars or wires

γ = reinforcement factor

   γ = 0.8 for #6 (metric #19) and smaller bars

   γ = 1.0 for #7 (metric #22) and larger bars

λ = lightweight concrete factor = 1.0 for normal concrete (lightweight concrete is not used in foundations)

c = spacing or cover dimension (in, mm) = the smaller of the distance from the center of the bar to the nearest concrete surface or one-half the center-to-center spacing of the bars

A_{tr} = total cross-sectional area of all transverse reinforcement that is within the spacing s and which crosses the potential plane of splitting through the reinforcement being developed (in^2, mm^2)—may conservatively be taken to be zero

s = maximum center-to-center spacing of transverse reinforcement within l_d (in, mm)

The term $(c + K_{tr})/d_b$ must be no greater than 2.5, and the product $\alpha\beta$ need not exceed 1.7. In addition, the development length must always be at least 300 mm (12 in).

Application to Spread Footings

Principles

A square footing bends in two perpendicular directions as shown in Figure 9.10a, and therefore might be designed as a *two-way slab* using methods similar to those that might be applied to a floor slab that is supported on all four sides. However, for practical purposes, it is customary to design footings as if they were a *one-way slab* as shown in Figure 9.10b. This conservative simplification is justified because of the following:

- The full-scale load tests on which this analysis method is based were interpreted this way.
- It is appropriate to design foundations more conservatively than the superstructure.
- The flexural stresses are low, so the amount of steel required is nominal and often governed by ρ_{min}.
- The additional construction cost due to this simplified approach is nominal.

Once we know the amount of steel needed to carry the applied load in one-way bending, we place the same steel area in the perpendicular direction. In essence the footing is reinforced twice, which provides more reinforcement than required by a more rigorous two-way analysis.

Steel Area

The usual procedure for designing flexural members is to prepare a moment diagram and select an appropriate amount of steel for each portion of the member. However, for simple spread footings, we again simplify the problem and design all the steel for the moment that occurs at the *critical section for bending*. The location of this section for various types of columns is shown in Figure 9.11.

We can simplify the computations by defining a distance l, measured from the critical section to the outside edge of the footing. In other words, l is the cantilever distance. It is computed using the formulas in Table 9.2.

The factored bending moment at the critical section, M_{uc}, is:

$$M_{uc} = \frac{P_u l^2}{2B} + \frac{2 M_u l}{B}$$

(9.16)

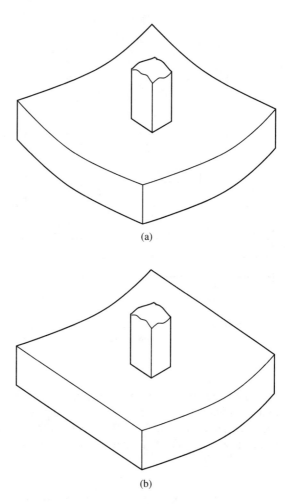

Figure 9.10   (a) A spread footing is actually a two-way slab, bending in both the "north-south" and "east-west" directions; (b) For purposes of analysis, engineers assume that the footing is a one-way slab that bends in one axis only.

(a)

(b)

Where:

M_{uc} = factored moment at critical section for bending

P_u = factored compressive load from column

M_u = factored moment load from column

l = cantilever distance (from Table 9.2)

B = footing width

The first term in Equation 9.16 is based on the assumption that P_u acts through the centroid of the footing. The second term is based on a soil bearing pressure with an assumed eccentricity of $B/3$, which is conservative (see Figure 5.15).

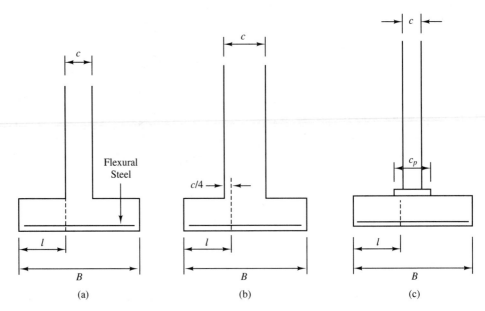

Figure 9.11 Location of critical section for bending: (a) with a concrete column; (b) with a masonry column; and (c) with a steel column.

After computing M_{uc}, find the steel area, A_s, and reinforcement ratio, ρ, using Equations 9.12 and 9.15. Check if the computed ρ is less than ρ_{min}. If so, then use ρ_{min}. Rarely will ρ be larger than 0.0040. This light reinforcement requirement develops because we made the effective depth d relatively large to avoid the need for stirrups.

TABLE 9.2 DESIGN CANTILEVER DISTANCE FOR USE IN DESIGNING REINFORCEMENT IN SPREAD FOOTINGS [15.4.2].

Type of Column	l
Concrete	$(B - c)/2$
Masonry	$(B - c/2)/2$
Steel	$(2B - (c + c_p))/4$

1. ACI does not specify the location of the critical section for timber columns, but in this context, it seems reasonable to treat them in the same way as concrete columns.
2. If the column has a circular, octagonal, or other similar shape, use a square with an equivalent cross-sectional area.
3. B = footing width; c = column width; c_p = base plate width. If column has a circular or regular polygon cross section, base the analysis on an equivalent square.

The required area of steel for each direction is:

$$\boxed{A_s = \rho\, B\, d} \tag{9.17}$$

Carry the flexural steel out to a point 70 mm (3 in) from the edge of the footing as shown in Figure 9.4.

Development Length

ACI [15.6] requires the flexural steel in spread footings meet standard development length requirements. This development length is measured from the critical section for bending to the end of the bars as defined in Figure 9.11 to the end of the bars, which is 70 mm (3 in) from the edge of the footing. Thus, the supplied development length, $(l_d)_{supplied}$ is:

$$\boxed{(l_d)_{supplied} = l - 70 \text{ mm (3 in)}} \tag{9.18}$$

Where:

$(l_d)_{supplied}$ = supplied development length

l = cantilever distance (per Table 9.2)

This supplied development length must be at least equal to the required development length, as computed using Equation 9.14 or 9.15. If this criteria is not satisfied, we do not enlarge the footing width, B. Instead, it is better to use smaller diameter bars, which have a correspondingly shorter required development length.

If the supplied development length is greater than the required development length, we still extend the bars to 70 mm (3 in) from the edge of the footing. Do not cut them off at a different location.

Example 9.1—Part B

Using the results from Part A, design the required flexural steel.

Solution

Find the required steel area

$$l = \frac{B - c}{2} = \frac{126 - 21}{2} = 52.5 \text{ in}$$

$$M_{uc} = \frac{P_u l^2}{2\,B} + 0 = \frac{(991{,}000)(52.5)^2}{(2)(126)} = 10{,}800{,}000 \text{ in-lb}$$

$$A_s = \left(\frac{f_c' b}{1.176 f_y}\right)\left(d - \sqrt{d^2 - \frac{2.353 \, M_u}{\phi f_c' b}}\right)$$

$$= \left(\frac{(4000 \text{ lb/in}^2)(126 \text{ in})}{(1.176)(60{,}000 \text{ lb/in}^2)}\right)\left(23 \text{ in} - \sqrt{(23 \text{ in})^2 - \frac{2.353 \,(10{,}800{,}000 \text{ in-lb})}{(0.9)(4000 \text{ lb/in}^2)(126 \text{ in})}}\right)$$

$$= 8.94 \text{ in}^2$$

Check minimum steel

$$A_s \geq 0.018 \,(27)\,(126)$$

$$\geq 6.12 \text{ in}^2$$

$$8.94 \geq 6.12 \quad \text{ok}$$

Use 12 #8 bars each way ($A_s = $ **9.42 in^2**)   ⇐ *Answer*

Clear space between bars $= 126/13\text{-}1 = 8.7 \text{ in} \quad$ OK

Check development length

$$(l_d)_{supplied} = l - 3 = 52.5 - 3 = 49.5 \text{ in}$$

$$\frac{c + K_{tr}}{d_b} = \frac{3.5 + 0}{1} = 3.5 > 2.5 \quad \text{use } 2.5$$

$$\frac{l_d}{d_b} = \frac{3}{40}\frac{f_y}{\sqrt{f_c'}}\frac{\alpha \, \beta \, \gamma \, \lambda}{\left(\frac{c + K_{tr}}{d_b}\right)} = \frac{3}{40}\frac{60{,}000}{\sqrt{4000}}\frac{(1)(1)(1)(1)}{2.5} = 28$$

$$l_d = 28 \, d_b = (28)(1) = 28 \text{ in}$$

$l_d < (l_d)_{supplied}$, so development length is OK.

The final design is shown in Figure 9.12.

QUESTIONS AND PRACTICE PROBLEMS

The ASD load for determining the required footing width should be computed using the largest of Equations 2.1, 2.2, 2.3a, or 2.4a, unless otherwise stated. The factored loads should be computed using the ACI load factors (Equations 2.7–2.17).

9.1 A column carries the following vertical downward loads: dead load = 400 kN, live load = 300 kN, wind load = 150 kN, and earthquake load = 250 kN. Compute the unfactored load, P, and the factored load, P_u, to be used in the design of a concrete footing.

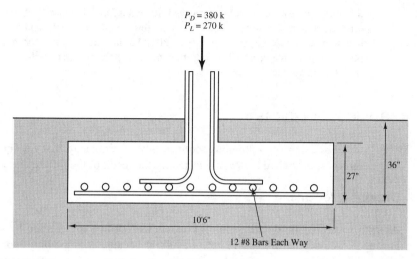

Figure 9.12   Footing design for Example 9.1.

9.2 A flexural member has a dead load moment of 200 ft-k and a live load moment of 150 ft-k. The computed nominal moment capacity, M_n, is 600 ft-k. Is the design of this member satisfactory? Use the ACI ultimate strength criterion.

9.3 Why are spread footings usually made of low-strength concrete?

9.4 Explain the difference between the shape of the actual shear failure surfaces in footings with those used for analysis and design.

9.5 A 400-mm square concrete column that carries a factored vertical downward load of 450 kN and a factored moment load of 100 kN-m is supported on a 1.5-m square footing. The effective depth of the concrete in this footing is 500 mm. Compute the ultimate shear force that acts on the most critical section for two-way shear failure in the footing.

9.6 A 16-in square concrete column carries vertical dead and live loads of 150 and 100 k, respectively. It is to be supported on a square footing with $f_c' = 3000$ lb/in^2 and $f_y = 60$ k/in^2. The soil has an allowable bearing pressure of 4500 lb/ft^2 and the groundwater table is at a great depth. Because of frost heave considerations, the bottom of this footing must be at least 30 inches below the ground surface. Determine the required footing thickness, size the flexural reinforcement, and show your design in a sketch.

9.7 A W16×50 steel column with a 22-inch square base plate is to be supported on a square spread footing. This column has a design dead load of 200 k and a design live load of 120 k. The footing will be made of concrete with $f_c' = 2500$ lb/in^2 and reinforcing steel with $f_y = 60$ k/in^2. The soil has an allowable bearing pressure of 3000 lb/ft^2 and the groundwater table is at a great depth. Determine the required footing thickness, size the flexural reinforcement, and show your design in a sketch.

9.8 A 500-mm square concrete column carries vertical dead and live loads of 500 and 280 kN, respectively. It is to be supported on a square footing with $f_c' = 17$ MPa and $f_y = 420$ MPa. The soil has an allowable bearing pressure of 200 kPa and the groundwater table is at a great depth. Determine the required footing thickness, size the flexural reinforcement, and show your design in a sketch.

9.6   CONTINUOUS FOOTINGS

The structural design of continuous footings is very similar to that for square footings. The differences, described below, are primarily the result of the differences in geometry.

Designing for Shear

As with square footings, the depth of continuous footings is governed by shear criteria. However, we only need to check one-way shear because it is the only type that has any physical significance. The critical surfaces for evaluating one-way shear are located a distance d from the face of the wall as shown in Figure 9.13.

The factored shear force acting on a unit length of the critical shear surface is:

$$V_{uc}/b = (P_u/b)\left(\frac{B - c - 2d}{B}\right) \tag{9.19}$$

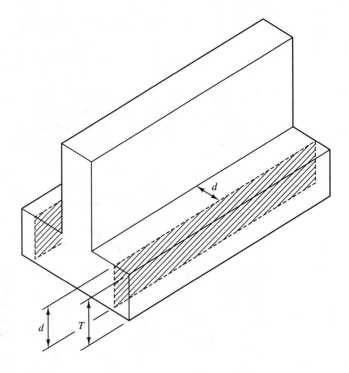

Figure 9.13 Location of idealized critical shear surface for one-way shear in a continuous footing.

Where:

V_{uc}/b = factored shear force on critical shear surface per unit length of footing

P_u/b = factored applied compressive load per unit length of footing

c = width of wall

b = unit length of footing (usually 1 ft or 1 m)

Setting $V_{uc} = \phi\, V_{nc}$, equating Equations 9.9 and 9.19, and solving for d gives:

$$d = \frac{(P_u/b)(B - c)}{48\,\phi\, B\,\sqrt{f'_c} + 2\,P_u/b} \qquad \text{(9.20 English)}$$

$$d = \frac{1500\,(P_u/b)(B - c)}{500\,\phi\, B\,\sqrt{f'_c} + 3\,P_u/b} \qquad \text{(9.20 SI)}$$

Where:

d = effective depth (in, mm)

P_u/b = applied vertical load per unit length of footing (lb/ft, kN/m)

b = footing width (in, mm)

c = wall width (in, mm)

ϕ = resistance factor = 0.85

f'_c = 28-day compressive strength of concrete (lb/in^2, MPa)

Then, compute the footing thickness, T, using the criterion described earlier.

Designing for Flexure

Nearly all continuous footings should have longitudinal reinforcing steel (i.e., running parallel to the wall). This steel helps the footing resist flexural stresses from non-uniform loading, soft spots in the soil, or other causes. Temperature and shrinkage stresses also are a concern. Therefore, place a nominal amount of longitudinal steel in the footing (0.0018 A_g to 0.0020 A_g) with at least two #4 bars (2 metric #13 bars). If large differential heaves or settlements are likely, we may need to use additional longitudinal reinforcement. Chapter 19 includes a discussion of this issue.

Transverse steel (that which runs perpendicular to the wall) is another issue. Most continuous footings are narrow enough so the entire base is within a 45° frustum, as shown in Figure 9.14. Thus, they do not need transverse steel. However, wider footings should include transverse steel designed to resist the flexural stresses at the critical section as defined in Table 9.2. The factored moment at this section is:

$$M_{uc}/b = \frac{(P_u/b)l^2}{2\,B} + \frac{2\,(M_u/b)l}{B} \qquad (9.21)$$

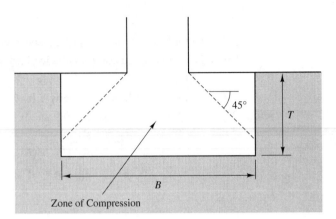

Figure 9.14   Zone of compression in lightly-loaded footings.

Where:

M_{uc}/b = factored moment at critical section per unit length of footing

P_u/b = factored applied compressive load per unit length of footing

M_u/b = factored applied moment load perpendicular to wall per unit length of footing

l = cantilever distance (from Figure 9.11 or Table 9.2)

Compute the required transverse steel area per unit length, A_s/b, using Equation 9.13, as demonstrated in Example 9.2.

Example 9.2

A 200-mm wide concrete block wall carries a vertical dead load of 120 kN/m and a vertical live load of 88 kN/m. It is to be supported on a continuous spread footing that is to be founded at a depth of at least 500 mm below the ground surface. The allowable bearing pressure of the soil beneath the footing is 200 kPa, and the groundwater table is at a depth of 10 m. Develop a structural design for this footing using $f_c' = 15$ MPa and $f_y = 300$ MPa.

Solution

Unfactored load—Equation 2.2 governs

$$P/b = (P/b)_D + (P/b)_L + \ldots = 120 \text{ kN/m} + 88 \text{ kN/m} + 0 = 208 \text{ kN/m}$$

Per Table 8.1, minimum $D = 400$ mm, but problem statement says use $D = 500$ mm

$$W_f/b = B D \gamma_c = B(0.5 \text{ m})(23.6 \text{ kN/m}^3) = 11.8 B$$

$$B = \frac{P/b + W_f/b}{q_A - u_D} = \frac{208 + 11.8 B}{200 - 0} = 1.1 \text{ m}$$

Factored load—Equation 2.7 governs

$$P_u/b = 1.4P_D/b + 1.7P_L/b$$

$$= 1.4(120) + 1.7(88)$$

$$= 318 \text{ kN/m}$$

Compute the required thickness using a shear analysis

$$d = \frac{1500\,(P_u/b)(B - c)}{500\,\phi\,B\sqrt{f_c'} + 3\,P_u/b}$$

$$= \frac{(1500)(318 \text{ kN/m})(1100 \text{ mm} - 200 \text{ mm})}{(500)(0.85)(1100 \text{ mm})\sqrt{15 \text{ MPA}} + (3)(318 \text{ kN/m})}$$

$$= 237 \text{ mm}$$

For ease of construction, place the longitudinal steel below the lateral steel. Assuming metric #13 bars (diameter = 12.7 mm), the footing thickness, T, is:

$$T = d + (1/2)(\text{diam. of lat. steel}) + \text{diam. of long steel} + 70 \text{ mm}$$

$$= 237 + 12.7/2 + 12.7 + 70$$

$$= 326 \text{ mm} \quad \rightarrow \quad \text{Use } 400 \text{ mm}$$

$$d = 400 - 12.7/2 - 12.7 - 70 = 311 \text{ mm}$$

In the square footing design of Example 9.1, we used an effective depth, d, as the distance from the top of the footing to the contact point of the two layers of reinforcing bars (as shown in Figure 9.4. We used this definition because square footings have two-way bending, this is the average d of the two sets of rebar. However, with continuous footings we are designing only the lateral steel, so d is measured from the top of the footing to the center of the lateral bars. The longitudinal bars will be designed separately.

Design the lateral steel

$$l = \frac{B - c/2}{2} = \frac{1.1 - 0.2/2}{2} = 0.50 \text{ m} = 500 \text{ mm}$$

$$M_{uc}/b = \frac{(P_u/b)l^2}{2B} + 0 = \frac{(318)(0.50)^2}{2(1.1)} = 36.1 \text{ kN-m/m}$$

$$A_s/b = \left(\frac{f_c'\,b}{1.176\,f_y}\right)\left(d - \sqrt{d^2 - \frac{2.353\,M_u}{\phi\,f_c'\,b}}\right)$$

$$= \left(\frac{(15 \text{ MPa})(1 \text{ m})}{(1.176)(300 \text{ MPa})}\right)\left(0.311 \text{ m} - \sqrt{(0.311 \text{ m})^2 - \frac{2.353\,(36.1 \text{ kN-m})}{(0.9)(15{,}000 \text{ kPa})(1 \text{ m})}}\right)\left(\frac{10^3 \text{ mm}}{\text{m}}\right)^2$$

$$= 437 \text{ mm}^2/\text{m}$$

Check minimum steel

$$A_s/b \geq 0.0020\,(460)(1000)$$

$$\geq 800 \text{ mm}^2/\text{m}$$

Use metric #13 bars @ 150 mm OC [$A_s = (129 \text{ mm}^2/\text{bar})(6.67 \text{ bars/m}) = 860 \text{ mm}^2/\text{m} > 800 \text{ mm}^2/\text{m}$]

Check development length

$$(l_d)_{supplied} = l - 75 = 500 - 75 = 425 \text{ mm}$$

$$\frac{c + K_{tr}}{d_b} = \frac{70 + 0}{12.7} = 5.5 > 2.5 \quad \text{use 2.5}$$

$$\frac{l_d}{d_b} = \frac{9}{10}\frac{f_y}{\sqrt{f_c'}}\frac{\alpha\,\beta\,\gamma\,\lambda}{\left(\dfrac{c + K_{tr}}{d_b}\right)} = \frac{9}{10}\frac{300}{\sqrt{15}}\frac{(1)(1)(1)(1)}{2.5} = 28$$

$$l_d = 28\,d_b = (28)(12.7) = 355 \text{ mm}$$

$l_d < (l_d)_{supplied}$, so development length is OK.
Design the longitudinal steel

$$A_s = \rho\,b\,d = (0.0020)(1100)(400) = 880 \text{ mm}^2$$

Use 7 metric #13 bars ($A_s = 903 \text{ mm}^2 \approx\, > 880 \text{ mm}^2$)

The final design is shown in Figure 9.15.

QUESTIONS AND PRACTICE PROBLEMS

9.9 A 12-in wide concrete block wall carries vertical dead and live loads of 13.0 and 12.1 k/ft, respectively. It is to be supported on a continuous footing made of 2000 lb/in² concrete and 40 k/in² steel. The soil has an allowable bearing pressure of 4000 lb/ft², and the groundwater table is at a great depth. The local building code requires that the bottom of this footing be at least 24 inches below the ground surface. Determine the required footing thickness, and design the lateral and longitudinal steel. Show your design in a sketch.

9.10 A 200-mm wide concrete block wall carries vertical dead and live loads of 50 and 70 kN/m, respectively. It is to be supported on a continuous footing made of 18 MPa concrete and

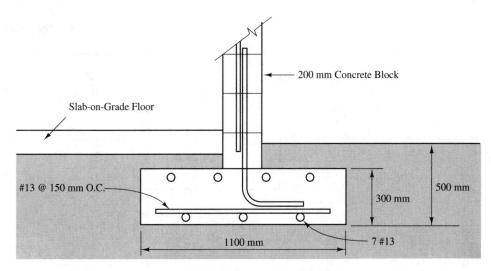

Figure 9.15   Final design for Example 9.2.

420 MPa steel. The soil has an allowable bearing pressure of 150 kPa, and the groundwater table is at a great depth. The local building code requires that the bottom of this footing be at least 500 mm below the ground surface. Determine the required footing thickness, and design the lateral and longitudinal steel. Show your design in a sketch.

9.7   RECTANGULAR FOOTINGS

Rectangular footings with width B and length L that support only one column are similar to square footings. Design them as follows:

1. Check both one-way shear (Equation 9.9) and two-way shear (Equation 9.6) using the critical shear surfaces shown in Figure 9.16a. Determine the minimum required d and T to satisfy both.
2. Design the long steel (see Figure 9.16b) by substituting L for B in Table 9.2 and Equation 9.16, and using Equation 9.17 with no modifications. Distribute this steel evenly across the footing as shown in Figure 9.16c.
3. Design the short steel (see Figure 9.16b) using Table 9.2 and Equation 9.16 with no modifications, and substituting L for B in Equation 9.17.
4. Since the central portion of the footing takes a larger portion of the short-direction flexural stresses, place more of the short steel in this zone [15.4.4]. To do so, divide the footing into inner and outer zones, as shown in Figure 9.16c. The portion of the total short steel area, A_s, to be placed in the inner zone is E:

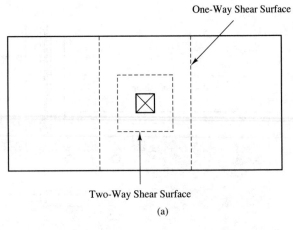

One-Way Shear Surface

Two-Way Shear Surface

(a)

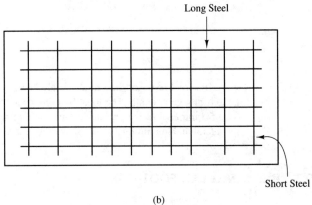

Long Steel

Short Steel

(b)

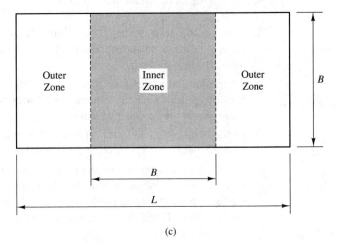

Outer
Zone

Inner
Zone

Outer
Zone

B

B

L

(c)

Figure 9.16   Structural design of rectangular footings; (a) critical shear surfaces; (b) long steel and short steel; (c) distribution of short steel.

$$E = \frac{2}{L/B + 1}$$ (9.22)

Distribute the balance of the steel evenly across the outer zones.

9.8 COMBINED FOOTINGS

Combined footings are those that carry more than one column. Their loading and geometry is more complex, so it is appropriate to conduct a more rigorous structural analysis. The rigid method, described in Chapter 10, is appropriate for most combined footings. It uses a soil bearing pressure that varies linearly across the footing, thus simplifying the computations. Once the soil pressure has been established, MacGregor (1996) suggests designing the longitudinal steel using idealized beam strips ABC, as shown in Figure 9.17. Then, design the transverse steel using idealized beam strips AD. See MacGregor (1996) for a complete design example.

Large or heavily loaded combined footings may justify a beam on elastic foundation analysis, as described in Chapter 10.

9.9 LIGHTLY-LOADED FOOTINGS

Although the principles described in Sections 9.5 to 9.8 apply to footings of all sizes, some footings are so lightly loaded that practical minimums begin to govern the design. For example, if P_u is less than about 400 kN (90 k) or P_u/b is less than about 150 kN/m (10 k/ft), the minimum d of 150 mm (6 in) [ACI 15.7] controls. Thus, there is no need to conduct a shear analysis, only to compute a T smaller than the minimum. In the same vein, if P_u is less than about 130 kN (30 k) or P_u/b is less than about 60 kN/m (4 k/ft), the minimum steel requirement ($\rho = 0.0018$) governs, so there is no need to conduct a flexural analysis. Often, these minimums also apply to footings that support larger loads.

In addition, if the entire base of the footing is within a 45° frustum, as shown in Figure 9.14, we can safely presume that very little or no tensile stresses will develop. This is often the case with lightly loaded footings. Technically, no reinforcement is required in such cases. However, some building codes [ICBO 1806.7] have minimum reinforcement requirements for certain footings, and it is good practice to include at least the following reinforcement in all footings:

Square footings

- If bottom of footing is completely within the zone of compression—no reinforcement required
- If bottom of footing extends beyond the zone of compression—as determined by a flexural analysis, but at least #4 @ 18 in o.c. each way (metric #13 @ 500 mm o.c. each way)

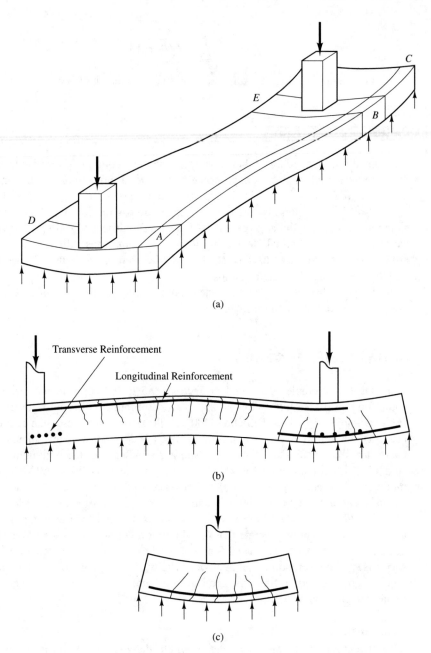

(a)

(b)

(c)

Figure 9.17   Structural design of combined footings: (a) idealized beam strips; (b) longitudinal beam strip; (c) transverse beam strip (Adapted from MacGregor, 1996).

Continuous footings
> Longitudinal reinforcement
> > • Minimum two #4 bars (metric #13)
> Lateral reinforcement
> > • If bottom of footing is completely within the zone of compression—no lateral reinforcement required
> > • If bottom of footing extends beyond the zone of compression—as determined by a flexural analysis, but at least #4 @ 18 in o.c. (metric #13 @ 500 mm o.c.)

This minimum reinforcement helps accommodate unanticipated stresses, temperature and shrinkage stresses, and other phenomena.

9.10 CONNECTIONS WITH THE SUPERSTRUCTURE

One last design feature that needs to be addressed is the connection between the footing and the superstructure. Connections are often the weak link in structures, so this portion of the design must be done carefully. A variety of connection types are available, each intended for particular construction materials and loading conditions. The design of proper connections is especially important when significant seismic or wind loads are present (Dowrick, 1987).

Connections are designed using either ASD (with the unfactored loads) or LRFD (with the factored loads) depending on the design method used in the superstructure.

Connections with Columns

Columns may be made of concrete, masonry, steel, or wood, and each has its own concerns when designing connections.

Concrete or Masonry Columns

Connect concrete or masonry columns to their footing [ACI 15.8] using *dowels*, as shown in Figure 9.18. These dowels are simply pieces of reinforcing bars that transmit axial, shear, and moment loads. Use at least four dowels with a total area of steel, A_s, at least equal to that of the column steel or 0.005 times the cross-sectional area of the column, whichever is greater. They may not be larger than #11 bars [ACI 15.8.2.3] and must have a 90° hook at the bottom. Normally, the number of dowels is equal to the number of vertical bars in the column.

Design for Compressive Loads

Check the bearing strength of the footing [ACI 10.17] to verify that it is able to support the axial column load. This is especially likely to be a concern if the column carries large compressive stresses that might cause something comparable to a bearing capacity failure

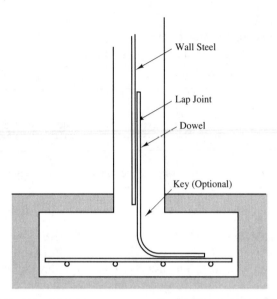

Figure 9.18   Use of dowels to connect a concrete or masonry column to its footing.

inside the footing. To check this possibility, compute the factored column load, P_u, and compare it to the nominal column bearing capacity, P_{nb}:

$$P_{nb} = 0.85 f_c' A_1 s$$ (9.23)

Then, determine whether the following statement is true:

$$P_u \leq \phi P_{nb}$$ (9.24)

Where:

P_u = factored column load

P_{nb} = nominal column bearing capacity (i.e., bearing of column on top of footing)

f_c' = 28-day compressive strength of concrete

$s = (A_2/A_1)^{0.5} \leq 2$ if the frustum in Figure 9.19 fits entirely within the footing (i.e., if $c + 4d \leq B$)

$s = 1$ if the frustum in Figure 9.19 does not fit entirely within the footing

A_1 = cross-sectional area of the column = c^2

$A_2 = (c + 4d)^2$ as shown in Figure 9.19

c = column width or diameter

ϕ = resistance factor = 0.7 [ACI 9.3.2.4]

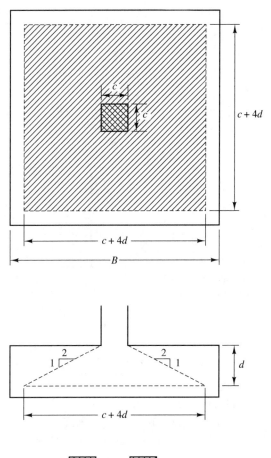

Figure 9.19 Application of a frustrum to find s and A_2.

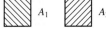 A_1 A_2

If Equation 9.24 is not satisfied, use a higher strength concrete (greater f_c') in the footing or design the dowels as compression steel.

Design for Moment Loads

If the column imparts moment loads onto the footing, then some of the dowels will be in tension. Therefore, the dowels must be embedded at least one development length into the footing, as shown in Figure 9.20 and defined by the following equations [ACI 12.5]:

$$l_{dh} = \frac{1200 \, d_b}{\sqrt{f_c'}}$$
(9.25 English)

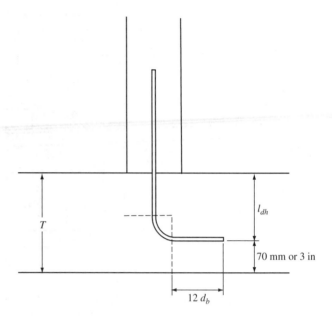

Figure 9.20   Minimum required embedment of dowels subjected to tension.

$$l_{dh} = \frac{100\, d_b}{\sqrt{f_c'}}$$
(9.25 SI)

Where:

T = footing thickness (in, mm)

l_{dh} = development length for 90° hooks, as defined in Figure 9.20 (in, mm)

d_b = bar diameter (in, mm)

f_c' = 28-day compressive strength of concrete (lb/in², MPa)

The development length computed from Equation 9.25 may be modified by the following factors [ACI 12.5.3][1] :

For standard reinforcing bars with yield strength other than 60,000 lb/in² : $f_y/60{,}000$

For metric reinforcing bars with yield strength other than 420 lb/in² : $f_y/420$

If at least 50 mm (2 in) of cover is present beyond the end of the hook: 0.7

Sometimes this development length requirement will dictate a footing thickness T greater than that required for shear (as computed earlier in this chapter).

---

[1]This list only includes modification factors that are applicable to anchorage of vertical steel in retaining wall footings. ACI 12.5.3 includes additional modification factors that apply to other situations.

As long as the number and size of dowels are at least as large as the vertical steel in the column, then they will have sufficient capacity to carry the moment loads.

Design for Shear Loads

If the column also imparts a shear load, V_u, onto the footing, the connection must be able to transmit this load. Since the footing and column are poured separately, there is a weak shear plane along the cold joint. Therefore, the dowels must transmit all of the applied shear load. The minimum required dowel steel area is:

$$A_s = \frac{V_u}{\phi f_y \mu}$$

(9.26)

Where:

A_s = minimum required dowel steel area

V_u = applied factored shear load

ϕ = 0.85 for shear

f_y = yield strength of reinforcing steel

μ = 0.6 if the cold joint not intentionally roughened or 1.0 if the cold joint is roughened by heavy raking or grooving [ACI 11.7.4.3]

However, the ultimate shear load, V_u, cannot exceed $0.2 \, \phi f_c' A_c$, where f_c' is the compressive strength of the column concrete, and A_c is the cross-sectional area of the column.

Splices

Most designs use a lap splice to connect the dowels and the vertical column steel. However, some columns have failed in the vicinity of these splices during earthquakes, as shown in Figure 9.21. Therefore, current codes require much more spiral reinforcement in columns subjected to seismic loads. In addition, some structures with large moment loads, such as certain highway bridges, require mechanical splices or welded splices to connect the dowels and the column steel.

Steel Columns

Steel columns are connected to their foundations using base plates and anchor bolts, as shown in Figure 9.22. The base plates are welded to the bottom of the columns when they are fabricated, and the anchor bolts are cast into the foundation when the concrete is placed. The column is then erected over the foundation, and the anchor bolts are fit through predrilled holes in the base plate.

The top of the footing is very rough and not necessarily level, so the contractor must use special construction methods to provide uniform support for the base plate and to make the column plumb. For traffic signal poles, light standards, and other lightweight columns, the most common method is to provide a nut above and below the base plate, as

Figure 9.21 Imperial County Services Building, El Centro, California. The bases of these columns failed during the 1979 El Centro earthquake, causing the building to sag about 300 mm. As a result, this six-story building had to be demolished. (U.S. Geological Survey photo)

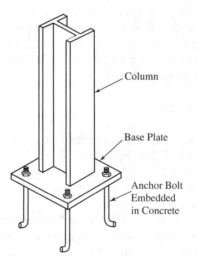

Column

Base Plate

Anchor Bolt Embedded in Concrete

Figure 9.22 Base plate and anchor bolts to connect a steel column to its foundations.

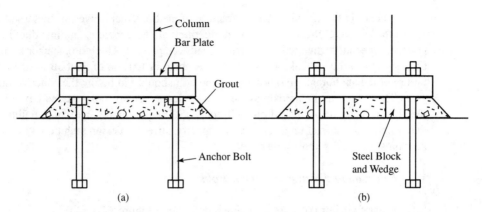

Figure 9.23   Methods of leveling the base plate: a) Double nuts, b) Blocks and shims.

shown in Figure 9.23a, and adjust these nuts as needed to make the column plumb. However, columns for buildings, bridges, and other large structures are generally too heavy for this method, so the contractor must temporarily support the base plate on steel blocks and shims, and clamp it down with a single nut on each anchor bolt, as shown in Figure 9.23b. These shims are carefully selected to produce a level base plate and a plumb column. Other construction methods also have been used.

Once the column is securely in place and the various members that frame into it have been erected, the contractor places a nonshrink grout between the base plate and the footing. This grout swells slightly when it cures (as compared to normal grout, which shrinks), thus maintaining continuous support for the base plate. The structural loads from the column are then transmitted to the footing as follows:

- Compressive loads are spread over the base plate and transmitted through the grout to the footing.
- Tensile loads pass through the base plate and are resisted by the anchor bolts.
- Moment loads are resisted by a combination of compression through the grout and tension in half of the bolts.
- Shear loads are transmitted through the anchor bolts, through sliding friction along the bottom of the base plate, or possibly both.

Design Principles

The base plate must be large enough to avoid exceeding the nominal bearing strength of the concrete (see earlier discussion under concrete and masonry columns). In addition, it must be thick enough to transmit the load from the column to the footing. The design of base plates is beyond the scope of this book, but it is covered in most steel design texts and in DeWolf and Ricker (1990).

Anchor bolts can fail either by fracture of the bolts themselves, or by loss of anchorage in the concrete. Steel is much more ductile than the concrete, and this ductility is important, especially when wind or seismic loads are present. Therefore, anchor bolts should be designed so the critical mode of failure is shear or tension of the bolt itself rather than failure of the anchorage. In other words, the bolt should fail before the concrete fails.

The following methods may be used to design anchor bolts that satisfy this principle. These methods are based on ACI and AISC requirements, but some building codes may impose additional requirements, or specify different design techniques, so the engineer must check the applicable code.

Selection and Sizing of Anchor Bolts

Five types of anchor bolts are available, as shown in Figure 9.24:

- *Standard structural steel bolts* may simply be embedded into the concrete to form anchor bolts. These bolts are similar to those used in bolted steel connections, except they are much longer. Unfortunately, these bolts may not be easily available in lengths greater than about 6 inches, so they often are not a practical choice.
- *Structural steel rods that have been cut to length and threaded* form anchor bolts that are nearly identical to a standard steel bolts and have the advantage of being more readily available. This is the most common type of anchor bolt for steel columns. If one nut is used at the bottom of each rod, it should be tack welded to prevent the rod from turning when the top nut is tightened. Alternatively, two nuts may be used.
- *Hooked bars* (also known as an *L-bolts* or a *J-bolts*) are specially fabricated fasteners made for this purpose. These are principally used for wood-frame structures, and are generally suitable for steel structures only when no tensile or shear loads are present.

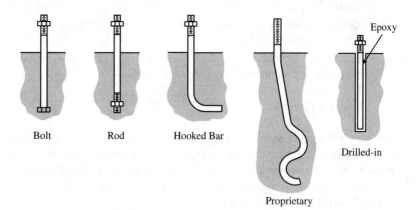

Figure 9.24   Types of anchor bolts.

- *Proprietary anchor bolts* are patented designs that often are intended for special applications, principally with wood-frame structures.
- *Drilled-in anchor bolts* are used when a cast-in-place anchor bolt was not installed during construction of the footing. They are constructed by drilling a hole in the concrete, then embedding a threaded rod into the hole and anchoring it using either epoxy grout or mechanical wedges. This is the most expensive of the five types and is usually required only to rectify mistakes in the placement of conventional anchor bolts.

Most anchor bolts are made of steel that satisfies ASTM A36 or ASTM A307, both of which have $F_y = 36$ k/in^2 (250 MPa). However, higher strength steel may be used, if needed. Each bolt must satisfy both of the following design criteria:

$$\boxed{P_u \leq \phi\, P_n} \tag{9.27}$$

$$\boxed{V_u \leq \phi\, V_n} \tag{9.28}$$

Where:

P_u = factored tensile force based on AISC load factors (Equations 2.18–2.23) expressed as a positive number

V_u = factored shear force based on AISC load factors (Equations 2.18–2.23)

ϕ = resistance factor

P_n = nominal tensile capacity

V_n = nominal shear capacity

In addition, the design must satisfy AISC requirements for interaction between shear and tensile stresses. Figure 9.25 presents the shear and tensile capacities for ASTM A36 and ASTM A307 bolts that satisfies Equations 9.27 and 9.28 and the interaction requirements, and may be used to select the required diameter.

Typically four anchor bolts are used for each column. It is best to place them in a square pattern to simplify construction and leave less opportunity for mistakes. Rectangular or hexagonal patterns are more likely to be accidentally built with the wrong orientation. More bolts and other patterns also may be used, if necessary.

If the design loads between the column and the footing consist solely of compression, then anchor bolts are required only to resist erection loads, accidental collisions during erection, and unanticipated shear or tensile loads. The engineer might attempt to estimate these loads and design accordingly, or simply select the bolts using engineering judgement. Often these columns simply use the same anchor bolt design as nearby columns, thus reducing the potential for mistakes during construction.

Anchorage

Once the bolt diameter has been selected, the engineer must determine the required depth of embedment into the concrete to provide the necessary anchorage. The required embedment depends on the type of anchor, the spacing between the anchors, the kind of loading,

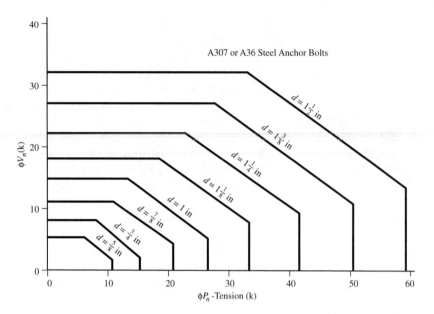

Figure 9.25   AISC tensile and shear capacities of A36 and A307 anchor bolts.

and the strength of the concrete. In addition, the bolt must be located at least a minimum horizontal distance from the edge of the concrete. Table 9.3 presents conservative design values for embedment depth and edge distance. Alternatively, the engineer may use the more refined procedure described by Marsh and Burdette (1985).

Shear Transfer

Shear forces may be transferred from the base plate to the foundation in two ways:

- Through shear in the anchor bolts
- By sliding friction along the bottom of the base plate

TABLE 9.3   ANCHORAGE REQUIREMENTS FOR BOLTS AND THREADED RODS (Shipp and Haninger, 1983) Copyright © American Institute of Steel Construction. Reprinted with permission.

Steel Grade	Minimum Embedment Depth	Minimum Edge Distance
A307, A36	12 d	5d or 100 mm (4 in), whichever is greater
A325, A449	17 d	7d or 100 mm (4 in), whichever is greater

d = nominal bolt diameter

Some engineers rely on both sources of shear resistance, while others rely only on one or the other.

When transferring shear loads through the anchor bolts, the engineer must recognize that the bolt holes in the base plate are oversized in order to simplify the erection of the column onto the footing. As a result, the resulting gap between the bolts and the base plate does not allow for efficient transfer of the shear loads. The base plate may need to move laterally before touching the bolts, and most likely only some of the bolts will become fully engaged with the plate. Therefore, when using this mode of shear transfer, it is probably prudent to assume the shear load is carried by only half of the bolts.

So long as the grout has been carefully installed between the base plate and the footing, and the base plate has not been installed using the double nut method as shown in Figure 9.23a, there will be sliding friction along the bottom of the base plate. AISC recommends using a coefficient of friction of 0.55 for conventional base plates, such as that shown in Figure 9.23b, and the resistance factor, ϕ, is 0.90. Thus, the available sliding friction resistance, ϕV_n, is:

$$
\begin{aligned}
\phi\, V_n &= \phi\, \mu\, P \\
&= (0.90)(0.55)\, P \\
&= 0.50\, P
\end{aligned}
\tag{9.29}
$$

The value of P in Equation 9.29 should be the lowest unfactored normal load obtained from Equations 2.1 to 2.4. Usually Equation 2.1 governs, except when uplift wind or seismic loads are present, as described in Equation 2.4. It is good practice to ignore any normal stress produced by live loads or the clamping forces from the nuts.

Sometimes short vertical fins are welded to the bottom of the base plate to improve shear transfer to the grout. These fins may justify raising the coefficient of friction to 0.7.

Example 9.3

A steel wide flange column with a steel base plate is to be supported on a spread footing. The AISC factored design loads are: $P_u = 270$ k compression and $M_u = 200$ ft-k. Design an anchor bolt system for this column using four bolts arranged in a 15×15 inch square.

Solution

Reduce the applied loads to a couple separated by 15 in:

$$
\begin{aligned}
P &= \frac{270\text{ k}}{2} \pm \frac{200\text{ ft-k}}{(15/12)\text{ ft}} \\
&= 135 \pm 160\text{ k}
\end{aligned}
$$

There are two bolts on each side, so the maximum tensile force in each bolt is:

$$P = \frac{135 - 160}{2} = 12.5 \text{ k} \quad \text{tension}$$

The shear force is zero.

Per Figure 9.25, use 3/4 inch bolts.

The depth of embedment should be $(12)(0.75) = 9$ in.

Use four 3/4 inch diameter x 13 inch long A36 threaded rods embedded 9 inches into the footing. Firmly tighten two nuts at the bottom of each rod. $\Leftarrow$ *Answer*

Wood Columns

Wood columns, often called *posts*, usually carry light loads and require relatively simple connections. The most common type is a metal bracket, as shown in Figure 9.26. These are set in the wet concrete immediately after it is poured. The manufacturers determine the allowable loads and tabulate them in their catalogs (for example, see www.strongtie.com).

It is poor practice to simply embed a wooden post into the footing. Although at first this would be a very strong connection, in time the wood will rot and become weakened.

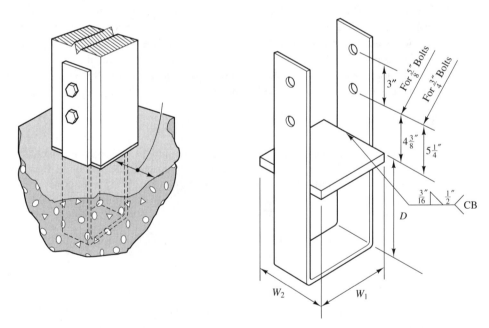

Figure 9.26 Steel post base for securing a wood post to a spread footing (Simpson Strong-Tie Co., Inc.).

Connections with Walls

The connection between a concrete or masonry wall and its footing is a simple one. Simply extend the vertical wall steel into the footing [ACI 15.8.2.2], as shown in Figure 9.27. For construction convenience, design this steel with a lap joint immediately above the footing.

The design of vertical steel in concrete retaining walls is discussed in Chapter 24. Design procedures for other walls are beyond the scope of this book.

Connect wood-frame walls to the footing using anchor bolts, as shown in Figure 9.28. Normally, standard building code criteria govern the size and spacing of these bolts. For example, the *Uniform Building Code* (ICBO, 1997) specifies 1/2 in (12 mm) nominal diameter bolts embedded at least 7 in (175 mm) into the concrete. It also specifies bolt spacings of no more than 6 ft (1.8 m) on center.

Sometimes we must supply a higher capacity connection between wood frame walls and footings, especially when large uplift loads are anticipated. Steel holdown brackets, such as that shown in Figure 9.29, are useful for these situations.

Many older wood-frame buildings have inadequate connections between the structure and its foundation. Figure 9.30 shows one such structure that literally fell off its foundation during the 1989 Loma Prieta Earthquake in Northern California.

Some wood-frame buildings have failed during earthquakes because the *cripple walls* collapsed. These are short wood-frame walls that connect the foundation to the floor. They may be retrofitted by installing plywood shear panels or by using diagonal steel bracing (Shepherd and Delos-Santos, 1991).

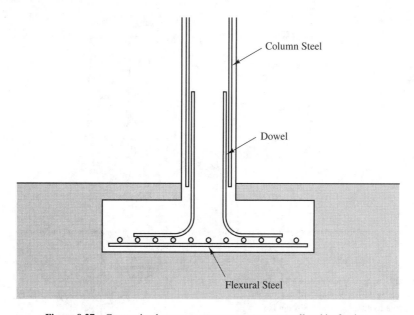

Column Steel

Dowel

Flexural Steel

Figure 9.27   Connection between a concrete or masonry wall and its footing.

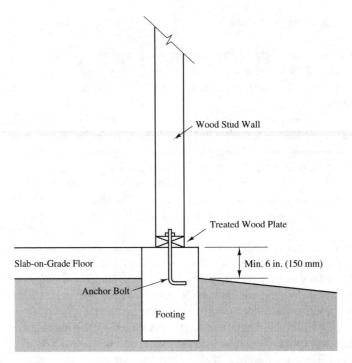

Wood Stud Wall

Treated Wood Plate

Slab-on-Grade Floor

Min. 6 in. (150 mm)

Anchor Bolt

Footing

Figure 9.28   Use of anchor bolts to connect a wood-frame wall to a continuous footing.

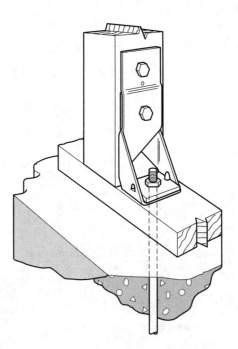

Figure 9.29   Use of steel holdown bracket
to connect a wood-frame wall with large uplift force to a footing (Simpson Strong-Tie
Co., Inc.).

Figure 9.30 House that separated from its foundation during the 1989 Loma Prieta Earthquake (Photo by C. Stover, U.S. Geological Survey).

QUESTIONS AND PRACTICE PROBLEMS

9.11 An 18-inch square concrete column carries dead and live compressive loads of 240 and 220 k, respectively. It is to be supported on a 8 ft wide 12 ft long rectangular spread footing. Select appropriate values for f_c' and f_y, then determine the required footing thickness and design the flexural reinforcing steel. Show the results of your design in a sketch.

9.12 The column described in Problem 9.11 is reinforced with 6 #8 bars. Design the dowels required to connect it with the footing, and show your design in a sketch.

9.13 A 400-mm diameter concrete column carrying a factored compressive load of 1500 kN is supported on a 600-mm thick, 2500-mm-wide square spread footing. It is reinforced with eight metric #19 bars. Using $f_c' = 18$ MPa and $f_y = 420$ MPa, design the dowels for this connection.

9.14 A 24-inch square concrete column carries a factored compressive load of 900 k and a factored shear load of 100 k. It is to be supported on a spread footing with $f_c' = 3000$ lb/ft^2 and $f_y = 60$ k/in^2. The column is reinforced with twelve #9 bars. Design the dowels for this connection.

9.15 A steel column with a square base plate is to be supported on a spread footing. The AISC factored design loads are: $P_u = 600$ k compression and $V_u = 105$ k. Design an anchor bolt system for this base plate and show your design in a sketch.

SUMMARY

Major Points

1. The plan dimensions and minimum embedment depth of a spread footing are governed by geotechnical concerns, and are determined using the unfactored loads.
2. The thickness and reinforcement of a spread footing are governed by structural concerns. Structural design is governed by the ACI code, which means these analyses are based on the factored load.
3. The structural design of spread footings must consider both shear and flexural failure modes. A shear failure consists of the column or wall punching through the footing, while a flexural failure occurs when the footing has insufficient cantilever strength.
4. Since we do not wish to use stirrups (shear reinforcement), we conduct the shear analysis first and select an effective depth, d, so the footing that provides enough shear resistance in the concrete to resist the shear force induced by the applied load. This analysis ignores the shear strength of the flexural steel.
5. Once the shear analysis is completed, we conduct a flexural analysis to determine the amount of steel required to provide the needed flexural strength. Since d is large, the required steel area will be small, and it is often governed by ρ_{min}.
6. For square footings, use the same flexural steel in both directions. Thus, the footing is reinforced twice.
7. For continuous footings, the lateral steel, if needed, is based on a flexural analysis. Use nominal longitudinal steel to resist nonuniformities in the load and to accommodate inconsistencies in the soil bearing pressure.
8. Design rectangular footings similar to square footings, but group a greater portion of the short steel near the center.
9. Practical minimum dimensions will often govern the design of lightly loaded footings.
10. The connection between the footing and the superstructure is very important. Use dowels to connect concrete or masonry structures. For steel columns and wood-frame walls, use anchor bolts. For wood posts, use specially manufactured brackets.

Vocabulary

Anchor bolts	Effective depth	Reinforcing bars
Critical section for bending	Factored load	Shear failure
Critical shear surface	Flexural failure	Two-way shear
Development length	Minimum steel	Unfactored load
Dowels	One-way shear	28-day compressive strength
	Post base	

COMPREHENSIVE QUESTIONS AND PRACTICE PROBLEMS

9.16 A 400-mm square concrete column reinforced with eight metric #19 bars carries vertical dead and live loads of 980 and 825 kN, respectively. It is to be supported on a 2.0 m × 3.5 m rectangular footing. The concrete in the footing will have $f_c' = 20$ MPa and $f_y = 400$ MPa. The building will have a slab-on-grade floor, so the top of the footing must be at least 150 mm below the finish floor elevation. Develop a complete structural design, including dowels, and show it in a sketch.

9.17 A 12-in wide masonry wall carries dead and live loads of 5 k/ft and 8 k/ft, respectively and is reinforced with #6 bars at 24 inches on center. This wall is to be supported on a continuous footing with $f_c' = 2000$ lb/in^2 and $f_y = 60$ k/in^2. The underlying soil has an allowable bearing pressure of 3000 lb/ft^2. Develop a complete structural design for this footing, including dowels, and show your design in a sketch.

10

Mats

The second type of shallow foundation is the *mat foundation*, as shown in Figure 10.1. A mat is essentially a very large spread footing that usually encompasses the entire footprint of the structure. They are also known as *raft foundations*.

Foundation engineers often consider mats when dealing with any of the following conditions:

- The structural loads are so high or the soil conditions so poor that spread footings would be exceptionally large. As a general rule of thumb, if spread footings would cover more than about one-third of the building footprint area, a mat or some type of deep foundation will probably be more economical.
- The soil is very erratic and prone to excessive differential settlements. The structural continuity and flexural strength of a mat will bridge over these irregularities. The same is true of mats on highly expansive soils prone to differential heaves.
- The structural loads are erratic, and thus increase the likelihood of excessive differential settlements. Again, the structural continuity and flexural strength of the mat will absorb these irregularities.

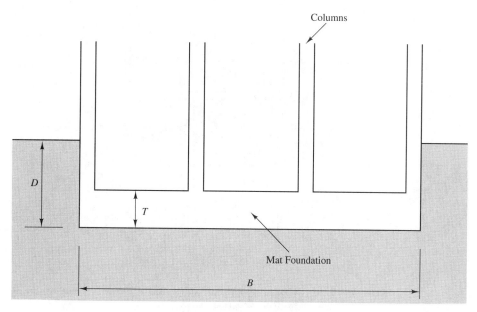

Figure 10.1   A mat foundation supported directly on soil.

• Lateral loads are not uniformly distributed through the structure and thus may cause differential horizontal movements in spread footings or pile caps. The continuity of a mat will resist such movements.

• The uplift loads are larger than spread footings can accommodate. The greater weight and continuity of a mat may provide sufficient resistance.

• The bottom of the structure is located below the groundwater table, so waterproofing is an important concern. Because mats are monolithic, they are much easier to waterproof. The weight of the mat also helps resist hydrostatic uplift forces from the groundwater.

Many buildings are supported on mat foundations, as are silos, chimneys, and other types of tower structures. Mats are also used to support storage tanks and large machines. Typically, the thickness, T, is 1 to 2 m (3–6 ft), so mats are massive structural elements. The seventy five story Texas Commerce Tower in Houston is one of the largest mat-supported structures in the world. Its mat is 3 m (9 ft 9 in) thick and is bottomed 19.2 m (63 ft) below the street level.

Although most mat foundations are directly supported on soil, sometimes engineers use pile- or shaft-supported mats, as shown in Figure 10.2. These foundations are often called *piled rafts*, and they are hybrid foundations that combine features of both mat and deep foundations. Pile- and shaft-supported mats are discussed in Section 11.9.

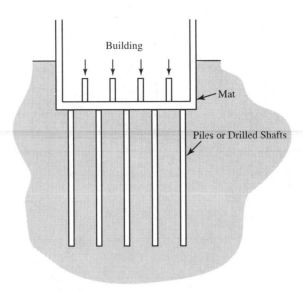

Figure 10.2 A pile- or shaft-supported mat foundation. The deep foundations are not necessarily distributed evenly across the mat.

Various methods have been used to design mat foundations. We will divide them into two categories: Rigid methods and nonrigid methods.

10.1 RIGID METHODS

The simplest approach to structural design of mats is the *rigid method* (also known as the *conventional method* or the *conventional method of static equilibrium*) (Teng, 1962). This method assumes the mat is much more rigid than the underlying soils, which means any distortions in the mat are too small to significantly impact the distribution of bearing pressure. Therefore, the magnitude and distribution of bearing pressure depends only on the applied loads and the weight of the mat, and is either uniform across the bottom of the mat (if the normal load acts through the centroid and no moment load is present) or varies linearly across the mat (if eccentric or moment loads are present) as shown in Figure 10.3. This is the same simplifying assumption used in the analysis of spread footings, as shown in Figure 5.10e.

This simple distribution makes it easy to compute the flexural stresses and deflections (differential settlements) in the mat. For analysis purposes, the mat becomes an inverted and simply loaded two-way slab, which means the shears, moments, and deflections may be easily computed using the principles of structural mechanics. The engineer can then select the appropriate mat thickness and reinforcement.

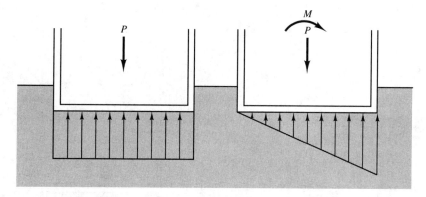

Figure 10.3   Bearing pressure distribution for rigid method.

Although this type of analysis is appropriate for spread footings, it does not accurately model mat foundations because the width-to-thickness ratio is much greater in mats, and the assumption of rigidity is no longer valid. Portions of a mat beneath columns and bearing walls settle more than the portions with less load, which means the bearing pressure will be greater beneath the heavily-loaded zones, as shown in Figure 10.4. This redistribution of bearing pressure is most pronounced when the ground is stiff compared to the mat, as shown in Figure 10.5, but is present to some degree in all soils.

Because the rigid method does not consider this redistribution of bearing pressure, it does not produce reliable estimates of the shears, moments, and deformations in the mat. In addition, even if the mat was perfectly rigid, the simplified bearing pressure distributions in Figure 10.3 are not correct—in reality, the bearing pressure is greater on the edges and smaller in the center than shown in this figure.

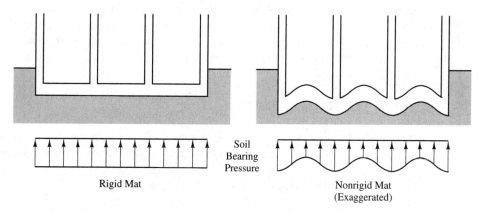

Figure 10.4   The rigid method assumes there are no flexural deflections in the mat, so the distribution of soil bearing pressure is simple to define. However, these deflections are important because they influence the bearing pressure distribution.

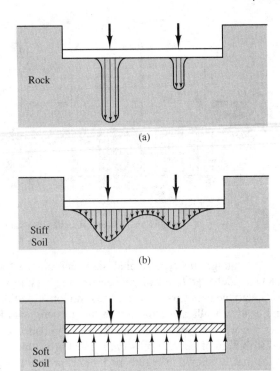

Figure 10.5 Distribution of bearing pressure under a mat foundation; (a) on bedrock or very hard soil; (b) on stiff soil; (c) on soft soil (Adapted from Teng, 1962).

10.2 NONRIGID METHODS

We overcome the inaccuracies of the rigid method by using analyses that consider deformations in the mat and their influence on the bearing pressure distribution. These are called *nonrigid methods*, and produce more accurate values of mat deformations and stresses. Unfortunately, nonrigid analyses also are more difficult to implement because they require consideration of *soil-structure interaction* and because the bearing pressure distribution is not as simple.

Coefficient of Subgrade Reaction

Because nonrigid methods consider the effects of local mat deformations on the distribution of bearing pressure, we need to define the relationship between settlement and bearing pressure. This is usually done using the *coefficient of subgrade reaction*, k_s (also known as the *modulus of subgrade reaction*, or the *subgrade modulus*):

$$k_s = \frac{q}{\delta}$$

(10.1)

Where:

k_s = coefficient of subgrade reaction

q = bearing pressure

δ = settlement

The coefficient k_s has units of force per length cubed. Although we use the same units to express unit weight, k_s is not the same as the unit weight and they are not numerically equal.

The interaction between the mat and the underlying soil may then be represented as a "bed of springs," each with a stiffness k_s per unit area, as shown in Figure 10.6. Portions of the mat that experience more settlement produce more compression in the "springs," which represents the higher bearing pressure, whereas portions that settle less do not compress the springs as far and thus have less bearing pressure. The sum of these spring forces must equal the applied structural loads plus the weight of the mat:

$$\Sigma P + W_f - u_D = \int q\,dA = \int \delta\,k_s\,dA \qquad (10.2)$$

Where:

ΣP = sum of structural loads acting on the mat

W_f = weight of the mat

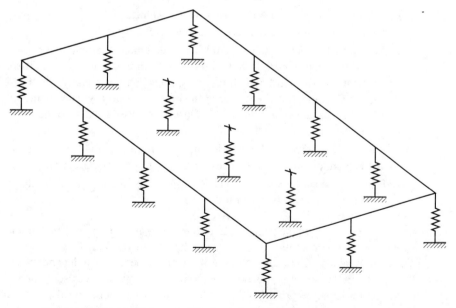

Figure 10.6 The coefficient of subgrade reaction forms the basis for a "bed of springs" analogy to model the soil-structure interaction in mat foundations.

u_D = pore water pressure along base of the mat

q = bearing pressure between mat and soil

A = mat–soil contact area

δ = settlement at a point on the mat

This method of describing bearing pressure is called a *soil-structure interaction analysis* because the bearing pressure depends on the mat deformations, and the mat deformations depend on the bearing pressure.

Winkler Method

The "bed of springs" model is used to compute the shears, moments, and deformations in the mat, which then become the basis for developing a structural design. The earliest use of these "springs" to represent the interaction between soil and foundations has been attributed to Winkler (1867), so the analytical model is sometimes called a *Winkler foundation* or the *Winkler method*. It also is known as a *beam on elastic foundation* analysis.

In its classical form the Winkler method assumes each "spring" is linear and acts independently from the others, and that all of the springs have the same k_s. This representation has the desired effect of increasing the bearing pressure beneath the columns, and thus is a significant improvement over the rigid method. However, it is still only a coarse representation of the true interaction between mats and soil (Hain and Lee, 1974; Horvath, 1983), and suffers from many problems, including the following:

1. The load-settlement behavior of soil is nonlinear, so the k_s value must represent some equivalent linear function, as shown in Figure 10.7.

2. According to this analysis, a uniformly loaded mat underlain by a perfectly uniform soil, as shown in Figure 10.8, will settle uniformly into the soil (i.e., there will be no differential settlement) and all of the "springs" will be equally compressed. In reality, the settlement at the center of such a mat will be greater than that along the edges, as discussed in Chapter 7. This is because the $\Delta\sigma_z$ values in the soil are greater beneath the center.

3. The "springs" should not act independently. In reality, the bearing pressure induced at one point on the mat influences more than just the nearest spring.

4. Primarily because of items 2 and 3, there is no single value of k_s that truly represents the interaction between soil and a mat.

Items 2 and 3 are the primary sources of error, and this error is potentially unconservative (i.e., the shears, moments, and deflections in the mat may be greater than those predicted by Winkler). The heart of these problems is the use of independent springs in the Winkler model. In reality, a load at one point on the mat induces settlement both at that point and in the adjacent parts of the mat, which is why a uniformly mat exhibits dish-shaped settlement, not the uniform settlement predicted by Winkler.

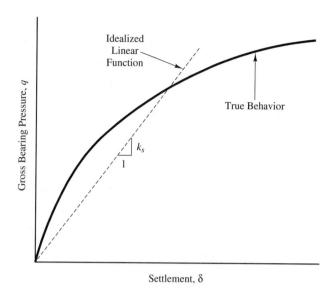

Figure 10.7 The q-δ relationship is nonlinear, so ks must represent some "equivalent" linear function.

Coupled Method

The next step up from a Winkler analysis is to use a *coupled method,* which uses additional springs as shown in Figure 10.9. This way the vertical springs no longer act independently, and the uniformly loaded mat of Figure 10.8 exhibits the desired dish shape. In principle, this approach is more accurate than the Winkler method, but it is not clear how to select the k_s values for the coupling springs, and it may be necessary to develop custom software to implement this analysis.

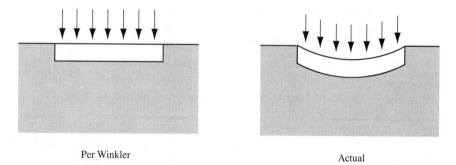

Per Winkler

Actual

Figure 10.8 Settlement of a uniformly-loaded mat on a uniform soil: (a) per Winkler analysis, (b) actual.

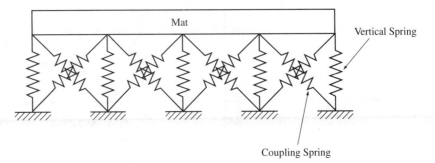

Figure 10.9   Modeling of soil-structure interaction using coupled springs.

Pseudo-Coupled Method

The *pseudo-coupled method* (Liao, 1991; Horvath, 1993) is an attempt to overcome the lack of coupling in the Winkler method while avoiding the difficulties of the coupled method. It does so by using "springs" that act independently, but have different k_s values depending on their location on the mat. To properly model the real response of a uniform soil, the "springs" along the perimeter of the mat should be stiffer than those in the center, thus producing the desired dish-shaped deformation in a uniformly-loaded mat. If concentrated loads, such as those from columns, also are present, the resulting mat deformations are automatically superimposed on the dish-shape.

Model studies indicate that reasonable results are obtained when k_s values along the perimeter of the mat are about twice those in the center (ACI, 1993). We can implement this in a variety of ways, including the following:

1. Divide the mat into two or more concentric zones, as shown in Figure 10.10. The innermost zone should be about half as wide and half as long as the mat.
2. Assign a k_s value to each zone. These values should progressively increase from the center such that the outermost zone has a k_s about twice as large at the innermost zone. Example 10.1 illustrates this technique.
3. Evaluate the shears, moments, and deformations in the mat using the Winkler "bed of springs" analysis, as discussed later in this chapter.
4. Adjust the mat thickness and reinforcement as needed to satisfy strength and serviceability requirements.

ACI (1993) found the pseudo-coupled method produced computed moments 18 to 25 percent higher than those determined from the Winkler method, which is an indication of how unconservative Winker can be.

Most commercial mat design software uses the Winkler method to represent the soil-structure interaction, and these software packages usually can accommodate the

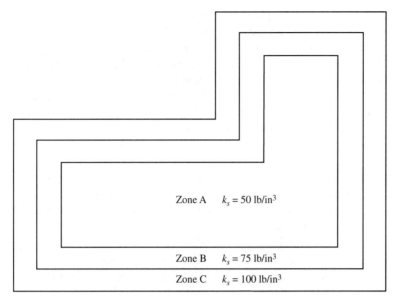

Figure 10.10   A typical mat divided into zones for a pseudo-coupled analysis. The coefficient of subgrade reaction, k_s, progressively increases from the innermost zone to the outermost zone.

pseudo-coupled method. Given the current state of technology and software availability, this is probably the most practical approach to designing most mat foundations.

Multiple-Parameter Method

Another way of representing soil-structure interaction is to use the *multiple parameter method* (Horvath, 1993). This method replaces the independently-acting linear springs of the Winkler method (a single-parameter model) with springs and other mechanical elements (a multiple-parameter model). These additional elements define the coupling effects.

The multiple-parameter method bypasses the guesswork involved in distributing the k_s values in the psuedo-coupled method because coupling effects are inherently incorporated into the model, and thus should be more accurate. However, it has not yet been implemented into readily-available software packages. Therefore, this method is not yet ready to be used on routine projects.

Finite Element Method

All of the methods discussed thus far attempt to model a three-dimensional soil by a series of one-dimensional springs. They do so in order to make the problem simple enough to perform the structural analysis. An alternative method would be to use a three-dimensional mathematical model of both the mat and the soil, or perhaps the mat, soil, and superstructure. This can be accomplished using the *finite element method.*

This analysis method divides the soil into a network of small elements, each with defined engineering properties and each connected to the adjacent elements in a specified way. The structural and gravitational loads are then applied and the elements are stressed and deformed accordingly. This, in principle, should be an accurate representation of the mat, and should facilitate a precise and economical design.

Unfortunately, such analyses are not yet practical for routine design problems because:

1. A three-dimensional finite element model requires tens of thousands or perhaps hundreds of thousands of elements, and thus place corresponding demands on computer resources. Few engineers have access to computers that can accommodate such intensive analyses.
2. It is difficult to determine the required soil properties with enough precision, especially at sites where the soils are highly variable. In other words, the analysis method far outweighs our ability to input accurate parameters.

Nevertheless, this approach may become more usable in the future, especially as increasingly powerful computers become more widely available.

This method should not be confused with structural analysis methods that use two-dimensional finite elements to model the mat and Winkler springs to model the soil. Such methods require far less computational resources, and are widely used. We will discuss this use of finite element analyses in Section 10.4.

10.3   DETERMINING THE COEFFICIENT OF SUBGRADE REACTION

Most mat foundation designs are currently developed using either the Winkler method or the pseudo-coupled method, both of which depend on our ability to define the coefficient of subgrade reaction, k_s. Unfortunately, this task is not as simple as it might first appear because k_s is not a fundamental soil property. Its magnitude also depends on many other factors, including the following:

- **The width of the loaded area**—A wide mat will settle more than a narrow one with the same q because it mobilizes the soil to a greater depth as shown in Figure 8.2. Therefore, each has a different k_s.
- **The shape of the loaded area**—The stresses below long narrow loaded areas are different from those below square loaded areas as shown in Figure 7.2. Therefore, k_s will differ.
- **The depth of the loaded area below the ground surface**—At greater depths, the change in stress in the soil due to q is a smaller percentage of the initial stress, so the settlement is also smaller and k_s is greater.

- **The position on the mat**—To model the soil accurately, k_s needs to be larger near the edges of the mat and smaller near the center.
- **Time**—Much of the settlement of mats on deep compressible soils will be due to consolidation and thus may occur over a period of several years. Therefore, it may be necessary to consider both short-term and long-term cases.

Actually, there is no single k_s value, even if we could define these factors because the q-δ relationship is nonlinear and because neither method accounts for interaction between the springs.

Engineers have tried various techniques of measuring or computing k_s. Some rely on plate load tests to measure k_s in situ. However, the test results must be adjusted to compensate for the differences in width, shape, and depth of the plate and the mat. Terzaghi (1955) proposed a series of correction factors, but the extrapolation from a small plate to a mat is so great that these factors are not very reliable. Plate load tests also include the dubious assumption that the soils within the shallow zone of influence below the plate are comparable to those in the much deeper zone below the mat. Therefore, plate load tests generally do not provide good estimates of k_s for mat foundation design.

Others have used derived relationships between k_s and the soil's modulus of elasticity, E (Vesić and Saxena, 1970; Scott, 1981). Although these relationships provide some insight, they too are limited.

Another method consists of computing the average mat settlement using the techniques described in Chapter 7 and expressing the results in the form of k_s using Equation 10.1. If using the pseudo-coupled method, use k_s values in the center of the mat that are less than the computed value, and k_s values along the perimeter that are greater. This should be done in such a way that the perimeter values are twice the central values, and the integral of all the values over the area of the mat is the same as the produce of the original k_s and the mat area. Example 10.1 describes this methodology.

Example 10.1

A structure is to be supported on a 30-m wide, 50-m long mat foundation. The average bearing pressure is 120 kPa. According to a settlement analysis conducted using the techniques described in Chapter 7, the average settlement, δ, will be 30 mm. Determine the design values of k_s to be used in a pseduo-coupled analysis.

Solution

Compute average k_s using Equation 10.1:

$$(k_s)_{avg} = \frac{q}{\delta} = \frac{120 \text{ kPa}}{0.030 \text{ m}} = 4000 \text{ kN/m}^3$$

Divide the mat into three zones, as shown in Figure 10.11, with $(k_s)_C = 2(k_s)_A$ and $(k_s)_B = 1.5(k_s)_A$

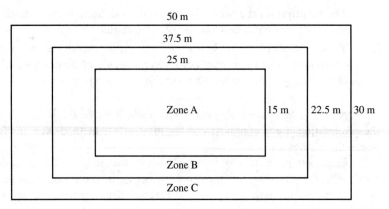

Figure 10.11   Mat foundation for Example 10.1.

Compute the area of each zone:

$$A_A = (25 \text{ m})(15 \text{ m}) = 375 \text{ m}^2$$

$$A_B = (37.5 \text{ m})(22.5 \text{ m}) - (25 \text{ m})(15 \text{ m}) = 469 \text{ m}^2$$

$$A_C = (50 \text{ m})(30 \text{ m}) - (37.5 \text{ m})(22.5 \text{ m}) = 656 \text{ m}^2$$

Compute the design k_s values:

$$A_A (k_s)_A + A_B (k_s)_B + A_C (k_s)_C = (A_A + A_B + A_C) (k_s)_{avg}$$

$$375 (k_s)_A + 469 (1.5)(k_s)_A + 656 (2)(k_s)_A = 1500 (k_s)_{avg}$$

$$2390 (k_s)_A = 1500 (k_s)_{avg}$$

$$(k_s)_A = 0.627 (k_s)_{avg}$$

$$(k_s)_A = (0.627)(4000 \text{ kN/m}^3) = \textbf{2510 kN/m}^3 \qquad \Leftarrow Answer$$

$$(k_s)_B = (1.5)(0.627)(4000 \text{ kN/m}^3) = \textbf{3765 kN/m}^3 \qquad \Leftarrow Answer$$

$$(k_s)_C = (2)(0.627)(4000 \text{ kN/m}^3) = \textbf{5020 kN/m}^3 \qquad \Leftarrow Answer$$

Because it is so difficult to develop accurate k_s values, it may be appropriate to conduct a parametric studies to evaluate its effect on the mat design. ACI (1993) suggests varying k_s from one-half the computed value to five or ten times the computed value, and basing the structural design on the worst case condition.

This wide range in k_s values will produce proportional changes in the computed total settlement. However, we ignore these total settlement computations because they are not reliable anyway, and compute it using the methods described in Chapter 7. These changes in k_s have much less impact on the shears, moments, and deflections in the mat, and thus have only a small impact on the structural design.

10.4   STRUCTURAL DESIGN

General Methodology

The structural design of mat foundations must satisfy both strength and serviceability requirements. This requires two separate analyses, as follows:

Step 1:   Evaluate the strength requirements using the factored loads (Equations 2.7–2.15) and LRFD design methods (which ACI calls *ultimate strength design*). The mat must have a sufficient thickness, T, and reinforcement to safely resist these loads. As with spread footings, T should be large enough that no shear reinforcement is needed.

Step 2:   Evaluate mat deformations (which is the primary serviceability requirement) using the unfactored loads (Equations 2.1–2.4). These deformations are the result of concentrated loading at the column locations, possible non-uniformities in the mat, and variations in the soil stiffness. In effect, these deformations are the equivalent of differential settlement. If they are excessive, then the mat must be made stiffer by increasing its thickness.

Closed-Form Solutions

When the Winkler method is used (i.e., when all "springs" have the same k_s) and the geometry of the problem can be represented in two-dimensions, it is possible to develop closed-form solutions using the principles of structural mechanics (Scott, 1981; Hetényi, 1974). These solutions produce values of shear, moment, and deflection at all points in the idealized foundation. When the loading is complex, the principle of superposition may be used to divide the problem into multiple simpler problems.

These closed-form solutions were once very popular, because they were the only practical means of solving this problem. However, the advent and widespread availability of powerful computers and the associated software now allows us to use other methods that are more precise and more flexible.

Finite Element Method

Today, most mat foundations are designed with the aid of a computer using the *finite element method* (*FEM*). This method divides the mat into hundreds or perhaps thousands of elements, as shown in Figure 10.12. Each element has certain defined dimensions, a specified stiffness and strength (which may be defined in terms of concrete and steel properties) and is connected to the adjacent elements in a specified way.

The mat elements are connected to the ground through a series of "springs," which are defined using the coefficient of subgrade reaction. Typically, one spring is located at each corner of each element.

The loads on the mat include the externally applied column loads, applied line loads, applied area loads, and the weight of the mat itself. These loads press the mat downward,

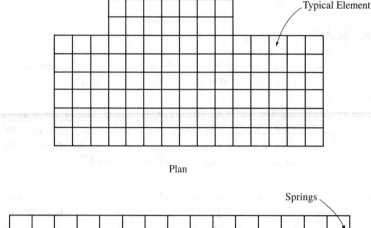

Plan

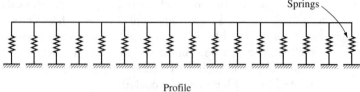

Profile

Figure 10.12   Use of the finite element method to analyze mat foundations. The mat is divided into a series of elements which are connected in a specified way. The elements are connected to the ground through a "bed of springs."

and this downward movement is resisted by the soil "springs." These opposing forces, along with the stiffness of the mat, can be evaluated simultaneously using matrix algebra, which allows us to compute the stresses, strains, and distortions in the mat. If the results of the analysis are not acceptable, the design is modified accordingly and reanalyzed.

This type of finite element analysis does not consider the stiffness of the superstructure. In other words, it assumes the superstructure is perfectly flexible and offers no resistance to deformations in the mat. This is conservative.

The finite element analysis can be extended to include the superstructure, the mat, and the underlying soil in a single three-dimensional finite element model. This method would, in principle, be a more accurate model of the soil-structure system, and thus may produce a more economical design. However, such analyses are substantially more complex and time-consuming, and it is very difficult to develop accurate soil properties for such models. Therefore, these extended finite element analyses are rarely performed in practice.

10.5   TOTAL SETTLEMENT

The bed of springs analyses produce a computed total settlement. However, this value is unreliable and should not be used for design. These analyses are useful only for computing shears, moments, and deformations (differential settlements) in the mat. Total settlement should be computed using the methods described in Chapter 7.

10.6   BEARING CAPACITY

Because of their large width, mat foundations on sands and gravels do not have bearing capacity problems. However, bearing capacity might be important in silts and clays, especially if undrained conditions prevail. The Fargo Grain Silo failure described in Chapter 6 is a notable example of a bearing capacity failure in a saturated clay.

We can evaluate bearing capacity using the analysis techniques described in Chapter 6. It is good practice to design the mat so the bearing pressure at all points is less than the allowable bearing capacity.

SUMMARY

Major Points

1. Mat foundations are essentially large spread footings that usually encompass the entire footprint of a structure. They are often an appropriate choice for structures that are too heavy for spread footings.
2. The analysis and design of mats must include an evaluation of the flexural stresses and must provide sufficient flexural strength to resist these stresses.
3. The oldest and simplest method of analyzing mats is the rigid method. It assumes that the mat is much more rigid than the underlying soil, which means the magnitude and distribution of bearing pressure is easy to determine. This means the shears, moment, and deformations in the mat are easily determined. However, this method is not an accurate representation because the assumption of rigidity is not correct.
4. Nonrigid analyses are superior because they consider the flexural deflections in the mat and the corresponding redistribution of the soil bearing pressure.
5. Nonrigid methods must include a definition of soil-structure interaction. This is usually done using a "bed of springs" analogy, with each spring having a linear force-displacement function as defined by the coefficient of subgrade reaction, k_s.
6. The simplest and oldest nonrigid method is the Winkler method, which uses independent springs, all of which have the same k_s. This method is an improvement over rigid analyses, but still does not accurately model soil–structure interaction, primarily because it does not consider coupling effects.
7. The coupled method is an extension of the Winkler method that considers coupling between the springs.
8. The pseudo-coupled method uses independent springs, but adjusts the k_s values to implicitly account for coupling effects.
9. The multiple parameter and finite element methods are more advanced ways of describing soil-structure interaction.
10. The coefficient of subgrade reaction is difficult to determine. Fortunately, the mat design is often not overly sensitive to global changes in k_s. Parametric studies are often appropriate.

11. If the Winkler method is used to describe soil–structure interaction, and the mat geometry is not too complex, the structural analysis may be performed using closed-form solutions. However, these methods are generally considered obsolete.

12. Most structural analyses are performed using numerical methods, especially the finite element method. This method uses finite elements to model the mat, and defines soil–structure interaction using the Winker or pseudo-coupled models. In principle, it also could use the multiple parameter model.

13. A design could be based entirely on a three-dimensional finite element analysis that includes the soil, mat, and superstructure. However, such analyses are beyond current practices, mostly because they are difficult to set up and require especially powerful computers.

14. The total settlement is best determined using the methods described in Chapter 7. Do not use the coefficient of subgrade reaction to determine total settlement.

15. Bearing capacity is not a problem with sands and gravels, but can be important in silts and clays. It should be checked using the methods described in Chapter 6.

Vocabulary

Beam on elastic foundation	Mat foundation	Rigid method
Bed of springs	Multiple parameter method	Shaft-supported mat
Coefficient of subgrade reaction	Nonrigid method	Soil–structure interaction
	Pile-supported mat	Winkler method
Coupled method	Pseudo-coupled method	
Finite element method	Raft foundation	

COMPREHENSIVE QUESTIONS AND PRACTICE PROBLEMS

10.1 Explain the reasoning behind the statement in Section 10.6: "Because of their large width, mat foundations on sands and gravels do not have bearing capacity problems."

10.2 How has the development of powerful and inexpensive digital computers affected the analysis and design of mat foundations? What changes do you expect in the future as this trend continues?

10.3 A mat foundation supports forty two columns for a building. These columns are spaced on a uniform grid pattern. How would the moments and differential settlements change if we used a nonrigid analysis with a constant k_s in lieu of a rigid analysis?

10.4 According to a settlement analysis conducted using the techniques described in Chapter 7, a certain mat will have a total settlement of 2.1 inches if the average bearing pressure is 5500 lb/ft^2. Compute the average k_s and express your answer in units of lb/in^3.

10.5 A 25-m diameter cylindrical water storage tank is to be supported on a mat foundation. The weight of the tank and its contents will be 50,000 kN and the weight of the mat will be 12,000 kN. According to a settlement analysis conducted using the techniques described in Chapter 7, the total settlement will be 40 mm. The groundwater table is at a depth of 5 m below the bottom of the mat. Using the pseudo-coupled method, divide the mat into zones and compute k_s for each zone. Then indicate the high-end and low-end values of k_s that should be used in the analysis.

10.6 An office building is to be supported on 150-ft $\times$ 300-ft mat foundation. The sum of the column loads plus the weight of the mat will be 90,000 k. According to a settlement analysis conducted using the techniques described in Chapter 7, the total settlement will be 1.8 inches. The groundwater table is at a depth of 10 ft below the bottom of the mat. Using the pseudo-coupled method, divide the mat into zones and composite each zone. Then indicate the high-end and low-end valves of k_s that should be used in the analysis.

Part C

Deep Foundation
Analysis and Design

11

Deep Foundations

For these reasons, Caesar determined to cross the Rhine, but a crossing by means of boats seemed to him both too risky and beneath his dignity as a Roman commander. Therefore, although construction of a bridge presented very great difficulties on account of the breadth, depth and swiftness of the stream, he decided that he must either attempt it or give up the idea of a crossing. The method he adopted in building the bridge was as follows. He took a pair of piles a foot and a half thick, slightly pointed at the lower ends and of a length adapted to the varying depth of the river, and fastened them together two feet apart. These he lowered into the river with appropriate tackle, placed them in position at right angles to the bank, and drove them home with pile drivers, not vertically as piles are generally fixed, but obliquely, inclined in the direction of the current. Opposite these, forty feet lower down the river, another pair of piles was planted, similarly fixed together, and inclined in the opposite direction to the current. The two pairs were then joined by a beam two feet wide, whose ends fitted exactly into the spaces between the two piles forming each pair . . . A series of these piles and transverse beams was carried right across the stream and connected by lengths of timber running in the direction of the bridge . . . Ten days after the collection of the timber had begun, the work was completed and the army crossed over.

From *The Conquest of Gaul*, translated by S.A. Handford, revised by Jane F. Gardner (Penguin Classics 1951, 1982) copyright © The Estate of S.A. Handford, 1951, revisions copyright © Jane F. Gardner, 1982). Used with permission.

Engineers prefer to use spread footings wherever possible, because they are simple and inexpensive to build. However, we often encounter situations where spread footings are not the best choice. Examples include the following:

- The upper soils are so weak and/or the structural loads so high that spread footings would be too large. A good rule-of-thumb for buildings is that spread footings cease to be economical when the total plan area of the footings exceeds about one-third of the building footprint area.
- The upper soils are subject to scour or undermining. This would be especially important with foundations for bridges.
- The foundation must penetrate through water, such as those for a pier.
- A large uplift capacity is required (the uplift capacity of a spread footing is limited to its dead weight).
- A large lateral load capacity is required.
- There will be a future excavation adjacent to the foundation, and this excavation would undermine shallow foundations.

In some of these circumstances, a mat foundation may be appropriate, but the most common alternative to spread footings is some type of *deep foundation.*

A deep foundation is one that transmits some or all of the applied load to soils well below the ground surface, as shown in Figure 11.1. These foundations typically extend to depths on the order of 15 m (50 ft) below the ground surface, but they can be much longer, perhaps extending as deep as 45 m (150 ft). Even greater lengths have been used in some offshore structures, such as oil drilling platforms. Since soils usually improve with depth, and this method mobilizes a larger volume of soil, deep foundations are often able to carry very large loads.

11.1   TYPES OF DEEP FOUNDATIONS AND DEFINITIONS

Engineers and contractors have developed many types of deep foundations, each of which is best suited to certain loading and soil conditions. Unfortunately, people use many different names to identify these designs. Different individuals often use the same terms to mean different things and different terms to mean the same thing. This confusion reigns in both verbal and written communications, and is often the source of misunderstanding, especially to the newcomer. This book uses terms that appear to be most commonly used and understood. This classification system is based primarily on the methods of construction, as follows:

- *Piles* are constructed by prefabricating slender prefabricated members and driving or otherwise forcing them into the ground.
- *Drilled shafts* are constructed by drilling a slender cylindrical hole into the ground, inserting reinforcing steel, and filling it with concrete.
- *Caissons* are prefabricated boxes or cylinders that are sunk into the ground to some desired depth, then filled with concrete. Some engineers use the term "caisson" to

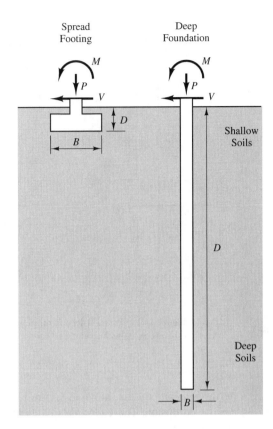

Figure 11.1 Deep foundations transfer most of the applied structural loads to deeper soil strata.

describe drilled shafts, so this is one of the more confusing terms in foundation engineering.

- *Mandrel-driven thin shells filled with concrete* consist of thin corrugated steel shells that are driven into the ground using a mandrel, then filled with concrete.
- *Auger-cast piles* are constructed by drilling a slender cylindrical hole into the ground using a hollow-stem auger, then pumping grout through the auger while it is slowly retracted.
- *Pressure-injected footings* use cast-in-place concrete that is rammed into the soil using a drop hammer.
- *Anchors* include several different kinds of deep foundations that are specifically designed to resist uplift loads.

The vast majority of deep foundation designs use one of these seven types. Other types also are available, but they are not often used and are beyond the scope of this book.

The various parts of deep foundations also have different names, which is another source of confusion (Fellenius, 1996). The upper end has many names, including "top," "butt," and "head," while those for the lower end include "tip," "toe," "base," "end," "point," and "bottom." Many of these terms can easily be misunderstood. We will use the

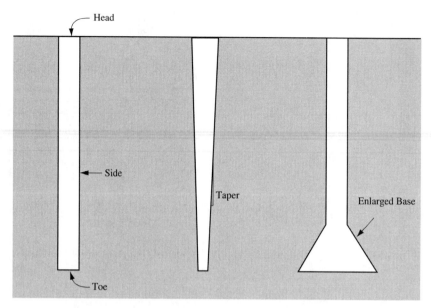

Figure 11.2   Parts of a deep foundation (a) straight foundations; (b) tapered foundations; (c) foundations with an enlarged base.

terms *head* and *toe*, as shown in Figure 11.2a, because they appear to have the least potential for confusion. These terms also are easy to remember by simply comparing the pile to a human body.

The exterior surface along the side of deep foundations is usually called the "side" or the "skin," either of which is generally acceptable. We will use the term *side*.

Most deep foundations are *straight*, which means the cross section is constant with depth, but some are *tapered*, as shown in Figure 11.2b. Others include an *enlarged base*, as shown in Figure 11.2c. The *axis* is a line through the center of the foundation and is parallel to its longest dimension.

11.2   LOAD TRANSFER

To properly select and design deep foundations, we need to understand how they transfer structural loads into the ground. These load-transfer mechanisms depend on the type of load being imposed, so it is convenient to divide structural loads into two categories:

- *Axial loads* are applied tensile or compressive loads that act parallel to the axis of the foundation, as shown in Figures 11.3a and 11.3b. These loads induce either tension or compression in the foundation, but no flexure. If the foundation is vertical, then the axial load is equal to the vertical applied load.

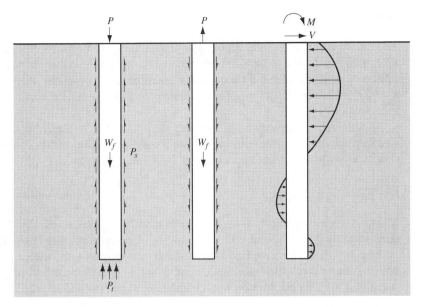

Figure 11.3   Transfer of structural loads from a deep foundation into the ground: (a) axial compressive loads; (b) axial tension loads; (c) lateral loads.

- *Lateral loads* are those that act perpendicular to the axis of the foundation, as shown in Figure 11.3c. This category includes shear and moment loads, V and M.

Axial Loads

An axial load may be either compressive (downward) or tensile (uplift). When it is compressive, deep foundations resist the load using both *side friction resistance* and *toe bearing resistance*, as shown in Figure 11.3a. However, when the load is tensile, the resistance is caused by side friction and the weight of the foundation, as shown in Figure 11.3b. In deep foundations with an enlarged base, uplift loads are also resisted by bearing along the ceiling of the enlarged base.

Lateral Loads

Lateral loads produce both shear and moment in a deep foundation, as shown in Figure 11.3c. These shears and moments produce lateral deflections in the foundation, which in turn mobilize lateral resistances in the adjacent soil. The magnitudes of these lateral deflections and resistances, and the corresponding load-bearing capacity of the foundation depend on the stiffness of both the soil and the foundation.

11.3   PILES

The first type of deep foundation is a *pile foundation*, which consists of long, slender, pre-fabricated structural members driven or otherwise inserted into the ground. Engineers use piles both on land and in the sea to support many kinds of structures. Piles are made from a variety of materials and in different diameters and lengths according to the needs of each project.

Although some engineers also use the word *pile* to describe certain types of cast-in-place deep foundations, we will use it only to describe prefabricated deep foundations. Cast-in-place methods are discussed later in this chapter.

History

Mankind has used pile foundations for more than 2000 years. Alexander the Great drove piles in the city of Tyre in 332 B.C., and the Romans used them extensively. Bridge builders in China during the Han Dynasty (200 B.C.–A.D. 200) also used piles. These early builders drove their piles into the ground using weights hoisted and dropped by hand (Chellis, 1961). By the Middle Ages, builders used pulleys and leverage to raise heavier weights.

Construction methods improved more quickly during the Industrial Revolution, especially when steam power became available. Larger and more powerful equipment was built, thus improving pile driving capabilities. These improvements continued throughout the twentieth century.

Pile materials also have become better. The early piles were always made of wood, and thus were limited in length and capacity. Fortunately, the advent of steel and reinforced concrete in the 1890s enabled the construction of larger and stronger piles, and better driving equipment made it possible to install them. Without these improved foundations, many of today's major structures would not have been possible.

Today, pile foundations can support very high loads, even in hostile environments. Perhaps the most impressive are those for offshore oil drilling platforms. These are as large as 10 ft (3 m) in diameter and must resist large lateral loads due to wind, wave, and earthquake forces.

Types of Piles

Most piles are now made from wood, concrete, or steel. Each material has its advantages and disadvantages and is best suited for certain applications. We must consider many factors when selecting a pile type, including the following:

- **The applied loads**—Some piles, such as timber, are best suited for low to medium loads, whereas others, such as steel, may be most cost-effective for heavy loads.
- **The required diameter**—Most pile types are available only in certain diameters.
- **The required length**—Highway shipping regulations and practical pile driver heights generally limit the length of pile segments to about 18 m (60 ft). Therefore,

longer piles must consist of multiple segments spliced together during driving. Some types of piles are easily spliced, whereas others are not.

- **The local availability of each pile type**—Some pile types may be abundant in certain geographic areas, whereas others may be scarce. This can significantly affect the cost of each type.
- **The durability of the pile material in a specific environment**—Certain environments may cause piles to deteriorate, as discussed in Chapter 2.
- **The anticipated driving conditions**—Some piles tolerate hard driving, while others are more likely to be damaged.

Timber Piles

Timber piles have been used for thousands of years and continue to be a good choice for many applications. They are made from the trunks of straight trees and resemble telephone poles, as shown in Figure 11.4. Because trees are naturally tapered, these piles are driven upside down, so the largest diameter is at the head, as shown in Figure 11.5.

Many different species of trees have been used to make timber piles. Today, most new timber piles driven in North America are either Southern pine or Douglas fir because these trees are tall and straight, and are abundant enough that the materials cost is low. They typically have head diameters in the range of 150 to 450 mm (6–18 in) and lengths between 6 and 20 m (20–60 ft), but greater lengths are sometimes available, up to 24 m (80 ft) in Southern pine and 38 m (125 ft) in Douglas fir. The branches and bark must be removed, and it is sometimes necessary to trim the pile slightly to give it a uniform taper. ASTM D25 gives detailed specifications.

Figure 11.4   Groups of timber piles. Those in the foreground have been cut to the final head elevation (National Timber Piling Council).

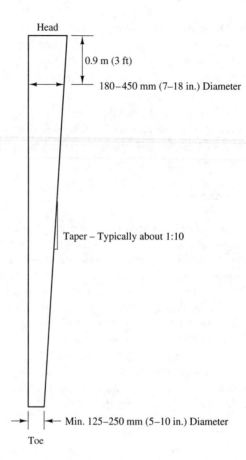

Head

0.9 m (3 ft)

180–450 mm (7–18 in.) Diameter

Taper – Typically about 1:10

Min. 125–250 mm (5–10 in.) Diameter

Figure 11.5   Typical timber pile.   Toe

Although it is possible to splice lengths of wood piling together to form longer piles, this is a slow and time-consuming process that makes the piles much more expensive. Therefore, if longer piles are necessary, use some other material.

Most timber piles are designed to carry downward axial loads of 100 to 400 kN (20–100 k). Their primary advantage is low construction cost, especially when suitable trees are available nearby. They are often used on waterfront structures because of their resistance to impact loads, such as those from ships.

When continually submerged, timber piles can have a very long life. For example, when the Campanile in Venice fell in 1902, its timber piles, which were driven in A.D. 900, were found in good condition and were reused (Chellis, 1962). However, when placed above the groundwater table, or in cyclic wetting conditions, timber piles are susceptible to decay, as discussed in Chapter 2. Therefore, they are nearly always treated with a preservative before installation.

When used in marine environments, timber piles are subject to attack from various marine organisms as well as abrasion from vessels and floating debris. In cooler waters (in North America, generally north of 40 degrees latitude), piles with heavy creosote treatment will usually remain serviceable for decades (ASCE, 1984). However, in warmer waters, biological attack is more of a problem and other chemical treatments are usually necessary. Even when treated, their usable life in such environments is often limited to about ten years.

These piles are also susceptible to damage during driving. Repeated hard hammer blows can cause *splitting* and *brooming* at the head and damage to the toe. It is often possible to control these problems by:

- Using lightweight hammers with appropriate cushions between the hammer and the pile
- Using steel bands near the butt (usually necessary only with Douglas fir)
- Using a steel shoe on the toe, as shown in Figure 11.6
- Predrilling (see discussion later in this section)

Nevertheless, even these measures are not always sufficient to prevent damage. Therefore, timber piles are best suited for light driving conditions, such as friction piles in loose sands and soft to medium clays. They are usually not good for dense or hard soils or as end bearing piles.

Steel Piles

By the 1890s, steel had become widely available and many structures were being built of this new material. The use of steel piling was a natural development. Today, steel piles are very common, especially on projects that require high-capacity foundations.

Figure 11.6   Use of steel toe points to reduce damage to timber piles during driving (Associated Pile and Fitting Corp.).

Because of their high strength and ductility, steel piles can be driven through hard soils and carry large loads. They also have the highest tensile strength of any major pile type, so they are especially attractive for applications with large applied tensile loads.

Steel piles are easy to splice, so they are often a good choice when the required length is greater than about 18 m (60 ft). The contractor simply drives the first section, then welds on the next one, and continues driving. Special steel splicers can make this operation faster and more efficient. Hunt (1987) reported the case of a spliced steel pile driven to the extraordinary depth of 210 m (700 ft). They are also easy to cut, which can be important with end-bearing piles driven to irregular rock surfaces.

Steel piles have the disadvantages of being expensive to purchase and noisy to drive. In certain environments, they may be subject to excessive corrosion, as discussed in Chapter 2.

H-Piles

Special rolled steel sections, known as HP sections, or simply *H-piles*, are made specifically to be used as piles. These sections are similar to WF (wide flange) shapes as shown in Figures 11.7 and 11.8. The primary difference is that the web is thinner than the flanges in wide flange members, while they have equal thicknesses in H-piles. Dimensions and

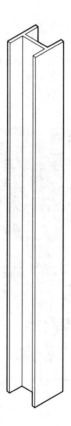

Figure 11.7   A steel H-pile.

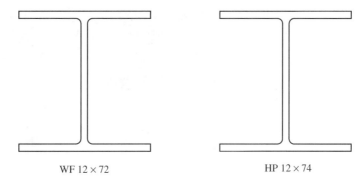

WF 12 × 72   HP 12 × 74

Figure 11.8   Comparison between typical wide flange (WF) and H-pile (HP) sections.

other relevant information for standard steel H-piles are listed in Table 12.1. These piles are typically 15 to 50 m (50–150 ft) long and carry working axial loads of 350 to 1800 kN 80–400 k.

H-piles are *small displacement piles* because they displace a relatively small volume of soil as they are driven. This, combined with their high strength, make them an excellent choice for hard driving conditions. They are often driven to bedrock and used as end bearing piles. If the pile will encounter hard driving, it may be necessary to use a hardened steel point to protect its toe, as shown in Figure 11.9.

Pipe Piles

Steel pipe sections are also commonly used as piles, as shown in Figure 11.10. They are typically 200 to 1000 mm (8–36 in) in diameter, 30 to 50 m (100–150 ft) long, and carry axial loads of 450 to 7000 kN (100–1500 k). A wide variety of diameters and wall thicknesses are available, and some engineers have even reclaimed used steel pipelines and

Figure 11.9   Hardened steel point attached to the toe of a steel H-pile to protect it during hard driving. (Associated Pile and Fitting Corp.)

Figure 11.10   A 16 inch (406 mm) diameter steel pipe pile.

used them as piles. Special sizes also can be manufactured as needed and pipe piles as large as 3 m (10 ft) in diameter with 75 mm (3 in) wall thickness have been used in offshore projects. Table 12.2 lists some of the more common sizes.

Pipe piles have a larger moment of inertia than H-piles, so they may be a better choice if large lateral loads are present.

Pipe piles may be driven with a *closed-end* or with an *open-end*. A closed-end pipe has a flat steel plate or a conical steel point welded to the toe. These are *large displacement piles* because they displace a large volume of soil. This increases the load capacity, but makes them more difficult to drive. Conversely, an open-end pipe has nothing blocking the toe and soil enters the pipe as it is being driven. The lower portion of small diameter open-end pipe piles usually becomes jammed with soil, thus forming a *soil plug*. Thus, an open-end pipe pile displaces less soil than a closed-end one, but more than an H-pile. Open-ended piles are primarily used in offshore construction.

Closed-end pipe piles can be inspected after driving because the inside is visible from the ground surface. Thus, it is possible to check for integrity and alignment.

Special steel pipe piles are also available, such as the *monotube pile*, which is tapered and has longitudinal flutes.

Figure 11.11 These steel forms are used to manufacture prestressed concrete piles. The prestressing cables, visible in the foreground, are in place and have been subjected to a tensile force. The spiral reinforcement also is in place. The next step will be to fill the forms with high quality concrete, cover them with a tarp, and steam-cure them overnight. The next day, the tension on the cables will be released, then the piles will be removed from the forms and allowed to cure.

Concrete Piles

Concrete piles are precast reinforced concrete members driven into the ground. This category does not include techniques that involve casting the concrete in the ground; They are covered later in this chapter.

Figure 11.11 shows steel molds used to manufacture precast prestressed concrete piles. This is usually done at special manufacturing facilities, then shipped to the construction site. Figure 11.12 shows completed piles ready to be driven.

Concrete piles usually have a square or octagonal cross section, as shown in Figure 11.13, although other shapes have been used (ACI, 1980). They are typically 250 to 600 mm (10–24 in) in diameter, 12 to 30 m (40–100 ft) long, and carry working axial loads of 450 to 3500 kN (100–800 k). A few nearshore projects have been built using much larger concrete piles.

Although conventionally-reinforced concrete piles were once common, prestressed piles have almost completely replaced them, at least in North America. These improved designs have much more flexural strength and are therefore less susceptible to damage during handling and driving. Prestressing is usually a better choice than post-tensioning because it allows piles to be cut, if necessary, without losing the prestress force.

Several methods are available to splice concrete piles, as shown in Figure 11.14. Although these techniques are generally more expensive than those for splicing steel

Figure 11.12   These 14-inch (356 mm) square prestressed concrete piles are stacked in a contractor's yard and are ready to be driven. The bars emerging from the end of these piles are conventional rebars that have been embedded in the end of the pile (they are not the reinforcing tendons). These rebars are used to structurally connect the pile with the pile cap.

piles, they can be cost-effective in some situations. However, unlike steel, concrete piles are difficult and expensive to cut. Therefore, they are best suited for use as friction piles that do not meet *refusal* during driving (refusal means that the pile cannot be driven any further, so it becomes necessary to cut off the upper portion) or as toe-bearing piles where the required length is uniform and predictable.

Concrete piles do not tolerate hard driving conditions as well as steel, and are more likely to be damaged during handling or driving. Nevertheless, concrete piles are very popular because they are often less expensive than steel piles, yet still have a large load capacity.

Composite Piles

A *composite pile* is one that uses two or more materials. For example, a steel pipe pile filled with concrete. Normal concrete piles are not considered to be composite piles even though they contain reinforcing steel.

Concrete-Filled Steel Pipe Piles

Sometimes steel pipe piles are filled with concrete after driving. These will have more up-lift capacity due to their greater weight, enhanced shear and moment capacity because of the strength of the concrete, and a longer useful life in corrosive environments. However,

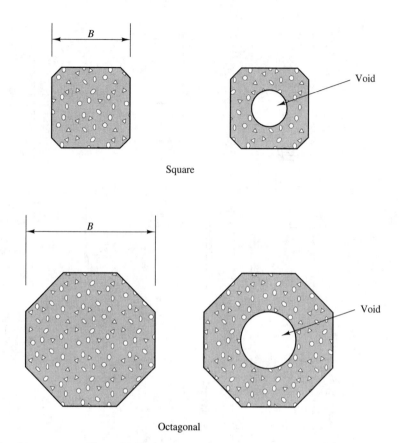

Square

Octagonal

Figure 11.13   Cross sections of typical concrete piles.

there is little, if any, usable increase in the downward load capacity because a pipe with sufficient wall thickness to withstand the driving stresses will probably have enough capacity to resist the applied downward loads. The net downward capacity may even be less because of the additional weight of the concrete in the pile.

Plastic-Steel Composite Piles

A plastic-steel composite pile consists of a steel pipe core surrounded by a plastic cover as shown in Figure 11.15, or steel rebars embedded in plastic. The plastic cover is typically made of recycled material, thus making this design attractive from a resource conservation perspective (Heinz, 1993).

   Plastic-steel composite piles have been used successfully in waterfront applications (see Figure 11.16), where their resistance to marine borers, decay, and abrasion along with their higher strength make them superior to timber piles. Although the materials cost for plastic-steel composites is higher, their longer life and resource conservation benefits make them an attractive alternative to timber piles.

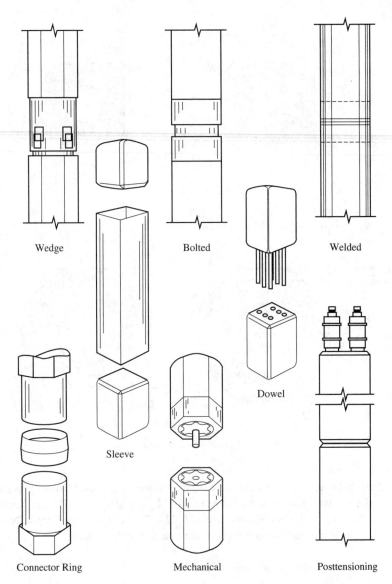

Wedge   Bolted   Welded

Dowel

Sleeve

Connector Ring   Mechanical   Posttensioning

Figure 11.14   Typical splices for concrete piles (Precast/Prestressed Concrete Institute).

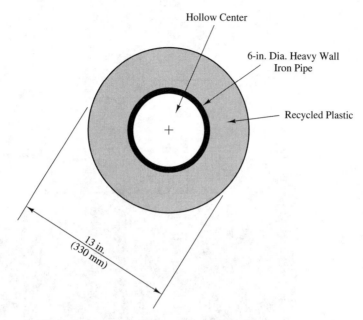

Figure 11.15   Cross section of a typical plastic-steel composite pile.

Construction Methods and Equipment

The construction of deep foundations is much more complex than that of shallow foundations, and the construction methods have a much greater impact on their performance.
Therefore, design engineers must understand how contractors build pile foundations.

Pile-Driving Rigs

Piles are installed using a *pile-driving rig* (or simply the *pile driver*). Its function is to
raise and temporarily support the pile while it is being driven and to support the pile hammer. Early rigs were relatively crude, but modern pile drivers, such as the ones in Figures
11.17 and 11.18, are much more powerful and flexible. Vertical tracks, called *leads*, guide
the hammer as the pile descends into the ground. Hydraulic or cable-operated actuators
allow the operator to move the leads into the desired alignment.

Hammers

The *pile hammer* is the device that provides the impacts necessary to drive the pile. Repeated blows are necessary, so the hammer must be capable of cycling quickly. It also
must deliver sufficient energy to advance the pile, while not being powerful enough to
break it. The selection of a proper hammer is one of the keys to efficient pile driving.

Figure 11.16 Installation of a plastic-steel composite pile in a waterfront application. The existing piles are timber (Photo courtesy of Plastic Pilings. Inc.).

Drop Hammers

The original type of pile hammer was the *drop hammer*. They consisted of a weight that was hoisted up, and then dropped directly onto the pile. These hammers became much larger and heavier during the late nineteenth century as described by Powell (1884):

> The usual method of driving piles is by a succession of blows given by a heavy block of wood or iron, called a ram, monkey or hammer, which is raised by a rope or chain, passed over a pulley fixed at the top of an upright frame, and allowed to fall freely on the head of the pile to be driven. The construction of a pile-driving machine is very simple. The guide frame

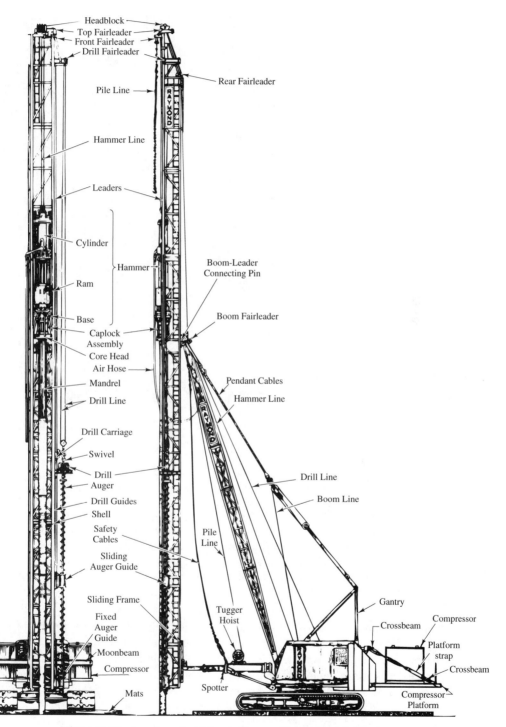

Figure 11.17   A modern pile-driving rig (Raymond International Builders).

Figure 11.18   A modern pile driver. The hammer is at the bottom of the leads, and the predrilling auger is attached to the side of the leads. The forklift on the left is used to transport the piles to the rig.

is about the same in all of them: the important parts are the two upright timbers, which guide the ram in its descent. The base of the framing is generally planked over and loaded with stone, iron, or ballast of some kind, to balance the weight of the ram. The ram is usually of cast-iron, with projecting tongue to fit the grooves of frame. Contractors have all sizes of frames, and of different construction, to use with hand or steam power, from ten feet to sixty feet in height. The height most in use is one of twenty feet, with about a twelve hundred pound ram. In some places the old hand-power method has to be used to avoid the danger of producing settling in adjoining buildings from jarring.

These hammers could deliver only about three to twelve blows per minute.

Drop hammers have since been replaced by more modern designs. They are now rarely used in North America for foundation piles, but they are sometimes used to install sheet piles. Drop hammers are effective for installing foundation piles in the very soft clays of Scandinavia, and thus are still used there (Broms, 1981). Drop hammers also form part of the *pressure-injected footing* process described in Section 11.8.

Steam, Pneumatic, and Hydraulic Hammers

New types of hammers began to appear in the late 1800s. These consisted of a self-contained unit with a ram, anvil, and raising mechanism, as shown in Figure 11.13. These hammers had slightly larger weights, but much shorter strokes than the drop hammers. For example, the "Nasmyth steam pile-drivers" of the 1880s had 1400 to 2300 kg (3000–5000 lb) rams with a stroke of about 900 mm (3 ft). Although these hammers delivered less energy per blow, they were more efficient because they cycled much more rapidly (about 60 blows/min for the Nasmyth hammer).

The early self-contained hammers used steam to raise the ram. This steam was produced by an on-site boiler. *Steam hammers* are still in use. Later, *pneumatic hammers* (powered by compressed air) and *hydraulic hammers* (powered by high-pressure hydraulic fluid) were introduced. The hydraulic hammers are becoming increasingly popular.

All three types can be built as a *single-acting hammer* or as a *double-acting hammer*. Single acting hammers raise the ram by applying pressure to a piston, as shown in Figure 11.19a. When the ram reaches the desired height, typically about 900 mm (3 ft), an exhaust valve opens and the hammer drops by gravity and impacts the anvil. When compared to other types, this design is characterized by a low impact velocity and heavy ram weights. These hammers have fixed strokes, which means each drop of the hammer delivers the same amount of energy to the pile.

A double-acting hammer, shown in Figure 11.19b, uses pressure for both the upward and downward strokes, thus delivering a greater impact than would be possible by gravity alone. The impact energy depends to some degree upon the applied pressure and therefore can be controlled by the operator. These hammers usually have shorter strokes and cycle more rapidly than single-acting hammers. Practical design limitations prevent these hammers from delivering as much energy as comparable single-acting hammers, so they are principally used for driving sheet piles.

A *differential hammer*, shown in Figure 11.19c, is similar to a double-acting hammer in that it uses air, steam, or hydraulic pressure to raise and lower the ram, but it differs in that it has two pistons with differing cross-sectional areas. This allows differential hammers to use the heavy rams of single-acting hammers and operate at the high speed and with the controllability of double-acting hammers.

Steam and pneumatic differential hammers cycle slowly under soft driving conditions and faster as the penetration resistance increases. The reverse is true of hydraulic hammers.

Diesel Hammers

A diesel hammer, shown in Figure 11.20, is similar to a diesel internal combustion engine. The ram falls from a high position and compresses the air in the cylinder below. At a certain point in the stroke, diesel fuel is injected (in either atomized or liquid form) and the air-fuel mixture is further compressed until the ram impacts the anvil. Combustion occurs about this time, forcing the ram up and allowing another cycle to begin.

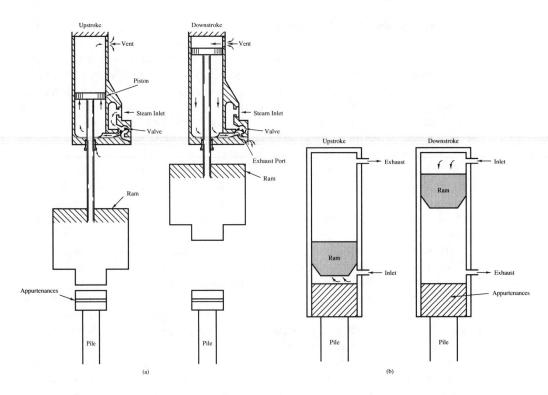

(a)

(b)

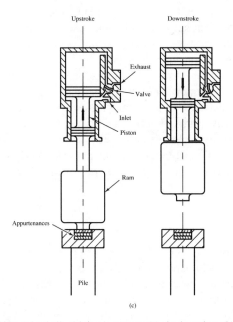

(c)

Figure 11.19   Self-contained pile-driving hammers: (a) single-acting; (b) double-acting; (c) differential.

(a)   (b)

Figure 11.20 Open-top diesel pile hammers. a) This hammer is at the bottom of the leads. The auger to the left is for predrilling, b) This hammer is in the process of driving a concrete pile. The ram is near the top of its stroke, and is visible at the top of the hammer (photo b courtesy of Goble, Rausche, Likins, and Associates).

Diesel hammers are either of the open-top (single acting) or closed-top (double acting) type. The closed-top hammer includes a bounce chamber above the ram that causes the hammer to operate with shorter strokes and at higher speeds than an open-top hammer with an equivalent energy output.

The operator and field engineer can monitor the energy output of a diesel hammer by noting the rise of the ram (in an open-top hammer) or the bounce chamber pressure (in a closed-top hammer). Diesel hammers develop their maximum energy under hard driving conditions and may be difficult to operate under soft conditions, which sometimes occur early in the driving sequence, because of a lack of proper combustion or insufficient hammer rebound. Once firm driving conditions are encountered, open-top hammers typically deliver forty to fifty five blows per minute, while closed-top hammers typically deliver about ninety blows per minute.

Although diesel hammers have been popular for many years, the exhaust is a source of air pollution, so air quality regulations may restrict their use in some areas.

Vibratory Hammers

A *vibratory hammer* (Warrington, 1992) is not a hammer in the same sense as those discussed earlier. It uses rotating eccentric weights to create vertical vibrations, as shown in Figure 11.21. When combined with a static weight, these vibrations force the pile into the ground. The operating frequency of vibratory hammers may be as high as 150 Hz and can be adjusted to resonate with the natural frequency of the pile.

Vibratory hammers are most effective when used with piles being driven into sandy soils. They operate more quickly and with less vibration and noise than conventional impact hammers. However, they are ineffective in clays or soils containing obstructions such as boulders.

Figure 11.21 A vibratory pile hammer. This hammer is extracting a steel pipe used as a casing for a drilled shaft foundation (ADSC: The International Association of Foundation Drilling).

Appurtenances

A pile-driving system also includes other components that are placed between the pile hammer and the pile, as shown in Figure 11.22. The ram hits a steel *striker plate*. It then transmits the impact energy through a *hammer cushion* (also known as a *capblock*) to a *drive head* (also known as a *drive cap*, *bonnet*, *hood*, or *helmet*). The drive head is placed directly on the pile except for a concrete pile where a *pile cushion* is inserted between them.

The cushions soften the sharp blow from the hammer by spreading it out over a longer time. Ideally, they should do this without absorbing too much energy. Hammer cushions do this to protect the hammer, and may consist of hardwood, or the more efficient man-made materials. Pile cushions, which are generally used only with concrete piles, are intended to protect the pile. They are usually made of plywood.

The optimal selection of the pile hammer and appurtenances is part of the key to efficient pile driving. Wave equation analyses, discussed in Chapter 14, can be very useful in this regard.

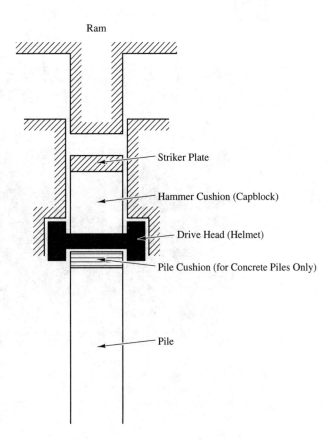

Figure 11.22   Pile-driving appurtenances.

Predrilling, Jetting, and Spudding

All piles are subject to damage during driving, especially in very hard ground or ground that contains boulders. Figure 11.23 shows an example of pile damage. One way to reduce the potential for damage and increase the contractor's production rate is to use predrilling, jetting, or spudding.

Predrilling means drilling a vertical hole, and then driving the pile into this hole. The diameter of the predrill hole must be less than that of the pile to assure firm contact with the soil. Predrilling also reduces the heave and lateral soil movement sometimes associated with pile driving. The predrill hole does not necessarily need to extend for the entire length of the pile.

To use *jetting*, the contractor pumps high-pressure water through a pipe to a nozzle located at the pile tip. This loosens the soil in front of the pile, thus allowing it to advance with very few or no hammer blows. Jetting is useful in sandy and gravelly soils, but is ineffective in clays. It is most often used to quickly penetrate through sandy layers to reach deeper bearing strata.

Spudding consists of driving hard metal points into the ground, and then removing them and driving the pile into the resulting hole. This method is much less common than predrilling or jetting, and is most often used to punch through thin layers of hard rock.

Figure 11.23   These steel pipe piles were used to support a temporary pier while the permanent pier (on the left) was under construction. Afterwards, they were extracted. The pile on the left experienced damage during driving, probably as a result of hitting an underground obstruction.

Pile Arrangements and Geometries

Usually each member of the superstructure that requires a foundation (for example, each column in a building) is supported on a group of three or more piles. *Pile groups* are used instead of single piles because:

- A single pile usually does not have enough capacity.
- Piles are *spotted* or located with a low degree of precision, and can easily be 150 mm (6 in) or more from the desired location, as shown in Figure 11.24. If a column for a building, which is located with a much greater degree of precision, were to be supported on a single pile, the centerlines would rarely coincide and the resulting eccentricity would generate unwanted moments and deflections in both the pile and the column. However, if the column is supported on three or more piles, any such eccentricities are much less significant.

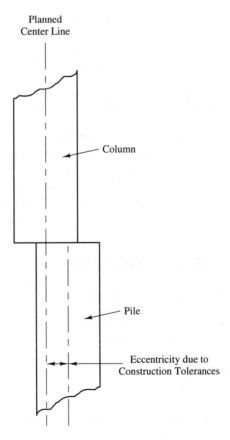

Figure 11.24   Unanticipated eccentricities between columns and single piles caused by construction tolerances.

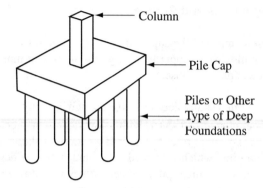

Figure 11.25   A pile cap is a structural member that connects the piles in a group.

- Multiple piles provide redundancy, and thus can continue to support the structure even if one pile is defective.
- The lateral soil compression during pile driving is greater, so the side friction capacity is greater than for a single isolated pile.

Each group of piles is connected with a *pile cap*, as shown in Figure 11.25, which is a reinforced concrete member that is similar to a spread footing. Its functions are to distribute the structural loads to the piles, and to tie the piles together so they act as a unit. The design of pile caps varies with the number of piles and the structural loads. Figure 11.26 shows typical pile cap layouts. Sometimes the individual pile caps are connected with *grade beams*, which are structural beams embedded in the ground. During construction, grade beams resemble continuous footings, but their purpose is significantly different.

QUESTIONS AND PRACTICE PROBLEMS

11.1 Explain the difference between axial loads and lateral loads.

11.2 Discuss some of the primary advantages and disadvantages of the following types of piles and suggest a potential application for each:
- Timber
- Steel
- Prestressed concrete

11.3 What is predrilling and when might it be used? What might happen if the predrill diameter or length was excessive?

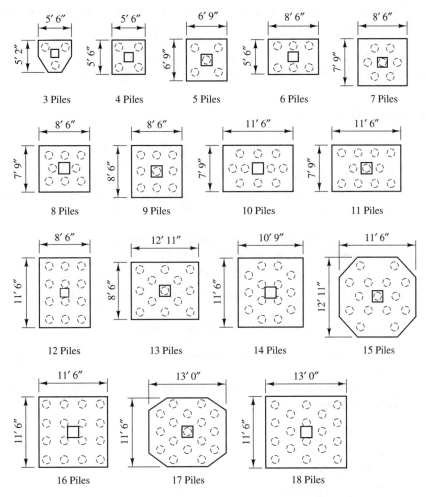

Figure 11.26 Typical configurations of pile caps (Adapted from CRSI, 1992).

11.4 What type or types of piles would be appropriate to support a heavy structure on an undulating bedrock surface located 25 to 40 m below the ground surface? Assume the side friction in the overlying soils provides less than 20 percent of the total axial load capacity. Explain the reasons for your choice.

11.5 Why are most concrete piles prestressed instead of being conventionally reinforced?

11.6 In the context of pile construction, what are cushions, when are they used, and what is their purpose?

11.7 Pile foundations that support buildings usually have at least three piles for each column. Why?

11.8 Pile driving in loose sands without predrilling tends to densify these soils. What effect does this densification have on the load bearing capacity of such piles?

11.4   DRILLED SHAFTS

Drilled shafts are another common type of deep foundation. The fundamental difference between piles and drilled shafts is that piles are prefabricated members driven into the ground, whereas drilled shafts are cast-in-place.

The construction procedure in competent soils, known as the *dry method*, is generally as follows:

1. Using a *drill rig*, excavate a cylindrical hole (the *shaft*) into the ground to the required depth, as shown in Figure 11.27a.
2. Fill the lower portion of the shaft with concrete as shown in Figure 11.27b.
3. Place a prefabricated reinforcing steel cage inside the shaft as shown in Figure 11.27c.
4. Fill the shaft with concrete as shown in Figure 11.27d.

Alternative construction procedures for use in difficult soils are discussed later in this chapter.

Engineers and contractors also use other terms to describe to this type of deep foundation, including the following:

- *Pier*
- *Drilled pier*
- *Bored pile*
- *Cast-in-place pile*
- *Caisson*
- *Drilled caisson*
- *Cast-in-drilled-hole (CIDH) foundation*

However, drilled shafts are not the same as certain other methods that also involve cast-in-place concrete, such as *auger-cast piles*, *pressure injected footings*, *step-taper piles*, and *grouted anchors*. They are covered later in this chapter.

History

The quality of soils usually improves with depth, so it often is helpful to excavate through weak surface soils to support structures on deeper bearing materials. Even the ancient Greeks understood the value of removing poor quality soils (Kerisel, 1987).

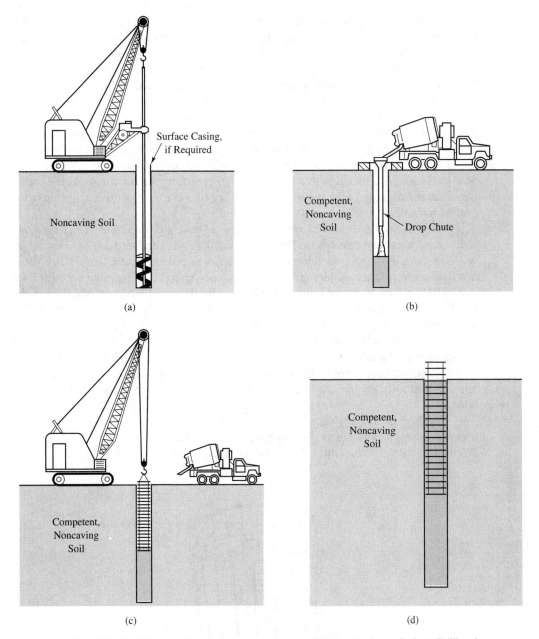

Figure 11.27   Drilled shaft construction in competent soils using the dry method: (a) Drilling the shaft; (b) Starting to place the concrete; (c) Placing the reinforcing steel cage; and (d) Finishing the concrete placement (Reese and O'Neill, 1988).

During the late nineteenth and early twentieth centuries, builders began to modify the traditional techniques of reaching good bearing soils. Many of the greatest advances occurred in the large cities of the Great Lakes region. As taller and heavier buildings began to appear, engineers realized that traditional spread footing foundations would no longer suffice. Following many years of problems with excessive settlements, they began to use foundation systems consisting of a single hand-dug shaft below each column.

General William Sooy-Smith was one of the pioneers in this new technology. His *Chicago Well Method*, developed in 1892, called for the contractor to hand-excavate a cylindrical hole about 1 m (3 ft) in diameter and 0.5 to 2 m (2–6 ft) deep, as shown in Figure 11.28. To prevent caving, they lined its wall with vertical wooden boards held in place with steel rings, and then repeated the process until reaching the bearing stratum. Finally, they enlarged the base of the excavation to form a bell and filled it with concrete up to the ground surface.

Hand excavation methods were slow and tedious, so the advent of machine-dug shafts was a natural improvement. The early equipment was similar to that used to drill oil wells, so much of the early development occurred in oil-producing areas, especially Texas and California. A few examples of horse- and engine-driven drills appeared between 1900 and 1930, but they had very limited capabilities. By the late 1920s, manufacturers were building practical truck-mounted engine-driven drill rigs, such as the one in Figure 11.29, thus bringing drilled shaft construction into its maturity.

During the next thirty five years, manufacturers and contractors developed larger and more powerful equipment along with a variety of cutting tools. They also borrowed construction techniques from the oil industry, such as casing and drilling mud, to deal with difficult soils. By the 1960s, drilled shafts had become a strong competitor to driven piles.

Today, drilled shafts support structures ranging from one story wood frame buildings to the largest skyscrapers. For example, the Sears Tower in Chicago is supported on 203 drilled shafts, some of them 30 m (100 ft) deep.

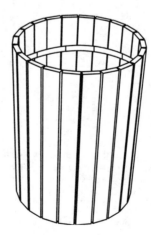

Figure 11.28 Early shaft construction using the Chicago Well Method. This is one set of wooden shores; the excavation would continue with additional sets until reaching the desired depth.

Figure 11.29   Early truck mounted drill rig and crew, circa 1925 (ADSC: The International Association of Foundation Drilling).

The advantages of drilled shaft foundations include the following:

- The costs of mobilizing and demobilizing a drill rig are often much less than those for a pile driver. This is especially important on small projects, where they represent a larger portion of the total costs.
- The construction process generates less noise and vibration, both of which are especially important when working near existing buildings.
- Engineers can observe and classify the soils excavated during drilling and compare them with the anticipated soil conditions.
- Contractors can easily change the diameter or length of the shaft during construction to compensate for unanticipated soil conditions.
- The foundation can penetrate through soils with cobbles or boulders, especially when the shaft diameter is large. It is also possible to penetrate many types of bedrock.
- It is usually possible to support each column with one large shaft instead of several piles, thus eliminating the need for a pile cap.

The disadvantages include the following:

- Successful construction is very dependent on the contractor's skills, much more so than with spread footings or even driven piles. Poor workmanship can produce weak foundations that may not be able to support the design load. Unfortunately, most of these defects are not visible. This is especially important because a single drilled shaft does not have the benefit of redundancy that is present in a group of driven piles.

- Driving piles pushes the soil aside, thus increasing the lateral stresses in the soil and generating more side friction capacity. However, shaft construction removes soil from the ground, so the lateral stresses remain constant or decrease. Thus, a shaft may have less side friction capacity than a pile of comparable dimensions. However, this effect is at least partially offset by rougher contact surface between the concrete and the soil and the correspondingly higher coefficient of friction.

- Pile driving densifies the soil beneath the tip, whereas shaft construction does not. Therefore, the unit end bearing capacity in shafts may be lower.

- Full-scale load tests are very expensive, so the only practical way to predict the axial load capacity is to use semiempirical methods based on soil properties. We typically have no independent check. However, the Osterberg load test device and high-strain dynamic impact tests, discussed in Chapters 12 and 15, may overcome this problem.

Figure 11.30   Typical drilling rig for constructing drilled shafts. This rig is able to drill shafts up to 1800 mm (72 inches) in diameter and 24 m (80 ft) deep. (ADSC: The International Association of Foundation Drilling).

Modern Construction Techniques

Contractors use different equipment and techniques depending on the requirements of each project (Greer and Gardner, 1986). The design engineer must be familiar with these methods to know when and where drilled shafts are appropriate. The construction method also influences the shaft's load capacity, so the engineer and contractor must cooperate to assure compatibility between the design and construction methods.

Drilling Rigs

Most drilled shafts are 500 to 1200 mm (18–48 inches) in diameter and 6 to 24 m (20–80 ft) deep. A typical modern truck-mounted drilling rig, such as the one shown in Figure 11.30, would typically be used to drill these shafts. Specialized rigs, such as those in Figures 11.31 and 11.32, are available for difficult or unusual projects. Some rigs are capable of drilling shafts as large as 8 m (26 ft) in diameter and up to 60 m (200 ft) deep.

Figure 11.31   Small track-mounted drilling rig capable of working on a hillside (ADSC: The International Association of Foundation Drilling).

Figure 11.32   Extremely large drill rig capable of drilling 8 m (26 ft) diameter holes to depths of 60 m (200 ft) (Anderson Drilling Co.)

Drilling Tools

Contractors have different drilling tools, each suited to a particular subsurface condition or drilling technique. The helix-shaped *flight auger*, shown in Figure 11.33, is most common.

The drill rig rotates the auger into the ground until it fills with soil. Then, it draws the auger out and spins it around to remove the cuttings, as shown in Figure 11.34. This process repeats until the shaft reaches the desired depth.

Conventional flight augers are effective in most soils and soft rocks. However, when encountering difficult conditions, the contractor has the option of switching to special augers or other tools. For example, augers with hardened teeth and pilot *stingers* are effective in hardpan or moderately hard rock. Spiral-shaped rooting tools help loosen cobbles and boulders, thus allowing the hole to advance under conditions that might cause refusal in a driven pile. Some of these special tools are shown in Figure 11.35.

Other drilling tools include:

- *Bucket augers* that collect the cuttings in a cylindrical bucket that is raised out of the hole and emptied. They are especially useful in running sands.
- *Belling buckets* that have extendable arms to enlarge the bottom of the shaft. These enlargements are called *bells* or *underreams*.
- *Core barrels* that cut a circular slot, creating a removable cylindrical core. They are especially useful in hard rock.
- *Multiroller percussion bits* to cut through hard rock.
- *Cleanout buckets* to remove the final cuttings from a hole and produce a clean bottom suitable for end bearing.

Figure 11.33 Typical flight auger (ADSC: The International Association of Foundation Drilling).

Figure 11.34   Spinning the auger to remove the cuttings (ADSC: The International Association of Foundation Drilling).

(a)

(b)

Figure 11.35   Special flight augers for difficult subsurface conditions (a) Auger with hardened teeth and a stinger; (b) Spiral-shaped rooting auger (ADSC: The International Association of Foundation Drilling).

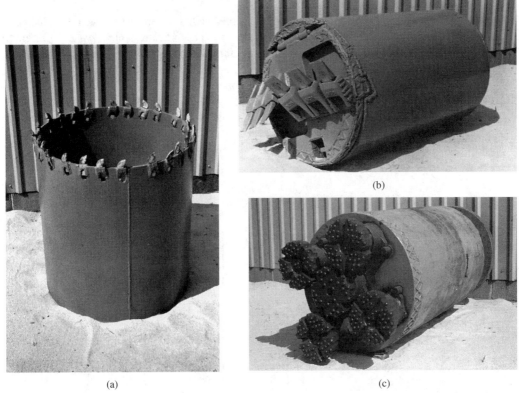

(b)

(a)   (c)

Figure 11.36   Special drilling tools (a) Core barrel; (b) Bucket auger; (c) Multiroller percussion bits (ADSC: The International Association of Foundation Drilling).

Some of these tools are shown in Figure 11.36.

Drilling Techniques in Firm Soils

In firm soils, contractors use the *dry method* (also known as the *open-hole method*) to build the shaft, as shown in Figure 11.27. These holes usually advance quickly using conventional flight augers and remain open without any special support. After checking the open hole for cleanliness and alignment, it is a simple matter to insert the steel reinforcing cage and dump concrete in from the top. Some contractors use a tremie or a concrete pump to deliver the concrete. Open-hole shafts in firm soils are very common because of their simplicity and economy of construction and their good reliability.

It also is possible to excavate stiff soils below the groundwater table using the open-hole method. Usually, the contractor simply pumps the water out as the hole advances and places the concrete in the dewatered shaft.

Drilling Techniques in Caving or Squeezing Soils

A hole is said to be *caving* when the sides collapse before or during concrete placement. This is especially likely in clean sands below the groundwater table. *Squeezing* refers to the sides of the hole bulging inward, either during or after drilling, and is most likely in soft clays and silts or highly organic soils. Either of these conditions could produce *necking* in the shaft (a local reduction in its diameter) or soil inclusions in the concrete, as shown in Figure 11.37, both of which could have disastrous consequences.

The two most common construction techniques for preventing these problems are the use of *casing* or the use of *drilling fluid*.

The casing method, shown in Figure 11.38, uses the following procedure:

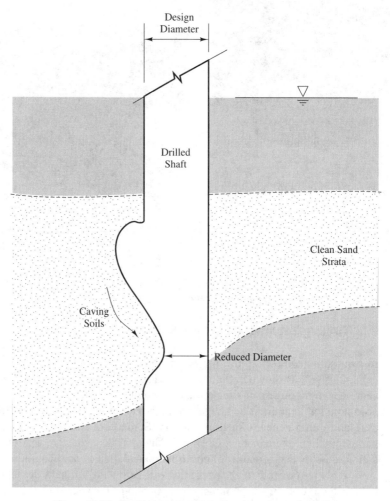

Figure 11.37   Possible consequences of caving or squeezing soils.

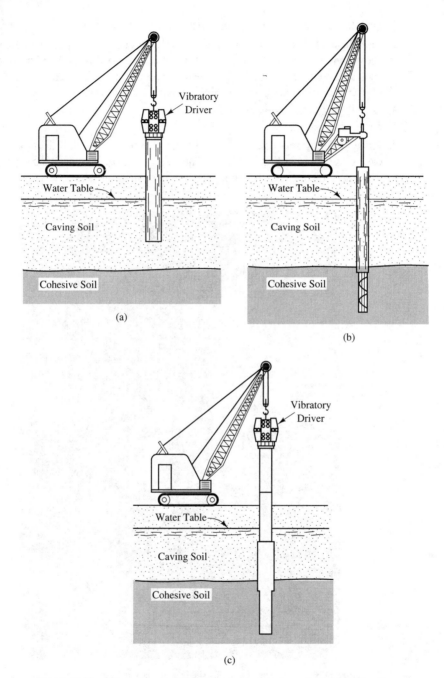

Figure 11.38 Using casing to deal with caving or sqeezing soils: (a) Installing the casing; (b) Drilling through and ahead of the casing; and (c) Placing the reinforcing steel and concrete, and removing the casing (Reese and O'Neill, 1988).

1. Drill the hole using conventional methods until encountering the caving strata.
2. Insert a steel pipe (the casing) into the hole and advance it past the caving strata as shown in Figure 11.38a and 11.39. Contractors do this using vibratory hammers such as the one in Figure 11.21. The diameter of this casing is usually 50 to 150 mm (2–6 in) less than the diameter of the upper part of the shaft.
3. Drill through the casing and into the non-caving soils below using a smaller diameter auger as shown in Figure 11.38b.
4. Place the reinforcing steel cage and the concrete through the casing and extract the casing as shown in Figure 11.38c. This is a very critical step, because premature extraction of the casing can produce soil inclusions in the shaft.

Figure 11.39   The contractor at this site is using casing. The first casing, visible at the bottom of the photograph, is already in place. However, its length is limited by the height of the rig. When casing must extend to greater depths, a second smaller casing is installed by passing it through the first casing, as shown here.

There are many variations to this method, including the option of leaving the casing in place and combining the casing and slurry methods.

The drilling fluid method (also known as the *slurry method*) is shown in Figure 11.40. It uses the following procedure:

1. Drill a *starter hole*, perhaps 3 m (10 ft) deep.
2. Fill the starter hole with a mixture of water and bentonite clay to form a *drilling mud* or *slurry*. When sea water is present, use attapulgite clay instead of bentonite. When properly mixed, the drilling mud has the appearance of very dirty water and keeps the hole open because of the hydrostatic pressure it applies to the soil.
3. Advance the hole by passing the drilling tools through the slurry as shown in Figure 11.39a. Continue to add water and bentonite as necessary.
4. Insert the reinforcing steel cage directly into the slurry as shown in Figure 11.39b.
5. Fill the hole with concrete using a *tremie pipe* that extends to the bottom as shown in Figure 11.39c. The concrete pushes the slurry to the ground surface, where it is captured.

Do not be concerned about the quality of the bond between the rebar and the concrete. Although the rebar is first immersed in slurry, research has shown that the bond is satisfactory. However, the slurry can form a cake on and in the surrounding soil, thus reducing the side friction resistance. Some specifications require the contractor to "scour" the sides of the holes to remove the slurry cake before placing the concrete.

Underreamed Shafts

An *underreamed shaft* (also known as a *belled shaft*) is one with an enlarged base, as shown in Figure 11.41. Usually, the ratio of the underream diameter to the shaft diameter (B_b/B_s) should be no greater than 3. Contractors build underreams using special belling buckets, such as the one shown in Figure 11.42.

The larger base area of underreamed shafts increases their end bearing capacity, and thus they are especially useful for shafts bottomed on strong soils or rock. However, the displacement required to mobilize the full end bearing is typically on the order of 10 percent of the base diameter, which may be more than the structure can tolerate. Underreamed shafts also have greater uplift capacities due to bearing between the ceiling of the underream and the soil above.

Unfortunately, the construction of underreamed shafts can be hazardous to the construction workers. In addition, the bottom of the underream must be cleaned of loose soil before placing concrete, and this process can be difficult and expensive.

Underreamed shafts are not being built as often as they were in the past, primarily because it is often more cost-effective to simply drill a deeper straight shaft and rely on the additional side friction. However, underreamed shafts are still built, especially when a firm bearing stratum is available.

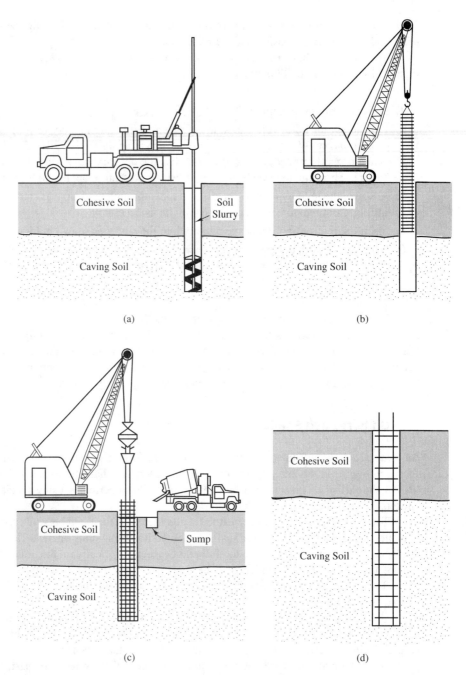

(a)

(b)

(c)

(d)

Figure 11.40 Using drilling fluid to deal with caving or squeezing soils: (a) Drilling the hole using slurry; (b) Installing the reinforcing steel cage through the slurry; (c) Placing the concrete using a tremie pipe and recovering slurry at the top; and (d) The completed foundation (Reese and O'Neill, 1988).

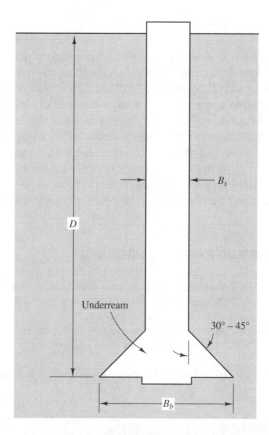

Figure 11.41 An underreamed drilled shaft.

Figure 11.42 A belling bucket used to produce a bell or underream at the bottom of a shaft (ADSC: The International Association of Foundation Drilling).

Concrete

Concrete for drilled shafts must have sufficient slump to flow properly and completely fill the hole. Using concrete that is too stiff creates voids that weaken the shaft. Typically, the slump should be between 100 and 220 mm (4–9 in), with the lower end of that range being most appropriate for large-diameter dry holes with widely spaced reinforcement and the high end for concrete placed under drilling fluid. Sometimes it is appropriate to include concrete admixtures to obtain a high slump while retaining sufficient strength.

Some people have experimented with expansive cements in drilled shaft concrete. These cements cause the concrete to expand slightly when it cures, thus increasing the lateral earth pressure and side friction resistance. So far, this has been only a research topic, but it may become an important part of future construction practice.

QUESTIONS AND PRACTICE PROBLEMS

11.9 Describe two situations where a drilled shaft would be preferable over a driven pile, then describe two situations where the reverse would be true.

11.10 In what circumstances would you expect caving or squeezing conditions to be a problem? What construction methods could a contractor use to overcome these problems?

11.11 The dry method of drilled shaft construction is most suitable for which types of soil conditions?

11.5 CAISSONS

The word *caisson* is derived from the French *caisse*, which means a chest or box. When applied to foundation engineering, it describes a prefabricated hollow box or cylinder that is sunk into the ground to some desired depth and then filled with concrete, thus forming a foundation.[1] Caissons have most often been used in the construction of bridge piers and other structures that require foundations beneath rivers and other bodies of water because the caissons can be floated to the job site and sunk into place.

Open Caissons

An *open caisson* is one that is open to the atmosphere, as shown in Figure 11.43. They may be made of steel or reinforced concrete, and normally have pointed edges at the bottom to facilitate penetration into the ground.

---

[1]The word "caisson" also is sometimes used to describe drilled shaft foundations because they were originally developed as "machine-dug caissons." However, this use of the term is confusing and should be avoided.

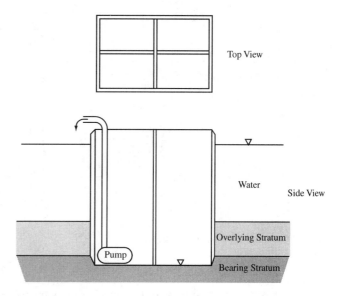

Figure 11.43   An open caisson.

Sometimes the site of the proposed foundation is dredged[2] before the caisson arrives on site. The dredging operation can be an economical way to remove some of the upper soils, thus reducing the quantities that must be excavated through the caisson. Then the caisson is floated into place and sunk into the soil. As it descends, the soil inside is removed and hauled out of the top and water that accumulates inside is pumped out. This process continues until the caisson sinks to the required depth and reaches the bearing stratum. It then is filled with concrete to form the foundation.

Caissons must be designed to resist the various loads imparted during construction, as well as the structural and hydrodynamic loads from the completed structure. In addition, it must have sufficient weight to overcome the side friction forces as it descends into the ground.

Pneumatic Caissons

When the excavation inside open caissons extends well below the surrounding water level, water flowing into the bottom can produce a quick condition in the soils as shown in Figure 11.44. This is most likely to occur in clean sands and is caused by the upward seepage forces of the flowing water.

One way to counteract this problem is to seal the bottom portion the caisson and fill it with compressed air, as shown in Figure 11.45. If the air pressure equals or exceeds the

---

[2]Dredging is the process of removing soil from the bottom of a body of water. This is normally done using specially-equipped ships, normally for the purpose of providing sufficient water depth for larger ships.

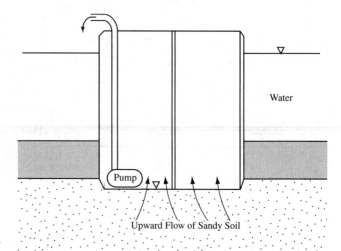

Figure 11.44 Development of a quick condition beneath an open caisson.

pore water pressure, very little water enters the excavation, thus eliminating the seepage forces and the potential for quick conditions. In addition, the required pumping costs then become minimal. This method was first used around 1850 by the British engineer Isambard Kingdom Brunel (1806–1859) during construction of the Chepstow Bridge across the Thames River in London. Many bridge foundations in North America also have been

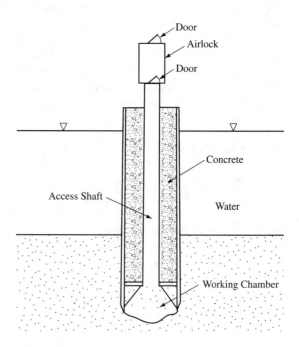

Figure 11.45 A pneumatic caisson uses compressed air to halt the flow of groundwater.

built using this method, including the Brooklyn and Williamsburg Bridges in New York City.

For example, an excavation 50 ft below the groundwater table would require a pressure, p, of about:

$$p \approx u = \gamma_w z_w = (62.4 \text{ lb/ft}^3)(50 \text{ ft})\left(\frac{1 \text{ ft}^2}{144 \text{ in}^2}\right) = 22 \text{ lb/in}^2$$

This is the same pressure a diver would encounter 50 ft below the water surface in a lake. Construction personnel, who are called *sandhogs*, can work for three-hour shifts under such pressures. In some cases, air pressures up to 48 lb/in^2 (35 kPa) may be used, but the shift time drops to only thirty minutes (White, 1962).

Workers enter these excavations by passing through an air lock, which is an intermediate chamber with doors connected to the outside and to the working chamber. The workers enter through the outside door, then both doors are closed and the chamber is slowly filled with compressed air. When the pressure reaches that in the working chamber, the workmen open the connecting door and enter the working chamber. The process is reversed when exiting.

If the air pressure in the air lock is lowered too quickly, nitrogen bubbles form inside the workers bodies, causing *caisson disease*, also known as *the bends*. It causes severe pains in the joints and chest, skin irritation, cramps, and paralysis. Fourteen men died of caisson disease during construction of the Eads Bridge in St. Louis. Divers can experience the same problem if they rise to the surface too quickly.

Because of the hazards of working under compressed air, the large expense of providing the necessary safety precautions for the workers, and the availability of other foundation types, pneumatic caissons are rarely used. However, in some circumstances they can be economically viable. For example, foundations for the proposed Great Belt Bridge in Denmark are to be built using pneumatic caissons (Prawit and Volmerding, 1995).

11.6   MANDREL-DRIVEN THIN-SHELLS FILLED WITH CONCRETE

One method of combining some of the best features of driven piles and cast-in-place drilled shafts is to use a mandrel-driven thin shell, as shown in Figure 11.46. Alfred Raymond developed an early version of this method in 1897 and later refined it to create the *step-taper pile*.

This type of foundation is built as follows:

1. Hold a steel mandrel in the leads of the pile driver. This mandrel is cylindrical and matches the inside of the thin shell. The purpose of the mandrel is to transmit the driving stresses from the hammer to the sides and bottom of the shell.
2. Pull the thin shell (which resembles corrugated steel drain pipe) onto the mandrel.
3. Drive the mandrel and shell into the ground using a pile hammer.

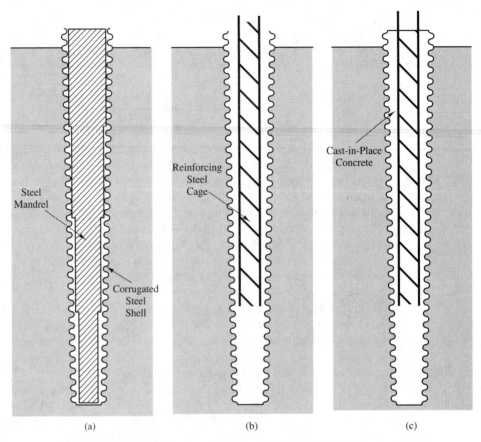

Steel
Mandrel

Reinforcing
Steel
Cage

Cast-in-Place
Concrete

Corrugated
Steel
Shell

(a)   (b)   (c)

Figure 11.46 Construction of mandel-driven thin-shell pile filled with concrete: (a) drive the thin shell into the ground using a steel mandrel; (b) Remove the mandrel and insert a cage of reinforcing steel; and (c) fill the shell with concrete.

4. Remove the mandrel and inspect the shell.

5. (optional) Place a cage of reinforcing steel into the shell.

6. Fill the shell with concrete.

The advantages of this design include:

- The shell provides a clean place to cast the concrete, so the structural integrity may be better than a drilled shaft.

- The displacement developed during driving and the corrugations on the shell produce high side friction resistance.

- The shells and mandrel can be shipped to the job site in pieces and assembled there, so it is possible to build long piles.

However, it also has disadvantages:

- It is necessary to mobilize a pile driver and other equipment, so the cost per pile will be at least as high as that for conventional driven piles.
- They cannot be spliced, so the total length is limited by the height of the pile driver.

11.7   AUGER-CAST PILES

The *auger-cast pile* is a type of cast-in-place pile developed in the United States during the late 1940s and early 1950s (Neate, 1989). They are known by many names (DFI, 1990), including:

- Augered pressure grouted (APG) pile
- Augered cast-in-place pile
- Continuous flight auger pile
- Intruded mortar piles
- Augerpress pile
- AugerPile
- Grouted bored pile
- Augered grout-injected pile

Specialty contractors build auger-cast piles as described below and as shown in Figure 11.47:

1. Using a hollow-stem auger with a temporary bottom plug, drill to the required depth. In the United States, 300, 350, or 400 mm (12, 14, or 16 inch) diameter augers are most common. Japanese contractors have built piles as large as 1 m (39 inches) in diameter. This equipment is similar to, but larger than, the hollow-stem augers used for soil exploration purposes. These augers are suspended from a crane and driven by a pneumatic or hydraulic motor. The depth of the pile may be as great as 27 m (90 ft), but lengths of 6 to 15 m (20–50 ft) are more typical.
2. Inject cement grout (sand, portland cement, and water) under high pressure through the middle of the auger. This grout forces the bottom plug out and then begins to flow out of the bottom of the auger.
3. While the grout is being injected, slowly and smoothly raise and rotate the auger to form the pile while bringing the soil cuttings to the ground surface.
4. Upon reaching the ground surface, remove the auger and insert reinforcing steel into the grouted hole. This may consist of a single centrally located bar or a prefab-

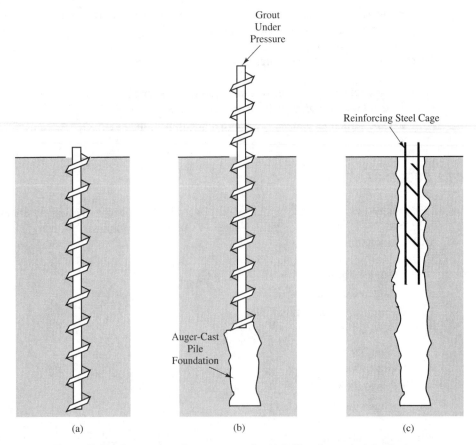

Figure 11.47   Construction of an auger-cast pile: (a) Drill to the required depth using a hollow-stem auger; (b) Withdraw the auger while injecting cement grout; (c) Install steel rebars (optional).

ricated steel cage. Because the grout has a very high slump and no gravel, it is possible to insert the steel directly into the newly grouted pile.

The advantages of this method include:

- The cost of construction is low, partly because the crane can be rented locally, thus reducing mobilization costs.
- The noise and vibration levels are much lower than those with driven piles.
- The auger protects the hole from caving, thus reducing the potential for ground movements during drilling.

- The grout is injected under pressure, so it penetrates the soil and provides and good bond. The pressure also provides some compaction of the soil.
- The technique is usable in a wide variety of soils, including some of those that might cause difficulties with driven piles or drilled shafts.

However, it also has disadvantages (Massarsch et al., 1988; Brons and Kool, 1988):

- The quality and integrity of the completed pile are very dependent on the skills of the contractor. For example, if the auger is raised too quickly or not rotated sufficiently, the concrete may become contaminated with soil. It is also difficult for the equipment operator to judge the correct grout pressure to use.
- In certain conditions, the augering process can draw up too much soil, thus causing a reduction in the lateral stresses in the ground. For example, if an auger passes through a loose sand, and then encounters a stronger strata, the auger will not advance as quickly, and continued rotation may bring too much of the sand to the surface.
- The construction process is very sensitive to equipment breakdowns. Once grouting has begun, any significant breakdown becomes cause for abandoning the foundation.
- Placement of reinforcing steel cages can be difficult, especially if heavy reinforcement is required. This may limit the pile's ability to resist lateral loads (resist uplift loads by placing a single large bar in the middle of the auger before grouting).
- These piles can not be used in soils that contain cobbles or boulders (the auger will not excavate them) or in thick deposits of highly organic soils (they compress under the grout pressure and thus require excessively large grout volumes).
- Unlike driven piles, there is no hammer blow count to use as an independent check on the pile capacity (the auger torque does not appear to be a good indicator).

Although auger-cast piles typically do not have as great a load capacity as conventional driven piles of comparable dimensions, they are often much less expensive. Because the equipment mobilization costs are much less than those for driven piles, and the technique works well in caving soils, auger-cast piles are most often used on small to medium-size structures on sandy soils.

11.8   PRESSURE-INJECTED FOOTINGS

Edgard Frankignoul developed the pressure-injected footing (PIF) foundation in Belgium before the First World War. This technique uses cast-in-place concrete that is rammed into the soil using a drop hammer. This ramming effect compacts the surrounding soil, thus increasing its load bearing capacity. PIF foundations are often called *Franki piles*. Other names include *bulb pile*, *expanded base pile*, *compacted concrete pile*, and *compacto pile*.

The construction techniques used to build PIFs are described below and shown in Figure 11.48.

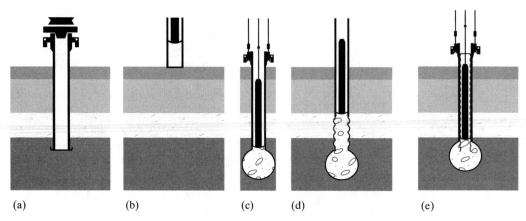

(a)   (b)   (c)   (d)   (e)

Figure 11.48   Construction of a PIF foundation: (a) top driving; (b) bottom driving; (c) finished base; (d) uncased shaft; and (e) cased shaft (Adapted from brochure by Franki Northwest Co., used with permission.)

Phase 1: Driving

The process begins by temporarily inserting the *drive tube* into the ground. This tube is a specially built 300–600 mm (12–24 in) diameter steel pipe. The contractor does so using one of the following methods:

- *Top driving method*: Install a temporary bottom plate on the drive tube, and then drive the tube to the required depth using a diesel pile hammer. The plate will later become detached when the concrete is pounded through the drive tube.
- *Bottom driving method*: Place a plug of low-slump concrete in the bottom of the tube and pack it in using the drop hammer. Then, continue to strike this plug, thus pulling the tube into the ground.

Phase 2: Forming the Base

Once the drive tube reaches the required depth, hold it in place using cables, place small charges of concrete inside the tube, and drive them into the ground with repeated blows of the drop hammer. This hammer has a weight of 1400 to 4500 kg (3–10 kips) and typically drops from a height of 6 m (20 ft). If the top driving method was used, this process will expel the temporary bottom plate. Thus, a bulb of concrete is formed in the soil, which increases the end-bearing area and compacts the surrounding soil. This process continues until a specified number of hammer blows is required to drive out a certain volume of concrete.

Phase 3: Building the Shaft

The shaft extends the PIF base to the ground surface. Two types of shafts are commonly used:

- To build a *compacted shaft,* raise the drive tube in increments while simultaneously driving in additional charges of concrete. This technique compacts the surrounding soil, thus increasing the side friction resistance. It also increases the end-bearing resistance by providing a stronger soil over the base.
- To build a *cased shaft,* insert a corrugated steel shell into the drive tube, place and compact a zero-slump concrete plug, and withdraw the tube. Then fill the shell with conventional concrete. Although this method does not develop as much load capacity, it is often more economical for piles that are longer than about 9 m (30 ft). A cased shaft may be mandatory if very soft soils, such as peat, are encountered because these soils do not provide the lateral support required for the compacted shaft method.

The contractor can reinforce either type of shaft to resist uplift or lateral loads. For the compacted shaft, the reinforcing cage fits between the drop hammer and the drive tube, thus allowing the hammer to fall freely.

PIF foundations may be installed individually or in a group of two or more and connected with a pile cap. Table 11.1 gives typical dimensions and typical capacities of PIF foundations. The actual design capacity must be determined using the techniques described in Chapter 14. Figure 11.49 shows a "mini" PIF that was extracted out of the ground.

TABLE 11.1   TYPICAL PIF DIMENSIONS AND CAPACITIES

| | Typical Allowable Downward Capacity | | Base Diameter[a] | | Nominal Shaft Diameter | | | |
| | | | | | Compacted | | Cased | |
PIF Type	(k)	(kN)	(in)	(mm)	(in)	(mm)	(in)	(mm)
Mini	100	450	24–30	600–750	n/a	n/a	10.6–11.1	270–280
Medium	200	900	34–40	850–1000	17	430	12.2–14	300–360
Standard	400	1800	34–40	850–1000	22	560	16–17.6	400–450
Large	500	2200	34–40	850–1000	23	580	19	480
Maxi	600	2700	34–40	850–1000	25	630	22	560

Adapted from a brochure by Franki Northwest Company. Used with permission.

[a]In very loose soils, the base diameter may be larger than listed here. Conversely, when PIFs are installed in groups, it may be slightly smaller.

Figure 11.49   This "mini" PIF was extracted from the ground. It had a base diameter of 600 mm (24 in). (Photo courtesy of William J. Neely.)

The advantages of PIF foundations include:

- The construction process compacts the soil, thus increasing its strength and load-bearing capacity. This benefit is most pronounced in sandy or gravelly soils with less than about 15 percent passing the #200 sieve, so PIFs are best suited for these kinds of soils.
- When compacted shafts are used, the construction process produces a rough interface between the shaft and the soil, thus further improving the side friction resistance.
- It is possible to build PIFs with large bases (thus gaining the additional end bearing area) in soils such as loose sands where belled drilled shafts would be difficult or impossible to build.

Disadvantages include:

- The side friction resistance for cased PIFs is unreliable because of the annular space between the casing and the soil. Although this space is filled with sand after the drive tube is lifted, we cannot be certain about the integrity of the connection between the shaft and the soil.

- The construction process generates large ground vibrations and thus may not be possible near sensitive structures. These vibrations also can damage wet concrete in nearby PIFs.
- The construction equipment is bulky and cumbersome, and thus requires large work areas.
- Compacted shafts cannot include large amounts of reinforcing steel.
- Although each PIF will have a higher load capacity than a pile or drilled shaft of comparable dimensions, it also is more expensive to build. Therefore, the engineer must evaluate the alternatives for each project individually to determine which type is most economical.
- They are generally economical only when the length is less than about 9 m (30 ft) for compacted PIFs or about 21 m (70 ft) for cased PIFs.

PIFs with Auger-Cast Shafts

Some contractors have combined a PIF base with an auger-cast shaft (Massarsch et al., 1988). The side friction resistance of this type of foundation will be much greater and more reliable than that of a conventional cased PIF, but not as large as that of a compacted shaft PIF. However, an auger-cast shaft could be built more quickly and at less cost than a compacted shaft PIF, thus providing reasonably high capacity at a moderate cost. Very few of these foundations have been built.

11.9   PILE-SUPPORTED AND PILE-ENHANCED MATS

Most mat foundations are supported directly on the underlying soils, as discussed in Chapter 10. However, when the net bearing pressure is too high or the soil is too compressible, such mats may experience excessive settlements. One option for such situations is to use a *pile-supported mat*, as shown in Figure 11.50. The piles are distributed across the mat, which then acts as a very large pile cap. "Pile"-supported mats also may be built using drilled shafts or other types of deep foundations.

Many pile-supported mats have been designed to transfer all of the structural loads to the deep foundations. However, others partially rely on the bearing pressure between the bottom of the mat and the underlying soil, and use the deep foundations to carry the balance of the load. This later design, which can be called a *pile-enhanced mat* can be much less expensive, and will probably be used more frequently in the future.

11.10   ANCHORS

The term *anchor* generally refers to a foundation designed primarily to resist uplift (tensile) loads. Although most foundations are able to resist some uplift, anchors are designed specifically for this task and are often able to do so more efficiently and at a lower cost.

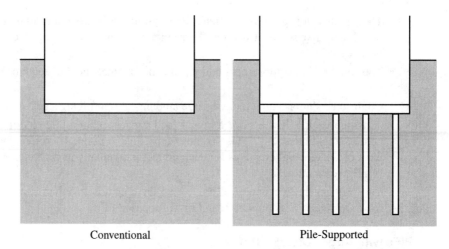

Conventional   Pile-Supported

Figure 11.50   A conventional mat foundation and a pile-supported (or pile-enhanced) mat foundation.

Lightweight structures often require anchors because the lateral wind and earthquake loads on the structure often produce uplift loads on some of the foundations. These structures include power transmission towers, radio antennas, and mobile homes. Some of these structures are stabilized with guy wires, which are then connected to the ground using anchors.

Anchors also can be installed horizontally (or nearly so) to provide lateral support to earth retaining structures. These are called *tieback anchors*. They eliminate the need for bracing outside the wall, and thus provide more space for construction and permanent structures.

Kulhawy (1985) divided anchors into three categories, as shown in Figure 11.51. These include:

- *Spread anchors* are specially designed structural members that are driven or inserted into the ground, then expanded or rotated to form an anchor.
- *Helical anchors* are steel shafts with helices that resemble large screws. They are screwed into the ground using specialized equipment, as shown in Figure 11.52.
- *Grouted anchors* are drilled holes filled with a steel tendon and grout. The tendon transmits the tensile loads into the anchor, then the grout transmits them to the surrounding ground through side friction.

The design load capacities may be computed from the geometry and soil type. For helical anchors, the torque required to install the anchor also can be an indicator of load capacity. In critical applications, such as tieback anchors, engineers often load test the installed anchors using hydraulic jacks.

Some anchors also can resist nominal downward, shear, and moment loads, and thus may be used in other foundation applications. For example, helical anchors similar to

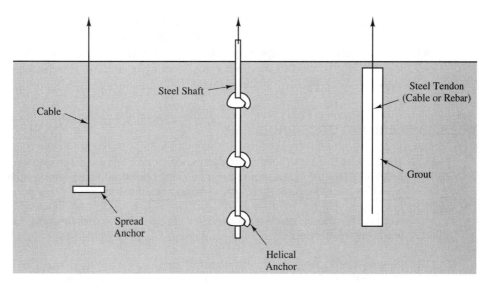

Figure 11.51 Types of anchors (Adapted from Kulhawy, 1985; Used with permission of ASCE).

Figure 11.52 Installation of a helical anchor to be used as a tieback for a sheet pile wall. Once these anchors are installed, the soil in the foreground will be excavated (Photo courtesy of A. B. Chance Company).

those in Figure 11.52 may be used to support streetlight standards, signs, cellphone antennas, and other similar structures. In some cases these foundations may be constructed without any concrete, which can be a significant advantage in remote locations. Helical anchors also may be used to underpin spread footings that have experienced excessive settlement.

QUESTIONS AND PRACTICE PROBLEMS

11.12 There are many ways to build midstream foundations for bridges that cross rivers and other bodies of water. One of them is to use a caisson, as described in this section. Another is to drive piles from a barge. Suggest some advantages and disadvantages of these two methods.

11.13 Suggest some critical items that a construction inspector should watch for during the construction of auger-cast piles.

11.14 Pressure-injected footings are best suited for sandy or gravelly soils with less than about 15 percent fines. Why would this construction method be less effective in a stiff saturated clay?

SUMMARY

Major Points

1. Deep foundations are those that transfer some or all of the structural loads to deeper soils.
2. We can classify deep foundations based on the method of construction. Major classifications include
   - piles
   - drilled shafts
   - caissons
   - mandrel-driven thin shells filled with concrete
   - auger-cast piles
   - pressure-injected footings
   - anchors
3. Load transfer analyses normally divide the applied loads into two categories: axial loads (tension and compression) and lateral loads (shear and moment).
4. Deep foundations transfer axial loads to the ground through two mechanisms: side friction and toe bearing.
5. Piles are prefabricated members that are driven or otherwise inserted into the ground. They are made of wood, concrete, steel, and other materials. Each has its advantages and disadvantages, and is best suited for particular applications.
6. Piles are installed using a pile-driving rig equipped with a pile hammer. Once again, a wide range of equipment is available to accommodate various field conditions.

7. Drilled shafts are constructed by drilling a cylindrical hole and casting the concrete in-place. Various methods are available for drilling the hole and for keeping it open until the concrete is placed.

8. An underreamed shaft is one that has an enlarged base. This design increases the allowable toe bearing load.

9. A caisson is a prefabricated box or cylinder that is sunk into the ground to form a foundation. They are most often used for bridges and other structures that require foundations beneath rivers or other bodies of water.

10. Mandrel-driven thin shells filled with concrete are a cross between piles and drilled shafts. These are also known as step-taper piles.

11. Auger-cast piles are constructed by injecting cement grout at high pressure through a hollow-stem auger. They generally have lower capacity than piles or drilled shafts, but also are generally less expensive, especially in squeezing or caving soil conditions.

12. Pressure-injected footings (also known as Franki piles) are made by pounding stiff concrete into the ground to form a bulb. These are high-capacity foundations and are generally used to support high loads in certain kinds of soil profiles.

13. Pile-supported or pile-enhanced mats consist of a mat foundation underlain by a deep foundation. These are often designed so the mat supports some of the load, and the deep foundations support the balance.

14. Anchors are special deep foundations designed primarily to resist tensile forces.

Vocabulary

Anchor	Head	Pressure-injected footing
Appurtenances	Helmet	Prestressed concrete
Auger-cast pile	Hydraulic hammer	Refusal
Axial load	Jetting	Side friction resistance
Caisson	Large displacement pile	Single-acting hammer
Casing method	Lateral load	Slurry method
Caving soil	Mandrel-driven thin shell	Small displacement pile
Closed-end pipe pile	filled with concrete	Soil plug
Composite pile	Open caisson	Spudding
Cushion	Open-end pipe pile	Squeezing soil
Deep foundation	Pile cap	Steam hammer
Diesel hammer	Pile hammer	Steel pipe pile
Double-acting hammer	Pile driver	Steel H-pile
Drilled shaft	Pile	Step-taper pile
Drilling rig	Pile-enhanced mat	Toe
Drop hammer	Pile-supported mat	Toe bearing resistance

Dry method   Pneumatic caisson   Underreamed shaft

Flight auger   Pneumatic hammer   Vibratory hammer

Franki pile   Predrilling

COMPREHENSIVE QUESTIONS AND PRACTICE PROBLEMS

11.15 Create a table with five columns labeled "foundation type," "typical applications," "best soil conditions," "advantages," and "disadvantages," along with rows labeled "piles," "drilled shafts," "caissons," "mandrel-driven thin shells filled with concrete," "auger-cast piles," and "pressure-injected footings." Fill in each cell of this table with the appropriate information for each foundation type.

11.16 A proposed ten-story office building is to be supported on a series of deep foundations embedded 60 ft below the ground surface. The soils at this site are loose to medium dense well-graded sands (SW) and silty sands (SM), and the groundwater table is at a depth of 12 ft. What type or types of deep foundations would be most appropriate for this project? What type or types would probably not be appropriate? Explain the reasons for your selections.

11.17 A new reinforced concrete pier is to be built in a major harbor area. This pier will service ocean-going cargo ships. The underlying soils are primarily low plasticity silts (ML) and clays (CL), with some interbedded sand layers. What type or types of deep foundations would be most appropriate for this project? What type or types would probably not be appropriate? Explain the reasons for your selections.

12

Deep Foundations— Structural Integrity

Amsterdam, die oude Stadt, is gebouwed op palen
Als die stad eens ommevelt, wie zal dat betalen?

An old Dutch nursery rhyme that translates to:

The old town of Amsterdam is built on piles
If it should fall down, who would pay for it?

Deep foundations must have sufficient structural integrity to safely transfer the applied loads from the structure to the ground. This means they must have sufficient strength to safely sustain the applied stresses, and sufficient stiffness to keep deformations within tolerable limits. This chapter covers the structural strength aspects of deep foundations, Chapter 16 covers the structural deformation aspects, Chapters 13 to 16 cover the geotechnical engineering aspects, and Chapter 17 shows how to synthesize this material into a complete design.

This chapter includes many references to code requirements, and uses both the model building codes (ICBO, BOCA, SBCCI, and ICC) as well as the AASHTO *Standard Specifications for Highway Bridges*. However, because of space and readability constraints, we will cover only selected code requirements. Each of these codes also includes many other detail requirements that are not covered here, and some local jurisdictions add their own requirements. Therefore, engineers should always refer to the applicable code when developing foundation designs.

12.1 DESIGN PHILOSOPHY

The technical literature provides very little information on the structural aspects of deep foundation design, which is a sharp contrast to the mountains of information on the geotechnical aspects. Building codes present design criteria, but they often are inconsistent with criteria for the superstructure, and sometimes are incomplete or ambiguous. In many ways this is an orphan topic that neither structural engineers nor geotechnical engineers have claimed as their own. However, in spite of these problems, deep foundations can be designed to safely carry the structural loads.

Buckling

Even the softest soils provide enough lateral support to prevent underground buckling in axially loaded deep foundations, especially when a pile cap is present. Bethlehem Steel Corporation (1979) quotes several load tests on steel H-piles, including one installed in soils so soft that the pile penetrated them without any hammer blows (i.e., it sank under its own weight). None of these piles buckled.

Slender deep foundations subjected to both axial and lateral loads might have problems with underground buckling if the upper soils are very soft. When this is a concern, it should be checked using a *p-y* analysis, as described in Chapter 16. If bucking proves to be a problem, it must be resolved by using a cross-section with a greater flexural rigidity, *EI*, as discussed in Chapter 16.

Above-ground buckling might be a problem in piles that extend above the ground surface, such as those for railroad trestles, or those driven through bodies of water. In these cases, the above-ground portion must be checked using standard structural analysis methods, and may be braced if necessary.

Buckling is a greater concern during pile driving (Fleming et al., 1985), especially in long, slender piles driven through water. Contractors can handle these cases by limiting the hammer size, using lower power settings during the initial stages of driving, or providing temporary lateral support.

Comparison with Superstructure Design

Because deep foundations are designed so that underground buckling is not a concern, the structural design is similar to that for short columns in the superstructure. However, there are some important differences:

- The construction tolerances for foundations are much wider and quality control is more difficult.
- Piles can be damaged during driving, so the as-built capacity may be less than anticipated.

- *Residual stresses* may be locked into piles during driving, so the actual stresses in the piles after the structure is completed may be greater than those generated by the applied structural loads.
- Concrete in drilled shafts and other cast-in-place foundations is not placed under ideal conditions, and thus may experience segregation of the aggregates, contamination from the soil, and other problems.

Therefore, we use more conservative design criteria for deep foundations. This extra conservatism primarily appears in the form of lower allowable stresses and conservative simplifications in the analysis methods.

However, sometimes the allowable stresses for working loads, as permitted by building codes, are lower than necessary. This is because some of these design values were not based on stresses from service loads, but are an indirect way of keeping driving stresses within tolerable limits. In other words, piles designed using these low allowable stresses have a larger cross section, and thus have correspondingly smaller driving stresses. However, this roundabout way of limiting driving stresses is no longer necessary, because they can now be computed more accurately using wave equation analyses, as discussed in Chapter 15. Therefore, some have suggested using larger allowable stresses for working loads, so long as driving stresses are independently evaluated using wave equation analyses. However, the applicable codes must change before this can be done in practice.

ASD vs. LRFD

Structural engineers use two different methods of designing structural members:

- The *allowable stress design (ASD) method* (also known as the *working stress design method*), which is based on the stresses induced in the structural member when subjected to the design loads. The engineer compares these working stresses with the allowable stress, which is the strength divided by a factor of safety, to determine if the design is satisfactory.
- The *load and resistance factor design (LRFD) method* (also known as the *ultimate strength design method*), which increases the design loads through the use of load factors, then compares these to the ultimate load-bearing capacity.

Chapters 2 and 21 discuss the differences between ASD and LRFD in more detail.

ASD is the older of these two methods. Before the 1960s, all structural designs were based on this method. However, since then the LRFD method has slowly been gaining favor, and it will eventually replace ASD. Currently, virtually all reinforced concrete superstructures are designed using LRFD (which the American Concrete Institute calls USD or *ultimate strength design*), and steel design is in the process of transition from the older ASD code to the newer LRFD code. However, timber and masonry structures are still designed using ASD.

Curiously, deep foundations have been exempted from these new LRFD codes. Building codes (ICBO, BOCA, SBCCI, IBC) currently specify allowable stresses in deep foundations using ASD, even those made of concrete, regardless of the method used to design members in the superstructure.

To complicate matters even further, caps and grade beams that connect deep foundations are designed using LRFD, much the same way as we used LRFD to design spread footings in Chapter 9. Thus, portions of the structural design are based on ASD, and portions on LRFD.

Hopefully this confusing state-of-affairs will eventually be resolved by universally adopting LRFD for all analyses. In the meantime, design engineers need to be very careful to use the correct loads and the correct method when designing each element of a deep foundation system. The discussions in this chapter define which load to use for each analysis.

12.2   LOADS AND STRESSES

The structural design must consider both axial (compression or tension) and lateral (shear and moment) loads in the foundation. Torsion loads are usually negligible, but might need to be considered in special circumstances.

When using ASD, the axial tension or compression stress at a depth z in a foundation subjected to an axial load is:

$$\boxed{f_a = \frac{P}{A}}$$

(12.1)

The maximum fiber stress caused by a moment M in the foundation is:

$$\boxed{f_b = \frac{M}{S}}$$

(12.2)

Where:

f_a = average normal stress caused by axial load

f_b = normal stress in extreme fiber caused by flexural load

P = axial tension or compression force in the foundation at depth z

M = moment in the foundation at depth z

A = cross-sectional area

S = elastic section modulus

Because this is an ASD analysis, the values of P and M should be the unfactored loads computed using Equations 2.1 to 2.4, or from similar equations presented in the governing code.

Section moduli for steel H-piles and common steel pipe piles are tabulated in Tables 12.1 and 12.2. For foundations with solid circular cross sections of diameter B, such as timber piles, use:

$$S = \frac{2I}{B} = \frac{\pi B^3}{32}$$

(12.3)

For square cross sections with side width B, use:

$$S = \frac{2I}{B} = \frac{B^3}{6}$$

(12.4)

For foundations with a constant cross-sectional area, the stress f_a becomes smaller with depth as some of the axial load shifts to the soil through side friction. However, for simplicity engineers usually compute f_a using the entire design load (i.e., neglecting any load transfer due to side friction). The stress f_b can increase or decrease with depth according to the moment. We are primarily interested in the f_b because of the maximum moment in the foundation, M_{max}, which may be computed using the methods described in Chapter 16.

Because we are using the ASD method, the allowable axial and flexural stresses are F_a and F_b, respectively, and the design must satisfy the following condition:

$$\frac{f_a}{F_a} + \frac{f_b}{F_b} \le 1$$

(12.5)

Therefore, the presence of moment loads in a foundation reduces its axial load capacity.

It also is possible to develop an *interaction diagram* that shows all possible combinations of axial and moment loads for a given cross section. The use of interaction diagrams will be illustrated later in this chapter.

For analysis purposes, neglect any interaction between the shear loads and the axial or moment loads. For a working stress analysis, the shear stress, f_v, must not exceed the shear capacity, F_v:

$$f_v = \frac{V}{A}$$

(12.6)

$$f_v \le F_v$$

(12.7)

Where:

f_v = shear stress in foundation at depth z

V = shear force in foundation at depth z

A = cross-sectional area of foundation at depth z

F_v = allowable shear stress in foundation at depth z

The greatest shear stress occurs at the top of the foundation, and is computed using the unfactored applied shear force, V.

12.3   PILES

The structural design of piles must consider each of the following loading conditions:

- *Handling loads* are those imposed on the pile between the time it is fabricated and the time it is in the pile driver leads and ready to be driven. They are generated by cranes, forklifts, and other construction equipment.
- *Driving loads* are produced by the pile hammer during driving.
- *Service loads* are the design loads from the completed structure.

The most critical handling loads often occur when the pile is suspended in a nearly horizontal position from only one or two support points, as shown in Figure 12.1. This can produce flexural stresses that are greater than those from the service loads. Concrete piles are especially prone to damage from these loads because of their greater weight and smaller tensile strength. PCI (1993a) recommends computing handling stress based on the

Figure 12.1   This prestressed concrete pile is being lifted into the pile leads by a single cable. This process induces handling stresses in the pile.

weight of the pile plus an additional 50 percent for inertial and impact effects. The pile designer must accommodate these handling loads by specifying pickup points along the length of the pile, as shown in Figure 12.2.

Driving loads also are important, especially if the contractor uses a large hammer. Timber and concrete piles are especially prone to such damage. Driving stresses are primarily compressive, but significant tensile stresses can develop in some circumstances. Use a wave equation analysis, as discussed in Chapter 15, to predict these driving stresses, and thus guide the selection of an appropriate hammer and pile driving appurtenances.

Most of the design effort focuses on the service loads. It is often necessary to consider various combinations of service loads to determine the most severe combination.

Timber Piles

ASTM D25-91 specifies minimum dimensions for timber piles. However, the available sizes in a particular region depend on the height and species of trees available locally. Typically timber piles have a head diameter of 200 to 450 mm (8–18 in) and a toe diame-

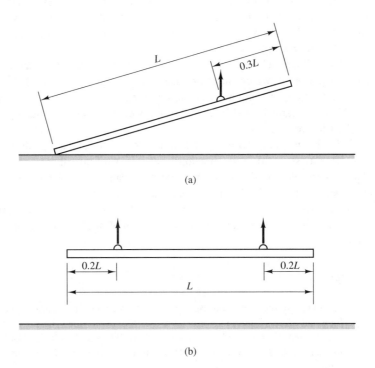

(a)

(b)

Figure 12.2 Using specified pickup points to keep handling stresses within tolerable limits: (a) single-point pickup; (b) double-point pickup.

ter of 125 to 250 mm (5–10 in). The length of timber piles is limited by the height of trees, and is typically 6 to 20 m (20–60 ft).

Because wood is a natural material, not a manufactured product, it is difficult to assign allowable design stresses. Design criteria for wood piles must consider many factors, including the following:

- The species of tree
- The quality of the wood (i.e., knots, straightness, etc.)
- The moisture content of the wood
- The extent of any damage incurred during driving
- The type and method of treatment (normally reduces strength)
- The number of piles in a pile group (redundancy if one pile is weak)

The vast majority of timber piles used in North America are either Southern pine or Douglas fir, and most building codes give an allowable axial stress under service loads, F_a, of about 8.3 MPa (1200 lb/in^2) for either type. This design value is probably satisfactory, as long as a wave equation analysis is used to check driving stresses. However, in the absence of a wave equation analysis, many engineers believe this allowable stress is too high (Armstrong, 1978; Davisson, 1989; Graham, 1985), especially when there is significant end bearing, and some have advocated values as low as 4.8 MPa (700 lb/in^2). This lower value for working loads results in a larger cross-section, which implicitly reduces the driving stresses.

The allowable extreme fiber stress caused by flexure (bending), F_b, is typically about $2F_a$. The allowable shear stress, F_v, is typically 0.09 F_a to 0.10 F_a. Because timber piles are tapered, they are not well suited to resist uplift loads. Therefore, we limit the allowable uplift capacity to 90 percent of the weight of the pile. Such piles will have minimal tensile stresses.

Timber piles are generally not subject to structural damage during handling, but the contractor must avoid large abrasions that might remove the preservative treatment and expose untreated wood. However, these piles can easily be damaged during driving, as described in Chapter 11. To avoid such damage, the maximum driving stresses should not exceed 20 MPa (3000 lb/in^2) (PDCA, 1998). This means that timber piles should only be driven with lightweight hammers and they should not be used at sites with hard driving conditions.

Example 12.1

A certain column load is to be supported on a group of 16 timber piles. Each pile is 30 ft long and has a 12-in diameter head and 7-in diameter toe. The top of each pile in the leading row will be subjected to an axial load of 40 k, and a shear load of 3.9 k. The maximum moment of 7.0 ft-k occurs at a depth of 8 ft. Is the structural design satisfactory? Use $f_a = 800$ lb/in^2.

Solution

1. Check the axial and flexural stresses at depth of maximum moment:

$$B \text{ at } 8 \text{ ft} = 12 - (8/30)(12 - 7) = 10.7 \text{ in.}$$

$$A = \frac{\pi B^2}{4} = \frac{\pi(10.7)^2}{4} = 89.9 \text{ in.}^2$$

Assume P at 8 ft = P at head (conservative).

$$f_a = \frac{P}{A} = \frac{40,000}{89.9} = 445 \text{ lb/in}^2$$

$$S = \frac{\pi B^3}{32} = \frac{\pi (10.7)^3}{32} = 120 \text{ in}^3$$

$$f_b = \frac{M}{S} = \frac{(7.0 \text{ ft-k})(12 \text{ in/ft})(1000 \text{ lb/k})}{120 \text{ in}^3} = 700 \text{ lb/in}^2$$

$$\frac{f_a}{F_a} + \frac{f_b}{F_b} = \frac{445}{800} + \frac{700}{1600} = 0.99 \leq 1 \quad \text{OK}$$

2. Check shear stress at the head of the pile (point of maximum shear):

$$F_v = 0.09 \, F_a = (0.09)(800) = 72 \text{ lb/in}^2$$

$$A = \frac{\pi B^2}{4} = \frac{\pi (12)^2}{4} = 113 \text{ in}^2$$

$$f_v = \frac{V}{A} = \frac{3900}{113} = 34.5 \text{ lb/in}^2 < 72 \quad \text{OK}$$

$$\therefore \textbf{Design is acceptable} \qquad \Leftarrow Answer$$

Steel Piles

Steel piles are usually made of mild steel that conforms to ASTM standard A36. This material has a yield strength, F_y, of 250 MPa (36,000 lb/in^2), and is adequate for most projects. Piles with F_y as high as 450 MPa (65,000 lb/in^2) are available, but they are not specified as often because most applications do not need the higher strength, and because they are more difficult to weld. In addition, some building codes do not permit the use of higher yield strengths in piles.

The allowable axial stress in steel piles, F_a, is typically $0.35F_y$ to $0.50F_y$ (87–124 MPa or 12,600–18,000 lb/in^2) for either tension or compression (Rempe, 1979). AASHTO is more conservative and uses $F_a = 0.25 \, F_y$ to $0.33 \, F_y$. The higher values in these two ranges are generally appropriate only when the driving conditions are favorable (i.e., where the pile will drive straight and not be deflected by boulders or other obstructions) and the driving stresses are checked with a wave equation analysis (as discussed in

Chapter 15). For comparison, engineers use an allowable stress of $0.60F_y$ to $0.66F_y$ for A36 steel in the superstructure.

Steel design methods used in the superstructure sometimes use a different allowable stress, F_b, for flexural (bending) stresses. However, for piles use $F_b = F_a$.

Structural engineers use an allowable shear stress of $F_v = 0.40 \, F_y$ in the superstructure. The greatest shear stress in piles occurs at the top, so the differences between the pile and the superstructure listed earlier are not as significant. Therefore, we may use the same allowable shear stress for piles. However, do not use the entire cross-sectional area for shear resistance. For H-piles, use only the area of the web; for pipe piles, use half of the total cross-sectional area.

Standard H-pile sections are listed in Table 12.1. Pipe piles are available in a wide variety of diameters and wall thicknesses; some of the more common sizes are listed in Table 12.2. Steel pipe piles have a constant moment of inertia, I, and section modulus, S,

TABLE 12.1   STANDARD STEEL H-PILE SECTIONS USED IN THE UNITED STATES[b] (*AISC Manual of Steel Construction*, American Institute of Steel Construction; Used with permission)

| | Area A (in^2) | | Depth B_1 (in) | Width B_2 (in) | $F_a A$ (k)[c] | X-X Axis | | Y-Y Axis | |
Designation[a]	Total	Web Only				I (in^4)	S (in^3)	I (in^4)	S (in^3)
HP 8×36	10.6	3.56	8.02	8.16	134	119	29.8	40	9.9
HP 10×42	12.4	4.05	9.70	10.07	156	210	43.4	72	14.2
HP 10×57	16.8	5.65	9.99	10.22	212	294	58.8	101	19.7
HP 12×53	15.5	5.11	11.78	12.04	195	393	66.8	127	21.1
HP 12×63	18.4	6.18	11.94	12.12	232	472	79.1	153	25.3
HP 12×74	21.8	7.34	12.13	12.21	275	569	93.8	186	30.4
HP 12×84	24.6	8.39	12.28	12.29	310	650	106	213	34.6
HP 13×60	17.5	5.75	12.54	12.90	221	503	80.3	165	25.5
HP 13×73	21.6	7.20	12.75	13.00	272	630	98.8	207	31.9
HP 13×87	25.5	8.65	12.95	13.10	321	755	117	250	38.1
HP 13×100	29.4	10.04	13.15	13.20	370	886	135	294	44.5
HP 14×73	21.4	6.88	13.61	14.58	270	729	107	261	35.8
HP 14×89	26.1	8.53	13.83	14.69	329	904	131	326	44.3
HP 14×102	30.0	9.87	14.01	14.78	378	1050	150	380	51.4
HP 14×117	34.4	11.47	14.21	14.88	433	1220	172	443	59.5

[a]The first number in the pile designation is the nominal section depth in inches. The second number is the weight in pounds per foot.

[b]Soft metric HP sections are the same size as those listed here.

[c]The allowable axial load when no moment is present ($F_a A$) is based on $F_a = 0.35 f_y$ and $f_y = 36$ k/in^2.

TABLE 12.2 COMMON STEEL PIPE PILE SECTIONS USED IN THE UNITED STATES

Designation[a]	Area, A (in^2)	Weight (lb/ft)	I (in^4)	S (in^3)	$F_a A$ (k)[b]
PP 12.75×0.250	9.82	33	192	30.1	124
PP 12.75×0.375	14.58	50	279	43.8	184
PP 12.75×0.500	19.24	65	362	56.7	242
PP 12.75×0.625	23.81	81	439	68.8	300
PP 12.75×0.750	28.27	96	511	80.1	356
PP 14×0.250	10.80	37	255	36.5	136
PP 14×0.375	16.05	55	373	53.3	202
PP 14×0.500	21.21	72	484	69.1	267
PP 14×0.625	26.26	89	589	84.1	331
PP 14×0.750	31.22	106	687	98.2	393
PP 14×1.000	40.84	139	868	124.0	515
PP 16×0.250	12.37	42	384	48.0	156
PP 16×0.375	18.41	63	562	70.3	232
PP 16×0.500	24.35	83	732	91.5	307
PP 16×0.625	30.19	103	894	111.7	380
PP 16×0.750	35.93	122	1047	130.9	453
PP 16×1.000	47.12	160	1331	166.4	594
PP 18×0.250	13.94	47	549	61.0	176
PP 18×0.375	20.76	71	807	89.6	262
PP 18×0.500	27.49	94	1053	117.0	346
PP 18×0.625	34.12	116	1289	143.2	430
PP 18×0.750	40.64	138	1515	168.3	512
PP 18×1.000	53.41	182	1936	215.1	673
PP 20×0.375	23.12	79	1113	111.3	291
PP 20×0.500	30.63	104	1457	145.7	386
PP 20×0.625	38.04	129	1787	178.7	479
PP 20×0.750	45.36	154	2104	210.4	571
PP 20×1.000	59.69	203	2701	270.1	752
PP 24×0.500	36.91	126	2549	212.4	465
PP 24×0.750	54.78	186	3705	308.8	690
PP 24×1.000	72.26	246	4787	398.9	910
PP 24×1.250	89.34	304	5797	483.1	1126
PP 24×1.500	106.03	361	6739	561.6	1336
PP 30×0.500	46.34	158	5042	336.1	584
PP 30×0.750	68.92	235	7375	491.7	868
PP 30×1.000	91.11	310	9589	639.3	1148
PP 30×1.250	112.90	384	11687	779.1	1423
PP 30×1.500	134.30	457	13674	911.6	1692

[a]The first number in the pile designation is the outside diameter; the second number is the wall thickness.
[b]The allowable axial load when no moment is present ($F_a A$) is based on $F_a = 0.35 f_y$ and $f_y = 36$ k/in^2.

regardless of the direction of the lateral load. However, I and S for H-piles depends on the direction of the load relative to the web. Usually the designer has no control over the as-built web orientation, so we must use I and S of the weak (Y-Y) axis, but sometimes the orientation is specified and the X-X axis properties may be used.

Because of their high strength and ductility, steel piles are normally not subject to damage during handling. However, they might be damaged during driving, especially if the contractor uses a large hammer. PDCA (1998) recommends limiting driving stresses to $0.9\,F_y$.

Example 12.2

A large sign is to be supported on a single, 400-mm diameter free-head steel pipe pile with 10-mm wall thickness. The sign will impose a vertical downward load of 20 kN, a shear load of 12 kN, and an overturning moment of 95 kN-m onto the top of the pile. The pile is made of A36 steel. Is this design adequate?

Solution

Because this is a free-head pile (as defined in Chapter 16), the maximum moment is equal to the applied moment.

Check axial and flexural stresses at the top of the pile.

$$A = \frac{\pi\, 0.400^2}{4} - \frac{\pi\, 0.380^2}{4} = 0.0123 \text{ m}^2$$

$$f_a = \frac{20}{0.0123} = 1{,}600 \text{ kPa} = 1.6 \text{ MPa}$$

$$I = \frac{\pi\, 0.400^4}{64} - \frac{\pi\, 0.380^4}{64} = 2.33 \times 10^{-4} \text{ m}^4$$

$$S = \frac{2I}{B} = \frac{2\,(2.33 \times 10^{-4} \text{ m}^4)}{0.400 \text{ m}} = 1.17 \times 10^{-3} \text{ m}^3$$

$$f_b = \frac{M}{S} = \frac{95}{1.17 \times 10^{-3}} = 81{,}500 \text{ kPa} = 81.5 \text{ MPa}$$

$$F_y = 250 \text{ MPa}$$

$$F_a = F_b = 0.35\,F_y = (0.35)(250) = 87 \text{ MPa}$$

$$\frac{f_a}{F_a} + \frac{f_b}{F_b} = \frac{1.6}{87} + \frac{81.5}{87} = 0.96 < 1.0 \quad \text{OK}$$

Check shear stresses using half of the cross-sectional area.

$$f_v = \frac{12}{0.0123/2} = 1950 \text{ kPa} = 1.95 \text{ MPa}$$

$$F_v = 0.4\,F_y = (0.4)(250) = 100 \text{ MPa}$$

$$f_v \ll F_v \quad \text{OK}$$

The design is satisfactory.

Note how the flexural stresses dominate this design.

Concrete-Filled Steel Pipe Piles

When empty steel pipe piles do not provide sufficient structural capacity, engineers sometimes fill them with concrete. The concrete increases both the axial and lateral structural load capacity, and provides some corrosion protection to the interior of the pipe. However, concrete infilling does not improve the geotechnical downward load capacity (because even open-end pipe piles generally become fully plugged). The geotechnical uplift capacity increases slightly, because of the added weight of the concrete.

Design concrete-filled steel pipe piles using a working stress analysis:

For axial compression loads:

$$F_a = \frac{0.35\, F_y A_s + 0.33\, f_c' A_c}{A} \tag{12.8}$$

For axial tension loads:

$$F_a = \frac{0.35\, F_y A_s}{A} \tag{12.9}$$

For flexural loads:

$$F_b = 0.35\, F_y \tag{12.10}$$

Where:

F_a = allowable axial stress based on total cross section

F_b = allowable flexural stress based on total cross section

F_y = yield stress of steel (usually 250 MPa or 36,000 lb/in²)

f_c' = 28-day compressive strength of concrete

A_s = cross-sectional area of steel

A_c = cross-sectional area of concrete

A = total cross-sectional area = $A_s + A_c$

The coefficient of the first term in Equations 12.8 to 12.10 may be increased to 0.50 if the driving conditions are especially favorable and a wave equation analysis is performed, as described earlier.

Once again, the AASHTO values are more conservative. They limit concrete stresses to $0.40\, f_c'$ and steel stresses to $0.25\, F_y$.

Prestressed Concrete Piles

The model building codes and the AASHTO code provide specific requirements on detailing prestressed concrete piles, and the following requirement for allowable compressive stress:

$$F_a = 0.33 f_c' - 0.27 f_{pc}$$ (12.11)

Where:

F_a = allowable compressive stress in concrete due to axial load

f_c' = 28-day compressive strength of concrete

f_{pc} = effective prestress stress on the gross section

This requirement is suitable for piles subjected only to axial compression, but provides no assistance for piles subjected to lateral or uplift loads.

The American Concrete Institute ACI 318-99 code (ACI, 1999) specifically excludes the design of concrete piles [1.1.5] except for some special seismic requirements [21.8.4]. The code commentary references ACI 543R-74 *Recommendations for Design, Manufacture, and Installation of Concrete Piles* (ACI, 1974), PCI (1993a), and the "general building code," which refers to other governing codes, such as the IBC. The provisions in ACI 543 use working stress (ASD) analyses with transformed sections based on now obsolete sections of ACI 318-63. Thus, at least from a code point-of-view, the structural design requirements for prestressed concrete piles are surprisingly vague and incomplete, especially when lateral loads are present.

The most workable solution to this dilemma is probably to design prestressed concrete piles using the code requirements for normal prestressed members, then check the design against Equation 12.11. The Precast/Prestressed Concrete Institute (PCI, 1993a) has developed standard designs for prestressed concrete piles based on this approach. Figure 12.3 and Tables 12.3 and 12.4 present the dimensions and other data, along with the allowable compressive load (based on ASD) as computed using Equation 12.11. Figure 12.4 is an *interaction diagram* for some of these piles. This interaction diagram shows the various combinations of factored normal and moment loads that may be sustained. PCA (1993) provides a much more detailed set of interaction diagrams for a wider range of pile sizes.

Use the following procedure to size a prestressed concrete pile:

1. Using Equations 2.1 to 2.4 (or comparable equations from the governing code) compute the unfactored compressive load acting on the pile.

2. Using the allowable compressive loads listed in the last four columns of Table 12.3 or 12.4, select a pile size. These allowable loads are based on Equation 12.11, and thus provide a pile that satisfies this code requirement. The optimal size is the smallest pile that has an allowable capacity $(F_a A)$ at least as large as the unfactored compressive load computed in Step 1. If no tensile or moment loads are present, then use this as the design pile size. Otherwise, continue with the remainder of this procedure.

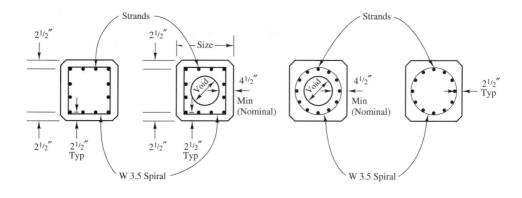

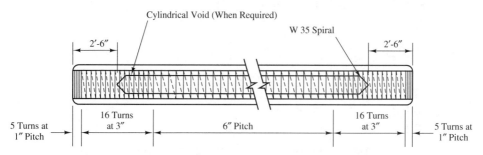

Figure 12.3   Standard prestressed pile designs (PCI, 1984).

3. Using Equations 2.7 to 2.17, compute the ACI factored compressive, tensile, and moment loads. The moment load should be based on the maximum moment, which is normally obtained using the techniques described in Chapter 16.

4. Using Figure 12.4 (or the more elaborate set of interaction diagrams provided in PCI, 1993a), select the required pile size. The optimal size is the smallest pile that satisfies both of the following conditions for both tension and compression:

$$P_u \leq \phi P_n$$

$$M_u \leq \phi M_n$$

5. For design, use the larger of the piles sizes obtained from Steps 2 and 4.

Prestressed concrete piles subjected to seismic loads may be prone to a brittle failure at the connection with the pile cap. To avoid this problem, many engineers embed mild steel bars into the top of the pile, as shown in Figure 11.12, and use them to connect it with the pile cap. These bars are more ductile than the prestressing strands.

Prestressed concrete piles also must be able to withstand handling and driving stresses. PCI (1993a) recommends designing for the maximum handling and driving stresses described in Table 12.5. These computations are based on the unfactored loads.

TABLE 12.3 STANDARD SQUARE PRESTRESSED CONCRETE PILES (Adapted from PCI, 1993a)

Pile Size (in)	Core Diameter (in)	Area A (in²)	Weight (lb/ft)	Moment of Inertia I (in⁴)	Section Modulus S (in³)	Perimeter (ft)	F_aA in Compression (k) $f_{pc} = 700$ lb/in² f'_c (lb/in²)			
							5000	6000	7000	8000
10	Solid	100	104	833	167	3.33	146	178	212	244
12	Solid	144	150	1,728	288	4.00	210	258	304	352
14	Solid	196	204	3,201	457	4.67	286	350	416	480
16	Solid	256	267	5,461	683	5.33	374	458	542	628
18	Solid	324	338	8,748	972	6.00	472	580	688	794
20	Solid	400	417	13,333	1,333	6.67	584	716	848	980
20	11	305	318	12,615	1,262	6.67	444	546	646	746
24	Solid	576	600	27,648	2,304	8.00	840	1,030	1,220	1,410
24	12	463	482	26,630	2,219	8.00	676	828	982	1,134
24	14	422	439	25,762	2,147	8.00	616	754	894	1,034
24	15	399	415	25,163	2,097	8.00	582	714	846	976
30	18	646	672	62,347	4,157	10.00	942	1,156	1,370	1,582
36	18	1,042	1,085	134,815	7,490	12.00	1,522	1,866	2,210	2,552

TABLE 12.4 STANDARD OCTAGONAL AND ROUND PRESTRESSED CONCRETE PILES (Adapted from PCI, 1993a)

Pile Size (in)	PileShape/ Core Diameter (in)	Area A (in²)	Weight (lb/ft)	Moment of Inertia I (in⁴)	Section Modulus S (in³)	Perimeter (ft)	$F_a A$ in Compression (k) $(f_{pc} = 700$ lb/in²) f'_c			
							5000	6000	7000	8000
10	Oct/Solid	83	85	555	111	2.76	120	148	176	202
12	Oct/Solid	119	125	1,134	189	3.31	172	212	252	290
14	Oct/Solid	162	169	2,105	301	3.87	236	290	344	396
16	Oct/Solid	212	220	3,592	449	4.42	308	378	448	518
18	Oct/Solid	268	280	5,705	639	4.97	390	480	568	656
20	Oct/Solid	331	345	8,770	877	5.52	482	592	702	810
20	Oct/11	236	245	8,050	805	5.52	342	422	500	578
22	Oct/Solid	401	420	12,837	1,167	6.08	584	718	850	982
22	Oct/13	268	280	11,440	1,040	6.08	390	480	566	656
24	Oct/Solid	477	495	18,180	1,515	6.63	696	854	1012	1168
24	Oct/15	300	315	15,696	1,318	6.63	438	536	636	736
36	Round/26	487	507	60,007	3,334	9.43	710	872	1032	1192
48	Round/38	675	703	158,222	6,592	12.57	986	1208	1430	1654

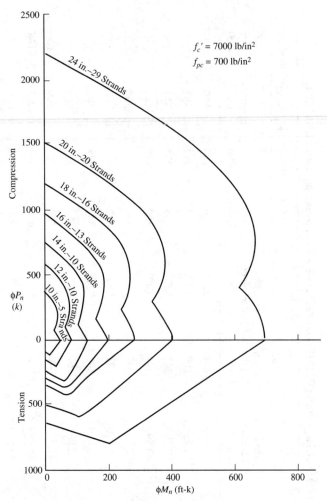

Figure 12.4   Interaction diagrams for selected PCI standard square prestressed concrete piles (Compiled from PCI, 1993a). These capacities are based on the ACI factored loads (Equations 2.7-2.17).

Example 12.3

A square prestressed concrete pile must sustain the following design service loads:
Axial compression
   Dead load = 100 k
   Live load = 120 k
Moment
   Dead load = 50 ft-k
   Live load = 20 ft-k

Select a pile size that will safely sustain these loads.

TABLE 12.5   ALLOWABLE HANDLING AND DRIVING STRESSES IN PRESTRESSED CONCRETE PILES (Adapted from PCI, 1993a; Used with permission)

Loading Condition	Allowable Stress on Gross Section	
	(MPa)	(lb/in^2)
Handling stresses[a]		
Flexure	$F_b = 0.50\,(f_c')^{0.5} + f_{pc}$	$F_b = 6\,(f_c')^{0.5} + f_{pc}$
Driving stresses		
Compression	$0.85\,f_c' - f_{pc}$	$0.85\,f_c' - f_{pc}$
Tension	$0.50\,(f_c')^{0.5} + f_{pc}$	$6\,(f_c')^{0.5} + f_{pc}$

[a]Computed using 1.5 times the weight of the pile

Solution

The unfactored compressive load is:

$$P = P_D + P_L = 100 + 120 = 220 \text{ k}$$

If we use $f_c' = 7000$ lb/in^2, the minimum pile size is 12-inch square (per Table 12.3)

Figure 12.4 is based on the factored loads, so it is necessary to apply the ACI load factors:

$$P_u = 1.4P_D + 1.7P_L = (1.4)(100) + (1.7)(120) = 344 \text{ k}$$

$$M_u = 1.4M_D + 1.7M_L = (1.4)(50) + (1.7)(20) = 104 \text{ ft-k}$$

Based on Figure 12.4 and the requirements that $P_u \le \phi P_n$ and $M_u \le \phi M_n$, we must use a 14-inch square pile.

    Use a 14-inch square prestressed concrete pile with
    $f_c' = $ **7000 lb/in^2 and 10 strands of prestressing steel**   ⇐ *Answer*

QUESTIONS AND PRACTICE PROBLEMS

12.1 An HP 13×60 pile is made of steel with $F_y = 50$ k/in^2 and is to be loaded in compression only (i.e., no moment or shear loads). Using $F_a = 0.35\,F_y$, compute the maximum allowable compressive load on this pile.

12.2 A pipe pile made of A36 steel ($F_y = 36$ k/in^2) must sustain a compressive load of 200 k and a moment load of 40 ft-k. Using $F_a = 0.35\,F_y$, select an appropriate pile size from the standard piles listed in Table 12.2.

12.3 A PP20×0.500 pile made of A36 steel ($F_y = 36$ k/in^2) is to be filled with concrete having a 28-day compressive strength of 3000 lb/in^2. This pile is to be loaded in compression only (i.e., no moment or shear loads). Compute the maximum allowable compressive load.

12.4 A group of twelve prestressed concrete piles is to support a building column. The total downward design load on the group is 4900 k. Assuming this load is evenly distributed across the 12 piles, and that there are no moment or shear loads, select a pile size and 28-day compressive strength of the concrete.

12.5 An 18-inch square prestressed concrete pile with $f_c' = 7000 \; lb/in^2$ is to sustain the following loads:

> compressive dead load = 400 k
>
> compressive live load = 200 k
>
> flexural dead load = 60 ft-k
>
> flexural live load = 75 ft-k

Is this design satisfactory? Justify the reason for your answer.

12.4 DRILLED SHAFTS

Drilled shaft foundations also suffer from a lack of well-defined code provisions for structural design. The model building codes all specify a maximum compressive stress for axial loads:

$$\boxed{F_a = 0.33 \, f_c'} \tag{12.12}$$

Combining Equations 12.1 and 12.12, and setting $f_a = F_a$ gives the code minimum diameter for drilled shafts:

$$\boxed{B = \sqrt{\frac{3.86 \, P}{f_c'}}} \tag{12.13}$$

Where:

> B = minimum shaft diameter to satisfy Equation 12.12
>
> P = unfactored compressive load
>
> f_c' = 28-day compressive strength of concrete

The design diameter should be a multiple of 200 mm or 6 in.

The American Concrete Institute ACI 318-99 code (ACI, 1999) specifically excludes deep foundations [1.1.5] (except for certain seismic provisions), while the AASHTO *Standard Specifications for Highway Bridges* (AASHTO, 1996) uses the same design criteria as for reinforced concrete columns [4.6.6.1], and give the designer the option of using ASD or LRFD.

Therefore, it is probably best to design drilled shafts so that they satisfy both the standard ACI design criteria for short columns and Equation 12.13. This requires evaluating the shaft using both factored loads (for the ACI design) and unfactored loads (for Equation 12.13).

Use of Interaction Diagrams

The interaction diagrams in Figures 12.6 to 12.9 are dimensionless plots that present the ACI design requirements for spirally reinforced circular short columns. These diagrams are based on the ratio of the longitudinal reinforcing steel circle diameter to the overall shaft diameter, as shown in Figure 12.5. These interaction diagrams are based on $f_c' = 4000$ lb/in^2 (27.6 MPa) and $f_y = 60$ k/in^2 (413.7 MPa). Additional interaction diagrams can be developed for other material strengths.

These interaction diagrams produce the required steel ratio, ρ to resist the combined axial and flexural loads:

$$\rho = \frac{A_s}{A_g} = \frac{4 A_s}{\pi B^2} \qquad (12.14)$$

Where:

ρ = steel ratio

A_s = cross-sectional area of longitudinal steel bars

A_g = gross cross-sectional area of drilled shaft

B = diameter of drilled shaft

The following procedure may be used to develop structural designs for drilled shafts based on these interaction diagrams and the model building code requirement:

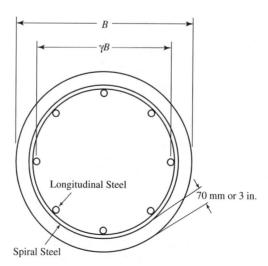

Figure 12.5 Definition of γ for the interaction diagrams in Figures 12.6 to 12.9.

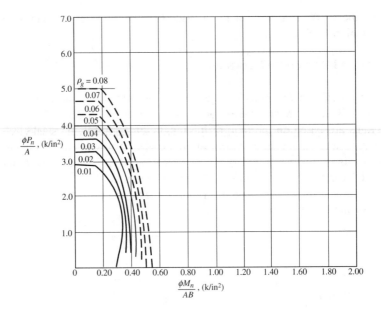

Figure 12.6 Interaction diagram for spirally-reinforced drilled shafts with $f'_c = 4000$ lb/in² (27.6 MPa), $f_y = 60$ k/in² (413.7 MPa) and $\gamma = 0.45$. (Adapted from McCormac, 1998, used with permission.)

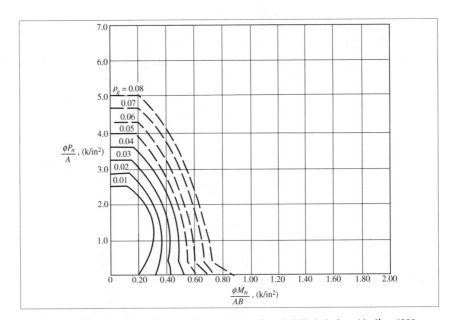

Figure 12.7 Interaction diagram for spirally-reinforced drilled shafts with $f'_c = 4000$ lb/in² (27.6 MPa), $f_y = 60$ k/in² (413.7 MPa) and $\gamma = 0.60$. (Adapted from McCormac, 1998, used with permission.)

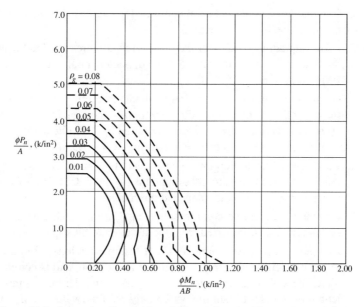

Figure 12.8 Interaction diagram for spirally-reinforced drilled shafts with $f'_c = 4000$ lb/in^2 (27.6 MPa), $f_y = 60$ k/in^2 (413.7 MPa) and $\gamma = 0.75$. (Adapted from McCormac, 1998, used with permission.)

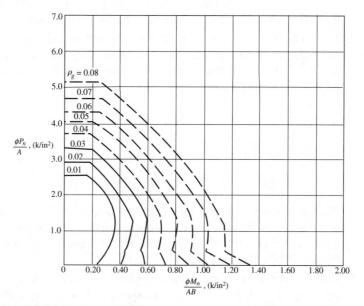

Figure 12.9 Interaction diagram for spirally-reinforced drilled shafts with $f'_c = 4000$ lb/in^2 (27.6 MPa), $f_y = 60$ k/in^2 (413.7 MPa) and $\gamma = 0.90$. (Adapted from McCormac, 1998, used with permission.)

1. Compute the unfactored axial compressive load using Equations 2.1 to 2.4 (or similar equations from a governing code).

2. Using Equation 12.13 with the unfactored compressive load from Step 1, compute the minimum shaft diameter, B. This diameter should be expressed as a multiple of 200 mm or 6 in.

3. Compute the diameter of the longitudinal reinforcing steel circle. Assuming #8 bars (metric #25) for the longitudinal steel, 1/2-inch diameter spiral reinforcement, and 70 mm (3 in) of concrete cover as required by ACI, the diameter of the longitudinal steel circle is approximately $B - 180$ mm or $B - 7.5$ in.

4. Use Figure 12.5 to compute γ and select the proper interaction diagram based on this value. This book provides diagrams for $\gamma = 0.45$, 0.60, 0.75, and 0.90. It may be necessary to use two diagrams and interpolate.

5. Compute the factored compressive load P_u using Equations 2.7 to 2.17. This is the compressive load imparted from the structure to the top of the foundation.

6. Compute the factored moment load M_u using Equations 2.7 to 2.17. This is the maximum moment load at any point in the foundation, and is computed using the techniques described in Chapter 16. If we wish to use the same reinforcing for the entire length of the shaft, then this is the only required value of M_u. Alternatively, if we wish to cut off part of the reinforcing at various depths (because of the reduction in moment with depth), then we need to compute M_u as a function of depth. Chapter 16 describes methods of developing this data.

7. Compute P_u/A and M_u/AB, where A is the gross cross-sectional area of the shaft (i.e., the combined area of the concrete and the reinforcing steel).

8. The interaction diagram(s) present the compressive and moment capacities, $\phi P_n/A$ and $\phi M_n/AB$ for various steel ratios, ρ. To use these diagrams, plot the values P_u/A and M_u/AB from Step 7 and select the required steel ratio, ρ, such that both of the following conditions are met:

$$\frac{P_u}{A} \leq \frac{\phi P_n}{A}$$

$$\frac{M_u}{AB} \leq \frac{\phi M_n}{AB}$$

The value of ρ must be between 0.01 and 0.08. If the γ value demands the use of two interaction diagrams, then determine ρ from each and interpolate. If $\rho > 0.08$, increase the diameter, B, and recompute ρ. However, do not use a B less than that computed from Equation 12.13.

9. Using Equation 12.14 and Table 9.1, select the size and number of longitudinal reinforcing bars. There must be at least six bars [ACI 10.9.2], and the clear space between bars must be at least equal to d_b, 25 mm (1 in), and 4/3 times the maximum aggregate size, whichever is greatest [ACI 3.3.2 and 7.6.1].

10. Size the spiral steel using the following equation [ACI 10.9.3]:

$$\rho_s = 0.45 \left(\frac{A}{A_c} - 1 \right) \frac{f_c'}{f_y}$$

(12.15)

Where:

ρ_s = ratio of volume of spiral reinforcement to total volume of core

A = gross cross-sectional area of drilled shaft = $\pi B^2/4$

A_c = cross-sectional area of core (i.e., area of circle bound by outside diameter of spiral reinforcement)

f_c' = 28-day compressive strength of concrete

f_y = yield strength of steel (maximum 420 MPa or 60,000 lb/in^2)

For design purposes, #3 or #4 bars (metric #10 or #13) might be sufficient spiral reinforcement, but construction procedures may dictate larger diameter rod (i.e., #6 or metric #19) in order to avoid excessive distortions of the cage. The pitch (i.e. the vertical distance between the spiral rods) is typically between 75 and 150 mm (3–6 in).

See O'Neill and Reese (1999) for more information on structural design of drilled shafts.

Example 12.4

The maximum compressive and moment loads in a drilled shaft foundation are as follows:
Compression
$P_D = 300$ k
$P_L = 260$ k
Moment
$M_D = 240$ ft-k
$M_L = 80$ ft-k

Using $f_c' = 4000$ lb/in^2 and $f_y = 60,000$ lb/in^2, determine the required diameter and reinforcement.

Solution

Unfactored load:

$$P = P_D + P_L = 300 + 260 = 560 \text{ k}$$

Required diameter:

$$B = \sqrt{\frac{3.86\,P}{f_c'}} = \sqrt{\frac{3.86(560{,}000 \text{ lb})}{4000 \text{ lb/in}^2}} = 23 \text{ in}$$

Use $B = 24$ in

$$\gamma = \frac{24 - 7.5}{24} = 0.69$$

Factored loads:

$$P_u = 1.4\,P_D + 1.7\,P_L = (1.4)(300) + (1.7)(260) = 862 \text{ k}$$
$$M_u = M_D + M_L = (1.4)(240) + (1.7)(80) = 472 \text{ ft-k}$$

Longitudinal reinforcement:

$$A = \frac{\pi(24)^2}{4} = 452 \text{ in}^2$$

$$\frac{P_u}{A} = \frac{862}{452} = 1.91 \text{ k/in}^2$$

$$\frac{M_u}{AB} = \frac{472 \text{ ft-k}}{(452 \text{ in}^2)(2 \text{ ft})} = 0.52 \text{ k/in}^2$$

Per Figure 12.7 ($\gamma = 0.60$): $\rho = 0.055$
Per Figure 12.8 ($\gamma = 0.75$): $\rho = 0.038$
By linear interpolation, use $\rho = 0.045$

$$(A_s)_{reqd} = \frac{\rho\,\pi\,B^2}{4} = \frac{0.045\,\pi\,(24)^2}{4} = 20.4 \text{ in}^2$$

Use sixteen #10 bars (satisfies required steel area, as well as minimum and maximum spacing requirements).

Spiral reinforcement:

$$A_c = \frac{\pi\,(24 - 6)^2}{4} = 254 \text{ in}^2$$

$$\rho_s = 0.45\left(\frac{A}{A_c} - 1\right)\frac{f'_c}{f_y}$$

$$= 0.45\left(\frac{452}{254} - 1\right)\frac{4000}{60000}$$

$$= 0.023$$

Let p = pitch

Volume of steel per turn = $18\,\pi\,A_s = 56.5\,A_s$

Volume of core per turn = $\pi\,18^2\,p\,/\,4 = 254\,p$

$$\rho_s = \frac{56.5\,A_s}{254\,p} = 0.023$$

For #6 bars ($A_s = 0.44 \text{ in}^2$), $p = 4.2$ in

Use #6 bars at 4 inches on center

Final Design:

**Use 24-inch diameter shaft with sixteen longitudinal #10 bars
and #6 spiral reinforcement at 4 inches on center**   ⇐ *Answer*

QUESTIONS AND PRACTICE PROBLEMS

12.6 A drilled shaft foundation will be subjected to the following design service loads: $P_D = 800$ k, $P_L = 350$ k, $M_D = 400$ ft-k, $M_L = 320$ ft-k. Using $f_c' = 4000$ lb/in^2 and $f_y = 60$ k/in^2, determine the required diameter (considering structural aspects only) and reinforcement.

12.7 A drilled shaft foundation will be subjected to the following design service loads: $P_D = 320$ k, $P_L = 150$ k, $M_D = 350$ ft-k, $M_L = 380$ ft-k. Using $f_c' = 4000$ lb/in^2 and $f_y = 60$ k/in^2, determine the required diameter (considering structural aspects only) and reinforcement.

12.5   CAPS

Pile foundations are nearly always placed in groups, so a single column is supported by multiple piles. Sometimes drilled shaft foundations also are placed in groups, although a single large-diameter shaft for each column is often more economical, especially in buildings. The reinforced concrete element that connects the column with its multiple deep foundations is called a *cap* or a *pile cap*. Figure 12.10 shows a cap under construction, and Figure 12.11 shows a completed cap.

The process of designing pile caps is very similar to that for spread footings. Both designs must distribute concentrated loads from the column across the bottom of the footing or cap. The primary differences are:

- The loads acting on caps are larger than those on most spread footings
- The loads on caps are distributed over a small portion of the bottom (i.e., the individual deep foundations), whereas those on spread footings are distributed across the entire base.

Although the ACI code does not govern the design of piles or drilled shafts, it does encompass the design of caps, which it calls "footings on piles" [15.2.3]. The design process for caps is very similar to that for spread footings, with the following additional requirements:

- The design must satisfy punching shear in the vicinity of the individual piles or shafts [ACI 15.5]
- The effective depth, d, must be at least 300 mm (12 in) [ACI 15.7]. This implies a minimum thickness T or 400 mm or 18 in.
- The bearing force between the individual piles or shafts and the cap must not exceed the capacity of either element [ACI 15.8.1]

12.6   GRADE BEAMS

Deep foundations are sometimes connected with grade beams, as shown in Figure 12.10. The Uniform Building Code requires grade beams for all deep foundations subjected to seismic loads [UBC 1807.2] and specifies they be designed to resist a horizontal compres-

Figure 12.10   This group of nine prestressed concrete piles (three rows of three piles) will support a single column for a building. They will be connected using a reinforced concrete cap. In addition, a grade beam will connect this cap with the adjacent caps. The vertical steel bars extending from the top of each pile can be seen in the photograph, and some of the cap and grade beam steel is now in place. Notice the series of stirrups in the grade beam. Once all of the steel is in place, the cap and the grade beam will be poured monolithically.

sive or tensile force equal to 10 percent of the column vertical load. However, the code does provide an exception if another suitable form of lateral restraint is provided.

Grade beams are designed the same as any member in the superstructure, and should not rely on any support from the underlying soil.

SUMMARY

Major Points

1. Deep foundations must have sufficient structural integrity to transfer the applied loads from the structure to the ground. This means they must have sufficient strength to safely sustain the applies stresses, and sufficient stiffness to keep the deformations within tolerable limits.

2. Underground buckling under service loads is usually not a problem with deep foundations.

Figure 12.11 This bridge bent is supported on a group of pile foundations. The reinforced concrete cap connects the bent and the foundations. Because of the clean sandy soils at this site, this cap was cast using plywood forms which produced the smooth sides. However, in more cohesive soils, caps usually are cast directly against the soil.

3. Deep foundations are usually designed more conservatively than similar members in the superstructure.

4. The design standards currently include both ASD and LRFD methods for structural design. This can be confusing. Timber and steel piles are normally designed exclusively using ASD, while prestressed concrete piles and drilled shafts are designed using a combination of ASD and LRFD.

5. Piles must be designed to accommodate handling loads, driving loads, and service loads. Drilled shafts are designed only for service loads.

6. The allowable stresses for various pile materials are defined in building codes, but these criteria are often incomplete.

7. Caps are designed using methods similar to those for spread footings, with some additional requirements.

8. Some codes require grade beams to connect deep foundations that are subjected to seismic loads.

Vocabulary

Allowable stress design (ASD)	Interaction diagram	Section modulus
Buckling	Load and Resistance Factor Design (LRFD)	Service loads
Driving loads	Model building code	Spiral reinforcement
Grade beam	Pile cap	Ultimate strength design (USD)
Handling loads	Residual stresses	Working stress design

COMPREHENSIVE QUESTIONS AND PRACTICE PROBLEMS

12.8 Why is it appropriate to use more conservative structural designs for foundations than for comparable superstructure members?

12.9 What type of pile is least prone to damage during handling?

12.10 A PP 18×0.500 steel pipe pile made of A36 steel carries a compressive load of 200 k. Compute the maximum allowable moment and shear forces that keep the pile stresses within acceptable limits.

12.11 Develop an interaction diagram (a plot of allowable moment load vs. allowable compressive load) for an HP 12×84 pile made of A36 steel. Assume bending occurs along the weak axis.

12.12 A 12-inch square, 60-ft long prestressed concrete pile with $f_c' = 6000$ lb/in^2 and $f_{pc} = 700$ lb/in^2 is to be lifted at a single point as shown in Figures 12.1 and 12.2a. Using the weight of the pile plus a 50 percent increase for inertial and impact effects, draw shear and moment diagrams for this pile. Then compute the resulting flexural stresses and determine if this pickup arrangement satisfies the PCI allowable stress criteria. If it does not, then suggest another method of pickup that is acceptable.

12.13 A 15-m long timber pile with a head diameter of 250 mm will be subjected to a compressive load of 500 kN. There are no shear or moment loads. Is this design satisfactory from a structural engineering perspective?

12.14 An H-pile made from A36 steel must support a compressive load of 250 k and a moment load of 50 ft-k. Select the most economical H-pile section. Assume the driving conditions are favorable, the driving stresses will be evaluated by a wave equation analysis, and the project is governed by one of the model building codes.

12.15 A prestressed concrete pile must sustain the following design loads: $P_D = 85$ k, $P_L = 70$ k, $M_D = 80$ ft-k, and $M_L = 70$ ft-k. Select a pile size that will safely sustain these loads.

12.16 Select the most inexpensive A36 steel pipe pile that will carry a compressive load of 375 k, a shear load 20 k, and a moment load of 200 ft-k simultaneously. Assume the price of steel per pound is the same for every pile size.

12.17 A drilled shaft foundation is to be made of concrete with $f_c' = 27.6$ MPa (4000 lb/in^2) and reinforced with grade 60 steel (metric grade 420). The design loads are $P_D = 1100$ kN, $P_L = 950$ kN, $M_D = 400$ kN-m, and $M_L = 250$ kN-m. Determine the required diameter and reinforcing that will sustain these loads.

13

Deep Foundations—Axial Load Capacity Based on Static Load Tests

In reality, soil mechanics is only one of the bodies of knowledge upon which the foundation engineer must draw. If studied to the exclusion of other aspects of the art, it leads to the erroneous and dangerous impression that all problems in foundation engineering are susceptible of direct scientific solution. Unfortunately, the vagaries of nature and the demands of economy combine to eliminate this possibility.

From *Foundation Engineering* by Peck, Hanson and Thornburn. Copyright ©1974 by John Wiley and Sons, Inc. Reprinted by permission.

The most important geotechnical design requirement for most deep foundations is that they have sufficient axial load capacity to support the applied loads. Therefore, geotechnical engineers have developed methods of evaluating axial load capacity and use these methods to properly size the foundations. We will divide these methods into three categories:

- Full-scale static load tests on prototype foundations
- Analytic methods, which are based on soil properties obtained from laboratory or in-situ tests
- Dynamic methods, which are based on the dynamics of pile driving or wave propagation

This chapter discusses full-scale static load tests; Chapters 14 and 15 discuss analytic and dynamic methods, respectively.

Many deep foundations also must support significant lateral loads, and these loads may control certain aspects of the design. Chapter 16 discusses methods of evaluating lateral load capacity. Finally, Chapter 17 draws this material together and discusses how to develop a deep foundation design.

It is important to be very careful when using the term *load capacity*. It can refer either to the *ultimate load capacity*, which is the load required to cause failure, or the *allowable load capacity*, which is the ultimate load capacity divided by a factor of safety. Confusion between these two definitions has often been a source of misunderstandings and litigation. Therefore, the term *load capacity* without any adjective should be used only in a generic sense, such as in the title of this chapter. However, when referring directly or indirectly to a numerical value, always include the adjective "ultimate" or "allowable."

13.1 LOAD TRANSFER

Deep foundations transfer applied axial loads to the ground via two mechanisms: *side friction* and *toe bearing*, as shown in Figure 13.1. The side-friction resistance (also known as *skin-friction resistance*) is the result of sliding friction along the side of the foundation and adhesion between the soil and the foundation. In contrast, the toe-bearing resistance (also known as *point-bearing resistance*, *tip-bearing resistance*, or *end-bearing resistance*) is the result of compressive loading between the bottom of the foundation and the soil, and thus is similar to load transfer in spread footings.

Downward (Compressive) Loads

Side friction and toe bearing are fundamentally different modes of resistance, so it is customary to evaluate each of them separately. Thus, the allowable downward (compressive) load capacity, P_a, is computed as follows:

$$P_a = \frac{P_{ult}}{F} = \frac{P_t + P_s - W_f}{F} \tag{13.1}$$

Where:

P_a = allowable downward load capacity

P_{ult} = ultimate downward load capacity

P_t = toe-bearing resistance

P_s = side-friction resistance

W_f = weight of foundation (considering buoyancy, if necessary)

F = factor of safety

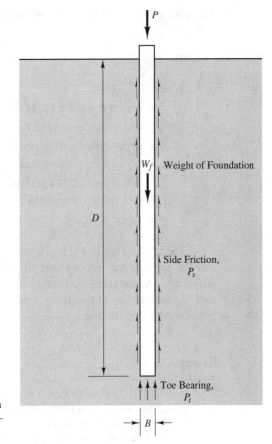

Figure 13.1   Transfer of axial loads from a deep foundation into the ground by side friction and toe bearing.

We can simplify this formula by using the net toe-bearing resistance, P_t':

$$P_t' = P_t - W_f \qquad (13.2)$$

$$P_a = \frac{P_t' + P_s}{F} \qquad (13.3)$$

Rewriting in terms of the unit toe-bearing and side-friction resistances gives:

$$P_a = \frac{q_t'A_t + \Sigma f_s A_s}{F} \qquad (13.4)$$

The foundation must satisfy the following design criterion:

$$\boxed{P \leq P_a}$$ (13.5)

Where:

P = design downward load (from Equations 2.1, 2.2, 2.3a, and 2.4a)

P_a = allowable downward load capacity

q_t' = net unit toe-bearing resistance

A_t = toe-bearing contact area

f_s = unit side-friction resistance

A_s = side-friction contact area

F = factor of safety

The toe-bearing and side-friction contact areas depend on the foundation geometry, and the net unit toe-bearing and side-friction resistances primarily depend on the soil properties. The net unit side-friction resistance typically varies with depth, so we must divide the foundation into sections, find the side friction in each section, and sum to compute the total side friction.

Example 13.1

An 800 kN compressive load is to be imposed on a 400-mm diameter, 15-m long steel pipe pile driven into the soil profile shown in Figure 13.2. The net unit toe-bearing and unit side-friction resistances are as shown. Compute the downward load capacity using a factor of safety of 3 and determine if the design is acceptable.

Solution

Toe bearing
$A_t = \pi(0.2)^2 = 0.126 \text{ m}^2$
$q_t' = 4000 \text{ kPa}$

Side friction—medium clay
$A_s = (0.4)\pi(4.0) = 5.03 \text{ m}^2$
$f_s = 25 \text{ kPa}$

Side friction—silty sand
$A_s = (0.4)\pi(10.0) = 12.57 \text{ m}^2$
$f_s = 100 \text{ kPa}$

Side friction—glacial till
$A_s = (0.4)\pi(1.0) = 1.26 \text{ m}^2$
$f_s = 800 \text{ kPa}$

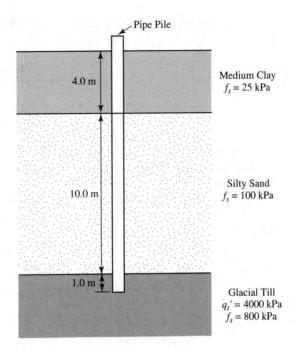

Figure 13.2 Proposed pipe pile for Example 13.1.

$$P_a = \frac{q_t' A_t + \Sigma f_s A_s}{F}$$

$$= \frac{(4000)(0.126) + (25)(5.03) + (100)(12.57) + (800)(1.28)}{3}$$

$$= \frac{504 + 126 + 1257 + 1024}{3}$$

$$= 970 \text{ kN} \quad \Leftarrow Answer$$

$$P = 800 \text{ kN} \leq P_a \quad \therefore \text{ The design is satisfactory} \quad \Leftarrow Answer$$

Upward (Tensile) Loads

The upward (tensile) load capacity of a shallow foundations is limited to its weight, so these foundations are not very effective in resisting such loads. However, deep foundations use both the weight and the side friction (which now acts in the opposite direction), and thus are more effective in resisting upward loads. Deep foundations with expanded bases can resist additional uplift loads through bearing on top of the base.

If upward loads are present, the foundation must satisfy the following design criterion:

$$\boxed{P_{upward} \leq (P_{upward})_a} \tag{13.6}$$

Where:

P_{upward} = applied upward (tensile) load, expressed as a positive number

$(P_{upward})_a$ = allowable upward (tensile) load capacity

Some foundations are subjected to both downward and uplift loads. For example, a downward load may be present under normal circumstances (Equations 2.1 and 2.2), but a net uplift load may exist when seismic loads are considered (Equations 2.3a and 2.4a). In such cases, the foundation must satisfy both Equations 13.5 and 13.6.

The allowable upward load capacity, $(P_{upward})_a$, for straight deep foundations with $D/B > 6$ is:

$$\boxed{(P_{upward})_a = \frac{W_f + \Sigma f_s A_s}{F}} \tag{13.7}$$

Where:

$(P_{upward})_a$ = allowable upward load capacity

W_f = weight of the foundation (considering buoyancy, if necessary)

f_s = unit side-friction resistance

A_s = side-friction contact area

F = factor of safety

The f_s value for upward loading is somewhat less than that for downward loading, as discussed in Section 14.4.

If some or all of the foundation is submerged below the groundwater table, then the computed value of W_f must consider buoyancy effects. This can be done by simply multiplying the submerged volume by the unit weight of water and subtracting this value from the dry weight.

If $D/B < 6$, a cone of soil may form around the foundation during an upward failure. This reduces its uplift capacity (Kulhawy, 1991). Fortunately, the vast majority of deep foundations are long enough to avoid this problem, so the side-friction resistance may be assumed to be equal to that for downward loading.

Cyclic up-and-down loads can be more troublesome than static loads of equal magnitude. Turner and Kulhawy (1990) studied deep foundations in sands and found that cyclic loads smaller than a certain critical level do not cause failure, but enough cycles larger than that level cause the foundation to fail in uplift. The greatest differences between static and cyclic capacities seem to occur in dense sands and in foundations with large depth-to-diameter ratios.

Deep foundations with enlarged bases, such as belled drilled shafts or pressure-injected footings, gain additional upward capacity from bearing on top of the base. Methods of evaluating this additional capacity are discussed in Chapter 14.

Contact Areas A_t and A_s

When using Equations 13.4 and 13.5, we must evaluate the toe-bearing and side-friction contact areas, A_t and A_s. The method of doing so depends on the shape and type of foundation.

Closed-Section Foundations

A *closed-section foundation* is a deep foundation in which the foundation-soil contact occurs along a well-defined surface around the perimeter of the foundation. These include virtually all deep foundations except H-piles and some open-end pipe piles. Because of their simple geometry, it is easy to compute the toe-bearing and side-friction contact areas, A_t and A_s, for these foundations. The design value of A_t is simply the toe area of the foundation (i.e., the solid cross-sectional area), while A_s for a particular stratum is the foundation surface area in contact with that stratum.

Closed-section foundations with enlarged bases, such as underreamed drilled shafts or pressure-injected footings, are slightly more complex. Typically, we use the full base area to compute A_t, but ignore the side of the enlargement when evaluating side friction.

Open-Section Foundations

Open-section foundations are deep foundations that have poorly defined foundation-soil contacts. These include open-end steel pipe piles and steel H-piles. These poorly defined contacts make it more difficult to compute A_t and A_s.

When open-end pipe piles are driven, they initially tend to "cookie cut" into the ground, and the toe-bearing area, A_t, is equal to the cross-sectional area of the steel. Soil enters the pipe interior as the pile advances downward. At some point, the soil inside the pile becomes rigidly embedded and begins moving downward with the pile. It has then become a *soil plug*, as shown in Figure 13.3, and the toe-bearing area then becomes the cross-sectional area of the pile and the soil plug. In other words, the pile now behaves the same as a closed-end pipe (i.e., one that has a circular steel plate welded to the bottom).

Many factors affect the formation of soil plugs (Paikowsky and Whitman, 1990; Miller and Lutenegger, 1997), including the soil type, soil consistency, in-situ stresses, pile diameter, pile penetration depth, method of installation, rate of penetration, and so forth. In open-end steel pipe piles, the soil plug may be considered rigidly embedded when the penetration-to-diameter ratio, D/B is greater than 10 to 20 (in clays) or 25 to 35 (in sands) (Paikowsky and Whitman, 1990). Many piles satisfy these criteria.

Once they become plugged, open-end pipe piles have the same side-friction area, A_s, as closed-end piles. Use only the outside of pipe piles when computing the side-friction area. Do not include the friction between the plug and the inside of the pile.

With H-piles, soil plugging affects both toe-bearing and side-friction contact areas. The space between the flanges of H-piles is much smaller than the space inside pipe piles, so less penetration is required to form a soil plug. For analysis purposes, we usually can compute A_t and A_s in H-piles based on the assumption they become fully plugged as shown in Figure 13.3.

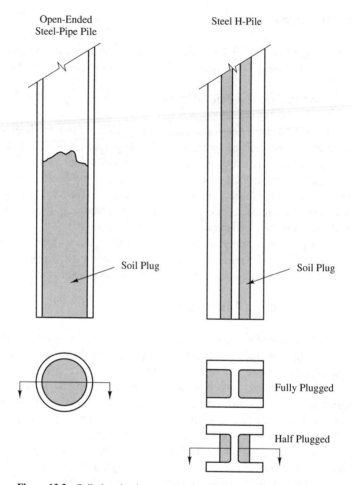

Figure 13.3   Soil plugging in open-ended steel pipe piles and steel H-piles.

If open-section foundations are driven to bedrock, the relative stiffnesses of the steel, soil plug, and bedrock are such that the toe-bearing probably occurs primarily between the steel and the rock. Therefore, in this case it is generally best to use A_t equal to the cross-sectional area of the steel, A_s, and ignore any plugging in the toe-bearing computations.

Factor of Safety

The design factor of safety, F, depends on many factors, including the following:

- **The type and importance of the structure and the consequences of failure**— Foundations for critical structures, such as major bridges, should have a higher factor of safety; those for minor uninhabited structures could use a lower factor of safety

- **The soil type**—Use a higher factor of safety in clays
- **The spatial variability of the soil**—Erratic soil profiles are more difficult to assess, and therefore justify use of a higher factor of safety
- **The thoroughness of the subsurface exploration program**—Intensive subsurface exploration programs provide more information on the subsurface conditions, and therefore can justify a lower factor of safety
- **The type and number of soil tests performed**—Extensive laboratory and/or in-situ tests also provide more information on the soil conditions and can justify a lower factor of safety
- **The availability of on-site or nearby full-scale static load test results**—These tests, which are described later in this chapter, are the most reliable way to determine load capacity, and thus provide a strong basis for using a lower factor of safety
- **The availability of on-site or nearby dynamic test results**—These tests, which are described in Chapter 15, also provide insights and can justify lower factors of safety
- **The anticipated level and methods of construction inspection and quality control**—Thorough methods can justify lower factors of safety
- **The probability of the design loads actually occurring during the life of the structure**—Some structures, such as office buildings, are unlikely to ever produce the design live loads, whereas others, such as tanks, probably will. Thus, the later might require a higher factor of safety.

Because of these many factors that must be considered, and the need for appropriate engineering judgement, the engineer usually has a great deal of discretion to determine the appropriate factor of safety for a given design.

It is good practice to use higher factors of safety for analyses of upward loads because uplift failures are much more sudden and catastrophic. For example, some pile foundations in Mexico City failed in uplift during the 1985 earthquake and heaved 3 m (10 ft) out of the ground. Therefore, most engineers use design values 1.5 to 2.0 times higher than those for downward loading.

Table 13.1 presents typical design factors of safety for pile foundations based on the type of construction control. These methods have not yet been described, but will be covered later in this chapter and in the following chapters. The values in this table are for piles used to support ordinary structures on typical soil profiles with average site characterization programs. Higher or lower values might be used for other conditions.

Except for full-scale load tests, the construction control methods described in Table 13.1 are based on the dynamics of pile driving, and thus are not applicable to drilled shaft foundations. Therefore, the factor of safety for drilled shafts depends primarily on the availability of static load test information, the uniformity of the soil conditions, and the thoroughness of the site characterization program. Table 13.2 presents typical factors of safety for design of drilled shafts that will support ordinary structures. Once again, the engineer might use higher or lower values for other conditions. For example, foundations supporting an especially critical structure might be designed using a higher value, while those supporting a temporary structure might use a lower value.

TABLE 13.1   TYPICAL FACTORS OF SAFETY FOR DESIGN OF PILES

	Factor of Safety, F	
Construction Control Method[a]	Downward Loading (Hannigan et al., 1997)	Upward Loading
Static load test with wave equation analysis	2.00[b]	3.00[b]
Dynamic testing with wave equation analysis	2.25	4.00
Indicator piles with wave equation analysis	2.50	5.00
Wave equation analysis	2.75	5.50
Pile driving formula[c]	3.50	6.00

Notes: a. Most of these terms have not yet been defined, but all of them will be discussed later in this chapter and in the following chapters.

   b. If the static load testing program is very extensive, the factors of safety for downward and uplift loads might be reduced to about 1.7 and 2.5, respectively.

   c. Hannigan, et al. refer specifically to the Gates formula.

Other types of deep foundations should generally be designed using factors of safety in the same range as those presented in Tables 13.1 and 13.2. The selection of these factors of safety depends on the various factors described earlier.

The actual factor of safety for both downward and upward loading (i.e., the real capacity divided by the real load) is usually much higher than the design F used in the formulas. This is because of the following:

TABLE 13.2   TYPICAL FACTORS OF SAFETY FOR DESIGN OF DRILLED SHAFTS

Design Information			Factor of Safety, F	
Static Load Test	Soil Conditions	Site Characterization Program	Downward Loading	Upward Loading
Yes	Uniform	Extensive	2.00[a]	3.00[a]
Yes	Erratic	Average	2.50	4.00
No	Uniform	Extensive	2.50	5.00
No	Uniform	Average	3.00	6.00
No	Erratic	Extensive	3.00	6.00
No	Erratic	Average	3.50	6.00

Note: [a] If the static load testing program is very extensive and the subsurface conditions are well-characterized, the factors of safety for downward and uplift loads might be reduced to about 1.7 and 2.5, respectively.

- We usually interpret the soil strength data conservatively.
- The actual service loads are probably less than the design loads, especially in buildings other than warehouses.
- The as-built dimensions of the foundation may be larger than planned.
- Some (but not all!) of the analysis methods are conservative.

Example 13.2

The steel H-pile foundation in Figure 13.4 has been designed with the benefit of a wave equation analysis (as discussed in Chapter 15), but without nearby static load test or dynamic test data. Compute the allowable uplift capacity of this foundation.

Solution

Reference: H-pile data in Table 12.1

$$W_f = \text{weight} - \text{buoyancy}$$

$$= (35 \text{ ft})(74 \text{ lb/ft}) - (35 - 12 \text{ ft})\left(\frac{21.8 \text{ in}^2}{144 \text{ in}^2/\text{ft}^2}\right)(62.4 \text{ lb/ft}^3)$$

$$= 2370 \text{ lb}$$

$$F = 5.5 \text{ (per Table 13.1)}$$

$$A_s = \left(2\frac{12.13}{12} + 2\frac{12.21}{12} \text{ ft}\right)(35 \text{ ft}) = 142 \text{ ft}^2$$

$$(P_{upward})_a = \frac{2370 + (900)(142)}{5.5} = \textbf{23.7k} \qquad \Leftarrow \textit{Answer}$$

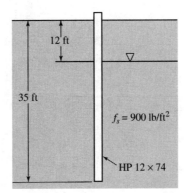

Figure 13.4   Proposed pile foundation for Example 2.2.

QUESTIONS AND PRACTICE PROBLEMS

13.1 The toe-bearing capacity in deep foundations is similar to the ultimate bearing capacity of spread footings. However, the side-friction capacity has no equivalent in spread footing design. Why do we ignore the friction acting on the side of spread footings?

13.2 When designing shallow foundations, we add the weight of the foundation to the applied column load (for example, see Equation 5.1). However, with downward loads on deep foundations, this weight is not explicitly computed (see Equation 13.4). Explain how we account for the weight of deep foundations subjected to downward loads.

13.3 A 500-mm diameter, 15-m deep drilled shaft foundation is to be constructed in the soil profile shown in Figure 13.5. The design axial loads are as follows: $P_D = 400$ kN, $P_L = 300$ kN, $P_E = \pm600$ kN. No static load test data is available, the soil conditions are comparatively uniform, and the site characterization program had average thoroughness. Compute the downward and upward design loads using Equations 2.1, 2.2, 2.3a, and 2.4a, then compute the allowable downward and upward load capacities and determine if this design is satisfactory.

13.2 CONVENTIONAL STATIC LOAD TESTS

The most precise way to determine the ultimate downward and upward load capacities for deep foundations is to build a full-size prototype foundation at the site of the proposed production foundations and slowly load it to failure. This method is known as a *conventional static load test*. All other methods determine the ultimate load capacities indirectly,

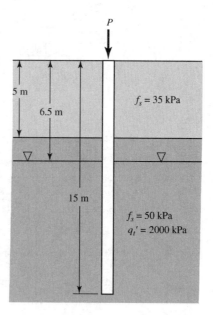

Figure 13.5 Proposed drilled shaft foundation for Problem 13.3.

and therefore are less precise. However, static load tests are much more expensive and time-consuming, and thus must be used more judiciously.

The objective of a static load test is to develop a load-settlement curve or, in the case of uplift tests, a load-heave curve. This curve is then used to determine the ultimate load capacity.

Equipment

To conduct a static load test, we must have a means of applying the desired loads to the foundation and measuring the resulting settlements. A wide variety of equipment has been used, with varying degrees of success (Crowther, 1988). Early tests often produced the test load by progressively stacking dead weights, such as precast concrete blocks or pig iron on top of the foundation, or by placing a tank on top of the foundation and progressively filling it with water. Unfortunately, these methods can be dangerous because it is difficult to place large weights without creating excessive eccentricities that can cause them to collapse. Therefore, these methods are rarely used today.

A safer alternative is to provide multiple supports for the weights, as shown in Figure 13.6, and use them as a reaction for a hydraulic jack, which is then used to provide

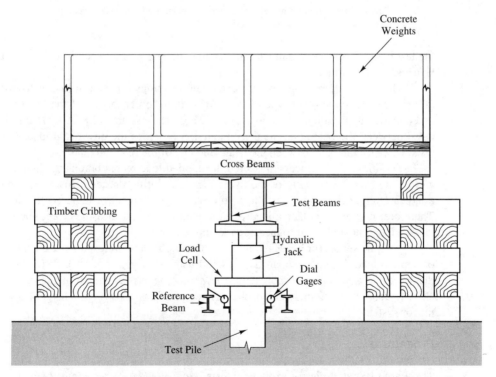

Figure 13.6   Use of a hydraulic jack reacting against dead weight to develop the test load in a static load test.

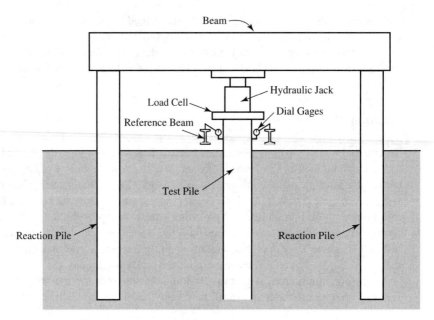

Figure 13.7   Use of a hydraulic jack reacting against a beam and reaction piles to develop the test load in a static load test.

the test load. This system is much more stable and less prone to collapse, but it is expensive and cumbersome.

The most common method of developing reactions for the test load is to install *reaction piles* (or some other type of deep foundation) on each side of the test foundation and connect them with a beam, as shown in Figures 13.7 and 13.8. A hydraulic jack located between the beam and the test foundation provides the downward load. For uplift tests, we place the jack between the beam and one of the reaction piles.

Traditionally, engineers measured the applied load by calibrating the hydraulic jack and monitoring the pressure of the hydraulic fluid during the test. However, even when done carefully, this method is subject to errors of 20 percent or more (Fellenius, 1980). Therefore, it is best to place a load cell (an instrument that measures force) between the jack and the pile and use it to measure the applied load.

The pile settles (or heaves in the case of an uplift test) in response to each increment of applied load. This settlement is measured using dial gages or LVDTs (linear voltage displacement transformers) mounted on *reference beams*, as shown in Figures 13.6 and 13.7. It also is wise to include a backup system, such as a surveyor's level or a wire and mirror system.

Procedure

There are two categories of static load tests: *controlled stress tests* (also known as *maintained load* or *ML tests*) and *controlled strain tests*. The former uses predetermined loads (the independent variable) and measured settlements (the dependent variable), while the

Figure 13.8   This hydraulic jack is loading the steel pipe pile in compression. The jack reacts against the beam above, which loads the reaction piles (background) in tension. Notice the reference beams and dial gages in the foreground. These are used to measure the foundation settlement.

latter uses the opposite approach. ASTM D1143 describes both procedures. The vast majority of tests use the controlled stress method, so it is the only one we will discuss.

Deep foundation construction, especially pile driving, alters the surrounding soil, as discussed in Chapter 14. This alteration often produces excess pore water pressures, which temporarily change the ultimate load capacity. Therefore, it is best to allow time for these excess pore water pressures to dissipate before conducting the test. This typically requires a delay of at least 2 days in sands and at least 30 days in clays.

When conducting controlled stress tests, the field crew applies the test load in increments and allows the foundation to move under each increment. They then record the corresponding movements and generate a load-displacement curve. Increments of 25, 50, 75, 100, 125, 150, 175, and 200 percent of the proposed design load are typically used when the test is intended to check a predetermined design. However, this load sequence is usually not sufficient to reach the ultimate load capacity, so the engineer cannot determine if the foundation is overdesigned and thus may lose an opportunity to economize the production foundations. Therefore, it is better to continue the test until reaching "failure" as defined in Section 13.3.

The traditional method, known as a *slow ML test*, maintains each of these load increments until the foundation stops moving or until the rate of movement is acceptably small. This criterion may require holding each load for at least 1 or 2 hours, sometimes more, so these tests may require 24 hours or more to complete.

More recently, the *quick ML test* is beginning to dominate. This method is similar to the slow test except that each load increment is held for a predetermined time interval regardless of the rate of pile movement at the end of that interval. Typically, each load increment is about 10 percent of the anticipated design load and is held for 2.5 to 15 minutes. Crowther (1988) suggests holding each increment for 5 minutes. This process continues until reaching about 200 percent of the anticipated design load or "failure" and generally requires 2 to 5 hours to complete. This may be the best method for most deep foundations.

Quality Control

Unfortunately, crews that conduct static load tests sometimes use less than satisfactory workmanship, so the test results are not always reliable. This is especially true with slow ML tests. Peck's (1958) review of static load tests demonstrates some of these problems, including the following:

- Cases where an increase in the downward load caused an upward deflection
- Lack of smoothness in the load-settlement curve
- Misleading load-settlement curves produced by unusual plotting scales

Therefore, it is important to have a formal quality control program. In addition, the test should be performed under the supervision of an professional engineer.

13.3   INTERPRETATION OF TEST RESULTS

Once we have obtained the load-settlement curve, it is necessary to determine the ultimate load capacity, which means we must define where "failure" occurs. For piles in soft clays, this is relatively straightforward. In these soils, the load-settlement curve has a distinct *plunge*, as shown by Curve A in Figure 13.9, and the ultimate capacity is the load that corresponds to this plunge. However, piles in sands, intermediate soils, and stiff clays have a sloped curve with no clear point of failure, as shown by Curve B.

Many different methods have been proposed to interpret the latter type of load-settlement curve (Fellenius, 1990), and these methods often produce significantly different results. For example, Fellenius (1980) used nine different methods to analyze the results of a static load test and found that the computed ultimate capacity varied from 362 k to 470 k. Another method, found in early editions of the AASHTO[1] specifications, produces an ultimate capacity of only 100 k!

---

[1]The American Association of State Highway and Transportation Officials.

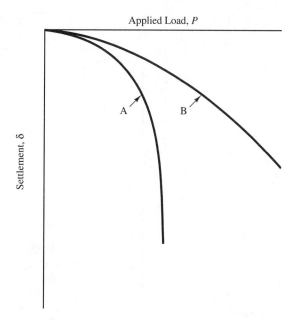

Figure 13.9 Typical load-settlement curves: Curve A is typical in soft clayey soils (note plunge); and Curve B is typical of intermediate, stiff clay and sandy soils (ever-increasing load).

Davisson's (1973) method is one of the most popular. It defines the ultimate capacity as that which occurs at a settlement of 4 mm (0.15 in) + $B/120$ + $PD/(AE)$ as shown in Figure 13.10. The last term in this formula is the elastic compression of a pile that has no side friction.

Davisson's method seems to work best with data from quick ML tests. It may produce overly conservative results when applied to data from slow ML tests.

Many engineers in the United States prefer to express pile capacities using tons (the 2000 lb variety), whereas others use kips. This book uses only kips to maintain consistency with the remainder of structural and geotechnical engineering practice.

Example 13.3

The load-settlement data shown in Figure 13.11 were obtained from a full-scale static load test on a 400-mm square, 17-m long concrete pile ($f_c' = 40$ MPa). Use Davisson's method to compute the ultimate downward load capacity.

Solution

$$E = 4700 \sqrt{f_c'} = 4700 \sqrt{40 \text{ MPa}} = 30,000 \text{ MPa}$$

$$4 \text{ mm} + \frac{B}{120} + \frac{PD}{AE} = 4 \text{ mm} + \frac{400 \text{ mm}}{120} + \frac{P(17,000 \text{ mm})}{(0.400^2)(30,000,000 \text{ kPa})}$$

$$= 7.3 \text{ mm} + 0.0035 \, P$$

Plotting this line on the load-displacement curve produces $P_{\text{ult}} = $ **1650 kN**.   ⇐ *Answer*

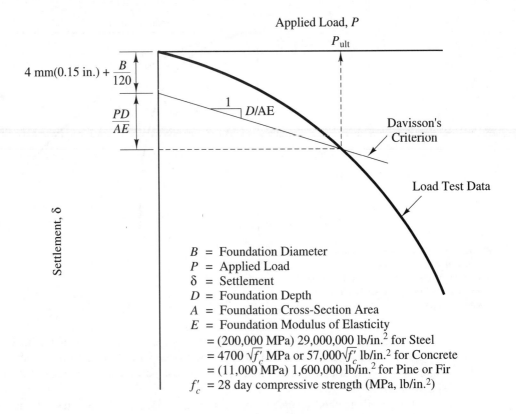

Figure 13.10   Davisson's method of interpreting pile static load test data.

QUESTIONS AND PRACTICE PROBLEMS

13.4 A typical deep foundation project may include several hundred piles, but only one or two static load tests. Thus, the information gained from these test pile must be projected to the production piles. Describe some of the factors that might cause the load capacity of the production piles to be different from that of the test pile.

13.5 Curve A in Figure 13.12 has been obtained from a static load test on a 40-ft long, 12-inch square solid concrete pile with $f_c' = 6000$ lb/in^2. Using Davisson's method, compute the ultimate load capacity.

13.6 Curve B in Figure 13.12 has been obtained from a static load test on a 60 ft long, PP18×0.375 pile. Using Davisson's method, compute the ultimate load capacity.

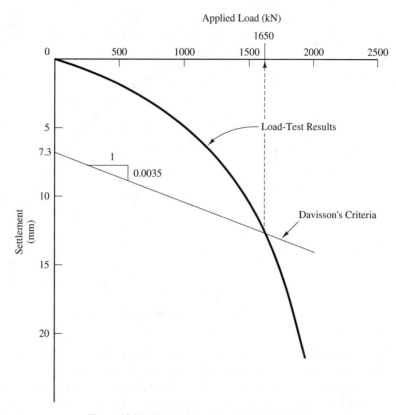

Figure 13.11   Static load test data for Example 13.3.

13.4   MOBILIZATION OF SOIL RESISTANCE

Stresses and strains always occur together. Whenever a stress is applied to a material, corresponding strains also develop. In other words, whenever we place a load on a material, it responds with a corresponding deformation. It is impossible to have one without the other.

This relationship between load and deformation helps us understand how deep foundations respond to applied loads. As discussed earlier in this chapter, downward loads acting on deep foundations are resisted by side-friction resistance and toe-bearing resistance. We compute each of these components separately, then combine them using Equation 13.4 to find the allowable downward load capacity. However, neither of these resistances can develop without a corresponding deformation. We call this deformation *settlement*. Therefore, in addition to evaluating the ultimate values of the side-friction and toe-bearing resistances, we also need to know something about the settlement required to

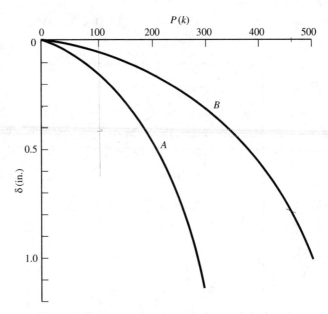

Figure 13.12   Load-settlement curves from static load tests.

mobilize these resistances. Some of this settlement is caused by elastic compression of the foundation, but most is the result of strains in the soil.

As shown in Figure 13.13, only 5 to 10 mm (0.2–0.4 in) of settlement is required to mobilize the full unit side-friction resistance, f_s. The load-settlement curve becomes quite steep, so there is a well-defined "ultimate side-friction resistance." However, the load-settlement curve for toe bearing is not nearly as steep, and often does not reach a well-defined ultimate value. Thus, the definition of "ultimate unit toe-bearing resistance"

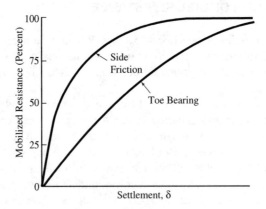

Figure 13.13   Load-displacement relationships for side-friction and toe bearing under downward loads.

becomes vague and subjective. In addition, much more settlement is required to mobilize the toe bearing, with the "ultimate" resistance corresponding to perhaps 10 percent of the foundation diameter. Even a small diameter foundations, perhaps with $B = 300$ mm, requires about 30 mm of settlement to develop the ultimate toe-bearing resistance. Larger foundations require even more settlement to reach this value.

This difference between side-friction and toe-bearing response has at least three important implications:

1. The load-settlement curve obtained at the head of the foundation during a conventional static load test is really a composite of the side-friction and toe-bearing curves.

2. Because of the shape of the load-settlement curve, the unit toe-bearing resistance, q_t', is usually a difficult number to define. This is a strong contrast to the ultimate bearing capacity, q_{ult}, in shallow foundations, which is based on a much better defined mode of failure. Engineers have used various methods of defining q_t', which is part of the reason different analysis methods often produce different results.

3. At a settlement of 5 to 10 mm, virtually all of the side friction will have been mobilized, but only a small fraction of the toe bearing will have been mobilized. Therefore, so long as sufficient side-friction resistance is available, it will carry nearly all of the service loads, and the toe bearing becomes the factor of safety.

   According to Tables 13.1 and 13.2, the design factors of safety for downward loads is usually between 2 and 3, so "sufficient side-friction resistance" should be available to resist all of the service load so long as the total side-friction resistance ($\Sigma f_s A_s$) comprises at least one-third to one-half of the ultimate capacity. This is the case for the vast majority of deep foundations, which means the settlement under service load should be less than that required to mobilize the side friction (5 to 10 mm) plus the elastic compression in the foundation (typically < 5 mm) for a total of no more than 15 mm. This is less than δ_a for virtually all structures.

   However, if the foundation relies primarily on toe bearing (i.e., $\Sigma f_s A_s$ comprises less than 1/3 to 1/2 of the ultimate capacity) then the settlement under service loads might be excessive, especially if the foundation diameter is large, and should be checked.

   Other conditions also might cause settlement problems in deep foundations, as discussed in Chapter 14.

13.5 INSTRUMENTED STATIC LOAD TESTS

Conventional static load tests only produce information on the load-settlement relationship at the top of the foundation. Although this data is very useful, we would also like to know how this load is transferred into the soil. How much side friction develops and how much toe bearing? How is the side friction distributed along the foundation? We can find the answers to these questions by installing additional instrumentation in the foundation, which allows us to perform an *instrumented static load test*.

Instrumentation Using Strain Gauges

Closed-end steel pipe piles may be instrumented by installing a series of strain gages on the inside of the pile, as shown in Figure 13.14. Using data from these strain gauges, we can compute the force in the pile at each depth, and thus can determine the distribution of side-friction resistance and the toe-bearing resistance.

    For example, if the applied load is transferred into the ground entirely through side friction, the plot of force vs. depth will be similar to Curve A in Figure 13.14 (i.e., the force in the foundation gradually dissipates with depth, reaching zero at the toe). Conversely, if the resistance is entirely toe bearing, the plot will be similar to Curve B (i.e., constant with depth). Most foundations have both side friction and toe bearing, and thus have plots similar to Curve C.

    Using this data, we can calculate the unit side-friction resistance, f_s, for each stratum, and the net unit toe-bearing resistance, q_t'. These values could then be used to design foundations with other lengths or diameters.

    Drilled shafts and concrete piles can be instrumented by installing load cells inside the concrete. These load cells measure the stress in the concrete at various depths.

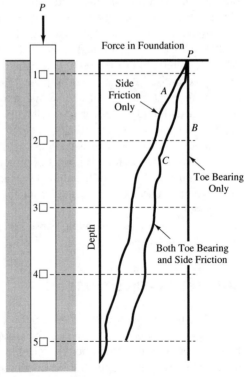

Figure 13.14   Typical instrumented static load test.

Instrumentation Using Telltale Rods

Another method of conducting instrumented static load tests is to install *telltale rods* inside the foundation. These rods extend from the top of the foundation to some specified depth, and are encased in a protective sleeve. By comparing the settlement of these rods with the settlement at the top of the foundation, we can compute the force in the foundation, which in turn may be used to compute f_s and q_t'.

Instrumented static load tests are more expensive than conventional static load tests, so they are generally used only on especially important projects, or for research. The results of such tests form the basis for some of the static design methods described in Chapter 14.

Example 13.4

Two telltale rods have been installed in a 45-ft long closed-end PP20×0.750 pile. This pile was subjected to a static load test, which produced the following results:

Test load at failure   250,040 lb

Dial gage settlement readings at failure:
   At pile head (Gauge #1)   0.572 in
   Tell-tale anchored at 20 ft depth (Gauge #2)   0.530 in
   Tell-tale anchored at 45 ft depth (Gauge #3)   0.503 in

Compute the force in the pile at 20 and 45 ft, and compute the average f_s values on the pile and the q_t' value.

Solution

Pile interval 1-2 (upper 20 ft)

$$\delta_1 - \delta_2 = \left(\frac{P_1 + P_2}{2}\right)\frac{L}{AE}$$

$$0.572 - 0.530 = \left(\frac{250,040 + P_2}{2}\right)\frac{(20)(12 \text{ in/ft})}{(45.36)(29 \times 10^6)}$$

$$\boldsymbol{P_2 = 210{,}360 \text{ lb}} \qquad \Leftarrow Answer$$

$$\bar{f}_s = \frac{\Delta P}{A}$$

$$= \frac{250,040 - 210,360}{\pi(20/12)(20)}$$

$$= \boldsymbol{379 \text{ lb/ft}^2} \qquad \Leftarrow Answer$$

Pile interval 2-3 (lower 25 ft)

$$\delta_2 - \delta_3 = \left(\frac{P_2 + P_3}{2}\right)\frac{L}{AE}$$

$$0.530 - 0.503 = \left(\frac{210,360 + P_3}{2}\right)\frac{(25)(12 \text{ in/ft})}{(45.36)(29 \times 10^6)}$$

$$P_3 = \mathbf{26,420 \ lb} \quad \Leftarrow Answer$$

$$\bar{f}_s = \frac{\Delta P}{A}$$

$$= \frac{210,360 - 26,420}{\pi(20/12)(25)}$$

$$= \mathbf{1410 \ lb/ft^2} \quad \Leftarrow Answer$$

$$q_t' = \frac{P_t}{A_t}$$

$$= \frac{26,420}{\pi(20/12)^2/4}$$

$$= \mathbf{12,100 \ lb/ft^2} \quad \Leftarrow Answer$$

Commentary

These values of f_s and q_t' may then be used to compute the load capacity of proposed piles at this site with other diameters and depths of embedment. Thus, this technique is a method of extrapolating the load test results.

This analysis has assumed the telltale dial gages were "zeroed" after the pile was driven but before the load was applied, and thus implicitly account for pile compression due to the weight of the pile.

13.6   OSTERBERG LOAD TESTS

Conventional static load tests on piles are nearly always conducted on a single pile, even though a pile group is normally needed to support each column. Therefore the load frame, jack, and other hardware only need to develop a fraction of the proposed column load. However, drilled shaft foundations are typically larger in diameter and have a correspondingly greater capacity. Many designs use a single high-capacity drilled shaft to support each column. This makes it more difficult and expensive to conduct conventional static load tests on drilled shafts because the hardware must be correspondingly larger. Often these additional costs are prohibitive.

Fortunately, Osterberg (1984) in the United States and DaSilva in Brazil have developed a method that reduces the cost of conducting high-capacity static load tests. It eliminates the expensive test frame and substitutes a hydraulic pancake jack at the bottom of the shaft, as shown in Figures 13.15 and 13.16. Once the concrete is in place, the opera-

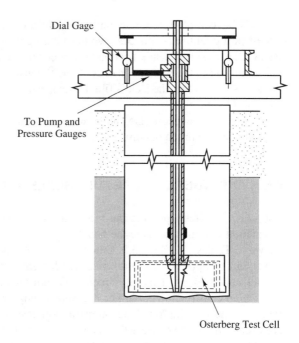

Figure 13.15 Osterberg load test device (Loadtest, Inc.).

tor pumps hydraulic fluid into the jack and keeps track of both pressure and volume. The jack expands and pushes up on the shaft (Osterberg, 1984). A dial gauge measures this movement. Thus, we obtain a plot of side-friction capacity vs. axial movement.

This device also includes a telltale rod that extends from the bottom of the pancake jack to the ground surface. It measures the downward movement at the bottom, and thus produces a plot of toe-bearing pressure vs. axial movement.

Figure 13.16 An Osterberg cell being installed at the bottom of a drilled shaft rebar cage (Photo courtesy of Loadtest, Inc.).

These two plots continue until the foundation fails in side friction or toe bearing. If a toe-bearing failure occurs first, we must extrapolate the data to obtain the ultimate skin friction. However, if a side-friction failure occurs first, it may be possible to add a nominal static load to the top, and then continue the test to find the toe-bearing capacity. Once the test is completed, the jack is filled with grout and the foundation may be used to carry structural loads.

The Osterberg test also can be used with piles. In this case, the pile is driven with the cell attached to the bottom.

13.7   WHEN AND WHERE TO USE FULL-SCALE STATIC LOAD TESTS

Full-scale load tests provide the most precise assessment of ultimate load capacity, so foundation designs that have the benefit of such tests may use a low factor of safety, as shown in Tables 13.1 and 13.2. Other methods of assessing ultimate load capacity are indirect and less precise, and thus must use higher factors of safety to compensate for the greater uncertainties. This difference in design factor of safety can significantly impact foundation construction costs. For example, changing the factor of safety from 2.0 to 3.0 increases most of the foundation construction costs by fifty percent. However, load tests are expensive, so the construction savings must be compared to the cost of performing the test. In many cases, a conservative design is less expensive than a load test. The decision whether or not to conduct static load tests, and how many tests to conduct at a particular site is primarily based on economics.

For example, extensive static load tests were performed for the I-880 freeway reconstruction project in Oakland, California (which was required because of the collapse of a viaduct during the 1989 Loma Prieta Earthquake). This $3.5 million test program produced $10 million in savings because they enabled a more precise design of the production foundations.

Full-scale static load tests are most likely to be cost-effective when one or more of the following conditions are present:

- Many foundations are to be installed, so even a small savings on each will significantly reduce the overall construction cost. Static load tests are more likely to be cost-effective if at least 3000 m (10,000 ft) of production piling is to be installed.
- The soil conditions are erratic or unusual, and thus difficult to assess with the analytic methods described in Chapter 14.
- The pile is supported in soils that are prone to dramatic failures (i.e., a steeply-falling load-settlement curve).
- The structure is especially important or especially sensitive to settlements.
- The engineer has little or no experience in the area.
- The foundations must resist uplift loads (the consequences of a failure are much greater).

Ideally, static load tests are conducted well before construction of the production piles. This allows time to finalize the design, develop a contract to build the foundation, and fabricate the production piles. Unfortunately, this also requires an extra mobilization and demobilization of the pile driver and crew, which is very expensive. Alternatively, the engineer could design and fabricate the piles without the benefit of a static load test, and then *proof test* some of the first production piles. This method is faster and saves the extra mobilization costs, but may generate last-minute changes in the design.

The state-of-practice for static load tests varies significantly from one geographic area to another. In some areas, they rarely are performed, even for very large structures, whereas elsewhere they are required for nearly all pile-supported structures. Sometimes these differences in practice are the result of local soil and geologic conditions or building code requirements. However, in other cases, the underutilization or overutilization of static load tests seems to be primarily the result of custom and habit.

SUMMARY

Major Points

1. The most important geotechnical design requirement for most deep foundations is that they have sufficient axial load capacity to support the applied loads. We can evaluate this capacity using static load tests, analytic methods, or dynamic methods.

2. Deep foundations transfer downward axial loads into the ground through side friction and toe bearing. Uplift loads are transferred through side friction, and possibly bearing on the ceiling of an enlarged base.

3. The design of open section foundations, such as steel H-piles and open-end steel pipe piles must consider soil plugging.

4. The design factor of safety must be based on many factors, and is somewhat subjective. A higher factor of safety is used for uplift loads because uplift failures are more catastrophic.

5. A conventional static load test consists of applying a series of loads to the top of a prototype deep foundation and measuring the corresponding settlements.

6. There is no universally accepted method of interpreting static load test results. However, Davisson's method is widely used.

7. A instrumented static load test is a conventional test that includes instruments inside the foundation to measure the distribution of side friction and toe bearing.

8. An Osterberg load test applies the test load using a hydraulic pancake jack located at the bottom of the foundation.

9. Static load tests are expensive to perform, and are usually cost-effective only when the project is large enough that the savings in production foundation costs is sufficient to offset the cost of the test. On smaller projects, a more conservative design without the benefit or cost of a load test may produce a lower total project cost.

Vocabulary

Closed-section foundation

Controlled stress test

Controlled strain test

Conventional static load test

Davisson's method

Downward load capacity

Enlarged base

Full-scale static load test

Instrumented static load test

Maintained load test

Open-section foundation

Osterberg load test

Plunge

Side friction

Side-friction resistance

Straight foundation

Temporary structure

Toe bearing

Toe-bearing resistance

Upward load capacity

COMPREHENSIVE QUESTIONS AND PRACTICE PROBLEMS

13.7 The results of pile load tests are usually considered to be the "correct" load capacity, and all other analysis methods are compared to this standard. However, there are many ways to conduct load tests, and many ways to interpret them. Therefore, can we truly establish a single "correct" capacity for a pile? Explain.

13.8 A 350-mm diameter closed-end steel pipe pile with 10-mm thick walls and a 20-mm thick bottom plate is to be driven to a depth of 18.5 m into a soil that has $f_s = 50$ kPa for downward loads and 40 kPa for upward loads, and $q_t' = 9000$ kPa. This pile will be constructed with the

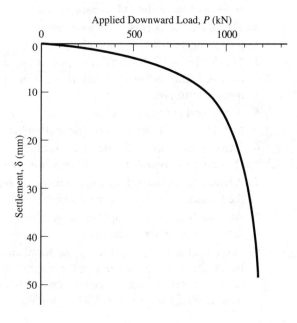

Figure 13.17   Static load test results for Problem 13.9.

benefit of a wave equation analysis and onsite dynamic testing. Compute the allowable downward and upward load capacities.

13.9 A 250-mm square, 15-m long prestressed concrete pile (f_c' = 40 MPa) was driven at a site in Amsterdam as described by Heijnen and Janse (1985). A conventional load test conducted 31 days later produced the load-settlement curve shown in Figure 13.17. Using Davisson's method, compute the ultimate load capacity of this pile.

13.10 A static load test has been conducted on a 60-ft long, 16-inch square reinforced concrete pile which has been driven from a barge through 20 ft of water, then 31 ft into the underlying soil. Telltale rods A and B have been embedded at points 30 ft and 59 ft from the top of the pile, respectively. The data recorded at failure was as follows: Load at head = 139,220 lb, settlement at head = 1.211 in, settlement of telltale rod A = 1.166 in, settlement of telltale rod B = 1.141 in. Use the data from tell tale rod A to compute the modulus of elasticity of the pile, then use this value and the remaining data to compute q_t' and the average f_s value in the soil.
Hint: Telltale rod A is anchored only 1 ft from the mud line (the top of the soil). There is essentially no side-friction resistance between the top of the pile and this point, so the force at a depth of 30 ft is essentially the same as that at the top of the pile.

14

Deep Foundations—Axial Load Capacity Based on Analytic Methods

There is no glory in the foundations.

Karl Terzaghi, 1951

The second way of determining the axial load capacity of deep foundations is to use *analytic methods*. These methods use certain properties of the soil, such as the effective friction angle, ϕ', or the CPT cone bearing, q_c, to compute the unit side-friction resistance, f_s, and the net unit toe-bearing resistance, q_t'. These values are then placed into Equations 13.4 and 13.5 to compute the allowable downward and upward load capacities. Additional analytic analyses are available to evaluate group effects and settlement.

Analytic methods are much less expensive than static load tests, and thus are very attractive. However, the computed capacities are not as precise, so designs based solely on analytic methods must be correspondingly more conservative. This additional conservatism is reflected in the factors of safety listed in Tables 13.1 and 13.2, which indicate that designs that do not include any load test data require a higher design factor of safety.

This chapter discusses various analytic methods for design of deep foundations to resist axial loads. Most of the chapter discusses specific ways to obtain q_t' and f_s through the use of various equations and charts, all of which are straightforward and easy to use. However, the proper application of these methods in practical engineering design problems also requires a thorough understanding of how side-friction and toe-bearing resistances develop in deep foundations, and the careful application of engineering judgment. The engineer also must understand the limitations of these methods. To help develop these skills, we will begin by discussing changes in the soil that occur during the construction of deep foundations, and how these changes affect analytic analysis methods.

14.1   CHANGES IN THE SOIL DURING CONSTRUCTION

When evaluating the load capacity of shallow foundations, engineers test the soil before construction and use the results of these tests in the analytic methods described in Chapters 5 to 10. The construction of shallow foundations does not significantly alter the underlying soils, so these pre-construction soil properties also reflect the post-construction conditions. Analytic methods of designing deep foundations use this same general approach, but there is an important difference: The process of constructing deep foundations changes the surrounding soils. For example, piles push the soil aside as they are driven into the ground, thus inducing large horizontal stresses in the soil. These changes are important because they alter the engineering properties of the soil, which means the pre-construction soil tests may not accurately reflect the post-construction conditions. Sometimes these changes are beneficial, while other times they are detrimental, but in either case they introduce another complexity into load capacity analyses. Therefore, in order to intelligently apply analytic methods of computing load capacity, the engineer must understand these changes.

Piles

Changes in Clays

Distortion

As a pile is driven into the ground, the soil below the toe must move out of the way. This motion causes both shear and compressive distortions. Additional distortions occur as a result of sliding friction along the side of the advancing pile. These distortions are greatest around large displacement piles, such as closed-end steel pipe piles. For example, Cooke and Price (1973) observed the distortion in London Clay as a result of driving a 168-mm (6.6 in) diameter closed-end pipe pile. The soil within a radius of 1.2 pile diameters from the edge of the pile was dragged down, while that between 1.2 and 9 diameters moved upward.

This remolding of the clay changes its structure and reduces its strength to a value near the residual strength. Nevertheless, current analysis techniques are based on the peak strength and implicitly consider the difference between peak and residual strength. An analysis based on the residual strength might be more reasonable, but no such method has yet been perfected.

Compression and Excess Pore Water Pressures

Pile driving also compresses the adjoining soils. If saturated clays are present, this compression generates excess pore water pressures, as discussed in Chapter 3. The ratio of the excess pore water pressure, u_e, to the original vertical effective stress, σ_z', may be as high as 1.5 to 2.0 near the pile, gradually diminishing to zero at a distance of 30 to 40 pile radii, as shown in Figure 14.1. The greatest compression occurs near the pile toe, so u_e/σ_z'

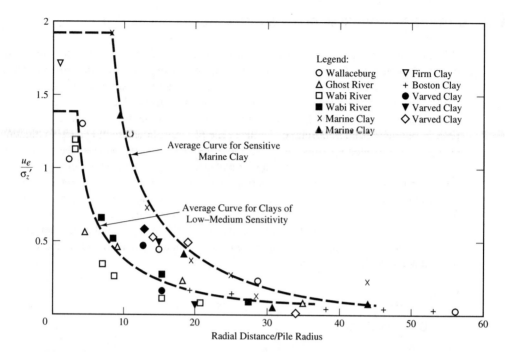

Figure 14.1   Summary of measured excess pore water pressures, uc, in the soil surrounding isolated piles driven into saturated clay (Adapted from Pile Foundation Analysis and Design by Poulos and Davis, Copyright c 1980 by John Wiley and Sons, Inc. Reprinted by permission of John Wiley & Sons, Inc.).

in that region may be as high as 3 to 4 (Airhart et al., 1969). These high pore water pressures dramatically decrease the shear strength of the soil, which makes it easier to install the pile, but temporarily decreases its load-bearing capacity.

The presence of excess pore water pressures is always a transient condition because the resulting hydraulic gradient causes some of the water to flow away; in this case, radially away from the pile. Thus, the pore water pressures eventually return to the hydrostatic condition. If the pile is timber, some of the water also might be absorbed into the wood. This, along with thixotropic effects[1] and consolidation, eventually restores or even increases the strength of the clay. We can observe this increase in strength by resetting the pile hammer a couple of days after driving the pile and noting the increase in the blow count, a phenomenon known as *setup* or *freeze*.

In most clays, the excess pore water pressures that develop around a single isolated pile completely dissipate in less than one month, with corresponding increases in load capacity, as shown in Figure 14.2. However, in pile groups, the excess pore water pressures

---

[1]Thixotropy is the hardening of a disturbed material (in this case, soil) that sometimes occurs as it is allowed to rest.

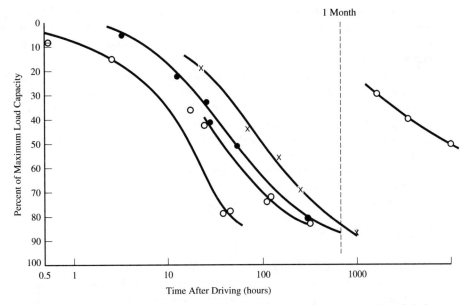

Figure 14.2   Increase in load capacity of an isolated pile with time (Adapted from Soderberg, 1962).

develop throughout a much larger zone of soil and may require a year or more to dissipate. This delay is significant because the rate of load capacity gain is now slower than the rate of construction, so the pile does not reach its full capacity until after the superstructure is built.

Loss of Contact Between Pile and Soil

Piles wobble during driving, thus creating gaps between them and the soil. Soft clays will probably flow back into this gap, but stiff clays will not. Tomlinson (1987) observed such gaps extending to a depth of 8 to 16 diameters below the ground surface. Piles subjected to applied lateral loads also can create gaps near the ground surface. Therefore, the side friction in this zone may be unreliable, especially in stiff clays.

Changes in Sands

Soil compression from the advancing pile also generates excess pore water pressures in loose saturated sands. However, sands have a much higher hydraulic conductivity (permeability) than do clays, so these excess pore water pressures dissipate very rapidly. Thus, the full pile capacity develops almost immediately.

Some local dilation (soil expansion) can occur when driving piles through very dense sands. This temporarily generates negative excess pore water pressures that increase the shear strength and make the pile more difficult to drive. This effect is especially evident when using hammers that cycle rapidly.

Usually, we are most interested in the permanent changes that occur in the sand during pile driving. The impact and vibrations from the pile will cause particle rearrangement, crushing, and densification. In loose sands, these effects are especially pronounced, and engineers sometimes use piles purely for densification. However, dense sands will likely require predrilling or jetting to install the pile, both of which loosen the sand and partially or fully negate this effect.

The sand in the center of pile groups is influenced by more than one pile, and therefore becomes denser than the sand near the edge of the group. This, in turn, probably causes the center piles to carry a larger share of the total downward load.

Drilled Shafts

The construction methods for drilled shaft foundations are completely different from those for piles. Piles push the soil aside as they are driven, thus increasing the lateral earth pressure, whereas drilled shaft construction produces a temporary hole in the ground that allows the soil to expand laterally, thus reducing the lateral earth pressure. In addition, pile driving compresses the soil below the toe, while drilled shaft construction does not. When the concrete is placed in the hole, it exerts an outward hydrostatic pressure on the soil, and this pressure may at least partially recompress it. However, this effect is probably not sufficient to return the soil to its original condition. Therefore, piles and drilled shafts may not develop the same side-friction and toe-bearing resistances, even when the foundation dimensions are identical.

The amount of net lateral expansion and its impact on the load capacity of drilled shaft foundations vary depending on how long the boring is left open before placing the concrete. If the boring is open for an extended period, too much expansion occurs and the load capacity can be significantly reduced. In addition, there is more opportunity for caving. Therefore, it is best to place the concrete as soon as possible, preferably within a couple of hours, and certainly within the same work day.

When working in clays, the process of drilling the hole also alters the soil properties because the auger smears and remolds the clay. This can reduce its shear strength and the side-friction resistance.

The use of drilling mud during construction also can affect the side-friction resistance because some of the mud may become embedded in the walls of the boring. This effect is very dependant on the contractor's construction methods and workmanship, as well as the soil type, and the corresponding reduction in load capacity can range from minimal to substantial. The best results are obtained when the slurry is carefully mixed and the slurry is in place for only a short time (i.e., the concrete is placed as soon as possible).

Sometimes drilled shafts have been intentionally or unintentionally built with a casing left in place. This method can seriously degrade the side-friction resistance because there is generally a gap between portions of the casing and the surrounding soil.

These detrimental effects can generally be kept under control by careful construction methods, and are often offset by other advantages of drilled shafts, such as the ability to drill into hard soil or soft rock that a pile could not penetrate.

Other Types of Deep Foundations

The changes in soil induced by construction of other types of deep foundations have not been studied as thoroughly. Nevertheless, these changes are still important and often are quite different than those due to construction of piles. For example, drilled shaft construction does not compress the soils like pile driving does, and thus does not induce significant positive excess pore water pressures. In fact, there may even be some relaxation in some soils, which could produce negative excess pore water pressures.

Because different types of deep foundation construction methods induce different changes in the soil, we often cannot use an analysis method developed for one type to design another type. It is important to use analysis methods developed specifically for the type of foundation being designed.

Impact on Analytic Design Methods

Most analytic methods of determining q_t' and f_s were developed empirically by comparing the results of static load tests with measured soil properties. Therefore, most of the soil changes discussed in this section are implicitly incorporated into the methods. For example, a research engineer could perform a series of static load tests on steel pipe piles driven into sands. These tests could be accompanied by measurements of the friction angle, ϕ', of each sand, and then could be used to develop an empirical formula for f_s as a function of ϕ' and the vertical effective stress, σ_z'. The remaining factors would be implicitly incorporated into the function.

Since most analytic methods were developed this way, it is important to apply these methods only to foundations that are comparable to those on which it was developed. For example, the equation described in the previous paragraph was developed for steel pipe piles, and thus is probably not applicable to auger cast piles even though they may have the same length and diameter. Therefore, most of the analytic methods described in this chapter are associated with a particular type of deep foundation.

QUESTIONS AND PRACTICE PROBLEMS

14.1 A prestressed concrete pile is being driven into a saturated clay with a hammer blow count of 17 blows per foot of pile penetration. Unfortunately, the pile driving rig breaks down before the pile reaches the required depth of penetration, resulting in a 15-hour delay. Once the rig is repaired, pile driving resumes but the blow count is now 25 blows per foot. Explain the primary reason for this change in blow count.

14.2 Sometimes pile driving contractors use predrilling when installing piles. This method consists of drilling a vertical hole that has a smaller diameter than the pile, then driving the pile into this hole. Could predrilling affect the side-friction resistance? Why? Is the diameter of the predrill hole important? Why?

14.3 Why is it important for drilled shaft contractors to place the concrete soon after drilling the shaft? What detrimental effects can occur if the contractor waits too long before placing the concrete?

14.2   TOE BEARING

The net unit toe-bearing resistance, q_t', for deep foundations appears to be similar to the ultimate bearing capacity, q_{ult}, for shallow foundations. Therefore, it might seem that we could compute it using Terzaghi's bearing capacity formulas (Equations 6.4 to 6.6). However, there are important differences:

- Most bearing-capacity analyses for shallow foundations, as discussed in Chapter 6, are based on a general shear failure. Thus, they are based on the assumptions that the soil is incompressible and the shear surfaces extend to the ground surface, as shown in Figure 14.3. However, toe-bearing failures in deep foundations usually occur as a result of local or punching shear, and thus involve shearing only near the pile toe, as shown in Figure 14.3. For example, this is what occurs beneath the toe of a pile as it is being driven into the ground. Therefore, q_t' depends on both the strength and the compressibility of the soil.

- As shown in Figure 6.2, the load-settlement curves for local and punching shear failures do not have a distinct point of failure. Thus, it is not always clear how to define "failure" or to know exactly when it occurs. In addition, failure criterion are often based on settlement, which blurs the distinction between strength and serviceability requirements.

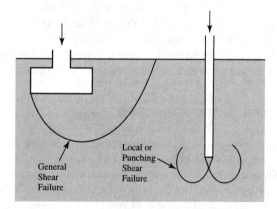

Figure 14.3   Comparison of general shear failure in spread footings with local and punching shear failures in deep foundations.

- In deep foundations, the depth of embedment, D, is very large and the diameter, B, is small. Therefore, the second term in Equations 6.4 to 6.6 dominates.

Therefore, we cannot compute toe-bearing resistance by simply applying the Terzaghi bearing-capacity formulas. It is necessary to have special methods developed specifically for deep foundations.

Sands

Any excess pore water pressures that might develop beneath deep foundations in sands dissipate very quickly, so the pore water pressure may be computed using Equation 3.9. Therefore we can evaluate the toe-bearing resistance using an effective stress analysis. Alternatively, we could use empirical correlations with other soil parameters, such as the CPT q_c value.

Piles

The toe-bearing resistance of piles in sandy soils may be expressed using the following formula (Kulhawy et al., 1983):

$$q_t' = B\, \gamma\, N_\gamma^* + \sigma_{zD}' N_q^* \tag{14.1}$$

Where:

q_t' = net unit toe-bearing resistance

B = pile diameter

N_γ^*, N_q^* = bearing capacity factors

γ = unit weight of soil immediately beneath pile toe (use $\gamma - \gamma_w$ if toe is located below the groundwater table)

σ_{zD}' = vertical effective stress at pile toe

When D/B exceeds 5, which is true for nearly all deep foundations, the first factor in Equation 14.1 becomes negligible and may be ignored.

The bearing capacity factors N_γ and N_q used in shallow foundations are functions only of the shear strength, as defined by the friction angle, ϕ', because general shear controls the failure. However, N_γ^* and N_q^* in Equation 14.1 depend on both shear strength and compressibility, because any of the three modes of failure (general, local, or punching shear) could govern. We define the compressibility effects using the *rigidity index*, I_r, of the soil (Vesić, 1977):

$$I_r = \frac{E}{2(1 + v)(\sigma_{zD}' \tan \phi')} \tag{14.2}$$

Where:

I_r = rigidity index

E = modulus of elasticity of soil in vicinity of toe

v = Poisson's ratio of soil in vicinity of toe

σ_{zD}' = vertical effective stress at the toe elevation

ϕ' = effective friction angle of soils in vicinity of toe

The magnitude of I_r is typically between about 10 and 400. Soils with high values of I_r fail in general shear, as modeled by conventional bearing capacity theory. However, low values of I_r suggest that soil compressibility is important and the local or punching shear failure modes govern.

For this analysis, we can assume the modulus of elasticity, E, is very close to the equivalent modulus of elasticity, E_s, used in the Schmertmann analysis described in Chapter 7. Therefore, we can estimate E from SPT or CPT data using Table 7.5 or Equation 7.17. If neither SPT nor CPT data is available, we could use an estimated N_{60}. The value of Poisson's ratio may be obtained from Table 14.1.

According to Vesić (1977), the bearing capacity factors N_γ^* and N_q^* are computed as follows:

$$N_\gamma^* = 0.6(N_q^* - 1)\tan \phi' \tag{14.3}$$

$$N_q^* = \frac{(1 + 2 K_0) N_\sigma}{3} \tag{14.4}$$

$$N_\sigma = \frac{3}{3 - \sin \phi'} e^{\frac{(90 - \phi')\pi}{180}} \tan^2 \left(45 + \frac{\phi'}{2} \right) I_r^{\frac{4 \sin \phi'}{3(1 + \sin \phi')}} \tag{14.5}$$

Where:

$N_\gamma^*, N_q^*, N_\sigma$ = bearing capacity factors

K_0 = coefficient of lateral earth pressure at rest (can estimate using Equation 3.15)

ϕ' = effective friction angle

TABLE 14.1 TYPICAL VALUES OF POISSON'S RATIO FOR SOILS AND ROCKS (Adapted from Kulhawy, et al., 1983)

Soil or Rock Type	Poisson's Ratio, v
Saturated clay, undrained conditions	0.50
Partially saturated clay	0.30–0.40
Dense sand, drained conditions	0.30–0.40
Loose sand, drained conditions	0.10–0.30
Sandstone	0.25–0.30
Granite	0.23–0.27

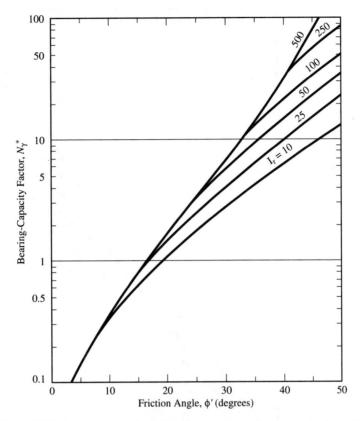

Figure 14.4 Bearing capacity factor N_γ^*. This plot is based on assumed values of K_0, and thus may produce results that differ from Equation 14.3. However, this difference represents a small part of the total toe-bearing resistance (Adapted from Kulhawy et al., 1983; Copyright © 1983 Electric Power Research Institute. Reprinted with permission).

The results of these equations are presented in graphical form in Figures 14.4 and 14.5 using typical values of K_0.

Example 14.1

A 400-mm square prestressed concrete pile is to be driven 16 m into the soil profile shown in Figure 14.6. Compute the net ultimate toe-bearing capacity.

Solution

$$\sigma'_{zD} = \Sigma \gamma H - u$$
$$= (17.8)(3) + (18.2)(13) - (9.8)(13)$$
$$= 163 \text{ kPa}$$

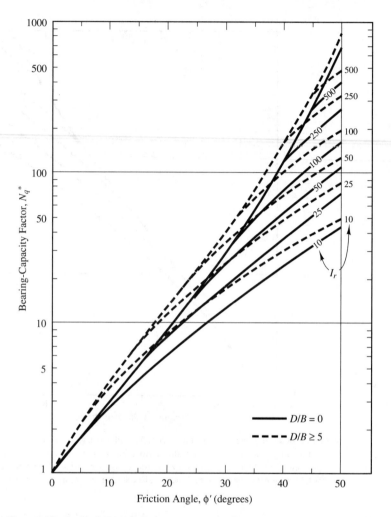

Figure 14.5   Bearing capacity factor N_q^* (Adapted from Kulhawy et al., 1983; Copyright © 1983 Electric Power Research Institute. Reprinted with permission).

Per Equation 7.17:

$$E_s = \beta_0 = \sqrt{OCR} + \beta_1 N_{60}$$
$$= (5000)\sqrt{1} + (1200)(25)$$
$$= 35{,}000 \text{ kPa}$$

$$I_r = \frac{E}{2(1 + \nu)\sigma'_{zD}\tan\phi'}$$
$$= \frac{35{,}000}{2(1 + 0.30)(163)\tan 36}$$
$$= 114$$

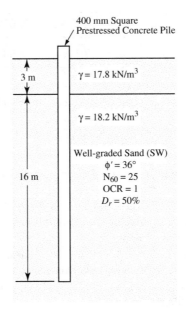

Figure 14.6   Proposed pile for Examples 14.1 and 14.2.

Per Figure 14.4: $N_\gamma^* = 14$
Per Figure 14.5: $N_q^* = 76$

$$q_t' = \beta \, \gamma \, N_\gamma^* + \sigma_{zD}' N_q^*$$

$$= (0.4)(18.2 - 9.8)(14) + (163)(76)$$

$$= 12,400 \text{ kPa}$$

$$q_t' A_t = (12,400)(0.4^2) = \mathbf{1990 \text{ kN}} \qquad \Leftarrow Answer$$

Drilled Shafts

The impacts and vibrations from pile driving compact the soil beneath the piles, which increases its toe-bearing resistance. Unfortunately, this does not occur beneath drilled shafts, where the augering process and the temporary stress relief may even loosen the soil. Therefore, the net unit toe-bearing resistance for drilled shafts in sands is less than that for piles.

Analytic design of drilled shafts is usually based on empirical formulas developed from instrumented full-scale static load tests (O'Neill and Reese, 1999). Various researchers have offered different interpretations of these load tests, but O'Neill and Reese defined the toe-bearing resistance as that which occurs at a settlement of 5 percent of the base diameter. Based on this criterion, they recommend computing the net unit toe-bearing resistance for drilled shafts in sands with $N_{60} \leq 50$ as follows:

$$q'_t = 1200 \, N_{60} \leq 60{,}000 \, \text{lb/ft}^2 \qquad \text{(14.6 English)}$$

$$q'_t = 57.5 \, N_{60} \leq 2900 \, \text{kPa} \qquad \text{(14.6 SI)}$$

Where:

q'_t = net unit toe-bearing resistance (lb/ft^2 or kPa)

N_{60} = mean SPT N-value between toe and a depth of $2B_b$ below the toe

B_b = base diameter of drilled shaft

If $N_{60} > 50$, the ground is classified as an intermediate geomaterial, as described later in this section.

Shafts with base diameters larger than about 1200 mm (50 in), require about 60 mm (2.5 in) of settlement to develop the "full" toe-bearing resistance as defined by O'Neill and Reese. With normal factors of safety, this translates to a settlement of about 25 mm (1 in) to develop the allowable toe-bearing resistance. Assuming this is equal to maximum allowable settlement, shafts with larger base diameters may experience excessive settlements, especially if toe bearing represents a large portion of the total capacity. There are two ways to deal with this problem:

1. Reduce q'_t to a lower value, q_{tr}' using Equations 14.7:

$$q'_{tr} = \frac{50 \text{ in}}{B_b} q'_t \qquad \text{(14.7 English)}$$

$$q'_{tr} = \frac{1200 \text{ mm}}{B_b} q'_t \qquad \text{(14.7 SI)}$$

The design is then based on q_{tr}'.

or **2.** Perform a settlement analysis, as described in Section 14.7, and adjust the design so the settlement under working loads is within tolerable limits.

If the allowable total settlement is greater or less than 25 mm, then the diameter at which excessive settlements may be a problem is correspondingly different.

Auger-Cast Piles

Neely (1991) developed the following empirical formula for toe-bearing resistance of auger-cast piles in sand. This formula is based on the results of sixty six load tests on piles with diameters between 300 and 600 mm (12–24 in) and depths of embedment between 4.5 and 26 m (15–85 ft). Grout consumption data was not always available, so he reduced the data in terms of the auger diameter, even though the actual pile diameter is somewhat larger. Therefore, capacity predictions using this formula also should be based on the auger diameter.

$$q'_t = 3800 \, N_{60} \leq 150{,}000 \, \text{lb/ft}^2 \qquad\qquad \text{(14.8 English)}$$

$$q'_t = 190 \, N_{60} \leq 7500 \, \text{kPa} \qquad\qquad \text{(14.8 SI)}$$

Where:

q'_t = net unit toe-bearing resistance (lb/ft^2, kPa)

N_{60} = SPT blow count between toe and a depth of about B below the toe

A comparison of Equations 14.6 and 14.8 suggests auger cast piles have substantially more toe-bearing resistance than drilled shafts. However, the differences between these formulas is probably caused by differences in interpretation of the load test data, not to any real difference in resistance for foundations of equal diameter. In any case, the small diameter of auger cast piles limits the $q'_t A_t$ value, so most of the load transfer occurs via side friction.

Pressure-Injected Footings

Neely (1989, 1990a, 1990b) developed empirical methods of computing the net unit toe-bearing resistance, q'_t, of pressure-injected footings (PIFs) from SPT N values. This method is based on the results of load tests conducted 93 PIFs with cased shafts and 41 PIFs with compacted shafts. All of these piles were bottomed in sands or gravels. His formula for cased PIFs bottomed in clean sands and gravels is:

$$q'_t = 560 \, (N_1)_{60} \frac{D}{B_b} \leq 5600 \, (N_1)_{60} \qquad\qquad \text{(14.9 English)}$$

$$q'_t = 28 \, (N_1)_{60} \frac{D}{B_b} \leq 280 \, (N_1)_{60} \qquad\qquad \text{(14.9 SI)}$$

Where:

q'_t = net unit toe-bearing resistance (lb/ft^2, kPa)

$(N_1)_{60}$ = corrected SPT N value between toe and a depth of about B below the toe

D = depth to bottom of PIF base

B_b = diameter of PIF base

For compacted shaft PIFs, Neely presented his design toe bearing recommendations in graphical form, as shown in Figure 14.7.

Neely's data suggest that silty sands and clayey sands have less toe-bearing resistance than clean sand with a comparable N value. Soils with a fines content of 25 percent may have a toe-bearing resistance of about half that predicted by Equation 14.9 or Figure 14.7.

Because of the soil compaction that occurs during PIF construction, their net unit toe-bearing resistance is much greater than that of a drilled shaft of comparable dimensions, especially in clean sands (compare Equations 14.6 and 14.9).

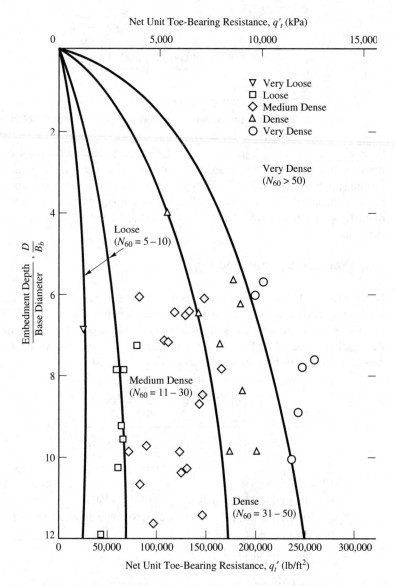

Figure 14.7   Net unit toe-bearing resistance, q_t', for compacted shaft PIFs in clean sands and gravels (Adapted from Neely, 1990b; Used with permission of ASCE).

Clays

Because of their low hydraulic conductivity, we assume undrained conditions exist in clays beneath the toe of deep foundations. Therefore, we compute q_t' using the undrained shear strength, s_u. For deep foundations with $D/B > 3$ in clays with $s_u \leq 250$ kPa (5000 lb/ft^2) use the following equation:

$$q_t' = N_c^* s_u$$

(14.10)

Where:

q_t' = net unit toe-bearing resistance

N_c^* = bearing capacity factor (O'Neill and Reese, 1999)

= 6.5 at s_u = 25 kPa (500 lb/ft^2)

= 8.0 at s_u = 50 kPa (1000 lb/ft^2)

= 9.0 at $s_u \geq 100$ kPa (2000 lb/ft^2)

s_u = undrained shear strength of soil between the toe and $2B$ below the toe

Clays with $s_u > 250$ kPa (5000 lb/ft^2) should be evaluated as intermediate geomaterials, as discussed later in this section.

The value of s_u is determined from laboratory tests, such as the unconfined compression or triaxial compression, or from in-situ tests. However, be especially cautious when working with fissured clays. The spacing of these fissures is often large compared to the size of the soil samples, so test results may represent the soil between the fissures rather than the entire soil mass. In such cases, reduce s_u accordingly (perhaps to a value of about 0.75 times the measured value).

As discussed earlier, large diameter foundations require correspondingly more settlement to achieve the ultimate toe-bearing resistance. If the base diameter, B_b, is greater than 1900 mm (75 in), the value of q_t' from Equation 14.10 could produce settlements greater than 1 in (25 mm), which would be unacceptable for most buildings. To keep settlements within tolerable limits, reduce the value of q_t' to q_{tr}', and use this value (O'Neill and Reese, 1999):

$$q_{tr}' = F_r q_t'$$

(14.11)

$$F_r = \frac{2.5}{\psi_1 B_b + 2.5\,\psi_2} \leq 1.0$$

(14.12)

$$\psi_1 = 0.0085\, B_b + 0.083\,(D/B_b)$$

(14.13 English)

$$\psi_1 = 0.28\, B_b + 0.083\,(D/B_b)$$

(14.13 SI)

$$\psi_2 = 0.014\,\sqrt{s_u}$$

(14.14 English)

$$\psi_2 = 0.065\,\sqrt{s_u}$$

(14.14 SI)

Where:

q_{tr}' = reduced net unit toe-bearing resistance

q_t' = net unit toe-bearing resistance

B_b = diameter at base of foundation (ft, m)

D = depth of embedment (ft, m)

s_u = undrained shear strength in the soil between the base of the foundation and a depth $2B_b$ below the base (lb/ft^2, kPa)

Intermediate Geomaterials and Rock

Intermediate geomaterial is a new term used to describe hard soils and soft rocks. O'Neill and Reese (1999) define them as cohesive or cemented materials, such as clay shales or mudstones, with 250 kPa (5000 lb/ft^2) < s_u < 2500 kPa (50,000 lb/ft^2) or noncohesive materials, such as glacial till, with N_{60} > 50. These materials can be difficult to evaluate because they have engineering properties between those of soil and rock.

We will define *rock* as a material with $s_u \geq 2500$ kPa (50,000 lb/ft^2), which corresponds to an unconfined compressive strength, $q_u \geq 5000$ kPa (100,000 lb/ft^2). This definition does not necessarily correspond to the geological definition of rock, which includes some materials that we would classify as intermediate geomaterials.

Some, but certainly not all, deep foundations extend down to bedrock. Such designs take advantage of the increased strength and stiffness of bedrock, and thus have more toe-bearing resistance than foundations bottomed in soil. However, the feasibility of such a design depends on many factors, including:

- The depth to bedrock, which ranges from zero (in which case a shallow foundation should be suitable) to thousands of meters below the ground surface (which is far below the technical and economic reach of deep foundations).
- The design loads, which may be too high to be supported on soil.
- The strength of the overlying soils, which may be sufficient to support the applied loads.
- The potential for scour, which may wash away some or all of the overlying soil.
- The potential for downdrag loads caused by settlement in the soils. This phenomenon is discussed in Chapter 18.

The toe-bearing resistance in intermediate geomaterials and bedrock is more difficult to evaluate than that for soil, mostly because it depends on the nature of the *discontinuities*, which can be difficult to assess. Fortunately, even conservative estimates of q_t' in these materials are often sufficiently high to produce an economical design. In very strong rock, q_t' may exceed the allowable stress in the foundation, especially with drilled shafts. In that case, the structural capacity may control the design. If the structural loads are very high, and toe bearing is the primary source of resistance, conservative design values may produce a foundation design that is too expensive. In these cases, full-scale static load tests may be appropriate. The Osterberg test has been especially useful in such circumstances.

O'Neill and Reese (1999) recommend the following methods of evaluating q_t' for drilled shaft foundations. These methods also should be appropriate for other types of deep foundations.

Cohesive Intermediate Geomaterial and Rock

The *rock quality designation* or *RQD* is a measure of the integrity of rock or intermediate geomaterial obtained from coring. It is defined as the percentage of the core that is recovered in pieces ≥ 100 mm in length. RQD values greater than 90 percent typically indicate excellent material, while those less than 50 percent reflect poor or very poor material. The *unconfined compressive strength, q_u,* is usually measured in the laboratory on core samples using a technique similar to that for measuring the compressive strength of concrete. This value also is equal to twice the undrained shear strength, s_u. O'Neill and Reese use the RQD and the unconfined compressive strength between the bottom of the foundation and a depth of about $2B$ below the bottom to evaluate toe-bearing resistance.

If RQD = 100% and the foundation extends to a depth of at least $1.5B$ into the intermediate geomaterial or rock:

$$q_t' = 2.5\, q_u \qquad (14.15)$$

If 70% < RQD < 100%, all joints are closed (i.e., not containing voids are soft infill material) and nearly horizontal, and $q_u > 500$ kPa (10,000 lb/ft^2):

$$q_t' = 7970\, (q_u)^{0.51} \qquad \text{(14.16 English)}$$

$$q_t' = 4830\, (q_u)^{0.51} \qquad \text{(14.16 SI)}$$

Where q_u is in lb/in^2 and MPa and q_t' is in lb/ft^2 and kPa.

If the material is jointed, the joints have random orientation, and the condition of the joints can be evaluated in the area or from test excavations:

$$q_t' = [t^{0.5} + (mt^{0.5} + t)^{0.5}]q_u \qquad (14.17)$$

Where m and t are determined from Tables 14.2 and 14.3.

Noncohesive Intermediate Geomaterial

For noncohesive intermediate geomaterials, O'Neill and Reese use:

$$q_t' = 0.59[(N_1)_{60}]^{0.8}\, \sigma_{zD}' \qquad (14.18)$$

TABLE 14.2 DESCRIPTION OF ROCK AND INTERMEDIATE GEOMATERIAL TYPES FOR USE IN TABLE 14.3 (O'Neill and Reese, 1999)

Rock or Intermediate Geomaterial Type	Description
A	Carbonate rocks with well-developed crystal cleavage (e.g., dolostone, limestone, marble)
B	Lithified argillacous rocks (e.g., mudstone, stiltstone, shale, slate)
C	Arenaceous rocks (e.g., sandstone, quartz)
D	Fine-grained igneous rocks (e.g., andesite, dolerite, diabase, rhyolite)
E	Coarse-grained igneous and metamorphic rocks (e.g., amphibole, gabbro, gneiss, granite, norite, quartz diorite)

Where:

q_t' = net unit toe-bearing resistance

$(N_1)_{60}$ = SPT N-value corrected for field procedures and overburden stress ≤ 100

σ_{zD}' = vertical effective stress at base of foundation

If the base diameter is larger than 1200 mm (50 in), this computed value should be reduced using Equation 14.7 in order to control settlement.

TABLE 14.3 VALUES OF m AND t FOR EQUATION 14.17 (O'Neill and Reese, 1999)

Quality of Rock or Intermediate Geomaterial	Joint Description	Joint Spacing	t	m				
				Type A	Type B	Type C	Type D	Type E
Excellent	Intact (closed)	> 3 m (> 10 ft)	1	7	10	25	17	25
Very Good	Interlocking	1–3 m (3–10 ft)	0.1	3.5	5	7.5	8.5	12.5
Good	Slightly weathered	1–3 m (3–10 ft)	4×10^{-2}	0.7	1	1.5	1.7	2.5
Fair	Moderately weathered	0.1–1 m (1–3 ft)	10^{-4}	0.14	0.2	0.3	0.34	0.5
Poor	Weathered with gouge	30–300 mm (1–12 in)	10^{-5}	0.04	0.05	0.08	0.09	0.13
Very Poor	Heavily weathered	< 50 mm (< 2 in)	0	0.007	0.01	0.015	0.017	0.025

14.3   SIDE FRICTION

The unit side-friction resistance, f_s, is more clearly defined than the toe bearing because of the sharp break in the load-settlement curve. The analysis of side friction is based on the principle of sliding friction, and is most accurately performed using effective stresses. However, side friction also may be evaluated using total stresses.

Effective Stress Analyses

Excess pore water pressures are often present along the sides of deep foundations during and immediately after construction, especially with piles in clays. However, these pressures dissipate rapidly, so hydrostatic conditions generally are present during the service life of the structure. If, during the service life of the structure, an additional load is placed on the foundation, little or no excess pore water pressure will be generated along the side of the foundation, because there is no additional compression of the soil. Therefore, we may compute the vertical effective stress using the hydrostatic pore water pressure and evaluate side-friction resistance using an effective stress analysis.

Principles

Side friction may be described using a simple sliding-friction model:

$$\boxed{f_s = \sigma_x' \tan \phi_f}$$

(14.19)

Where:

f_s = unit side-friction resistance

σ_x' = horizontal effective stress (i.e., perpendicular to the foundation axis)

$\tan \phi_f = \mu$ = coefficient of friction between the soil and the foundation

ϕ_f = soil-foundation interface friction angle

Researchers have used laboratory tests to measure the coefficient of friction, $\tan \phi_f$, and have correlated it with the effective friction angle of the soil, ϕ'. Table 14.4 gives typical values of ϕ_f / ϕ'. Saturated dense soils have ϕ_f / ϕ' ratios on the high end of these ranges, whereas dry loose soils tend to the low end.

As discussed in Chapter 3, the ratio between the horizontal and vertical effective stresses is defined as the coefficient of lateral earth pressure, K:

$$K = \frac{\sigma_x'}{\sigma_z'}$$

(14.20)

Deep foundation construction induces significant changes in the surrounding soils, as discussed in Section 14.1. Therefore, the coefficient of lateral earth pressure, K, is generally not equal to the coefficient of lateral earth pressure in the ground before construction, K_0. The ratio K/K_0 depends on many factors, including the following:

TABLE 14.4   APPROXIMATE ϕ_f/ϕ' VALUES FOR THE
INTERFACE BETWEEN DEEP FOUNDATIONS AND SOIL
(Adapted from Kulhawy et al., 1983 and Kulhawy, 1991)

Foundation Type	ϕ_f/ϕ'
Rough concrete	1.0
Smooth concrete (i.e., precast pile)	0.8–1.0
Rough steel (i.e., step-taper pile)	0.7–0.9
Smooth steel (i.e., pipe pile or H-pile)	0.5–0.7
Wood (i.e., timber pile)	0.8–0.9
Drilled shaft built using dry method or with temporary casing and good construction techniques	1.0
Drilled shaft built with slurry method (higher values correspond to more careful construction methods)	0.8–1.0

- **Type of deep foundation**—Pile driving causes compression of the surrounding soils, while drilled shaft construction may result in some stress relaxation.
- **High displacement vs. low displacement piles**—High displacement piles, such as closed-end steel pipes, displace much more soil than do low displacement piles, such as a steel H-piles, and therefore have a much higher K/K_0 ratio. However, some of this gain may be lost over time as creep effects tend to relax the locally high horizontal stresses.
- **Soil consistency**—Dense soils provide more resistance to distortion, which results in a greater K/K_0 ratio.
- **Special construction techniques**—Predrilling or jetting loosens the soil, thus reducing the K/K_0 ratio.

The largest possible value of K is K_p, the coefficient of passive earth pressure,[2] which is equal to $\tan^2 (45 + \phi'/2)$. Kulhawy et al. (1983) suggested the K/K_0 ratios in Table 14.5.

Combining Equations 14.19 and 14.20 with the factors in Tables 14.4 and 14.5 gives:

$$ f_s = K_0\, \sigma'_z \left(\frac{K}{K_0} \right) \left(\frac{\phi_f}{\phi'} \right) \tag{14.21} $$

The most difficult factor to assess in Equation 14.21 is K_0. It can be estimated using Equation 3.15, which in turn requires an estimate of the overconsolidation ratio (OCR). In clays this can be done by performing consolidation tests, as described in Chapter 3, and

---

[2]Passive earth pressures are discussed in more detail in Chapter 23.

TABLE 14.5   APPROXIMATE RATIO OF COEFFICIENT OF LATERAL EARTH
PRESSURE AFTER CONSTRUCTION TO THAT BEFORE CONSTRUCTION
(Adapted from Kulhawy et al., 1983 and Kulhawy, 1991)

Foundation Type and Method of Construction	K/K_0
Pile—jetted	0.5–0.7
Pile—small displacement, driven	0.7–1.2
Pile—large displacement, driven	1.0–2.0
Drilled shaft—built using dry method with minimal sidewall disturbance and prompt concreting	0.9–1.0
Drilled shaft—slurry construction with good workmanship	0.9–1.0
Drilled shaft—slurry construction with poor workmanship	0.6–0.7
Drilled shaft—casing method below water table	0.7–0.9

using Equation 3.24. However, even soils that are supposedly normally consolidated usually have an overconsolidated crust as shown in Figure 14.8. This crust often extends from the ground surface to a depth of 3 to 6 m (10–20 ft) or more. As a result, K_0 in "normally consolidated" soil deposits is typically quite high near the ground surface and gradually becomes smaller with depth (Kulhawy, 1984, 1991).

In more classical overconsolidated clays (i.e., those precompressed by removal of overburden), the overconsolidation ratio, OCR, decreases with depth, so K_0 also decreases with depth.

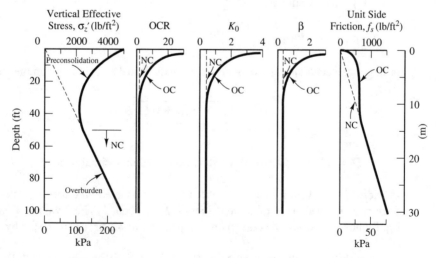

Figure 14.8   Typical distributions of parameters that influence side-friction resistance (Adapted from Kulhawy, 1991; Used with permission of Van Nostrand Publishers). NC = normally consolidated, OC = overconsolidated.

Unfortunately, it is much more difficult to evaluate the OCR in sands because we cannot obtain suitable undisturbed samples. This is true even when interbedded clays are present because the OCR of these strata may be different, especially if desiccation has occurred.

Some in-situ tests, such as the pressuremeter (PMT), dilatometer (DMT), or K_0 stepped blade, might be used to evaluate lateral earth pressures. None of these tests is widely available in North America, but their use may increase in the future.

β Method

Because of the difficulties in evaluating some of the factors in Equation 14.21, engineers sometimes rewrite this equation as follows (Burland, 1973):

$$\boxed{f_s = \beta\,\sigma'_z} \tag{14.22}$$

$$\beta = K_0 \left(\frac{K}{K_0}\right) \tan\left[\phi'\left(\frac{\delta}{\phi'}\right)\right] \tag{14.23}$$

This technique of computing side-friction resistance is known as the *beta method*. In practice, design β values have been obtained by backcalculating them from full-scale static load tests and correlating these values with soil properties and foundation type. We then use these correlations to apply these values to new foundations.

When using the β method, divide the soil into layers with layer boundaries located at the soil strata boundaries and at the groundwater table. Then compute f_s at the midpoint of each layer using Equation 14.22.

Sands

For large-displacement pile foundations in sands, Bhushan (1982) recommends:

$$\boxed{\beta = 0.18 + 0.65\,D_r} \tag{14.24}$$

Where:

D_r = relative density of the sand expressed in decimal form

Equation 14.24 is based on a series of load tests conducted on closed-end steel pipe piles which probably had ϕ_f/ϕ' of about 0.7 and K/K_0 of about 1. The β value for other types of deep foundations would be correspondingly higher or lower, as suggested by Tables 14.4 and 14.5.

For drilled shafts in sand with $N_{60} \geq 15$, O'Neill and Reese (1999) recommend assessing the following equation for β, and limit f_s to a maximum value of 190 kPa (4000 lb/ft²):

$$\boxed{\beta = 1.5 - 0.135 \sqrt{z} \qquad 0.25 \leq \beta \leq 1.20}$$ (14.25 English)

$$\boxed{\beta = 1.5 - 0.245 \sqrt{z} \qquad 0.25 \leq \beta \leq 1.20}$$ (14.25 SI)

Where:

z = depth to midpoint of soil layer (ft, m)

If N_{60} <15, multiply the β values obtained from Equation 14.25 by the ratio $N_{60}/15$.

This β function reaches its limits at depths of 1.5 m (5 ft) and 26 m (86 ft), so layer boundaries should be placed at these two depths. Another boundary should be placed at the groundwater table. Additional boundaries should be placed every 6 m (20 ft) and where the sand strata end and it becomes necessary to begin using clay or rock analyses.

For auger-cast piles, Neely (1991) computes the average unit side-friction resistance using the β method:

$$\boxed{\bar{f}_s = \bar{\beta}\,\bar{\sigma}'_z \leq 140 \text{ kPa } (2800 \text{ lb/ft}^2)}$$ (14.26)

Where:

$\bar{f}_s$ = average unit side-friction resistance

$\bar{\beta}$ = average beta factor (from Figure 14.9)

$\bar{\sigma}'_z$ = average vertical effective stress along length of pile

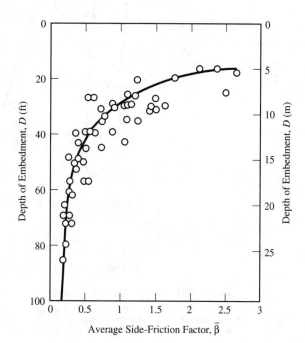

Figure 14.9 Average beta factor, $\bar{\beta}$ for use in Equation 14.26. (Adapted from Neely, 1991; used by permission of ASCE.)

The beta factor β varies nonlinearly with depth. Figure 14.9 gives the average value, $\overline{\beta}$, for various pile lengths. Therefore, when using this chart, determine only one $\overline{\beta}$ value based on the length of the pile and use it in Equation 14.26. Do not divide the soil into layers to determine the β value for each layer.

For pressure injected footings (PIFs), the side-friction resistance should be neglected if a cased shaft is used because of the annular space between the casing and the soil. However, PIFs with compacted shafts have excellent shaft-soil contact with a high lateral stress and a rough surface, and therefore can develop high side-friction resistance. Neely (1990b) computes it using Equation 14.26 and Figure 14.10.

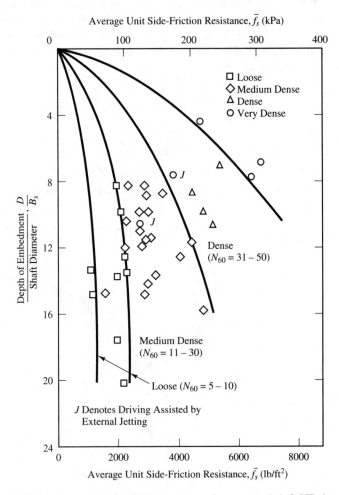

Figure 14.10 Average unit side-friction resistance for compacted shaft PIFs in sandy soils (Adapted from Neely, 1990b; used with permission of ASCE). These values are appropriate only when the drive tube is driven into the ground in the conventional fashion. The use of jetting to assist this process reduces the side-friction resistance.

Gravels

Rollins, Clayton, and Mikesell (1997) used a series of full-scale static load tests to develop a revised version of Equation 14.25 for drilled shaft foundations in gravels (> 50 percent gravel size):

$$\beta = 3.4 \, e^{-0.026 \, z} \qquad 0.25 \le \beta \le 3.00 \qquad \text{(14.27 English)}$$

$$\beta = 3.4 \, e^{-0.085 \, z} \qquad 0.25 \le \beta \le 3.00 \qquad \text{(14.27 SI)}$$

and for gravelly sands (25–50 percent gravel size):

$$\beta = 2.0 - 0.061 \, z^{0.75} \qquad 0.25 \le \beta \le 1.80 \qquad \text{(14.28 English)}$$

$$\beta = 2.0 - 0.15 \, z^{0.75} \qquad 0.25 \le \beta \le 1.80 \qquad \text{(14.28 SI)}$$

Where:

z = depth to midpoint of soil layer (ft, m)

e = base of natural logarithms = 2.718

β for piles in gravelly soils also should be 20 to 30 percent higher than that in sandy soils, so Equation 14.24 should be conservative.

Silts and Clays

Normally consolidated silts and clays typically have β = 0.27 to 0.50 and β = 0.25 to 0.35, respectively (Fellenius, 1999). Soft, normally consolidated soils are usually toward the lower end of these ranges, while lightly overconsolidated soils are toward the high end. If the soil is heavily overconsolidated, β could be much higher, as shown in Figure 14.11.

In clays, the side-friction resistance within 1.5 m (5 ft) of the ground surface should be ignored because of clay shrinkage caused by drying, foundation movement produced by lateral loads, pile wobble during driving, and other factors.

Example 14.2

Using a factor of safety of 2.75, compute the allowable downward capacity of the pile described in Example 14.1. This pile is to be driven without the use of jetting.

Solution

$$\beta = 0.18 + 0.65(0.50) = 0.50 \quad \text{(Based on } \phi_f/\phi' = 0.7 \text{ and } K/K_0 = 1)$$

Per Table 14.4, $\phi_f/\phi' \approx 0.9$

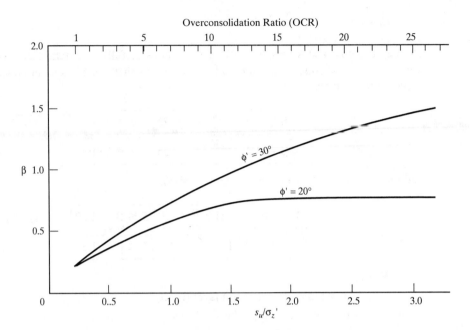

Figure 14.11 β values for clays (Adapted from Randolph and Wroth, 1982, and Equation 3.15). For normally consolidated or slightly overconsolidated soils, these values are comparable to those recommended by Fellenius. However, values for heavily overconsolidated soils are much higher. When working with low displacement piles, reduce these values by about 20 percent.

Per Table 14.5, $K/K_0 \approx 1$

$$\therefore \text{ use } \beta = 0.50\left(\frac{0.9}{0.7}\right)\left(\frac{1}{1}\right) = 0.64$$

Layer	Depth (m)	σ_z' (kPa)	f_s (kPa)	A_s (m²)	$f_s A_s$ (kN)
1	0–3	26.7	17.1	4.8	82
2	3–10	82.8	53.0	11.2	594
3	10–16	137.4	87.9	9.6	844
					1520

$$P_a = \frac{q_t' A_t + \Sigma f_s A_s}{F} = \frac{1990 + 1520}{2.75} = \textbf{1276 kN} \quad \Leftarrow Answer$$

Example 14.3

The drilled shaft shown in Figure 14.12 is to be designed without the benefit of any onsite static load tests. The soil conditions are uniform and the site characterization program was average. Compute the allowable downward load capacity.

Solution

The unit weights of these soils, γ, have not been given (probably because it was not possible to obtain a suitably undisturbed sample of these sandy soils). We can't compute the load capacity without this information, so we must estimate γ for each strata using typical values from Table 3.2:

Silty sand above groundwater table: $\gamma \approx 17 \text{ kN/m}^3$
Silty sand below groundwater table: $\gamma \approx 20 \text{ kN/m}^3$
Sand below groundwater table: $\gamma \approx 20 \text{ kN/m}^3$

Using Equation 14.25 to compute β:

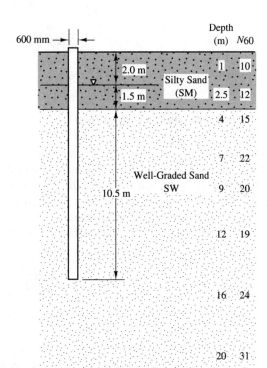

Figure 14.12   Drilled shaft foundation for Example 14.3.

Strata	z (m)	β	σ_z' (kPa)	f_s (kPa)	A_s (m^2)	P_s (kN)
Silty sand above GWT	1.00	1.20	17.0	20.4	3.77	77
Silty sand below GWT	2.75	1.09	41.5	45.2	2.83	128
Sand (3.5–9 m)	6.25	0.88	77.3	68.0	10.37	705
Sand (9–16 m)	11.15	0.68	130.9	89.0	9.42	838
						1748

Note how we compute z and σ_z' at the midpoint of each strata.

Use Equation 14.6 to compute the net unit toe-bearing resistance. Although no N_{60} values are available within a depth of $2B$ below the bottom of the shaft, it appears that $N_{60} = 22$ would be a reasonable value for design.

$$q_t' = 57.5\, N_{60} = (57.5)(22) = 1265 \text{ kPa}$$

$$A_t = \frac{\pi(0.6)^2}{4} = 0.283 \text{ m}^2$$

Using a factor of safety of 3 (per Table 13.2):

$$P_a = \frac{q_t' A_t + \Sigma f_s A_s}{F}$$

$$= \frac{(1265)(0.283) + 1748}{3}$$

$$= 702 \text{ kN} \qquad \Leftarrow Answer$$

Total Stress Analyses (α Method)

Even though effective stress analyses are easily implemented and technically more correct, engineers often evaluate side-friction resistance in clays soils using total stress analyses based on the undrained strength. The *alpha method*, as defined in Equation 14.29, is the most common way of formulating this approach. Although the alpha method is less precise than the beta method, it has been used much more widely and thus has the benefit of a more extensive experience base. The alpha method equation is:

$$\boxed{f_s = \alpha\, s_u} \qquad (14.29)$$

Where:

f_s = unit side-friction resistance

α = adhesion factor

s_u = undrained shear strength in soil adjacent to the foundation

The formulation of the alpha method and the term *adhesion factor* give the mistaken impression that side-friction resistance is due to a "gluing" effect between the soil and the pile. This is misleading. It is more accurate to think of side-friction resistance being a classical sliding friction problem.

The adhesion factor, α, is determined empirically from pile load test results. Ideally, this could be performed on a site-specific basis, which allows us to extend load test results to assist in the design of piles with other diameters or lengths at that site. In the absence of site-specific data, we can use generic α values obtained from load tests at a variety of sites. Figure 14.13 shows back-calculated α values from full-scale static load tests, along with several suggested functions. Clearly, there is a large scatter in this data.

Of the various α functions in Figure 14.13, the API function is probably the most commonly used for piles:

For $s_u < 25$ kPa (500 lb/ft^2):

$$\boxed{\alpha = 1.0} \tag{14.30}$$

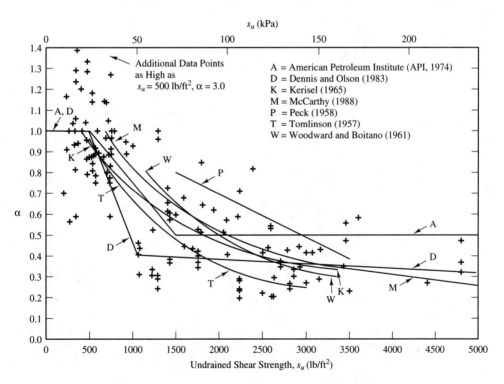

Figure 14.13 Measured values of α as backcalculated from full-scale static load tests compared with several proposed functions for α (Load test data adapted from Vesić, 1977).

For 25 kPa (500 lb/ft^2) $< s_u <$ 75 kPa (1500 lb/ft^2):

$$\alpha = 1.0 - 0.5\left(\frac{s_u - 500 \text{ lb/ft}^2}{1000 \text{ lb/ft}^2}\right)$$ (14.31 English)

$$\alpha = 1.0 - 0.5\left(\frac{s_u - 25 \text{ kPa}}{50 \text{ kPa}}\right)$$ (14.31 SI)

For $s_u >$ 75 kPa (1500 lb/ft^2):

$$\alpha = 0.5$$ (14.32)

Figure 14.14 shows backcalculated values of α for drilled shafts obtained from instrumented load tests. Once again, there is a wide scatter in the results. We can use the α curve shown in this plot, or the Reese and O'Neill (1989) approach as shown in Figure 14.15. Reese and O'Neill limit f_s to a maximum value of 260 kPa or 5500 lb/ft^2.

All of the α factors presented in this section are for insensitive clays ($S_t <$ 4). In sensitive clays, full-scale static load tests, special lab tests, or some other method of verification are appropriate (O'Neill and Reese, 1999).

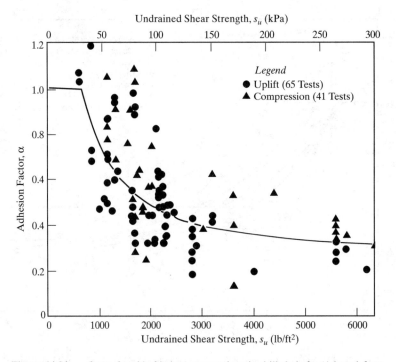

Figure 14.14 α factor for side-friction computations in drilled shafts (Adapted from Kulhawy and Jackson, 1989; used with permission of ASCE).

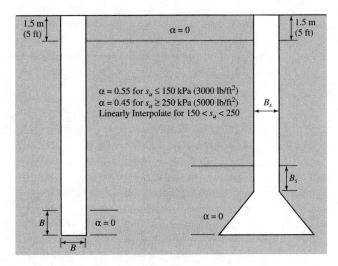

Figure 14.15 α factor for side-friction computations in drilled shafts (Adapted from O'Neill and Reese, 1999).

Example 14.4

A 60-ft long drilled shaft is to be built using the open hole method in the soil shown in Figure 14.16. The shaft will be 24 inches in diameter and the bottom will be belled to a diameter of 60 inches. This foundation is to be designed without the benefit of an onsite static load test, the soil conditions are uniform, and the subsurface investigation program was extensive. Compute the allowable downward load capacity.

Solution

$$q'_t = 9\,s_u = (9)(4000) = 36{,}000 \text{ lb/ft}^2$$

$$A_t = \frac{\pi B_b^2}{4} = \frac{\pi\,5^2}{4} = 19.6 \text{ ft}^2$$

Layer	Depth (ft)	Thickness (ft)	s_u (lb/ft^2)	α	f_s (lb/ft^2)	A_s (ft^2)	$f_s A_s$ (k)
1	0–5	5.0	—	0	—	—	0
2	5–12	7.0	1600	0.55	880	44	39
3	12–37	25.0	1400	0.55	770	157	121
4	37–56.5	19.5	4000	0.48	1920	122	234
5	56.5–60	3.5	—	0	—	—	0
							394

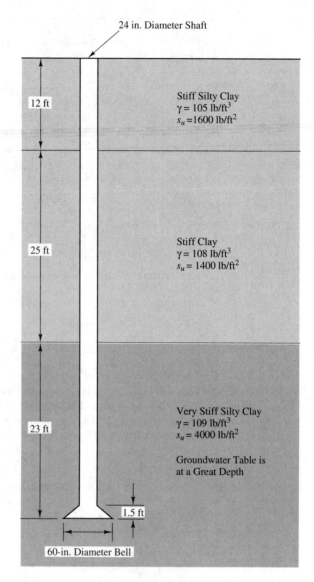

Figure 14.16 Drilled shaft foundation for Example 14.4.

Use $F = 2.50$ (per Table 13.2)

$$P_a = \frac{q'_t A_t + \Sigma f_s A_s}{F}$$

$$= \frac{(36)(19.6) + 394}{2.5}$$

$$= \textbf{440 k} \quad \Leftarrow \textit{Answer}$$

Intermediate Geomaterial and Rock

One of the advantages of drilled shaft foundations is that they can be drilled through intermediate geomaterials and rock that a pile could not penetrate. These strata can develop large side-friction and toe-bearing resistances, and thus can support high-capacity foundations. O'Neill and Reese (1999) present methods of computing these resistances for drilled shafts.

14.4   UPWARD LOAD CAPACITY

When a deep foundation is subjected to downward loads, it experiences some elastic compression and, because of the Poisson effect, a small increase in diameter. However, the opposite occurs when they are subjected to upward loads: The foundation "stretches" and the diameter becomes slightly smaller. As a result, the unit side-friction resistance, f_s, for upward loading (Equation 13.7) may be smaller than that for downward loading. For design, the f_s values for upward loading may conservatively be set to 75 percent of those for downward loading unless static load test data justifies the use of higher values. See O'Neill and Reese (1999) for additional information.

Deep foundations with enlarged bases, such as belled drilled shafts or pressure-injected footings, gain upward capacity through bearing on top of the base, as shown in Figure 14.17. This capacity can be large, but it is difficult to quantify. O'Neill (1987) and O'Neill and Reese (1999) suggested the following relationship for clays:

$$(P_{upward})_a = \frac{(s_u N_u + \sigma_{zD})(\pi/4)(B_b^2 - B_s^2)}{F} \tag{14.33}$$

For unfissured clay:

$$N_u = 3.5 \, D_b/B_b \leq 9 \tag{14.34}$$

For fissured clay:

$$N_u = 0.7 \, D_b/B_b \leq 9 \tag{14.35}$$

Where:

$(P_{upward})_a$ = allowable upward load capacity

s_u = undrained shear strength in soil above the base

σ_{zD} = total stress at the bottom of the base

B_b = diameter of the enlarged base

B_s = diameter of the shaft

D_b = depth of embedment of enlarged base into bearing stratum

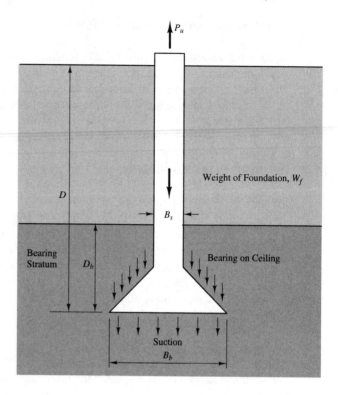

Figure 14.17 Additional upward capacity in deep foundations with enlarged bases.

The capacity computed from Equation 14.33 may be added to that computed in Equation 13.7. However, the upward pressure from the enlarged base interacts with the side-friction resistance of the lower portion of the shaft, so O'Neill recommends neglecting the side friction between the bottom of the foundation and a distance $2B_b$ above the bottom.

If the bottom is below the groundwater table, suction forces might produce additional upward resistance. Although they might be large, especially for short-term loading (i.e., less than 1 minute), it is best to ignore them until additional research better defines their character and magnitude.

For pressure injected footings in sands, we use an effective stress analysis:

$$(P_{upward})_a = \frac{N_u \sigma'_{zD} A_b}{F} \tag{14.36}$$

Where:

$(P_{upward})_a$ = allowable upward load capacity

N_u = breakout factor (from Figure 14.18)

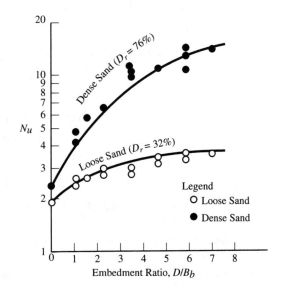

Figure 14.18   Breakout factor, N_u, for foundations with enlarged bases in sandy soils (Adapted from Dickin and Leung, 1990).

σ_{zD}' = vertical effective stress at bottom of foundation

A_b = area of bottom of enlarged base

F = factor of safety

Once again, this capacity may be added to that computed from Equation 13.7.

Belled drilled shafts are not practical in sands, so this equation is not applicable to drilled shaft foundations.

Example 14.5

Compute the allowable upward load capacity of the drilled shaft described in Example 14.4.

Solution

Use $F = 5.0$ (per Table 13.2)

Neglect side friction below a point $2B_b$ above the bottom of the bell

$$2\,B_b = (2)(5) = 10 \text{ ft}$$

Thus, layer 4 changes to; Depth = 37–50 ft; $A_s = 82$ ft^2; $f_s A_s = 157$ k

$$\Sigma f_s A_s = 39 + 121 + 157 = 317 \text{ k}$$

The soil near the bell is described as "very stiff clay" and thus is probably fissured.

$$N_u = 0.7 \frac{D_b}{B_b} = 0.7 \left(\frac{60}{5}\right) = 8.4$$

$$\sigma_{zD} = \Sigma \gamma H = (105)(12) + (108)(25) + (109)(23) = 6467 \text{ lb/ft}^2$$

$$(P_{upward})_a = \frac{(s_u N_u + \sigma_{zD})(\pi/4)(B_b^2 - B_s^2)}{F}$$

$$= \frac{[(4000)(8.4) + 6467](\pi/4)(5^2 - 2^2)}{(5)(1000 \text{ lb/k})}$$

$$= 132 \text{ k}$$

$$W_f = \left[\frac{\pi(2)^2}{4}(58.5) + \frac{\pi[(2 + 5)/2]^2}{4}(1.5)\right](0.150 \text{ k/ft}^3) = 30 \text{ k}$$

No adjustment for buoyancy is necessary because the groundwater table is below the bottom of the shaft.

Compute the uplift capacity exclusive of the bell using Equation 13.7 with a 0.75 factor applied to the side-friction resistance.

$$(P_{upward})_a = \frac{W_f + \Sigma f_s A_s}{F} = \frac{30 + (317)(0.75)}{5} = 54 \text{ k}$$

The total upward load capacity is:

$$(P_{upward})_a = 132 + 54 = \mathbf{186 \ k} \qquad \Leftarrow Answer$$

QUESTIONS AND PRACTICE PROBLEMS

14.4 An office building is to be supported on a series of 700-mm diameter, 12.0-m long drilled shafts that will be built using the open hole method. The soil profile at this site is as follows:

Depth (m)	Soil Classification	Undrained Shear Strength, s_u (kPa)
0–2.2	Stiff clayey silt (CL)	70
2.2–6.1	Stiff silty clay (CL)	85
6.1–11.5	Very stiff sandy clay (CL)	120
11.5–30.0	Very stiff sandy clay (CL)	180

The groundwater table is at a depth of 50 m. No onsite static load test data is available, the soil conditions are uniform, and the site characterization program was average. Compute the allowable downward and upward capacities of each shaft.

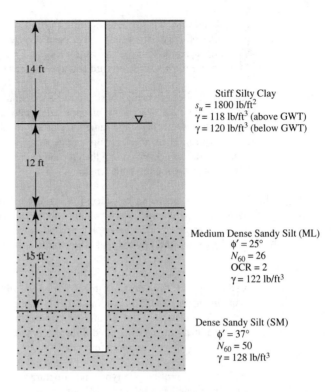

Stiff Silty Clay
$s_u = 1800 \text{ lb/ft}^2$
$\gamma = 118 \text{ lb/ft}^3 \text{ (above GWT)}$
$\gamma = 120 \text{ lb/ft}^3 \text{ (below GWT)}$

Medium Dense Sandy Silt (ML)
$\phi' = 25°$
$N_{60} = 26$
$OCR = 2$
$\gamma = 122 \text{ lb/ft}^3$

Dense Sandy Silt (SM)
$\phi' = 37°$
$N_{60} = 50$
$\gamma = 128 \text{ lb/ft}^3$

Figure 14.19   Soil profile for Problem 14.5.

14.5 A closed-end PP16×0.500 pile with a 1-in thick bottom plate will be driven 45 ft into the soil profile shown in Figure 14.19. This pile will then be filled with concrete. It has been designed without the benefit of any onsite static load tests, but onsite dynamic testing and wave equation analyses have been performed. Compute the allowable downward and upward load capacities.

14.6 An industrial building is to be supported on a series of 14-in square prestressed concrete piles which will be driven 45 ft into the following soil profile:

Depth (ft)	Soil Classification	Friction Angle ϕ' (deg)	Relative Density D_r	N_{60}	Unit Weight γ (lb/ft³)
0–10.0	Silty sand (SM)	33	40%	12	100
10.0–16.5	Sandy silt (ML)	31	50%	16	110
16.5–35.0	Fine to medium sand (SW)	35	62%	30	125
35.0–65.0	Well-graded sand (SW)	37	68%	35	126

The groundwater table is at a depth of 15 ft and all of the strata may be assumed to be normally consolidated. These piles have been designed without the benefit of any onsite static load tests, dynamic tests, or indicator piles, but a wave equation analysis has been performed. Compute the allowable downward and upward load capacities of each pile.

14.7 Auger-cast piles are being considered as an alternative design for the industrial building described in Problem 14.6. Select an appropriate pile diameter and determine the required depth of embedment to achieve the same allowable downward capacity per pile as provided by the concrete piles. Then, check the upward capacity and compare with the upward capacity of the concrete piles. If necessary, lengthen the auger-cast pile to provide sufficient upward capacity. Finally, state the required pile length.

14.8 Drilled shafts are being considered as an alternative design for the industrial building described in Problem 14.6. Each column will be supported on a single drilled shaft, which will replace a group of piles. Select an appropriate shaft diameter, then determine the length required to resist the upward and downward loads that would otherwise have required four piles.

14.9 A series of 20-ft deep pressure injected footings with compacted shafts are being considered as an alternative design for the industrial building described in Problem 14.6. Each PIF will replace four concrete piles. Using the information in Table 11.1, select an appropriate PIF type, then determine the required dimensions to resist the downward and upward loads that would otherwise have required four piles.

14.10 Using the soil profile, pile type, and structure described in Problem 14.6, determine the required pile length to support a 95-k compressive load.

14.11 According to Equations 14.6 and 14.9, how does the net unit toe-bearing resistance of PIF foundations in sand compare to drilled shafts in sand? Explain the reason for this difference. In what kinds of sand would you expect this difference to be greatest?

14.5   ANALYSES BASED ON CPT RESULTS

Engineers also have developed analytic methods based on cone penetration test (CPT) results. These methods are very attractive because of the similarities between the CPT and the load transfer mechanisms in deep foundations. The cone resistance, q_c, is very similar to the net unit toe-bearing resistance, q_t', and the cone side friction, f_{sc}, is very similar to the unit side-friction resistance, f_s. The CPT is essentially a miniature pile load test, and was originally developed partially as a tool for predicting pile capacities. Although we still must use empirical correlations to develop design values of q_t' and f_s from CPT data, these correlations should be more precise than those based on parameters that have more indirect relationships to deep foundations.

CPT-based methods include the following:

- *Dutch method* (also known as the *European method*) (Heijnen, 1974; DeRuiter and Beringen, 1979)

- *Nottingham and Schmertmann method* (Nottingham, 1975; Nottingham and Schmertmann, 1975; Schmertmann, 1978)
- *LCPC method* developed by the Laboratoire Central des Ponts et Chaussées (also known as the *French Method*) (Bustamante and Gianeselli, 1982; Briaud and Miran, 1991)
- *Meyerhof method* (Meyerhof, 1956, 1976, 1983)
- *Tumay and Fakhroo method* (Tumay and Fakhroo, 1981)
- *Eslami and Fellenius method* (Eslami and Fellenius, 1997)

We will discuss only the Eslami and Fellenius method.

Basis

Eslami and Fellenius used the results of 102 full-scale static load tests from a total of forty sites in thirteen countries, along with nearby CPT soundings to develop a method for evaluating axial load capacity. All of the load tests were conducted on piles, and the soil conditions ranged from soft clay to gravelly sand.

This method takes advantage of the additional data gained through the use of a *piezocone* (also known as a CPTU test) which is a standard CPT probe equipped with a piezometer to measure the pore water pressure near the cone tip while the test is in progress. This pore water pressure is the sum of the hydrostatic pore water pressure (such as could be measured using a conventional stationary piezometer) and any excess pore water pressures induced by the advancing cone. In sandy soils, the excess pore water pressure is usually very small, but in clays it can be large.

When using piezocones, the pore pressure data is combined with the measured q_c values to obtain the *corrected cone resistance*, q_T. The correction factors depend on the details of the piezocone, so this correction is normally applied when reducing the original CPT data, and a plot of q_T vs. depth is provided to the engineer.

The Eslami and Fellenius method requires application of an additional pore water pressure correction to the q_T values as follows:

$$\boxed{q_E = q_T - u_2} \tag{14.37}$$

Where:

  q_E = effective cone resistance

  q_T = corrected cone resistance

  u_2 = pore water pressure measured behind the cone point

This correction is intended to more closely align the analysis with the effective stresses. In sands, u_2 should be approximately equal to the hydrostatic pore water pressure. Therefore, this method could still be used in sands even if only conventional CPT data (i.e., no pore pressure data) is available, so long as the position of the groundwater table is known and no artesian conditions are present.

Toe Bearing

This method correlates the net unit toe-bearing resistance, q_t', with the effective cone resistance, q_E. Toe bearing failures occur as a result of punching and local shear, and thus affect only the soils in the vicinity of the toe. Therefore, the analysis considers only the q_E values in the following zones:

> For piles installed through a weak soil and into a dense soil: $8B$ above the pile toe to $4B$ below the pile toe
>
> For piles installed through a dense soil and into a weak soil: $2B$ above the pile toe to $4B$ below the pile toe

In both cases, B is the pile diameter. The geometric average, q_{Eg}, of the n measured q_E values within the defined depth range is then computed using:

$$q_{Eg} = \frac{(q_E)_1(q_E)_2(q_E)_3 \cdots (q_E)_n}{n} \qquad (14.38)$$

In general, odd spikes or troughs in the q_E data should be included in the computation of q_{Eg}. However, extraordinary peaks or troughs might be "smoothed over" if they do not appear to be representative of the soil profile. For example, occasional gravel in the soil can produce false spikes.

The net unit toe-bearing resistance has then been empirically correlated with q_{Eg} using the load test results:

$$q_t' = C_t q_{Eg} \qquad (14.39)$$

Where:

q_t' = net unit toe-bearing resistance

C_t = toe bearing coefficient

q_{Eg} = geometric average effective cone resistance

Eslami and Fellenius recommend using $C_t = 1$ for pile foundations in any soil type. In addition, unlike some other methods, they do not place any upper limit on q_t'.

Side Friction

The procedure for computing the unit side-friction resistance, f_s, is similar to the method used to compute q_t'. If the analysis is being performed by a computer (which is often the case, because CPT data can be provided in electronic form), a side-friction analysis is performed for each CPT data point using the following equation:

TABLE 14.6   SIDE-FRICTION COEFFICIENT, C_S (Eslami and Fellenius, 1997)

Soil Type	C_s	
	Range	Typical Design Value
Soft sensitive soils	0.0737–0.0864	0.08
Clay	0.0462–0.0556	0.05
Stiff clay or mixture of clay and silt	0.0206–0.0280	0.025
Mixture of silt and sand	0.0087–0.0134	0.01
Sand	0.0034–0.0060	0.004

$$\boxed{f_s = C_s q_E}$$ (14.40)

Where:

f_s = unit side-friction resistance

C_s = side-friction coefficient (from Table 14.6)

q_E = effective cone resistance

The C_s value depends on the soil type, and should be selected using Table 14.6. this soil classification may be determined directly from the CPT data using Figure 14.20.

Because CPT data is typically provided at depth intervals of 100 to 200 mm, this procedure is too tedious to use at every data point when performing computations by

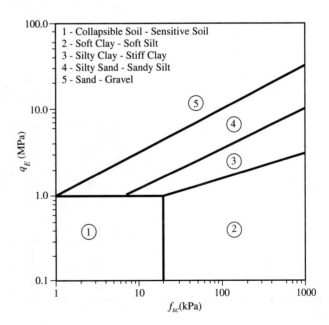

Figure 14.20   Soil classification from CPT data (Eslami and Fellenius, 1997).

hand. Therefore, hand computations usually divide the soil between the ground surface to the pile tip into layers according to the CPT results, with a representative q_E for each layer. For most soil profiles, five to ten layers are sufficient.

Accuracy

Finally, Eslami and Fellenius applied this method to independent load test data (i.e., not the data used to develop the method). This comparison indicates the average ultimate capacity prediction using this method is within about 2 percent of the measured ultimate capacity (i.e., there is no systematic bias), and that 95 percent of the predictions are within about 30 percent of the measured ultimate capacity. This is very good accuracy (e.g., compare with Figure 14.13), and certainly well within the range implied by the factors of safety in Tables 13.1 and 13.2. However, attaining this accuracy in practice requires careful selection of the C_s values from Table 14.6, which may be difficult for some soils.

Example 14.6

A 400-mm square prestressed concrete pile is to be driven 12.0 m into the soil profile shown in Figure 14.21. Using the Eslami and Fellenius method with a factor of safety of 2.5, compute the allowable downward load capacity.

Solution

Toe bearing

$$q'_t = C_t q_{Eg} = (1)(2500) = 2500 \text{ kPa}$$

$$A_t = (0.4)^2 = 0.16 \text{ m}^2$$

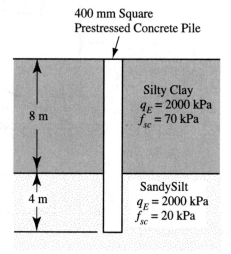

400 mm Square
Prestressed Concrete Pile

Silty Clay
$q_E = 2000$ kPa
$f_{sc} = 70$ kPa

8 m

SandySilt
$q_E = 2000$ kPa
$f_{sc} = 20$ kPa

4 m

Figure 14.21   Soil profile for Example 14.6.

Side friction
    Silty clay stratum (ignore upper 1.5 m)

$$f_s = C_s q_E = (0.025)(2000) = 50 \text{ kPa}$$
$$A_s = (4)(0.4)(8 - 1.5) = 10.4 \text{ m}^2$$

Sandy silt stratum

$$f_s = C_s q_E = (0.01)(2500) = 25 \text{ kPa}$$
$$A_s = (4)(0.4)(4) = 6.4 \text{ m}^2$$

Downward load capacity

$$P_a = \frac{q_t' A_t + \Sigma f_s A_s}{F}$$

$$= \frac{(2500)(0.16) + (50)(10.4) + (25)(6.4)}{2.5}$$

$$= \textbf{432 kN} \quad \Leftarrow Answer$$

QUESTIONS AND PRACTICE PROBLEMS

14.12 Using the Eslami and Fellenius method, compute the allowable downward load capacity of an 18-inch diameter closed-end steel pipe pile driven 60 ft into the soil profile shown in Figure 14.22. Use a factor of safety of 2.5.

14.6   GROUP EFFECTS

Pile foundations are usually installed in groups of three or more, as discussed in Chapter 11. Other types of deep foundations, such as auger-cast piles, also can be installed in groups.

The proper spacing of piles in the group is important. If they are too close [i.e., less than 2.0–2.5 diameters or 600 mm (2 ft) on center], there may not be enough room for errors in positioning and alignment. Conversely, if the spacing is too wide, the pile cap will be very large and expensive. Therefore, piles are usually spaced 2.5 to 3.0 diameters on-center.

Group Efficiency

The interactions between piles in a group and the adjacent soil are very complex, and the ultimate capacity of the group is not necessarily equal to the ultimate capacity of a single isolated pile multiplied by the number of piles. The effect of these interactions on the axial load capacity is called the *group efficiency*, which depends on several factors, including the following:

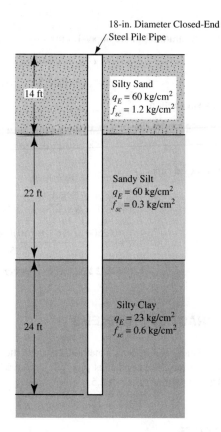

18-in. Diameter Closed-End
Steel Pile Pipe

Silty Sand
$q_E = 60$ kg/cm²
$f_{sc} = 1.2$ kg/cm²

14 ft

Sandy Silt
$q_E = 60$ kg/cm²
$f_{sc} = 0.3$ kg/cm²

22 ft

Silty Clay
$q_E = 23$ kg/cm²
$f_{sc} = 0.6$ kg/cm²

24 ft

Figure 14.22 CPT results and soil profile for Problem 14.12.

- The number, length, diameter, arrangement, and spacing of the piles
- The load transfer mode (side friction vs. end bearing)
- The construction procedures used to install the piles
- The sequence of installation of the piles
- The soil type
- The elapsed time since the piles were driven
- The interaction, if any, between the pile cap and the soil
- The direction of the applied load

Engineers compute the allowable downward load capacity of pile groups using a *group efficiency factor*, η, as follows[3]:

$$\boxed{P_{ag} = \eta\, N\, P_a} \qquad (14.41)$$

---

[3]Some authorities suggest applying the efficiency factor to the entire pile capacity, as shown here, while others apply it only to the side-friction component.

Where:

P_{ag} = allowable downward or upward capacity of pile group

η = group efficiency factor

N = number of piles in group

P_a = allowable downward or upward capacity of a single isolated pile

When computing the allowable load capacity, we always ignore the bearing pressure acting between the bottom of the cap and the underlying soil.

Converse and Labarre were among the first to address group efficiency, but they had little or no test data available to develop a formula for computing η. Therefore, they were forced to rely on assumed relationships between the pile group geometry and group efficiency, thus forming the basis for the *Converse-Labarre Formula* (Bolin, 1941):

$$\eta = 1 - \theta\,\frac{(n - 1)m + (m - 1)n}{90\, m\, n}$$

(14.42)

Where:

m = number of rows of piles

n = number of piles per row

$\theta = \tan^{-1}(B/s)$ (expressed in degrees)

B = diameter of a single pile

s = center-to-center spacing of piles (not the clear space between piles)

Another approach is to compare *individual failure* with *block failure*, as shown in Figure 14.23. Individual failure means the soil between the piles remains stationary and the individual piles punch through it, whereas block failure means the soil moves with the piles, thus failing as a large single unit. Presumably block failure governs if the sum of the perimeters of the piles is greater than the circumference of the pile group, and the group efficiency factor is assumed to be the ratio of these two perimeters:

$$\eta = \frac{2s(m + n) + 4B}{\pi\, m\, n\, B} \leq 1$$

(14.43)

Although pile group efficiency formulas, such as Equations 14.42 and 14.43, have been widely used, they were based primarily on intuition and speculation, and had little or no hard data to substantiate them. To overcome this problem, some researchers have conducted model load tests to study the performance of pile groups. Although these tests provide some insight, scale effects, especially the lower effective stresses in the model compared to the real soil, make the results difficult to interpret. We can alleviate some of these problems by performing model tests inside of a centrifuge, which has the net effect of increasing the apparent unit weight of the soil and thereby increases the effective stresses, but few such tests have yet been performed.

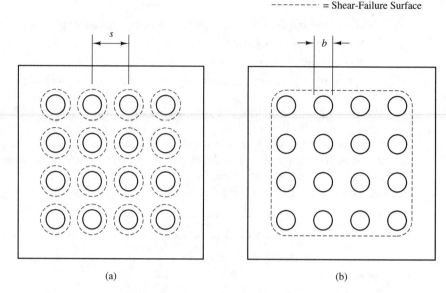

Figure 14.23   Types of failures in pile groups: (a) individual failure occurs along the perimeter of each pile, (b) block failure occurs along the perimeter of the pile group.

Full-scale static load tests on pile groups overcome the problems with model tests, and are the best method for studying group behavior. Unfortunately, these are very expensive because of the large load frames, jacks, and other equipment required to conduct the test. They also lack the versatility of model tests. Therefore, very few full-scale static tests have been done (DiMillio et al., 1987a, 1987b; O'Neill, 1983).

Neither model tests nor full-scale static tests are used as routine design tools, except possibly for exceptional projects. However, the results of such tests done in a research environment may be used to develop criteria for evaluating group effects in routine pile designs.

Tests in Sands

Group efficiency factors from model and full-scale static load tests of piles in sands, as reported by O'Neill (1983) suggest the following:

- In loose sands, η is always greater than 1 and reaches a peak at $s/B \approx 2$. It also seems to increase with the number of piles in the group.
- In dense sands with $2 < s/B < 4$ (the normal range), η is usually slightly greater than 1 so long as the pile was installed without predrilling or jetting.
- Piles installed by predrilling or jetting, and drilled shafts, have lower group efficiency factors, perhaps as low as about 0.7.

The high values of η in sands seem to be primarily due to the radial consolidation that occurs during driving and the resulting increase in lateral stress. Less consolidation occurs if predrilling or jetting is used.

Tests in Clays

The results from model and full-scale static tests in clays are quite different from those in sand (O'Neill, 1983). The group efficiency factor, η, is generally less than one, and becomes smaller as the number of piles in the group increases. Some of the measured η values were as low as 0.5.

Another important difference is that η for groups in clays increases with time. This is because the efficiency of pile groups in clays is largely governed by excess pore water pressures induced by pile driving. Although the magnitude of these pressures is not significantly higher than those near single piles, they encompass a larger volume of soil and therefore dissipate much more slowly, as shown in Figure 14.24. For single piles, nearly all of the excess pore water pressures dissipate within days or weeks of pile driving, whereas in groups they may persist for a year or more.

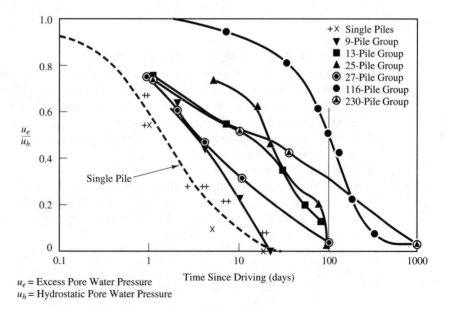

u_e = Excess Pore Water Pressure
u_h = Hydrostatic Pore Water Pressure

Figure 14.24   Measured excess pore water pressures in soil surrounding full-scale static pile groups (Adapted from O'Neill, 1983; used with permission of ASCE).

Guidelines for Practice

Although research conducted thus far has provided many insights, the behavior of pile groups is still somewhat mysterious, and no comprehensive method of assessing group action has yet been developed. Therefore, we must use the available information, along with engineering judgment and conservative design methods to develop design values of the group efficiency factor, η. Hannigan et al. (1997) recommend the following guidelines for pile groups:

In sands:

- So long as no predrilling or jetting is used, the piles are at least 3 diameters on center, and the group is not underlain by weak soils, use $\eta = 1$.
- Avoid predrilling or jetting whenever possible, because these methods can significantly reduce the load capacity. If these methods must be used, they should be carefully controlled.
- If a pile group founded on an firm bearing stratum of limited thickness is underlain by a weak deposit, then the ultimate group capacity is the smaller of either the sum of the ultimate capacities of the individual piles, or the group capacity against block failure of an equivalent foundation consisting of the pile group and the enclosed soil mass punching through the underlying weak soil.
- Piles should be installed at center-to-center spacings of at least 3 diameters.

In clays:

- Use a center-to-center spacing of at least 3 pile diameters.
- Use the following procedure to estimate the allowable capacity of the pile group:
  1. If the undrained shear strength, s_u, is less than 95 kPa (2000 lb/ft^2) and the pile cap is not in firm contact with the ground, use Equation 14.41 with $\eta = 0.7$ for groups with center-to-center spacings of 3 diameters, and $\eta = 1.0$ with center-to-center spacings of 6 diameters or more. For intermediate spacings, linearly interpolate between these two values.
  2. If the undrained shear strength, s_u, is less than 95 kPa (2000 lb/ft^2) and the pile cap is in firm contact with the ground, use Equation 14.41 with $\eta = 1.0$.
  3. If the undrained shear strength, s_u, is greater than 95 kPa (2000 lb/ft^2), use Equation 14.41 with $\eta = 1.0$ regardless of whether or not the cap is on contact with the soil.
  4. Compute the group capacity against block failure using the following formula:

$$P_{ag} = 2D(B_g + L_g)s_{u1} + B_g L_g s_{u2} N_c^* \qquad (14.44)$$

$$N_c^* = 5\left(1 + \frac{D}{5B}\right)\left(1 + \frac{B}{5L}\right) \le 9 \qquad (14.45)$$

Where:

P_{ag} = allowable downward load capacity

B_g = width of pile group

L_g = length of pile group

D = depth of embedment of pile group

s_{u1} = weighted average of undrained shear strength in clays over depth of embedment

s_{u2} = average undrained shear strength between the bottom of the pile group and a depth $2B_g$ below the bottom

N_c^* = bearing capacity factor

5. Use the lowest of the applicable values from Steps 1 to 4.

- Because of the excess pore water pressures produced by pile driving, the short-term ultimate capacity of pile groups in saturated clays will be reduced to about 0.4 to 0.8 times the ultimate value. However, as these excess pore water pressures dissipate, the ultimate capacity will increase. The rate at which it rises depends primarily on the dissipation of excess pore water pressures. Small groups will probably reach the long-term η within 1 to 2 months, which may be faster than the rate of loading, whereas larger groups may require a year or more. If the group will be subjected to the full design load before the excess pore water pressures fully dissipate, then a more detailed analysis may be warranted. In some cases, it may be appropriate to install piezometers to monitor the dissipation of excess pore water pressures.

14.7   SETTLEMENT

Most deep foundations designed using the methods described in Chapters 12 to 17 will have total settlements of no more than about 12 mm (0.5 in), which is acceptable for nearly all structures. Therefore, engineers often do not perform any settlement computations for deep foundations. However, certain conditions can produce excessive settlements, so the engineer must be able to recognize and evaluate them. These include the following:

- The structure is especially sensitive to settlement.
- The foundation has a large diameter and a large portion of the allowable capacity is due to toe bearing.
- One or more highly compressible strata are present, especially if these strata are below the toe.
- Downdrag loads might develop during the life of the structure.
- The engineer must express the pile response in terms of an equivalent "spring" located at the bottom of the column. This analytical model is used in some sophisticated structural analyses.

Load-Settlement Response

The load-settlement response of deep foundations is approximately described by the following relationships (adapted from Fellenius, 1999):

$$\frac{(q_t')_m}{q_t'} = \left(\frac{\delta}{\delta_u}\right)^g \tag{14.46}$$

$$\frac{(f_s)_m}{f_s} = \left(\frac{\delta}{\delta_u}\right)^h \leq 1 \tag{14.47}$$

Where:

q_t' = unit toe-bearing resistance

$(q_t')_m$ = mobilized net unit toe-bearing resistance

f_s = unit side-friction resistance

$(f_s)_m$ = mobilized unit side-friction resistance

δ = settlement

δ_u = settlement required to mobilize ultimate resistance

= $B/10$ for toe bearing

= 10 mm (0.4 in) for side friction

g = 0.5 (clay) – 1.0 (sand)

h = 0.02–0.5

Deep foundations also experience elastic compression, which is another source of apparent "settlement." It can be computed using:

$$\delta_e = \frac{P z_c}{A E} \tag{14.48}$$

Where:

δ_e = settlement due to elastic compression of foundation

P = downward load on each foundation

z_c = depth to centroid of soil resistance (typically about 0.75 D)

D = depth of embedment

A = cross-sectional area of a single foundation

E = modulus of elasticity of the foundation

= 29,000,000 lb/in² (200,000 MPa) for steel

= 57,000 $\sqrt{f_c'}$ lb/in²   (4700 $\sqrt{f_c'}$ MPa)   for concrete

= 1,600,000 lb/in² (11,000 MPa) for Southern pine or Douglas fir

Equations 14.46 to 14.48 may be used to develop approximate load-settlement curves.

Example 14.7

A 40-ft long HP14×73 pile driven into a sandy clay has a computed ultimate side-friction capacity ($\Sigma f_s A_s$) of 170 k and an ultimate toe-bearing capacity ($q_t' A_t$) of 90 k. Develop a load-settlement curve using Equations 14.46 to 14.48, then determine the settlement when the foundation is subjected to the allowable load (with a factor of safety of 3).

Solution

Use $\delta_u = 14/10 = 1.4$ in for toe bearing and 0.4 in for side friction
Use $g = 0.75$ and $h = 0.10$

δ (in)	Side Friction			Toe Bearing			P (k)	δ_e (in)	Adj δ (in)
	δ/δ_u	$(f_s)_m/f_s$	$(f_s A_s)_m$ (k)	δ/δ_u	$(q_t')_m/q_t'$	$(q_t' A_t)_m$ (k)			
0	0	0	0	0	0	0	0	0	0
0.05	0.12	0.81	138	0.04	0.08	7	145	0.08	0.13
0.10	0.25	0.87	148	0.07	0.14	12	160	0.09	0.19
0.20	0.50	0.93	159	0.14	0.23	21	180	0.10	0.30
0.40	1.00	1.00	170	0.29	0.39	35	205	0.12	0.52
0.60	1.50	1.00	170	0.43	0.53	48	218	0.13	0.73
1.00	2.50	1.00	170	0.71	0.78	70	240	0.14	1.14
1.40	3.50	1.00	170	1.00	1.00	90	260	0.15	1.55

The P vs. adjusted δ values are plotted in Figure 14.25.

The allowable downward load is:

$$P_a = \frac{q_t' + \Sigma f_s A_s}{F} = \frac{90 + 170}{3} = 87 \text{ k}$$

δ at $P = 87$ k, $\delta = \mathbf{0.08 \text{ in}}$ $\Leftarrow$ *Answer*

Commentary

Sixty five percent of the ultimate capacity is side friction, only half of which is mobilized under the allowable load. Therefore, the corresponding settlement is very small.

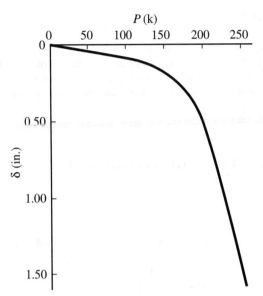

Figure 14.25   Results of load-settlement computations for Example 14.7.

O'Neill and Reese's Method

O'Neill and Reese (1999) developed the charts in Figures 14.26 to 14.29 to estimate the settlement of drilled shafts under service loads. These charts express the settlement in terms of the ratio of the mobilized resistance to the actual resistance. They were developed from full-scale load tests on drilled shafts, and thus should be more accurate than Equations 14.46 and 14.47. If the computed settlement is too large, use these charts to modify the design accordingly.

Drilled shafts in hard or dense soils tend to have load-settlement curves toward the upper end of the ranges shown in Figures 14.26 to 14.29 (i.e., less settlement is required to reach their ultimate capacities). Conversely, those in soft or loose soils tend toward the lower end of these ranges, and require greater settlement.

If the drilled shafts are underlain by a compressible soil, the settlement will be greater than indicated by Figures 14.26 to 14.29. This additional settlement should be evaluated using the equivalent footing method, discussed below.

Example 14.8

Compute the settlement of the shaft in Example 14.4 when it supports the allowable downward load. Use $f_c' = 3000 \text{ lb/in}^2$.

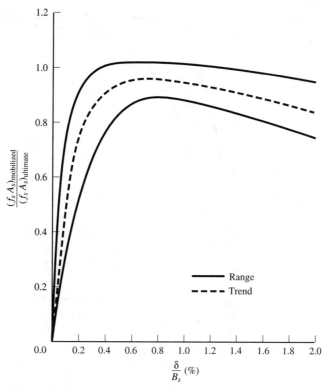

Figure 14.26   Normalized curves showing load transfer in side friction vs. settlement for drilled shafts in clays (O'Neill and Reese, 1999).

Solution:

From Example 14.4:

$B = 24$ in

$B_b = 60$ in

$P_s = 394$ k

$P_t' = 706$ k

$P_a = 440$ k

$$E = 57{,}000 \sqrt{f_c'} = 57{,}000 \sqrt{3000} = 3{,}100{,}000 \text{ lb/in}^2$$

$$A = \pi \, 12^2 = 452 \text{ in}^2$$

Try $\delta = 0.20$ in

$\delta/B = 0.2/24 = 0.8\%$. From Fig. 14.26 → working $P_s = (1.0)(394) = 394$ k

$\delta/B_b = 0.2/60 = 0.3\%$. From Fig. 14.27 → working $P_t' = (0.25)(706) = 176$ k

$\overline{}$

570 k > 440

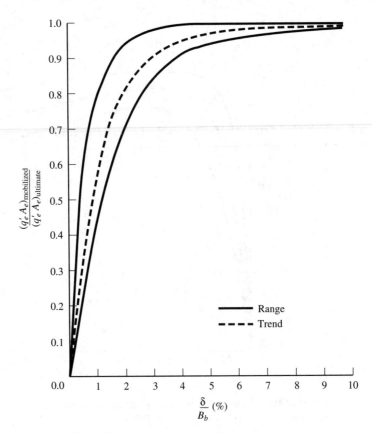

Figure 14.27   Normalized curves showing load transfer in toe bearing vs. settlement for drilled shafts in clays (O'Neill and Reese, 1999).

$$\delta_{adj} = \delta + \frac{P z_c}{A E}$$

$$= 0.20 + \frac{(570{,}000)(0.75)(60)(12 \text{ in/ft})}{(452)(3{,}100{,}000)}$$

$$= 0.42 \text{ in}$$

Try $\delta = 0.10$ in

$\delta/B = 0.1/24 = 0.4\%$. From Fig. 14.26 $\rightarrow$ working $P_s = (0.92)(394) = 362$ k

$\delta/B_b = 0.1/60 = 0.2\%$. From Fig. 14.27 $\rightarrow$ working $P_t' = (0.1)(706) = \underline{\quad 71 \text{ k}}$

$$\overline{\quad\quad\quad 433 \text{ k} < 440 \quad}$$

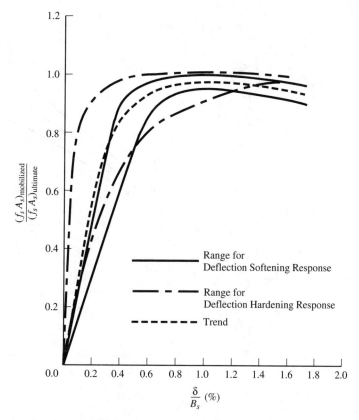

Figure 14.28 Normalized curves showing load transfer in side friction vs. settlement for drilled shafts in sand (O'Neill and Reese, 1999).

$$\delta_{adj} = \delta + \frac{Pz_c}{AE}$$

$$= 0.10 + \frac{(433,000)(0.75)(60)(12 \text{ in/ft})}{(452)(3,100,000)}$$

$$= 0.27 \text{ in}$$

$$\delta = 0.27 + (0.42 - 0.27)\left(\frac{440 - 433}{570 - 433}\right) = \mathbf{0.28 \ in} \qquad \Leftarrow \textit{Answer}$$

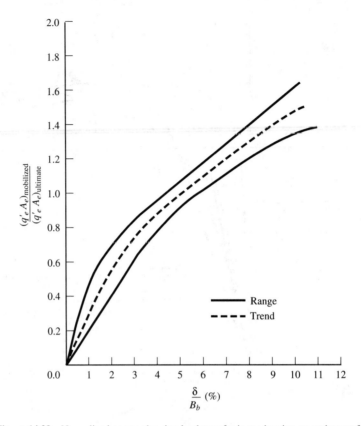

Figure 14.29   Normalized curves showing load transfer in toe bearing vs settlement for drilled shafts in sand (O'Neill and Reese, 1999).

Commentary

The side-friction resistance provides only 36 percent of the ultimate capacity, yet it carries 82 percent of the working load. This is because the shaft will mobilize the full side friction after only a small settlement (in this case, about 0.2 in), whereas mobilization of the full toe bearing requires much more settlement (in this case, about 3 in). Therefore, under working loads, this shaft has mobilized nearly all of the side friction, but only a small portion of the toe bearing. Thus, the remaining toe-bearing resistance becomes the reserve that forms the factor of safety. It would need to settle about 3 inches to attain the ultimate capacity.

t-z Method

The *t-z method* uses the numerical model shown in Figure 14.30 to compute the settlement of a deep foundation. This model divides the foundation into a series of elements, each of which has a certain modulus of elasticity. The side-friction resistance acting on

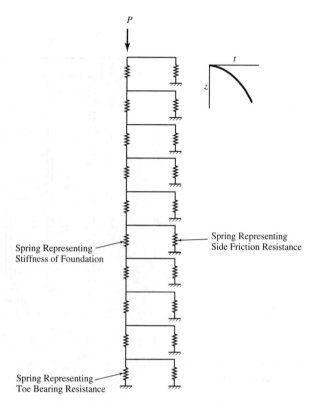

Figure 14.30   Numerical model for *t-z* method.

each element is modeled using a nonlinear "spring," as is the toe bearing acting on the bottom element. The load-displacement characteristics of these springs are defined using *t-z curves* (Kraft, Ray and Kagawa, 1981), where *t* is the load and *z* is the settlement of that pile segment. A load is then applied to the top of this model, and the foundation moves downward until it reaches static equilibrium. The corresponding settlement is then recorded.

This model explicitly considers the axial compression of the foundation, as well as potential variations in the soil properties along its length, and thus should be more precise than methods that consider these factors implicitly. However, the actual precision of this method depends on our ability to properly define the *t-z* curves.

Experimental *t-z* curves have been developed by backcalculating them from instrumented load tests. They have been correlated with soil properties, such as undrained shear strength or effective friction angle, which allows them to be applied to other foundations. Commercial software is available to conduct *t-z* analyses (for example, see www.ensoft-inc.com).

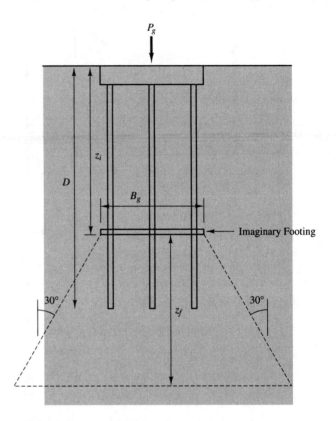

Figure 14.31 Use of an imaginary footing to compute the settlement of a deep foundation group.

Imaginary Footing Method

The *imaginary footing method* computes the settlement of a deep foundation group by replacing it with an imaginary footing, as shown in Figure 14.31. This method is especially useful when the design consists of a group of deep foundations that are underlain by compressible soils such that the compression of these soils is more significant than the settlements required to mobilize the side friction and toe bearing.

For foundations that rely primarily on side friction, place the imaginary footing at a depth of 0.67 D (where D is the depth of embedment). For foundations that rely principally on toe bearing, place it at the toe elevation. When both side friction and toe bearing are significant, use linear interpolation to place the imaginary footing between these two positions.

Then use the techniques described in Chapter 7 to compute the settlement of the imaginary footing, and add the elastic compression of the foundations, δ_e, using Equation 14.48, with $z_c =$ the depth z_i to the imaginary footing.

This method also can consider settlements produced by causes other than the structural loads on the piles. For example, the construction process may include lowering the groundwater table, which increases the effective stress in the soil and thus create settlement.

Downdrag Loads

If the soils surrounding a deep foundation settle, they induce a downward side-friction load known as *downdrag*. This load can be very large, and has been a source of excessive settlement. This phenomenon is discussed in Chapter 18.

QUESTIONS AND PRACTICE PROBLEMS

14.13 What is "block failure" in a group of piles?

14.14 What is a typical group efficiency factor for piles driven into loose sand without predrilling or jetting? How does predrilling and jetting affect this factor? Why?

14.15 The group efficiency factor for piles in saturated clay is low soon after driving, and increases with time. Why does this occur?

14.16 Based on the results of a static load test, a certain pile has a net toe bearing capacity of 200 k and a side-friction capacity of 500 k. A series of these piles are to be driven in a 3×4 group, which will have an estimated group efficiency factor of 1.15. Compute the allowable downward load capacity of this pile group using a factor of safety of 3.0.

14.17 A 500-mm square prestressed concrete pile ($f_c' = 40$ MPa) is to be driven 20 m into a clay. The ultimate side-friction capacity, $\Sigma f_s A_s$, is 1450 kN and the ultimate net toe bearing capacity, $q_t' A_t$ is 300 kN. Using Equations 14.46 to 14.48, develop a load-settlement curve, then determine the allowable load for a factor of safety of 2.5 and the corresponding settlement.

14.18 A 1500-mm diameter, 22-m long drilled shaft is to be constructed in a silty sand. The ultimate side-friction capacity, $\Sigma f_s A_s$, is 6200 kN and the ultimate net toe-bearing capacity, $q_t' A_t$, is 4000 kN. This toe bearing capacity was computed using Equation 14.6 without applying the reduction factor in Equation 14.7. Using O'Neill and Reese's method, develop a load-settlement curve, then determine the allowable load for a factor of safety of 2.5 and the corresponding settlement.

14.19 A 3.5-m square, 5×5 pile group supports a downward load of 12,000 kN. Each pile in the group is 300-mm square, 20-m long concrete ($f_c' = 40$ MPa). The load transfer consists of 75 percent side friction and 25 percent toe bearing. The soil profile at this site is as follows:

> 0–25 m: Overconsolidatd stiff clay: $\gamma = 18.6$ kN/m^3,
>
> $\gamma_{sat} = 19.7$ kN/m^3, $C_r/(1+e) = 0.04$
>
> >25 m: Dense sand and gravel

The groundwater table is at a depth of 10 m. Using the imaginary footing method, compute the settlement of this pile group.

SUMMARY

Major Points

1. Analytic methods of evaluating axial load capacity are those based on soil properties. These methods evaluate the values of $q_t{}'$ and f_s, which are then combined with the foundation geometry to compute the axial load capacity.

2. Analytic methods are less accurate than full-scale static load tests, but have the advantages of being less expensive to analyze and more flexible. However, because of this loss in accuracy, they require the use of higher factors of safety, which increases construction costs.

3. The process of constructing deep foundations induces significant changes in the soil. These changes include distortion, compression, excess pore water pressures, loss of contact, and loosening. It is difficult to quantify these changes, which is why all analytic methods must be calibrated from static load tests.

4. The load-settlement curve for side friction has a well-defined capacity, but that for toe bearing is much more poorly defined. Therefore the definition of toe bearing "capacity" is subject to interpretation.

5. The full side-friction resistance is mobilized after only 5 to 10 mm of settlement. However, much more settlement is required to mobilize the full toe-bearing capacity. Therefore, side friction usually carries most of the service loads.

6. Unlike bearing-capacity failures in shallow foundations, toe-bearing failures in deep foundations are usually local or punching shear.

7. Side-friction analyses are best performed using effective stresses, because drained conditions are normally present. However, total stress methods also are available.

8. The upward load capacity of straight foundations is less than the downward load capacity because of the Poisson effect. The presence of an enlarged base can significantly increase the upward load capacity.

9. The cone penetration test forms the basis for a good analytic analysis because it is essentially a miniature pile load test.

10. When deep foundations are installed in groups, we must consider the group interaction effects. These effects are described by the group efficiency factor.

11. Settlement analysis methods are available for deep foundations. These analyses are especially important when the diameter is large, the foundation is underlain by soft soils, or downdrag loads are present.

Vocabulary

Alpha method	Elastic compression	Presumptive capacity
Analytic method	Excess pore water pressure	Rigidity index
Beta method	Freeze	Set up

Block failure	Group efficiency factor	Side friction
Compression	Group efficiency	t-z method
Distortion	Imaginary footing	Toe bearing
Effective stress analysis	Individual failure	Total stress analysis

COMPREHENSIVE QUESTIONS AND PRACTICE PROBLEMS

14.20 Why is the cone penetration test a good source of data for pile designs?

14.21 The soil profile beneath a proposed construction site is as follows:

Depth (m)	Soil Classification	Undrained Shear Strength, s_u (kPa)
0–2.5	Stiff silty clay (CL)	80
2.5–6.7	Soft clay (CL)	15
6.7–15.1	Medium clay (CL)	30
15.1–23.0	Stiff clay (CL)	100

Develop a plot of allowable downward load capacity vs. depth for a 350-mm square concrete pile. Consider pile embedment depths between 5 and 20 m and use a factor of safety of 3.0.

14.22 An HP 13×87 pile is embedded 45 ft into a clay. The unit weight of this soil is 100 lb/ft³ above the groundwater table (which is 12 ft below the ground surface) and 112 lb/ft³ below. The soil in the vicinity of the pile tip has an undrained shear strength of 2800 lb/ft². According to a static load test, the ultimate downward load capacity is 143 k.

You wish to compute a site-specific β factor for HP 13×87 piles to be used in the design of other piles at this site. Based on these test results, what is that β factor?

Hint: Compute β based on the average σ_z' and the average measured f_s.

14.23 Using the information in Problem 14.22, compute a site-specific α factor.

14.24 A 12-inch square prestressed concrete pile is to be driven 45 ft into the soil described by the CPT results in Figure 4.14 in Chapter 4. Compute the allowable downward load capacity using a factor of safety of 2.8 and each of the following methods:

14.25 A group of five closed-end steel pipe piles were driven into a sandy hydraulic fill at Hunter's Point in San Francisco, California (DiMillio, et al., 1987a). A single isolated pile also was driven nearby. Each pile had an outside diameter of 10.75 in and a length of 30 ft. The group piles were placed 3 to 4 ft on-center, and their pile cap was elevated above the ground surface.

The upper 4.5 ft of the soil was predrilled to a diameter larger than the piles, and the top of the completed piles extended 5 ft above the ground surface. Therefore, only 20.5 ft of each pile was in contact with the soil. No other predrilling or jetting was done.

An extensive subsurface investigation was conducted before these piles were installed. This included SPT, CPT, DMT and other tests. The CPT results are shown in Figure 14.32.

a. Using this CPT data, compute the ultimate downward load capacity of the single pile.

b. Based on a pile load test, the ultimate downward load capacity of the single pile was 80 k (based on Davisson's method). Other methods of reducing the load test data gave ultimate load capacities of 80 to 117 k. How accurate was your prediction?

c. Using this CPT data, compute the ultimate downward load capacity of the pile group.

d. Based on a group pile load test, the ultimate downward load capacity of the pile group was 432 to 573 k, depending on the method of reducing the load test data. How accurate was your prediction?

14.26 Compute the settlement of the shaft in Example 14.3 when it is subjected to the allowable downward load.

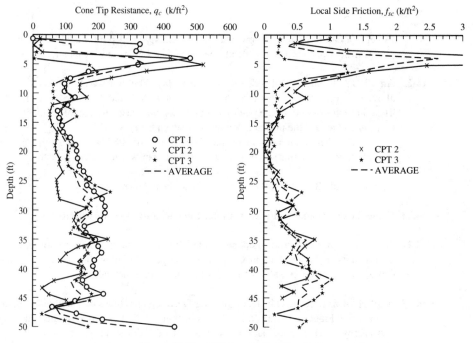

Figure 14.32   CPT data for Problem 14.25.

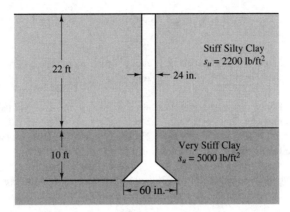

Figure 14.33 Proposed drilled shaft for Problem 14.27.

14.27 A highway bridge is to be supported on a series of belled drilled shaft foundations as shown in Figure 14.33. Although no static load test data is available, the soil conditions are uniform and an extensive site characterization program has been completed. The groundwater table is at a great depth. The shaft will be drilled using the open-hole method. Compute the allowable downward and upward capacities of each shaft.

14.28 Compute the settlement of the shaft in Problem 14.27 when it is subjected to the allowable downward load.

14.29 A full-scale load test has been conducted on a 24-in diameter, 40-ft long instrumented drilled shaft similar to the one shown in Figure 13.14. The test crew maintained records of the load-settlement data and the forces in each of the five load cells. The applied load at failure (using Davisson's method as described in Chapter 13) was 739,600 lb. The corresponding forces in the load cells were as follows:

Load Cell Number	Depth (ft)	Force (lb)
1	3.0	719,360
2	12.0	636,120
3	21.0	487,500
4	30.0	304,320
5	39.0	135,400

There are two soil strata: the first extends from the ground surface to a depth of 15 ft and has a unit weight of 117 lb/ft^3; the second extends from 15 ft to 60 ft and has a unit weight of 120 lb/ft^3 above the groundwater table and 127 lb/ft^3 below. The groundwater table is at a depth of 17 ft.

Compute the average β factor in each of the two soil strata, and the net unit toe-bearing resistance, q_t'.

Note: Once these site-specific β and q_t' values have been computed, they could be used to design shafts of other diameters or lengths at this site.

14.30 An 18-inch diameter auger-cast pile is to be built in a well graded silty sand and must support a compressive load of 150 k. The soil has a unit weight of 120 lb/ft^3 above the groundwater table, 128 lb/ft^3 below and $N_{60} = 25$. The groundwater table is at a depth of 45 ft. Compute the required depth of embedment using a factor of safety of 3.0.

14.31 Using the data in Problem 14.30 and a 30-ft deep cased-shaft pressure-injected footing instead of an auger-cast pile, determine the required base diameter.

15

Deep Foundations—Axial Load Capacity Based on Dynamic Methods

I read some of the papers last night where some of these pile driving formulas were derived, and the result was that my sleep was very much disturbed.

Pioneer Foundation Engineer Lazarus White (1936)

The third way of determining the axial load capacity of deep foundations is to use *dynamic methods*, which are based on the foundation's response to dynamic loads, such as those from a pile hammer or some other impact source. By monitoring the response to these dynamic loads, engineers can develop predictions of the static load capacity. Some dynamic methods also provide information on driveability and structural integrity.

15.1 PILE-DRIVING FORMULAS

When driving piles, it is very easy to monitor the *blow count*, which is the number of hammer blows required to drive the pile a specified distance. In English units, blow count is normally expressed as blows/ft; with SI units it is expressed as blows/250 mm or perhaps with some other units. Blow count records are normally maintained for the entire driving process, but the most important value is the blow count for the last foot (or 250 mm), because it represents the completed pile's resistance to driving.

Intuitively, we would expect piles that are difficult to drive (i.e., those with a high blow count) will have a greater downward load capacity than those that drive more easily. Thus, there should be some correlation between blow count and load capacity. Engineers have attempted to define this relationship by developing empirical correlations between

559

hammer weight, blow count, and other factors, with the static load capacity. These relationships are collectively known as *pile-driving formulas*.

Hundreds of pile-driving formulas have been proposed, some of them as early as the 1850s. Although these formulas have different formats, all share a common methodology of computing the pile capacity based on the driving energy delivered by the hammer. They use the principle of conservation of energy to compute the work performed during driving, and attempt to consider the various losses and inefficiencies in the driving system using empirical coefficients.

Engineers have used pile-driving formulas as follows:

- At sites where full-scale load test data is not available, standard pile-driving formulas have been used to assess the static load capacity of the piles. In practice, each pile has a required load capacity which corresponds to a certain minimum acceptable blow count. Therefore, each pile is driven until it reaches the specified blow count.

- At sites were full-scale load test data is available, the engineer modifies one of the standard pile driving formulas to match the load test results. For example, if a certain formula overpredicted the test pile capacity by 20 percent, then it is modified with a site-specific correction factor of 1/1.20. This custom formula is then applied to other piles at this site, and thus is a means of extrapolating the load test results.

Pile-driving formulas are convenient because the engineer can compute the capacity of each pile as it is driven by simply determining the final blow count. Thus, these formulas have often served as a means of construction control.

Typical Pile-Driving Formulas

The basic relationship common to all pile driving formulas is:

$$P_a = \frac{W_r h}{s\, F}$$

Where:

P_a = allowable downward load capacity

W_r = hammer ram weight

h = hammer stroke (the distance the hammer falls)

s = pile set (penetration) per blow at the end of driving = 1/blow count

F = factor of safety

The *Sanders formula* of 1851 uses Equation 15.1 with a factor of safety as high as 8. This relatively high value is partially a true factor of safety and partially a method of accounting for energy losses.

One of the most popular pile driving formulas is the one first published over one hundred years ago in the journal *Engineering News* (Wellington, 1888). It has since become known as the *Engineering News Formula*:

$$P_a = \frac{W_r h}{F(s + c)}$$ (15.2)

Based on load test data, Wellington recommended using a c coefficient of 1 in (25 mm) to account for the difference between the theoretical set and the actual set. However, his database included only timber piles driven with drop hammers. Some engineers use $c = 0.1$ in (2 mm) for single-acting hammers, although this was not part of the original formula. He also recommended using a factor of safety of 6.

Wellington apparently had much confidence in his work when he stated that his formula was:

> . . . first deduced as the correct form for a theoretically perfect equation of the bearing power of piles, barring some trifling and negligible elements to be noted; and I claim in regard to that general form that it includes in proper relation to each other every constant which ought to enter into such a theoretically perfect practical formula, and that it cannot be modified by making it more complex . . . (Wellington, 1892)

The Engineering News Formula has been used quite extensively since then and has routinely been extrapolated to other types of piles and hammers. Other pile-driving formulas include the *Modified Engineering News Formula*, the *Hiley Formula*, the *Gates Formula*, and many others.

Precision

Pile-driving formulas are attractive, and they continue to be widely used in practice. Unfortunately, the accuracy of these methods is less than impressive. Cummings (1940) was one of the first to describe their weaknesses. Since then, many engineers have objected to the use of these formulas and many lively discussions have ensued. Terzaghi's (1942) comments are typical:

> In spite of their obvious deficiencies and their unreliability, the pile formulas still enjoy a great popularity among practicing engineers, because the use of these formulas reduces the design of pile foundations to a very simple procedure. The price one pays for this artificial simplification is very high. In some cases the factor of safety of a foundation designed on the basis of the results obtained by means of the pile formula is excessive and in other cases significant settlements have been experienced.
> . . . on account of their inherent defects, all the existing formulas are utterly misleading as to the influence of vital conditions, such as the ratio between the weight of the pile and the hammer, on the result of the pile driving operations. In order to obtain reliable information

concerning the effect of the impact of the hammer on the penetration of the piles it is necessary to take into consideration the vibrations that are produced by the impact.

. . . Newton himself warned against the application of his theory to problems involving the impact produced "by the stroke of a hammer." (Used with permission of ASCE)

In another article, Peck (1942) suggested marking various pile capacities on a set of poker chips, selecting a chip at random, and using that capacity for design. His data suggest that even this method would be more accurate than pile driving formulas.

Not everyone agreed with Cummings, Terzaghi, and Peck, so this topic was the subject of heated discussions, especially during the 1940s. However, comparisons between pile load tests and capacities predicted by pile driving formulas have clearly demonstrated the inaccuracies in these formulas. Some of this data are presented in Figure 15.1 in the form of 90 percent confidence intervals. All of these piles were driven into soils that were

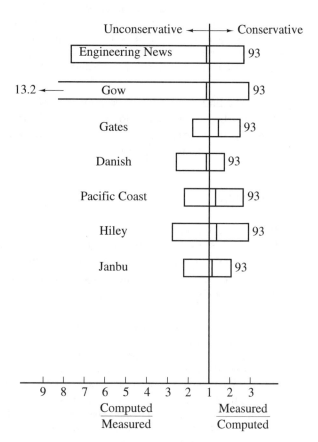

Figure 15.1 Ratio of measured pile capacity (from pile load tests) to capacities computed using various pile-driving formulas. The bars represent the 90 percent confidence intervals, and the line near the middle of each bar mean. The number to the right of each bar is the number of data points (based on data from Olson and Flaate, 1967).

primarily or exclusively sand. Predictions of piles driven into clay would be much worse because of freeze effects.

Although the principle of conservation of energy is certainly valid, pile-driving formulas suffer because it is very difficult to accurately account for all of the energy losses in a real pile-driving situation. The sources of these uncertainties include the following:

- The pile, hammer, and soil types used to generate the formula may not be the same as those at the site where it is being used. This is probably one of the major reasons for the inaccuracies in the original Engineering News Formula.
- The formulas do not account for freeze effects.
- The hammers do not always operate at their rated efficiencies.
- The energy absorption properties of cushions can vary significantly.
- The formulas do not account for flexibility in the pile.
- There is no simple relationship between the static and dynamic strength of soils.

Because of these many difficulties, Davisson (1979) stated ". . . it is hoped that such formulas have been purged from practice." Wave equation analyses, as discussed in the next section, provide much better results. Therefore, there is little need to continue using pile-driving formulas.

15.2 WAVE EQUATION ANALYSES

It is unfortunate that pile driving formulas are so unreliable, because it is very useful to define pile capacity in terms of hammer blow counts. Such functions provide a convenient means of construction control that can easily be applied to every pile installed at a given site. Therefore, it would be very useful to have a reliable dynamic method.

To satisfy this need, engineers studied the dynamics of pile driving in more detail and eventually developed another type of dynamic analysis: the *wave equation method*. This method provides a more accurate function of capacity vs. blow count, helps optimize the driving equipment, and computes driving stresses.

Pile-Driving Dynamics

Pile-driving formulas consider the pile to be a rigid body subject to classical Newtonian physics. In other words, they assume the entire pile moves downward as a unit. In reality, the impact load provided by the pile hammer is very short compared to the time required for the resulting stress wave to reach the bottom of the pile, so portions of the pile may be moving downward, while other portions are stationary or even moving upward (in response to a reflected wave). Therefore, it is much better to consider stress wave propagation effects when evaluating the pile driving process. Isaacs (1931) appears to have been the first to suggest the advantages of evaluating piles based on wave propagation.

The one-dimensional propagation of stress waves in a long slender rod, such as a pile, is described by the *one-dimensional wave equation*:

$$\frac{\partial^2 u}{\partial t^2} = \frac{E}{\rho}\frac{\partial^2 u}{\partial z^2}$$

(15.3)

Where:

   z = depth below the ground surface

   t = time

   u = displacement of the pile at depth z

   E = modulus of elasticity of the pile

   ρ = mass density of the pile

This formula is manageable when the boundary conditions are simple, but it becomes much more difficult with the complex boundary conditions associated with pile foundations. A closed-form solution is available (Warrington, 1997), but virtually all practical problems are solved using numerical methods, and this was not possible until digital computers became available. Smith experimented with numerical solutions soon after the Second World War (Smith, 1951) in what appears to have been one of the first civilian applications of digital computers. He later refined this work (Smith, 1960, 1962), thus forming the basis for modern wave equation analyses of piles.

Smith's numerical model divides the pile, hammer and driving accessories into discrete elements, as shown in Figure 15.2. Each element has a mass that equals that of the corresponding portion of the real system. These elements are connected with springs that have the same stiffness as the corresponding element. For example, if the pile is divided into 1-foot long elements, the corresponding masses and springs in the numerical model pile would correspond to the mass and stiffness of 1 foot of piling.

Some portions of the pile system are more difficult to model than others. For example, the hammer is more difficult to model than the pile. This is especially true of diesel hammers because their energy output varies with the hardness of the driving. Fortunately, some of the newer programs include improved hammer models.

The method also models the interface between the pile and the soil using a series of springs and dashpots along the sides of the elements and on the bottom of the lowest element to model the side-friction and toe-bearing resistances. The springs model resistance to driving as a function of displacement; the dashpots model resistance as a function of velocity.

Smith proposed using a bilinear elastic-plastic spring and a linear dashpot as shown in Figure 15.3. The spring resistance increases until it reaches a displacement q, known as the *quake*. At that point, it reaches the *ultimate resistance*, R_u, and becomes completely plastic. The dashpot resistance is a linear function of the velocity and is defined by the *Smith damping factor*, J_s. Although this is a simplified view of the forces that act between the pile and the soil, it seems to work well for practical problems. More elaborate spring and dashpot functions also may be used.

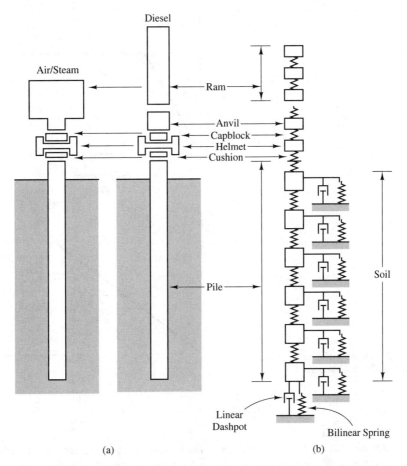

Figure 15.2   The numerical model used in wave equation analyses: (a) actual system;
(b) numerical model (adapted from Goble, 1986).

The values of q, R_u, and J_s for each element are based on the soil type and other factors. Researchers have developed recommended values by comparing wave equation analyses with pile load test results. Therefore, it is best to think of them as experimental calibration coefficients, not physical soil properties.

Smith suggested using a quake value of 2.5 mm (0.1 in) in all cases. Others have modified his suggestion slightly and used a quake value of 0.008 mm per mm of pile diameter (0.1 in per foot). Although these guidelines often produce acceptable results, there are situations where the quake is much larger, perhaps as much as 150 mm (0.5 in). Failure to recognize these special situations, and the use of a quake value that is too small can result in pile breakage and/or premature refusal.

Once assembled, the numerical model simulates the pile driving by imparting a downward velocity to the hammer ram elements equal to the impact velocity. This gener-

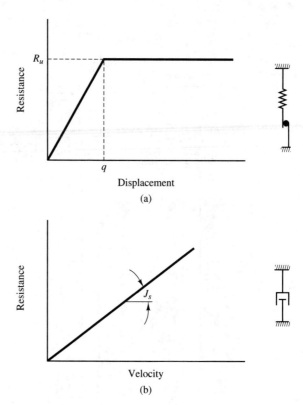

Figure 15.3   Smith's model of the soil-pile interface: (a) bilinear springs and (b) linear dashpots.

ates a stress impulse that travels to the bottom of the pile, generating displacements and resistances along the way. Upon reaching the bottom, the impulse reflects back upward and travels back to the top of the pile. Eventually the wave dissipates due to energy losses. In the real pile, this process occurs very quickly, and the stress waves are fully dissipated before the next hammer blow occurs.

Usually the reflected waves are compressive, so the pile is subjected only to compressive stresses. However, if the pile penetrates through a hard stratum with the tip founded in a much softer stratum, the reflected wave could be tensile. This is especially important in concrete piles, because their tensile strength is much less than the compressive strength, so the tensile wave can cause breakage in the pile.

Once the stresses have dissipated, the pile will have advanced some distance, known as the *set*. The model predicts the magnitude of the set, but often expresses it as the inverse of the set, which is the *blow count* (i.e., hammer blows/ft). A *bearing graph*, as shown in Figure 15.4, is a plot of ultimate load bearing capacity vs. blow count for a certain pile type and size driven with a certain hammer at a certain site. It may be necessary to develop several bearing graphs to represent piles of different diameters, lengths, or types, or for different hammers.

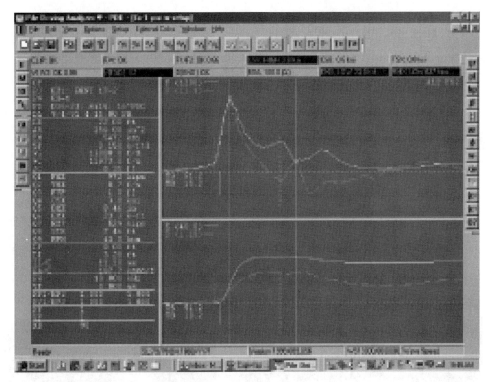

Figure 15.4   A typical bearing graph obtained from a wave equation analysis. (Courtesy
of Goble Rausche Likins and Associates, Inc.).

Once the bearing graph (or graphs) have been obtained, they may be used as a
means of construction control. For example, if a certain pile has a design ultimate capacity of 100 k, the field engineer simply refers to the appropriate bearing graph and determines the necessary blow count, or given an observed blow count, the graph indicates the
corresponding load capacity.

Freeze (Setup) Effects

As discussed in Section 14.1, piles driven into saturated clays produce excess pore water
pressures that temporarily decrease their load capacity. The capacity returns when these
pressures dissipate, a process known as *freeze* or *setup*. Thixotropic effects also contribute
to pile freeze.

We are primarily interested in the load capacity after freeze has occurred, yet dynamic analyses based on data obtained during driving only give the prefreeze capacity.
Therefore, engineers use *retap* blow counts in these soils. These are obtained by bringing
the pile driver back to the pile sometime after it has been driven (perhaps a few days) to

drive it an additional couple of inches and monitor the blow count (Hussein, Likins, and Hannigan, 1993).

Analyses of Driving Stresses and Selection of Optimal Driving Equipment

Pile driving hammers are available in many different sizes and energy ratings. For example, the rated energy of a small hammer might by only 10 kN-m, whereas that for a large hammer could be over 300 kN-m. Thus, it is very important to select the proper hammer for each project. If the hammer is too small, it will not efficiently drive the pile and it may be possible to reach the design embedment; if it is too large, the pile may be overstressed and break, as shown in Figure 15.5. In addition, other parts of the pile-driving system, such as cushions, should be optimized for each project. Fortunately, wave equation analyses can be used to select the optimal hammer and driving accessories because these analyses evaluate the driving stresses in the pile, which can be compared to the allowable stresses listed in Chapter 12. Through trial and error, a suitable hammer and accessories can be selected. This process is known as a *driveability analysis*.

Figure 15.5   These piles were damaged during driving because the contractor selected a hammer that was too large. A driveability analysis could have predicted these problems, and would have helped the contractor select a more appropriate hammer. (Courtesy of Goble Rausche Likins and Associates, Inc.)

For example, an engineer might propose driving a pile into a very hard stratum. An analytic analysis conducted using the techniques described in Chapter 14 might indicate a very high allowable capacity for this pile. However, it may be difficult or impossible to drive the pile into this hard soil without using an exceptionally large hammer, and that hammer might damage the pile. In this case it may be necessary to use a stronger pile (i.e., greater wall thickness) to accommodate the increase driving stresses, or redesign the foundation to better accommodate the soil conditions. Such problems can be identified and rectified by conducting a drivability analysis before construction.

Wave Equation Analysis Software

The first publicly available wave equation software was the TTI program developed at Texas A&M University (Edwards, 1967). In 1976, researchers at the Case Institute of Technology[1] developed the WEAP (Wave Equation Analysis of Piles) program. It has been revised on several occasions and is in the public domain (see http://uftrc.ce.ufl.edu/info-cen/info-cen.htm). The WEAP program has since formed the basis for other more advanced proprietary programs, such as GRLWEAP (see www.pile.com).

Wave equation analysis software is widely available, and can easily run on personal computers. Thus, there is no need to continue using pile driving formulas.

15.3 HIGH-STRAIN DYNAMIC TESTING

Another dynamic method of evaluating the static load capacity of deep foundations is to install instruments on the foundation and use them to monitor load and settlement data obtained while the foundation is subjected to a dynamic impact load. This measured response to dynamic loads can then be used to develop design static load capacities.

The most common source of dynamic loading is a pile hammer, because it is already on site and thus represents little or no additional cost. Therefore, these tests are most commonly performed on pile foundations. However, dynamic loads also can be obtained with drop hammers or with explosives, which enables testing of drilled shafts and other types of deep foundations.

This method requires sufficient strain in the foundation to mobilize the side friction and toe bearing, and thus is called *high-strain dynamic testing*. The next section discusses low strain dynamic testing, which is used to evaluate structural integrity.

The Case Method

High-strain dynamic tests were developed during the 1960s and early 1970s at the Case Institute of Technology in Cleveland, Ohio (now known as Case Western Reserve University). The *Case Method* is based on an analysis of dynamic forces and accelerations measured in the field while the pile is being driven (Rausche, Goble, and Likins, 1985; Hannigan, 1990). The Case Method provides real-time information on pile capacity, pile-driving stresses, structural integrity, and hammer/driving system performance.

---

[1]Now known as Case Western Reserve University.

Pile Driving Analyzer

Field equipment for measuring the forces and accelerations in a pile during driving was developed during the 1960s and became commercially available in 1972. The methodology is now standardized and is described in ASTM standard D4945.

This equipment includes three components:

- A pair of *strain transducers* mounted near the top of the pile
- A pair of *accelerometers* mounted near the top of the pile
- A *pile driving analyzer* (PDA)

The strain transducers and accelerometers and a pile driving analyzer is shown in Figure 15.6.

The pile driving analyzer monitors the output from the strain transducers and accelerometers as the pile is being driven, and evaluates this data as follows:

- The strain data, combined with the modulus of elasticity and cross-sectional area of the pile, gives the axial force in the pile.
- The acceleration data integrated with time produces the particle velocity of the waves travelling through the pile.
- The acceleration data, double integrated with time produces the pile displacement during the hammer blow.

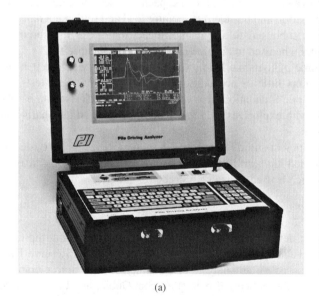
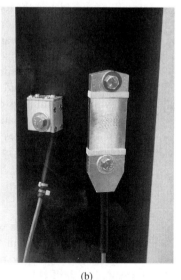

(a)   (b)

Figure 15.6   (a) Pile-driving analyzer; (b) Accelerometer (left) and strain gage (right) mounted near the top of a pile to provide input to the pile-driving analyzer (Pile Dynamics, Inc.).

Using this data, the PDA computes the Case method capacity, using the method described later, and displays the results immediately. It also can store the field data on a floppy disk to provide input for a CAPWAP analysis, also described later in this chapter.

Wave Propagation

The hammer impact creates a compressive wave pulse that travels down the pile. As it travels, the pulse induces a downward (positive) particle velocity. If the pile has only toe-bearing resistance (no side friction), the pulse reflects off the bottom and travels back up as another compression wave. This reflected wave produces an upward moving (negative) particle velocity.

The time required for the wave to travel to the bottom of the pile and return is $2D_2/c$, where:

D_2 = distance from strain transducers and accelerometers to the pile tip

c = wave velocity in the pile

The time $2D_2/c$ is very short compared to the interval between hammer blows. Therefore, the PDA can observe the effects of a single blow.

The plots of force and particle velocity near the top of this toe bearing pile (as measured by the PDA) are similar to those in Figure 15.7. Note the arrival of the return pulse at time $2D_2/c$. These plots are called *wave traces*.

Now, consider another pile that also has no side-friction resistance and much less toe-bearing resistance than the previous one. In this case, more of the energy in the downward moving wave is expended in advancing the pile, so the reflected wave has a smaller

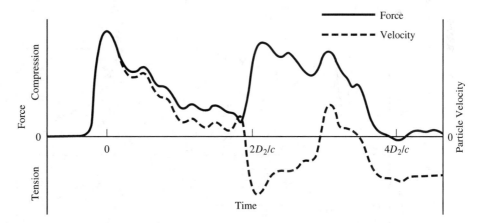

Figure 15.7 Typical plots of force and particle velocity near the top of the pile vs. time for a toe-bearing pile (Hannigan, 1990; Used with permission of Deep Foundations Institute).

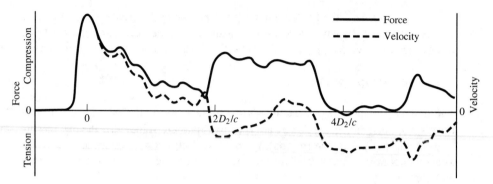

Figure 15.8   Typical plots of force and particle velocity near the top of the pile vs. time for a toe-bearing pile with less toe-bearing resistance than that shown in Figure 15.7 (Hannigan, 1990; Used with permission of Deep Foundations Institute).

force and a smaller velocity, as shown in the wave trace in Figure 15.8. Therefore, the shape of the wave trace at time $2D_2/c$ reflects the bearing resistance.

Finally, consider a friction pile with very little toe-bearing resistance. As the compression wave pulse travels down the pile, it encounters the side-friction resistance. Each increment of resistance generates a reflected wave that travels back up the pile, so the wave trace measured near the top will be similar to that in Figure 15.9. The time scale corresponds to the depth below the instruments, and the vertical distance between the force and velocity plots reflects the soil resistance at various depths.

The wave trace also provides pile integrity data. For example, if the pile breaks during driving, the fracture produces a reflected wave that changes the wave trace recorded by the pile driving analyzer (Rausche and Goble, 1978).

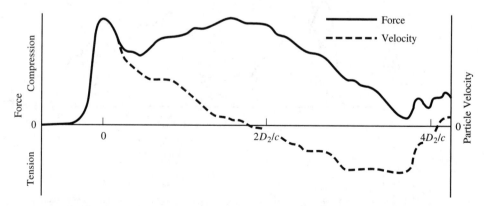

Figure 15.9   Typical plots of force and particle velocity near the top of the pile vs. time for a friction pile (Hannigan, 1990; Used with permission of Deep Foundations Institute).

Case Method Analyses

The Case method is an analytical technique for determining the static pile capacity from wave trace data (Hannigan, 1990). The PDA is programmed to solve for pile capacity using this method and gives the results of this computation in real time in the field.

The Case method computations include an empirical correlation factor, j_c, that can be determined from an on-site static load test. Thus, engineers can use this method to extend static load test results to indicator piles or selected production piles. However, for most projects, it would not be cost-effective to obtain PDA measurements on all of the production piles.

It is also possible to use the Case method without an on-site pile load test by using j_c values from other similar soils. This approach is less accurate, but still very valuable.

CAPWAP

The Case method, while useful, is a simplification of the true dynamics of pile driving and the associated response of the adjacent soil. The empirically obtained damping factor, j_c, calibrates the analysis, so the final results are no better than the engineer's ability to select the proper value. In contrast, a wave equation analysis utilizes a much more precise numerical model, but suffers from weak estimates of the actual energy delivered by the hammer. Fortunately, the strengths and weaknesses of these two methods are complimentary, so we can combine them to form an improved analysis (Rausche, Moses, and Goble, 1972). This combined analysis is known as CAPWAP (CAse Pile Wave Analysis Program).

The numerical model used in CAPWAP is essentially the same as that in Figure 15.3 except that the hammer and accessories are removed and replaced with force-time and velocity-time data obtained from the pile driving analyzer. The analysis produces values of R_u (the ultimate resistance in the soil "springs"), q (the quake), and j_c (the Case method damping factor).

A CAPWAP analysis performed on PDA data could be used as follows:

- To provide more accurate input parameters for a wave equation analysis that could then be used to select the optimal driving equipment as well as to produce a bearing graph.
- To provide a site-specific Case method damping factor, j_c, for use in PDA analyses of selected production piles.
- To obtain quantitative measurements of pile setup (Fellenius et al., 1989).
- To produce simulated static load test results.

Figure 15.10 shows a sample output from a CAPWAP analysis.

CAPWAP analyses can be used to reduce the required number of static load tests, or used where load tests are not cost-effective.

Below is the PDA screen; it is what the measurement engineer sees during dynamic pile testing.

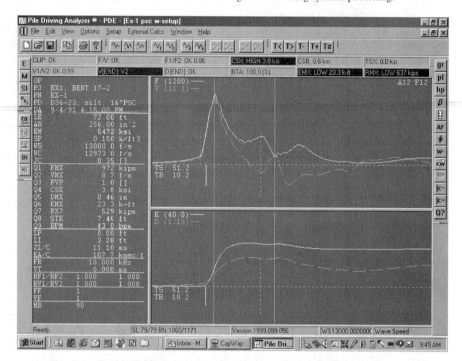

Figure 15.10 Sample output from a CAPWAP analysis. The upper plots reflect the computational match between the CAPWAP model and the PDA data obtained from the field. The plot in the lower left is a computed load-settlement curve, and that in the lower right shows the computed distribution of soil resistance (Courtesy of Goble Rausche Likins and Associates, Inc.)

Tests Using Drop Hammers

Case Method and CAPWAP analyses were originally developed for use with pile foundations, because they rely on acceleration and stress data obtained while the pile is being driven. More recently, engineers have applied these methods to drilled shaft foundations, using drop hammers to generate the required strains. This method can provide dynamic verification of the load capacity determined from analytic methods, while being less costly than a static load test.

Figure 15.11 shows a typical drop hammer used to perform these tests. Typically the test uses a weight equal to about 1.5 percent of the required static test load, and a drop height of about 8.5 percent of the shaft length or 2 m, whichever is greater. However, the actual equipment needs should be determined in advance using a wave equation analysis (Hussein, Likins, and Rausche, 1996). The instrumentation and data analysis are essentially the same as for piles.

Figure 15.11 This drop hammer is being used to conduct a high-strain dynamic load test on a drilled shaft foundation (ADSC: The International Association of Foundation Drilling).

Statnamic Test

The *statnamic test* is another high strain dynamic testing method for evaluating the static load capacity of deep foundations. This method loads the foundation by detonating slow-burning explosives located inside a pressure chamber placed between the foundation and a mass, as shown in Figures 15.12 and 15.13. The force from the explosion generates a downward movement in the foundation, which is monitored using load and displacement instruments, and the data obtained from these instruments is plotted directly as a load-displacement curve similar to that obtained from a static load test. The mass provides a reaction for this force.

The loading from a statnamic test is of much longer duration (0.1–0.2 s) than that from a pile hammer or drop hammer, so the foundation moves downward almost as a single unit and may be analyzed as if it were a rigid body. Therefore, the statnamic test does not rely on a numerical model of the foundation's dynamic response. Instead, it directly develops a load-displacement curve, and thus is conceptually similar to a static load test, except that it is performed very quickly. This feature also allows the test to be performed without a knowledge of the cross-sectional area or modulus of elasticity of the foundation, which is especially helpful when testing drilled shafts.

The load-displacement curve obtained from the test is then adjusted to account for inertial and damping forces in an attempt to produce a "derived static" or "equivalent static" load-displacement curve. For most foundations, this analysis can assume the foundation acts as a rigid body

Data reduction methods attempt to account for inertial and damping forces. Typically, the foundation settles about 5 mm (0.2 in) during the test, which is probably suffi-

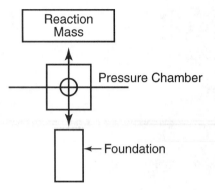

Figure 15.12   Schematic of the statnamic test.

cient to develop most of the side-friction resistance, but less than the settlement required to fully mobilize the toe-bearing resistance. Therefore, the test should, in principle, be conservative. However, the static response of soil differs from the dynamic response, so the load-settlement curve obtained from statnamic can differ from that obtained from a static load test. The data reduction attempts to account for this difference, along with the inertial effects in the foundation itself, but the proper method of doing so is not always clear. As a result, the capacities obtained from statnamic tests can be higher or lower than those obtained from static load tests. In addition, the rigid body assumption may not be valid for long slender foundations.

Statnamic tests also have been used to evaluate lateral load capacity of deep foundations, and are an alternative to static lateral load tests. These statnamic tests are performed using a horizontal pressure chamber. Since design lateral loads are usually dynamic (i.e., wind or earthquake), the statnamic test may be a better representation of the actual service loads, and thus may provide more realistic capacities than those obtained from static lateral load tests.

Application of High-Strain Dynamic Tests

All of these high-strain dynamic test methods help an engineer evaluate the downward load capacity of a deep foundation, and thus serve as a independent check on capacities determined using analytic methods. The Case Method provides this data in real time as the pile is being driven, and thus may be used to determine the required embedment depth. Often this depth is different from that indicated by the analytic methods described in Chapter 14, which produces an as-built foundation that is longer or shorter than that indicated on the design drawings.

CAPWAP analyses, statnamic tests, and drop hammer tests are conducted after the foundation has been built, and thus cannot be used to control the as-built depth of the test shaft. However, the test results are still valuable in that they provide an independent con-

(a)

(b)

Figure 15.13 A statnamic test on a pile driven over water. a) Usually the reaction mass is located above or around the statnamic pressure chamber. However, in this case, the reaction mass consisted of water inside submerged containers. b) The statnamic pressure chamber is located above the test pile, and in this case is connected to the submerged mass through tendons. In most tests this pressure chamber is hidden by the reaction mass. c) The explosion occurs inside the pressure chamber, which then imparts a downward force on the pile. (Photos courtesy of Berminghammer Foundation Equipment).

(c)

Figure 15.14 Conducting a low-strain dynamic test to assess the integrity of a concrete-filled pipe pile (Photo courtesy of Pile Dynamics, Inc.).

firmation of the downward load capacity which verifies the suitability of the test foundation and assists in the design of future foundations at that site.

15.4 LOW-STRAIN DYNAMIC TESTING

Low-strain dynamic testing consists of striking the top of the foundation with a small load and monitoring the resulting waves using one or more accelerometers. These tests typically use a carpenter's mallet to generate the load, as shown in Figure 15.14, so the induced strains are very small and the settlement is not nearly sufficient to fully mobilize the side-friction or toe-bearing resistance. Therefore, low-strain dynamic tests do not provide an indication of static load capacity. However, they are very useful for evaluating structural integrity and for determining the as-built length. These methods are discussed in Section 17.10.

15.5 CONCLUSIONS

Dynamic methods have progressed from the crude pile driving formulas of the late nineteenth century to sophisticated numerical models with extensive calibrations to static load tests. These methods are very attractive to foundation engineers, because they promise to provide reliable information on load capacities without the necessity of investing in expensive static load tests. In addition, some dynamic methods provide information on driveability and structural integrity.

It is unlikely that static load tests will ever become obsolete, but dynamic methods have substantially reduced the need for them. As dynamic methods continue to be refined, they will probably enjoy increased use in a wider range of projects.

SUMMARY

Major Points

1. Dynamic methods use evaluations of a foundation's response to applied dynamic loads, such as those from a pile hammer or other impact source, to determine its static load capacity. Some dynamic methods also provide information on driveability and structural integrity.

2. Pile-driving formulas are the oldest type of dynamic method. They attempt to correlate static load capacity with blow count, hammer type, and other factors. Unfortunately, these formulas oversimplify the pile-driving process, and thus are not very precise, especially if they have not been calibrated with an onsite load test.

3. Wave equation analyses have generally replaced pile-driving formulas. Instead of relying solely on empirical correlations, the wave equation analysis uses a detailed analytical model of the pile and its driving system. The results of these analyses include a bearing graph (which is a plot of static capacity vs. blow count), and the pile driving stresses, which form the basis for a driveability analysis.

4. High-strain dynamic tests use instruments mounted on the foundation to record its response to applied dynamic loads. These loads usually are from the pile hammer, but also may be generated by a drop hammer or by a controlled explosion. The data obtained from these instruments may be used to determine the static load capacity.

5. Low-strain dynamic testing uses much smaller impact loads (typically imparted by a carpenter's hammer) and is used to evaluate structural integrity.

Vocabulary

Bearing graph	Engineering News Formula	Pile-driving formula
Blow count	Freeze (setup)	Pile driving analyzer
CAPWAP	High-strain dynamic	Quake
Case method	testing	Smith damping factor
Driveability analysis	Low-strain dynamic	Statnamic test
Dynamic methods	testing	Wave equation

COMPREHENSIVE QUESTIONS AND PRACTICE PROBLEMS

15.1 Explain why pile-driving formulas are not reliable, and why a wave equation analysis is a better choice.

15.2 How can analyses based on the wave equation account for increased pile capacity caused by freeze or setup?

15.3 A 15-m long concrete pile has a wave velocity of 4000 m/s and is being driven at a rate of 25 blows per minute. Will the wave from the first hammer blow reach the bottom of the pile and return to the top before the next hammer blow? Justify your answer with appropriate computations and comment on the results.

15.4 Why does the statnamic test use slow-burning explosives?

15.5 An HP 10 × 53 pile is to be driven with a Delmag D-12 hammer in a soil profile described in Figure 15.4. The required allowable downward load capacity is 500 kN. Using the bearing graph in Figure 15.4 and the appropriate factor of safety from Table 13.1, determine the minimum required blow count.

15.6 The pile described in Problem 15.5 is to be made of A36 steel, which has $F_y = 250$ MPa. Determine the maximum allowable driving stress using the criterion described in Section 12.3, then use the wave equation analysis results in Figure 15.4 to determine if the driving stresses are acceptable.

15.7 According to a static analysis, the pile described in Problem 15.5 will develop the required downward load capacity if it is driven 12.2 m into the ground. The pile has now been driven to that depth, and the final blow count was 200 blows/m. Is this acceptable? Why or why not? Does any remedial action need to be taken? Explain.

15.8 A 14-inch square prestressed concrete pile is to be driven with a certain hammer. According to a wave equation analysis, the driving stresses will exceed the maximum allowable values described in Table 12.5. What can be done to resolve this problem? Provide at least two possible solutions.

15.9 A series of large-diameter steel pipe piles are to be driven at an offshore site. These piles will support a major marine structure. Unfortunately, because of the location and size of these piles, a full-scale static load test is not economically feasible. Suggest an appropriate dynamic method for evaluating the load capacity of these piles and indicate the reason for your selection.

16

Deep Foundations— Lateral Load Capacity

Some say the cup is half empty, while others say it is half full. However, in my opinion both are wrong. The real problem is the cup is too big.

Sometimes all we need is a new perspective on an old problem.

The discussions in Chapters 13 to 15 considered only axial loads. However, many deep foundations also must support *lateral loads*, which are applied shear and/or moment loads as shown in Figure 16.1. Sources of lateral loads include the following:

- Earth pressures on the back of retaining walls
- Wind loads
- Seismic loads
- Berthing loads from ships as they contact piers or other harbor structures
- Downhill movements of earth slopes
- Vehicle acceleration and braking forces on bridges
- Eccentric vertical loads on columns
- Ocean wave forces on offshore structures
- River current forces on bridge piers
- Cable forces from electrical transmission towers or electric railway wire support towers
- Structural loads on abutments for arch or suspension bridges

Deep foundations transfer these loads into the ground through lateral bearing, as shown in Figure 16.1, which is quite different from the side friction and toe bearing associated with

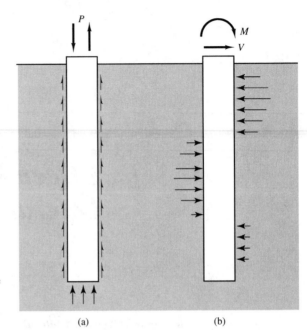

Figure 16.1   Load transfer from a deep foundation to the ground: (a) Axial loads are transferred through the side friction and toe bearing; (b) lateral loads are transferred through lateral bearing.

axial loads. Thus, when evaluating load transfer from the foundation to the ground, axial and lateral load capacities require two separate analyses. This chapter discusses methods of performing lateral load analyses, and shows how to compute the resulting stresses and lateral deflections in the foundation.

16.1   BATTER PILES

Until the middle twentieth century, engineers did not know how to evaluate laterally loaded deep foundations, so they assumed deep foundations were only able to resist axial loads. Therefore, when horizontal loads were present, engineers installed some of the foundations at an angle from the vertical, as shown in Figure 16.2. Thus, each foundation supposedly is subjected to axial loads only.

Although it is possible to build most types of deep foundations on an angle, driven piles are the most common choice. Piles driven in this fashion are known as *batter piles* or *raker piles*. Contractors with the proper equipment can usually install them at a batter of up to 4 vertical to 1 horizontal by simply tilting the pile-driver leads. However, these operations are not as efficient as driving vertical piles.

Although batter piles have generally performed well, there are at least two situations that cause problems:

- Large offshore structures, such as drilling platforms, are subjected to very large lateral loads from water currents, wind loads, and seismic loads, but it is not practical to drive piles at a sufficiently large angle to accommodate these loads.

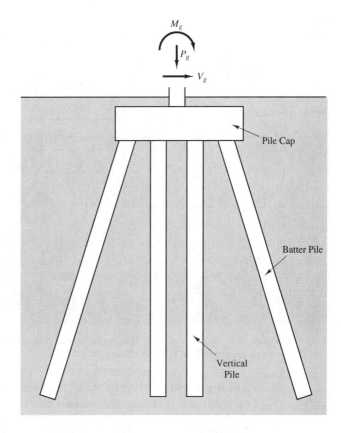

Figure 16.2   Batter piles being used to resist combined vertical and horizontal loads.

- Batter piles form a very stiff foundation system. This is suitable when only static loads are present, but can cause problems when dynamic loads are applied, such as those imposed by an earthquake. Figure 16.3 shows the kinds of damage that can occur as a result of the high loads that develop in such a stiff system. A more flexible system would perform better because it would allow inertial effects to absorb some of the load.

Engineers also recognized that groups of batter piles are not subject to axial loads only. In reality, shear and moment loads also are present because:

- Only certain combinations of applied vertical and horizontal loads on the pile cap produce only axial loads in all of the piles. Because the horizontal loads often are temporary (i.e., wind or seismic loads), the actual working loads often are not properly combined. For example, if the horizontal load is sometimes zero, then the vertical load generates flexural stresses in the batter piles.
- The soil may consolidate after the piles are driven (see discussion of downdrag loads in Section 18.1). This downward movement of the soil produces lateral loads on the batter piles, and might even cause them to fail in flexure. Unfortunately, such

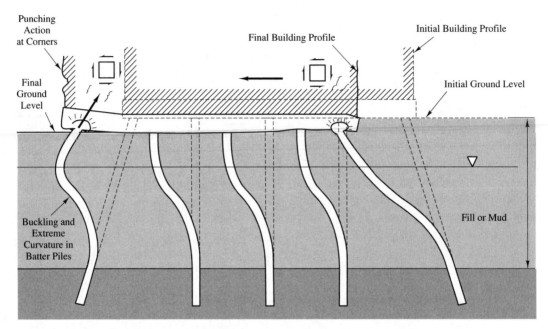

Figure 16.3   Behavior of batter piles during seismic loading (Adapted from Dowrick, 1987; Used with permission of John Wiley and Sons).

a failure would probably be hidden, and would not be evident until an earthquake or other extreme event occurred and the batter piles failed to perform.

In spite of these problems, batter piles are still an appropriate solution for some foundation problems. However, research and development in designing deep foundations to resist lateral loads has opened new possibilities, as discussed in the next section.

16.2   RESPONSE TO LATERAL LOADS

Because of the difficulties with batter piles and the need to produce more economical and reliable designs, engineers reconsidered the assumption that deep foundations are only able to resist axial loads. If a deep foundation system could be designed to resist both axial and lateral loads, then it could be built using only vertical members, which would be more economical and more flexible. These characteristics are especially important in off-shore structures, so the petroleum industry led much of the early effort to develop lateral load analysis methods.

Short vs. Long Foundations

When evaluating lateral loads, we divide deep foundations into two categories: *short* and *long*, as shown in Figure 16.4. A short foundation is one that does not have enough embedment to anchor the toe against rotations, whereas a long foundation is one in which the

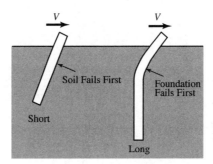

Figure 16.4   Short vs. long foundations.

toe is essentially fixed. The minimum length to be considered "long" depends both on the stiffness of the foundation and the lateral resistance provided by the soil. In general, flexible foundations, such as timber piles, are long if D/B greater than about 20, while stiffer foundations, such as those made of steel or concrete, typically require D/B greater than about 35.

The ultimate lateral capacity of short foundations is controlled primarily by the soil. In other words, the soil fails before the foundation reaches its flexural capacity. Conversely, the ultimate lateral capacity of long foundations is controlled primarily by the flexural strength of the foundation because it will fail structurally before the soil fails.

Soil–Structure Interaction

The transfer of lateral loads from deep foundations to the ground is a *soil–structure interaction* problem, and is very similar to the interaction between mat foundations and the underlying soil as discussed in Chapter 10. In other words, the movements and flexural stresses in the foundation depend on the soil resistance, while the soil resistance depends on the movements of the foundation. Thus, we cannot artificially separate the geotechnical and structural aspects of the analysis. Both must be evaluated concurrently.

Figure 16.5 shows diagrams of lateral deflection, slope, moment, and shear in a long foundation, and the lateral soil reaction, all as a function of depth. The applied shear and/or moment loads produce a lateral deflection at the top of the foundation. This deflection induces lateral resistance in the soil, and this resistance is a function of the deflection as shown in Figure 16.6. Near the ground surface, this soil resistance opposes the applied loads, so the deflection decreases with depth, eventually reaching zero deflection at some depth. However, the shear and moment at this depth are not zero, so the foundation deflects in the opposite direction and induces soil reactions that also are in the opposite direction. This interaction continues with depth until all of the parameters are essentially zero.

The shapes and magnitudes of these plots depend on many factors, including:

- The type (shear and/or moment) and magnitude of the applied loads
- The resistance-deflection relationship in the soil (known as the *p-y* curve)
- The flexural rigidity of the foundation, which is the product of its modulus of elasticity, E, and moment of inertia, I

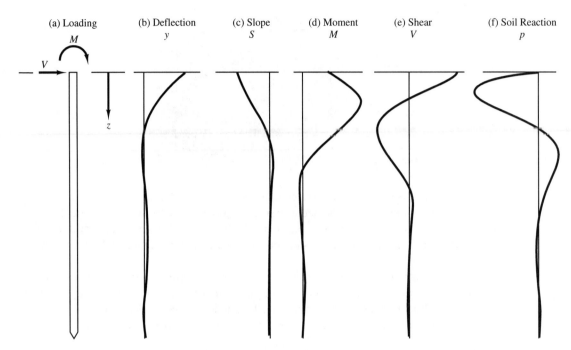

Figure 16.5   Forces and deflections in a long deep foundation subjected to lateral loads (Adapted from Matlock and Reese, 1960; Used with permission of ASCE).

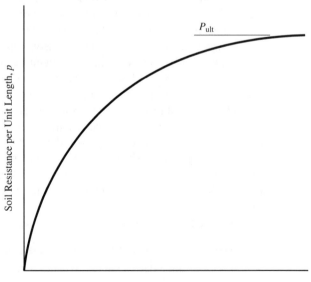

Figure 16.6   Soil resistance per unit length, p, as a function of lateral deflection, y. This is known as a p-y curve.

The change in each of these parameters with depth are defined by the principles of structural mechanics as follows:

$$S = \frac{dy}{dz}$$

(16.1)

$$M = EI\frac{dS}{dz} = EI\frac{d^2y}{dz^2}$$

(16.2)

$$V = \frac{dM}{dz} = EI\frac{d^3y}{dz^3}$$

(16.3)

$$p = \frac{dV}{dz} = EI\frac{d^4y}{dz^4}$$

(16.4)

Where:

S = slope of foundation

M = bending moment in foundation

V = shear force in foundation

p = lateral soil resistance per unit length of the foundation

E = modulus of elasticity of foundation

I = moment of inertia of foundation in the direction of bending

y = lateral deflection

z = depth below ground surface

If the shape of one of these curves is known, either through computation or field measurements, the others may be computed through progressive integration or differentiation with appropriate boundary conditions.

Example 16.1

According to measurements made during a full-scale lateral load test, the deflection of a certain deep foundation between the ground surface and a depth of 2.0 m is defined by the equation:

$$y = 0.035 - 0.010z^{1.8}$$

Develop equations for slope, moment, shear, and soil reaction vs. depth for $0 \leq z \leq 2$ m.

Solution

$$S = \frac{dy}{dz} = -0.018z^{0.8} \qquad \Leftarrow Answer$$

$$M = EI\frac{dS}{dz} = -0.0144z^{-0.2}EI \qquad \Leftarrow Answer$$

$$V = \frac{dM}{dz} = 0.288z^{-1.2}EI \qquad \Leftarrow Answer$$

$$p = \frac{dV}{dz} = -0.346z^{-2.2}EI \qquad \Leftarrow Answer$$

Commentary

In practice, the shapes of these curves are not easily defined by simple polynomials, so the integration or differentiation must be performed by numerical methods.

The type of connection between the foundation and the structure also is important because it determines the kinds of restraint, if any, acting on the foundation. These define the boundary conditions needed to perform integrations on Equations 16.1 to 16.4. Engineers usually assume that one of the following restraint conditions prevail (although others are also possible):

- The *free-head condition*, shown in Figure 16.7a, means that the top of the pile may freely move laterally and rotate when subjected to shear and/or moment loads. A single drilled shaft supporting a highway sign is an example of the free-head condition. In this case, the applied shear and moment loads, V and M, at the top are known but the slope and deflection, S and y, are unknown.

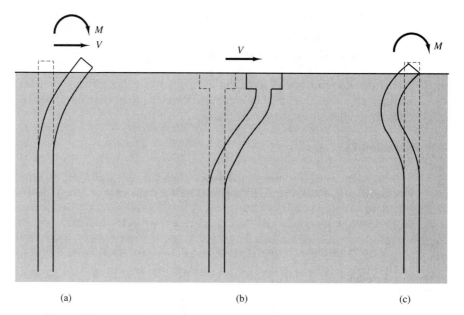

(a)   (b)   (c)

Figure 16.7 Types of connections between the pile and the structure: (a) free-head; (b) restrained-head; and (c) pure moment.

- The *restrained-head condition* (also known as the *fixed-head condition*), shown in Figure 16.7b, means that the top of the pile may move laterally, but is not permitted to rotate. A group of piles connected with a pile cap closely approximate this condition, because there will be very little rotation in the cap. In this case, *V* and *S* at the top are known, but *M* and *y* are unknown.
- The *pure moment condition*, shown in Figure 16.7c, occurs when there is an applied moment load, but no applied shear load. It results in rotation of the top of the pile, but no lateral movement. This is the least common of the three conditions, and is not often used in practice. In this case, *V, S, M,* and *y* at the top are all known.

Thus, selecting one of these alternatives defines the boundary conditions at the top of the pile. Although real piles often have connections that are intermediate between these three ideal conditions, one of these assumptions is usually reasonable for design purposes.

So long as the foundation is "long," the shear, moment, rotation, and deflection at the toe are all zero, thus forming four additional boundary conditions.

The foundation transmits most of the applied lateral loads to the upper soils, so the soil properties within about 10 diameters of the ground surface are the most important. Therefore, we must be especially careful to consider scour or other phenomena that might eliminate some of the upper soils.

16.3   METHODS OF EVALUATING LATERAL LOAD CAPACITY

Methods of evaluating lateral load capacity include both experimental and analytical methods, and range from simple to complex. The results of extensive research, and the widespread availability of powerful computers have greatly improved our ability to analyze these foundations.

The objectives of lateral load capacity analyses generally include one or more of the following:

- Determine the minimum required depth of embedment to transfer the lateral loads into the ground. When lateral loads are the primary loads acting on the foundation, this criteria may control the required depth. However, when significant axial loads are present, such as with foundations that support buildings or bridges, the depth of embedment is usually controlled by axial load considerations.
- Determine the lateral deflection under the design lateral loads. Buildings, bridges, and other similar structures typically can tolerate no more than 5 to 20 mm (0.25–0.75 in) of lateral movement.
- Determine the shears and moments induced in the foundation by the lateral loads. Sometimes we need only the maximum values and design the entire foundation to resist them, while other times we need shear and moment diagrams and design the foundation accordingly.

Full-Scale Lateral Load Tests

One way to evaluate lateral load capacity is to conduct a full-scale load test. These tests involve installing one or more prototype foundations at the project site, applying a series of lateral loads, and measuring the resulting lateral deformation and slope at the top of the foundation (Reese, 1984; ASTM D3966). Thus, we can directly determine load-deformation characteristics. In addition, if strain gages are installed in the foundation and/or deformations are measured at different depths (such as with an inclinometer), we can compute the shears and moments in the foundation using Equations 16.1 to 16.4.

Full-scale lateral load tests are less common than full-scale axial load tests, but they are performed on occasion, especially for very large projects or when the soil conditions are unusual. Figure 16.8 shows a typical test setup. Most load tests are performed using a free-head condition with applied shear loads, because this is the easiest test condition to produce in the field.

The results of a lateral load test (deflections, shears, and moments) can be directly used to design the production foundations. Alternatively, they may be used to backcalculate site-specific *p-y* curves (Kramer, 1991). These curves may then be used with the *p-y* analyses described later in this chapter to design a wider variety of foundations with different head restraint conditions.

The primary disadvantage of full-scale load tests is cost. They are very expensive, and thus can be economically justified only on large projects.

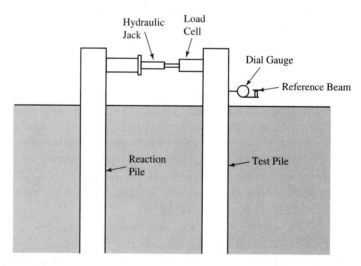

Figure 16.8 Typical full-scale lateral load test on a prototype deep foundation. Unlike vertical load tests, both foundations are "reaction" and both are "tested" because both move about the same distance during the test.

Figure 16.9   Model load tests conducted at Utah State University to evaluate the response of pile groups subjected to cyclic lateral loads. Each of the five model piles is made of 1.3 in. (33 mm) diameter aluminum tubing and instrumented with fourteen pairs of strain gauges (photo courtesy of Professor Joseph Caliendo). See Moss, Rawlings, Caliendo, and Anderson (1998) for more details.

Model Lateral Load Tests

The cost of load tests can be substantially reduced by using scale model foundations and conducting the tests in a laboratory. Figure 16.9 shows the apparatus used for a model load test. In addition to their lower cost, these tests have the advantages of more flexibility in loading, restraint conditions, soil conditions, etc. However, it can be difficult to extrapolate model tests to full-scale foundations because the various parameters (linear dimensions, stresses, masses, etc.) have different scaling ratios.

Some engineers have attempted to overcome some of the scaling problems inherent in model load tests by conducting the tests inside a centrifuge. The additional "gravity" in a centrifuge model helps bring the scaling ratio for mass and stress closer to that for linear dimensions. However, centrifuge tests introduce other complexities, and are limited to very small scale models.

Model lateral load tests are most effective when evaluating undrained loading conditions in a clay, because the undrained shear strength, s_u, is not dependant on the effective stress and thus is not affected by scaling.

Lateral Statnamic Test

Another option is to conduct a lateral statnamic test. This technique uses the same principles as the vertical statnamic test described in Chapter 15, except the load is applied horizontally as shown in Figure 16.10. Unlike conventional lateral load tests, lateral statnamic

Figure 16.10   Lateral statnamic test being conducted from a barge. The test foundation is on the far left side of the photograph. (Courtesy of Berminghammer Foundation Equipment).

tests require only one test foundation because the statnamic equipment provides its own reaction.

Rigid Analyses

The earliest analytical solutions to lateral load problems assumed the foundation was perfectly rigid (i.e., a very high EI). This is the same assumption used in early mat foundation analyses, as discussed in Chapter 10, and it was introduced for the same reason: to simplify the computation of soil reaction forces. Broms (1964a, 1964b, 1965) and others used this method to evaluate lateral load capacity.

These rigid analyses provide information on the minimum required depth of embedment and the maximum moment. However, because these methods neglect flexural rotations in the foundation and use simplified descriptions of the soil resistance, they are not as accurate as the nonrigid methods discussed later.

Rigid analyses are now used primarily for lightweight "short" laterally loaded foundations, such as those supporting streetlights or small highway signs, where the required depth of embedment is controlled by the lateral loads. The Uniform Building Code [UBC 1806.8.2] and International Building Code [IBC 1805.7.2] include the following rigid analysis formulas for "posts or poles" embedded in the ground:

For the free-head condition:

$$D_{\min} = \frac{A}{2} \left(1 + \sqrt{\frac{4.36\,h}{A}} \right) \tag{16.5}$$

$$A = \frac{2.34\,V}{S_1 B} \tag{16.6}$$

For the restrained-head condition:

$$D_{\min} = \sqrt{\frac{4.25\,V\,h}{S_3 B}} \tag{16.7}$$

Where:

$D_{\min}$ = minimum required depth of embedment (ft, m)

V = applied shear load (lb, kN)

h = vertical distance from ground surface to point of application of V (ft, m)

B = diameter of foundation (ft, m)

S_1 = allowable lateral soil pressure at $z = D_{\min}/3$ (lb/ft^2, kPa)

S_3 = allowable lateral soil pressure at $z = D_{\min}$ (lb/ft^2, kPa)

z = depth below the ground surface (ft, m)

The values of S_1 and S_3 are presented in Table 16.1. Both UBC and IBC permit the use of twice the values shown in this table for short-term loads (presumably wind or seismic loads), so long as a lateral deflection of 13 mm (0.5 in) is acceptable.

TABLE 16.1   ALLOWABLE LATERAL SOIL PRESSURE (ICBO, 1997 and ICC, 2000; used with permission)

Soil Classification	Allowable Lateral Bearing Pressure	
	(lb/ft^2)	(kPa)
Massive crystalline bedrock	1200 z	190 z
Sedimentary and foliated rock	400 z	63 z
Sandy gravel and/or gravel (GW and GP)	200 z	31 z
Sand, silty sand, clayey sand, silty gravel, and clayey gravel (SW, SP, SM, SC, GM, and GC)	150 z	24 z
Clay, sandy clay, silty clay, and clayey silt (CL, ML, MH, and CH)	100 z	16 z

z = depth below the ground surface to a maximum of 15 ft (4.6 m). Below that depth, use an allowable lateral bearing pressure equal to the value at 15 ft (4.6 m).

Note: The UBC and IBC imply the maximum z is 15 ft or 15 m, which is an incorrect conversion to metric units. Since this table was originally developed using English units, we can assume they represent the correct value.

Example 16.2

A sign is to be supported on an 18-inch diameter drilled shaft foundation, as shown in Figure 16.11. The design wind load is 800 lb, and the center of the sign is 20 ft above the ground. Compute the minimum required depth of embedment, D_{min}.

Solution

This is a free-head condition where a 0.5 inch lateral deflection would be acceptable, so use twice the value shown in Table 16.1.

$$S_1 = 150(2)(D_{min}/3) = 100D_{min}$$

$$A = \frac{2.34(800)}{100(D_{min})(1.5)} = \frac{12.5}{D_{min}}$$

$$D_{min} = \frac{12.5}{2\,D_{min}}\left(1 + \sqrt{1 + \frac{4.36\,(20)}{12.5/D_{min}}}\right)$$

$$\boldsymbol{D_{min} = 7\ ft} \quad \Leftarrow Answer$$

Depth to Fixity Analyses

During the 1960s and 1970s, many geotechnical engineers described the results of lateral load analyses in terms of the *depth to fixity*, as shown in Figure 16.12 (Davisson, 1970; Tomlinson, 1987). This method ignored the soil between the ground surface and the depth to fixity, and considered the soil beneath this depth to be infinitely strong and rigid. They provided this depth to the structural engineer, who simply computed the stresses and deflections in the foundation as if there was a rigid connection at the depth of fixity and no soil resistance above that point. This method was especially convenient for the structural engineer because the structural computations were straightforward and appeared to provide all of the needed information.

The depth to fixity method does consider the flexural rigidity of the foundation, and in that respect is an improvement over the rigid analyses. However, it uses a simplistic representation of the soil resistance, and thus is not as precise as the *p-y* method discussed later in this chapter.

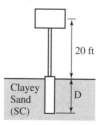

20 ft

Clayey Sand (SC)

D

Figure 16.11   Proposed sign for Example 16.2.

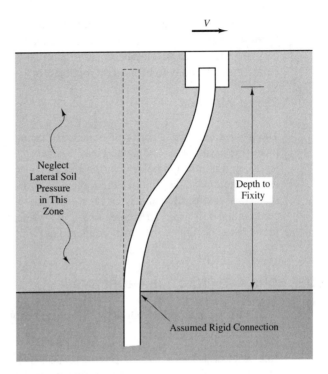

Figure 16.12 Depth to fixity method of describing the behavior of laterally-loaded deep foundations.

Nonrigid Soil-Structure Interaction Analyses

Because of the shortcomings of the methods described thus far, engineers have developed more thorough lateral load analysis methods that consider the flexural rigidity of the foundation, the soil's response to lateral loads, and soil-structure interaction effects. This can be done using either of two methods: the finite element method or the *p-y* method.

Finite Element Method

A *finite element method* (FEM) analysis consists of dividing both the foundation and the soil into a series of small elements and assigning appropriate stress–strain properties to each element. The analysis then considers the response of these elements to applied loads, and uses this response to evaluate shears, moments, and lateral deflections in the foundation. Finite element analyses may be performed using either two-dimensional or three-dimensional elements.

The accuracy of finite element analyses depends on our ability to assign correct engineering properties to the elements. This is easy to do for the foundation because the properties of structural materials are well-defined, but very difficult to do for the soil because it is more complex. For example, the stress-strain properties in the soil are definitely nonlinear. In addition, three-dimensional FEM analyses, which are more accurate, require more extensive computer resources.

Finite element analyses have been used on specialized projects, and ultimately may become the preferred method of evaluating laterally-loaded deep foundations. However, they are still under development and require additional calibration with load test results before they are likely to be used on routine projects.

p-y Method

The *p-y method* uses a series of nonlinear "springs" to model the soil–structure interaction. This is similar to the method used to analyze mat foundations, as discussed in Chapter 10, and is much simpler than the finite element method. Although the *p-y* method is not as rigorous as the finite element method, it has been extensively calibrated with full-scale load test results, and is easier to implement due to the widespread availability of commercial software. Therefore, this is the preferred method for most practical design problems, especially with "long" foundations.

QUESTIONS AND PRACTICE PROBLEMS

16.1 What are the primary advantages of using laterally-loaded vertical piles instead of batter piles?

16.2 A full-scale lateral load test has been conducted on an HP 10×57 pile made of A36 steel. The lateral load was applied so that bending occurred in the X-X axis. According to measurements made during this test, the lateral deflection between the ground surface and a depth of 10 ft is defined by the equation $y = 1.52 - 0.001131\ z^{1.5}$, where both y and z are expressed in inches. Develop plots of shear, moment, and lateral soil pressure vs. depth.

16.3 A group of cell-phone antennas are to be installed on a steel pole that will be embedded into sandstone bedrock. The embedded portion of this pole will be placed in a 2-ft diameter boring that will then be backfilled with concrete. The resultant of the lateral wind load will be 600 lb and will act at a point 50 ft above the ground surface. Using a rigid analysis and a free-head condition, compute the required depth of embedment.

16.4 A satellite dish is to be supported on a steel pole that will be embedded through a concrete slab and into the underlying ground. The embedded portion of the pole will be placed in a 6-inch diameter hole that will be backfilled with concrete. The lateral wind load on this dish will be 200 lb, and the center of the dish will be 15 ft above the concrete slab. The soils are clayey sands and sandy clays (SC and CL). Using a rigid analysis and a restrained-head condition (because of the slab), compute the required depth of embedment.

16.5 A sign is to be supported on two steel poles, each of which will be embedded into a stiff silty clay. The lateral wind load will be 5.4 kN and will act at a height of 8 m above the ground surface. This load will be equally distributed to the two poles. The embedded portions of the poles will be in 300-mm diameter holes that will be backfilled with concrete. Using a rigid analysis and a free-head condition, compute the required depth of embedment.

16.4   *p-y* METHOD

Numerical Model

The *p-y method* models the foundation using a two-dimensional finite difference analysis. It divides the foundation into *n* intervals with a node at the end of each interval, and the soil as a series of nonlinear "springs" located at each node, as shown in Figure 16.13. The flexural stiffness of each interval is defined by the appropriate *EI*, and the load-deformation properties of each spring is defined by a *p-y curve* such as those in Figure 16.14. It also is necessary to apply appropriate boundary conditions, as described earlier.

Using this information and applying the structural loads in increments, the software finds a condition of static equilibrium and computes the shear, moment, and lateral deflection at each interval.

Experiments with this method began in the late 1950s (McClelland and Focht, 1958; Matlock and Reese, 1960). However, the full development of this method required development of new software and calibration from full-scale load tests. Much of this work was performed during the 1960s and 1970s, and the method was well established by 1980. Nevertheless, this method continues to be refined through continued research and experience.

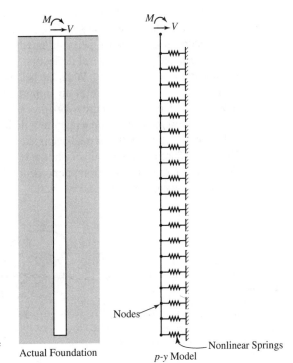

Figure 16.13   Analytical model used in the *p-y* method.

Actual Foundation   *p-y* Model

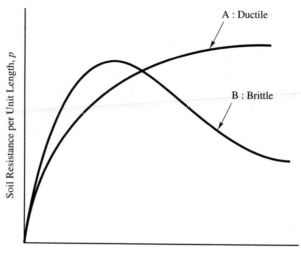

Figure 16.14   Typical *p-y* curves.

p-y Curves

The heart of the *p-y* method is the definition of the lateral load-deflection relationships between the foundation and the soil. These are expressed in the form of *p-y curves*, where *p* is the lateral soil resistance per unit length of the foundation (expressed in units of force per length), and *y* is the lateral deflection. The *p-y* relationship might first appear to be a nonlinear extension of the Winkler beam-on-elastic-foundation concept described in Chapter 10. However, there is an important difference between the two: The Winkler model considers only compressive forces between the foundation and the soil, whereas the lateral soil load acting on a deep foundation is the result of compression on the leading side, shear friction on the two adjacent sides, and possibly some small compression on the back side. These components are shown in Figure 16.15. Thus, it is misleading to think of the *p-y* curve as a compression phenomenon only (Smith, 1989), even though the numerical model appears to treat them as such.

Figure 16.15   A foundation with no lateral load is subjected to a uniform lateral earth pressure. Once the load is applied and the foundation moves laterally a distance *y*, the lateral pressure becomes greater on one side. The resultant of this pressure distribution is *p*.

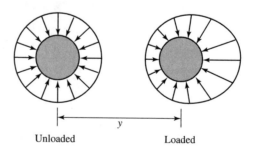

The ultimate compression resistance is probably much greater than the ultimate side shear resistance. However, mobilization of the side shear requires much less deflection, so it may be an important part of the total resistance at the small deflections generally associated with the working loads.

The p-y curve for a particular point on a foundation depends on many factors, including the following:

- Soil type
- Type of loading (i.e., short-term static, sustained static, repeated, or dynamic)
- Foundation diameter and cross-sectional shape
- Coefficient of friction between foundation and soil
- Depth below the ground surface
- Foundation construction methods
- Group interaction effects

The influence of these factors is not well established, so it has been necessary to develop p-y curves empirically by back-calculating them from full-scale load tests. Most of this data were obtained from 250 to 600 mm (10–24 in) diameter steel pipe piles because these are easier to instrument. These tests have been performed on various soil types, and the results have been correlated with standard soil properties, such as effective friction angle or undrained shear strength. Thus, they may be extended to other sites by measuring the appropriate soil properties and using the correlations. Reese (1984, 1986) summarizes many of the tests conducted thus far and provides recommended p-y curves for analysis and design. These curves and correlations also have been incorporated into p-y software.

Some curves are *ductile*, as shown in Curve A in Figure 16.14. These curves reach the maximum resistance, p_{max}, at a certain deflection, and then maintain this resistance at greater deflections. Other curves are *brittle*, as shown in Curve B in Figure 16.14, and have a decreasing p at large deflections. Brittle curves can occur in some clays, especially if they are stiff or if the loading is repeated or dynamic. Soft clays under static loading and sands appear to have ductile curves. Brittle curves are potentially more troublesome because of their potential for producing large foundation movements.

Although the empirical p-y data collected thus far have been an essential part of making the p-y method a practical engineering tool, we continue to need more data. Additional instrumented load tests would further our understanding of this relationship and thus provide more accurate data for analysis and design. For large projects, it may be appropriate to conduct full-scale lateral load tests and develop site-specific p-y curves.

Some engineers have attempted to develop p-y curves from in-situ pressuremeter or dilatometer tests (Baguelin et al., 1978; Robertson et al., 1989). Although these tests directly measure something similar to the compression component of lateral pile resistance, they do not address the side shear component. This may be at least a partial explanation of Baguelin's assessment that this approach produces pessimistic results. There also are scale effects to consider. Further research may improve the reliability of these methods.

Software

The first widely used *p-y* analysis software was a program called COM624, which was developed at the University of Texas at Austin. This program is in the public domain (see http://uftrc.ce.ufl.edu/info-cen/info-cen.htm), but it has since been superceded by other software that is more up-to-date and easier to use. Another public domain program, *Florida Pier*, is available from http://www.dot.state.fl.us/Structures/proglib.htm. A popular commercial Windows-based program is called LPILE (see http://www.ensoftinc.com).

Accuracy

Reese and Wang (1986) compared the results of twenty two static full-scale lateral load tests with the predicted deflections. They also compared the measured and predicted maximum moments for twelve of these tests. These data, expressed in Figure 16.16 using confidence intervals, suggest that this method provides good predictions of deflections and very good predictions of moments.

16.5 EVANS AND DUNCAN'S CHARTS

Evans and Duncan (1982) developed a convenient method of expressing the lateral load-deflection behavior in chart form. They compiled these charts from a series of *p-y* method computer analyses using the computer program COM624.

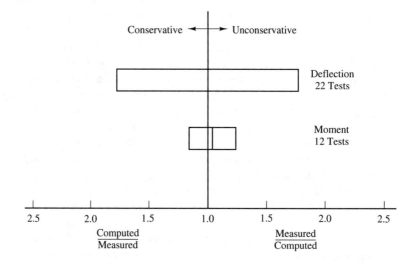

Figure 16.16 90 percent confidence interval for computed lateral deflection and bending moment predictions from *p-y* analysis (based on data from Reese and Wang, 1986). The line in the middle of each bar represents the average prediction, and the number to the right is the number of data points.

The advantages of these charts include the following:

- The analyses can be performed more quickly, and they do not require the use of a computer.
- The load corresponding to a given pile deflection can be determined directly, rather than by trial.
- The load corresponding to a given maximum moment in the pile can be determined directly, rather than by trial.

These charts are also a useful way to check computer output from more sophisticated analyses.

The charts presented here are a subset of the original method and apply only to deep foundations that satisfy the following criteria:

- The stiffness, EI, is constant over the length of the foundation.
- The shear strength of the soil, expressed either as s_u or ϕ', and the unit weight, γ, are constant with depth.
- The foundation is long enough to be considered fixed at the bottom. For relatively flexible foundations, such as timber piles, this corresponds to a length of at least 20 diameters. For relatively stiff foundations, such as those made of steel or concrete, the length must be at least 35 diameters.

We can idealize deep foundations that deviate slightly from these criteria, such as tapered piles, by averaging the EI, s_u, ϕ', or γ values from the ground surface to a depth of 8 pile diameters.

Characteristic Load and Moment

Evans and Duncan defined the *characteristic shear load*, V_c, and *characteristic moment load*, M_c, as follows:

$$V_c = \lambda\, B^2\, ER_I \left(\frac{\sigma_p}{ER_I} \right)^m (\epsilon_{50})^n \tag{16.8}$$

$$M_c = \lambda\, B^3\, ER_I \left(\frac{\sigma_p}{ER_I} \right)^m (\epsilon_{50})^n \tag{16.9}$$

$$R_I = \frac{I}{\pi\, B^4/64} \tag{16.10}$$

$$= 1.00 \text{ for solid circular cross sections}$$

$$= 1.70 \text{ for solid square cross sections}$$

For plastic clay and sand:

$$\lambda = 1.00 \tag{16.11}$$

For brittle clay[1]:

$$\lambda = (0.14)^n \tag{16.12}$$

For clay:

$$\sigma_p = 4.2\, s_u \tag{16.13}$$

For sand:

$$\sigma_p = 2\, C_{p\phi}\, \gamma\, B\, \tan^2 (45 + \phi'/2) \tag{16.14}$$

Where:

V_c = characteristic shear load

M_c = characteristic moment load

λ = a dimensionless parameter dependent on the soil's stress-strain behavior

B = diameter of foundation

E = modulus of elasticity of foundation

     = 29,000,000 lb/in² (200,000 MPa) for steel

     = 57,000 $\sqrt{f_c'}$ lb/in² (4700 $\sqrt{f_c'}$ MPa) for concrete

     = 1,600,000 lb/in² (11,000 MPa) for Southern pine or Douglas fir

f_c' = 28-day compressive strength of concrete (lb/in², MPa)

R_I = moment of inertia ratio (dimensionless)

σ_p = representative passive pressure of soil

ϵ_{50} = axial strain at which 50 percent of the soil strength is mobilized (see Table 16.2)

m, n = exponents from Table 16.3

I = moment of inertia of foundation

     = $\pi B^4/64$ for solid circular cross sections

     = $B^4/12$ for square cross sections

     Also see tabulated values in Chapter 12

s_u = undrained shear strength of soil from the ground surface to a depth of 8 pile diameters

ϕ' = effective friction angle of soil (deg) from ground surface to a depth of 8 pile diameters

$C_{p\phi}$ = passive pressure factor = $\phi'/10$

γ = unit weight of soil from ground surface to a depth of 8 pile diameters. If the groundwater table is in this zone, use a weighted average of γ and γ_b, where $\gamma_b = \gamma - \gamma_w$ (the buoyant unit weight in the zone below the groundwater table.

---

[1]A brittle clay is one with a residual strength that is much less than the peak strength.

TABLE 16.2   TYPICAL ε_{50} VALUES
(Reese and Wang, 1997)

Soil Type	ε_{50}
Soft clay	0.020
Medium clay	0.010
Stiff clay	0.005
Medium dense sand with little or no mica	0.002

The value of ε_{50} could be determined from triaxial compression tests. However, when such test data is not available, use the typical values in Table 16.2.

Using the Charts

The charts in Figures 16.17 to 16.26 express the relationships between the actual shear, moment and deflection, where:

V = applied shear at top of foundation

M = applied moment at top of foundation

M_{max} = maximum moment in foundation

y_t = lateral deflection at top of foundation

Some foundations are subjected to both shear and moment loads. As a first approximation, compute the lateral deflections and moments from each component separately and add them. Alternatively, use the nonlinear superposition procedure described in Evans and Duncan (1982).

Example 16.3

A 20-k shear load will be applied to a 12-inch square, 60-ft long restrained-head concrete pile. The soil is a sand with $\phi' = 36°$ and $\gamma = 120$ lb/ft^3. The groundwater table is at a depth of 40 ft. The pile is made of concrete with a 28-day compressive strength of 6000 lb/in^2. Compute the lateral deflection at the top of this pile and the maximum moment.

TABLE 16.3   VALUES OF EXPONENTS m
AND n (Evans and Duncan, 1982)

Soil Type	For V_c		For M_c	
	m	n	m	n
Clay	0.683	−0.22	0.46	−0.15
Sand	0.57	−0.22	0.40	−0.15

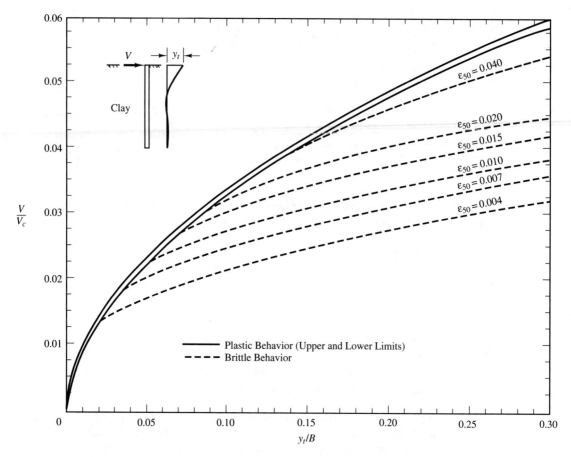

Figure 16.17   Shear load vs. lateral deflection curves for free-head condition in clay (Evans and Duncan, 1982).

Solution

Use units of pounds and inches. $\lambda = 1.00$; $R_I = 1.70$:

$$C_{p\phi} = \frac{\phi'}{10} = \frac{36°}{10} = 3.6$$

$$\sigma_p = 2\,C_{p\phi}\,\gamma\,B\,\tan^2(45 + \phi'/2)$$

$$= (2)(3.6)\left(\frac{120\ \text{lb/ft}^3}{(12\ \text{in/ft})^3}\right)(12\ \text{in})\tan^2(45 + 36°/2)$$

$$= 23.1\ \text{lb/in}^2$$

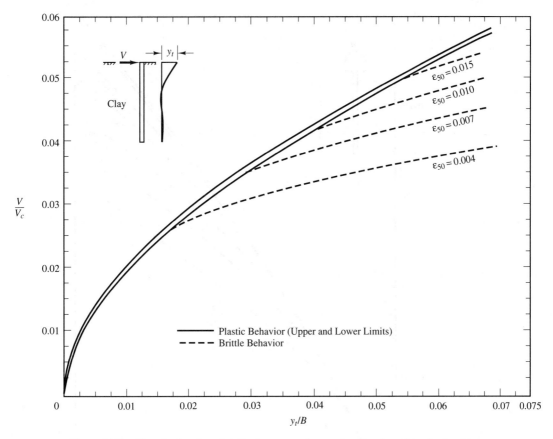

Figure 16.18   Shear load vs. lateral deflection curves for restrained-head condition in clay (Evans and Duncan, 1982).

$$E = 57,000 \sqrt{f_c'}$$
$$= 57,000 \sqrt{6000}$$
$$= 4,400,000 \text{ lb/in}^2$$

Using Equation 16.8:

$$V_c = \lambda B^2 E R_1 \left(\frac{\sigma_p}{ER_1} \right)^m (\epsilon_{50})^n$$

$$= (1.00)(12)^2(4,400,000)(1.70) \left(\frac{23.1}{(4,400,000)(1.70)} \right)^{0.57} (0.002)^{-0.22}$$

$$= 3,056,000 \text{ lb}$$

$$V/V_c = 20,000/3,056,000 = 0.0065$$

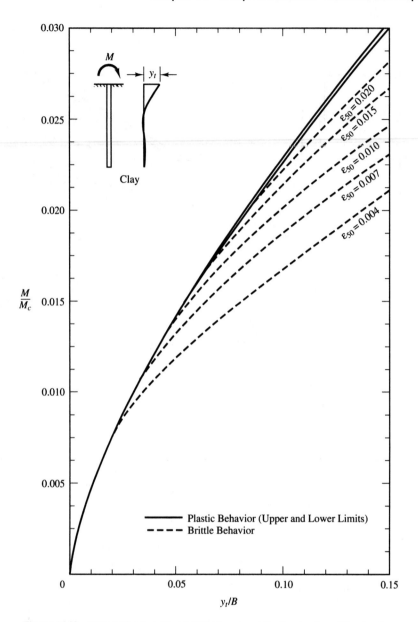

Figure 16.19　Moment load vs. lateral deflection curves for free-head condition in clays (Evan and Duncan, 1982).

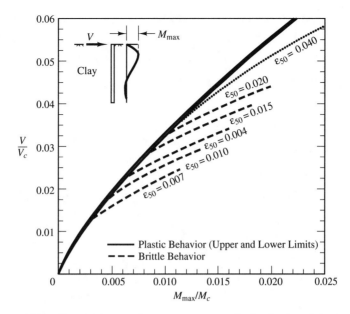

Figure 16.20   Shear load vs. maximum moment curves for free-head condition in clay (Evans and Duncan, 1982).

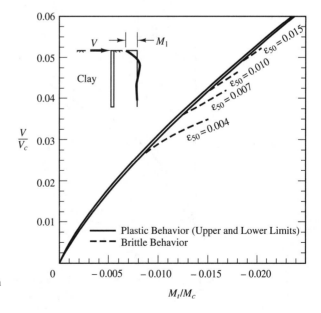

Figure 16.21   Shear load vs. maximum moment curves for restrained-head condition in clay (Evans and Duncan, 1982).

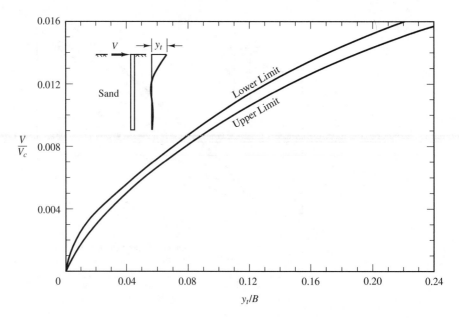

Figure 16.22   Shear load vs. lateral deflection curves for free-head condition in sand (Evans and Duncan, 1982).

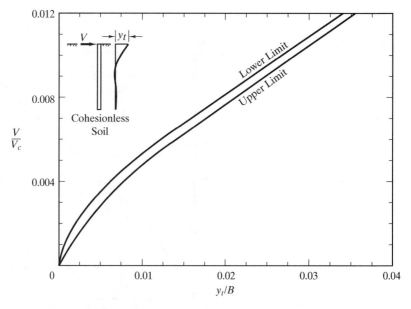

Figure 16.23   Shear load vs. lateral deflection curves for restrained-head condition in sand (Evans and Duncan, 1982).

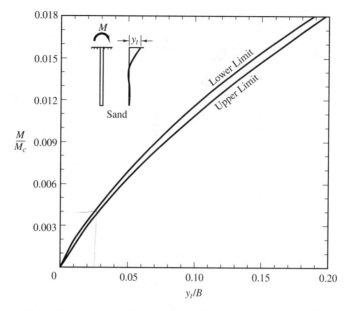

Figure 16.24 Moment load vs. lateral deflection curves for free-head condition in sand (Evans and Duncan, 1982).

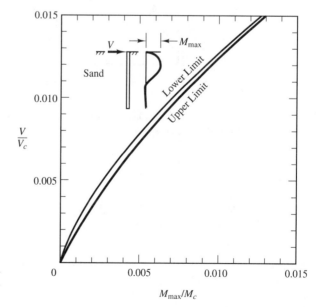

Figure 16.25 Shear load vs. maximum moment curves for free-head condition in sand (Evans and Duncan, 1982).

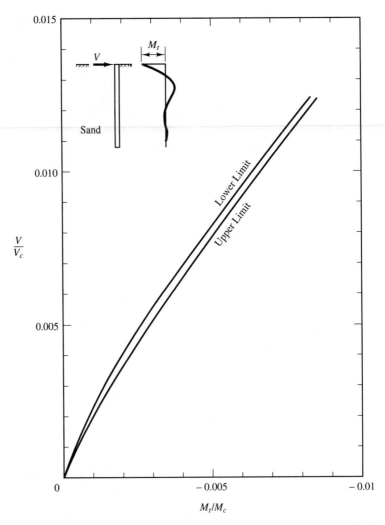

Figure 16.26   Shear load vs. maximum moment curves for restrained-head condition in sand (Evans and Duncan, 1982).

From Figure 16.23: $y_t/B = 0.0150$

$$y_t = (0.0150)(12) = \textbf{0.18 in} \quad \Leftarrow Answer$$

Using Equation 16.9:

$$M_c = \lambda\, B^3\, E\, R_I \left(\frac{\sigma_p}{E\, R_I}\right)^m (\epsilon_{50})^n$$

$$= (1.00)(12)^3(4{,}400{,}000)(1.70)\left(\frac{23.1}{(4{,}400{,}000)(1.70)}\right)^{0.40}(0.002)^{-0.15}$$

$$= 205{,}200{,}000 \text{ in-lb}$$

From Figure 16.26: $M_{max}/M_c = 0.0041$

$$M_{max} = (0.0041)(205{,}200{,}000) = \textbf{841{,}000 in-lb} \quad \Leftarrow Answer$$

Example 16.4

Solve Example 16.3 using p-y method software.

Solution

The plots in Figures 16.27 and 16.28 were obtained using LPILE PLUS 3.0, which is a commercial software package.

The computed deflection of 0.17 in and maximum moment of 790,000 in-lb are very close to the 0.18 in and 841,000 in-lb obtained in Example 16.3. However, the computer solution provides moment and deflection as functions of depth, while Evans and Duncan gives only the top deflection and maximum moment. The software also can produce shear plots.

QUESTIONS AND PRACTICE PROBLEMS

16.6 What are the advantages of using a p-y method computer program to evaluate laterally-loaded deep foundations instead of using Evans and Duncan's charts?

16.7 A 30 kN shear load will be applied to a 400-mm diameter restrained-head steel pipe pile with 10-mm thick walls. This pile will be embedded to a sufficient depth to be considered "long." The soil is a brittle stiff clay with $s_u = 250$ kPa. Using Evans and Duncan's method, compute the lateral deflection at the top of this pile and the maximum moment in the pile.

16.8 A 120 ft-k moment load will be applied to a 14-inch square prestressed "long" concrete pile. The top of this pile will have a free-head condition. The underlying soil is a silty sand with $\phi' = 29°$ and $\gamma = 119$ lb/ft^3. The pile has a 28-day compressive strength of 7000 lb/in^2. Using Evans and Duncan's method, compute the lateral deflection at the top of this pile. Note: for this pile, the maximum moment is equal to the applied moment.

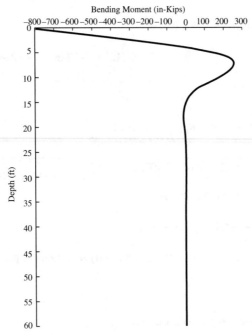

Figure 16.27 Moment diagram for Example 16.4.

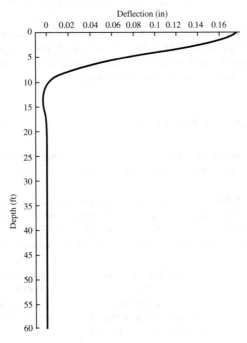

Figure 16.28 Deflection diagram for Example 16.4.

16.9 A restrained-head HP 13 × 60 pile is embedded 50 ft into a stiff plastic clay with $s_u = 3500$ lb/ft^2. A shear load is to be applied to the top of this pile such that bending will occur in the Y-Y axis. Using Evans and Duncan's method, determine the maximum allowable shear load such that the lateral deflection does not exceed 0.5 in.

16.6 GROUP EFFECTS

The analysis of lateral loads becomes more complex when we consider deep foundations installed in groups. There are two basic questions:

- How are the applied loads distributed among the individual foundations in the group?
- How does the ultimate capacity and load-deflection behavior of the group compare to that of a single isolated foundation?

Shear loads are distributed one way, while moment loads are distributed differently, so we must consider each type of loading separately.

Response to Applied Shear Loads

Shear loads are difficult to evaluate, largely because of the many factors that influence group behavior (O'Neill, 1983). These factors include the following:

- The number, size, spacing, orientation, and arrangement of the foundation members
- The soil type
- The type of connection at the top (fortunately, the foundations in a group are nearly always connected with a cap, so only the restrained-head condition need be considered)
- The interaction between the cap and the individual foundations
- The vertical contact force between the cap and the soil
- The lateral resistance developed between the side of the cap and the soil
- The differences between the *p-y* curves for the inner foundations and those for the leading row of foundations
- The method and sequence of installation
- The as-built inclination of the foundations (although the design drawings may show them perfectly plumb, in reality they will have some accidental batter)

Evaluations of shear loads on deep foundations must explicitly or implicitly consider *pile–soil–pile interaction (PSPI)* (O'Neill, 1983). PSPI occurs because the lateral movement of a pile (or other type of deep foundation) relieves some of the stress on the soil behind it. This soil, in turn, provides less resistance to lateral movement of the next pile. Thus, different piles may have different *p-y* curves. This has also been called the *shadow effect*.

PSPI is most pronounced when the foundations are closely spaced. Because of this effect, the leading row of foundations carries more than a proportionate share of the load. Some load test data confirm this behavior (Holloway et al., 1982; Brown et al., 1988; McVay et al., 1995). The net result is that the lateral deflection of a group will be greater than that of a single isolated foundation subjected to proportionate share of the group load. For conventionally spaced onshore groups, this ratio may be on the order of 2 to 3.

One way to evaluate the group response to applied shear loads is to modify the *p-y* curves to reflect PSPI effects. This may be done by applying *p-multipliers* as follows (Brown and Bollman, 1993; Hannigan et al., 1997):

1. Develop the *p-y* curves for a single isolated foundation using the techniques described earlier in this chapter.

2. Apply the following *p*-multiplier factors to the *p* values in the *p-y* curves for each row of foundations in the group:

    Leading row: 0.8

    Second row: 0.4

    Third and subsequent rows: 0.3

    These values are based on center-to-center spacing of 3 diameters and lateral deflections of 0.10 to 0.15 diameter, and should be conservative for larger spacings or smaller deflections. Since all of the p-multipliers are less than one, the *p-y* curves for foundations in a group are softer than those for an single isolated foundation.

3. Using the *p-y* method with the revised *p-y* curves developed in Step 2, determine the shear load vs. lateral deflection behavior for a single foundation in each row.

4. Consider a small lateral deflection (e.g., 5 mm) applied to the group and, using the functions developed in Step 3, determine the corresponding shear load for each pile in the group. Sum these loads to find the total shear load acting on the group. Repeat this analysis with other deflection values to develop a shear load-deflection plot for the group.

5. Using the plot developed in Step 4 and the design shear load, determine the lateral deflection. If this deflection is not acceptable, then revise the design and return to Step 1. Otherwise, proceed to Step 6.

6. Using a series of *p-y* analyses with the *p-y* curves developed in Step 2, develop plots of maximum moment vs. lateral deflection for a single foundation in each row.

7. Using the plot from Step 4, determine the lateral deflection of the group when subjected to the design shear load.

8. Using the deflection from Step 7 and the plot from Step 5, determine the maximum moment in each foundation.

Normally, the same structural design is used for all of the foundations in a group. Therefore, this design must be based on the shears and moments in the leading row, because they are the largest values.

The p-multiplier method is most appropriate for deep foundations that support major structures, such as large bridges. A simpler but less precise method may be used for smaller structures. This alternative method assigns two shares of the shear load to each foundation in the leading row and one share to all of the other foundations. The lateral load design for all foundations is then based on the load acting on those in the leading row. For example, a 200 kN shear load applied to a 4(4 group of piles has a total of 20 shares and would be evaluated using a design load of $(200/20)(2) = 20$ k per pile.

A complete analysis also might consider the lateral resistance on the side of a buried pile cap, which could be computed using the second term of Equation 8.8. The sliding friction component, $(P + W_f)$ in Equation 8.8 should be ignored because we cannot rely upon any normal force between the base of the cap and the underlying soil.

Response to Applied Moment Loads

When a moment load is applied to a group of deep foundations, the pile cap acts like a beam as shown in Figure 16.29. The deep foundations respond with upward or downward reactions, as shown. To compute these reactions, assume pinned connections exist be-

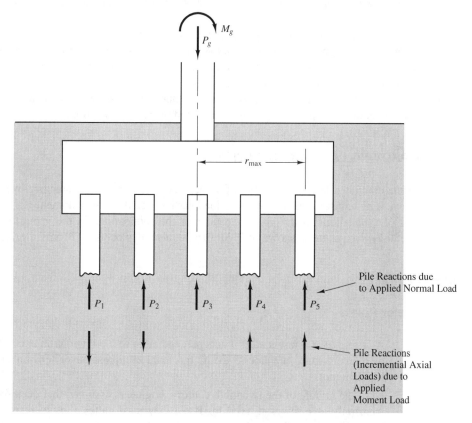

Figure 16.29   Additional axial loads in a group of deep foundations subjected to an applied moment load.

tween the foundations and the cap, and the reaction in each foundation is proportional to its distance from the centerline of the cap. Thus:

$$M_g = \Sigma P_i r_i \qquad\qquad (16.15)$$

$$P_i = P_{max} \left(\frac{r_i}{r_{max}} \right) \qquad\qquad (16.16)$$

Where:

M_g = applied moment on pile group

P_i = axial reaction force in pile i resulting from the applied moment

r_i = distance from centerline of cap to pile i

P_{max} = axial reaction force in one of the piles farthest from centerline

r_{max} = distance from centerline of cap to farthest pile

These reaction forces are then added or subtracted from the reactions due to the applied downward load.

Pile caps subjected to applied moment loads experience very little rotation, and thus do not induce significant moments in the foundations. In other words, if the structure imparts only downward and moments loads to the cap (i.e., no applied shear load), then the foundations are subjected to axial loads only.

If normal, shear, and moment loads are applied to the cap, then each may be evaluated separately and the resulting stresses added using superposition.

16.7 IMPROVING LATERAL CAPACITY

If the computed lateral resistance of a deep foundation is not satisfactory, we could improve it by adjusting the factors described earlier (i.e., diameter, moment of inertia, number of foundations in the group, spacing of the foundations in the group, etc.). Other methods, such as those shown in Figure 16.30, also may be used (Broms, 1972).

SUMMARY

Major Points

1. A *lateral load* is any load that acts perpendicular to the foundation axis. Thus, shear or moment loads are lateral loads, but axial compression or tension or torsional loads are not.

2. Until the middle of the twentieth century, engineers assumed that deep foundations were only able to resist axial loads, so they used batter piles to resist horizontal loads. More recently, we have reconsidered that assumption and now rely on both axial and lateral capacities.

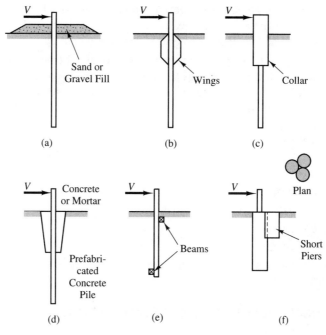

Figure 16.30   Methods of increasing the lateral resistance in a deep foundation (Adapted from Broms, 1972).

3. The utilization of lateral capacities in design often produces foundations that are more economical to build, more efficient in resisting seismic loads, and possibly more reliable.

4. The analysis of laterally-loaded deep foundations is a soil–structure interaction problem that requires simultaneous consideration of structural and geotechnical aspects.

5. When conducting lateral load analyses, engineers usually assume one of the following boundary conditions at the top of the foundation: The *free-head* condition, the *restrained-head* condition, or the *pure moment* condition.

6. Lateral load capacity may be evaluated using full-scale load tests, model load tests, rigid analyses, depth to fixity analyses, or nonrigid soil–structure interaction analyses. Load tests are sometimes used on larger projects, and rigid analyses may be appropriate for some lightly-loaded foundations, but most lateral load problems are analyzed using the *p-y* method, which is a type of nonrigid soil–structure interaction analysis.

7. The *p-y method* uses nonlinear *p-y curves* to describe the lateral soil pressure acting on the foundation and a finite difference analysis to compute the deflections, shears, and moments. Design *p-y curves* are based on empirical data obtained from full-scale lateral load tests.

8. Evans and Duncan have developed chart solutions based on *p-y* method analyses. These charts are not as flexible as a computer solution, but are useful when a computer is not available or to check computer output.

9. The behavior of pile groups subjected to lateral loads is only partially understood. With typical pile spacings (i.e., about 3 diameters on center), the lateral deflections will be greater than that of a single isolated pile subjected to a proportionate share of the applied load.

Vocabulary

Batter pile	Lateral load	Pure moment condition
Brittle soil response	Model load test	Restrained-head condition
Depth to fixity analysis	Nonrigid analysis	Rigid analysis
Ductile soil response	*p-y* curve	Shadow effect
Free-head condition	*p-y* method	Soil–structure interaction
Full-scale load test	Pile–soil–pile interaction	

COMPREHENSIVE QUESTIONS AND PRACTICE PROBLEMS

16.10 Give two examples of design situations when each of the following methods would be appropriate:

    a. Rigid analyses as presented in the International Building Code (Equations 16.5–16.7)

    b. *p-y* method analyses using appropriate computer software

    c. Full-scale lateral load tests

16.11 Explain how *p-y* curves could be back-calculated from a full-scale lateral load test, and how these curves could then be used to design other deep foundations.

16.12 A 75-ft long restrained-head H-pile must resist an applied shear load of 36 k while deflecting no more than 0.40 in. The soil is a silty sand with $c' = 0$, $\phi' = 34°$ and $\gamma = 127$ lb/ft^3, and the groundwater table is at a great depth. Using Evans and Duncan's method, determine the required H-pile section (see Table 12.1). Assume the bending occurs in the Y-Y axis.

16.13 The H-pile described in Problem 16.12 also carries an axial load of 50 k. Using Evans and Duncan's method and the techniques described in Chapter 12, determine if the stresses in the pile are acceptable. The pile is made of A36 steel, and the driving conditions are only moderately favorable.

16.14 A downward load of 100 k and a shear load of 20 k will be applied to the top of a 50-ft long restrained-head steel pipe pile. This pile will be embedded in a well-graded sand with occasional cobbles. This soil has $c' = 0$, $\phi' = 34°$ and $\gamma = 127$ lb/ft^3. The groundwater table is at a depth of 30 ft. The pile will be made of A36 steel and has a maximum allowable lateral deflection of 0.75 in. Using Evans and Duncan's method, select a standard pile section from Table 12.2 that satisfies this deflection criterion and satisfies the allowable stress criteria described in Chapter 12.

17

Deep Foundations—Design

Let the foundations of those works be dug from a solid site and to a solid base if it can be found. But if a solid foundation is not found, and the site is loose earth right down, or marshy, then it is to be excavated and cleared and remade with piles of alder or of olive or charred oak, and the piles are to be driven close together by machinery, and the intervals between are to be filled with charcoal . . . even the heaviest foundations may be laid on such a base.

Marcus Vitruvius, Roman Architect and Engineer
1st century B.C.
(as translated by Morgan, 1914)

Chapters 11 to 16 covered various structural, geotechnical, and construction aspects of deep foundation design. This chapter shows how to synthesize this information and develop a comprehensive design.

The process of designing and building deep foundations differs from those used for other structural elements, such as girders or columns, because the final design depends on the soil conditions encountered during construction. As a result, the design shown on the project drawings is tentative, and often changes during construction. For example, the drawings may indicate a certain pile is to be embedded 15 m into the ground, but the soil conditions may be harder or softer than anticipated, and thus require a different "as-built" pile length. Such changes are common, even routine, in deep foundation construction.

In addition, deep foundation design requires careful coordination between geotechnical, structural, and construction engineers. Each of these parties has expertise to contribute, but none can work in isolation from the others.

17.1   DESIGN SERVICE LOADS AND ALLOWABLE DEFORMATIONS

The design process begins with defining the service loads. These are the loads the structure imparts onto the foundation, and are determined using the techniques described in Chapter 2. They can include downward, uplift, moment, and/or shear loads, and should be expressed both as the unfactored loads (i.e., the largest of Equations 2.1, 2.2, 2.3a, and 2.4a) and as the ACI factored loads (i.e., the largest of Equations 2.7–2.17).

This also is the time to determine the allowable total and differential settlements, which may be evaluated using the techniques described in Section 2.3 or by some other suitable analysis. If lateral loads are present, the allowable lateral deflection also must be defined. Buildings, bridges, and other similar structures typically can tolerate no more than 5 to 20 mm (0.25–0.75 in) of lateral deflection in the foundation.

These tasks are clearly within the domain of the structural engineer, who will use the results to design the individual foundations. The structural engineer also must transmit the allowable settlements and lateral deflections and the range of design loads to the geotechnical engineer who will use this information to guide the subsurface characterization program and develop the geotechnical design parameters.

17.2   SUBSURFACE CHARACTERIZATION

Next, the geotechnical engineer uses the techniques described in Chapter 4 to characterize the subsurface conditions at the site. Because deep foundations generally carry heavier loads and often support more important structures, the subsurface investigation program is usually more intensive than those performed for shallow foundations. At least some of the exploratory borings must extend well below the toe elevation of the proposed foundations, so the geotechnical engineer must have at least a preliminary notion of the foundation depths even before drilling the borings.

In addition to providing the necessary soil and rock parameters for computing axial and lateral load capacities and driveability, the subsurface investigation program also provides much of the information needed to select the optimal type of deep foundation.

If possible, the subsurface investigation program also should include research on deep foundations that have already been built in the vicinity, and the nature of any construction difficulties that may have occurred. For example, reports of pile foundations reaching refusal at shallow depths may be cause for concern, especially if certain minimum embedment depths are required to achieve the required uplift capacity. This information might come from the geotechnical engineer's personal experience, or from consultation with other engineers or contractors who have worked in the area.

17.3   FOUNDATION TYPE

Next, the geotechnical engineer must decide whether a deep foundation is truly necessary. This is an important decision with far-reaching consequences, and thus should not be taken lightly. Sometimes even large structures may be safely supported on another type of foundation, such as a mat. Alternatively, it may be feasible to improve the soil conditions using techniques described in Section 18.4 so they are able to support the structure using a shallow foundation system.

If deep foundations appear to be most appropriate, the engineer must then determine which type to use (piles, drilled shafts, etc.). Chapter 11 discussed the various types of deep foundations, and described their advantages and disadvantages. Sometimes one type is clearly the best choice for a particular application, but more often there are many potential designs, all of which may be feasible. The ultimate selection then depends on many factors, including:

- *Design Loads*—Structures with heavy loads, such as major bridges or large buildings, require different types of foundations than those with lighter loads. In addition, the type of load (compression, uplift, shear, moment) also influences the selection.

- *Subsurface Conditions*—Sites underlain by good soil and rock require different foundations than those with poor subsurface conditions.

- *Constructibility*—The foundation must be buildable using reasonable construction methods and equipment. For example, piles or caissons would probably be the best choice for foundations intended to support the center pier of a bridge constructed over a major river. Either type could be built from a barge. However, drilled shaft foundations would be virtually impossible to build at such a site.

- *Reliability*—The as-built foundations must be able to reliably support the applied loads. Certain foundation types might not be good choices at certain sites, because the subsurface conditions might not be conducive to construction of reliable foundations.

- *Cost*—The cost of construction must be kept to a minimum, commensurate with the other requirements.

- *Availability of materials, equipment, and expertise*—Both piles and drilled shafts are viable options in all but the most remote locations because of the large number of contractors who are able to build these foundations. However, some of the other types may not be readily available in certain areas. For example, auger-cast piles are very commonly used in the southeastern United States, and can be specified without concern about the availability of qualified contractors. However, this type is rarely used in California, so suitable equipment and expertise would probably need to be brought in from far away, and thus be more expensive.

- *Local experience and precedent*—Most projects are near other structures with similar foundation requirements. Methods that have been successful on these past projects are often favored over untried methods.

Although this task is primarily within the geotechnical engineer's domain, it should be performed in consultation with the structural engineer. In some cases, it is also appropriate to consult with a contractor.

17.4   LATERAL LOAD CAPACITY

Deep foundations are very slender structural members, and thus do not transmit lateral loads as efficiently as axial loads. When lateral loads are present, they often dictate the required foundation diameter, so it is usually most efficient to begin the design process by addressing the lateral load capacity.

Begin by considering the column with the greatest design lateral load and estimating the size and number of foundations required to support this load. For example, the design engineer might estimate a group of nine 16-inch diameter steel pipe piles will be required. This initial design is based primarily on engineering judgement, and serves as a starting point for the design process. If a group of foundations is being considered, the design loads should be distributed to the individual foundations as discussed in Section 16.6, then the analysis process should continue by examining a single foundation in the leading row.

The lateral load capacity analysis should usually be performed using the *p-y* method, as described in Chapter 16, and should assume the foundations are embedded at a sufficient depth to be considered fixed at the bottom. For relatively flexible foundations, such as timber piles, this corresponds to a depth of at least 20 diameters. For relatively stiff foundations, such as those made of steel or concrete, the depth should be at least 35 diameters. The actual design depths will be determined later.

In the case of steel pipe piles, the objective of this analysis is to determine the required wall thickness to satisfy the strength and serviceability requirements, as discussed in Chapter 16. Based on the results of this analysis, the initial design may need to be modified. For example, if the required wall thickness is excessive, the engineer may choose to use larger diameter pipes, or more piles per group. Other foundation types have different parameters that can be adjusted.

In major structures with relatively few foundations, such as large bridges, the engineer repeats this design process for the remaining foundations, thus developing a custom design for each. However, for structures with more foundations, such as most buildings, it is usually more convenient to develop lateral load design charts, such as those shown in Figure 17.1, which then can be used to design the remaining foundations. These charts facilitate the evaluation of various design alternatives, and thus simplify the design process. Typically the geotechnical engineer develops these charts, then the structural engineer uses them to design specific foundations.

Example 17.1

The design charts in Figure 17.1 have been developed for steel pipe piles at a certain site. One of the proposed columns at this site will impart a downward load of 1300 k and a shear load of 110 k. The allowable lateral deflection is 0.5 in. The upper soils have an allowable equivalent passive fluid density of 300 lb/ft^3. Determine the required number of piles, along with their diameter and wall thickness.

Solution

Try a group of nine 16-in diameter piles arranged in a 3×3 group and connected with a concrete pile cap. If these piles are arranged 3 diameters on-center, the cap will be about 12 ft

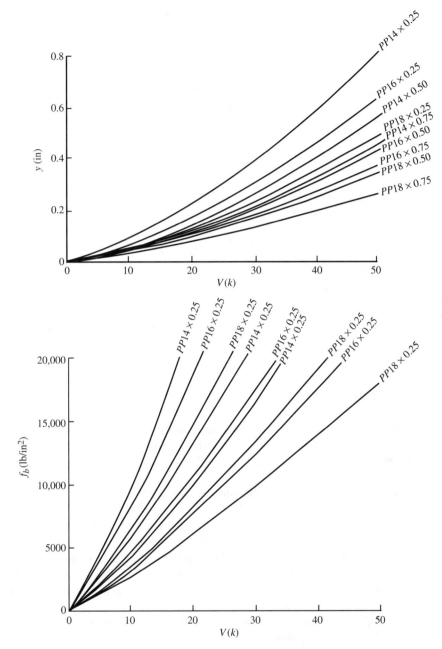

Figure 17.1 Typical format for presenting allowable lateral load capacities for a specific site. These charts were developed for restrained-head steel pipe piles subjected to shear loads. Other kinds of charts may be more appropriate for different conditions. Note: This chart was developed using *p-y* analysis software and is based on specific soil conditions. It is not a generic chart for use on all sites.

wide and 3 ft deep. Therefore, the lateral resistance acting on the side of the pile cap would be:

$$V_{fa} = 0.5 \, \lambda_a BD^2 = (0.5)(300)(12)(3^2) = 16{,}200 \text{ lb}$$

Therefore, the piles must resist a $110 - 16 = 94$ k shear load. Using the alternative simplified method of distributing shear loads, as discussed in Section 16.6, there are 12 shares and the shear load acting on one of the piles in the leading row is $(94/12)(2) = 16$ k. The downward load is evenly distributed, and is $1300/9 = 144$ k/pile.

Try PP 16×0.75 piles $(A - 35.93 \text{ in}^2)$

$$f_a = \frac{P}{A} = \frac{144{,}000 \text{ lb}}{35.93 \text{ in}^2} = 4008 \text{ lb/in}^2$$

If we use A36 steel with $F_a = F_b = 0.35 \, F_y = 12{,}600 \text{ lb/in}^2$, then we can use Equation 12.5 to compute the maximum allowable f_b:

$$\frac{f_a}{F_a} + \frac{f_b}{F_b} \le 1 \rightarrow \frac{4008}{12{,}600} + \frac{f_b}{12{,}600} \le 1 \rightarrow f_b \le 8592 \text{ lb/in}^2$$

According to Figure 17.1, a 16 k shear load on a PP 16×0.75 pile at this site will produce $f_b = 6000 \text{ lb/in}^2$ and a lateral deflection of only 0.08 in. Further trials will demonstrate that a PP 16×0.50 pile would be overstressed.

Therefore, **use nine PP 16×0.75 piles**   ⟸ *Answer*

This design could be further refined using the p-multiplier method described in Section 16.6.

17.5   AXIAL LOAD CAPACITY

The next step in the design process is to evaluate the geotechnical axial load capacity. The objective is to determine the required number of deep foundations and their depth of embedment.

Pre-Construction Full-Scale Static Load Tests

The most reliable method of designing deep foundations is to conduct one or more full-scale static load tests at the site of the proposed structure, as discussed in Chapter 13. Ideally, these tests should be conducted during the design stage so the results can be used to develop the design shown on the contract drawings. This method should produce a design that has the lowest construction cost, and the least number of changes during construction.

Unfortunately, pre-construction load tests are very expensive because they require an extra mobilization and demobilization of the construction equipment. Typically such tests also require a special construction contract because they are performed months before the production foundations are installed. As a result, pre-construction load tests are probably cost-effective only on large projects with expensive foundation systems, as discussed in Section 13.7. For smaller projects, the cost of conducting pre-construction load tests often exceeds the benefit gained from the test. In other words, it may be less expensive to use other design methods on smaller projects, even if they result in more conservative designs.

For high-capacity foundations, statnamic tests or Osterberg tests are often more economical than conventional static load tests. Another alternative is to drive pre-construction prototype indicator piles, monitor them using a pile driving analyzer, and perform a CAPWAP analysis. All three of these methods are often cost-effective, especially when multiple foundations need to be built, and can save time.

Analytic Methods

Regardless of whether or not a pre-construction load test is performed, the geotechnical engineer virtually always develops foundation designs using the analytic methods described in Chapter 14. These methods can easily consider various diameters and lengths, and thus can be used to design a wide range of foundations. If load test data is available, it can be used to calibrate the analytic methods, thus producing a design with a lower factor of safety. The analysis is almost identical if load test data is not available, except the factor of safety is higher to reflect the greater uncertainty.

The results of these analyses are typically presented as a design chart, such as the one in Figure 17.2. Such charts should consider a range of foundation diameters consistent with the results of the lateral load capacity analysis, and a range of lengths consistent with the design loads. Typically the geotechnical engineer develops this chart, then the structural engineer uses it to design the individual foundations.

If settlement is a concern, it also should be addressed at this stage. Settlement analyses may be performed using the techniques described in Section 14.7. If necessary, the design chart may need to be modified to satisfy settlement requirements.

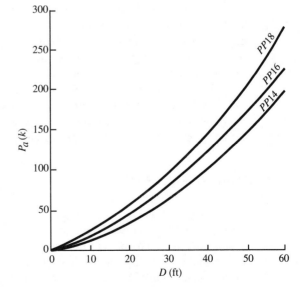

Figure 17.2 Typical design chart for allowable downward axial load capacities. This chart was developed using a particular soil profile, and is applicable only to that profile. It is not a generic chart for use on other sites. The allowable uplift capacity may be presented as a separate design chart, or as a stated percentage of the downward capacity.

Example 17.2

The axial load capacity design chart in Figure 17.2 has been developed for the site described in Example 17.1. Using this chart, determine the design embedment depth.

Solution

According to this chart, 40-ft deep piles are needed to resist the design load of 144 k.

Therefore, **use nine PP 16 × 0.75 piles driven to a depth of 40 ft**   ← *Answer*

17.6   DRIVEABILITY

If pile foundations are being designed, the geotechnical engineer also should perform a *driveability analysis* using the wave equation method, as discussed in Chapter 15. The purpose of this analysis is to determine if the proposed piles can be driven using typical construction equipment and techniques without overstressing the pile and without requiring an excessive blow count. Normally the contractor has not yet been selected, so the engineer must estimate the hammer size and other parameters for the analysis. This is one of the many reasons why design engineers must be familiar with construction methods! Even though the actual driving equipment to be used is not yet known, a driveability analysis at this stage should produce useful results, and it is an important check to help avoid construction problems. If this analysis indicates potential driveability problems, the design must be revised accordingly. For example, in the case of the steel pipe piles, it may be necessary to use a greater wall thickness than required to resist the service loads.

Driveability problems may indicate an overdesigned foundation. For example, foundations designed using a factor of safety that is too large will have a larger diameter and/or greater depth of embedment than necessary to support the structural loads, and such a foundation may be correspondingly difficult to build. In this case, increasing the structural capacity (such as increasing the wall thickness) may make the piles driveable, but often at a premium price. Figure 17.3 shows the relationship between capacity problems, driveability problems, and cost problems, and the optimum design window that avoids all three.

If the engineer finds it difficult to achieve adequate capacity and driveability without creating cost problems, it may be necessary to conduct full-scale load tests, CAPWAP analyses, or other means of justifying a lower factor of safety. Alternatively, the best solution may be to use some other type of foundation, such as drilled shafts.

17.7   STRUCTURAL DESIGN

At this stage, the geotechnical engineer's design recommendations have been prepared in a formal report and presented to the structural engineer, who then uses them to design the foundations for each column. In other words, this stage of the design process combines the design loads with graphs such as those in Figures 17.1 and 17.2 to determine the size, number, and depth of foundations needed to support each part of the structure.

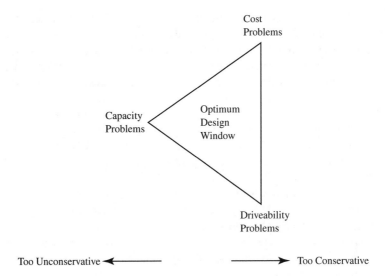

Figure 17.3 Improperly designed pile foundations can have capacity problems (they cannot properly support the design loads), driveability problems (they are too difficult to install) or cost problems (they are too expensive). An overly conservative design normally has cost and/or driveability problems, whereas an unconservative design has capacity problems. The optimum design falls within the triangular window.

The structural engineer also must be sure the foundations have sufficient structural integrity, and does so using the principles described in Chapter 12. Pile caps and grade beams also are designed at this stage.

The results of this effort are presented in the form of design drawings and specifications. The drawings indicate the number, size, spacing, and anticipated depth of foundations for each column, and the specifications provide other design information, such as material strengths, special procedures, and so forth. The completed design drawings should be sent back to the geotechnical engineer for review, to be sure the geotechnical design recommendations have been properly implemented. On difficult projects, it may be helpful to also have the drawings reviewed by a construction engineer.

Once they have been finalized, these drawings and specifications become part of the complete design package, which is presented to the contractor. It is important that these documents be correct and complete, because they become a key element in the contract between the owner and the contractor.

17.8   SEISMIC DESIGN

The design of deep foundations in seismically active regions must consider the earthquake loads imparted by the structure, the ability of the soil to absorb these loads, and any seismically-induced degradation of the soil. In some cases, these considerations control the foundation design, and thus can be very important.

The vast majority of earthquake-related deep foundation failures have been the result of *seismic liquefaction* of the soils around or beneath the foundation (Martin and Lam, 1995). If liquefaction occurs beside the foundations, the resulting loss of support can cause excessive lateral movements and flexural failure. If liquefaction occurs below the foundations, excessive settlement or rotation of the pile cap may occur.

If the ground surface is not level, such as in the vicinity of bridge abutments, the liquefied soil often moves laterally creating a phenomenon known as a *lateral spread*. This moving soil typically pushes deep foundations out of position, which can produce catastrophic failures (Bartlett and Youd, 1992). Figure 17.4 shows one such failure. Lateral spreads have occurred even on gently sloping ground, and therefore must be carefully evaluated.

If the liquefiable soils are confined to shallow depths and do not pose a lateral spread hazard, it might be possible to ignore their load-bearing capacity and design deep foundations accordingly. However, extensive zones of liquefied soil are very difficult to accommodate in the design, especially if they extend to significant depths. In those cases, it may be necessary to employ soil improvement methods to remediate the liquefaction hazard.

Very few seismically-induced deep foundation failures have occurred in non-liquefiable soils, probably because of the high factors of safety typically used in deep foundation design. Nevertheless, foundation designs in these soils must still provide sufficient strength, stiffness, and ductility to resist seismic loads, and these considerations can

Figure 17.4   The Showa Bridge in Niigata, Japan, collapsed due to the formation of lateral spreads in the underling soils during the 1964 earthquake. These lateral spreads pushed the pile foundations out of position, thus removing support from the simply-supported deck (Earthquake Engineering Research Center Library, University of California, Berkeley, Steinbrugge Collection).

have a significant impact on the final design. Seismic considerations are especially important when existing structures are being upgraded to meet new seismic design criteria.

Much of the emphasis in seismic design is on lateral loads, which are typically evaluated using *p-y* analyses as described in Chapter 16. These analyses usually treat the seismic load as an equivalent static load, but adjust the soil properties to account for dynamic effects. Two important considerations must be included in these analyses:

1. The cyclic lateral loading produced by earthquakes can cause a softening of the *p-y* curves, even if no liquefaction occurs, which results in additional lateral deflection.
2. Sometimes cyclic loading and the resulting lateral movements produce a tapered gap in the soil surrounding the upper portion of the foundation, as shown in Figure 17.5. This phenomenon is especially important in clays, and can be modeled in a *p-y* analysis.

The axial load capacity, both downward and upward, also can be an important part of seismic design, and controls the moment capacity of pile groups. In addition, excessive settlement or heave can cause significant rotations of pile caps, which can damage the superstructure. In some cases, especially with bridges, these considerations can be more important than the lateral load capacity (Martin and Lam, 1995). The axial capacity and movement are typically evaluated using the techniques for static loads, as discussed in the

Figure 17.5   This pile, which supported a bridge near Watsonville, California, moved laterally during the 1989 Loma Prieta Earthquake, creating a gap between the pile and the soil. The size of this gap is an indication of the excessive lateral deflections that occurred during the earthquake (Loma Prieta Collection, Earthquake Engineering Research Center, University of California, Berkeley).

preceding chapters, with appropriate consideration for dynamic effects, liquefied zones, and soil gaps.

The structural design of concrete piles and drilled shafts must consider special code requirements as outlined in ACI Section 21.8. These requirements are intended to provide continuous load paths for seismic forces and appropriate detailing in the reinforcing. It is especially important to provide a ductile connection between these foundations and their cap.

17.9   SPECIAL DESIGN CONSIDERATIONS

Sometimes deep foundations must be designed to accommodate special considerations. Two of the most common are scour and downdrag.

Scour

Scour is the loss of soil because of erosion in river bottoms or in waterfront areas. This is an important consideration for the design of foundations for bridges, piers, docks, and other structures because the soils around and beneath the foundations could be washed away. Section 8.8 discusses the consequences of scour in more detail.

A hydraulic evaluation forms the basis for estimating the scour zone, which can extend several meters below the river bottom. All foundations must extend below any potential scour zone, so deep foundations (usually piles) are usually necessary. These foundations must be designed to safely support the design loads even if the upper soils are lost to scour.

Piles are normally driven at such sites before the scour occurs, so driveability is governed by the unscoured soil profile, while design capacity is governed by the post-scour soil profile. This difference narrows the optimum design window of Figure 7.3, making it difficult to simultaneously meet both driveability and capacity requirements.

Scour also reduces the lateral load capacity of piles because the upper soils are no longer present to provide lateral support. In other words, the distance from the structure to the ground surface becomes greater, so the moments in the piles increases. Therefore, it becomes necessary to use larger pile sections or battered piles to support the lateral loads.

Downdrag

Engineers often need to build structures on fills underlain by soft soils. The weight of these fills causes the underlying soils to consolidate, thus producing settlement at the ground surface (see Chapter 11 in Coduto, 1999). If the structure is supported on deep foundations that penetrate through the fill and soft soil and into underlying stiff soil, these foundations will remain roughly in place while the upper soils settle around it. This downward soil movement creates a corresponding side-friction force known as *downdrag*. Unfortunately, this force is now a load rather than a resistance. In addition, it can be large, and thus becomes a significant design issue. Section 18.1 discusses downdrag loads in more detail.

17.10   VERIFICATION AND REDESIGN DURING CONSTRUCTION

Deep foundation design drawings are normally bound with the remaining drawings, and may appear to be no different than floor plans, structural connection details, or the many other structural and architectural drawings. It may seem that completing these drawings and the related specifications signals the end of the design process, and that the contractor must simply build the deep foundation as shown on the drawings. However, this is not the case. The design of deep foundations continues throughout construction, not because there was any deficiency in the original design, but because all of the information needed for design was not yet available.

The as-built foundations may be significantly different from those shown on the design drawings. These differences are primarily due to the differences between the sub-surface conditions anticipated in the design and those actually encountered during construction. For example, the foundation may be designed to develop toe bearing in a certain stratum, but the depth to that stratum inevitably varies across the site. Thus, the as-built depth of embedment cannot be determined until the foundation is actually being built. In the case of piles, this might be based on blow counts, while with drilled shafts it is based on the soil types extracted by the auger. This is why the geotechnical engineer (and to a lesser degree the structural engineer) continue to play an active role during construction, and why the as-built foundations are often substantially different from those shown on the design drawings. In fact, deep foundations built exactly as designed are the exception rather than the rule.

Wave Equation Analysis

If pile foundations are being used, a second wave equation analysis should be performed before construction begins. This analysis should use the characteristics of the contractor's driving equipment, and is intended to serve the following purposes:

- To help the contractor select the optimal driving equipment
- To identify and resolve potential driveability problems
- To develop a bearing graph to be used in construction control

Construction Observation and Testing

During construction, the geotechnical engineer has a representative on site to observe the construction process. This field engineer has two primary responsibilities: to confirm that the as-built foundations satisfy the intent of the design drawings and specifications, and to note the need for changes in the design as a result of changed subsurface conditions. Small design changes are typically made immediately, while more substantive changes require consultation with the geotechnical engineer and/or the structural engineer.

If no full-scale load tests were performed prior to construction, the engineer may chose to conduct one or more tests immediately before construction of the production foundations. Such tests are significantly less expensive because they do not require an extra mobilization and demobilization of the pile driving equipment and crew, yet they still provide the data needed to justify a lower factor of safety. However, if the test results are not as anticipated, it is more difficult to change the design at this stage because the production piles have already been ordered. Such changes are especially difficult with concrete piles because they are much more difficult to splice (if too short) or cutoff (if too long).

If piles are being used, another alternative is to drive special *indicator piles* at the beginning of construction. These are full-scale piles installed using the production driving equipment and monitored using high strain dynamic testing methods as discussed in Section 15.3. The Case Method is used most frequently because it is much less expensive than a load test, yet provides good information on pile capacity. The data obtained from these indicator piles is then used to reevaluate the design and to provide driving criteria for the production piles.

The field engineer uses the driving criteria (which is primarily in the form of a bearing graph) with a knowledge of the design parameters to determine the final penetration depth for each production pile. Sometimes the piles must be driven until a certain blow count is achieved, while other times it must be driven a certain distance into a specific stratum. In clays, it may be necessary to occasionally obtain *retap* blow counts by redriving piles after the excess pore water pressures have dissipated and *setup* has occurred. Sometimes additional Case Method analyses are performed on some of the piles as another check on capacity.

For drilled shaft foundations, the field engineer observes the soils being excavated and compares them to the anticipated soil conditions. If necessary, the depth of the shaft may be changed to accommodate the actual subsurface conditions. The field engineer also checks for a variety of construction procedures, especially those that might impact caving or other similar problems. With end bearing shafts, the contractor must confirm that the bottom is clean and free of debris before placing the concrete. In dry holes, it may be possible to shine a light down the hole and observe the bottom from the ground surface. A variety of remote techniques have also been used, especially down-hole video cameras. Holden (1988) describes a down-hole camera equipped with water and air jets that is capable of observing the bottom of holes drilled under bentonite slurry.

17.11 INTEGRITY TESTING

An increasingly common part of deep foundation construction, especially with drilled shafts, is post-construction integrity testing. The purpose of these integrity tests is to confirm that the as-built foundation is structurally sound, and does not contain any significant defects.

The best defense against construction defects is to use only experienced and conscientious contractors, and to use sufficiently thorough and competent construction inspection methods implemented by qualified personnel. Unfortunately, even with these precautions, construction defects can occur, especially when working in difficult soil conditions. Such defects are especially important in drilled shaft foundations that use only one shaft per column, because they lack the redundancy of a group of foundations.

Integrity tests include a variety of nondestructive methods for evaluating the quality of a completed foundation (Litke, 1986; Hertlein, 1992). These methods include the following:

- *Sonic logging*: Install two vertical tubes (usually PVC pipe) in the drilled shaft before placing the concrete. Once the shaft is completed, lower a compression wave source down one tube and a receiver down the other while taking readings of the wave propagation through the shaft. Any voids will show up as anomalies in the wave propagation pattern.

- *Nuclear logging*: This method, also known as gamma-gamma logging, is similar to sonic logging except that it uses radiation instead of compression waves. Christopher, Baker, and Wellington (1989) describe a case history.

- *Vibration analysis*: This consists of placing a mechanical vibrator at the top of the completed foundation and varying the frequency to determine the natural frequencies and impedance of the shaft.

- *Stress-wave propagation*: This may be the most common method of integrity testing because it is reasonably reliable (except in marginal cases), quick, and relatively inexpensive. It consists of tapping the top of the completed foundation with a mallet and using accelerometers to monitor the stress waves it creates at the top of the foundation. The required equipment is shown in Figure 17.6. The accelerometers

Figure 17.6 Equipment used in the stress propagation method of integrity testing (Courtesy of Pile Dynamics, Inc.).

can be located at the top of the shaft or embedded within it (Hearne, Stokoe, and Reese, 1981 and Olson and Wright, 1989).

This method is able to detect large defects, say, those that are on the order of twenty percent of the cross-sectional area. However, it will probably not reveal smaller defects because they do not generate a sufficiently strong reflection.

- *Tomography*: This technique is being used for medical purposes to produce three-dimensional images of the human body. Some researchers have suggested this method might be applied to drilled shafts, thus producing a three-dimensional image of the entire shaft and showing even the smallest defects. However, major technical hurdles must be overcome before this method becomes usable.

Other methods of integrity testing are also possible, such as full scale load tests (too expensive to use in quantity) or coring the foundation top-to-bottom (expensive and easy to miss defects), or dynamic load tests.

SUMMARY

Major Points

1. The process of designing deep foundations differs from that of most structural elements in that it does not end with completion of the design drawings. The design often changes during construction to reflect the subsurface conditions actually encountered in the field.

2. The proper design of deep foundations requires careful coordination between geotechnical, structural, and construction engineers.

3. The geotechnical engineer has the primary responsibility for evaluating the axial and lateral load capacity of the foundations. The results of the geotechnical analyses are usually presented as design charts

4. The structural engineer is usually responsible for designing the individual foundations, and does so using the design charts provided by the geotechnical engineer. In addition, the structural engineer develops a design that has sufficient structural integrity.

5. The design of deep foundations subjected to seismic loads must consider both the transfer of these loads into the ground and the earthquake effects on the surrounding soil. Most seismically-induced failures have been caused by soil liquefaction.

6. Some deep foundation designs must consider scour, downdrag, or other special issues.

7. A field engineer should be present when deep foundations are constructed. He or she has authority to revise the design as needed to achieve the design objectives.

8. In some cases, completed foundations are checked for stuctural integrity using a variety of non-destructive testing methods.

Vocabulary

Constructibility	Driveability	Scour
Design chart	Indicator pile	Seismic liquefaction
Downdrag	Integrity testing	

COMPREHENSIVE QUESTIONS AND PRACTICE PROBLEMS

17.1 You are the foundation engineer for a new baseball stadium that is now under construction. The design drawings indicate the stadium will be supported on a series of steel H-piles, and provides the "estimated toe elevation" for each pile. However, during construction it became necessary to drive many of the piles to greater depths, which resulted in a significant increase in the construction cost. The owner (who knows a great deal about baseball, but little about foundations) is very unhappy about this additional cost. His construction manager has advised the owner to pay the invoice, but so far the owner has refused. Therefore, the construction manager has asked you to write a 300- to 500-word letter to the owner explaining the situation and encouraging him to pay for the extra pile expense.

Keep in mind that the owner is a layperson who should be trusting the judgement of his construction manager but is too strong-willed to stay in the background. At the moment he does not intend to approve the extra payment to the foundation contractor, which could eventually lead to an expensive and time-consuming lawsuit. A well-written letter by you could diffuse the situation and avoid litigation.

17.2 What are indicator piles, and how do they benefit the foundation design process?

17.3 Many engineers believe the only disadvantage to overdesigning pile foundations is the additional cost of the pile materials. However, there is another important consequence of overdesign which can significantly affect construction cost and may even make the foundation impossible to build. What is this consequence and how can it be avoided?

17.4 Under what circumstances would you most likely require integrity testing of newly-constructed drilled shaft foundations?

17.5 Redo Examples 17.1 and 17.2 using nine 18-inch diameter piles, then using the following assumptions determine which design would be less expensive to build:

    1. The pile caps represent 20 percent of the total construction cost, and is the same for both designs. The remaining 80 percent is the pile construction cost, which is different for each design.

    2. Half of the pile construction cost is materials and half is installation.

    3. The materials cost is proportional to the total weight of steel.

    4. Mobilization and demobilization represent 20 percent of the pile installation cost, and will be independent of the amount of piling to be driven.

    5. The remaining portion of the pile installation cost is proportional to the total feet of piles to be driven.

17.6 A set of cell-phone antennas are to be placed on a 12-inch diameter "monopole" steel tower, which will be supported on a single drilled shaft foundation. The unfactored design loads acting on the top of this foundation are as follows: $P_D = 47$ k, $V_W = 72$ k, $M_W = 7100$ in-k. The underlying soil is a stiff plastic sandy clay with $s_u = 3000$ lb/ft^2 and an allowable lateral bearing pressure of 300 lb/ft^2 to a maximum of 4500 lb/ft^2. The groundwater table is at a depth of 60 ft. Select appropriate values of f_c' and f_y, and develop a design for this drilled shaft. Use the IBC rigid analysis for the lateral load design. Your design should show all relevant dimensions and reinforcement, and should include the anchor bolts required to connect the monopole tower. You may use any appropriate number and spacing of anchor bolts. There is no restriction on lateral deflection, so the lateral design needs to be based only on strength considerations.

17.7 Solve Problem 17.6 using p-y analysis software instead of the IBC rigid analysis formulas. Develop a moment diagram and determine appropriate cutoff points for the reinforcing steel. Discuss the differences between this design and that developed in Problem 17.6.

Part D

Special Topics

18

Foundations on Weak and Compressible Soils

As the correct solution of any problem depends primarily on a true understanding of what the problem really is, and wherein lies its difficulty, we may profitably pause upon the threshold of our subject to consider first, in a more general way, its real nature; the causes which impede sound practice; the conditions on which success or failure depends; the directions in which error is most to be feared.

A. M. Wellington, 1887

As cities grow and land becomes more scarce, it often becomes necessary to erect buildings and other structures on sites underlain by poor soils. These sites are potentially troublesome, so the work of the foundation engineer becomes even more important.

The most common of these problematic soils are the soft, saturated clays and silts often found near the mouths of rivers, along the perimeter of bays, and beneath wetlands. These soils are very weak and compressible, and thus are subject to bearing capacity and settlement problems. They frequently include organic material, which further aggravates these problems.

Areas underlain by these soft soils frequently are subject to flooding, so it often becomes necessary to raise the ground surface by placing fill. Unfortunately, the weight of these fills frequently causes large settlements. For example, Scheil (1979) described a building constructed on fill underlain by varved clay in the Hackensack Meadowlands of New Jersey. About 10 in (250 mm) of settlement occurred during placement of the fill, 0.5 in (12 mm) during construction of the building, and an additional 4 in (100 mm) over the following ten years.

In seismic areas, loose saturated sands can become weak through the process of *liquefaction*. Moderate to strong ground shaking can create large excess pore water pressures in these soils, which temporarily decrease the effective stress and shear strength. One of the most dramatic illustrations of this phenomenon occurred in Niigata, Japan, during the 1964 earthquake. Many buildings suddenly settled more than 1 m (3 ft), and these settlements were often accompanied by severe tilting (Seed, 1970). One apartment building tilted to an angle of 80 degrees from the vertical!

Fortunately, engineers and contractors have developed methods of coping with these problematic soils, and have successfully built many large structures on very poor sites. These methods include the following, either individually or in combination:

- Support the structure on deep foundations that penetrate through the weak soils.
- Support the structure on shallow foundations and design them to accommodate the weak soils.
- Use a floating foundation, either deep or shallow.
- Improve the engineering properties of the soils.

18.1   DEEP FOUNDATIONS

One of the most common methods of dealing with poor soils is to use deep foundations that penetrate down to stronger soils or bedrock, as shown in Figure 18.1. Thus, the structural loads bypass the troublesome soils. Although deep foundations often perform well, the engineer must be conscious of two potential problems: downdrag loads and the interface between the structure and the adjacent ground.

Downdrag Loads

Construction on sites underlain by soft soils often requires placement of fills to raise the ground surface elevation. The reasons for placing these fills include:

- Providing protection against flooding from nearby bodies of water
- Providing sufficient surface drainage to accommodate rainfall and snowmelt
- Accommodating grade changes, such as the approach fills to a bridge
- Providing a firmer layer of soil to support pavements and other lightweight improvements

Structures to be placed on these fills might still be supported on deep foundations, which are typically installed after the fill is placed. These foundations normally extend through the fill and soft soils, and into underlying firmer soils.

The weight of these fills causes consolidation in the underlying soils, so the fill and the soft soil move downward. However, if the deep foundations are suitably supported by the underlying soils, they experience very little settlement. Because of this downward movement of the soil with respect to the foundations, the side friction force in the upper zone now acts downward instead of upward, and becomes a load instead of a resistance, as shown in Figure 18.1. This load is known as the *downdrag load*, or the *negative skin*

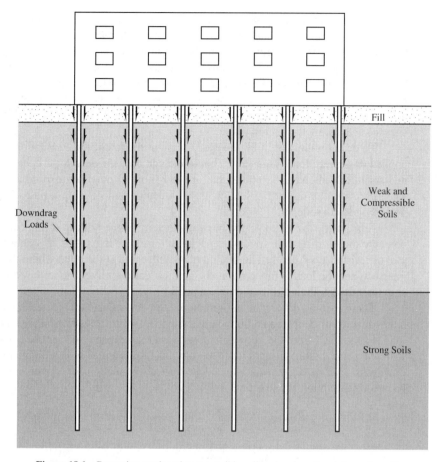

Figure 18.1   Bypassing weak and compressible soils using deep foundations; formation of downdrag loads.

friction load. It can be very large and may cause excessive settlements in the foundations (Bozozuk, 1972a, 1981).

Because of the variations in soil properties, fill thickness, and other factors, down-drag loads are usually greater on some foundations beneath a structure and less on others. This can cause differential settlements, and, in severe cases, these loads may pull some of the piles out of their caps. Although downdrag loads are most pronounced in soft soils and the overlying fills, they will develop whenever the soils are moving downward in re-lation to the pile. Only a few millimeters of relative movement is required to mobilize the full side friction resistance (as discussed in Chapter 13), so downdrag can even occur in moderately stiff soils if the toe of the pile is bottomed in bedrock.

Downdrag loads also can develop if the groundwater table drops to a lower eleva-tion, even if no fill is placed, because this reduction in pore water pressure increases the effective stress and causes consolidation.

To compute the magnitude of downdrag loads, use the β method as described in Chapter 14, along with Equation 14.22. The β values are the same as those for side friction, except the force acts downward instead of upward.

To design the foundation, determine the location of the *neutral plane* as shown in Figure 18.2. This plane is located at the point on the foundation where there is no relative movement between the soil and the foundation (Fellenius, 1999). In other words, the downdrag load acts from the ground surface to the neutral plane, and the side-friction resistance acts below the neutral plane.

To develop the plot in Figure 18.2, begin at the top of the foundation and mark the applied downward load. This value should include only the dead load, because live, wind, and seismic loads have durations that are too short in duration to impact this analysis. Then compute the downdrag load at various depths and develop a plot of applied plus downdrag load vs. depth.

Next, mark the ultimate toe bearing capacity at the bottom of the foundation, and compute the ultimate side friction capacity at various depths. Use this data to develop a plot of resistance vs. depth. The neutral plane is located at the point where these two plots intersect, and the load at this point is the greatest compressive load in the foundation. The structural design should be based on this load.

There is some difference in opinion on how to evaluate the geotechnical downward load capacity of deep foundations subjected to downdrag loads. Usually the downdrag load is subtracted from the allowable downward load capacity determined from Equation 13.4. In some cases, this approach results in a substantial reduction in the allowable downward load capacity, which means additional foundations or longer foundations will be needed to support the structure.

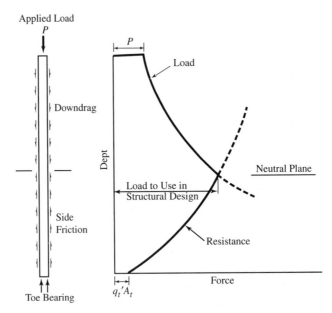

Figure 18.2   Method of locating the neutral plane (Adapted from Fellenius, 1999).

An alternative approach (Fellenius, 1998) argues the downdrag load is detrimental to the geotechnical load capacity of a deep foundation only if it induces excessive settlement. Therefore, he recommends conducting a settlement analysis. If the toe of the foundation is supported by firm soil or bedrock, the downdrag load will not induce significant foundation settlement and thus should not be subtracted from the geotechnical load capacity. In addition, Fellenius argues that live loads cause enough settlement to temporarily reverse the direction of relative movement between the foundation and the soil, so downdrag load and the live load will never act simultaneously.

If the design downdrag loads are large enough to significantly impact the foundation design, engineers often consider implementing one or more of the following downdrag reduction techniques:

- Coat the pile with bitumen, thus reducing the coefficient of friction (Bjerrum et al., 1969; Tawfiq and Caliendo, 1994). This method is very effective, so long as the pile is not driven through an abrasive soil, such as sand, that might scrape off the bitumen coating.
- Drive the piles before placing the fill, wrap the exposed portions with lubricated polyethylene sheets or some other low-friction material as shown in Figure 18.3, and place the fill around the piles.
- Use a large diameter predrill hole, possibly filled with bentonite, thus reducing K.
- Use a pile tip larger in diameter than the pile, thus making a larger hole as the pile is driven.

Figure 18.3 These prestressed concrete piles will support a bridge abutment. They extend above the ground surface because the abutment fill has not yet been placed. Since the underlying soils are compressible, and will settle due to the weight of the fill, the piles have been wrapped with lubricated polyethylene sheets to reduce the downdrag forces between the piles and the fill. The next stage of construction will consist of building a mechanically stabilized earth (MSE) wall around these piles and backfilling it to approximately the height of the piles (Photo by Dr. Kamal Tawfiq).

- Drive an open-end steel pipe pile through the consolidating soils, remove the soil plug, then drive a smaller diameter load-bearing pile through the pipe and into the lower bearing strata. This isolates the inner pile from the downdrag loads.
- Accelerate the settlement using surcharge fills or other techniques, and then install the foundations after the settlement is complete.

Interface with the Adjacent Ground

Even if the problem with downdrag loads is adequately addressed, deep foundations that extend through fills and soft soils can have another problem: As the ground settles, the structure remains at the same elevation. Thus, a gap forms between the structure and the ground. Settlements of 0.5 m (2 ft) or more are not unusual at a soft ground site, so the structure could eventually be suspended well above the ground surface. This creates serious serviceability problems, including access difficulties, stretched and broken utility lines, and poor aesthetics. This problem is especially common where approach fills meet pile-supported bridge abutments, and results in an abrupt bump in the pavement at the edge of the bridge.

Figure 18.4 shows this type of settlement problem in a supermarket building which was built on a pile foundation. The ground surface around this building settled about 400 mm in twenty five years, which required continual maintenance. This change in grade required continual placement of new asphalt pavement around the building.

Figure 18.4   The ground around this pile-supported supermarket building settled about 400 mm in twenty-five years, but the building remained at essentially the same elevation. This required continual maintenance around the perimeter of the building. Note the steep slope to reach the front door and the spacer blocks required at the top of the footing-supported posts (which settled with the surrounding ground).

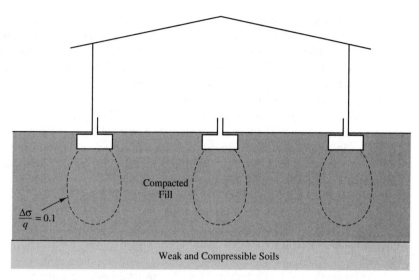

Figure 18.5   Use of spread footings to support structures on fills underlain by weak soils.

Thus, properly designed deep foundations can safely support a structure on a soft ground site, but they also introduce new problems of downdrag loads and structure/ground separation. Such foundations can also be very expensive, especially for light-weight structures, because the downdrag loads may be greater than the structural loads.

18.2   SHALLOW FOUNDATIONS

To avoid the problems with downdrag loads and structure/ground separation, some engineers have used shallow foundation systems (either spread footings or mats), as shown in Figure 18.5. These are especially suitable for lightweight structures.

Lightly loaded foundations located in a fill probably will not have bearing capacity problems. However, they may be subjected to large total and differential settlements. Therefore, the engineer must provide a way to accommodate or avoid these settlement problems.

Coping with Settlement Problems

One way to cope with settlement problems is to place the fill, and then delay construction of the structure until most of the settlement has occurred. This method is reliable and inexpensive, but may require many years of waiting. However, it also is possible to accelerate the settlement process using surcharge fills, as described in Section 18.4.

Another method is to design the structure to accommodate large settlements. An airline terminal building at LaGuardia Airport in New York City (originally built to serve

Eastern Airlines) is a good example (York and Suros, 1989). The building site is underlain by a 24-m (80 ft) deep deposit of soft organic clay that was covered with 6 to 12 m (20–38 ft) of incinerated refuse fill in the late 1930s. The organic clay has a compressibility, $C_c/(\text{ite})$, of 0.29 to 0.33. When the construction of this building began in 1979, the ground surface had already settled more than 2 m (7 ft) and was expected to settle an additional 450 mm (18 in) during the following twenty years.

Other buildings at the airport are supported on pile foundations and require continual maintenance to preserve access for aircraft, motor vehicles, and people. Therefore, the Eastern Airlines terminal building was built on a floating foundation with spread footings. It was also designed to be very tolerant of differential settlements and included provisions for leveling jacks between the footings and the building.

By 1988, the building had settled as much as 315 mm (12.4 in), with differential settlements of up to 56 mm (2.2 in). However, because of the tolerant design, the structure was performing well. The first leveling operation was planned for 1989. In this case, a pile foundation would have been much more expensive, yet it would not have performed as well.

This method also has been used on the terminal building at the Kansai International Airport in Japan. This building is located on a man-made island that is underlain by very soft marine soils. The differential settlements between the various columns are periodically monitored, and extra shims are placed between the base plate and the pile cap as needed.

Buildings with heavily loaded slab-on-grade floors, such as warehouses, also can have problems because the floor may settle more than the footings. Thus, we must either structurally support the floor or provide construction details that permit this differential movement.

Another alternative is to use lightweight fill materials. The weight of the fill is usually the primary cause of settlement, so it is helpful to use fill materials that have a very low unit weight. These can consist of natural materials, such as lightweight aggregates, or synthetics. The most common synthetic is *expanded polystyrene* (EPS), a plastic foam with a unit weight of 1.2 lb/ft^3 (0.19 kN/m^3). This is the same material used to make disposable coffee cups. In the United States, it is available in $2 \times 4 \times 8$ ft ($610 \times 1220 \times 2440$ mm) blocks that can be stacked to form a fill (Horvath, 1992).

Some projects have involved excavating some of the upper soft soils, placing EPS, and then covering the foam with a thin soil layer. Thus, the net increase in stress on the natural soils can be very small. The foam can support light to moderately loaded spread footings.

18.3 FLOATING FOUNDATIONS

Another way to reduce the settlement of a structure is to build a *floating foundation* (also known as a *compensated foundation*) (Golder, 1975). This consists of excavating a volume of soil with a weight nearly equal to that of the proposed structure, and then building the structure in this excavation, as shown in Figure 18.6. Thus, the increase in vertical ef-

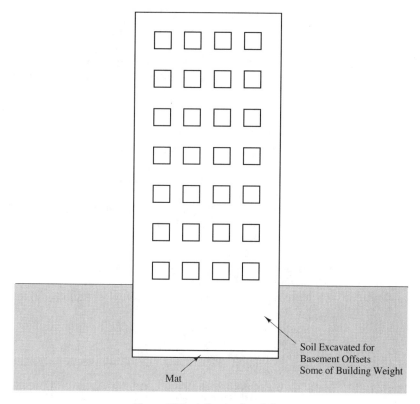

Figure 18.6   A floating foundation.

fective stress in the soil, $\Delta\sigma_z'$, is very small. Nearly all floating foundations are mats or pile supported mats.

The earliest documented floating foundation was for the Albion Mill, which was built on a soft soil in London around 1780. It had about 50 percent flotation (i.e., the excavation reduced $\Delta\sigma_z'$ by 50 percent) and, according to Farley (1827), "the whole building would have floated upon it, as a ship floats in water." In spite of this pioneering effort, floating foundations did not become common until the early twentieth century. Early examples include the Empress Hotel in Victoria, British Columbia, 1912; the Ohio Bell Telephone Company Building in Cleveland, 1925; and the Albany Telephone Building in Albany, New York, 1929.

If the dead and live loads from the structure were constant, we theoretically could select a depth of excavation that would produce $\Delta\sigma_z' = 0$. However, the live load varies over time, so the weights are not always perfectly balanced. Fluctuations in the groundwater table elevation are also important because the groundwater produces an upward buoyancy force on submerged and dewatered basements, thus further reducing $\Delta\sigma_z'$. Therefore, we must determine the worst possible loading condition and select a depth of excavation such that $\Delta\sigma_z'$ is always greater than zero.

Torre Latino Americana

The construction of the 43-story Torre Latino Americana in Mexico City was an important milestone in floating foundation technology (Zeevaert, 1957). It is significant because of the exceptionally difficult soils there.

The soil profile is generally as follows:

0–18 ft (0–5.5 m) depth
> Old fill that includes Aztec artifacts. Groundwater table at 7 ft (2 m).

18–30 ft (5.5–9.1 m) depth
> *Becarra sediments*—Interbedded sands, silts, and clays.

30–110 ft (9.1–33.5 m) depth
> *Tacubaya clays*—Soft volcanic clay; moisture content = 100–400%, $C \approx 0.80$; $s_u = 700$–1400 lb/ft^2 (35–70 kPa).

110–230 ft (33.5–70.0 m) depth
> *Tarango sands and clays*—Harder and stronger deposits; much less compressible than the Tacubaya clays.

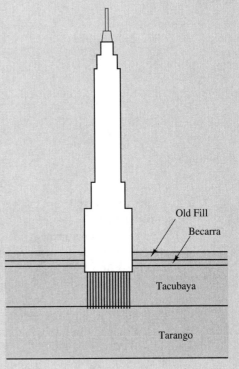

The highly compressible Tacubaya clays have caused dramatic settlements in other structures. For example, the Palace of Fine Arts, located across the street from the Tower, settled over 10 ft (3 m) from 1904 to 1962 and continues to settle at a rate of 0.5 in (12 mm) per year (White, 1962)!

To avoid these large settlements, engineers excavated to a depth of 43 ft (13.0 m) and built a mat supported on piles driven to the Tarango sands. The removal of this soil compensated for about half the building weight. Thus, the building does load the deeper soils and has settled. However, this is by design, and the settlement has approximately matched that of the surrounding ground.

Figure 18.7   The Torre Latino Americana.

When considering floating foundations, it is often useful to conduct parametric studies that evaluate the effect of various parameters on $\Delta\sigma_z'$. Sometimes small changes in loads, groundwater level, or other factors can produce a much greater change in $\Delta\sigma_z'$. The greatest problems with floating foundations usually occur during construction. The most critical time is when the excavation is open, but construction of the structure has not yet begun. These excavations may not be stable, and thus may require special precautions.

18.4   SOIL IMPROVEMENT

Another way to cope with weak or compressible soils is to improve them. Some soil improvement techniques have been used for many years, whereas others are relatively new (Mitchell et al., 1978; Mitchell and Katti, 1981). This section discusses some of the more common methods.

Removal and Replacement, or Removal and Recompaction

Sometimes poor soils can simply be *removed and replaced* with good quality compacted fill. This alternative is especially attractive if the thickness of the deposit is small, the groundwater table is deep, and good quality fill material is readily available.

If the soil is inorganic and not too wet, then it probably is not necessary to haul it away. Such soils can be improved by simply compacting them. In this case, the contractor excavates the soil until firm ground is exposed, then places the excavated soil back in its original location, compacting it in lifts. This technique is often called *removal and recompaction*.

Surcharge Fills

Covering poor soils with a temporary surcharge fill, as shown in Figure 18.8, causes them to consolidate more rapidly. When the temporary fill is removed, some or all of the soil is now overconsolidated, and thus stronger and less compressible. This process is known as *preloading* or *precompression* (Stamatopoulos and Kotzias, 1985).

Engineers have primarily used preloading to improve saturated silts and clays because these soils are most conducive to consolidation under static loads. Sandy and gravelly soils respond better to vibratory loads.

If the soil is saturated, the time required for it to consolidate depends on the ability of the excess pore water to move out of the soil voids (see the discussion of consolidation theory in Chapter 3). This depends on the thickness of the soil deposit, its coefficient of permeability, and other factors, and can be estimated using the principles of soil mechanics. The time required could range from only a few weeks to thirty years or more.

The consolidation process can be accelerated by an order of magnitude or more by installing *vertical drains* in the natural soil, as shown in Figure 18.8. These drains provide a pathway for the excess water to escape more easily. The most common design is a *wick*

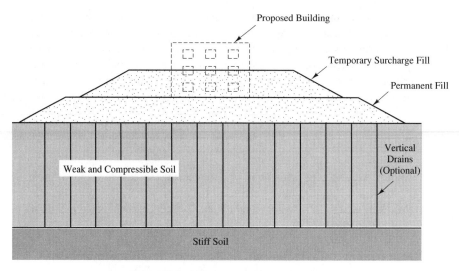

Figure 18.8   Soil improvement by preloading.

drain, which is inserted into the ground using a device that resembles a giant sewing machine.

Preloading is less expensive than some other soil improvement techniques, especially when the surcharge soils can be moved from place to place, thus preloading the site in sections. Vertical drains, if needed, may double the cost.

Vibro-Compaction and Vibro-Replacement

Sandy and gravelly soils consolidate most effectively when subjected to vibratory loads. This is especially true at depths of about 3 m (10 ft) or more because of the greater overburden stress.

Engineers in Germany recognized this behavior and developed depth vibrators in the 1930s. These methods were refined during the 1970s and now are frequently used throughout the world.

Some of these depth vibrators are essentially vibratory pile hammers attached to a length of steel pipe pile (Brown and Glenn, 1976). Known as a *terra probe*, this device is vibrated into the ground, densifying the adjacent soils, and then retracted. This process is repeated on a grid pattern across the site until all of the soil has been densified.

Another technique uses a probe with a built-in vibrator that is lowered into the ground, also on a grid pattern, as shown in Figure 18.9. The probe is known as a *vibroflot* and the technique is called *vibroflotation*. The vibroflots are also equipped with water jets to aid penetration and improve soil compaction.

Figure 18.9   This crane is lowering a vibroflot into the ground (Photo courtesy of GKN Hayward Baker, Inc.).

Both of these techniques may be classified as *vibro-compaction* methods because they compact the soils in-situ using vibration. In either case, additional sandy soil is added to assist the compaction process. Another closely related method is *vibro-replacement*, which uses the vibrator to create a shaft, and then the shaft is backfilled with gravel to form a *stone column* (Mitchell and Huber, 1985). This technique may be used in cohesive soils and is primarily intended to provide load bearing members that extend through the weak soils. The stone column also acts as a vertical drain, thus helping to accelerate consolidation settlements.

Dynamic Consolidation

The Soviets tried dropping heavy weights to stabilize loess in the 1930s. However, this method did not become widely accepted until the French developed *dynamic consolidation* (also known as *heavy tamping*) in 1970. This technique consists of dropping 4 to 36 Mg (5–40 ton) weights, called *pounders*, from heights of 6 to 30 m (20–100 ft) (Mitchell et al., 1978). This equipment is shown in Figure 18.10. The impact of the falling weight compacts the soil to significant depths. This tamping process is repeated on a grid pattern, and then the upper soils are smoothed and compacted using conventional earthmoving and compaction equipment.

Dynamic consolidation has been effectively used with a wide range of soil types. It is relatively inexpensive, but it also generates large shock waves and therefore cannot be used close to existing structures.

Reinforcement Using Geosynthetics

Soil is similar to concrete in that both materials are strong in compression but weak in tension. Fortunately, both can be improved by introducing tensile members to form a composite material. Steel bars are used to form reinforced concrete, which is vastly supe-

Figure 18.10   This crane has just dropped a large weight, thus producing dynamic consolidation in the underlying ground (Photo courtesy of GKN Hayward Baker, Inc.).

rior to plain concrete. Likewise, various metallic and nonmetallic materials can be used to reinforce soil.

Engineers most frequently use materials known as *geosynthetics* to reinforce soils (Koerner, 1990). Although most of these applications have been in the context of retaining structures and earth slopes, strategically placed tensile reinforcement also can provide flexural strength in soil beneath spread footings, which increases the bearing capacity. In addition, such reinforcement spreads the applied load over a larger area, thus reducing the change in effective stress and reducing the consolidation settlement. This technique has been used as a means of using spread footings at sites that ordinarily would have required deep foundations.

SUMMARY

Major Points

1. Many areas have a scarcity of good building sites, thus forcing development on sites with poor soil conditions. These soils are often weak and compressible, and create special problems for the foundation engineer.

2. Structures can be supported on deep foundations that extend through the weak soils and into deeper, stronger strata. However, these foundations are often subjected to large downdrag loads. In addition, a gap may form between the ground and the structure.

3. Shallow foundations avoid the downdrag load and structure/ground separation problems, but can have problems with excessive settlements. These can be accommodated by delaying construction or by providing jacks between the structure and the foundation.

4. Lightweight fills can be used to achieve the desired ground surface elevation without excessively loading the soil, thus reducing settlements.

5. Floating foundations compensate for some of the structural load by excavating a certain weight of soil. These can be a very effective way of controlling settlements.

6. Some poor soil conditions can be remedied by improving the soil. Many techniques are available, and they are often a cost-effective solution.

Vocabulary

Downdrag	Negative skin friction	Surcharge Fill
Dynamic consolidation	Preloading	Terra probe
Expanded polystyrene	Removal and	Vertical drains
Floating foundation	recompaction	Vibro-compaction
Geosynthetics	Removal and replacement	Vibroflotation
Liquefaction	Soil improvement	

COMPREHENSIVE QUESTIONS AND PRACTICE PROBLEMS

18.1 A 300-mm diameter, 25-m long steel pipe pile must penetrate through 2 m of recently placed compacted fill and 16 m of soft clay before reaching firm bearing soils. The unit weights of the fill clay, and firm soil are 20.0 kN/m³, 13.0 kN/m³, and 18.5 kN/m³, respectively, and β values are 0.35, 0.25, and 0.38, respectively. The groundwater table is at a depth of 2 m below the top of the fill. The net unit toe bearing resistance is 4000 kPa. Compute the downdrag load.

18.2 Two soft clay deposits are identical except that one has a higher coefficient of lateral earth pressure, K_0. A 12-inch square reinforced concrete pile is to be driven into each deposit. Will the downdrag load be the same for both piles? Explain.

18.3 A certain structure is supported on a floating foundation located below the groundwater table. What would happen if $\Delta\sigma_z'$ in the soil below this foundation became less than zero?

18.4 The most common remediation for soils prone to seismic liquefaction is to densify them. Which soil improvement methods might be suitable for this task?

18.5 A proposed construction site is underlain by 50 ft of soft clays that have a unit weight of 80 lb/ft³. The ground surface elevation is +2.0 ft and the groundwater table is at elevation +1.0 ft.

The proposed site development will consist of excavating some of the soft clay (with temporary dewatering), placing expanded polystyrene blocks to elevation +6.0 ft, and then covering the blocks with 2.0 ft of 120 lb/ft³ compacted fill. The proposed building will be supported on spread footings in the compacted fill. The total weight of this building divided by its footprint area will be 100 lb/ft².

Compute the required elevation of the bottom of the temporary excavation so that the net increase in effective stress in the soft clay will be 75 lb/ft². Be sure to consider buoyancy effects in your computations.

19

Foundations on Expansive Soils

Question: *What causes more property damage in the United States than all the earthquakes, floods, tornados, and hurricanes combined?*
Answer: *Expansive soils!*

According to a 1987 study, expansive soils in the United States inflict about $9 billion in damages per year to buildings, roads, airports, pipelines, and other facilities—more than twice the combined damage from the disasters listed above (Jones and Holtz, 1973; Jones and Jones, 1987). The distribution of these damages is approximately as shown in Table 19.1. Many other countries also suffer from expansive soils. Although it is difficult to estimate the global losses, this is clearly an international problem.

Sometimes the damages from expansive soils are minor maintenance and aesthetic concerns, but often they are much worse, even causing major structural distress, as illustrated in Figures 19.1 through 19.3. According to Holtz and Hart (1978), 60 percent of the 250,000 new homes built on expansive soils each year in the United States experience minor damage and 10 percent experience significant damage, some beyond repair.

In spite of these facts, we do not expect to see newspaper headlines "Expansive Soils Waste Billions" and certainly not "Expansive Soil Kills Local Family." Expansive soils are not as dramatic as hurricanes or earthquakes and they cause only property damage, not loss of life. In addition, they act more slowly and the damage is spread over wide areas rather than being concentrated in a small locality. Nevertheless, the economic loss is large and much of it could be avoided by proper recognition of the problem and incorporating appropriate preventive measures into the design, construction, and maintenance of new facilities.

TABLE 19.1   ANNUAL DAMAGE IN THE UNITED
STATES FROM EXPANSIVE SOILS (Adapted from
Jones and Holtz 1973; Jones and Jones, 1987; Used
with permission of ASCE)

Category	Annual Damage
Highways and streets	$4,550,000,000
Commercial buildings	1,440,000,000
Single family homes	1,200,000,000
Walks, drives and parking areas	440,000,000
Buried utilities and services	400,000,000
Multi-story buildings	320,000,000
Airport installations	160,000,000
Involved in urban landslides	100,000,000
Other	390,000,000
Total annual damages (1987)	$9,000,000,000

Foundation engineers must be aware of this potential problem and be ready to take appropriate action when encountering such soils.

19.1   THE NATURE, ORIGIN, AND OCCURRENCE OF EXPANSIVE SOILS

When geotechnical engineers refer to expansive soils, we usually are thinking about clays or sedimentary rocks derived from clays, and the volume changes that occur as a result of changes in moisture content. This is the most common expansion phenomenon, and thus is the primary focus of this chapter. Other less common mechanisms of soil expansion are discussed in Section 19.6. Expansion due to frost heave, which is an entirely different phenomenon, is discussed in Chapter 8.

Clays are fundamentally very different from gravels, sands, and silts. All of the latter consist of relatively inert bulky particles and their engineering properties depend primarily on the size, shape, and texture of these particles. In contrast, clays are made up of very small particles that are usually plate-shaped. The engineering properties of clays are strongly influenced by the very small size and large surface area of these particles and their inherent electrical charges.

What Causes a Clay to Expand?

Several different *clay minerals* occur in nature, the differences being defined by their chemical makeup and structural configuration. Three of the most common clay minerals are *kaolinite*, *illite*, and *montmorillonite* (part of the smectite group). The different chemi-

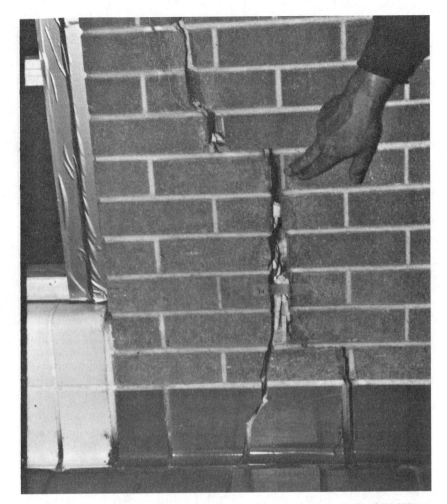

Figure 19.1   Heaving of an expansive soil caused this brick wall to crack. The $490,000 spent to repair this and other walls, ceilings, doors, and windows represented nearly one-third of the original cost of the six-year-old building (Photo courtesy of the Colorado Geological Survey).

cal compositions and crystalline structures of these minerals give each a different susceptibility to swelling, as shown in Table 19.2.

Swelling occurs when water infiltrates between and within the clay particles, causing them to separate. Kaolinite is essentially nonexpansive because of the presence of strong hydrogen bonds that hold the individual clay particles together. Illite contains weaker potassium bonds that allow limited expansion, and montmorillonite particles are only weakly linked. Thus, water can easily flow into montmorillonite clays and separate the particles. Field observations have confirmed that the greatest problems occur in soils with a high montmorillonite content.

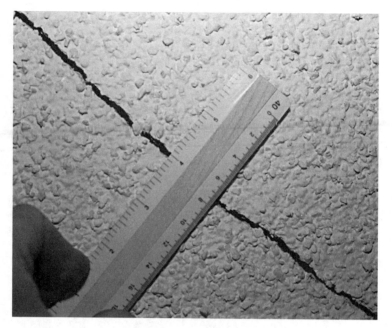

Figure 19.2   Heaving of expansive soils caused this 0.1-inch wide crack in the ceiling of a one-story wood-frame house.

Several other forces also act on clay particles, including the following:

- Surface tension in the menisci of water contained between the particles (tends to pull the particles together, compressing the soil).
- Osmotic pressures (tend to bring water in, thus pressing the particles further apart and expanding the soil).
- Pressures in entrapped air bubbles (tend to expand the soil).
- Effective stresses due to external loads (tend to compress the soil).
- London-Van Der Waals intermolecular forces (tend to compress the soil).

Expansive clays swell or shrink in response to changes in these forces. For example, consider the effects of changes in surface tension and osmotic forces by imagining a montmorillonite clay that is initially saturated, as shown in Figure 19.4a. If this soil dries,

Figure 19.3 Expansive soil caused this brick building to crack (Photo courtesy of the Colorado Geological Survey).

TABLE 19.2 SWELL POTENTIAL OF PURE CLAY MINERALS
(Adapted from Budge et al., 1964)

Surcharge Load		Swell Potential (%)		
(lb/ft^2)	(kPa)	Kaolinite	Illite	Montmorillonite
200	9.6	Negligible	350	1500
400	19.1	Negligible	150	350

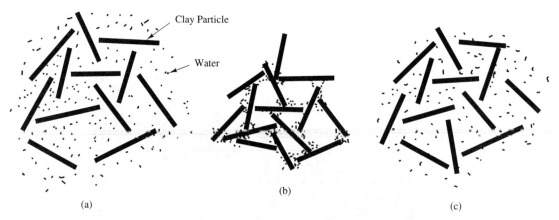

Figure 19.4   Shrinkage and swelling of an expansive clay.

the remaining moisture congregates near the particle interfaces, forming menisci, as shown in Figure 19.4b, and the resulting surface tension forces pull the particles closer together causing the soil to shrink. We could compare the soil in this stage to a compressed spring; both would expand if it were not for forces keeping them compressed.

The soil in Figure 19.4b has a great affinity for water and will draw in available water using osmosis. We would say that it has a very high *soil suction* at this stage. If water becomes available, the suction will draw it into the spaces between the particles and the soil will swell, as shown in Figure 19.4c. Returning to our analogy, the spring has been released and perhaps is now being forced outward.

What Factors Control the Amount of Expansion?

The portion of a soil's potential expansion that will actually occur in the field depends on many factors. One of these factors is the percentage of expansive clays in the soil. For example, a pure montmorillonite could swell more than fifteen times its original volume (definitely with disastrous results!), but clay minerals are rarely found in such a pure form. Usually, the expansive clay minerals are mixed with more stable clays and with sands or silts. A typical "montmorillonite" (really a mixed soil) would probably not expand more than 35 to 50 percent, even under the worst laboratory conditions, much less in the field.

There are two types of montmorillonite clay: calcium montmorillonite and sodium montmorillonite (also known as bentonite). The latter is much more expansive, but less common.

Two of the most important variables to consider are the initial moisture content and the surcharge pressure. If the soil is initially moist, then there is much less potential for additional expansion than if it were dry. Likewise, even a moderate surcharge pressure restrains much of the swell potential (although large loads are typically required to completely restrain the soil). Figure 19.5 illustrates a typical relationship between swell potential, initial moisture content, and surcharge pressure.

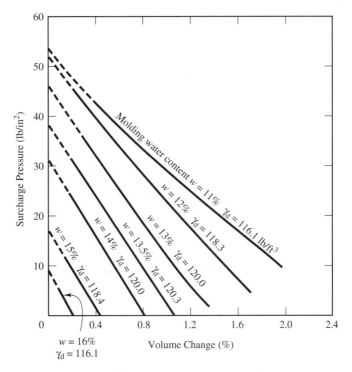

Figure 19.5   Swell potential as a function of initial moisture content and surcharge load (typical). (Adapted from Seed, Mitchell, and Chan, 1962).

This relationship demonstrates why pavements and slabs-on-grade are so suscepti-ble to damage from expansive soils (see Table 19.1). They provide such a small surcharge load that there is little to resist the soil expansion. However, it also demonstrates how even a modest increase in surcharge, such as 300 mm (12 in) or so of subbase, signifi-cantly decreases the potential heave.[1]

Remolding a soil into a compacted fill may make it more expansive (O'Neill and Poormoayed, 1980), probably because this process breaks up cementation in the soil and produces high negative pore water pressures that later dissipate. Many other factors also affect the expansive properties of fills, especially the methods used to compact the fill (kneading vs. static) and the as-compacted moisture content and dry unit weight (Seed and Chan, 1959).

Figure 19.6 illustrates how compacting a soil wet of the optimum moisture content reduces its potential for expansion. It also illustrates that compacting the soil to a lower

---

[1]The use of a subbase for this purpose is somewhat controversial. Although it provides additional surcharge pressure, which is good, it can also become an avenue for additional water to enter the expansive soils, which is bad. Many engineers feel that the risk of additional water infiltration is too great and therefore do not use this method.

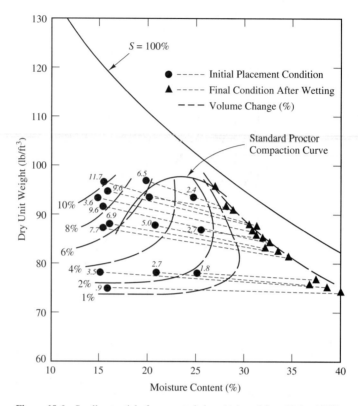

Figure 19.6 Swell potential of compacted clays (Adapted from Holtz, 1969).

dry unit weight also reduces its swell potential (although this also will have detrimental effects, such as reduced shear strength and increased compressibility).

Although laboratory tests are useful, they may not accurately predict the behavior of expansive soils in the field. This is partly because the soil in the lab is generally inundated with water, whereas the soil in the field may have only limited access to water. The flow of water into a soil in the field depends on many factors, including the following:

- The supply of water (depends on rainfall, irrigation, and surface drainage).
- Evaporation and transpiration (depends on climate and vegetation; large trees can extract large quantities of water from the soil through their roots).
- The presence of fissures in the soil (water will flow through the fissures much more easily than through the soil).
- The presence of sand or gravel lenses (helps water penetrate the soil).
- The soil's affinity for water (its suction).

Because of these factors, Jones and Jones (1987) suggested that soils in the field typically swell between 10 and 80 percent of the total possible swell.

Occurrence of Expansive Clays

Chemical weathering of materials such as feldspars, micas, and limestones can form clay minerals. The particular mineral formed depends on the makeup of the parent rock, topography, climate, neighboring vegetation, duration of weathering, and other factors.

Montmorillonite clays often form as a result of the weathering of ferromagnesian minerals, calcic feldspars, and volcanic materials. They are most likely to form in an alkaline environment with a supply of magnesium ions and a lack of leaching. Such conditions would most likely be present in semi-arid regions. *Bentonite* (sodium montmorillonite) is formed by chemical weathering of volcanic ash. Figure 19.7 shows the approximate geographical distribution of major montmorillonite deposits in the United States.

Expansive clays are also common in the Canadian prairie provinces, Israel, South Africa, Australia, Morocco, India, Sudan, Peru, Spain, and many other places in the world.

Figure 19.7 illustrates regional trends only. Not all the soils in the shaded areas are expansive and not all the soils outside are nonexpansive. The local occurrence of expansive soils can vary widely, as illustrated in Figure 19.8. However, even maps of this scale are only an aid, not a substitute for site-specific field investigations.

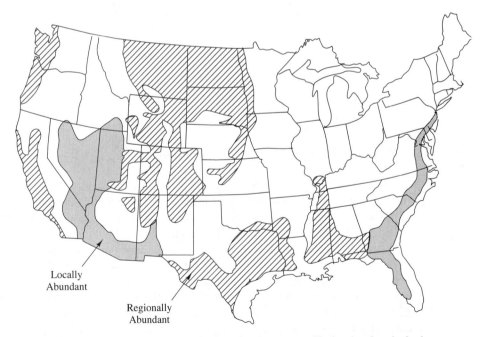

Locally
Abundant

Regionally
Abundant

Figure 19.7   Approximate distribution of major montmorillonite clay deposits in the United States (Adapted from Tourtelot, 1973). Erosion, glacial action, and other geologic processes have carried some of these soils outside the zones shown here. Thus, the unshaded areas are not immune to expansive soils problems.

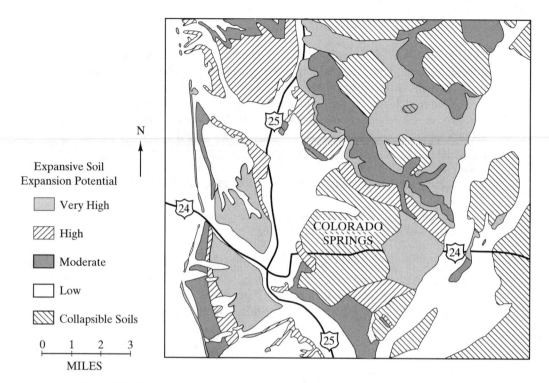

Figure 19.8   Distribution of expansive soils in the Colorado Springs, Colorado area (Adapted from Hart, 1974).

Influence of Climate on Expansion Potential

As discussed earlier, any expansive soil could potentially shrink and swell, but in practice this occurs only if its moisture content changes. The likelihood of such changes depends on the balance between water entering a soil (such as by precipitation or irrigation) and water leaving the soil (often by evaporation and transpiration).

In humid climates, the soil is moist or wet and tends to remain so throughout the year. This is because the periods of greatest evaporation and transpiration (the summer months) also coincide with the greatest rainfall. The climate in North Carolina, as shown in Figure 19.9b, is typical of this pattern. Because the variations in moisture content are small, very little shrinkage or swelling will occur. However, some problems have been reported during periods of extended drought when the soil dries and shrinks (Hodgkinson, 1986; Sowers and Kennedy, 1967).

Most of the problems with expansive soils occur in arid, semi-arid, and monsoonal areas because the seasonal distribution of precipitation and evaporation/transpiration

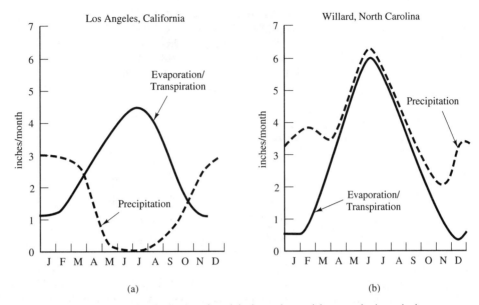

Figure 19.9 Annual distribution of precipitation and potential evaporation/transpiration in (a) Los Angeles, California, and (b) Willard, North Carolina (Adapted from Thornthwaite, 1948). Note how the wet Los Angeles winters are followed by very dry summers. In contrast, the total annual precipitation in Willard is much higher and most of it occurs during the summer. Used with permission of the American Geographical Society.

causes wide fluctuations in the soil's moisture content. Most of the precipitation in arid and semi-arid areas occurs during the winter and spring when evaporation and transpiration rates are low. Thus, the moisture content of the soil increases. Then, during the summer, precipitation is minimal and evaporation/transpiration is greatest, so the soil dries. Thus, the soil expands in the winter and shrinks in the summer. The climate in Los Angeles, shown in Figure 19.9a, displays this pattern.

A useful measure of precipitation and evaporation/transpiration as it affects expansive soil problems is the Thornthwaite Moisture Index (TMI) (Thornthwaite, 1948). This index is a function of the difference between the mean annual precipitation and the amount of water that could be returned to the atmosphere by evaporation and transpiration. A positive value indicates a net surplus of soil moisture whereas a negative value indicates a net deficit. Using this index, Thornthwaite classified climates as shown in Table 19.3.

Because expansive soils are most troublesome in areas where there the moisture content varies during the year, and this is most likely to occur in arid climates, regions with the lowest TMI values should have the greatest potential for problems. Researchers have observed that expansive soils are most prone to cause problems in areas where the TMI is no greater than +20. However, this is not an absolute upper limit. For example, some expansive soil problems have occurred in Alabama and Mississippi (TMI ≈ 40).

TABLE 19.3   CLASSIFICATION
OF CLIMATE BASED ON THORNTH-
WAITE MOISTURE INDEX (TMI)
(Adapted from Thornthwaite (1948);
Used with permission of the American
Geographical Society)

TMI	Climate Type
−60 to −40	Arid
−40 to −20	Semiarid
−20 to 0	Dry subhumid
0 to 20	Moist subhumid
20 to 100	Humid
> 100	Perhumid

Figure 19.10 shows Thornthwaite contours for the United States. Combining this information with Figure 19.7 shows that the areas most likely to have expansive soils problems include the following:

- Central and southern Texas
- Portions of Colorado outside the Rocky Mountains
- Much of California south of Sacramento
- The northern plains states
- Portions of the great basin area (Arizona, Nevada, Utah)

Man-made improvements can change the TMI at a given site, as discussed later in this section.

Depth of the Active Zone

In its natural state, the moisture content of a soil fluctuates more near the ground surface than at depth. This is because these upper soils respond more rapidly to variations in precipitation and evaporation/transpiration.

An important criterion when evaluating expansive soils problems is the *depth of the active zone*, which is the greatest depth of moisture content fluctuations (see Figure 19.11). Presumably, the moisture content is reasonably constant below that depth, so no expansion occurs there. From Figure 19.9, the soil moisture content in Los Angeles varies dramatically from summer to winter, so we would expect the active zone there to extend much deeper than in Willard, where the moisture content is less variable. As a result, expansive soils should be more of a problem in Los Angeles (and they are!).

O'Neill and Poormoayed (1980) presented active zone depths for selected cities as shown in Table 19.4. According to this table, a soil profile in San Antonio would generate

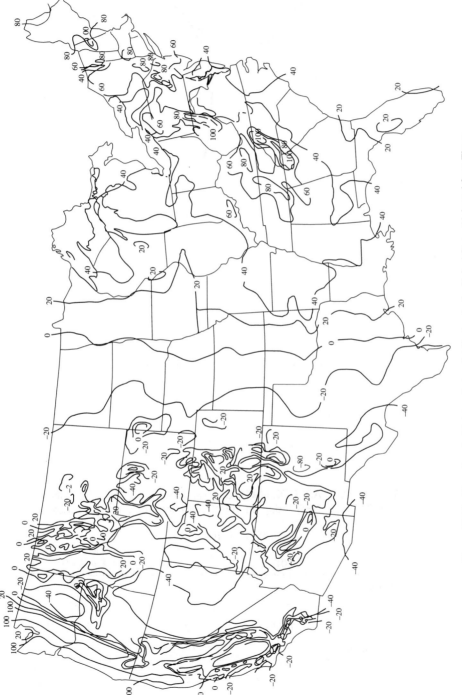

Figure 19.10 Thornthwaite Moisture Index distribution in the United States (Adapted from Thornthwaite 1948, and PTI, 1980). Used with permission of the American Geographical Society and the Post-Tensioning Institute.

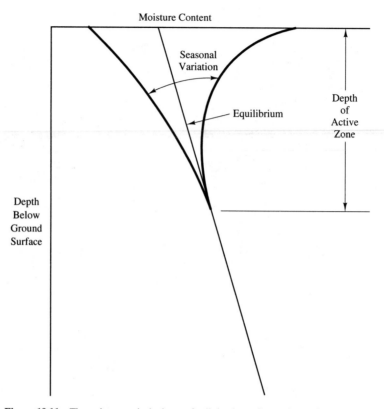

Figure 19.11   The active zone is the layer of soil that has a fluctuating moisture content.

TABLE 19.4   APPROXIMATE DEPTH OF THE ACTIVE ZONE IN
SELECTED CITIES (Adapted from O'Neill and Poormoayed, 1980)

| City | Depth of the Active Zone | | TMI (from Fig. 19.10) |
	(ft)	(m)	
Houston	5–10	1.5–3.0	18
Dallas	7–15	2.1–4.2	−1
Denver	10–15	3.0–4.2	−15
San Antonio	10–30	3.0–9.0	−14

more heave than an identical profile 320 km (200 miles) away in Houston because the active zone is deeper in San Antonio. As a result, the appropriate preventive design measures for identical structures would also be different.

Clays that are heavily fissured will typically have deeper active zones because the fissures transmit the water to greater depths. For example, field studies conducted in Colorado indicate the active zone in some locations may extend as deep as 16 m (50 ft) below the ground surface (Holtz, 1969).

The depth of the active zone at a given site is difficult to determine, and this is a major source of uncertainty in heave analyses.

Influence of Human Activities

Engineers also need to consider how human activities, especially new construction, change the moisture conditions at a particular site. For example:

- Removal of vegetation brings an end to transpiration.
- Placement of slab-on-grade floors, pavements, or other impervious materials on the ground stops both evaporation and the direct infiltration of rain water.
- Irrigation of landscaping introduces much more water into the ground. In southern California, some have estimated that irrigation in residential areas is the equivalent of raising the annual precipitation from a natural 375 mm (15 in) to an inflated 1500 mm (60 in) or more.
- Placement of aggressive trees or heated buildings can enhance desiccation.

These changes are difficult to quantify, but the net effect in arid and semi-arid areas is normally to increase the moisture content under structures. This results in more swelling and more structural damage.

19.2   IDENTIFYING, TESTING, AND EVALUATING EXPANSIVE SOILS

When working in an area where expansive soils can cause problems, geotechnical engineers must have a systematic method of identifying, testing, and evaluating the swelling potential of troublesome soils. The ultimate goal is to determine which preventive design measures, if any, are needed to successfully complete a proposed project.

An experienced geotechnical engineer is usually able to visually identify potentially expansive soils. To be expansive, a soil must have a significant clay content, probably falling within the unified symbols CL or CH (although some ML, MH, and SC soils also can be expansive). A dry expansive soil will often have fissures, slickensides, or shattering, all of which are signs of previous swelling and shrinking. When dry, these soils usually have cracks at the ground surface. However, any such visual identification is only a first step; we must obtain more information before we can develop specific design recommendations.

TABLE 19.5   CORRELATIONS WITH COMMON SOIL TESTS (Adapted from Holtz, 1969; and Gibbs, 1969)

Percent Colloids	Plasticity Index	Shrinkage Limit	Liquid Limit	Swelling Potential
<15	<18	<15	<39	Low
13–23	15–28	10–16	39–50	Medium
20–31	25 41	7–12	50–63	High
>28	>35	>11	>63	Very high

The next stage of the process—determining the degree of expansiveness—is more difficult. A wide variety of testing and evaluation methods have been proposed, but none of them is universally or even widely accepted. Engineers who work in certain geographical areas often use similar techniques, which may be quite different from those used elsewhere. This lack of consistency continues to be a stumbling block.

We can classify these methods into three groups. The first group consists of purely *qualitative methods* that classify the expansiveness of the soil with terms such as "low," "medium," or "high" and form the basis for empirically based preventive measures. The second group includes *semiquantitative methods*. They generate numerical results, but engineers consider them to be an *index* of expansiveness, not a fundamental physical property. The implication here is that the design methods will also be empirically based. The final group includes methods that provide *quantitative* results that are measurements of fundamental physical properties and become the basis for a rational or semirational design procedure.

Qualitative Evaluations

This category of evaluations is usually based on correlations with common soil tests, such as the Atterberg limits or the percent colloids.[2] Such correlations are approximate, but they can be useful, especially for preliminary evaluations.

The U.S. Bureau of Reclamation developed the correlations in Table 19.5. An engineer could use any or all of them to classify the swelling potential of a soil, but the plasticity index and liquid limit correlations are probably the most reliable. Montmorillonite particles are generally smaller than illite or kaolinite, so expansiveness roughly correlates with the percent colloids. Engineers rarely perform the shrinkage limit test, and some have questioned the validity of its correlation with expansiveness.

Chen (1988) proposed the correlations in Table 19.6 based on his experience in the Rocky Mountain area.

---

[2]Colloids are usually defined as all particles smaller than 0.001 mm; clay-size particles are sometimes defined as those smaller than 0.002 mm. A hydrometer test is an easy way to measure the percentage of colloids or clay-size particles in a soil, which can be a rough indicator of its potential expansiveness.

TABLE 19.6   CORRELATIONS WITH COMMON SOIL TESTS (Adapted from Chen, 1988; Used with permission of Elsevier Science Publishers)

Laboratory and Field Data			Degree of Expansiveness			
Percent Passing #200 Sieve	Liquid Limit	SPT N Value	Probable Expansion (%)[a]	Swelling Pressure (k/ft^2)	Swelling Pressure (kPa)	Swelling Potential
<30	<30	<10	<1	1	50	Low
30–60	30–40	10–20	1–5	3–5	150–250	Medium
60–95	40–60	20–30	3–10	5–20	250–1000	High
>95	>60	>30	>10	>20	>1000	Very High

[a]Percent volume change when subjected to a total stress of 1000 lb/ft^2 (50 kPa).

Semiquantitative Evaluations

Loaded Swell Tests

The most common semiquantitative method of describing expansive soils is in terms of its *swell potential*, which engineers usually measure in some kind of *loaded swell test*. Unfortunately, these are very ambiguous terms because there are many different definitions of swell potential and an even wider range of test methods.

Loaded swell tests usually utilize a laterally confined cylindrical sample, as shown in Figure 19.12. The initially dry sample is loaded with a surcharge, and then soaked. The sample swells vertically, and this displacement divided by the initial height is the swell potential, usually expressed as a percentage.

This methodology is attractive because it measures the desired characteristics directly, is relatively easy to perform, and does not require exotic test equipment (the test can be performed in a conventional consolidometer). However, because there is no universally accepted standard test procedure, the specifics of the test vary and results from different tests are not always comparable. The typical ranges of test criteria are as follows:

- **Sample size**: What is its diameter and height? Typically, the sample is 50 to 112 mm (2.0–4.5 inches) in diameter and 12 to 37 mm (0.5–1.5 inches) tall. Larger diameter samples are less susceptible to side friction and therefore tend to swell more.

- **Method of preparation**: Is the sample undisturbed or remolded? If it is undisturbed, how was it sampled and prepared? If it is remolded, how was it compacted, to what density and moisture content, and what curing, if any, was allowed?

- **Initial moisture content**: What is the moisture content at the beginning of the test? Some possibilities include:
  - In-situ moisture content
  - Optimum moisture content
  - Air dried moisture content

  Other options are also possible.

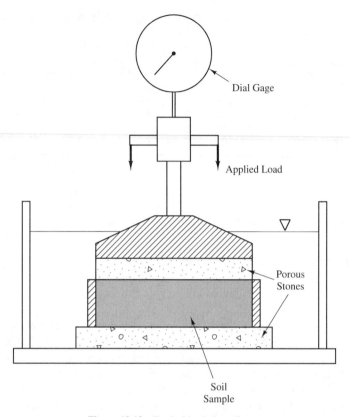

Figure 19.12   Typical loaded swell test.

- **Surcharge load**: How large is the surcharge load? It is usually between 2.9 and 71.8 kPa (60–1500 lb/ft^2). Some engineers prefer to use a surcharge equal to the in-situ or anticipated as-built overburden stress.
- **Duration**: Expansive soils do not swell immediately upon application of water. It takes time for the water to seep into the soil. This raises the question of how long to allow the test to run. Some engineers conduct the test for a specified time (i.e., 24 hr) whereas others continue until a specified rate of expansion is reached (such as no more than 0.03 mm/hr). The latter could take several days in some soils.

Snethen (1984) suggested the following definition of potential swell:

> Potential swell is the equilibrium vertical volume change or deformation from an oedometer-type[3] test (i.e., total lateral confinement), expressed as a percent of original height, of an undisturbed specimen from its natural water content and density to a state of saturation under an applied load equivalent to the in-situ overburden pressure.

---

[3]The terms *oedometer* and *consolidometer* are synonymous.

TABLE 19.7   TYPICAL CLASSIFICATION
OF SOIL EXPANSIVENESS BASED
ON LOADED SWELL TEST RESULTS
AT IN-SITU OVERBURDEN STRESS
(Adapted from Snethen, 1984)

Swell Potential (%)	Swell Classification
<0.5	Low
0.5–1.5	Marginal
>1.5	High

Snethen also suggested that the applied load should consider any applied external loads, such as those from foundations.

Using Snethen's test criteria, we could classify the expansiveness of the soil, as shown in Table 19.7.

Expansion Index Test

The *expansion index test* [ASTM D4829] (ICBO, 1997; Anderson and Lade, 1981) is an attempt to standardize the loaded swell test. In this test a soil sample is remolded into a standard 102-mm (4.01 in) diameter, 25-mm (1 in) tall ring at a degree of saturation of about 50 percent. A surcharge load of 6.9 kPa (1 lb/in^2) is applied, and then the sample is saturated and allowed to stand until the rate of swelling reaches a certain value or 24 hours, whichever is longer. The amount of swell is expressed in terms of the *expansion index*, or EI, which is defined as follows:

$$EI = 1000 \, h \, F \tag{19.1}$$

Where:

EI = expansion index

h = expansion of the soil (in)

F = percentage of the sample by weight that passes through a #4 sieve

Table 19.8 gives the interpretation of EI test results.

If the EI varies with depth, the test procedure also includes a series of weighting factors that emphasize the shallow soils and deemphasize the deeper soils. Although this concept is sound in principle, the stated factors imply an active zone depth of only 1.2 m (4 ft), which is far too shallow. Therefore, it may be best to ignore this portion of the test procedure.

Because the expansion index test is conducted on a remolded sample, it may mask certain soil fabric effects that may be present in the field.

TABLE 19.8   INTERPRETATION OF EXPAN-
SION INDEX TEST RESULTS (Reproduced from
the 1997 Edition of the *Uniform Building Code,*
©1997, with permission of the publisher, the International Conference of Building Officials)

EI	Potential Expansion
0–20	Very Low
21–50	Low
51–90	Medium
91–130	High
>130	Very High

Correlations

Several researchers have developed empirical correlations between swell potential and basic engineering properties. Vijayvergiya and Ghazzaly (1973) developed relationships for undisturbed soils, as shown in Figure 19.13. They use moisture content, liquid limit, and dry unit weight as independent variables and define the swell potential at a surcharge load of 9.6 kPa (200 lb/ft^2) and an initial moisture content equal to the in-situ moisture content.

Expansion Pressure Tests

As discussed earlier in this chapter, the surcharge pressure affects the swelling of a soil. Higher pressures provide more restraint, and a certain pressure, called the *expansion pressure*, or *swell pressure*, σ_s, prevents all swell. Some engineers elect to measure σ_s and use it as a measure of expansiveness.

Engineers can measure the expansion pressure at the end of a loaded swell test by simply increasing the normal load in increments until the sample returns to its original volume, as shown in Figure 19.14. Another method is to use a modified consolidometer that allows no vertical strain, as shown in Figure 19.15. The first method tends to produce larger swelling pressure. However, neither test precisely duplicates the actual sequence of loading and wetting in the field.

Some engineers believe the expansion pressure is independent of the initial moisture content, initial degree of saturation, and strata thickness of the soil and varies only with the dry unit weight and is therefore a fundamental physical property of an expansive soil (Chen, 1988). Others disagree with this evaluation and claim that it varies.

When testing undisturbed samples, Chen (1988) recommends defining the swelling pressure as that required to keep the soil at its in-situ dry unit weight. He also recom-

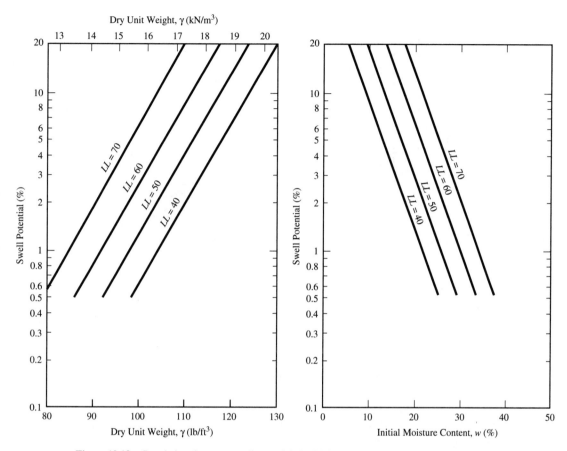

Figure 19.13 Correlations between swell potential, liquid limit, LL; initial moisture content, w; and dry unit weight, γ_d. Adapted from Vijayvergiya and Ghazzaly (1973), used with permission of the Israel Geotechnical Society.

mends testing remolded samples at 100 percent relative compaction and defining the swelling pressure as that required to maintain this dry unit weight.

Quantitative Evaluations

There is increasing emphasis on developing more rational analysis techniques to deal with expansive soils problems, a concept discussed later in this chapter. These methods require tests that evaluate the soil on a more fundamental basis. This approach has not yet been developed in detail and is not commonly used in practice. However, it will probably become much more dominant in the future.

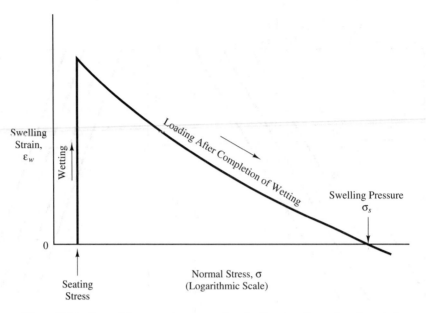

Figure 19.14   Determining expansion pressure by reloading the soil sample at the end of a loaded swell test.

Variation of Swell with Normal Stress

The swell potential varies with the normal stress acting on the soil. Shallow, lightly loaded soils will swell more than those that are deeper and more heavily loaded. Therefore, any rational method must consider this relationship.

One way to assess this relationship is to obtain undisturbed soil samples at various depths and perform a loaded swell test on each. The *constant volume swell* (CVS) test

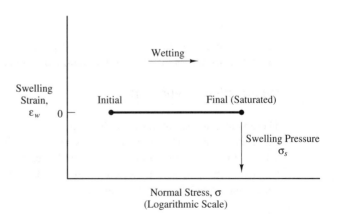

Figure 19.15   Determining expansion pressure using a zero-strain consolidometer.

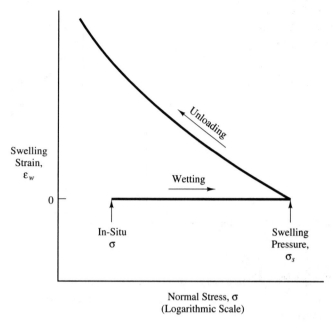

Figure 19.16 Constant volume swell (CVS) test results.

(Johnson and Stroman, 1976) is one such test. The procedure is generally as follows[4] and as shown in Figure 19.16:

1. Place an undisturbed soil sample in a consolidometer and apply a normal load equal to the in-situ overburden stress.

2. Inundate the sample and begin increasing the normal load in increments as necessary to restrain any swelling. Continue until the swelling pressure is fully developed.

3. Unload the soil in increments to obtain the swell curve. Continue until the load is less than the in-situ overburden stress.

Another similar procedure is the *modified swell overburden* (MSO) test (Johnson and Stroman, 1976), which follows and is shown in Figure 19.17:

1. Place an undisturbed soil sample in a consolidometer and apply a normal load equal to the design overburden stress (i.e., the stress at the sample location after the foundation is in place).

2. Inundate the sample and allow it to swell under the design overburden stress.

---

[4]See Johnson and Stroman (1976) for complete details on the CVS and MSO test procedures.

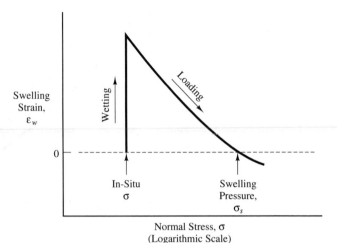

Figure 19.17   Modified swell overburden
(MSO) test results.

3. After the swelling is complete, load the sample in increments until the soil returns
    to its original volume. The pressure that corresponds to the original volume is the
    swelling pressure.

4. Unload the sample in increments to a stress less than the in-situ overburden.

Johnson and Stroman (1976) recommend the MSO test when the design overburden
pressures are known in advance, and the CVS test when they are not known in advance.

The strain measured in these tests is the *potential swell strain*, ϵ_w, for each normal
stress. Note that the test results are expressed in terms of total stress, σ, not effective
stress, σ'. Thus, this information will be used in a *total stress analysis*, not an *effective
stress analysis*.

Wetting Processes

The swell strain that occurs in the field is not necessarily equal to that measured in the laboratory because the soil in the field may not become completely saturated. The ratio of the
actual swell strain to the potential swell strain is the *wetting coefficient*, α. If the soil remains at its in-situ moisture content, then $\alpha = 0$; if it becomes completely saturated, $\alpha = 1$.

Chen (1988) suggests that α is approximately proportional to the change in the degree of saturation. Thus:

$$\boxed{\alpha \approx \frac{S - S_0}{1 - S_0}} \tag{19.2}$$

Where:

S_0 = degree of saturation before wetting (in decimal form)

S = degree of saturation after wetting (in decimal form)

For example, if the soil in the field is initially at a degree of saturation of 40 percent, and is wetted until it reaches $S = 80$ percent, then $\alpha = (0.80 - 0.40)/(1 - 0.40) = 0.67$. In other words, the swelling in the field will be only 67 percent of that in the lab.

Unfortunately, it is very difficult to predict the degree of saturation that will occur in the field. It depends on many factors, including the following:

- The rate and duration of water inflow (wetting) and outflow (evaporation and transpiration).
- The rate at which water flows through the soil.
- Stratification of the soil.

Flow in unsaturated soils is driven by soil suction, not by conventional hydraulic heads, so some engineers have attempted to use measurements of soil suction to predict the degree and depth of wetting. See Nelson and Miller (1992) for more details.

19.3   ESTIMATING POTENTIAL HEAVE

The current state-of-practice in most areas is to move directly from laboratory test results to the recommended design measures with no quantitative analyses to connect the two. Such leaps are possible only when the engineer is able to rely on local experience obtained from trial-and-error. For example, we may know that in a certain geologic formation, slab-on-grade floors have performed adequately only if the expansion index is less than some certain value. If the EI at a new project site in that formation is less than the specified value, then the engineer will recommend using a slab-on-grade floor; if not, then some other floor must be used.

This kind of methodology implicitly incorporates such factors as climate, depth of the active zone, hydraulic conductivity (especially the presence of fissures), and structural tolerance of differential heave, so they are limited only to the geologic formations, geographic areas and types of structures that correspond to those from which the method was derived. They generally work well as long as these restrictions are observed, but can be disastrous when extrapolated to new conditions.

We would prefer to have a more rational method of designing structures to resist the effects of expansive soils; one that *explicitly* considers these factors. Ideally, such a method would predict potential heave and differential heave. Just as engineers design spread footings based on their potential for settlement, it would be reasonable to design structures on expansive soils based on the potential for heave.

Laboratory Testing

Heave analyses are normally based on laboratory swell tests, such as the MSO or CVS tests described earlier. Conduct these tests on undistarbed samples from different depths within the active zone to establish the expansive properties of each strata. Typically, the moisture content of the soil at the beginning of each test is equal to the in-situ moisture content. Thus, the laboratory tests represent the swelling that would occur if the soil becomes wetter than the in-situ moisture. Sometimes, engineers will first dry the samples to a lower moisture content, thus modeling a worse condition.

Because the laboratory swell tests are laterally confined, they model a field condition in which the swell occurs only in the vertical direction. This may be a suitable model when the ground surface is level, but a poor one when it is sloped or when a retaining wall is present. In the latter cases, the horizontal swell is often very important.

In the field, some of the swell may be consumed by the filling of fissures in the clay. This is not reflected in the laboratory tests because the samples normally will not include fissures. However, this error should be small and conservative, and thus can be ignored.

Analysis

The heave caused by soil expansion is:

$$\boxed{\delta_w = \Sigma\, \alpha\, H\, \epsilon_w}$$

(19.3)

Where:

δ_w = heave caused by soil expansion

α = wetting coefficient

H = thickness of layer

ϵ_w = potential swell strain

We implement this analysis as follows:

Step 1: Divide the active zone of soil beneath the foundation into layers in a fashion similar to that used for settlement analyses. These layers should be relatively thin near the bottom of the footing (perhaps 1 ft or 0.25 m thick) and become thicker with depth. The bottom of the last layer should coincide with the bottom of the active zone.

Step 2: Compute the vertical total stress, σ_z, at the midpoint of each layer. This stress should be the sum of the geostatic and induced stresses (i.e., it must consider both the weight of the soil and load from the footing.

Step 3: Using the results of the laboratory swell tests, compute the potential swell strain, ϵ_w, at the midpoint of each layer.

Step 4: Determine the initial profile of degree of saturation vs. depth. This would normally be based on the results of moisture content tests from soil samples recovered from an exploratory boring.

Step 5: Estimate the final profile of degree of saturation vs. depth. As discussed earlier, this profile is difficult to predict. It is the greatest source of uncertainty in the analysis. Techniques for developing this profile include the following:

Option 1 Use empirical estimates based on observations of past projects. Studies of the equilibrium moisture conditions under covered areas, including buildings and pavements, suggest that the final moisture content is usually in the range of 1.1 to 1.3 times the plastic limit.

Option 2   Assume that the soil becomes saturated, but the pore water pressure above the original groundwater table remains equal to zero. This assumption is common for many foundation engineering problems and may be appropriate for many expansive soils problems, especially if extra water, such as that from irrigation or poor surface drainage, might enter the soil.

Option 3   Assume that a suction profile will develop such that a negative hydrostatic head is present. This scenario is based on a soil suction that diminishes with depth at the rate of 9.8 kPa/m of depth (62.4 lb/ft^2 per foot of depth).

Option 4   Assume $S = 100\%$ at the ground surface, and tapers to the natural S at the bottom of the active zone.

The second and third options are shown graphically in Figure 19.18.

Step 6:   Compute the heave for each layer and sum them using Equation 19.3.

For additional information on heave estimates, see McDowell (1956), Lambe and Whitman (1959), Richards (1967), Lytton and Watt (1970), Johnson and Stroman (1976), Snethen (1980), Mitchell and Avalle (1984), and Nelson and Miller (1992).

Example 19.1

A compressive column load of 140 kN is to be supported on a 0.50-m deep square footing. The allowable bearing pressure is 150 kPa. The soils beneath this proposed footing are expansive clays that currently have a degree of saturation of 25%. This soil has a unit weight of 17.0 kN/m^3, and the depth of the active zone is 3.5 m. The results of laboratory swell tests are shown in the figure.

   Compute the potential heave of this footing due to wetting of the expansive soils.

Solution

$$W_f = 0.50\, B^2\, (23.6\ \text{kN/m}^3) = 11.8\, B^2$$

$$B = \sqrt{\frac{P + W_f}{q_A}}$$

$$= \sqrt{\frac{140 + 11.8\, B^2}{150}}$$

$$= 1.00\ \text{m}$$

$$\sigma'_{zD} = \gamma D - u = (17.0\ \text{kN/m}^3)(0.5\ \text{m}) - 0 = 8\ \text{kPa}$$

Assume S after wetting varies from 100% at the ground surface to 25% at the bottom of the active zone. Compute $\Delta\sigma_z$ using Equation 7.7 and add it to σ_{z0} (the geostatic stress) to compute σ_z. Find ϵ_s using the lab data, α using Equation 19.2, and δ_w using Equation 19.3.

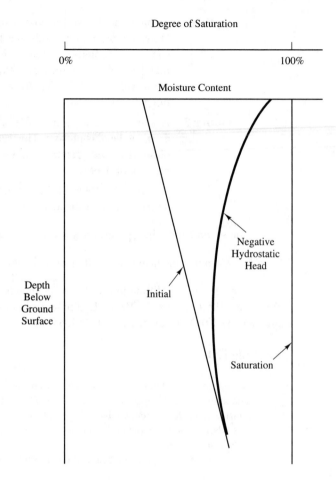

Figure 19.18 Final moisture content profiles.

			At Midpoint of Soil Layer							
Depth (m)	H (mm)	z_f (m)	σ_{z0} (kPa)	$\Delta\sigma_z$ (kPa)	σ_z (kPa)	ϵ_w (%)	S_0 (%)	S (%)	α	δ_w (mm)
0.50–0.75	250	0.12	11	141	152	2.0	25	90	0.87	4.3
0.75–1.00	250	0.32	15	126	141	2.1	25	80	0.73	3.8
1.00–1.50	500	0.75	21	68	89	3.5	25	70	0.60	10.5
1.50–2.00	500	1.25	30	33	63	3.9	25	50	0.33	6.4
2.00–3.00	1000	2.00	42	14	56	4.5	25	30	0.07	3.1
									Total	28

The estimated heave is **28 mm.**   ⇐ *Answer*

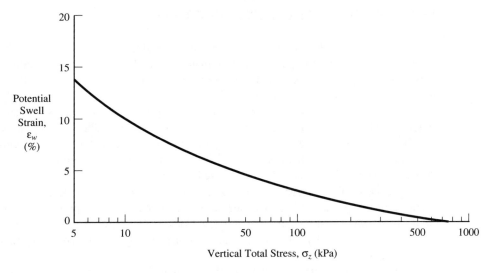

Figure 19.19   Laboratory test results for Example 19.1.

Differential Heave

Just as differential settlements often control the design of normal building foundations, differential heaves control the design of foundations on expansive soils. The differential heave can range from zero to the total heave, but is typically between one-quarter and one-half of the total heave (Johnson and Stroman, 1976). The greatest differential heaves are most likely to occur when swelling is due to such extraneous influences as broken water lines, poor surface drainage, or aggressive tree roots. Soil profiles with numerous fissures are also more likely to have higher differential heaves because of their higher hydraulic conductivity.

Donaldson (1973) recommended designing for the following heave factors (where heave factor = differential heave/total heave):

Profile with high hydraulic conductivity (i.e., fissures) in upper 3 m (10 ft)
>   Without extraneous influences   0.50
>   With extraneous influences   0.75

Profile with low hydraulic conductivity in upper 3 m (10 ft)
>   Without extraneous influences   0.25
>   With extraneous influences   0.40

19.4   TYPICAL STRUCTURAL DISTRESS PATTERNS

It is difficult to describe a typical distress pattern in buildings on expansive soils because the exact pattern of heaving depends on so many factors. However, in broad terms, buildings in arid areas tend to experience an *edge lift*, as shown in Figure 19.20a, that causes them to distort in a concave-up fashion (Simons, 1991). Conversely, in humid climates, the expansive soil may shrink when it dries, causing the edges to depress, as shown in Figure 19.20b. Heated buildings with slab-on-grade floors in colder climates sometimes experience a center depression caused by drying and shrinkage of the underlying clay soils. In the Dallas area, heaves of 125 to 150 mm (5–6 in) are common, and heaves of 200 to 300 mm (8–12 in) have been measured (Greenfield and Shen, 1992).

Special local conditions will often modify this pattern. For example, poor surface drainage or a leaky water line near one corner of the building will probably cause additional local heave, as shown in Figure 19.20c. Conversely, aggressive tree roots at another location may locally dry the soil and cause local shrinkage, as shown in Figure 19.20d (Byrn, 1991).

In arid climates, the heaving usually responds to seasonal moisture changes, producing annual shrink-swell cycles, as shown in Figure 19.21. However, during the first 4 to 6 years, the cumulative heave will usually exceed the cumulative shrinkage, so there will be a general heaving trend, as shown in Figure 19.21. As a result, expansive soil problems in dry climates will usually become evident during the first six years after construction.

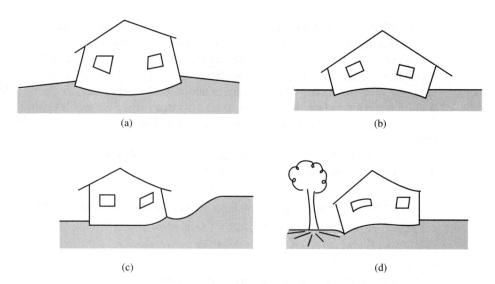

(a)   (b)

(c)   (d)

Figure 19.20   Typical distress patterns resulting from heave of expansive soils: (a) edge lift; (b) center lift; (c) localized heave due to drainage problems; and (d) localized shrinkage caused by aggressive tree roots.

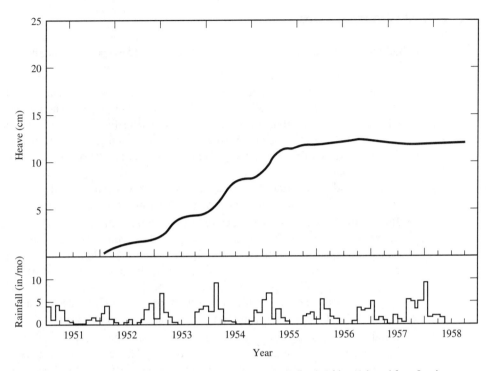

Figure 19.21   Heave record from single story brick house in South Africa (Adapted from Jennings, 1969).

Although this is a typical scenario, it does not mean that expansive soils will always behave this way. It is also very possible that unprecedented swelling may occur later in the life of the building. For example, an exceptionally wet year may invoke larger heaves. Likewise, changes in surface topography could cause ponding near the building and generate heave.

19.5   PREVENTIVE DESIGN AND CONSTRUCTION MEASURES

The next step is to develop appropriate design and construction methods to minimize (but not eliminate!) the potential for damage from expansive soils. As with most engineering problems, dealing with expansive soils ultimately boils down to a matter of risk vs. cost. A geotechnical engineer cannot guarantee that a structure will have no problems with expansive soils, but can recommend the use of certain preventive measures that seem to be an appropriate compromise between reducing the risk of damage (especially major damage) and keeping construction costs to a minimum.

Basic Preventive Measures

Any building site on an expansive soil should include at least the following features:

- **Surface Drainage**—Although good surface drainage is important at all building sites, it is especially critical where expansive soils are present. The ground surface should slope away from the structure, as shown in Figure 19.22. Bare or paved areas should have a slope of at least 2 percent, vegetated ground should have at least 5 percent. If possible, slope the ground within 3 m (10 ft) of the structure at a 10 percent grade.

    It is also important to install gutters or other means of collecting rainwater from the roof and discharging it away from the foundation.

- **Basement Backfills**—If the structure has a basement, the backfill should consist of nonexpansive soils. It should be well compacted to avoid subsequent settlement that would adversely affect the surface drainage patterns. In addition, a well-compacted backfill is less pervious, so water will be less likely to infiltrate the soil.

    Install a drain pipe at the bottom of the backfill to capture any water that might enter and carry it outside or to a sump pump. Carefully design such drains to avoid having them act as a conduit to bring water *into* the backfill.

- **Landscaping**—Irrigation near the structure can introduce large quantities of water into the soil and is a common cause of swelling. This can be an especially troublesome source of problems because irrigation systems are usually installed by homeowners or others who are not sufficiently conscious of expansive soil concerns. Specific preventative measures include:
    - Avoid placing plants and irrigation systems immediately adjacent to the structure.
    - Avoid placing irrigation pipes near the structure (to prevent problems from leaks).
    - Direct all spray heads away from the structure.

    As discussed earlier, large trees near the structure are often troublesome, especially those with shallow root systems. These trees can draw large quantities of water out of the soil, thus causing it to shrink. Therefore, it is best to avoid planting large trees near the structure.

- **Underground Utilities**—Utility lines often become distorted because of differential swelling of expansive soils. With water, sewer, or storm drain pipes, these distortions can create leaks that in turn cause more expansion. This progression is likely to occur where the pipes enter the building, and could cause large heaves and serious structural distress.

    The risk of this potential problem can be reduced by:
    - Using flexible pipe materials (i.e., PVC or ABS instead of clay or concrete pipe).
    - Installing the pipe such that large shear or flexural stresses will not develop. In some cases, this may require the use of flexible joints.

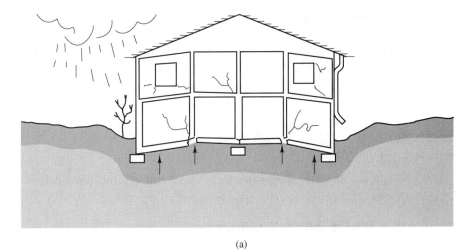

(a)

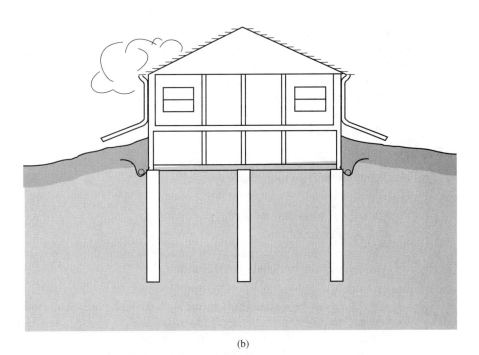

(b)

Figure 19.22   Surface drainage details: (a) poor drainage, wet expanded clay; and (b) good drainage; dry stable clay (Colorado Geological Survey).

Additional Preventive Measures

Beyond these basics, it also is possible to incorporate more extensive measures. O'Neill and Poormoayed (1980) divided them into three basic categories:

- Alter the expansive clay to reduce or eliminate its swelling potential.
- Bypass the expansive clay by isolating the foundation from its effect.
- Mitigate the movements in the superstructure.

Each of these approaches includes several specific methodologies, as discussed below.

Altering the Expansive Clay

Replacement

Perhaps the most obvious method of overcoming expansive soil problems is to remove the soil and bring in a nonexpansive soil as a replacement. When done carefully, this can be a very effective, although expensive, method. However, be careful not to use a highly permeable soil that could provide an avenue for water to infiltrate the natural soils below (which are probably expansive), thus increasing the depth of the active zone.

Lime Treatment

When hydrated lime is mixed with an expansive clay, a chemical reaction occurs and the clay is improved in the following ways:

- Swelling potential is reduced.
- Shear strength is increased.
- Moisture content is decreased (helpful when working during the wet season because it increases the workability of the soil).

The lime can be mechanically mixed with the soil at a rate of about 2 to 8 percent by weight. Special equipment is needed to assure adequate mixing and the process generally limited to shallow depths (i.e., 300 mm or 12 in).

Another method of treating soil with lime is to inject it in a slurry form using a technique known as *pressure-injected lime* (PIL). The lime slurry is forced into the soil under high pressure using equipment similar to that shown in Figure 19.23. This method is capable of treating soils to depths of up to about 2.5 m (8 ft).

The PIL technique is most effective in highly fissured soils because the fissures provide pathways for the slurry to disperse. In addition to the chemical effects, the filling of the fissures with lime also retards moisture migration in the soil.

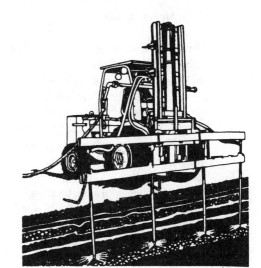

Figure 19.23   Pressure-injected lime (PIL) system (courtesy of GKN Hayward Baker, Inc.)

Lime treatment is most commonly used on canals, highways, and other projects that have no foundation. The pressure injected lime technique is also useful when making repairs to an existing structure that has suffered damage from expansive soils.

Prewetting

This technique, also known as *ponding*, *presoaking*, or *presaturation*, consists of covering the site with water before construction in an attempt to increase the moisture content of the soil, thus preswelling it. When used with a project that will include a slab-on-grade, the moisture will remain reasonably constant, especially if the perimeter of the site is landscaped and irrigated or if moisture barriers are installed around the structure. The idea here is to cause the soil to expand before building the structure, and then maintain it at a high moisture content.

In some areas, such as southern California, this technique works well and generally requires a soaking time of a few days or weeks. However, in other areas, the required soaking time is unacceptably long (i.e., many months). These differences in soaking times may be due to differing required depths of soaking (a function of the depth of the active zone) and the presence or absence of fissures in the clay.

The soaking process can be accelerated by first drilling a grid of vertical sand drains (borings filled with sand) to help the water percolate into the soil. The use of wetting agents in the water also can accelerate this process.

After the prewetting is completed, it is usually necessary to treat the ground surface to provide a working platform. This could consist of lime treating the upper soils or placing a 100 to 150 mm (4–6 in) thick layer of sand or gravel on the surface.

Moisture Barriers

Impermeable moisture barriers, either horizontal or vertical, can be an effective means of stabilizing the moisture content of the soil under a structure. These barriers may be located on the ground surface in the form of sidewalks or other paved areas, or they can be buried. The latter could consist of underground polyethylene or asphalt membranes. Moisture barriers are especially helpful under irrigated landscape areas where they can take the form of sealed planter boxes.

The primary advantage of barriers is that they promote more uniform moisture conditions below the structure, thus reducing the differential heave. They may or may not affect the total heave.

Although moisture barriers can be very helpful, never consider them to be completely impervious. They are generally supplementary measures that work in conjunction with other techniques.

Bypassing the Expansive Clay

Because the moisture content of a soil will fluctuate more near the ground surface than it will at depth (the active zone concept described earlier), one method of mitigating swelling effects is to support the structure on deeper soils, thus bypassing some or all of the active zone. This method is also useful when the expansive soil strata is relatively thin and is underlain by a nonexpansive soil.

Deepened Footings

When working with mildly expansive soils it often is possible to retain a spread footing foundation system by simply deepening the footings, perhaps to about 0.5 m (3 ft) below grade. This will generally be less than the depth of the active zone, so some heave would still be possible, but its magnitude will be much less.

This method also has the advantage of increasing the rigidity of the footing (which is usually supplemented by additional reinforcement—say, one or two #4 bars top and bottom), thus spreading any heave over a greater distance and improving the structure's tolerance of heave.

When spread footings are used, they should be designed using as high a bearing pressure as practicable to restrain the heaving. A bearing pressure equal to the swelling pressure would be ideal, but is generally possible only in very mildly expansive soils.

Drilled Shafts

In a highly expansive soil, deepened spread footings cease to be practical and a drilled shaft foundation often becomes the preferred system. In the Denver, Colorado area, shafts for lightweight structures are typically 250 to 300 mm (10–12 inches) in diameter and 4.5 to 6 m (15–20 ft) deep (Greenfield and Shen, 1992). The shafts must extend well below the active zone. The individual shafts are connected with grade beams that are cast on top of collapsible cardboard or foam forms, as shown in Figure 19.24. The purpose of

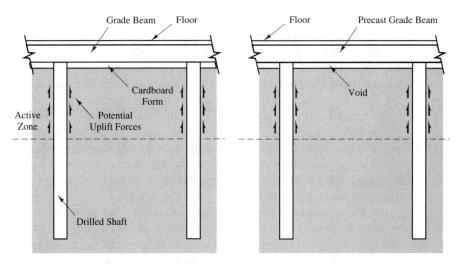

Figure 19.24   Typical lightweight building foundation consisting of drilled shafts and raised grade beams: (a) cast-in-place grade beam with cardboard forms are designed to collapse at pressures slightly greater than those from the wet concrete; (b) precast concrete grade beam. Note the uplift forces acting on the shaft due to the heave of the soil in the active zone.

these forms is to permit the soil to freely expand without pressing against the grade beam. Another alternative is to use precast grade beams.

This system also works well on larger structures, although much larger shafts are then required.

One of the special concerns of using drilled shafts in expansive soils is the development of uplift forces along the sides of the shafts within the active zone, as shown in Figure 19.24. The shafts must be designed to accommodate these forces, both in terms of load transfer and the need for tension steel extending through the active zone.

This matter is further aggravated because the soil also attempts to swell horizontally, which translates to an increased normal stress between the soil and the shaft. This, in turn, permits more side friction to be mobilized and increases the uplift force.

Reese and O'Neill (1988) recommend computing the unit uplift side friction in the active zone as follows:

$$f_s = a\,\sigma_{hs}\tan\phi_r \qquad\qquad (19.4)$$

Where:

f_s = unit uplift side friction

a = an empirical coefficient

σ_{hs} = horizontal swelling pressure

ϕ_r = effective residual friction angle of the clay

For design purposes, consider the horizontal swelling pressure, σ_{hs}, to be equal to the vertical swelling pressure obtained from a swelling pressure test.

Unfortunately, we cannot yet define the correlation coefficient, a, with any great precision. O'Neill and Poormoayed (1980) back-calculated a value of 1.3 from a single instrumented shaft in San Antonio. This value agrees very closely with Chen (1988) whose work was based on model tests and has been successfully used in practice. This is probably a reasonable number to use until we obtain more experimental data.

Obtain the effective residual friction angle from laboratory tests. Chen (1988) suggests that for stiff clay and clay shale, it is usually on the order of 5 to 10 degrees.

Consider the possibility that parts of the shaft might develop a net tensile force if the soil were to swell. Therefore, the design must include steel reinforcing bars and they should extend to the bottom of the shaft. In Denver, shafts that support lightweight structures typically have at least two full-length #5 grade 40 rebars, whereas in Dallas, two #6 bars are common (Greenfield and Shen, 1992).

An alternative to designing for uplift loads is to isolate the shaft from the soil in the active zone. One way to do so is by forming the shaft with a cardboard tube inside a permanent steel casing and filling the annular space with a weak but impervious material.

For design purposes, the resistance to either upward or downward movements is considered to begin at the bottom of the active zone. This resistance can be generated by a straight shaft, one with a bell, or one with shear rings. The latter two are commonly preferred when large uplift loads are anticipated.

Structurally-Supported Floors

When the computed total heave exceeds about 25 to 50 mm (1–2 in), conventional slab-on-grade floors cease to perform well. In such cases, the most common design is to use a raised floor supported on a drilled shaft foundation that penetrates through the active zone, as shown in Figure 19.25. Not only does this design isolate the floor from direct heaving, it also provides ventilation of the ground surface while still shielding it from precipitation. This keeps the soil under the building much drier than it would be with a slab-on-grade floor or a mat floor.

Mitigating Movements in the Structure

Another method of addressing differential settlement and heave problems is to make the structure more tolerant of these movements. There are many ways to accomplish this, and these measures can be used alone or in conjunction with the methods described earlier.

Flexible Construction

Some structures will tolerate a large differential movement and still perform acceptably. Lightweight industrial buildings with steel siding are an example of this type of construction. See the discussion in Chapter 2 for more details on this concept.

Another way of adding flexibility to a structure is to use *floating slabs*, as shown in Figure 19.26. Casting these slabs separate from the foundation and providing a slip joint between the slab and the wall allow it to move vertically when the soil swells.

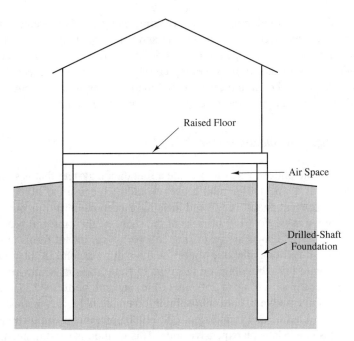

Figure 19.25   Bypassing an expansive clay with a raised floor and a drilled shaft foundation.

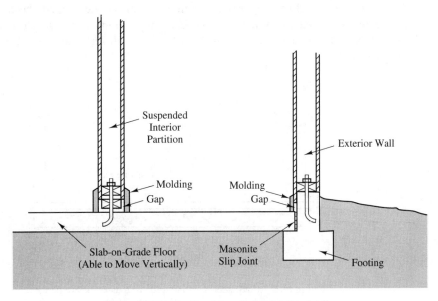

Figure 19.26   Floating slab and related design details.

Any construction resting on a floating slab must also be able to move. For example, furnaces would need a flexible plenum and vent pipe. Partition walls could be suspended from the ceiling with a gap at the bottom covered with flexible molding. Stairways also could be suspended from the ceiling and wall and not be connected to the slab. Each of these details is shown in Figure 19.26. Floating slabs are most commonly used in garages and basements because these design details are easier to implement.

Rigid Foundation System

The opposite philosophy is to provide a foundation system that is so rigid and strong that it moves as a unit. Differential heaves would then cause the structure to tilt without distorting. Conventionally reinforced mats have been used for this purpose. These mats are also known as *waffle slabs* because of the shape of their integral beams, as shown in Figure 19.27. These slabs are cast using collapsible cardboard forms to provide void spaces under the slab, thus allowing space for the soil to expand. Kantey (1980) noted routine success with brick buildings on this type of foundation in South Africa, where some had experienced heaves of up to 250 mm (10 in) and still performed adequately.

An alternative to conventional reinforcement is to use prestressed or post-tensioned slabs as a kind of mat foundation. PTI (1980) presents a complete design procedure for post-tensioned slabs on expansive soils. This method is gaining popularity, especially for residential projects, and has been used successfully on highly expansive soils in California, Texas, and elsewhere.

We can apply this concept more economically to mildly expansive soils by using conventional spread footings in such a way that no footing is isolated from the others. This can be accomplished by using continuous footings and/or grade beams, as shown in Figure 19.28. Such a system does not have the rigidity of a mat, but is much more rigid than isolated footings and will help to spread differential heaves over a longer distance.

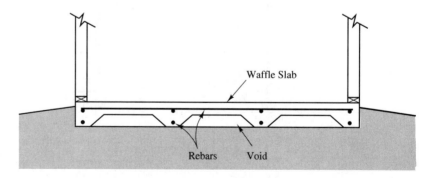

Figure 19.27   Conventionally reinforced mat foundation "waffle slab."

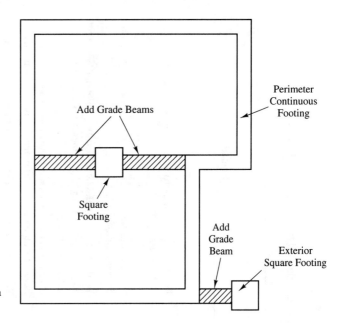

Figure 19.28 Use of continuous footings and grade beams to improve the rigidity of a spread footing system.

Determining Which Methods to Use

Gromko (1974) suggested the criteria in Table 19.9 to guide the selection of preventive design measures based on the estimated heave and the length-to-height ratio, L/H, of the walls.

19.6 OTHER SOURCES OF HEAVE

Although expansive clays are the most common source of heave, other mechanisms also have been observed.

Expansive Rocks

Sedimentary rocks formed from clays, such as claystone, and shale, are often expansive (Lindner, 1976). The physical mechanisms are similar to those for clay soils, but the swelling pressures and potential heave are often very high because of the high unit weight of rock. However, these rocks do not transmit water as easily, so the potential heave may be more difficult to attain in the field.

Some other rocks can expand as a result of chemical processes, such as oxidation or carbonation. These processes often create by-products that occupy a larger volume than the original materials (Lindner, 1976).

TABLE 19.9 PREVENTIVE DESIGN MEASURES BASED ON COMPUTED HEAVE (Adapted from Gromko, 1974. Used with permission of ASCE).

Total Predicted Heave		Recommended Construction	Method	Remarks
L/H = 1.25	L/H = 2.50			
< 6 mm <0.25 in	< 12 mm <0.5 in	No precautions		
6–12 mm 0.25–0.5 in	12–50 mm 0.5–2 in	Rigid building tolerating movement (steel reinforcement as necessary)	Foundations: Spread footings, mats, or waffle	Footings should be narrow and deep, consistent with the soil bearing capacity.
			Floor slabs: Waffle, tile	Slabs should be designed to resist bending and should be independent of grade beams.
			Walls	Walls on a mat should be as flexible as the mat. No rigid connections vertically. Strengthen brickwork with tiebars or bands.
12–50 mm 0.5–2 in	50–100 mm 2–4 in	Building damping movement	Joints: Clear, flexible	Contacts between structural units should be avoided, or a flexible waterproof material may be inserted in the joint.
			Walls: Flexible, unit construction, steel frame	Walls or rectangular building units should heave as a unit.
			Foundations: Three-point, cellular or jacks	Cellular foundations allow slight soil expansion to reduce the swelling pressure. Three-point loading allows motion without duress.
> 50 mm > 2 in	> 100 mm > 4 in	Building independent of movement	Drilled shafts: Straight, belled	Use smallest diameter and most widely spaced shafts compatible with load. Allow clearance under grade beams.
			Suspended floor	Suspend floors on grade beams 300–450 m (12–18 in) above the soil.

Steelmaking Slag

The process of making steel from iron ore produces two principal types of solid wastes: *blast furnace slag* and *steelmaking slag* (Lankford et al., 1985). Blast furnace slag is the waste produced when iron ore and limestone are combined in a blast furnace to produce iron. This material has very favorable engineering properties and has been used for concrete aggregate (with some problems), base material beneath pavements, and many other applications (Lee, 1974). In contrast, steelmaking slag is produced when iron is made into steel, and it is a much more troublesome material.

The primary problem with steelmaking slag is that it can expand in volume after being put in place, thus producing problems similar to those caused by expansive clays. For example, Crawford and Burn (1968) described a case history of a floor slab built over sand-sized steelmaking slag that heaved up to 75 mm (3 in) in 5 years. Uppot (1980) described another case history of an industrial building that experienced heaves of up to 200 mm (8 in) in columns and up to 250 mm (10 in) in floor slabs within six years of construction.

This expansion is the result of hydration of unslaked lime, magnesium oxides, and other materials, and may exceed 20 percent of the original volume of the slag. Spanovich and Fewell (1968) observed that the expansion potential is reduced by more than 50 percent if the slag is allowed to age for 30 days before being used. This aging process requires exposure to oxygen and water. However, they also observed that slag that had been buried for more than 30 years (and thus had not been cured) was still very expansive.

Some engineers have used cured steelmaking slag for applications that have a high tolerance of movements, such as open fills, unpaved roads, and railroad ballast. Engineers in Japan have used mixtures of steelmaking slag and blast furnace slag to produce aggregate base material for roadway pavements (Nagao, et al., 1989). However, because of its expansive properties, steelmaking slag should not be used beneath structural foundations.

Salt Heave

Soils in arid areas sometimes contain high concentrations of water-soluble salts that can crystallize out of solution at night when the temperature is low and return to solution during the day as the temperature rises. The formation and dissipation of these salt crystals can cause daily cycles of heave and shrinkage in the soil, especially in the late fall and early spring when the difference between day and night air temperatures may be 60° F (33° C) or more. Although some moisture must be present for crystallization to occur, this process is driven by changes in temperature. This phenomenon, known as *salt heave*, or *chemical heave*, normally occurs only in the upper 0.3 to 0.6 m (1–2 ft) of the soil because the salt concentrations and temperature fluctuations are greatest in this zone.

Based on studies of soils in the Las Vegas, Nevada area, Cibor (1983) suggested that salt heave can generate swelling pressures of 10 to 15 kPa (200–300 lb/ft²). The bearing pressure beneath most spread footings is much larger than this swelling pressure, so these footings are able to resist the heaving forces without moving. In addition, the bottoms of footings are typically below the zone of greatest heave potential. However, heave has been observed in very shallow and lightly loaded footings.

Damage from salt heave is most often seen in slab-on-grade floors, sidewalks, and other shallow, very lightly loaded areas. For example, Blaser and Scherer (1969) observed heaves of 100 mm (4 in) in exterior concrete slabs.

Evaluate these soils by measuring the percentage of salts in solution. Typically, concentrations of greater than 0.5 percent soluble salts in soils with more than 15 percent fines may be of concern. The heave potential can be measured using a thermal swell test (Blaser and Scherer, 1969). This test is similar to the swell tests discussed earlier in this chapter, except that the soil expansion is induced by cooling the sample instead of adding water.

A common method of avoiding salt heave problems is to excavate the salt-laden natural soils to a depth of 0.3 to 0.6 m (1–2 ft) below the proposed ground surface and backfill with open graded gravel. Because the gravel provides thermal insulation, the temperature below will not vary as widely as on the ground surface, so salt heave will be less likely to occur. In addition, the weight of the gravel will resist any swelling pressures that might develop.

SUMMARY

Major Points

1. Expansive soils cause more property damage in the United States than all the earthquakes, floods, tornados, and hurricanes combined.

2. Most of this damage is inflicted on lightweight improvements, such as houses, small commercial buildings, and pavements.

3. Soil swelling is the result of the infiltration of water into certain clay minerals, most notably montmorillonite. Swelling will occur only if the moisture content of the soil changes.

4. Expansive soils are most likely to cause problems in subhumid, semiarid, and arid climates. The patterns of precipitation and evaporation/transpiration in these areas usually cause the moisture content of the soil to fluctuate during the year, which creates cycles of shrinking and swelling. New construction in these areas normally promotes an increased moisture content in the soil that further aggravates the problem.

5. The potential heave at a given site is a function of the soil profile, variations in soil moisture, overburden loads, and superimposed loads.

6. A wide variety of test methods are available to evaluate the degree of expansiveness. Some of these methods are primarily qualitative, and others provide quantitative results.

7. Preventive measures fall into three categories:
   a. Alter the condition of the expansive clay.
   b. Bypass the expansive clay by isolating the foundation from its effect.

    c. Provide a shallow foundation capable of withstanding differential movements and mitigating their effect in the superstructure.

8. The current state-of-practice in most places is to select the appropriate preventive measures based on qualitative or semiquantitative test results and local experience. These methods are generally applicable only to limited geographic areas and for certain types of structures.

9. The profession is moving toward the use of heave calculations as a basis for determining preventive measures. Such calculations can be based on either loaded swell tests or soil suction tests. Hopefully, these methods will help the engineer more accurately characterize the soil and soil-structure interaction.

Vocabulary

Active zone	Expansive soil	Montmorillonite
Bentonite	Heave	Prewetting
Clay minerals	Illite	Slag
Constant volume swell test	Kaolinite	Swell potential
Differential heave	Lime treatment	Thornthwaite moisture index
Edge lift	Loaded swell test	Waffle slab
Expansion index	Modified swell overburden test	Wetting coefficient
Expansion pressure	Moisture barrier	

COMPREHENSIVE QUESTIONS AND PRACTICE PROBLEMS

19.1 Why are lightweight structures usually more susceptible to damage from expansive soils?

19.2 What types of climates are most prone to cause problems with expansive clays?

19.3 How do human activities often aggravate expansive soil problems?

19.4 What is the "active zone"?

19.5 What is the "swelling pressure"?

19.6 What are the primary sources of uncertainty in heave analyses?

19.7 Why do the soils beneath buildings with raised floors tend to be drier than those beneath slab-on-grade floors?

19.8 The foundation described in Example 19.1 is to be redesigned using a net allowable bearing pressure of 75 kPa. Compute the potential heave and compare it with that computed in the example. Discuss.

19.9 An 8 inch diameter drilled shaft will penetrate through an expansive stiff clay to a depth well below the active zone. It will carry a downward load of 5200 lb. The horizontal swelling pressure in this soil is approximately 5000 lb/ft^2 and the active zone extends to a depth of 10 ft. No residual strength data are available. Determine the following:

a. What is the uplift skin friction load?

b. Is a tensile failure in the shaft is possible (do not forget to consider the weight of the shaft)?

c. What is the required reinforcing steel to prevent a tensile failure (if necessary)? Use $f_y = 40$ k/in^2 and a load factor of 1.7.

20

Foundations on Collapsible Soils

It is better to fail while attempting to do something worthwhile
than to succeed at doing something that is not.

Foundation engineers who work in arid and semiarid areas of the world often encounter deposits of *collapsible soils*. These soils are dry and strong in their natural state and appear to provide good support for foundations. However, if they become wet, these soils quickly consolidate, thus generating unexpected settlements. Sometimes these settlements are quite dramatic, and many buildings and other improvements have been damaged as a result. These soils are stable only as long as they remain dry, so they are sometimes called *metastable soils*, and the process of collapse is sometimes called *hydroconsolidation*, *hydrocompression*, or *hydrocollapse*.

To avoid these kinds of settlements, the foundation engineer must be able to identify collapsible soils, assess the potential settlements, and employ appropriate mitigation measures when necessary (Clemence and Finbarr, 1981).

20.1 ORIGIN AND OCCURRENCE OF COLLAPSIBLE SOILS

Collapsible soils consist predominantly of sand and silt size particles arranged in a loose "honeycomb" structure, as shown in Figure 20.1. Sometimes gravel is also present. This loose structure is held together by small amounts of water-softening cementing agents, such as clay or calcium carbonate (Barden et al., 1973). As long as the soil remains dry, these cements produce a strong soil that is able to carry large loads. However, if the soil becomes wet, these cementing agents soften and the honeycomb structure collapses.

Various geologic processes can produce collapsible soils. By understanding their geologic origins, we are better prepared to anticipate where they might be found.

701

Figure 20.1 Microscopic view of a collapsible soil. In their natural state, these soils have a honeycomb structure that is held together by water soluble bonds. However, if the soil becomes wet, these bonds soften and the soil consolidates (Adapted from Houston, et al., 1988; Used with permission of ASCE).

Loaded Soil Structure
Before Inundation

Loaded Soil Structure
After Inundation

Collapsible Alluvial and Colluvial Soils

Some *alluvial soils* (i.e., soils transported by water) and some *colluvial soils* (i.e., soils transported by gravity) can be highly collapsible. These collapsible soils are frequently found in the southwestern United States as well as other regions of the world with similar climates. In these areas, short bursts of intense precipitation often induces rapid downslope movements of soil known as *flows*. While they are moving, these soils are nearly saturated and have a high void ratio. Upon reaching their destination, they dry quickly by evaporation, and capillary tension draws the pore water toward the particle contact points, bringing clay and silt particles and soluble salts with it, as shown in Figure 20.2. Once the

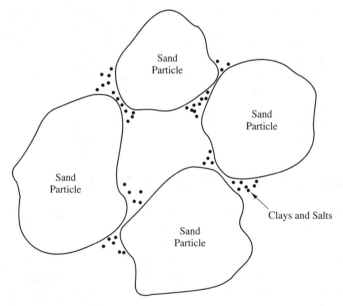

Figure 20.2 When flow deposits dry by evaporation, the retreating water draws the suspended clay particles and dissolved salts toward the particle contact points.

Figure 20.3   Aerial view of flow topography in the desert near Palm Springs, California. Both surface water and windstorms in this area move from the right side of the photo to the left side. Thus, the alluvial and aeolian soil deposits form long stripes on the ground. These soils are often highly collapsible, so settlement problems can occur in the heavily irrigated developments, such as the golf course community shown in the bottom of this photograph.

soil becomes dry, these materials bond the sand particles together, thus forming the honeycomb structure.

When the next flow occurs, more honeycomb-structured soil forms. It, too, dries rapidly by evaporation, so the previously deposited soil remains dry. Thus, deep deposits of this soil can form. The remains of repeated flows are often evident from the topography of the desert, as shown in Figure 20.3. These deposits are often very erratic, and may include interbedded strata of collapsible and noncollapsible soils.

In some areas, only the upper 1 to 3 m (3–10 ft) of these soils is collapsible, whereas elsewhere the collapse-prone soils may extend 60 m (200 ft) or more below the ground surface. An example of the latter is the San Joaquin Valley in Central California. Irrigation canals are especially prone to damage in that area because even small leaks that persist for a long time may wet the soil to a great depth. Settlements of 600 to 900 mm (2–3 ft) are common, and some cases of up to 4.7 m (15 ft) have been reported (Dudley, 1970).

Collapsible Aeolian Soils

Soils deposited by winds are known as *aeolian soils*. These include windblown sand dunes, loess, volcanic dust deposits, as well as other forms. *Loess* (an aeolian silt or sandy silt) is the most common aeolian soil and it covers much of the earth's surface. It is found in the United States, central Europe, China, Africa, Australia, the former Soviet Union, India, Argentina, and elsewhere (Pye, 1987). The locations of major loess deposits in the United States are shown in Figure 20.4 along with locations of other reported collapsible soil deposits.

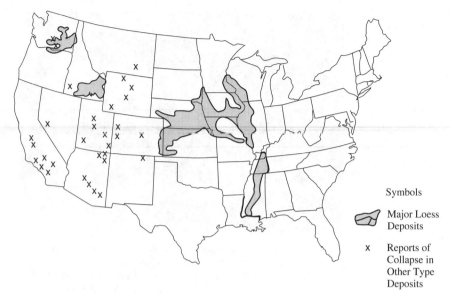

Figure 20.4 Locations of major loess deposits in the United States along with other sites of reported collapsible soils (Adapted from Dudley, 1970; Used with permission of ASCE).

Collapsible loess has a very high porosity (typically on the order of 50 percent) and a correspondingly low unit weight (typically 11–14 kN/m^3 or 70–90 lb/ft^3). The individual particles are usually coated with clay, which acts as a cementing agent to maintain the loose structure. This cementation is often not as strong as that in many alluvial soils, so collapse can occur either by wetting under a moderate normal stress or by subjecting the soil to higher normal stresses without wetting it.

Loess deposits are generally much less erratic than other types of collapsible soils, but they are often much thicker. Deposits 60 m (200 ft) thick are not unusual.

Collapsible Residual Soils

Residual soils are those formed in-place by the weathering of rock. Sometimes this process involves decomposition of rock minerals into clay minerals that may be removed by leaching, leaving a honeycomb structure and a high void ratio. When this structure develops, the soil is prone to collapse. For example, Brink and Kantey (1961) reported that residual decomposed granites in South Africa often collapse upon wetting, leading to a 7 to 10 percent increase in unit weight. Residual soils derived from sandstones and basalts in Brazil also are collapsible (Hunt, 1984).

Dudley (1970) reported test results from a residual soil from Lancaster, California, that showed nearly zero consolidation when loaded dry to a stress of 670 kPa (14,000 lb/ft^2) over the natural overburden stress, yet collapsed by 10 percent of its volume when soaked.

Residual soils are likely to have the greatest amount of spatial variation, thus making it more difficult to predict the collapse potential.

20.2   IDENTIFICATION, SAMPLING, AND TESTING

Engineers have used many different techniques to identify and evaluate collapsible soils. They may be divided into two categories: indirect methods and direct methods.

The indirect methods assess collapse potential by correlating it with other engineering properties such as unit weight, Atterberg limits, or percent clay particles. The result of such efforts is usually a qualitative classification of collapsibility, such as "highly collapsible." Although these classifications can be useful, they provide little, if any, quantitative estimates of the potential settlements (Lutenegger and Saber, 1988). In addition, most of them have been developed for certain types of soil, such as loess, and cannot necessarily be used for other types, such as alluvial soils.

As a result of these difficulties, many engineers prefer to use direct methods of evaluating potentially collapsible soils. These methods involve actually wetting the soil, either in the laboratory or in-situ, and measuring the corresponding strain. We can then extrapolate the results of such tests to the entire soil deposit and predict the potential settlements.

Obtaining Samples of Collapsible Soils

Laboratory tests are more useful than in-situ tests because of the greater degree of control that is possible in the lab. However, to perform such tests, we must obtain relatively undisturbed samples of the soil and bring them to the lab.

Collapsible soils that are moderately to well cemented and do not contain much gravel usually can be sampled for purposes of performing collapse tests without undue difficulty. This includes many of the alluvial collapsible soils. As with any sampling operation, the engineer must strive to obtain samples that are as undisturbed as possible and representative of the soil deposit. Unfortunately, collapsible soils, especially those of alluvial or residual origin, are often very erratic. It is difficult to obtain representative samples of erratic soil deposits, so we must obtain many samples to accurately characterize the collapse potential. Considering the usual limitations of funding for soil sampling, it is probably wiser to obtain many good samples rather than only a couple of extremely high quality (but expensive) samples.

Sometimes, conventional thin-wall Shelby tube samplers can be pressed into the soil. It is best to use short tubes to avoid compressing the samples. Unfortunately, because collapsible soils are strong when dry (it is important not to artificially wet the soil during sampling), it often is necessary to hammer the tube in place. Fortunately, studies by Houston and El-Ehwany (1991) suggest that hammering does not significantly alter the results of laboratory collapse tests for cemented soils.

Although Shelby tubes work well in silty and sandy soils, they are easily bent when used in soils that contain even a small quantity of gravel. This is often the case in collapsible soils, so we may be forced to use a sampler with heavier walls, such as the ring-lined barrel sampler shown in Figure 4.5 in Chapter 4. Although these thick-wall samplers generate more sample disturbance, their soil samples may still be suitable for laboratory collapse tests (Houston and El-Ehwany, 1991).

Lightly cemented collapsible soils, such as loess, are much more difficult to sample and require more careful sampling techniques. Fortunately, loess is usually much more homogeneous, so fewer samples are needed.

Gravelly soils are more difficult to sample and therefore more difficult to evaluate. Special in-situ wetting tests are probably appropriate for these soils (Mahmoud, 1991).

Laboratory Soil Collapse Tests

Once samples have been obtained, they may be tested in the laboratory by conducting *collapse tests*. These are conducted in a conventional oedometer (consolidometer) and directly measure the strain (collapse) that occurs upon wetting. Although the details of the various test procedures vary, we can divide them into two groups: double oedometer tests and single oedometer tests.

Double Oedometer Method

Jennings and Knight (1956, 1957, 1975) developed the *double oedometer method* while investigating collapsible soils in South Africa. This method uses two parallel consolidation (oedometer) tests on "identical" soil samples. The first test is performed on a sample at its in-situ moisture content; the second on a soaked sample. The test results are plotted together, as shown in Figure 20.5. The vertical distance between the test results represents the *potential hydrocollapse strain*, ϵ_w, as a function of normal stress.

Single Oedometer Method

Many engineers prefer to use the single oedometer method to assess collapse potential. With this method, each test requires only one soil sample.

The single oedometer test is usually conducted as follows (Houston et al., 1988):

1. Place an undisturbed soil sample in an oedometer and maintain the in-situ moisture content.

2. Apply a seating load of 5 kPa (100 lb/ft^2) and zero the dial gauge.

3. Increase the vertical stress in increments, allowing the soil to consolidate each time. Normally the load may be changed when the rate of consolidation becomes less than 0.1 percent per hour. Continue this process until the vertical stress is equal to or slightly higher than that which will occur in the field.

4. Inundate the soil sample and monitor the resulting hydrocompression. This is the potential hydrocollapse strain, ϵ_w, for this overburden stress.

5. Once the hydroconsolidation has ceased, apply an additional stress increment and allow the soil to consolidate.

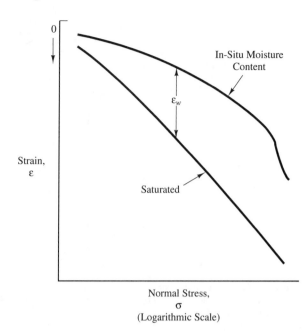

Figure 20.5   Typical results from double
oedometer tests.

This procedure can normally be completed in 24 to 36 hours. The results of such a test are shown in Figure 20.6.

The single oedometer method is faster and easier and it more closely simulates the actual loading and wetting sequence that occurs in the field. It also overcomes the problem of obtaining the two identical samples needed for the double oedometer test. However, this method provides less information because it only gives the hydrocollapse strain at one normal stress. Therefore, the soil should be wetted at a normal stress that is as close as possible to that which will be present in the field.

Evaluation of Laboratory Collapse Test Results

The most common practice has been to evaluate collapse test results using criteria such as that shown in Table 20.1. When using this approach, engineers specify the required remedial measures, if any, based on a qualitative assessment of the test results. For example, soils with an ϵ_w greater than some specified value might always be excavated and compacted back in place.

Although such methods often are sufficient, they do not provide an estimate of the potential settlement due to hydrocollapse. For example, a thick stratum of "moderate trouble" soil that becomes wetted to a great depth may cause more settlement than a "severe trouble" soil that is either thinner or does not become wetted to as great a depth. Therefore, it would be more useful to perform a *quantitative* evaluation of the test data to estimate the potential settlements due to collapse and develop any preventive design measures accordingly.

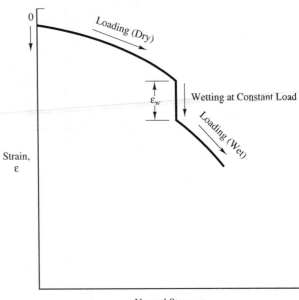

Figure 20.6 Typical results from a collapse test using the single oedometer method (Adapted from Houston et al., 1988; Used with permission of ASCE).

One of the soil characteristics to identify before conducting a settlement analysis is the relationship between the potential hydrocollapse strain, ϵ_w, and the normal stress, σ. Although additional research needs to be conducted to further understand this relationship, it appears that there is some *threshold collapse stress*, σ_t, below which no collapse will occur, and that ϵ_w becomes progressively larger at stresses above σ_t. This trend probably continues until the stress is large enough to break down the dry honeycomb structure, as shown in Figure 20.7. This latter stress is probably quite high in most collapsible soils, but in some lightly cemented soils, such as loess, it may be within the range of stresses that might be found beneath a foundation. For example, Peck, Hanson and Thornburn

TABLE 20.1 CLASSIFICATION OF SOIL COLLAPSIBILITY

Potential Hydrocollapse Strain, ϵ_w	Severity of Problem
0–0.01	No problem
0.01–0.05	Moderate trouble
0.05–0.10	Trouble
0.10–0.20	Severe trouble
> 0.20	Very severe trouble

Adapted from Jennings and Knight, 1975; Proceedings 6th Regional Conference for Africa; ©A.A. Balkema Publishers, Brookfield, VT; Used with permission).

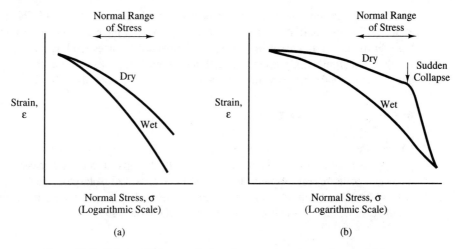

Figure 20.7   Relationship between hydrocollapse strain and normal stress: (a) for most collapsible soils; and (b) for loess.

(1974) noted a sudden and dramatic collapse of certain dry loessial soils in Iowa when the normal stress was increased to about 260 kPa (5500 lb/ft^2).

For collapsible soils other than loess, it appears that the relationship between ϵ_w and σ_t within the typical range of stresses beneath foundations may be expressed by the following formula:

$$\begin{aligned} \epsilon_w &= C_w \left(\log \sigma - \log \sigma_t \right) \\ &= C_w \log \left(\frac{\sigma}{\sigma_t} \right) \end{aligned} \tag{20.1}$$

$$C_w = \frac{d\,\epsilon_w}{d\,(\log \sigma)} \tag{20.2}$$

Where:

ϵ_w = potential hydrocollapse strain

C_w = hydroconsolidation coefficient

σ_t = threshold collapse stress

σ = normal stress at which hydroconsolidation occurs

The definition of C_w, as shown in Equation 20.2, also could be expressed as the additional hydrocollapse strain that occurs for every tenfold increase in normal stress.

We could determine C_w and σ_t by conducting one or more double oedometer tests. Alternatively, we could perform a series of single oedometer tests at a variety of normal stresses, and plot the results, as illustrated in Example 20.1. In either case, conduct the tests at stresses that are comparable to those in the field.

The magnitude of σ_t can often be taken to be about 5 kPa (100 lb/ft^2) (Houston et al., 1988), although a higher value would be likely if the clay content or unit weight is high. Fortunately, any error in determining σ_t, and the corresponding error in computing C_w, does not significantly affect the settlement computations as long as the samples are inundated at normal stresses within the range of those that exist in the field.

In-Situ Soil Collapse Tests

Gravelly soils pose special problems because they are very difficult to sample and test; yet they still may be collapsible. To evaluate these soils, it may be necessary to conduct an in-situ collapse test. Some of these tests have consisted of large-scale artificial wetting with associated monitoring of settlements (Curtin, 1973), and others have consisted of small-scale wetting in the bottom of borings (Mahmoud, 1991).

20.3 WETTING PROCESSES

To assess potential settlements caused by soil collapse, it is necessary to understand the processes by which the soil becomes wetted. We must identify the potential sources of water and understand how it infiltrates into the ground.

Usually, the water that generates the collapse comes from artificial sources, such as the following:

- Infiltration from irrigation of landscaping or crops
- Leakage from lined or unlined canals
- Leakage from pipelines and storage tanks
- Leakage from swimming pools
- Leakage from reservoirs
- Seepage from septic tank leach fields
- Infiltration of rainwater as a result of unfavorable changes in surface drainage

Although the flow rate from most of these sources may be slow, the duration is long. Therefore, the water often infiltrates to a great depth and wets soils that would otherwise remain dry.

As water penetrates the soil, a *wetting front* forms, as shown in Figure 20.8. This process is driven primarily by soil suction, so the wetting front will be very distinct. The distance it advances depends on the rate and duration of the water inflow as well as the hydraulic conductivity of the soil, the magnitude of the soil suction, and other factors. These are difficult to quantify, so this may be the greatest source of uncertainty in estimating the settlement due to collapse. Hopefully additional research and experience will generate more insight.

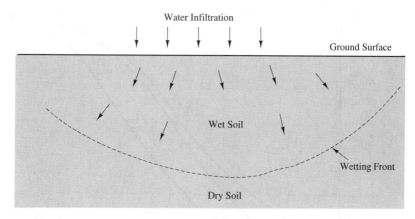

Figure 20.8 As water infiltrates into the soil, a distinct wetting front advances downward.

Curtin (1973) gives an interesting illustration of wetting processes in collapsible soils. He conducted large-scale wetting collapse tests in a deposit of collapsible alluvial soil in California's San Joaquin Valley. This soil is collapsible to a depth of at least 75 m (250 ft). After applying water continuously for 484 days, the wetting front advanced to a depth of at least 45 m (150 ft). The resulting collapse caused a settlement of 4.1 m (13.5 ft) at the ground surface.

In another case, irrigation of lawns and landscaping, and poor surface drainage around a building in New Mexico caused the wetting front to extend more than 30 m (100 ft) into the ground, which resulted in 25 to 50 mm (1–2 in) of settlement (Houston, 1991). In this case, the water inflow was slow and gradual, so the soil did not become completely saturated. If the rate of inflow had been greater, and the soil became wetter, the settlements would have been larger.

The New Mexico case illustrates the importance of defining the degree of saturation that might occur in the field. Laboratory collapse tests wet the soil to nearly 100 percent saturation, which may be a much worse situation than that in the field. This is why we refer to ϵ_w as the *potential* hydrocollapse strain; the actual strain depends on the degree of wetting.

Tests conducted on collapsible soils in Arizona and New Mexico indicate that these soils typically become only 40 to 60 percent saturated, even after extended periods of wetting (Houston, 1991). It appears that long-term intensive wetting, such as that obtained in Curtin's tests, is necessary to obtain greater degrees of saturation.

Obtain the relationship between the percentage of potential collapse and the degree of saturation for a given soil by conducting a series of single oedometer collapse tests with different degrees of wetting. Typical results from such tests are shown in Figure 20.9. The *wetting coefficient*, α, is the ratio between the collapse that occurs when the soil is partially wetted to that which would occur if it were completely saturated. Although

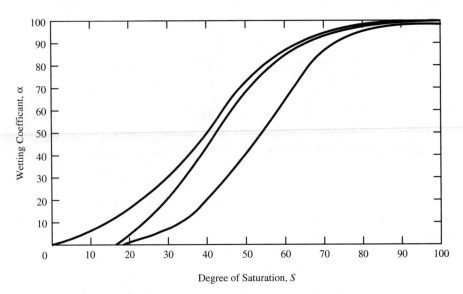

Figure 20.9 Experimental relationships between collapse and degree of saturation during wetting. These functions are for particular soils from Arizona, and must be determined experimentally for other soils (Adapted from Houston, 1992).

very few such tests have been conducted, it appears that α values in the field will typically
be on the order of 0.5 to 0.8. Once again, additional research and experience probably will
give more insight on selecting α values for design.

20.4 SETTLEMENT COMPUTATIONS

The settlement that results from collapse of the soil may be computed as follows:

$$\delta_w = \int \alpha \, C_w \log \left(\frac{\sigma_z}{\sigma_t} \right) dz \qquad (20.3)$$

Where:

δ_w = settlement due to hydroconsolidation

α = wetting coefficient

C_w = hydrocollapse coefficient

σ_z = vertical total stress

σ_t = threshold stress

z = depth

For practical analyses, evaluate this formula using several finite layers of soil as follows:

$$\boxed{\delta_w = \Sigma\ \alpha\ C_w\ H \log\left(\frac{\sigma_z}{\sigma_t}\right)} \qquad\qquad (20.4)$$

Where:

$H =$ thickness of soil layer

Compute the total stress, σ, at the midpoint of each layer using a unit weight that corresponds to about 50 percent saturation. The values of α, σ_t, and C_w may vary with depth and should be assigned to each layer accordingly. Continue the analysis down to the maximum anticipated depth of the wetted front or to the bottom of the collapsible soil, whichever comes first.

Notice that this is a *total stress analysis*, which means that both the laboratory and field data are evaluated in terms of total stress. This differs from the *effective stress analyses* used to evaluate consolidation settlement, as discussed in Chapter 7. An effective stress analysis would need to consider soil suction (negative pore water pressures).

Example 20.1

A 6 kip load is to be applied to a 2-ft square, 1.5-ft deep footing that will be supported on a collapsible soil. This soil extends to a depth of 12 ft below the ground surface and has a natural unit weight of 100 lb/ft³. At 50 percent saturation, it has a unit weight of 120 lb/ft³. Seven single oedometer collapse tests have been performed, the results of which are plotted in the following figure.

Solution:

For design purposes, the depth of wetting will probably be at least 12 ft. Therefore, base the analysis on the assumption that the entire depth of this collapsible soil will become wetted.

The line in the figure that follows is a conservative interpretation of the collapse test data. It represents $\sigma_t = 200$ lb/ft² and $C_w = 0.078$.

The limited data available suggests that α varies between 0.5 and 0.8. Therefore, it seems appropriate to use α values ranging from 0.9 near the bottom of the footing to 0.5 at a depth of 12 ft.

$$W_f = (2\ \text{ft})(2\ \text{ft})(1.5\ \text{ft})(150\ \text{lb/ft}^3) = 900\ \text{lb}$$

$$q = \frac{P + W_f}{A} = \frac{6000 + 900}{2^2} = 1725\ \text{lb/ft}^2$$

$$\sigma_{zD} = \gamma D = (100\ \text{lb/ft}^3)(2\ \text{ft}) = 200\ \text{lb/ft}^2$$

A tabular format, such as that which follows, is a convenient way to compute the settlement. The parameter σ_{z0} is the geostatic stress, $\Delta\sigma_z$ is the stress caused by the footing load, and $\sigma_z = \sigma_{z0} + \Delta\sigma_z$. Compute $\Delta\sigma_z$ using Equation 7.7 and δ_w using Equation 20.4.

We also could compute the settlement that would occur if the footing was not present by using σ_{z0} instead of σ_z. Such an analysis produces a settlement of 3.3 in. Thus, even if the wetting and the soil were perfectly uniform, the differential settlement would be $4.8 - 3.3 =$

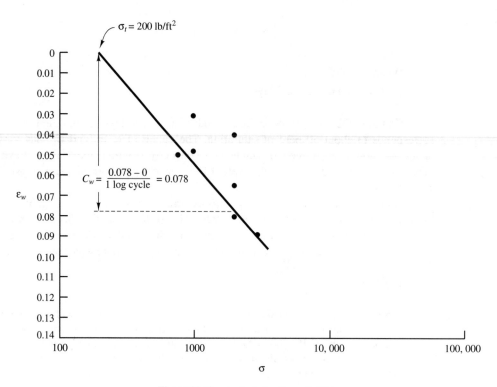

Figure 20.10　Analysis for Example 20.1.

1.5 in. In reality, both the wetting and the soil would be erratic, so the differential settlement could be even greater.

Depth Increment (ft)	H (ft)	z_f (ft)	σ_{z0} (lb/ft^2)	$\Delta\sigma_z$ (lb/ft^2)	σ_z (lb/ft^2)	σ_t (lb/ft^2)	C_w	α	δ_w (ft)
1.5–2.0	0.5	0.25	210	1515	1725	200	0.078	0.9	0.033
2.0–3.0	1.0	1.00	300	1075	1375	200	0.078	0.8	0.052
3.0–5.0	2.0	2.50	480	351	831	200	0.078	0.8	0.077
5.0–7.0	2.0	4.50	720	124	844	200	0.078	0.7	0.068
7.0–9.0	2.0	6.50	960	62	1022	200	0.078	0.6	0.066
9.0–12.0	3.0	9.00	1260	33	1293	200	0.078	0.5	0.095
> 12.0							0		0
								Total	0.39 ft
					Answer $\Rightarrow$				**= 4.7 in**

20.5   COLLAPSE IN DEEP COMPACTED FILLS

Although most collapsing soil problems are associated with natural soil deposits, engineers have observed a similar phenomenon in deep compacted fills (Lawton et al., 1992). Some deep fills can collapse even when they have been compacted to traditional standards (Lawton et al., 1989, 1991; Brandon et al., 1990). For example, Brandon reported settlements of as much as 450 mm (18 in) occurred in 30 m (100 ft) deep compacted fills near San Diego that became wet sometime after construction.

This suggests that the hydroconsolidation potential is dependent both on the void ratio and the normal stress. Very loose soils will collapse upon wetting even at low normal stresses, but denser soils will be collapsible only at higher stresses. In other words, these soils have a very high σ_t, but the normal stresses near the bottom of deep fills are greater than σ_t. At shallower depths (i.e., $\sigma_z < \sigma_t$), fills with a significant clay content may expand when wetted.

It appears that this phenomenon is most likely to occur in soils that are naturally dry and compacted at moisture contents equal to or less than the optimum moisture content. We can reduce the collapse potential by compacting the fill to a higher dry unit weight at a moisture content greater than the optimum moisture content.

20.6   PREVENTIVE AND REMEDIAL MEASURES

In general, collapsible soils are easier to deal with than are expansive soils because collapse is a one-way process, whereas expansive soils can shrink and swell again and again. Many mitigation measures are available, several of which consist of densifying the soil, thus forming a stable and strong material.

Houston and Houston (1989) identified the following methods of mitigating collapsible soil problems:

1. **Removal of the collapsible soil:** Sometimes, the collapsible soil can simply be excavated and the structure then may be supported directly on the exposed non-collapsible soils. We could accomplish this either by lowering the grade of the building site or by using a basement. This method would be most attractive when the collapsible soil extends only to a shallow depth.

2. **Avoidance or minimization of wetting:** Collapse will not occur if the soil is never wetted. Therefore, when working with collapsible soils, always take extra measures to minimize the infiltration of water into the ground. This should include maintaining excellent surface drainage, directing the outflow from roof drains and other sources of water away from the building, avoiding excessive irrigation of landscaping, and taking extra care to assure the water-tightness of underground pipelines.

    For some structures, such as electrical transmission towers, simple measures such as this will often be sufficient. If collapse-induced settlement did occur, the tower could be releveled without undue expense. However, the probability of success would be much less when dealing with foundations for buildings because there are many more opportunities for wetting and the consequences of settlement are more expensive to repair. Therefore, in most cases, we would probably combine these techniques with other preventive measures.

3. **Transfer of load through the collapsible soils to the stable soils below:** If the collapsible soil deposit is thin, it may be feasible to extend spread footing foundations through this stratum as shown in Figure 20.11. When the deposit is thick, we could use deep foundations for the same purpose. In either case, the ground floor would need to be structurally supported.

   When using this method, consider the possibility of negative skin friction acting on the upper part of the foundation.

4. **Injection of chemical stabilizers or grout:** Many types of soils, including collapsible soils, can be stabilized by injecting special chemicals or grout. These techniques strengthen the soil structure so future wetting will not cause it to collapse. These methods are generally too expensive to use over large volumes of soil, but they can be very useful to stabilize small areas or as a remedial measure beneath existing structures.

   A variation of this method, known as *compaction grouting*, involves injecting stiff grout that forms hard inclusions in the soil. This is often used to remediate settlement problems.

5. **Prewetting:** If the collapsible soils are identified before construction begins, they can often be remedied by artificially wetting the soils before construction (Knodel, 1981). This can be accomplished by sprinkling or ponding water at the ground surface, or by using trenches or wells. This method is especially effective when attempting to stabilize deep deposits, but may not be completely satisfactory for shallow soils where loads from the proposed foundations may significantly increase the normal stress.

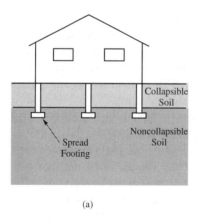

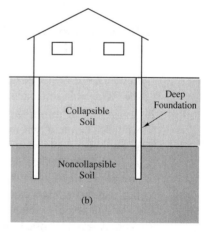

Figure 20.11 Transferring structural loads through collapsible soils to deeper, more stable soils: (a) with deepened spread footings; and (b) with deep foundations.

If the soil has strong horizontal stratification, as is the case with many alluvial soils, then the injected water may flow horizontally more than it does vertically. Therefore, be cautious when using this method near existing structures that are underlain by collapsible soils.

It is important to monitor prewetting operations to confirm that the water penetrates to the required depth and lateral extent.

This method can also be combined with a temporary surcharge load. The increased normal stress beneath such loads will intensify the collapse process and produce greater settlements.

6. **Compaction with rollers or vehicles:** Collapsible soils can be converted into excellent bearing material with little or no collapse potential by simply compacting them (Basma and Tuncer, 1992). Sometimes, this compaction may consist simply of passing heavy vibratory sheepsfoot rollers across the ground surface, preferably after first prewetting the soil.

   More frequently, this procedure includes excavating and stockpiling the soil, and then placing it back in layers using conventional earthmoving techniques. If the collapsible stratum is thin, say, less than 3 m (10 ft), this method can be used to completely eradicate the problem. It is often the preferred method when minimum risk is necessary and the collapsible soil deposit is shallow.

   If the collapsible stratum is thick, then we may choose to estimate the depth of the wetting front and extend the removal and recompaction to that elevation. This method also reduces the likelihood that the lower soils will become wet because the recompacted soil has a reduced hydraulic conductivity (permeability). Also, if the lower soils should collapse, the compacted fill will spread any settlements over a larger area, thus reducing differential settlements.

   Some collapsible soils have sufficient clay content to become slightly expansive when compacted to a higher unit weight.

7. **Compaction with displacement piles:** Large displacement piles, such as closed-end steel pipe piles, can be driven into the ground, compacting the soil around the pile. It may then be extracted and the hole backfilled with sandy gravel or some other soil. Repeating this process on a grid pattern across the site will reduce the collapse potential both by soil compaction and by the column action of the backfill.

8. **Compaction by heavy tamping:** This technique, which is discussed in more detail in Chapter 18, consists of dropping heavy weights (several tons) from large heights (several meters) to compact the soil. This process is continued on a grid pattern across the site, leaving craters that are later backfilled. This technique can be very effective, especially when combined with prewetting.

9. **Vibroflotation:** This technique, also discussed in more detail in Chapter 18, consists of penetrating the soil with a vibrating probe equipped with water jets (known as a vibroflot). The water softens the soil and the vibrations help the collapse process. The vertical hole formed by the vibroflot is also filled with gravel, thus reinforcing the soil and adding bearing capacity. This process is repeated on a grid pattern across the site.

10. **Deep blasting combined with prewetting:** Engineers in the former Soviet Union have experimented with stabilizing collapsible loess by detonating buried explosives. In some cases, the ground is first thoroughly prewetted and then the collapse is induced by vibrations from the detonations. In other cases, the explosives are detonated while the soil is still dry, and the voids created are filled first with water and then with sand and gravel.

11. **Controlled wetting:** This method is similar to method 5 in that it involves injecting water into the soil through trenches or wells. However, it differs in that the wetting is much more controlled and often concentrated in certain areas. This would be used most often as a remedial measure to correct differential settlements that have accidently occurred as a result of localized wetting. When used with careful monitoring, this method can be an inexpensive yet effective way of stabilizing soils below existing buildings that have already settled.

12. **Design structure to be tolerant of differential settlements:** As discussed in Chapter 2, some types of structures are much more tolerant of differential settlements than are others. Therefore, if the potential for collapse-induced settlement is not too large, we may be able to use a more tolerant structure. For example, a steel storage tank would be more tolerant than a concrete one.

The selection of the best method or methods to use at a given site depends on many factors, including the following:

- How deep do the collapsible soils extend?
- How deep would the wetting front extend if the soil was accidentally wetted?
- How much settlement is likely to occur if the soil is accidentally wetted?
- What portion of the total stress is due to overburden and what portion is due to applied loads?
- Is the building or other structure already in place?
- Has any artificial wetting already occurred?

SUMMARY

Major Points

1. Collapsible soils have a loose honeycomb structure and are dry in their natural state. If they are later wetted, this structure will collapse and settlements will occur.

2. Collapsible soils are usually encountered only in arid or semi-arid climates and are continually dry in their natural state. The water that causes them to collapse is normally introduced artificially (as compared to natural infiltration of rainfall). However, collapsible loess is also found in more humid climates.

3. Collapsible soils can be formed by various geologic processes. They may be alluvial soils (especially debris flow deposits), aeolian soils (especially loess), or residual soils.

4. Collapsible soils usually can be effectively sampled, as long as they are moderately to well cemented and do not contain too much gravel. Lightly cemented soils can be sampled with more difficulty, whereas gravelly soils are very difficult or impossible to sample.

5. Laboratory collapse tests may be conducted to measure the collapse potential as a function of overburden stress. Either the double oedometer test or the single oedometer test may be used.

6. In the field, collapsible soils are normally not wetted to 100 percent saturation and therefore do not strain as much as those tested in the laboratory. Typically, the strain in the field is 50 to 80 percent of that observed in a thoroughly wetted laboratory test specimen.

7. The settlement also depends on the depth of wetting, but this is difficult to evaluate in advance.

8. We can estimate the amount of settlement by projecting the laboratory collapse tests back to the field conditions while making appropriate corrections for overburden stress and degree of saturation.

9. Many remedial and preventive measures are available to prevent or repair structural damage caused by collapsible soils.

Vocabulary

Aeolian soil	Double oedometer method	Loess
Alluvial soil	Honeycomb structure	Residual soil
Collapse test	Hydrocollapse	Single oedometer method
Collapsible soil	Hydrocompression	Threshold stress
Colluvial soil	Hydroconsolidation	Wetting coefficient

COMPREHENSIVE QUESTIONS AND PRACTICE PROBLEMS

20.1 What is a "honeycomb" structure?

20.2 Why are collapsing soils rarely found in areas with very wet climates?

20.3 What are the advantages of evaluating collapse potential using direct methods instead of indirect methods?

20.4 Why is it difficult to obtain samples of gravelly soils and use them in laboratory collapse tests?

20.5 The analysis method presented in this chapter includes the simplifying assumption that ϵ_w increases with the log of σ. Although this appears to be a reasonable assumption within the usual range of stresses beneath a spread footing, would you expect this relationship to hold true at very high stresses? Explain.

20.6 Why is σ_t usually larger in soils that have a greater clay content?

20.7 A proposed industrial building will include an 800-mm wide, 400-mm deep continuous foot-ing that will carry a load of 90 kN/m. The soil beneath this proposed footing is collapsible to a depth of 2.5 m, and noncollapsible below that depth. The upper 2.5 m has a natural unit weight of 16.0 kN/m^3, $\sigma_t = 7.2$ kPa and $C_w = 0.085$. At 50 percent saturation, this soil has a unit weight of 18.0 kN/m^3. Compute the potential settlement due to wetting of this collapsible soil. Assume the depth of wetting is greater than 2.5 m.

20.8 What preventive measures would you use to avoid settlement problems in the footing de-scribed in Problem 20.7?

20.9 Assume you have several undisturbed samples of a collapsible soil and need to develop a plot of α vs. S similar to those in Figure 20.9. How would you obtain these data?

20.10 Is it possible for some soils to be expansive when the normal stress is low, and collapsible when it is high? Explain.

Reliability-Based Design

You can know a lot and not really understand anything.

Charles Kettering
American Inventor

Reliability has been defined as "the probability of an object (item or system) performing its required function adequately for a specified period of time under stated conditions" (Harr, 1996). Thus, a reliable foundation is one that has a high probability of supporting the structure (the required function) without collapse or excessive deformation (adequately) for its design life (the specified period) at the project location (the condition). Achieving a sufficient reliability is one of the most fundamental and important goals of foundation design and construction.

In one sense this is an easy goal to achieve: we could simply "overdesign" the foundation so that its load-bearing capacity and anticipated settlement surpass design requirements by a large margin (i.e., use high factors of safety) and use extremely durable materials that will survive well beyond the design life. Unfortunately, this solution conflicts with another important requirement: that the foundation be economical. We cannot afford to grossly overdesign foundations, especially with large structures. Thus, reliability and economy are often conflicting goals.

Traditionally, foundation engineers strike a balance between reliability and economy using a combination of precedent, judgement, and analysis, and express this balance primarily in terms of a factor of safety. We typically use higher factors of safety when reliability is especially important or when there are large uncertainties in the analysis, and lower factors of safety when the opposite conditions are present. For example, Figure 6.11

presents various factors to consider when selecting a factor of safety for bearing capacity analyses of shallow foundations. This approach is called a *deterministic design method*.

Unfortunately, deterministic methods are very subjective and are generally not based on a systematic assessment of reliability, especially when we consider their use in the entire structure (not only the foundations). These methods can produce structures with some "overdesigned" components and perhaps some "underdesigned" components. The additional expense incurred in constructing the overdesigned components probably does not contribute to the overall reliability of the structure, so this is not a very cost-effective way to produce reliable structures. In other words, it would be better to redirect some of the money used to build the overdesigned components toward strengthening the underdesigned ones.

Therefore, there is increasing interest in adopting *reliability-based design methods* in civil engineering. These methods are intended to quantify reliability, and thus may be used to develop balanced designs that are both more reliable and less expensive. Another objective of reliability-based design is to better evaluate the various sources of failure and use this information to develop design and construction methods that are both more reliable and more *robust*. A robust design is one that it is insensitive to variations in materials, construction technique, and environment.

Reliability-based design methods have been used extensively in manufacturing, where they have produced significant improvements in reliability and economy. However, these methods are only beginning to be used in civil engineering. This chapter discusses the application of reliability-based design methods to foundation engineering.

21.1   METHODS

Reliability-based design methods could be used to address many different aspects of foundation design and construction. However, most of these efforts to date have focused on geotechnical and structural strength requirements, such as the bearing capacity of shallow foundations, the side friction and toe-bearing capacity of deep foundations, and the stresses in deep foundations. All of these are based on the difference between load and capacity, so we can use a more specific definition of reliability as being the probability of the load being less than the capacity for the entire design life of the foundation.

Various methods are available to develop reliability-based design of foundations, most notably stocastic methods, the first-order second-moment method, and the load and resistance factor design method. Each has its advantages and disadvantages.

Stochastic Methods

Stochastic methods differ from deterministic methods in that they define a range of possible values for load and ultimate capacity rather than a single value for each. These ranges are determined using the principles of probability and statistics, and expressed as *probability density functions* such as those shown in Figure 21.1. These functions describe the

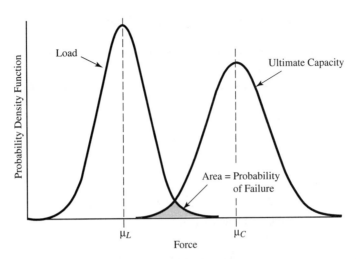

Figure 21.1   Probability density functions for load and ultimate capacity. The area under each of the probability density functions is equal to 1, and the probability of failure is equal to the shaded area.

probability of various loads and capacities occurring in the foundation. Thus, the high point of each curve occurs at the most likely value, and the width of the curve indicates the variation or uncertainty. Since each curve encompasses the entire range of potential values, the area under the curve is 1, which represents a probability of 1 (or 100%). The probability of a load or ultimate capacity being within a specified range is equal to the area under the curve within that range.

Both of these probability density functions are generally based on other probability density functions for various factors. For example, the function describing the load on a spread footing would depend on the distribution for each component of load (dead, live, wind, etc.), and that for its bearing capacity would depend on the distribution of shear strength, groundwater location, as-built foundation dimensions, and possibly other factors. The assessment of each factor and compilation of the probability density functions can be a complex and time-consuming task.

This analysis defines "failure" as the load being greater than the ultimate capacity. Thus, the probability of failure is equal to the shaded area in Figure 21.1. The engineer then compares this probability to some maximum acceptable probability and thus determines if the design is acceptable. If necessary, the design is then modified. For example, in the case of a bearing capacity analysis for a spread footing, increasing the footing width B will move the capacity curve to the right, thus decreasing the probability of failure.

Stochastic methods are potentially very powerful, but they require sufficient data to define the probability density functions. They also require expertise in probability and statistics.

First-Order Second-Moment (FOSM) Method

The first-order second-moment (FOSM) method (Cornell, 1969) is a simplification of the stochastic method that still defines probability density functions. It assumes the load and ultimate capacity are independent variables, and defines them using their means and standard deviations. It then uses the *reliability index*, β, which is defined as:

$$\beta = \frac{\mu_C - \mu_L}{\sqrt{\sigma_C^2 - \sigma_L^2}}$$ (21.1)

Where:

   β = reliability index (also known as safety index)

   μ_C = mean ultimate capacity

   μ_L = mean load

   σ_C = standard deviation of ultimate capacity

   σ_L = standard deviation of load

High values of β correspond to lower probabilites of failure, so the design methodology is based on achieving a certain minimum β.

Load and Resistance Factor Design (LRFD) Method

Load and resistance factor design (LRFD) is the third method of implementing reliability-based design. It is also known as *ultimate strength design* (USD) or *limit states design*. It applies *load factors*, γ, most of which are greater than one, to the nominal loads to obtain the *factored load*, U. In the case of normal loads, the factored load P_u is:

$$P_u = \gamma_1 P_D + \gamma_2 P_L + \ldots$$ (21.2)

Where:

   P_u = factored normal load

   γ = load factor

   P_D = normal dead load

   P_L = normal live load

These load factors reflect the bias and variability of the various load types, and are based on extensive statistical analyses. Design codes present a series of equations in the form of Equation 21.2, each with a different load combination, and define the factored load as the largest load computed from these equations.

   The LRFD method also applies a *resistance factor*, ϕ (also known as a *strength reduction factor*) to the ultimate capacity from a strength limit analysis. Finally, the design must satisfy the following criteria:

$$\boxed{P_u \leq \phi P_n} \tag{21.3}$$

Where:

P_u = factored normal load

ϕ = resistance factor

P_n = nominal normal load capacity

Similar equations also are used for shear and moment loads.

Research engineers who develop LRFD codes choose load and resistance factors that force a certain offset between the probability density functions for load and resistance, thus providing for a specified maximum probability of failure. These relationships are shown in Figure 21.2.

The primary advantage of LRFD is that the design engineer does not need to perform any probability computations because the probabilistic content is implicit within the load and resistance factors. Thus, unlike the stochastic and FOSM methods, the application of LRFD is very similar to the traditional allowable stress design (ASD) methods, and thus is more readily accepted. Thus, LRFD is the most widely accepted method of reliability-based design in structural engineering.

The LRFD method was first developed in the 1960s for reinforced concrete design, and has almost completely replaced the previous ASD methods for this material. In 1986, the American Institute for Steel Construction (AISC) released its first LRFD code for the design of steel structures, and it is gradually replacing the ASD code. Widely accepted

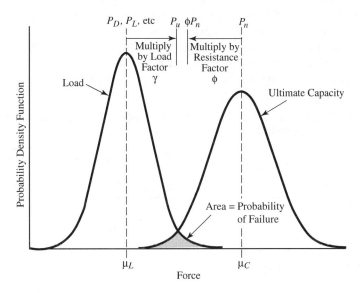

Figure 21.2   Probability density functions for load and resistance and LRFD criteria. The area under each of the probability density functions is equal to 1, and the probability of failure is equal to the shaded area.

codes, including the *Building Code Requirements for Structural Concrete (ACI 318-99)* (ACI, 1999) and the *Manual of Steel Construction, Load and Resistance Factor Design* (AISC, 1995). LRFD methods for timber design are currently under development. Eventually, LRFD methods will completely replace ASD methods in all structural engineering strength analyses.

Unfortunately, progress in implementing LRFD design of foundations has been much slower. Except for the structural design of spread footings and pile caps, nearly all structural and geotechnical aspects of foundation design still use ASD. As superstructure design moves toward universal use of LRFD, it becomes increasingly important to develop similar methods for foundation design. The use of LRFD for the entire structure, including the foundation, should provide for more optimized designs, and leave less opportunity for mistakes due to the use of two different design methodologies. In addition, LRFD makes a clearer distinction between strength and serviceability requirements, and thus helps clarify the design process.

21.2   LRFD FOR STRUCTURAL STRENGTH REQUIREMENTS

Although LRFD design procedures are well established for concrete and steel members in the superstructure, and are becoming established for wood and masonry, their applicability to foundations has thus far been limited to spread footings and pile caps. Codes still specify the use of ASD for the structural design of deep foundations (see discussion in Chapter 12).

Unfortunately, because of construction tolerances, residual stresses, and other factors described in Chapter 12, we cannot design deep foundations using the resistance factors developed for the superstructure. Just as we used lower allowable stresses for ASD, we also need to use lower resistance factors for LRFD. Table 21.1 presents a set of proposed resistance factors for pile foundations. However, no North American code has yet adopted these or any other structural resistance factors specifically for deep foundations.

TABLE 21.1   PROPOSED STRUCTURAL RESISTANCE FACTORS
FOR PILE FOUNDATIONS (PDCA, 1998)

Pile Material	Structural Resistance Factor, ϕ	
	Using ACI Load Factors	Using ANSI Load Factors
Timber	0.60	0.55
Steel	0.95	0.85
Prestressed concrete with spiral reinforcement that satisfies the ACI code	0.75	0.70
Prestressed concrete with other types of transverse reinforcement	0.70	0.65

21.3 LRFD FOR GEOTECHNICAL STRENGTH REQUIREMENTS

The following codes and suggested codes include resistance factors for geotechnical strength design of foundations:

- The American Association of State Highway and Transportation Officials (AASHTO) *Standard Specifications for Highway Bridges* (AASHTO, 1996; Barker et al., 1991) deals only with highway bridges, and is used by state and provincial transportation departments. It includes LRFD criteria for all major types of foundations.
- The Ontario Highway Bridge Design Code (MTO, 1991) governs the design of highway bridges in the Province of Ontario, and includes LRFD criteria for foundations.
- The Federal Highway Administration (FHWA) has sponsored research on LRFD design of drilled shaft foundations (O'Neill and Reese, 1999).
- The Electric Power Research Institute (EPRI) has developed LRFD guidelines for electric power transmission structure foundations (Phoon, Kulhawy, and Grigoriu, 1995).
- The Pile Driving Contractor's Association (PDCA, 1998) has developed a proposed LRFD code for the design and construction of pile foundations.

Thus far, the most prominent LRFD code for foundation design in North America is the AASHTO *Standard Specifications for Highway Bridges*. Therefore, we will examine it as an example. Tables 21.2 to 21.4 present resistance factors (AASHTO calls them *per-*

TABLE 21.2 AASHTO RESISTANCE FACTORS FOR BEARING CAPACITY OF SHALLOW FOUNDATIONS (AASHTO, 1996)

Soil or Rock Type	Analysis Method	Resistance Factor, ϕ
Sand	Semi-empirical procedure using SPT data	0.45
	Semi-empirical procedure using CPT data	0.55
	Rational method using ϕ estimated from SPT data[1]	0.35
	Rational method using ϕ estimated from CPT data[1]	0.45
Clay	Semi-empirical procedure using CPT data	0.50
	Rational method using shear strength measured in laboratory tests[1]	0.60
	Rational method using shear strength measured in field vane tests[1]	0.60
	Rational method using shear strength estimated from CPT data[1]	0.50
Rock	Semi-empirical procedure (Carter and Kulhawy)	0.60

[1]In this context, AASHTO's reference to the "rational method" refers to Terzaghi's or Vesić's bearing capacity equations, as presented in Chapter 6.

TABLE 21.3   AASHTO RESISTANCE FACTORS FOR AXIALLY-LOADED PILES
(AASHTO, 1996)

	Analysis		Resistance Factor, ϕ
Capacity of single piles	Side friction	α method	0.70
		β method	0.50
		λ method	0.55
	Toe bearing	Clay (Skempton method)	0.70
		Sand (Kulhawy method)	
		ϕ from CPT	0.45
		ϕ from SPT	0.35
		Rock (CGS method)	0.50
	Side friction and toe bearing	SPT method	0.45
		CPT method	0.55
		Load test	0.80
		Pile-driving analyzer	0.70
Block failure	Clay		0.65
Uplift capacity of single piles	α method		0.60
	β method		0.40
	λ method		0.45
	SPT method		0.35
	CPT method		0.45
	Load test		0.80
Group uplift capacity			0.55

formance factors) for geotechnical design. These resistance factors are applicable only to foundations used to support structures designed using the AASHTO code (most notably bridges). They may or may not be applicable to buildings and other structures designed using the ACI or AISC codes, because they use different load factors.

The LRFD design process for geotechnical load capacity uses load and resistance factors instead of the more traditional factor of safety. For example, with shallow foundations this process is as follows:

1. Compute the ultimate bearing capacity, q_{ult}, using Terzaghi's method (Equations 6.4–6.6) or Vesić's method (Equation 6.13).
2. Compute the nominal downward load capacity, P_n, using:

$$P_n = q_{ult}A \qquad (21.4)$$

TABLE 21.4   AASHTO RESISTANCE FACTORS FOR AXIALLY-LOADED DRILLED SHAFTS (AASHTO, 1996)

Analysis	Method		Resistance Factor, ϕ
Downward load capacity of single drilled shaft	Side friction in clay	α method (Reese and O'Neill)	0.65
	Toe bearing in clay	Total stress method (Reese and O'Neill)	0.55
	Side friction and toe bearing in sand	1. Thomas and Reese 2. Meyerhof 3. Quiros and Reese 4. Reese and Wright 5. Reese and O'Neill	Based on experience and judgement
	Side friction in rock	1. Carter and Kulhawy 2. Horvath and Kenney	0.55 0.65
	Toe bearing in rock	1. Canadian Geotechnical Society 2. CGS pressuremeter method	0.50 0.50
	Side friction and toe bearing	Load test	0.80
Block failure	Clay		0.65
Uplift load capacity of single drilled shaft	Clay	α method (Reese and O'Neill) Belled shafts (Reese and O'Neill)	0.55 0.50
	Sand	1. Touma and Reese 2. Meyerhof 3. Quiros and Reese 4. Reese and Wright 5. Reese and O'Neill	Based on experience and judgement
	Rock	1. Carter and Kulhawy 2. Horvath and Kenney	0.45 0.55
	Any	Load test	0.80
Uplift capacity of a group of drilled shafts	Sand		0.55
	Clay		0.55

Where:

P_n = nominal downward load capacity

q_{ult} = ultimate bearing capacity

A = base area of foundation

In theory, P_n is the downward load required to produce a bearing capacity failure.

3. Obtain the appropriate resistance factor, ϕ, from the appropriate code.

4. Compute the factored load, P_u, using Equations 2.7 to 2.17, 2.18 to 2.23, or some other standard.

5. Design the foundation so that the following condition is satisfied:

$$\boxed{P_u \le \phi P_n} \tag{21.5}$$

Where P_u includes both the factored column load and the factored weight of the foundation, which is treated as a dead load.

A similar process is used for shear loads and for deep foundations.

Example 21.1

A single-column highway bridge bent is to be supported on a square spread footing. The bottom of this footing will be 1.8 m below the adjacent ground surface. The factored vertical compressive load, P_u, is 4500 kN (including the weight of the foundation) when computed using the AASHTO load factors. The soil has $c' = 0$ and $\phi'=31°$ (based on empirical correlations with SPT data) and $\gamma = 17.5$ kN/m³. The groundwater table is at a great depth. Using LRFD, determine the required footing width, B to satisfy bearing-capacity requirements.

Solution

Use Terzaghi's method:

$$\text{For } \phi' = 31° \quad N_c = 40.4 \quad N_q = 25.3 \quad N_\gamma = 23.7$$

$$\sigma'_{zD} = \gamma D - u = (17.5 \text{ kN/m}^3)(1.8 \text{ m}) - 0 = 31.5 \text{ kPa}$$

$$q_{ult} = 1.3cN_c + \sigma'_{zD}N_q + 0.4\gamma BN_\gamma$$
$$= 0 + (31.5)(25.3) + 0.4(17.5)B(23.7)$$
$$= 797 + 166B$$

Per Table 21.2, the resistance factor $\phi = 0.35$

$$P_u \le \phi q_{ult}A$$
$$4500 \le 0.35(797 + 166B)B^2$$
$$B \ge 3.2 \text{ m} \quad \Leftarrow Answer$$

The footing width must be at least 3.2 m to satisfy bearing capacity requirements.

21.4 SERVICEABILITY REQUIREMENTS

In principle, reliability-based design methods are just as applicable to serviceability requirements as they are to strength requirements. For example, a reliability-based approach to settlement analysis would be especially useful in the design of shallow foundations, because settlement often controls the design. However, research and development efforts have thus far focused primarily on strength requirements, and no codes have any criteria for reliability-based serviceability requirements.

Therefore, even when load capacity is determined using LRFD (or some other reliability-based method), settlement analyses should still be performed using the traditional methods with the unfactored loads. This is parallel to the structural engineering practice of evaluating beam deflections and other serviceability requirements using the unfactored loads.

21.5   THE ROLE OF ENGINEERING JUDGMENT

Engineering is both an art and a science, and cannot be reduced to a set of formulas and procedures. This is why engineering judgement plays such an important role in the design process, especially in geotechnical engineering. *Engineering judgment* is one of the ways we incorporate the engineer's experience and subjective assessments into a design.

Reliability-based design methods should be viewed as a supplement to, but not a replacement of, engineering judgment. It is a tool we can use to more effectively implement our judgement, and to strengthen the ties between our judgment and the experience of other engineers. Therefore, it is important to implement reliability-based design methods in a way that compliments engineering judgment rather than destroys it. This will require some changes in the way we implement judgment (Kulhawy and Phoon, 1996).

21.6   TRANSITION TO LRFD

The vast majority of geotechnical designs are still based on ASD methods, and these methods will continue to dominate for the foreseeable future. That is why the geotechnical analyses in this book use ASD. However, the growing dominance of LRFD in structural engineering is driving the need for new LRFD based geotechnical analysis and design methods.

This usage of two different design methods introduces needless confusion into the design process. Sometimes the superstructure is designed using one method, then the foundation is designed using another. In the case of shallow foundations, the geotechnical analysis is performed using ASD, but the structural design uses LRFD. This confusion makes the design process more difficult, and increases the potential for mistakes. In addition, ASD-based designs do not have the reliability-based design benefits of LRFD.

The state and provincial transportation departments (the primary users of the AASHTO code) are leading this transition. Few if any non-transportation structures now use LRFD, or are likely to do so in the near future. However, LRFD will eventually become the method of choice for all structures.

In the meantime, foundation engineers need to be aware of the differences between these two methods, and be sure they know which is being used. It also is very important for geotechnical design criteria to clearly state whether they are intended for use with the factored or unfactored loads, and to clearly state what is a strength limit and what is a serviceability limit.

SUMMARY

Major Points

1. Structures, including their foundations, should be both reliable and economical. Traditionally, engineers attempt to meet this objective using deterministic design methods and assigning factors of safety to various aspects of the design.

2. Engineers are beginning to replace deterministic methods with reliability-based design methods, which should result in designs that are both more reliable and less costly.

3. Reliability-based design methods use the principles of probability and statistics to describe the various parameters that affect reliability. When applied to foundation design, this can be done using stochastic methods, the first-order second moment method, or the load and resistance factor design (LRFD) method.

4. LRFD is the most popular method of reliability-based design for foundations. It uses load factors and resistance factors instead of a factor of safety.

5. The load factors for LRFD analyses are the same as those used in the superstructure. However, the resistance factors, both for structural and geotechnical design, are unique to foundations. Some of these factors have been developed and published, but they are best considered preliminary.

6. Presently, LRFD analyses are used only to evaluate strength requirements. Serviceability requirements, such as settlement and lateral deformation, must still be evaluated using conventional methods with the unfactored load.

7. Reliability-based design will change the role of engineering judgement, but will not replace it.

Vocabulary

Deterministic design
method
Engineering judgment
First-order second-
moment method
Limit states design
Load and resistance factor
design (LRFD)

Load factor
Nominal load capacity
Probability density
function
Reliability
Reliability-based design
method
Resistance factor

Reliability index
Robust
Stochastic method
Ultimate strength design
(USD)

COMPREHENSIVE QUESTIONS AND PRACTICE PROBLEMS

21.1 A spread footing foundation supporting a portion of a highway bridge is subjected to an AASHTO downward ultimate column load, P_U, of 2500 kN. This load includes the weight of the foundation. The underlying soils are silty sands with $c' = 5$ kPa, $\phi' = 34°$, and $\gamma = 18.1$ kN/m^3. The groundwater table is at a depth of 5 m. Using LRFD and Terzaghi's bearing capacity formula, determine the required width of an 800 mm deep square footing.

21.2 The following unfactored downward loads act on a 3-ft diameter drilled shaft foundation: $P_D = 200$ k, $P_L = 150$ k. The soil has $f_s = 2000$ lb/ft^2 and $q_t' = 80,000$ lb/ft^2. The geotechnical analysis for this foundation is to be performed using a resistance factor of 0.65. Using the ANSI load factors, compute the required depth of embedment, D.

21.3 Referring to the resistance factors for side friction and toe bearing, as presented in Table 21.3, why may we use a higher ϕ when using load test data than when using SPT data?

21.4 Explain the difference between structural resistance factors, such as those listed in Table 21.1, with geotechnical resistance factors, such as those listed in Table 21.3, and describe when each would be used in the design process.

Part E

Earth-Retaining Structure
Analysis and Design

Earth-Retaining Structures

It would be well if engineering were less generally thought of . . . as the art of constructing. In a certain important sense it is the art of not constructing . . . of doing that well with one dollar which any bungler can do with two after a fashion.

Arthur M. Wellington (1887)

Many civil engineering projects require sharp transitions between one ground surface elevation and another. Often these transitions are made by short sections of sloping ground, as shown in Figure 22.1. However, when space is at a premium, such slopes can be replaced with *earth-retaining structures*, which are vertical or near-vertical facilities that maintain the ground surface at two different elevations. Practical applications of earth-retaining structures include:

- Highway and railroad projects where the required grade is significantly above or below the adjacent ground and the right-of-way is not large enough to accommodate a slope.
- Bridge abutments.
- Building sites on sloping ground where earth-retaining structures are used to create level building pads.
- Waterfront facilities where earth-retaining structures are built to accommodate the berthing of ships
- Flood control facilities.
- Unstable ground, where the earth-retaining structure provides the needed resistance to prevent landslides.

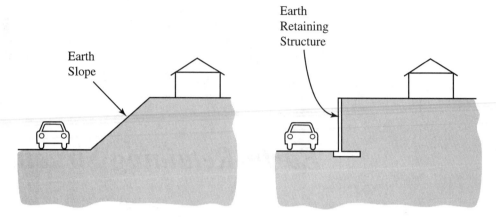

Figure 22.1 Earth slopes and earth retaining structures are used to maintain two different ground surface elevations.

Many kinds of retaining structures are available, each of which is best suited for particular applications. O'Rourke and Jones (1990) classified earth-retaining structures into two broad categories: *externally stabilized systems* and *internally stabilized systems*, as shown in Figure 22.2. Some hybrid methods combine features from both systems.

22.1 EXTERNALLY STABILIZED SYSTEMS

Externally stabilized systems are those that resist the applied earth loads by virtue of their weight and stiffness. This was the only type of retaining structure available before 1960, and they are still very common. O'Rourke and Jones subdivided these structures into two categories: *gravity walls* and *in-situ walls*.

Gravity Walls

Massive Gravity Walls

The earliest retaining structures were *massive gravity walls*, as shown in Figures 22.3 and 22.4. They were often made of mortared stones, masonry, or unreinforced concrete and resisted the lateral forces from the backfill by virtue of their large mass. In addition, these walls are very thick, so the flexural stresses are minimal and no tensile reinforcement is needed.

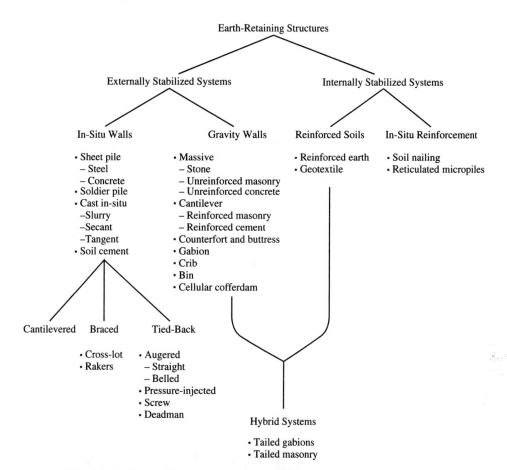

Figure 22.2   Classification of earth-retaining structures (Adapted from O'Rourke and Jones, 1990; Used with permission of ASCE).

Construction of massive gravity walls requires only simple materials and moderately skilled labor, but the required volume of materials is very large and the construction process is very labor-intensive. Therefore, these walls are rarely used today except if the required height is very short.

Cantilever Gravity Walls

The *cantilever gravity wall*, shown in Figure 22.5, is a refinement of the massive gravity wall concept. These walls have a much thinner stem, and utilize the weight of the backfill soil to provide most of the resistance to sliding and overturning. Since the cross section is much smaller, these walls require much less construction material. However, they have large flexural stresses, and thus are typically made of reinforced concrete or reinforced

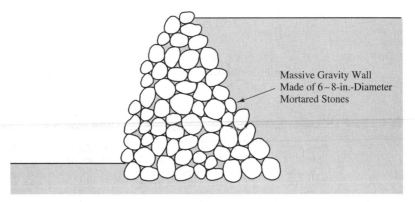

Figure 22.3   A massive gravity wall.

masonry. Such walls must be carefully constructed, and thus requires skilled labor. Nevertheless, they are generally much less expensive than massive gravity walls, and thus are the most common type of earth-retaining structure. Chapter 24 discusses these walls in more detail.

Crib Walls

A *crib wall*, shown in Figure 22.6, is another type of gravity retaining structure. It consists of precast concrete members linked together to form a crib. These members resemble a child's Lincoln Log toy. The zone between the members is filled with compacted soil.

Figure 22.4   Short massive gravity walls like this one may be economically viable. However, this design is prohibitively expensive for tall walls because of the large required width and the large material requirements and labor costs.

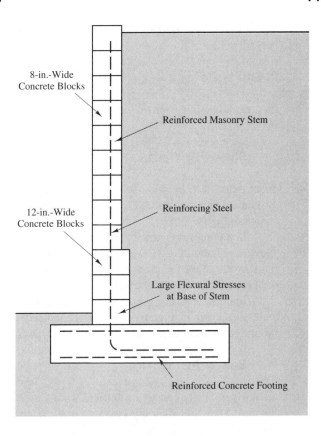

8-in.-Wide
Concrete Blocks

Reinforced Masonry Stem

Reinforcing Steel

12-in.-Wide
Concrete Blocks

Large Flexural Stresses
at Base of Stem

Reinforced Concrete Footing

Figure 22.5   A cantilever gravity wall.

Figure 22.6   This crib wall produced enough level ground to build the condominium buildings in the background.

The soil supplies most of the weight required to resist the lateral loads imposed by the backfill soils.

In-Situ Walls

In-situ walls differ from gravity walls in that they rely primarily on their flexural strength, not their mass.

Sheet Pile Walls

Sheet piles are thin, wide steel piles as shown in Figure 22.7. They are driven into the ground using a pile hammer. A series of sheet piles in a row form a *sheet pile wall*, as shown in Figure 22.8.

Sometimes it is possible to simply cantilever a short sheet pile out of the ground, as shown in Figure 22.9. However, it is usually necessary to provide lateral support at one or more levels above the ground. This may be accomplished in either of two ways: by *internal braces* or by *tieback anchors*.

Internal braces are horizontal or diagonal compression members that support the wall, as shown in Figure 22.9. Tieback anchors are tension members drilled into the ground behind the wall. The most common type is a grouted anchor with a steel tendon.

Soldier Pile Walls

Soldier pile walls consist of vertical wide flange steel members with horizontal timber lagging. They are often used as temporary retaining structures for construction excavations, as shown in Figures 22.10 and 22.11.

Figure 22.7   These sheet piles are being stored in a contractors yard, and are ready to be driven into the ground. The interlocking joints at each edge are used to join adjacent piles to form a continuous wall.

Figure 22.8   This sheet pile wall has been installed along a riverfront.

Slurry Walls

Slurry walls are cast-in-place concrete walls built using bentonite slurry. The contractor digs a trench along the proposed wall alignment and keeps it open using the slurry. Then, the reinforcing steel is inserted and the concrete is placed using tremie pipes or pumps. As the concrete fills the trench, the slurry exits at the ground surface.

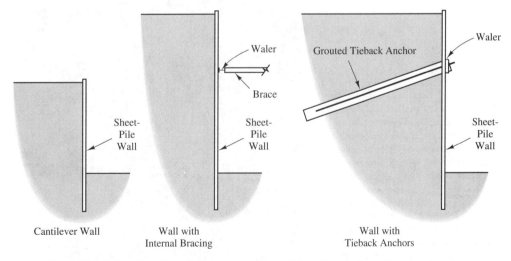

Figure 22.9   Short sheet pile walls can often cantilever, taller walls usually require internal bracing or tieback anchors.

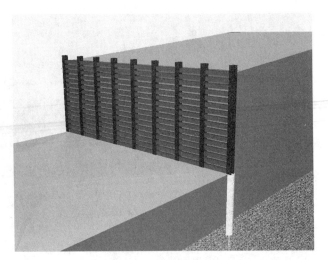

Figure 22.10   A soldier pile wall.

Figure 22.11   The ground surface at this
site was originally level with the street.
However, the proposed building, shown
under construction, is to have a basement, so
the site was excavated to a lower elevation.
A soldier pile wall is being used to provide
temporary support for the street. The permanent basement wall will be located directly
above the grade beam shown in the foreground, and the columns above the pile caps
will be embedded in the exterior wall. Once
this permanent wall is in place, the zone between it and the street will be backfilled and
the timber lagging will be removed. The
steel sheet piles may or may not be removed.

Slurry walls have been used as basement walls in large urban construction, and often eliminate the need for temporary walls.

22.2   INTERNALLY STABILIZED SYSTEMS

Internally stabilized systems reinforce the soil to provide the necessary stability. Various schemes are available, all of which have been developed since 1960. They can be subdivided into two categories: *reinforced soils* and *in-situ reinforcement*.

Reinforced Soils

Soil is strong in compression, but has virtually no tensile strength. Therefore, the inclusion of tensile reinforcing members in a soil can significantly increase is strength and load-bearing capacity, much the same way that placing rebars in concrete increases its strength. The resulting reinforcing soil is called *mechanically stabilized earth (MSE)*.

Often MSE is used so that slopes may be made steeper than would otherwise be possible. Thus, this method forms an intermediate alternative between earth slopes and retaining walls.

MSE also may be used with vertical or near-vertical faces, thus forming a type of retaining wall. In this case, it becomes necessary to place some type of facing panels on the vertical surface, even though the primary soil support comes from the reinforcement, not the panels. Such structures are called *MSE walls*. The earliest MSE walls were developed by Henri Vidal in the early 1960, using the trade name *reinforced earth*. This design uses strips of galvanized steel for the reinforcement and precast concrete panels for the facing, as shown in Figures 22.12 and 22.13.

Many other similar methods also are used to build MSE walls. The reinforcement can consist of steel strips, polymer geogrids, wire mesh, geosynthetic fabric, or other materials. The facing can consist of precast concrete panels, precast concrete blocks, rock filled cages called *gabions*, or other materials. Sometimes the reinforcement is simply curved around to form a type of facing. Figure 22.14 shows an MSE wall being built using wire mesh reinforcement and gabion facing.

MSE walls are becoming very popular for many applications, especially for highway projects. Their advantages include low cost and high tolerance of differential settlements.

In-Situ Reinforcement

In-situ reinforcement methods differ from reinforced soils in that the tensile members are inserted into a soil mass rather than being embedded during placement of fill.

One type of in-situ reinforcement is called *soil nailing*. It consists of drilling near-horizontal holes into the ground, inserting steel tendons, and grouting. The face of the wall is typically covered with shotcrete, as shown in Figure 22.15.

These walls do not require a construction excavation, and thus are useful when space is limited.

Figure 22.12   Reinforced earth walls consist of precast concrete facing panels and steel or polymer reinforcing strips that extend into the retained soil. This wall is under construction (The Reinforced Earth Company).

Figure 22.13   A completed reinforced earth wall (The Reinforced Earth Company).

Figure 22.14   An MSE wall under construction using galvanized wire mesh as the tensile reinforcement and rock-filled cages called gabions for the facing (Federal Highway Administration).

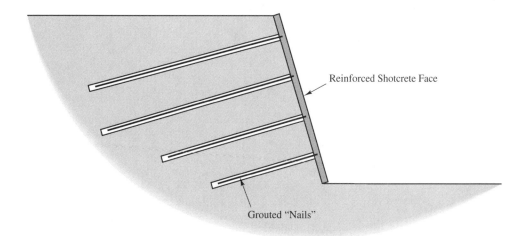

Reinforced Shotcrete Face

Grouted "Nails"

Figure 22.15   A soil nail wall.

SUMMARY

Major Points

1. An earth-retaining structure is a structural system designed to maintain a sharp transition between two ground surface elevations.
2. Externally stabilized earth-retaining systems resist the applied earth pressures by virtue of their weight and stiffness. These include gravity walls and in-situ walls.
3. Gravity walls include massive gravity walls, which are made of mortared stones, masonry, or unreinforced concrete, cantilever gravity walls, which are made of reinforced concrete or masonry, and crib walls, which are made of precast concrete members with a soil infill.
4. In-situ walls include sheet pile walls, soldier pile walls, and slurry walls.
5. Internally stabilized earth-retaining systems reinforce the soil to provide the necessary stability. These include reinforced soils and in-situ reinforcement.

Vocabulary

Cantilever gravity wall	In-situ reinforcement	Reinforced soil
Crib wall	Internal brace	Reinforced earth
Earth-retaining structure	Internally stabilized	Sheet pile wall
Externally stabilized	systems	Slurry wall
systems	Massive gravity wall	Soil nailing
Gabion	Mechanically stabilized	Soldier pile wall
Gravity wall	earth	Tieback anchor
In-situ wall	MSE wall	

23

Lateral Earth Pressures

*For every complex problem, there is a
solution that is simple, neat, and wrong.*

H.L. Mencken

One of the first steps in the design of earth-retaining structures is to determine the magnitude and direction of the forces and pressures acting between the structure and the adjacent ground, as shown in Figure 23.1. The most important of these is the pressure between the retained earth and the back of the earth-retaining structure. We call this a *lateral earth pressure* because its primary component is horizontal. Another lateral earth pressure acts between the front of the foundation and the adjacent ground. These pressures are the subject of this chapter.

The terms *pressure* and *stress* are nearly synonymous. In the context of this discussion, let us define *pressure* as the contact force per unit area between a structure and the adjacent ground, and *stress* as the force per unit area within the soil or the structure. However, because these two terms are so closely related, and the pressure at a point on the structure is equal to the stress in soil immediately adjacent to the structure, we will use the same symbols to represent both stress and pressure. The symbol σ represents normal stresses and pressures, while τ represents shear stresses and pressures.

23.1 HORIZONTAL STRESSES IN SOIL

Lateral earth pressures are the direct result of horizontal stresses in the soil. In Chapter 3 we defined the ratio of the horizontal effective stress to the vertical effective stress at any point in a soil as the *coefficient of lateral earth pressure, K*:

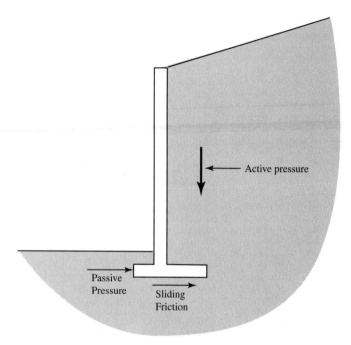

Figure 23.1   Forces and pressures acting between an earth retaining structure and the adjacent ground.

$$K = \frac{\sigma'_x}{\sigma'_z}$$ (23.1)

Where:

     K = coefficient of lateral earth pressure

     σ'_x = horizontal effective stress

     σ'_z = vertical effective stress

In the context of this chapter, K is important because it influences the lateral earth pressures acting on an earth retaining structure.

For purposes of describing lateral earth pressures, engineers have defined three important soil conditions: the *at-rest condition*, the *active condition*, and the *passive condition*.

The At-Rest Condition

Let us assume a certain retaining wall is both *rigid* and *unyielding*. In this context, a rigid wall is one that does not experience any significant flexural movements. The opposite would be a *flexible* wall—one that has no resistance to flexure. The term *unyielding* means the wall does not translate or rotate, as compared to a *yielding* wall that can do either or both. Let us also assume this wall is built so that no lateral strains occur in the ground. Therefore, the lateral stresses in the ground are the same as they were in its natural undisturbed state.

The most accurate way to evaluate K_0 would be to measure σ_x' in-situ using a dilatometer, pressuremeter, or some other test, compute σ_z' using the techniques described in Chapter 3, then compute K_0 using Equation 23.1. However, these in-situ tests are not often used in engineering practice, so we usually must rely on empirical correlations with other soil properties. Several such correlations have been developed, including the following by Mayne and Kulhawy (1982), which is based on laboratory tests on 170 soils that ranged from clay to gravel. This formula is applicable only when the ground surface is level:

$$K_0 = (1 - \sin \phi')OCR^{\sin \phi'}$$ (23.2)

Where:

    K_0 = coefficient of lateral earth pressure at rest

    ϕ' = effective friction angle of soil

    OCR = overconsolidation ratio of soil

For gravity walls that are backfilled with sandy soil and have footings founded on bedrock, Duncan et al. (1990) recommend using $K_0 = 0.45$ if the backfill is compacted, or $K_0 = 0.55$ if the backfill is not compacted.

If no groundwater table is present, the lateral earth pressure, σ, acting on a rigid and unyielding wall is, in theory, equal to the horizontal stress in the soil:

$$\sigma = \sigma_z'K_0$$ (23.3)

As long as no settlement is occurring, there are no shear forces acting on the back of the wall in the at-rest condition. Lateral earth pressures with shallow groundwater are discussed later in this chapter.

In a homogeneous soil, K_0 is a constant and σ_z' varies linearly with depth. Therefore, in theory, σ also varies linearly with depth, forming a triangular pressure distribution, as shown in Figure 23.2. Thus, if at-rest conditions are present, the horizontal force acting on a unit length of a vertical wall is the area of this triangle:

$$P_0/b = \frac{\gamma H^2 K_0}{2}$$ (23.4)

Where:

    P_0/b = normal force acting between soil and wall per unit length of wall

    b = unit length of wall (usually 1 ft or 1 m)

    γ = unit weight of soil

    H = height of wall

    K_0 = coefficient of lateral earth pressure at rest

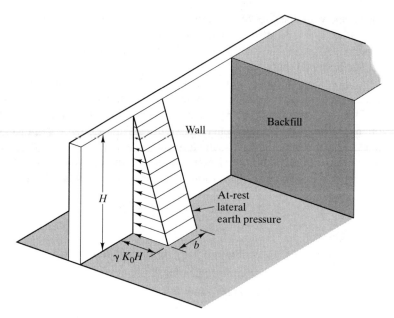

Figure 23.2   At-rest pressure acting on a retaining wall.

Example 23.1

An 8-ft tall basement wall retains a soil that has the following properties: $\phi' = 35°$, $\gamma = 127 \text{ lb/ft}^3$, OCR = 2. The ground surface is horizontal and level with the top of the wall. The groundwater table is well below the bottom of the wall. Consider the soil to be in the at-rest condition and compute the force that acts between the wall and the soil.

Solution

$$K_0 = (1 - \sin \phi')\text{OCR}^{\sin \phi'}$$

$$= (1 - \sin 35°)2^{\sin 35°}$$

$$= 0.635$$

$$P_0/b = \frac{\gamma H^2 K_0}{2}$$

$$= \frac{(127 \text{ lb/ft}^2)(8 \text{ ft})^2(0.635)}{2}$$

$$= \mathbf{2580\ lb/ft} \qquad \Leftarrow Answer$$

Because the theoretical pressure distribution is triangular, this resultant force acts at the lower third-point on the wall.

The Active Condition

The at-rest condition is present only if the wall does not move. Although this may seem to be a criterion that all walls should meet, even very small movements alter the lateral earth pressure.

Suppose Mohr's circle A in Figure 23.3 represents the state of stress at a point in the soil behind the wall in Figure 23.4, and suppose this soil is in the at-rest condition. The inclined lines represent the Mohr-Coulomb failure envelope. Because the Mohr's circle does not touch the failure envelope, the shear stress, τ, is less than the shear strength, s.

Now, permit the wall to move outward a short distance. This movement may be either translational or rotational about the bottom of the wall. It relieves part of the horizontal stress, causing the Mohr's circle to expand to the left. Continue this process until the circle reaches the failure envelope and the soil fails in shear (circle B). This shear failure will occur along the planes shown in Figure 23.4, which are inclined at an angle of $45 + \phi/2$ degrees from the horizontal. A soil that has completed this process is said to be in the *active condition*. The value of K in a $c = 0$ soil in the active condition is known as K_a, the *coefficient of active earth pressure*.

Once the soil attains the active condition, the horizontal stress in the soil (and thus the pressure acting on the wall) will have reached its lower bound, as shown in Figure 23.5. The amount of movement required to reach the active condition depends on the soil type and the wall height, as shown in Table 23.1. For example, in loose sands, the active condition is achieved if the wall moves outward from the backfill a distance of only $0.004\,H$ (about 12 mm for a 3-m tall wall). Although basement walls, being braced at the top, cannot move even that distance, a cantilever wall (one in which the top is not connected to a building or other structure) could very easily move 12 mm outward, and such

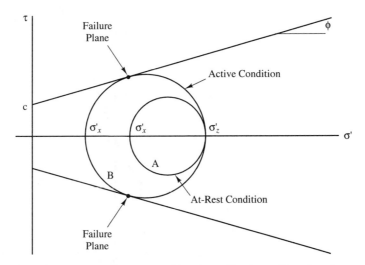

Figure 23.3   Changes in the stress conditions in a soil as it transitions from the at-rest condition to the active condition.

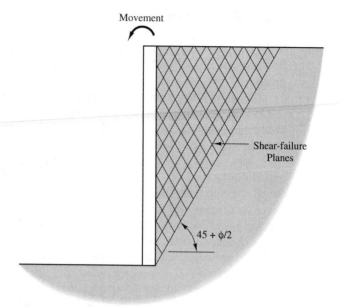

Figure 23.4 Development of shear failure planes in the soil behind a wall as it transitions from the at-rest condition to the active conditon.

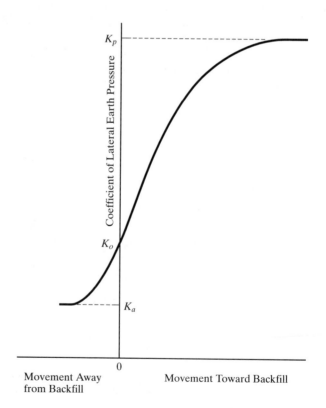

Figure 23.5 Effect of wall movement on lateral earth pressure in sand.

TABLE 23.1 WALL MOVEMENT REQUIRED TO REACH THE ACTIVE CONDITION (Adapted from CGS, 1992).

Soil Type	Horizontal Movement Required to Reach the Active Condition
Dense sand	$0.001\,H$
Loose sand	$0.004\,H$
Stiff clay	$0.010\,H$
Soft clay	$0.020\,H$

H = Wall height

a movement would usually be acceptable. Thus, a basement wall may need to be designed to resist the at-rest pressure, whereas the design of a free-standing cantilever wall could use the active pressure. Since the active pressure is smaller, the design of free-standing walls will be more economical.

The Passive Condition

The *passive condition* is the opposite of the active condition. In this case, the wall moves *into* the backfill, as shown in Figure 23.6, and the Mohr's circle changes, as shown in Figure 23.7. Notice how the vertical stress remains constant whereas the horizontal stress changes in response to the induced horizontal strains.

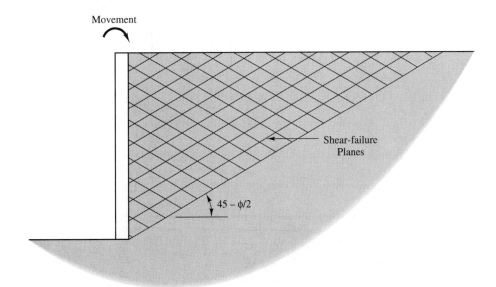

Figure 23.6 Development of shear failure planes in the soil behind a wall as it transitions from the at-rest condition to the passive condition.

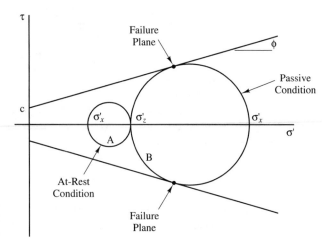

Figure 23.7 Changes in the stress condition in a soil as it transitions from the at-rest condition to the passive condition.

In a homogeneous soil, the shear failure planes in the passive case are inclined at an angle of 45 − ϕ/2 degrees from the horizontal. The value of K in a cohesionless soil in the passive condition is known as K_p, the *passive coefficient of lateral earth pressure*. This is the upper bound of K and produces the upper bound of pressure that can act on the wall.

Engineers often use the passive pressure that develops along the toe of a retaining wall footing to help resist sliding, as shown in Figure 23.8. In this case, the "wall" is the side of the footing.

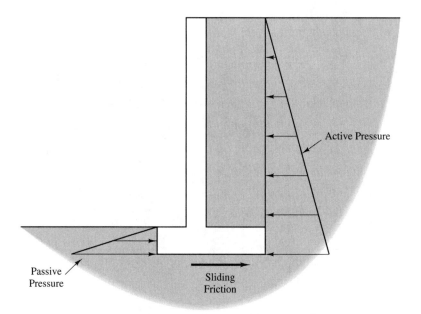

Figure 23.8 Active and passive pressures acting on a cantilever retaining wall.

TABLE 23.2   WALL MOVEMENT REQUIRED TO REACH
THE PASSIVE CONDITION (Adapted from CGS, 1992)

Soil Type	Horizontal Movement Required to Reach the Passive Condition
Dense sand	0.020 H
Loose sand	0.060 H
Stiff clay	0.020 H
Soft clay	0.040 H

H = Wall height

More movement must occur to attain the passive condition than for the active condition. Typical required movements for various soils are shown in Table 23.2.

Although movements on the order of those listed in Tables 23.1 and 23.2 are necessary to reach the full active and passive states, respectively, much smaller movements also cause significant changes in the lateral earth pressure. While conducting a series of full-scale tests on retaining walls, Terzaghi (1934b) observed:

> With compacted sand backfill, a movement of the wall over an insignificant distance (equal to one-tenthousanth of the depth of the backfill) decreases the [coefficient of lateral earth pressure] to 0.20 or increases it up to 1.00.

This effect is not as dramatic in other soils, but even with those soils only the most rigid and unyielding structures are truly subjected to at-rest pressures.

23.2   CLASSICAL LATERAL EARTH PRESSURE THEORIES

The solution of lateral earth pressure problems was among the first applications of the scientific method to the design of structures. Two of the pioneers in this effort were the Frenchman Charles Augustin Coulomb and the Scotsman W. J. M. Rankine. Although many others have since made significant contributions to our knowledge of earth pressures, the contributions of these two men were so fundamental that they still form the basis for earth pressure calculations today. More than fifty earth pressure theories are now available; all of them have their roots in Coulomb and Rankine's work.

Coulomb presented his theory in 1773 and published it three years later (Coulomb, 1776). Rankine developed his theory more than eighty years after Coulomb (Rankine, 1857). In spite of this chronology, it is conceptually easier for us to discuss Rankine's theory first.

For clarity, we will begin our discussion of these theories by considering only soils with $c = 0$ and $\phi \geq 0$. These are sometimes called *cohesionless soils*. Once we have established the basic concepts, we will then expand the discussion to include soils with cohesion ($c \geq 0$, $\phi \geq 0$). Lateral earth pressure theories may be used with either effective stress analyses (c', ϕ') or total stress analyses (c_T, ϕ_T). However, effective stress analyses are usually more appropriate, and are the only type we will consider in this chapter.

Rankine's Theory for Soils with $c = 0$ and $\phi \geq 0$

Assumptions

Rankine approached the lateral earth pressure problem with the following assumptions:

1. The soil is homogeneous and isotropic, which means c, ϕ, and γ have the same values everywhere, and they have the same values in all directions at every point (i.e., the strength on a vertical plane is the same as that on a horizontal plane). Section 23.9 extends this discussion to consider layered soils, where each layer has different values of c, ϕ, and γ.

2. The most critical shear surface is a plane. In reality, it is slightly concave up, but this is a reasonable assumption (especially for the active case) and it simplifies the analysis.

3. The ground surface is a plane (although it does not necessarily need to be level).

4. The wall is infinitely long so that the problem may be analyzed in only two dimensions. Geotechnical engineers refer to this as a *plane strain* condition.

5. The wall moves sufficiently to develop the active or passive condition.

6. The resultant of the normal and shear forces that act on the back of the wall is inclined at an angle parallel to the ground surface (Coulomb's theory provides a more accurate model of shear forces acting on the wall).

Active Condition

With these assumptions, we can treat the wedge of soil behind the wall as a free body and evaluate the problem using the principles of statics, as shown in Figure 23.9a. This is similar to methods used to analyze the stability of earth slopes, and is known as a *limit equilibrium analysis*, which means that we consider the conditions that would exist if the soil along the base of the failure wedge was about to fail in shear.

Weak seams or other nonuniformities in the soil may control the inclination of the critical shear surface. However, if the soil is homogeneous, P_a/b is greatest when this surface is inclined at an angle of $45 + \phi/2$ degrees from the horizontal, as shown in the Mohr's circle in Figure 23.3. Thus, this is the most critical angle.

Solving this free body diagram for P_a/b and V_a/b gives:

$$P_a/b = \frac{\gamma H^2 K_a \cos\beta}{2} \qquad (23.5)$$

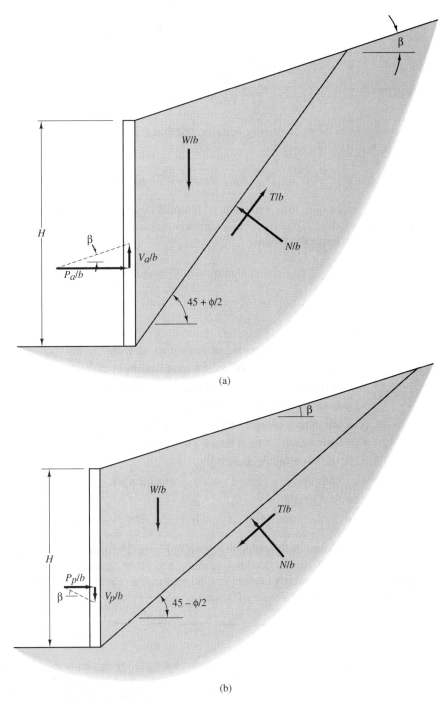

(a)

(b)

Figure 23.9 Free body diagram of soil behind a retaining wall using Rankine's solution: (a) active case; and (b) passive case.

$$V_a/b = \frac{\gamma H^2 K_a \sin\beta}{2} \tag{23.6}$$

$$K_a = \frac{\cos\beta - \sqrt{\cos^2\beta - \cos^2\phi}}{\cos\beta + \sqrt{\cos^2\beta - \cos^2\phi}} \tag{23.7}$$

The magnitude of K_a is usually between 0.2 and 0.9. Equation 23.7 is valid only when $\beta \le \phi$. If $\beta = 0$, it reduces to:

$$K_a = \tan^2(45° - \phi/2) \tag{23.8}$$

A solution of P_a/b as a function of H would show that the theoretical pressure distribution is triangular. Therefore, the theoretical pressure and shear stress acting against the wall, σ and τ, respectively, are:

$$\sigma = \sigma_z' K_a \cos\beta \tag{23.9}$$

$$\tau = \sigma_z' K_a \sin\beta \tag{23.10}$$

Where:

σ = soil pressure imparted on retaining wall from the soil

τ = shear stress imparted on retaining wall from the soil

P_a/b = normal force between soil and wall per unit length of wall

V_a/b = shear force between soil and wall per unit length of wall

b = unit length of wall (usually 1 ft or 1 m)

K_a = active coefficient of lateral earth pressure

σ_z' = vertical effective stress

β = inclination of ground surface above the wall

H = wall height

However, observations and measurements from real retaining structures indicate that the true pressure distribution, as shown in Figure 23.10, is not triangular. This difference is because of wall deflections, arching, and other factors. The magnitudes of P_a/b and V_a/b are approximately correct, but the resultant acts at about $0.40H$ from the bottom, not $0.33H$ as predicted by theory (Duncan et al., 1990).

Example 23.2

A 6-m tall cantilever wall retains a soil that has the following properties: $c' = 0$, $\phi' = 30°$, and $\gamma = 19.2$ kN/m³. The ground surface behind the wall is inclined at a slope of 3 horizontal to 1 vertical, and the wall has moved sufficiently to develop the active condition. Determine the normal and shear forces acting on the back of this wall using Rankine's theory.

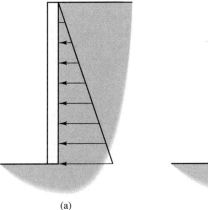
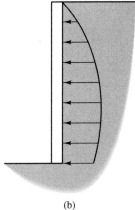

Figure 23.10 Comparison between (a) theoretical and (b) observed distributions of earth pressures acting behind retaining structures.

(a)   (b)

Solution

$$\beta = \tan^{-1}(1/3)$$
$$= 18°$$

$$K_a = \frac{\cos\beta - \sqrt{\cos^2\beta - \cos^2\phi}}{\cos\beta + \sqrt{\cos^2\beta - \cos^2\phi}}$$

$$= \frac{\cos18° - \sqrt{\cos^2 18° - \cos^2 30°}}{\cos18° + \sqrt{\cos^2 18° - \cos^2 30°}}$$

$$= 0.415$$

$$P_a/b = \frac{\gamma H^2 K_a\cos\beta}{2} = \frac{(19.2\text{kN/m}^2)(6\text{m})^2(0.415)\cos18°}{2} = \textbf{136 kN/m} \qquad \Leftarrow Answer$$

$$V_a/b = \frac{\gamma H^2 K_a\sin\beta}{2} = \frac{(19.2\text{kN/m}^2)(6\text{m})^2(0.415)\sin18°}{2} = \textbf{44 kN/m} \qquad \Leftarrow Answer$$

These results are shown in Figure 23.11.

Passive Condition

Rankine analyzed the passive condition in a fashion similar to the active condition except that the shear force acting along the base of the wedge now acts in the opposite direction (it always opposes the movement of the wedge) and the free body diagram becomes as shown in Figure 23.9b. Notice that the failure wedge is much flatter than it was in the active case and the critical angle is now $45 - \phi/2$ degrees from the horizontal.

The normal and shear forces, P_p/b and V_p/b, respectively, acting on the wall in the passive case are:

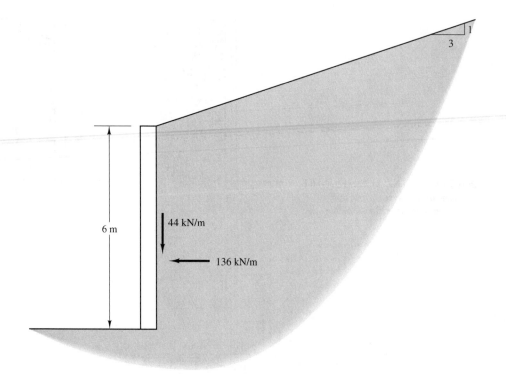

Figure 23.11   Results from Example 23.2.

$$P_p/b = \frac{\gamma H^2 K_p \cos\beta}{2} \tag{23.11}$$

$$V_p/b = \frac{\gamma H^2 K_p \sin\beta}{2} \tag{23.12}$$

$$K_p = \frac{\cos\beta + \sqrt{\cos^2\beta - \cos^2\phi}}{\cos\beta - \sqrt{\cos^2\beta - \cos^2\phi}} \tag{23.13}$$

Equation 23.13 is valid only when $\beta \leq \phi$. If $\beta = 0$, it reduces to:

$$K_p = \tan^2(45° + \phi/2) \tag{23.14}$$

The theoretical pressure and shear acting against the wall, σ and τ, respectively, are:

$$\sigma = \sigma'_z K_p \cos\beta \tag{23.15}$$

$$\tau = \sigma'_z K_p \sin\beta \tag{23.16}$$

Example 23.3

A six-story building with plan dimensions of 150 ft × 150 ft has a 12-ft deep basement. This building is subjected to horizontal wind loads, and the structural engineer wishes to transfer these loads into the ground through the basement walls. The maximum horizontal force acting on the basement wall is limited by the passive pressure in the soil. Using Rankine's theory, compute the maximum force between one of the basement walls and the adjacent soil assuming full passive conditions develop, then convert it to an allowable force using a factor of safety of 3. The soil is a silty sand with $c' = 0$, $\phi' = 30°$, and $\gamma = 119$ lb/ft³, and the ground surface surrounding the building is essentially level.

Solution

$$K_p = \tan^2(45° + \phi'/2) = \tan^2(45° + 30°/2) = 3.00$$

$$P_p = \frac{(25{,}700 \text{ lb/ft})(150 \text{ ft})}{1000 \text{ lb/k}} = 3860 \text{ k}$$

$$P_p/b = \frac{\gamma H^2 K_p \cos\beta}{2}$$

$$= \frac{(119 \text{ lb/ft}^3)(12 \text{ ft})^2(3.00)\cos 0}{2}$$

$$= 25{,}700 \text{ lb/ft}$$

The allowable passive force, $(P_p)_a$, is:

$$(P_p)_a = \frac{P_p}{F} = \frac{3860\text{k}}{3} = \mathbf{1290 \text{ k}}   \Leftarrow Answer$$

Note: The actual design computations for this problem would be more complex because they must consider the active pressure acting on the opposite wall, sliding friction along the basement floor, lateral resistance in the foundations, and other factors. In addition, the horizontal displacement required to develop the full passive resistance may be excessive, so the design value may need to be reduced accordingly. Finally, to take advantage of this resistance, the wall would need to be structurally designed to accommodate this large load, which is much greater than that due to the active or at-rest pressure.

Coulomb's Theory for Soils With $c = 0$ and $\phi \geq 0$

Coulomb's theory differs from Rankine's in that the resultant of the normal and shear forces acting on the wall is inclined at an angle ϕ_w from a perpendicular to the wall, where $\tan \phi_w$ is the coefficient of friction between the wall and the soil, as shown in Figure 23.12. This is a more realistic model, and thus produces more precise values of the active earth pressure.

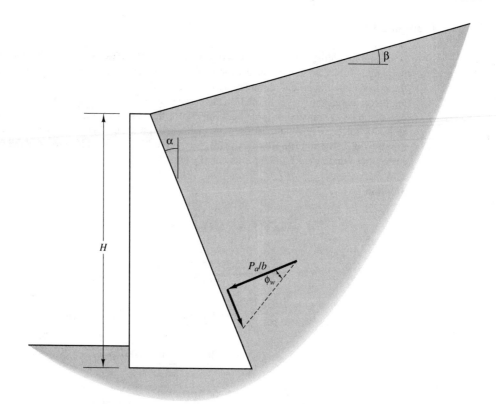

Figure 23.12   Parameters for Coulomb's active earth pressure equation. Walls inclined in the opposite direction have a negative α. V_a/b normally acts in the direction shown, thus producing a positive ϕ_w.

Coulomb presented his earth pressure formula in a difficult form, so others have rewritten it in a more convenient fashion, as follows (Müller Breslau, 1906; Tschebotari-off, 1951):

$$P_a/b = \frac{\gamma H^2 K_a \cos\phi_w}{2} \qquad (23.17)$$

$$V_a/b = \frac{\gamma H^2 K_a \sin\phi_w}{2} \qquad (23.18)$$

$$K_a = \frac{\cos^2(\phi - \alpha)}{\cos^2\alpha \cos(\phi_w + \alpha)\left[1 + \sqrt{\dfrac{\sin(\phi + \phi_w)\sin(\phi - \beta)}{\cos(\phi_w + \alpha)\cos(\alpha - \beta)}}\right]^2} \qquad (23.19)$$

Where:

σ = normal pressure imparted on retaining wall from the soil

τ = shear stress imparted on retaining wall from the soil

P_a/b = normal force between soil and wall per unit length of wall

V_a/b = shear force between soil and wall per unit length of wall

b = unit length of wall (usually 1 ft or 1 m)

K_a = active coefficient of lateral earth pressure

σ_z' = vertical effective stress

α = inclination of wall from vertical

β = inclination of ground surface above the wall

ϕ_w = wall-soil interface friction angle

Equation 23.19 is valid only for $\beta \leq \phi$. When designing concrete or masonry walls it is common practice to use $\phi_w = 0.67\,\phi'$. Steel walls have less sliding friction, perhaps on the order of $\phi_w = 0.33\,\phi'$.

Coulomb did not develop a formula for passive earth pressure, although others have used his theory to do so. However, the addition of wall friction can substantially increase the computed passive pressure, possibly to values that are too high. Therefore, engineers normally neglect wall friction in passive pressure computations ($\phi_w = 0$) and use Rankine's method to compute passive earth pressures (Equations 23.11–23.16).

Example 23.4

Using Coulomb's method, compute the active pressure acting on the reinforced concrete retaining wall shown in Figure 23.13.

Solution

$$\beta = \tan^{-1}\left(\frac{1}{2}\right) = 27°$$

$$\phi_w = 0.67\phi' = 0.67(32°) = 21°$$

$$K_a = \frac{\cos^2(\phi - \alpha)}{\cos^2\alpha \, \cos(\phi_w + \alpha)\left[1 + \sqrt{\dfrac{\sin(\phi + \phi_w)\,\sin(\phi - \beta)}{\cos(\phi_w + \alpha)\,\cos(\alpha - \beta)}}\right]^2}$$

$$= \frac{\cos^2(32° - 2°)}{\cos^2 2° \, \cos(21° + 2°)\left[1 + \sqrt{\dfrac{\sin(32° + 21°)\,\sin(32° - 27°)}{\cos(21° + 2°)\,\cos(2° - 27°)}}\right]^2}$$

$$= 0.491$$

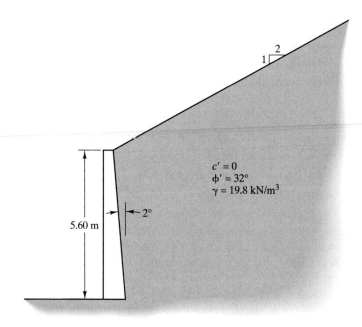

Figure 23.13   Retaining wall for Example 23.4.

$$P_a/b = \frac{\gamma\, H^2 K_a \cos\phi_w}{2}$$

$$= \frac{(19.8\ \text{kN/m}^2)(5.60\ \text{m})^2(0.491)\cos 21°}{2}$$

$$= 142\ \text{kN/m} \qquad \Leftarrow Answer$$

$$V_a/b = \frac{\gamma\, H^2 K_a \sin\phi_w}{2}$$

$$= \frac{(19.8\ \text{kN/m}^2)(5.60\ \text{m})^2(0.491)\sin 21°}{2}$$

$$= 55\ \text{kN/m} \qquad \Leftarrow Answer$$

QUESTIONS AND PRACTICE PROBLEMS

23.1 Explain the difference between the active, at-rest, and passive earth pressure conditions.

23.2 Which of the three earth pressure conditions should be used to design a rigid basement wall? Why?

23.3 A basement is to be built using 2.5-m tall masonry walls. These walls will be backfilled with a silty sand that has $c' = 0$, $\phi' = 35°$, and $\gamma = 19.7$ kN/m³. Assuming the at-rest conditions will exist and using an overconsolidation ratio of 2, compute the normal force per meter acting on the back of this wall. Also, draw a pressure diagram and indicate the lateral earth pressure acting at the bottom of the wall.

23.4 A 10-ft tall concrete wall with a vertical back is to be backfilled with a silty sand that has a unit weight of 122 lb/ft³, an effective cohesion of 0, and an effective friction angle of 32°. The ground behind the wall will be level. Using Rankine's method, compute the normal force per foot acting on the back of the wall. Assume the wall moves sufficiently to develop the active condition in the soil.

23.5 The wall described in Problem 23.4 has a foundation that extends from the ground surface to a depth of 2 ft. As the wall moves slightly away from the backfill soils to create the active condition, the footing moves into the soils below the wall, creating the passive condition as shown in Figure 23.9. Compute the ultimate passive pressure acting on the front of the foundation.

23.6 A 12-ft tall concrete wall with a vertical back is to be backfilled with a clean sand that has a unit weight of 126 lb/ft³, an effective cohesion of 0, and an effective friction angle of 36°. The ground behind the wall will be inclined at a slope of 2 horizontal to 1 vertical. Using Rankine's method, compute the normal and shear forces per foot acting on the back of the wall. Assume the wall moves sufficiently to develop the active condition in the soil.

23.7 Repeat Problem 23.6 using Coulomb's method.

23.3   LATERAL EARTH PRESSURES IN SOILS WITH $c \geq 0$ AND $\phi \geq 0$

Rankine did not address lateral earth pressures in soils with cohesion ($c \geq 0$ and $\phi \geq 0$) and Coulomb did not address passive pressures. However, later investigators developed complete formulas for cohesive soils. Bell (1915) was among those who contributed.

Theoretical Behavior

Soils with cohesion can stand vertically to a height of no more than the *critical height*, H_c:

$$H_c = \frac{2c}{\gamma \sqrt{K_a}}$$

(23.20)

In other words, if $H < H_c$ the earth will stand vertically without a wall. In practice we would apply some factor of safety to H_c (perhaps 1.5 to 2) before deciding not to build a wall. An engineer also would want to consider the potential for surface erosion and other modes of failure.

If $H > H_c$, the theoretical pressure distribution is as shown in Figure 23.14.

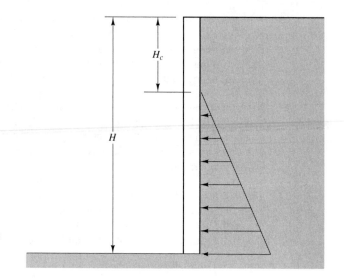

Figure 23.14 Theoretical active pressure distribution in soils with cohesion ($c \geq 0$, $\phi \geq 0$).

Equations 23.5 and 23.6 then become:

$$P_a/b = \left(\frac{\gamma H^2 K_a}{2} - 2cH\sqrt{K_a} + \frac{2c^2}{\gamma} \right) \cos\beta \geq 0 \tag{23.21}$$

$$V_a/b = \left(\frac{\gamma H^2 K_a}{2} - 2cH\sqrt{K_a} + \frac{2c^2}{\gamma} \right) \sin\beta \geq 0 \tag{23.22}$$

These formulas often are incorrectly stated without the $2c^2/\gamma$ term. This term must be present to account for the lack of tensile forces between the wall and the soil at depths shallower than H_c.

The Rankine equations for passive conditions in soils with cohesion are as follows:

$$P_p/b = \left(\frac{\gamma H^2 K_p}{2} + 2cH\sqrt{K_p} \right) \cos\beta \tag{23.23}$$

$$V_p/b = \left(\frac{\gamma H^2 K_p}{2} + 2cH\sqrt{K_p} \right) \sin\beta \tag{23.24}$$

The theoretical shape of the passive pressure distribution is shown in Figure 23.15.

Actual Behavior

The theoretical behavior may be approximately correct for sandy soils that develop cohesive strength from cementing agents, such as calcium carbonate or iron oxide. However, it is not a good indicator of lateral earth pressures in clays, and may produce unconservative

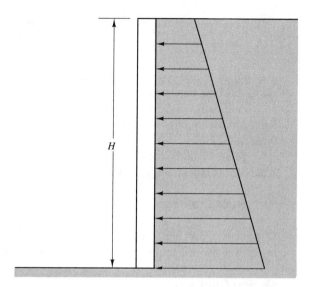

Figure 23.15   Theoretical distribution of passive earth pressure in soils with cohesion $(c \geq 0, \phi \geq 0)$.

designs. This discrepancy occurs because the earth pressure theories do not consider the following aspects of clay behavior:

Creep

When the shear stresses in a clayey soil are a large percentage of the shear strength, a phenomenon known as *creep* can occur. The soil slowly shears and never reaches complete equilibrium unless the shear stresses become sufficiently small. When this occurs behind a retaining wall, the failure wedge slowly moves toward the wall, so it is impossible to maintain the active condition (which demands high shear stresses) for long periods. Therefore, design cantilever walls backfilled with clay using something higher than the active pressure.

Expansiveness

Another potentially more significant problem is that clayey soils may be expansive (see Chapter 19). An expanding backfill places very large loads on the wall, far greater than the computed active or at-rest pressure. The exact magnitude of lateral pressures caused by expansive soils is difficult to predict, but the passive pressure would be an upper bound.

The cyclic expansion and contraction cycles in expansive soils further aggravate this situation and could cause the following scenario:

1. The soil expands and the retaining wall moves outward.
2. The soil shrinks and moves slightly downhill, thus remaining in intimate contact with the wall. Cracks form.

3. Loose debris falls into the cracks, preventing them from closing.

4. The soil expands again and moves the wall farther out.

This process could continue indefinitely, moving the wall much farther than would a single cycle of expansion.

A slightly cohesive soil, such as an SC, would probably not pose any serious problems, but highly cohesive soils, such as a CH, could produce large movements or even failure.

Many of the preventive measures described in Chapter 19 are also appropriate for reducing the potential for expansion behind a retaining wall.

Poor Drainage

Clays have a low hydraulic conductivity, and thus obstruct drainage. Therefore, groundwater may become trapped behind the wall, producing hydrostatic pressures.

Design Guidelines

The best procedure is to avoid backfilling any wall with clay, especially in regions with adverse climates. It is often possible to bring in other soils from on-site or off-site to use for backfill material. However, if clay is used, the design must reflect their presence. Jennings (1973) described the South African practice of designing such walls using at-rest pressures with a K_0 of 0.8 to 1.0 and making them as flexible as possible. Terzaghi and Peck's design values, described in Section 23.5, also provide lateral earth pressures for walls supporting clay.

23.4   EQUIVALENT FLUID METHOD

As discussed earlier, the theoretical distribution of lateral earth pressure acting on a wall backfilled with a cohesionless soil is triangular. This is the same shape as the pressure distribution that would be imposed if the wall was backfilled with a fluid instead of soil. Furthermore, if this fluid had the proper unit weight, the magnitude of the lateral pressure also would be equal to that from the soil. Engineers often use this similarity when expressing lateral earth pressures for design purposes. Instead of quoting K values, we can describe the lateral earth pressure using the *equivalent fluid density*, G_h, which is the unit weight of a fictitious fluid that would impose the same horizontal pressures on the wall as the soil. The vertical component of the lateral earth pressure also can be expressed using a similar value, G_v. Thus, the normal and shear forces acting on the wall may then be expressed as:

$$P_a/b = \frac{G_h H^2}{2}$$

(23.25)

$$V_a/b = \frac{G_v H^2}{2}$$ (23.26)

Where:

P_a/b = normal force between soil and wall per unit length of wall

V_a/b = shear force between soil and wall per unit length of wall

b = unit length of wall (usually 1 ft or 1 m)

G_h, G_v = horizontal and vertical equivalent fluid densities

H = height of wall

By setting Equations 23.25 and 23.26 equal to the Equations 23.5 and 23.6, we can derive the following Rankine equations for G_h and G_v in soils with $c = 0$:

$$G_h = \gamma K_a \cos\beta$$ (23.27)

$$G_v = \gamma K_a \sin\beta$$ (23.28)

Where:

γ = unit weight of soil

K_a = Rankine's coefficient of active earth pressure (Equations 23.7 or 23.8)

β = inclination of ground surface above the wall

If the earth pressures are to be computed using Coulomb's method, then the equations for G_h and G_v in soils with $c = 0$ are:

$$G_h = \gamma K_a \cos\phi_w$$ (23.29)

$$G_v = \gamma K_a \sin\phi_w$$ (23.30)

Where:

γ = unit weight of soil

K_a = Coulomb's coefficient of active earth pressure (Equation 23.19)

ϕ_w = wall-soil interface friction angle

The equivalent fluid method also may be used to express at-rest earth pressures, as follows:

$$G_h = \gamma K_0$$ (23.31)

$$G_v = 0$$ (23.32)

Where:

γ = unit weight of soil

K_0 = coefficient of lateral earth pressure at rest (Equation 23.2)

Geotechnical engineers develop design values of the equivalent fluid density and present them to the civil engineer or structural engineer who then design the wall using the principles of fluid statics. This method is popular because it reduces the potential for confusion and mistakes. All engineers understand fluid statics (or at least they should!), so these design parameters are easy to use.

Example 23.5

A 12-ft tall cantilever retaining wall will be supported on a 2-ft deep continuous footing and will retain a sandy soil with $c' = 0$, $\phi' = 35°$, and $\gamma = 124$ lb/ft^3. The ground surface above the wall will be inclined at a 2:1 slope ($\beta = 27°$) and there will be no surcharge loads. Compute the design values of G_h and G_v, then use them to compute the total normal and shear forces imposed by the backfill on the back of the wall and footing.

Solution

Use Coulomb's method

$$\phi_w = 0.67(35°) = 23°$$

$$K_a = \frac{\cos^2(\phi - \alpha)}{\cos^2\alpha \cos(\phi_w + \alpha)\left[1 + \sqrt{\frac{\sin(\phi + \phi_w)\sin(\phi - \beta)}{\cos(\phi_w + \alpha)\cos(\alpha - \beta)}}\right]^2}$$

$$= \frac{\cos^2(35 - 0)}{\cos^2 0 \cos(23 + 0)\left[1 + \sqrt{\frac{\sin(35 + 23)\sin(35 - 27)}{\cos(23 + 0)\cos(0 - 27)}}\right]^2}$$

$$= 0.383$$

$$G_h = \gamma K_a \cos\phi_w = (124)(0.383)\cos23 = \textbf{44 lb/ft}^3 \qquad \Leftarrow Answer$$

$$G_v = \gamma K_a \sin\phi_w = (124)(0.383)\sin23 = \textbf{19 lb/ft}^3 \qquad \Leftarrow Answer$$

$$P_a/b = \frac{G_h H^2}{2}$$

$$= \frac{(44 \text{ lb/ft}^3)(14 \text{ ft})^2}{2}$$

$$= \textbf{4310 lb/ft} \qquad \Leftarrow Answer$$

$$V_a/b = \frac{G_v H^2}{2}$$

$$= \frac{(19 \text{ lb/ft}^3)(14 \text{ ft})^2}{2}$$

$$= \mathbf{1860 \text{ lb/ft}} \quad \Leftarrow Answer$$

23.5   PRESUMPTIVE LATERAL EARTH PRESSURES

The computation of lateral earth pressures, as discussed thus far in this chapter, requires assessments of the shear strength parameters c and ϕ. For large walls, the cost of obtaining these strength parameters (i.e., soil exploration, testing, and analysis) is normally justified. However, for small walls it may be less expensive to eliminate these costs and use conservative assumed earth pressures. These *presumptive lateral earth pressures* are similar to the presumptive allowable bearing pressures discussed in Section 8.3, which are often used to design spread footings for lightweight structures. Presumptive earth pressure values also are useful for walls backfilled with clay, because classical lateral earth pressures do not work well with cohesive soils.

Terzaghi and Peck (1967) developed presumptive earth pressures based on soil classification and the observed performance of real walls. This method classifies the soil into five types, as described in Table 23.3. The normal and shear forces acting on the wall are then computed using Equations 23.25 and 23.26 with Figures 23.16 and 23.17.

TABLE 23.3   CLASSIFICATION OF SOIL TYPES FOR TERZAGHI & PECK'S METHOD (Adapted from Terzaghi and Peck, 1967).

Soil Type	Description
1	Coarse grained soil without admixture of fine soil particles, very permeable (i.e., clean sand or gravel).
2	Coarse-grained soil of low permeability due to admixture of particles of silt size.
3	Residual soil with stones, fine silty sand, and granular materials with conspicuous clay content.
4	Very soft clay, organic silts, or silty clays.
5	Medium or stiff clay, deposited in chunks and protected in such a way that a negligible amount of water enters the spaces between the chunks during floods or heavy rains. If this condition cannot be satisfied, the clay should not be used as backfill material. With increasing stiffness of the clay, danger to the wall due to infiltration of water increases rapidly.

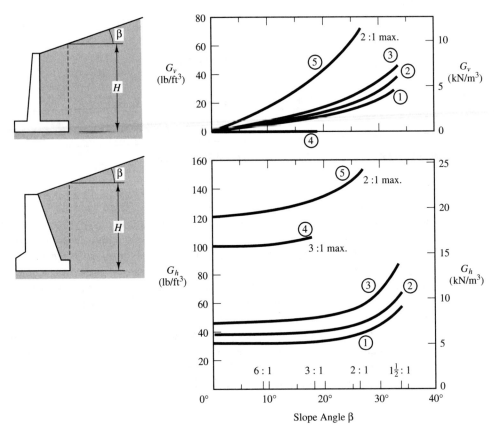

Figure 23.16 Charts for estimating the loads acting against a retaining wall beneath a planar ground surface (Adapted from Terzaghi and Peck, 1967).

Alternatively, Table 23.4 presents presumptive lateral earth pressures for retaining walls with essentially level backfills (i.e., where the ground surface above the wall has a slope of less than 5 percent). These are expressed as the equivalent fluid density, G_h. The design value of G_v should be taken to be zero in all cases. The active values could be used for cantilever walls, and the at-rest values for braced walls. Basement walls should be evaluated as braced walls. However, this table should not be used for walls more than 2.5 m (8 ft) tall.

QUESTIONS AND PRACTICE PROBLEMS

23.8 A 4-m tall cantilever wall is to be backfilled with a stiff clay. How far must this wall move to attain the active condition in the soil behind it? Is it appropriate to use the active pressure for design? Explain.

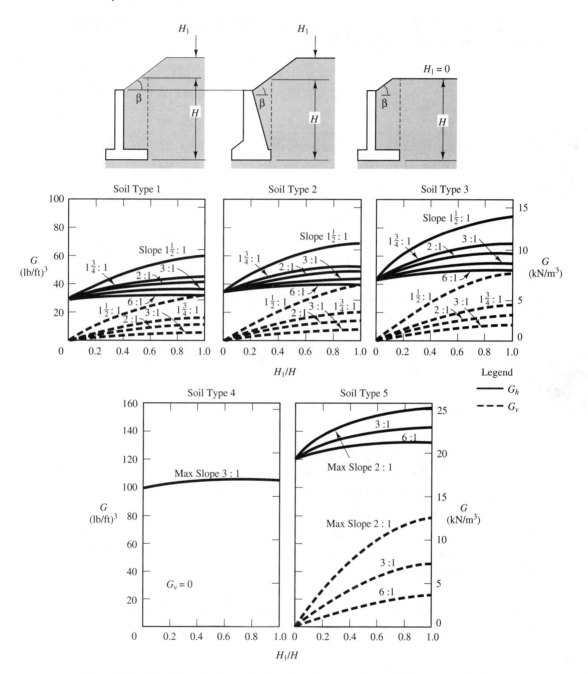

Figure 23.17 Charts for estimating the loads acting against a retaining wall below a ground surface that is sloped and then becomes level. For soil type 5, use an H value 4 ft (1 m) less than the actual height (Adapted from Terzaghi and Peck, 1967).

TABLE 23.4   PRESUMPTIVE LATERAL EARTH PRESSURES FOR WALLS WITH LEVEL BACKFILLS (Adapted from ASCE, 1996a and ICC, 2000)

| | | Equivalent Fluid Density, G_h | | | |
| | | Active Condition | | At-Rest Condition | |
	Unified Classification and Description of Retained Soils[a,b]	(lb/ft³)	(kN/m³)	(lb/ft³)	(kN/m³)
GW	Well-graded clean gravels and sandy gravels	35	5.5	60	9.4
GP	Poorly-graded clean gravels and sandy gravels	35	5.5	60	9.4
GM GW-GM GP-GM	Silty gravels	35	5.5	60	9.4
GC GW-GC GP-GC GC-GM	Clayey gravels	45	7.1	60	9.4
SW	Well-graded clean sands and gravelly sands	35	5.5	60	9.5
SP	Poorly-graded clean sands and gravelly sands	35	5.5	60	9.5
SM SW-SM SP-SM	Silty sands	45	7.1	60	9.5
SC SW-SC SP-SC SC-SM	Clayey sands	85	13.4	100	15.7
ML	Inorganic silts with low plasticity	85	13.4	100	15.7
CL-ML	Inorganic silty clays with low plasticity	85	13.4	100	15.7
CL	Inorganic clays with low plasticity	100	15.7	100	15.7
OL	Organic silts and clays with low plasticity				
MH	Inorganic silts with low plasticity		Not suitable unless evaluated		
CH	Inorganic clays with high plasticity		by a geotechnical engineer.		
OH	Organic silts and clays with high plasticity				

[a]The retained soils are those in the zone above a plane inclined upward at a slope of 1:1 from the bottom rear of the footing. If more than one soil type is present in this zone, the soil lateral load shall be the highest for any of those soils.

[b]The use of this table is limited to walls no more than 2.5 m (8 ft) tall.

23.9 A 3-m tall cantilever retaining wall with a vertical back is to be backfilled with a soil that has $G_h = 6.0$ kN/m^3 and $G_v = 0$. Compute the lateral force per meter acting on the back of this wall.

23.10 A 13-ft tall cantilever retaining wall supports a residual fine silty sand (SM) and the ground surface behind the wall is level. Using Terzaghi and Peck's charts, determine G_h and G_v, then use these values to compute the normal and shear earth pressures acting on the back of this wall.

23.11 Repeat Problem 23.10 using the presumptive lateral earth pressures in Table 23.4.

23.12 A 4.0-m tall cantilever concrete retaining wall is to be backfilled with a clayey sand that has $c'=100$ lb/ft^2, $\phi'=30°$, and $\gamma=120$ lb/ft^3. Using Coulomb's equations, compute G_h and G_v, then use these values to compute the normal and shear earth pressures acting on the back of this wall. In this case, it is probably best to ignore the cohesive strength of the backfill soil.

23.13 A 2.5-m tall cantilever retaining wall is to be backfilled with a clayey sand (SC). Using any appropriate method, compute G_h and G_v, then use these values to compute the normal and shear earth pressures acting on the back of this wall. Then, consider the possibility of backfilling this wall with an imported poorly-graded sand (SP) and recompute the values. Discuss the difference and list the major factors that would need to be considered when deciding whether or not to use the imported backfill.

23.6   LATERAL EARTH PRESSURES FROM SURCHARGE LOADS

Surcharge loads often are present along the ground surface above a retaining structure, as shown in Figure 23.18. These loads might be the result of structural loads on shallow footings, vehicles, above-ground storage, backfill compaction equipment, or other causes. If a surcharge load occurs within a horizontal distance of about H from the wall (where H equals the height of the wall), then it may impose significant additional lateral pressures on the wall and therefore is of interest.

   If the surcharge load is uniformly distributed, or if it can be assumed so, then the additional lateral earth pressure also is uniformly distributed, as shown in Figure 23.18. Its magnitude is:

$$\boxed{\sigma = Kq} \tag{23.33}$$

Where:

   $\sigma =$ additional lateral earth pressure due to surcharge load

   $K = K_a$, K_0, or K_p, depending on whether active, at-rest, or passive conditions prevail, respectively

   $q =$ surcharge pressure

   If the surcharge load occurs along a point, line, or limited area, the resulting additional lateral earth pressure may be computed using the techniques described in Chapter

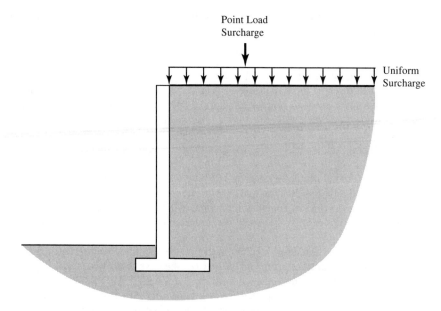

Figure 23.18   Typical surcharge loads near a retaining wall.

10 of *Geotechnical Engineering: Principles and Practices* (Coduto, 1999). The effects of multiple surcharge loads may be combined using superposition.

Example 23.6

The ground surface above a 4.0-m tall retaining wall will be subjected to a uniform surcharge load of 13.0 kPa and a vertical point load of 1000 kN, as shown in Figure 23.19. Compute the additional lateral earth pressures due to these surcharge loads.

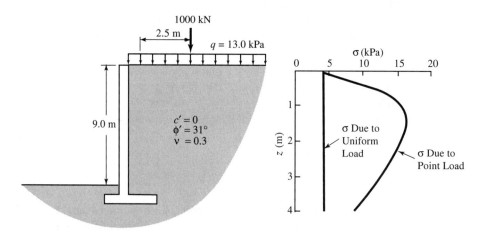

Figure 23.19   Proposed retaining wall with surcharge load for Example 23.6.

Solution

Lateral earth pressure caused by uniform load

$$K_a = \tan^2(45° - \phi/2) = \tan^2(45° - 31°/2) = 0.320$$

$$\sigma = K_a q = (0.320)(13.0 \text{ kPa}) = \textbf{4.2 kPa}$$

Lateral earth pressure caused by concentrated load
Using Equation 10.11 from *Geotechnical Engineering: Principles and Practices*:

Depth below top of wall (m)	R (m)	σ (kPa)
0.00	2.50	3.8
0.40	2.53	10.2
0.80	2.62	14.6
1.20	2.77	16.4
1.60	2.97	16.2
2.00	3.20	15.1
2.40	3.47	13.6
2.80	3.75	12.3
3.20	4.06	11.2
3.60	4.38	10.3
4.00	4.72	9.6

These results are plotted in Figure 23.19, and are in addition to the active earth pressure acting on this wall.

An often-neglected surcharge load is that imposed by the equipment that places and compacts the backfill. Although these loads are trivial if the contractor uses only light-weight hand-operated compaction equipment, they can be significant if larger equipment, such as loaders, dozers, or sheepsfoot rollers, operate behind the wall. Unfortunately, design engineers sometimes fail to consider construction procedures, so it is not unusual for walls to tilt excessively during backfilling because they were not designed to support heavy equipment loads.

One way to predict the effect of these loads on the wall is to combine a knowledge of the weight and position of the compaction equipment with the surcharge charts in this chapter. Another way is the method proposed by Ingold (1979).

Compacted backfills also induce large horizontal stresses in the ground (i.e., the K_0 value is large). This also increases the lateral earth pressure acting on the wall (Duncan et al., 1991).

23.7 GROUNDWATER EFFECTS

The discussions in this chapter have thus far assumed that the groundwater table is located below the base of the wall. If the groundwater table rises to a level above the base of the wall, three important changes occur:

- The effective stress in the soil below the groundwater table will decrease, which decreases the active, passive, and at-rest pressures.
- Horizontal hydrostatic pressures will develop against the wall and must be superimposed onto the lateral earth pressures.
- The effective stress between the bottom of the footing and the soil becomes smaller, so there is less sliding friction.

The net effect of the first two changes is a large increase in the total horizontal pressure acting on the wall (i.e., the increased hydrostatic pressures more than offset the decreased effective stress). The resulting pressure diagram is shown in Figure 23.20.

Example 23.7

This cantilever wall has moved sufficiently to create the active condition. Compute the lateral pressure distribution acting on this wall with the groundwater table at locations **a** and **b**, as shown in Figure 23.21.
    The soil properties are: $c' = 0$, $\phi' = 30°$, $\gamma = 20.4$ kN/m³, and $\gamma_{sat} = 22.0$ kN/m³

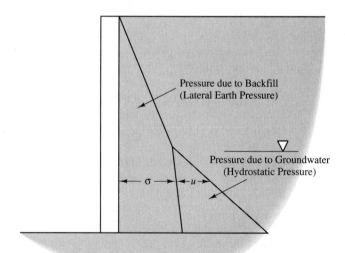

Figure 23.20 Theoretical lateral pressure distribution with shallow groundwater table.

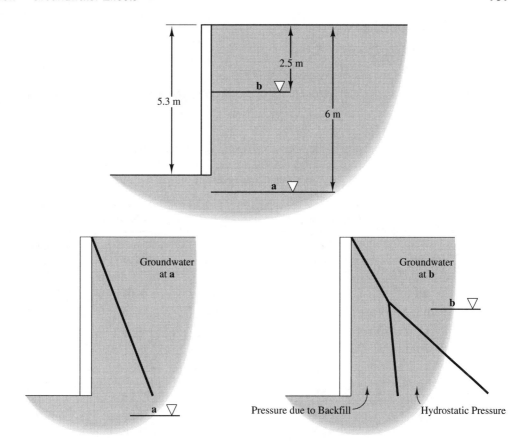

Figure 23.21   Retaining wall for Example 23.7.

Solution

Use Rankine's Method

$$K_a = \tan^2(45° - \phi/2) = \tan^2(45° - 30°/2) = 0.333$$

With the groundwater table at **a**:

$$\sigma = \sigma_z' K_a \cos\beta$$
$$= \gamma z K_a \cos\beta$$
$$= 20.4\, z(0.333)\cos0$$
$$= 6.79\, z$$

Where z = depth below the top of the wall

$$\beta = 0 \therefore V_a = 0 \text{ (per Rankine)}$$

With the groundwater table at **b**:

$$\sigma @ z \geq 2.5 \text{ m} = \sigma'_z K_a \cos\beta$$

$$= (\Sigma \gamma H - u)K_a \cos\beta$$

$$= [(20.4 \text{ kN/m}^3)(2.5 \text{ m}) + (22.0 \text{ kN/m}^3)(z - 2.5 \text{ m}) - u](0.333)\cos 0$$

$$= 7.33 z - 0.33u - 1.33$$

$$u - 9.80 \text{ kN/m}^3(z - 2.5 \text{ m}) \geq 0$$

Total horizontal pressure on wall $= \sigma + u$

z (m)	Groundwater at **a** σ (kPa)	Groundwater at **b** u (kPa)	σ (kPa)	Total Pressure (kPa)
0.0	0.00	0.00	0.00	0.00
0.5	3.40	0.00	3.40	3.40
1.0	6.79	0.00	6.79	6.79
1.5	10.19	0.00	10.19	10.19
2.0	13.58	0.00	13.58	13.58
2.5	16.98	0.00	16.98	16.98
3.0	20.37	4.90	19.04	23.94
3.5	23.77	9.80	21.09	30.89
4.0	27.16	14.70	23.14	37.84
4.5	30.56	19.60	25.19	44.79
5.0	33.95	24.50	27.24	51.74
5.3	35.99	27.44	28.46	55.90

Example 23.7 demonstrates the profound impact of groundwater on retaining walls. If the groundwater table rises from **a** to **b**, the total horizontal force acting on the wall increases by about 30 percent. Therefore, the factor of safety against sliding and overturning could drop from 1.5 to about 1.0, and the flexural stresses in the stem would be about 30 percent larger than anticipated. There are two ways to avoid these problems:

1. Design the wall for the highest probable groundwater table. This can be very expensive, but it may be the only available option.
2. Install drains to prevent the groundwater from rising above a certain level. These could consist of *weep holes* drilled in the face of the wall or a *perforated pipe drain*

installed behind the wall. Drains such as these are the most common method of designing for groundwater.

Further problems can occur if the groundwater becomes frozen and ice lenses form. The same processes that cause frost heave at the ground surface also produce large horizontal pressures on retaining walls. This is another good reason to provide good surface and subsurface drainage around retaining walls.

QUESTIONS AND PRACTICE PROBLEMS

23.14 A 20-ft tall cantilever retaining wall supports a soil with $c' = 0$, $\phi' = 36°$, and $\gamma = 126$ lb/ft^3. The ground surface above the wall is essentially level, and will be used for storage of 55 gallon drums filled with oil. These drums will impart a uniform surcharge load of 300 lb/ft^2. Assuming active conditions will exist, determine the Rankine active pressure and the lateral earth pressure caused by the surcharge load. Show both of these pressure distributions on a cross-section of the wall.

23.15 A 6-m tall cantilever retaining wall on a 0.5-m thick footing will support a backfill with $G_h = 5.5$ lb/ft^3 and $G_v = 0$. The ground above this wall will be essentially level and used as a parking lot. The vehicles in this parking lot can be represented by a uniform surcharge pressure of 15 kPa. Determine the earth pressures acting on this wall and show these pressures in a sketch.

23.16 An 18-ft tall cantilever retaining wall on a 2-ft thick footing will support a sandy backfill with $c' = 0$, $\phi' = 38°$, $\gamma = 125$ lb/ft^3, and $\gamma_{sat} = 128$ lb/ft^3. The wall has been designed with weep holes so the groundwater table can rise no higher than the top of the footing. Using Rankine's method, compute the total horizontal force per foot acting on the back of the wall and footing. Unfortunately, the weep holes were inadvertently omitted from the construction, so the as-built wall had no drainage provisions. As a result, the water table rose to a level 7 ft below the top of the wall. Compute the new total horizontal force per foot acting on the back of the wall and footing. Discuss your findings.

23.8   PRACTICAL APPLICATION

Even the most thorough analytical method is a simplification of the truth. Traditional lateral earth pressure calculations ignore many real-world effects, such as temperature fluctuations, readjustment of soil particles due to creep, and vibrations from traffic. Add the ever-present uncertainties in defining soil profiles and obtaining truly representative samples for testing and we should not be surprised when measured earth pressures are often quite different from those that the equations predict. See Gould (1970) for interesting comparisons of predicted and measured earth pressures on real structures.

This does not mean that the analyses are not useful. However, be careful to apply them carefully and with judgment. Specific practical guidelines include the following:

1. Use extra caution in clays

Although these methods model sands reasonably well, they are less reliable when used with clays because of their more complex behavior. Therefore, an extra note of caution is appropriate when working with clays.

One important characteristic of clays is their tendency to creep, which is a continued deformation under a constant applied load. We normally expect materials subjected to a given stress to develop a corresponding deformation and then stabilize, but materials that creep will continue to deform and not reach equilibrium until the stress is relieved. Although creep can occur in clays at any stress level, it will usually be most pronounced when the shear stress is at least half of the shear strength.

In the context of lateral earth pressures, creep can be a problem when we design clays for the active or passive condition, which, by definition, stress the soil to a level near its shear strength. This means that once a wall has moved sufficiently to develop the active or passive condition, the soil will begin to creep and the lateral earth pressures will begin to migrate back toward the at-rest condition.

Therefore, do not design cantilever retaining walls backfilled with clay using the active pressure. A design pressure between the active and at-rest values would be most appropriate. Likewise, it is best not to rely on the full passive pressure, at least not for long-term loads. Terzaghi and Peck's charts could be used to estimate the active pressure, and Rankine's method (with a high factor of safety) for the passive.

2. Select appropriate strength parameters

Retaining wall designs should nearly always be based on the saturated strength of the soil. Even though the soil may be relatively dry during construction, the designer must assume that it may one day become wet.

If the design incorporates a drainage system, such as weep holes, then designers can safely assume that the groundwater will not rise above the level of the drain. In such cases, use zero pore water pressure in the soil and zero hydrostatic pressure against the wall.

It is also important to remember that walls are usually built with construction excavations that are smaller than the active wedge. Any analysis of such walls should be based on the soil strength along the base of the wedge, not those from the narrow backfill. In all cases, use plane strain strengths.

3. Use an appropriate method of analysis

Morgenstern and Eisenstein (1970) compared the earth pressure coefficients from Coulomb and four other theories with those of Rankine. They found that the computed values of K_a are within 10 percent of Rankine's K_a for all values of ϕ, which we would consider to be very close agreement. Coulomb's theory produces K_a values that are slightly more accurate than Rankine's, but the difference is so small that for practical problems an engineer could achieve satisfactory results with either method. However, as discussed earlier, passive pressures are best computed using Rankine's K_p.

SUMMARY

Major Points

1. The lateral earth pressures acting on a structure can vary from a lower bound, known as the active condition, to an upper bound, known as the passive condition. The magnitude of the active pressure could be as low as zero, whereas the passive pressure could be as high as several times the vertical stress.

2. The lateral stress condition in a soil depends on the lateral strain. If the soil expands laterally a sufficient distance, then the active condition will exist. If it contracts laterally a sufficient distance, then the passive condition will exist. If no lateral strains occur, then the at-rest condition exists. Intermediate states are also possible.

3. The ratio of the horizontal to vertical effective stress at a point is known as the coefficient of lateral earth pressure, K. The value of K in the active, passive, and at-rest conditions is represented by K_a, K_p, and K_0, respectively.

4. The two classical theories for computing active and passive pressures are Coulomb's theory and Rankine's theory. Either theory is acceptable for computing active pressures, but only Rankine's theory accurately predicts passive pressures.

5. Use special care when computing lateral earth pressures in clayey soils because of the potential for creep and the potential for expansion upon wetting.

6. Lateral earth pressures are often expressed in terms of the equivalent fluid density, which is the unit weight of a fictitious fluid that imposes the same lateral earth pressure as the real backfill soil.

7. Surcharge loads that act within a distance H behind the wall, where H is the height of the wall, will increase the lateral pressures.

8. The presence of groundwater behind the wall can significantly increase the lateral earth pressure.

Vocabulary

Active condition	Coulomb theory	Presumptive lateral earth pressure
At-rest condition	Creep	
Coefficient of lateral earth pressure	Equivalent fluid density	Rankine theory
	Lateral earth pressure	Rigid
Coefficient of passive earth pressure	Limit equilibrium analysis	Surcharge load
	Passive condition	Unyielding
Coefficient of active earth pressure	Plane strain condition	

COMPREHENSIVE QUESTIONS AND PRACTICE PROBLEMS

23.17 A basement wall will be built of concrete masonry and supported on a concrete footing. The bottom of the footing will be 2.5 m below the adjacent ground surface. The backfill soils will consist of silty sand (SM) and well-graded sand (SW), and the groundwater will be controlled through the use of subdrains. Determine the horizontal force per meter acting on the back of this wall and footing.

23.18 An 8.0-m tall cantilever concrete retaining wall on a 700-mm thick footing is to be backfilled with a silty sand having $c' = 0$, $\phi' = 35°$, and $\gamma = 122$ lb/ft^3. The ground surface above this wall will be inclined at a slope of 3 horizontal to 1 vertical, and the groundwater will be controlled with subdrains. The back of the wall is tapered at an angle of 5 degrees. Using Coulomb's equations, compute G_h and G_v, then use these values to compute the normal and shear force acting on the back of this wall and footing.

24

Cantilever Retaining Walls

Theories require assumptions which often are true only to a limited degree. Thus any new theoretical ideas have questionable points which can be removed only by checks under actual conditions. This statement holds especially true for theories pertaining to soil mechanics because assumptions relative to soil action are always more or less questionable. Such theories may sometimes be checked to a limited degree by laboratory tests on small samples, but often the only final and satisfactory verification requires observations under actual field conditions.

MIT Professor Donald W. Taylor (1948)

Cantilever retaining walls are the most common type of earth-retaining structure because they are often the most economical, especially when the wall height is less than about 5 m (16 ft). This chapter discusses the analysis and design of such walls.

The design of these walls must satisfy two major requirements: First, the wall must have adequate *external stability*, which means it must remain fixed in the desired location (except for small movements required to mobilize the active or passive pressures). Second, it must have sufficient *internal stability* (or *structural integrity*) so it is able to carry the necessary internal stresses without rupturing. Both of these requirements are shown in Figure 24.1. Walls that have insufficient external stability experience failure in the soil, while those that have insufficient internal stability experience structural failure in the wall itself. These are two separate requirements, and each must be satisfied independently. Extra effort in one does not compensate for a shortcoming in the other. For example, adding more rebars (improving the internal stability) does not compensate for a footing that is too short (a deficiency in external stability).

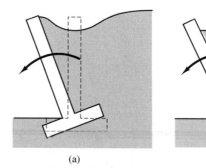

Figure 24.1   (a) A wall that lacks sufficient external stability moves away from its desired location because the soil fails; (b) A wall with inadequate internal stability (structural integrity) is unable to carry the necessary internal stresses and experiences a structural failure.

(a)   (b)

24.1   EXTERNAL STABILITY

A cantilever retaining wall must be externally stable in all the following ways:

- It must not *slide* horizontally, as shown in Figure 24.2a.
- It must not *overturn*, as shown in Figure 24.2b and Figure 24.3.
- The resultant of the normal force that acts on the base of the footing must be within the middle third of the footing, as shown in Figure 24.2c.
- The foundation must not experience a *bearing-capacity failure*, as shown in Figure 24.2d.
- It must not undergo a *deep-seated shear failure*, as shown in Figure 24.2e.
- It must not *settle* excessively, as shown in Figure 24.2f.

The external stability of a wall in each of these modes is dependent on its dimensions and on the forces between the wall and the ground. Figure 24.3 shows a wall that does not satisfy the overturning criteria and is slowly rotating outward.

When evaluating external stability, engineers consider the wall and the soil above the footing as a unit, as shown in Figure 24.4. We will refer to it as the *wall-soil unit* and evaluate its external stability using the principles of statics.

We can evaluate the external stability of a wall-soil unit only after its dimensions are known. Therefore, first develop a trial design using the guidelines in Figure 24.5, then check its external stability, and progressively refine the design. Continue this converging trial-and-error process until an optimal design is obtained (one that minimizes costs while satisfying all external stability criteria).

Sliding

We evaluate the sliding stability using a *limit equilibrium* approach by considering the forces acting on the wall-soil unit if it were about to fail. The factor of safety is the ratio of the forces required to cause the wall to fail to those that actually act on it.

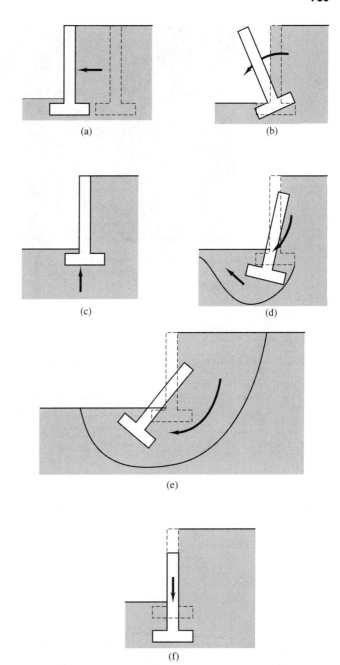

Figure 24.2 Potential external stability problems in a cantilever retaining wall: (a) sliding failure, (b) overturning failure, (c) normal force acting on the base of the footing not within the middle third, (d) bearing capacity failure, (e) deep-seated shear failure, and (f) excessive settlement.

Figure 24.3   This retaining wall, which was backfilled with an expansive clay, is slowly failing by overturning and will eventually collapse. This photograph was taken approximately thirty years after construction.

The forces tending to cause sliding (known as the *driving forces*) are as follows:

- The horizontal component of the lateral earth pressures acting on the back of the wall-soil unit.
- Hydrostatic forces, if any, acting on the back of the wall-soil unit.
- Seismic forces from the backfill (these are beyond the scope of this chapter).

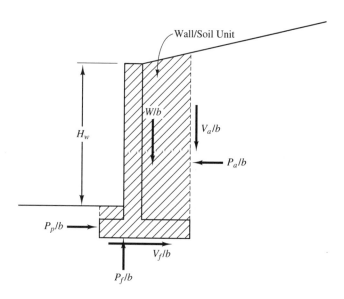

Figure 24.4   Forces acting between a cantilever retaining wall and the ground. The wall, footing, and backfill soil immediately above the footing form the wall-soil unit, which is used to perform external stability analyses.

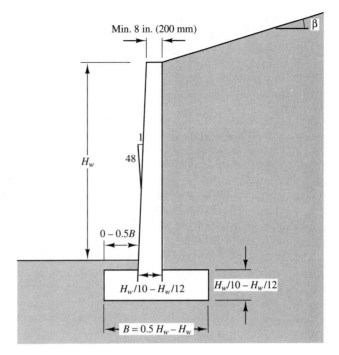

Figure 24.5 Suggested first trial dimensions for cantilever retaining walls backfilled with sandy soils. For short walls with strong soils and level backfill, the toe extension will be about 0.5B. For weaker soils or inclined backfill, the toe extension will be less (with a corresponding increase in the heel extension).

These are countered by the following *resisting forces*:

- Lateral earth pressures acting on the front of the wall-soil unit.
- Sliding friction along the bottom of the footing.
- Hydrostatic forces, if any, acting on the front of the wall-soil unit.

The factor of safety, F, is defined as the ratio of the horizontal resisting forces, $\Sigma\,(P_R/b)$ to the horizontal driving forces, $\Sigma\,(P_D/b)$:

$$F = \frac{\Sigma(P_R/b)}{\Sigma(P_D/b)} \tag{24.1}$$

The unit length, b, is usually 1 ft or 1 m.

IBC 1610.2 requires a factor of safety against sliding of at least 1.5. This criterion is suitable for walls backfilled and supported on sands, sandy silts, gravels, or rock. However, if the backfill or the underlying soil is a clay or clayey silt, design for a factor of safety of at least 2.0. This higher factor of safety is justified because of the historically poorer performance of such walls (Duncan et al., 1990), and because the shear strength of clayey soils is less reliable.

The resisting force is typically computed using Equation 8.8, which uses allowable coefficient of friction and passive pressure values (i.e., a factor of safety has already been applied to these values). Thus, this design procedure produces a compound factor of safety against sliding: the first is applied by the geotechnical engineer when developing design values of μ_a and λ_a, and the second by the structural engineer when performing the sliding analysis. This compounding typically produces a "true" factor of safety between 2.5 and 3.5, which is probably appropriate.

The use of Equation 8.8 in this analysis also implicitly double-counts part of the active earth pressure. This is conservative, and helps keep the lateral movements under control.

Example 24.1—Part A

A 12-ft tall retaining wall supports a backfill inclined at a slope of 4:1 as shown in Figure 24.6. The soil behind the wall is a fine to medium sand with the following properties: $c' = 0$, $\phi' = 35°$, and $\gamma = 122$ lb/ft^3. The soil below the footing is also a fine to medium sand: Its properties are $c' = 0$, $\phi' = 38°$, $\gamma = 125$ lb/ft^3, and the allowable bearing pressure, q_A, is 5000 lb/ft^2. Determine the required footing dimensions to satisfy external stability criteria.

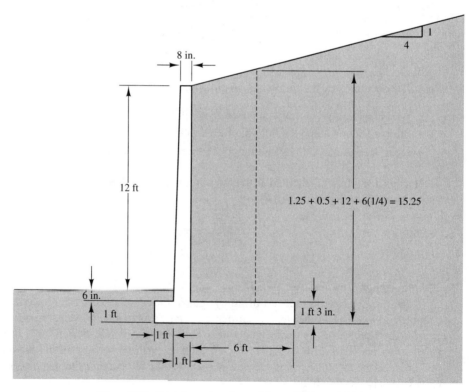

Figure 24.6   Trial dimensions for retaining wall, Example 24.1.

Solution

Select the trial dimensions using Figure 24.5. These trial dimensions are shown in Figure 24.6.

Develop the geotechnical design parameters

Active earth pressure—use Coulomb's method (Equation 23.19). Rankine's method also would be acceptable.

$$\beta = \tan^{-1}(1/4) = 14°$$

$$\phi_w = (2/3)(35°) = 23°$$

$$K_a = \frac{\cos^2(\phi - \alpha)}{\cos^2\alpha \cos(\phi_w + \alpha)\left[1 + \sqrt{\dfrac{\sin(\phi + \phi_w)\sin(\phi - \beta)}{\cos(\phi_w + \alpha)\cos(\alpha - \beta)}}\right]^2}$$

$$= \frac{\cos^2(35 - 0)}{\cos^2 0 \cos(23 + 0)\left[1 + \sqrt{\dfrac{\sin(35 + 23)\sin(35 - 14)}{\cos(23 + 0)\cos(0 - 14)}}\right]^2}$$

$$= 0.291$$

$$G_h = \gamma K_a \cos\phi_w = 122(0.291)\cos 23° = \mathbf{33\ lb/ft^3} \qquad \Leftarrow Answer$$

$$G_v = \gamma K_a \sin\phi_w = 122(0.291)\sin 23° = \mathbf{14\ lb/ft^3} \qquad \Leftarrow Answer$$

Per Table 8.3: $\mu = 0.45–0.55$

Per Equation 8.12: $\mu = \tan[0.7(38)] = 0.50$

Use $\mu = 0.50$

$$\mu_a = \frac{\mu}{F} = \frac{0.50}{1.5} = \mathbf{0.33} \qquad \Leftarrow Answer$$

$$\lambda_a = \frac{\gamma[\tan^2(45° + \phi/2) - \tan^2(45° - \phi/2)]}{F}$$

$$= \frac{125[\tan^2(45° + 38°/2) - \tan^2(45° - 38°/2)]}{2}$$

$$= 248\ lb/ft^3 \qquad \text{use } \mathbf{250\ lb/ft^3} \qquad \Leftarrow Answer$$

These geotechnical design parameters, along with the allowable bearing pressure given in the problem statement, are normally developed by the geotechnical engineer.

Evaluate sliding stability of trial design

$$H = 1.25 + 0.5 + 12 + 6(1/4)$$

$$= 15.25\ ft$$

$$P_a/b = \frac{G_h H^2}{2} = \frac{(33)(15.25^2)}{2} = 3,837 \text{ lb/ft}$$

$$V_a/b = \frac{G_v H^2}{2} = \frac{(14)(15.25^2)}{2} = 1,628 \text{ lb/ft}$$

Weight on bottom of footing:

Stem: [(8/12 + 12/12)/2] (12.5)(150 lb/ft³)	=	1,562 lb/ft
Footing: $W_f/b = (1.25)(8)(150)$	=	1,500 lb/ft
Soil behind wall: (6)[12.5 + (1/2) (6) (1/4)](122 lb/ft³)	=	9,699 lb/ft
V_a/b	=	1,628 lb/ft
$P/b + W_f/b$	=	14,389 lb/ft

Using a modified version of Equation 8.8:

$$V_f/b = (P/b + W_f/b)\mu_a + 0.5\lambda_a D^2$$

$$= (14,389)(0.33) + (0.5)(250)(1.75^2)$$

$$= 5,131 \text{ lb/ft}$$

$$F = \frac{\Sigma(P_R/b)}{\Sigma(P_D/b)} = \frac{5,131}{3,837} = 1.34 < 1.5$$

If the sliding stability criterion has not been satisfied, as in this example, the engineer must modify the trial design using one or more of the following methods (see Figure 24.7):

- **Extend the heel of the footing:** This increases the weight acting on the footing, thus increasing the sliding resistance. Unfortunately, it also increases construction costs because it requires a larger construction excavation. If the wall is near a property line or some other limit to construction, this method may not be feasible.

- **Add a key beneath the footing:** This improves the sliding stability by increasing the passive pressure. Unfortunately, the active pressure also increases. However, the increase in the passive is greater than the increase in the active, so there is a net gain. This method is most effective in soils with a relatively high friction angle because the ratio K_p/K_a is greatest in such soils and the net increase in resisting force is greatest.

    Some engineers do not favor the use of keys, especially in clean sands (where they are theoretically most effective) because they feel that the benefits of the key are more than offset by the soil disturbance during construction.

- **Use a stronger backfill soil:** This soil must extend at least to the line shown in Figure 24.7 to assure that the critical failure surface passes through it. Usually this requires a larger construction excavation and often requires the use of imported soil. Therefore, this method is usually expensive.

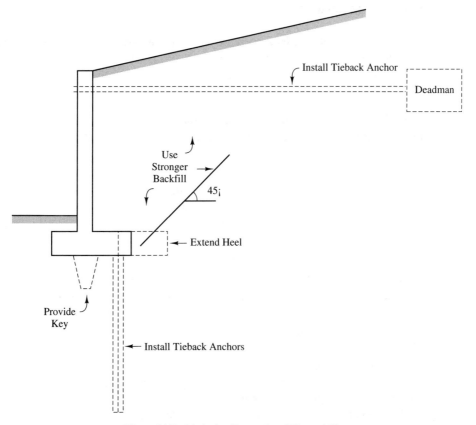

Figure 24.7 Methods of improving sliding stability.

- **Install tiedown anchors:** This increases the normal force acting on the footing, P/b, thus increasing the sliding friction. This method is most effective when ϕ_f is large.
- **Install a tieback anchor:** This increases the total resisting force, $\Sigma\ (P_R/b)$. Tieback anchors might be in the form of a *deadman* as shown here, or they may develop resistance through friction along an augered hole. Screw-type anchors also are available.

Conversely, if the sliding factor of safety was excessive, the engineer would reduce the heel extension and/or remove or shorten the key. Adjusting the toe extension has very little effect on the sliding stability.

We express the final dimensions of the footing and the key (if used) as a multiple of 3 in or 100 mm. Therefore, the trial-and-error design process ends when the required dimensions are known within this tolerance.

Example 24.1—Part B

Additional trials will show that a 7 ft 6 in heel extension will produce $F = 1.52$ against sliding. This corresponds to a total footing width of 9 ft 6 in.

Overturning

Once the trial design satisfies the sliding stability requirements, begin evaluating its overturning stability using a limit equilibrium approach with the wall-soil unit shown in Figure 24.8. The factor of safety is now defined in terms of moments:

$$F = \frac{\Sigma(M_R/b)}{\Sigma(M_D/b)} \tag{24.2}$$

Where:

$F =$ factor of safety against overturning

$\Sigma M_R/b =$ sum of the resisting moments per unit length of wall

$\Sigma M_D/b =$ sum of the driving moments (also known as overturning moments) per unit length of wall

$b =$ unit length (usually 1 ft or 1 m)

This is not a $\Sigma M = 0$ analysis. We are taking the sum of all moments acting in one direction (the resisting moments) and dividing them by the sum of all moments acting in the other direction (the driving moments). Thus, the computed factor of safety depends on the location of the point about which we compute the moments.

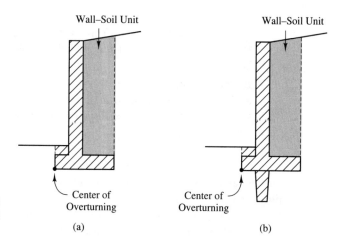

Figure 24.8   Wall-soil unit for overturning analysis: (a) for walls without a key; and (b) for walls with a key.

A rigorous overturning analysis would compute the factor of safety about a range of points and search for the point that produces the lowest F. This is the most critical point, and the factor of safety about this point is the "true" F for overturning. It also is the point about which the wall would rotate if it were to fail by overturning. This kind of searching is similar to the process used to evaluate the stability of earth slopes.

However, customary engineering practice does not use such a rigorous analysis. Engineers normally evaluate the overturning stability by computing the moments about only one point: the toe of the footing, as shown in Figure 24.8. Since this is probably not the critical point, the computed F is almost always greater than the "true" F, sometimes by as much as 40 percent, and the analysis is unconservative. Fortunately, the rest of the design process, especially the requirements for "sliding stability" and "location of normal force on base of footing," implicitly catch any such problems and prevent the final design from being unconservative. In other words, whenever the overturning analysis produces misleading results, the other analyses control the design. Therefore, the customary practice of computing moments about the toe is acceptable, even though it sometimes produces incorrect results.

Because this is a limit equilibrium analysis, we consider the forces that would act on the wall-soil unit if an overturning failure was occurring. The following forces generate driving moments:

- The horizontal component of the lateral earth pressures acting on the back of the wall-soil unit. In theory, the resultant of this force acts at a height of $0.33H$ from the bottom of the footing, but field measurements indicate the actual location is at a height of about $0.4H$ (Duncan et al., 1990). In other words, the actual pressure distribution is not triangular. Nevertheless, it is customary practice to base the analysis on the theoretical height, even though this procedure is somewhat unconservative.
- Hydrostatic forces acting behind the wall-soil unit. The resultant of these forces may be determined using the principles of fluid statics.
- Seismic forces from the backfill (which are beyond the scope of this chapter).

These moments are resisted by the following:

- The vertical component of the lateral earth pressures acting on the back of the wall-soil unit.
- The weight of the wall-soil unit.
- Surcharge loads acting above the wall-soil unit.
- Hydrostatic pressure acting on the front of the footing.

The bottom portion of the footing may move slightly rearward if an overturning failure were to occur (i.e., the center of rotation may actually be slightly higher than the bottom of the footing). Therefore, the resistance offered by the passive pressure acting on the front of the footing is not reliable, and should be neglected in the overturning analysis.

Since the analysis assumes the wall is in the process of overturning about the toe, the normal force between the bottom of the footing and the ground, $P/b + W_f/b$, acts through the center of overturning and thus has no moment arm. The same is true of the sliding friction force acting along the bottom of the footing, V_f/b. Therefore, neither of these forces need be considered in the moment computations.

The minimum required factor of safety is the same as for sliding: at least 1.5 for walls backfilled and supported on sands, sandy silts, gravels, or rock [ICC 1610.2], or at least 2.0 if the backfill or the underlying soil is a clay or clayey silt.

Location of Normal Force Along Base of Footing

During in-service conditions (compared to the limit equilibrium condition used in the overturning analysis), the force $P/b + W_f/b$ must be located within the middle third of the footing (i.e., the eccentricity, e, must be no greater than $B/6$) to maintain a compressive stress along the entire base of the footing. This avoids excessive rotation of the footing and wall, as discussed in Section 5.3. For walls supported on bedrock, the rotation will be small, so $P/b + W_f/b$ should be located within the middle half of the footing (i.e., e must be no greater than $B/4$) (Duncan et al., 1990).

For the in-service condition, the wall is in static equilibrium, so the sum of the moments about any point must be equal to zero. We may use this fact to compute the location of $P/b + W_f/b$, as demonstrated in the next example.

Example 24.1—Part C

Check the overturning stability of the wall described in Part B. This analysis uses revised weights and forces that reflect the 7 ft 6 in heel extension.

Source	Force (lb)	Arm (ft)	M_D/b (ft-lb/ft)	M_R/b (ft-lb/ft)
P_a/b	4,028	0.333 (15.62) = 5.20	20,946	
V_a/b	1,708	9.5		16,226
Stem	1,562	1.0 + 0.5 = 1.5		2,343
Footing (W_f/b)	1,781	9.5/2 = 4.75		8,460
Soil	12,293	1 + 1 + 3.75 = 5.75		70,685
			20,946	97,714

$$F = \frac{\Sigma(M_R/b)}{\Sigma(M_D/b)} = \frac{97,714}{20,946} = 4.67 > 1.5 \quad \text{OK}$$

Find the location of the resultant force acting on the footing.

This analysis differs from the overturning analysis in that it considers the in-service conditions. The objective is to find the location of the resultant force acting on the bottom of the footing and determine whether or not it is within the middle third. Because the wall is in static equilibrium, find the location of this force by taking the sum of moments about any point and setting it equal to zero. We already have computed moments about the center of overturning, so it is easiest to continue using this point.

The magnitude of $P/b + W_f/b$ with the 7 ft 6 in heel extension is 17,348 lb/ft.

Let x = horizontal distance from the center of overturning to the resultant force.

$$\Sigma M/b = 20{,}946 + 17{,}348x - 97{,}714 = 0$$

$$x = 4.43 \text{ ft}$$

Let e = eccentricity = distance from the center of footing to the resultant force.

$$e = \frac{B}{2} - x = \frac{9.5 \text{ ft}}{2} - 4.43 \text{ ft} = 0.32 \text{ ft}$$

$$\frac{B}{6} = \frac{9.5 \text{ ft}}{6} = 1.58 \text{ ft}$$

$$e < B/6 \quad \therefore \text{ Resultant is within the middle third} \quad \textbf{OK}$$

If the overturning stability of the trial design is not satisfactory, or if the resultant is not in the middle third of the footing, modify the design in one or more of the following ways:

- **Extend the toe of the footing:** This is almost always the most cost-effective method because it increases the moment arms of all the resisting moments and increases the bearing capacity of the footing without increasing the driving moments or significantly increasing the volume of the construction excavation.
- **Extend the heel of the footing**, as shown in Figure 24.7: This method is much more expensive because it also increases the volume of the construction excavation. An engineer would most likely choose this approach only if a property line or some other restriction prevented extending the toe.
- **Use a stronger backfill soil**, as shown in Figure 24.7: This, too, would be expensive for the reasons described earlier.
- **Use tiedown or tieback anchors**, as shown in Figure 24.7: This is very effective, but may be expensive.

Example 24.1—Part D

In this case, both the overturning stability and the resultant in the middle third criteria have been exceeded by wide margins. Therefore, the wall is overdesigned. Try removing all of the toe extension. The resisting moment arms all will be reduced by 1 ft.

Source	Force (lb)	Arm (ft)	M_D/b (ft-lb/ft)	M_R/b (ft-lb/ft)
P_a/b	4,028	0.333 (15.62) = 5.20	20,951	
V_a/b	1,708	8.5		14,518
Stem	1,562	0.5		781
Footing (W_f/b)	1,594	4.25		6,774
Soil	12,293	4.75		58,392
			20,951	80,645

$$F = \frac{\Sigma(M_R/b)}{\Sigma(M_D/b)} = \frac{80,465}{20,951} = 3.84 > 1.5 \quad \text{OK}$$

Find the location of the resultant force acting on the footing.

The magnitude of $P/b + W_f/b$ with no toe extension is 17,160 lb/ft.

Let x = horizontal distance from the center of overturning to the resultant force.

$$\Sigma M/b = 20,951 + 17,160x - 80,465 = 0$$

$$x = 3.47 \text{ ft}$$

Let e = eccentricity = distance from the center of footing to the resultant force.

$$e = \frac{B}{2} - x = \frac{8.5 \text{ ft}}{2} - 3.47 \text{ ft} = 0.78 \text{ ft}$$

$$\frac{B}{6} = \frac{8.5 \text{ ft}}{6} = 1.42 \text{ ft}$$

$$e < B/6 \quad \therefore \text{ Resultant is within the middle third} \quad \textbf{OK}$$

The stability criteria are still exceeded by a significant margin, but there is no room for further economizing. The reduction in footing weight is minimal, so there is no need to recheck the sliding stability.

Bearing Capacity and Settlement

The footing that supports the wall may be subject to a bearing capacity failure, as shown in Figure 24.2d, or excessive settlement, as shown in Figure 24.2f. Check this possibility using the techniques for footings with moment loads, as described in Section 8.2 of Chapter 8.

Example 24.1—Part E

Check the bearing capacity and settlement.

$$B' = B - 2e = 8.5 - 2(0.78) = 6.94 \text{ ft}$$

$$q_{equiv} = \frac{P + W_f}{B'L'} - u_D = \frac{17{,}160 \text{ lb/ft}}{(6.94 \text{ ft})(1 \text{ ft})} - 0 = 2473 \text{ lb/ft}^2$$

Per problem statement, $q_A = 5000 \text{ lb/ft}^2$. This value reflects both bearing capacity and settlement requirements, as discussed in Chapter 8.

$q_{equiv} \le q_A$, so the design is acceptable.

Note: This footing is subject to both downward and shear loads, so technically the bearing capacity analyses used to develop q_A should use Vesić's method and take advantage of his load inclination factors. However, in this case there is such a large margin between q_{equiv} and q_A that it does not matter if this refinement was considered.

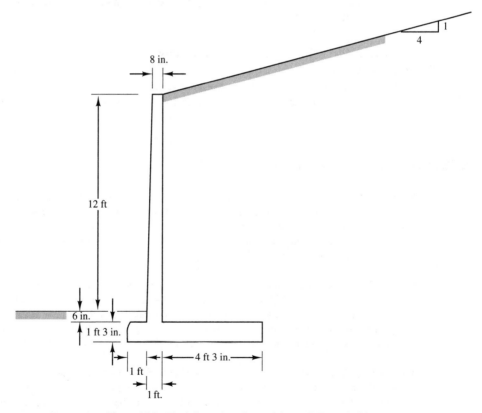

Figure 24.9   Final dimensions for retaining wall, Example 24.1.

Conclusion

For reasons discussed below, we will not check for deep seated shear failure or excessive settlement. Therefore, the design shown in Figure 24.9 satisfies all stability criteria.

Deep-Seated Shear Failure

A deep-seated shear failure would be a catastrophic event that could be much larger in scope than any of the modes described earlier. Be concerned about this mode of failure if any of the following conditions is present:

- The soil is a soft or medium clay and is subjected to undrained conditions. Generally, this is a concern only if $\gamma H_w/s_u > 6$ where H_w is the height of the wall as defined in Figure 24.5, s_u is the undrained shear strength, and γ is the unit weight of the soil.
- Adversely oriented weak seams or bedding planes are present.
- Some or all of the soil is prone to liquefaction. This phenomenon occurs during earthquakes in loose, saturated sands and silty sands and causes the soil to lose most or all of its strength.

If so, evaluate the deep-seated shear stability using a slope stability analysis. Fortunately, the vast majority of walls do not satisfy any of these conditions, so we usually can conclude that a deep-seated failure is not a concern. The wall in Example 24.1 does not satisfy any of the above criteria, so we do not need to check for deep-seated failure.

QUESTIONS AND PRACTICE PROBLEMS

24.1 Using Coulomb's active earth pressure formula to compute G_h and G_v for the retaining wall in Figure 24.10. Also compute μ_a and λ_a. The allowable bearing pressure, q_A, is 200 kPa. Then, use these geotechnical design parameters to evaluate the external stability of the wall. Does it satisfy normal design requirements? If not, what changes would you recommend? The groundwater table is at a great depth.

24.2 A 2.2-m tall cantilever retaining wall will be backfilled with a well graded sand that has $G_h = 5$ kN/m^3 and $G_v = 0$. The soils beneath the footing are similar and have $\mu_a = 0.35$, $\lambda_a = 45$ kN/m^3, and $q_A = 250$ kPa. The ground surface above and below the wall will be level, and the groundwater table is at a great depth. Determine the required footing dimensions to satisfy external stability requirements. Show your final design in a sketch.

24.3 A 10-ft tall cantilever retaining wall will be backfilled with a silty sand that has $G_h = 40$ lb/ft^3 and $G_v = 10$ lb/ft^3. The soils beneath the footing are similar and have $\mu_a = 0.30$, $\lambda_a = 200$ lb/ft^3, and $q_A = 3500$ lb/ft^2. The ground surface above and below the wall will be level, and the groundwater table is at a great depth. Determine the required footing dimensions to satisfy external stability requirements. Show your final design in a sketch.

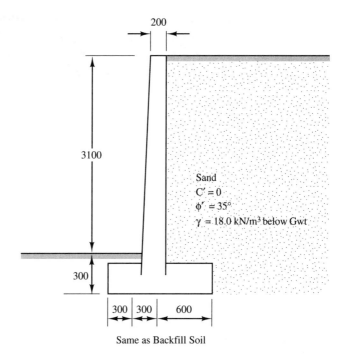

Figure 24.10   Proposed cantilever retaining wall for Problem 24.1

Same as Backfill Soil

24.2   RETWALL SPREADSHEET

External stability analyses, as presented in Section 24.1, can be performed using a spreadsheet, such as Microsoft® Excel. Such spreadsheet solutions are useful because they reduce the potential for mistakes, and make the trial-and-error design process much faster.

A Microsoft® Excel spreadsheet RETWALL.XLS has been developed in conjunction with this book and may be downloaded from the Prentice Hall website. Downloading instructions are presented in Appendix B. A typical screen is shown in Figure 24.11. Usually, the most effective way to use this spreadsheet is to follow the sequence shown in Example 24.1 (i.e., sliding analysis first, etc.).

QUESTIONS AND PRACTICE PROBLEMS—SPREADSHEET ANALYSES

24.4 Use the retwall spreadsheet to solve Problem 24.1.

24.5 Use the retwall spreadsheet to solve Problem 24.2.

24.6 Use the retwall spreadsheet to solve Problem 24.3.

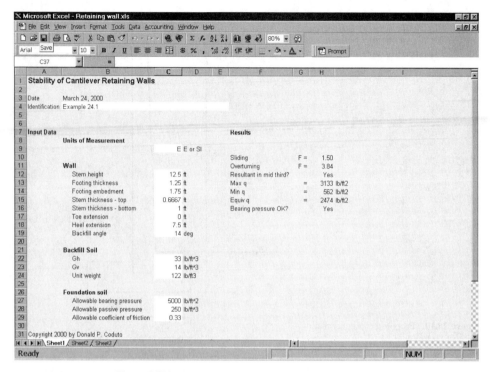

Figure 24.11   Typical screen from RETWALL.XLS spreadsheet.

24.3   INTERNAL STABILITY (STRUCTURAL DESIGN)

Once the wall has been sized to satisfy external stability requirements, we must check its internal stability. We will satisfy internal stability requirements by developing a structural design with sufficient structural integrity to safely resist the applied loads.

The footing is nearly always made of reinforced concrete, but the stem can be either reinforced concrete or reinforced masonry. Concrete stems have greater flexural strength, and greater flexibility in the design dimensions, and thus seem to be most cost-effective for tall walls. However, masonry stems do not require formwork, which is very expensive, do this type of stem is generally preferred for shorter walls.

The internal stability assessment and structural design begins with the stem, then proceeds to the footing.

Formed Concrete Stems

The design of formed concrete stems is governed by the *Building Code Requirements for Structural Concrete* ACI 318-99 and ACI 318M-99 (ACI, 1999). Local building codes sometimes have additional requirements.

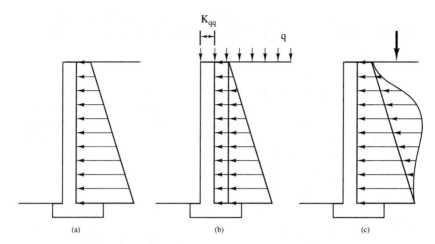

Figure 24.12   Typical loading diagrams for stem design: (a) with no surcharge loads; (b) with uniform surcharge load; (c) with point surcharge load.

Developing Shear and Moment Diagrams

The lateral loads acting on the stem should be computed using the techniques described in Chapter 23, and should consider surcharge loads, if any, as shown in Figure 24.12. The ACI code uses LRFD design (ACI calls it ultimate strength design) so we must multiply the computed loads by the ACI load factors defined by Equations 2.7 to 2.17. The load factor for earth pressures is 1.7 (per Equation 2.11). We then use this factored earth pres-

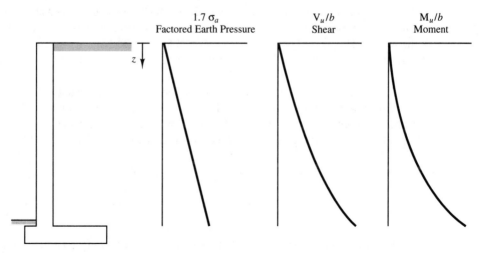

Figure 24.13   Earth pressure acting on the stem and typical shear and moment diagrams.

sure load to develop the shear and moment diagrams for the stem. Figure 24.13 shows typical shear and moment diagrams in the stem of a cantilever retaining wall.

Example 24.2—Part A

Evaluate the internal stability and develop a structural design for the wall in Example 24.1.

Develop an equation for moment in the stem. In this case, no surcharge loads are present so the loading diagram is triangular, the shear diagram is a second-order curve (a parabola), and the moment diagram is a third-order curve (a cubic). Therefore, we can develop the moment diagram by computing the moment at the bottom of the stem and backcalculating the coefficient in the moment equation.

At the bottom of the stem:

$$P_a/b = \frac{G_h H^2}{2} = \frac{(33)(12.5)^2}{2} = 2578 \text{ lb/ft}$$

$$P_u/b = 1.7 \, P_a/b = (1.7)(2578) = 4383 \text{ lb/ft}$$

$$M_u/b = (4383 \text{ lb/ft})(12.5 \text{ ft})(0.33) = 18,079 \text{ ft-lb/ft} = 217,000 \text{ in-lb/ft}$$

Set z equal to the distance in inches below the top of the stem, and m equal to the coefficient in the moment equation.

$$M_u/b = mz^3$$

$$217,000 \text{ in-lb/ft} = m[(12.5 \text{ ft})(12 \text{ in/ft})]^3$$

$$m = 0.0643$$

$$M_u/b = 0.0643z^3$$

Determining Stem Thickness and Reinforcement

The thickness of the stem should be at least 200 mm (8 in) and expressed as a multiple of 20 mm or 1 in. For short walls, it may be possible to have a uniform thickness, but taller walls are usually tapered or stepped so the thickness at the bottom is greater than at the top. A greater thickness and more reinforcing steel are required at the bottom because the stresses there are larger.

The design process begins with an evaluation of the shear capacity of the stem. The most critical section for shear is at the bottom. Since there are no stirrups, the concrete must have sufficient capacity to resist the shear force (i.e., we neglect any shear resistance provided by the flexural steel. The nominal shear capacity is [ACI 11.3.1.1, 11.3.2.1] is:

$$\boxed{V_n/b = V_c/b = 2b_w d\sqrt{f'_c}} \qquad \text{(24.3 English)}$$

$$\boxed{V_n/b = V_c/b = \frac{1}{6} b_w d\sqrt{f'_c}} \qquad \text{(24.3 SI)}$$

The stem must have sufficient thickness so the following condition is satisfied [ACI 11.5.5.1]:

$$\boxed{V_u/b \leq 0.5\phi V_n/b}$$ (24.4)

Where:

V_u/b = factored shear force per unit length of wall (lb/ft, kN/m)

V_n/b = nominal shear capacity per unit length of wall (lb/ft, kN/m)

V_c/b = nominal shear capacity of concrete (lb/ft, kN/m)

b_w = width of shear surface = 12 in/ft or 1000 mm/m

d = effective depth (in, mm)

f'_c = 28-day compressive strength of concrete (lb/in^2, MPa)

ϕ = resistance factor = 0.85

Using Equations 24.3 and 24.4, we can determine the minimum required thickness at the bottom of the stem to satisfy shear requirements.

The shear force also must pass from the stem to the footing. This must be addressed separately, because the stem and footing are cast in two separate concrete pours, and there is a cold joint between them. The shear capacity of this joint is based on sliding friction, with the normal force provided by the reinforcing steel. The required steel area per unit length of wall, A_{vf}/b, is [ACI 11.7.4.1]:

$$\boxed{A_{vf}/b = \frac{V_u/b}{\phi f_y \mu}}$$ (24.5)

Where:

A_{vf}/b = required steel area per unit length of wall

b = unit length of wall (usually 1 ft or 1 m)

ϕ = resistance factor = 0.85

V_u/b = factored shear force per unit length of wall

f_y = yield strength of steel

μ = coefficient of friction = 0.6

Rather than using special bars to provide A_{vf}/b, we simply increase the size of the vertical flexural steel so that it serves double duty.

Although not required by ACI, it is good practice to provide a key between the footing and the stem, as shown in Figure 24.14. This key is usually made with tapered 2×6 or 2×8 lumber, and thus has a depth of 1.5 in (37 mm). It provides additional shear transfer between the stem and the footing.

The next step is to evaluate the flexural stresses at the bottom of the stem and providing sufficient stem thickness and reinforcing steel to safely resist these flexural stresses.

ACI permits reinforcement up to $0.75\rho_b$, where ρ_b is the balanced steel ratio [ACI 10.3.3]. However, because of difficulties in accurately placing rebars in the field, it is best to use steel ratios of $0.35\rho_b$ to $0.50\rho_b$ (MacGregor, 1992). Tables 24.1 and 24.2 present these steel ratios for various combinations of f'_c and f_y. The minimum required steel area is as follows [ACI 14.3.2]:

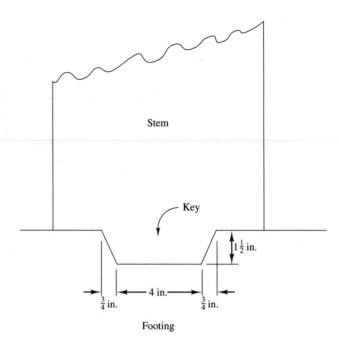

Figure 24.14 Use of a key between the footing and the stem.

For bars not larger than #5 (metric #16) and $f_y \geq 420$ MPa (60 k/in²): $A_s \geq 0.0012\, A_g$
For all other cases: $A_s \geq 0.0015\, A_g$

Where:

A_s = flexural steel area
A_g = gross concrete area
f_y = yield strength of steel

TABLE 24.1   MAXIMUM STEEL RATIO, ρ — ENGLISH UNITS
(Adapted from MacGregor, 1992)

f_y (lb/in²)		f_c' (lb/in²)			
		3000	4000	5000	6000
40,000	ρ_b	0.0371	0.0495	0.0582	0.0655
	0.75 ρ_b	0.0278	0.0371	0.0437	0.0491
	0.50 ρ_b	0.0186	0.0247	0.0291	0.0328
	0.35 ρ_b	0.0130	0.0173	0.0204	0.0229
60,000	ρ_b	0.0214	0.0285	0.0335	0.0377
	0.75 ρ_b	0.0161	0.0214	0.0251	0.0283
	0.50 ρ_b	0.0107	0.0143	0.0168	0.0189
	0.35 ρ_b	0.0075	0.0100	0.0117	0.0132

TABLE 24.2 MAXIMUM STEEL RATIO, ρ — SI UNITS
(Adapted from MacGregor, 1992)

f_y (MPa)		f_c' (MPa)				
		20	25	30	35	40
300	ρ_b	0.0321	0.0401	0.0482	0.0536	0.0582
	0.75 ρ_b	0.0241	0.0301	0.0361	0.0402	0.0436
	0.50 ρ_b	0.0161	0.0201	0.0241	0.0268	0.0291
	0.35 ρ_b	0.0112	0.0140	0.0169	0.0187	0.0204
420	ρ_b	0.0202	0.0253	0.0304	0.0338	0.0367
	0.75 ρ_b	0.0152	0.0190	0.0228	0.0253	0.0275
	0.50 ρ_b	0.0101	0.0126	0.0152	0.0169	0.0183
	0.35 ρ_b	0.0071	0.0089	0.0106	0.0118	0.0128

The maximum center-to-center steel spacing is three times the stem thickness, or 500 mm (18 in), whichever is less [ACI 7.6.5].

The next step is to determine if some of the vertical flexural steel can be cut off part-way up the stem, as shown in Figure 24.15. This is often possible, because the moment equation is cubic, so the flexural stresses in the stem decrease quickly. Example 24.2 illustrates a method of determining cutoff heights.

Finally, we design the longitudinal steel in the stem. Although in theory there are no flexural stresses in that direction, such stresses may develop if the soil is not perfectly uniform or if isolated surcharge loads are present. This steel accommodates these loads as well as temperature and shrinkage stresses. ACI 14.3.3 requires the following minimum longitudinal steel:

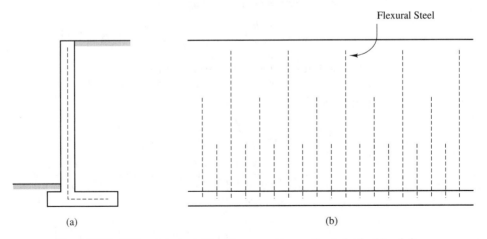

Figure 24.15 Different steel cutoff lengths are used to vary the steel ratio: (a) end view; (b) side view.

- If the bar size is no greater than #5 (metric #16) and $f_y \geq 420$ MPa (60 k/in^2): use $A_s \geq 0.0020\, A_g$
- For all other conditions: use $A_s \geq 0.0025\, A_g$

Where A_g is the gross concrete area.

It is not necessary to conduct a longitudinal flexural analysis unless large surcharge point loads are present.

Example 24.2—Part B

Determine the required wall thickness and flexural steel at the bottom of the stem.
Use $f_c' = 4000$ lb/in^2 and $f_y = 60$ k/in^2

Check shear at bottom of stem

The most critical section for shear occurs at the bottom of the stem. Since there are no stirrups, the concrete must be sufficient to resist this shear force (i.e., we neglect any shear resistance provided by the flexural steel).

$$V_u/b = P_u/b = 4383 \text{ lb/ft}$$

$$V_n/b = V_c/b = 2\, b_w d\sqrt{f_c'}$$
$$= 2(12 \text{ in/ft})d\sqrt{4000 \text{ lb/in}^2}$$
$$= 1518\, d \text{ lb/ft}$$

Combining with Equation 24.4 and solving produces $d \geq 6.79$ in

$$T \geq d + d_b + \text{cover} = 6.79 + 0.5 + 3 = 10.29 \text{ in}$$

Use $T = 11$ in, $d = 7.5$ in

Check transfer of shear from stem to footing

$$A_{vf}/b = \frac{V_u/b}{\phi f_y \mu} = \frac{(4383 \text{ lb/ft})}{(0.85)(60,000 \text{ lb/in}^2)(0.6)} = 0.143 \text{ in}^2/\text{ft}$$

This steel area will need to be added to the computed flexural steel area.
Evaluate flexural stresses at base of stem and design flexural steel.
Using Equation 9.13:

$$A_s/b = \left(\frac{f_c' b}{1.176 f_y}\right)\left(d - \sqrt{d^2 - \frac{2.353\, M_u/b}{\phi f_c' b}}\right)$$
$$= \left(\frac{(4000)(12)}{(1.176)(60,000)}\right)\left(7.5 - \sqrt{7.5^2 - \frac{(2.353)(215,000)}{(0.9)(4000)(12)}}\right)$$
$$= 0.561 \text{ in}^2/\text{ft}$$

$$\rho = \frac{A_s}{bd} = \frac{0.561}{(7.5)(12)} = 0.0062$$

Per Table 24.1, the steel ratio must be no greater than 0.0214, and preferably between about 0.0100 and 0.0143. The minimum acceptable steel area is $(0.0012)(11)(12) = 0.16$ in^2/ft. Therefore, the computed steel ratio is acceptable.

The total required vertical steel area, A_s, is $0.561 + 0.143 = 0.704$ in^2/ft.

Use #7 bars at 9 in OC

$$A_s/b = \left(\frac{0.60 \text{ in}^2}{9 \text{ in}}\right)\left(\frac{12 \text{ in}}{1 \text{ ft}}\right) = 0.800 \text{ in}^2/\text{ft}$$

$$\rho = \frac{A_s}{bd} = \frac{0.800 \text{ in}^2}{(12 \text{ in})(7.5 \text{ in})} = 0.0089$$

This steel area is sufficient to provide both flexural reinforcement and the normal force required for shear transfer.

Design remainder of stem steel

We will use a tapered stem, with a thickness of 8 in ($d = 4.5$ in) at the top and 11 in ($d = 7.5$ in) at the bottom. The steel at the bottom of the stem will be spaced at 9 inches on center, but at some height we will cut off every other bar, leaving steel at 18 inches on center. The maximum permissible steel spacing is $3T$ or 18 in, whichever is less [ACI 7.6.5], so we cannot cut off any more bars.

Compute ρ at the top and bottom of the stem for each steel spacing using Equation 9.12, then use Equations 9.10 and 9.11 with $\phi = 0.9$ to determine the corresponding moment capacities.

Sample computation:

$$a = \frac{\rho d f_y}{0.85 f'_c} = \frac{(0.0089)(7.5)(60,000)}{(0.85)(4000)} = 1.18 \text{ in}$$

$$\phi M_n = \phi A_s f_y\left(d - \frac{a}{2}\right) = (0.90)(0.800)\left(7.5 - \frac{1.18}{2}\right) = 299,000 \text{ in-lb/ft}$$

Steel	A_s/b (in^2/ft)	$\phi M_n/b$ at bottom ($d = 7.5$ in) (in-lb/ft)	$\phi M_n/b$ at top ($d = 4.5$ in) (in-lb/ft)
#7 at 9 in OC	0.800	299,000	169,000
#7 at 18 in OC	0.400	156,000	91,000

The factored moment and nominal moment capacities along the height of the stem are shown in Figure 24.16. We use this diagram to determine the cutoff location, which is one development length higher than the intersection of the #7 @ 18-in curve with the factored moment curve. In other words, this intersection is where the wider steel spacing provides enough flexural capacity, but we need to extend the steel one development length past this point to provide sufficient anchorage.

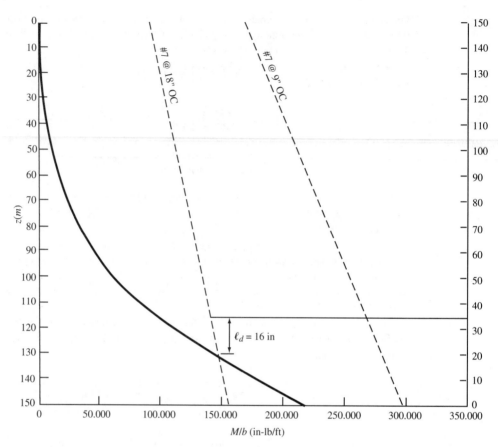

Figure 24.16   Factored moments and flexural capacity along stem for Example 24.2.

Per Equation 9.14, the required development length, l_d is:

$$\frac{l_d}{d_b} = \frac{3}{40}\frac{f_y}{\sqrt{f_c'}}\frac{\alpha\beta\gamma\lambda}{\left(\dfrac{c + K_{tr}}{d_b}\right)} = \frac{3}{40}\frac{60{,}000}{\sqrt{4000}}\frac{(1)(1)(1)(1)}{\left(\dfrac{3.5 + 0}{0.875}\right)} = 17.8$$

$$l_d = 17.8\,d_b = (17.8)(0.875\text{ in}) = 16\text{ in}$$

Based on this information, use the following steel in the stem:

Vertical steel:

Distance from top of footing	Steel
0–36 in	#7 at 9 in OC
36 in–Top	#7 at 18 in OC

Longitudinal steel:

$$A_s = 0.0020A_g = (0.0020)(12.5 \text{ ft} \times 12 \text{ in/ft})\left(\frac{11 \text{ in} + 8 \text{ in}}{2}\right) = 2.70 \text{ in}^2$$

Use 14 #4 bars ($A_s = 2.80 \text{ in}^2$).

Masonry Stems

The design of masonry stems is similar to that for concrete stems. However, masonry is a slightly different material, and has some unique design provisions and code requirements. See Schneider and Dickey (1987) for more information.

Footing

Regardless of the material used for the stem, the footing is almost always made of reinforced concrete. The structural design process for this footing is slightly different than that for columns or non-retaining walls, as described in Chapter 9, because the loading from retaining walls is different.

The minimum required footing thickness, T, may be governed by the required development length of the vertical steel from the stem. This steel normally has a standard 90° hook, as shown in Figure 9.20, which requires a footing thickness of at least:

$$\boxed{T \geq l_{dh} + 3 \text{ in}} \qquad \text{(24.6 English)}$$

$$\boxed{T \geq l_{dh} + 70 \text{ mm}} \qquad \text{(24.6 SI)}$$

The total footing thickness, T, should be a multiple of 100 mm or 3 in, and never less than 300 mm or 12 in.

The required development length for hooked bars made of grade 60 (metric grade 420) steel is [ACI 12.5]:

$$\boxed{l_{dh} = \frac{1200 \, d_b}{\sqrt{f'_c}}} \qquad \text{(24.7 English)}$$

$$\boxed{l_{dh} = \frac{100 \, d_b}{\sqrt{f'_c}}} \qquad \text{(24.7 SI)}$$

Where:

T = footing thickness (in, mm)

l_{dh} = development length, as defined in Figure 9.20 (in, mm)

d_b = bar diameter (in, mm)

f'_c = 28-day compressive strength of concrete (lb/in², MPa)

The development length computed from Equation 24.7 may be modified by the following factors [ACI 12.5.3][1] :

For standard reinforcing bars with yield strength other than 60,000 lb/in^2 : $f_y/60,000$

For metric reinforcing bars with yield strength other than 420 lb/in^2 : $f_y/420$

With at least 50 mm (2 in) of cover beyond the end of the hook: 0.7

Next we design the heel extension, which is the portion of the footing beneath the backfill. The shear and flexural stresses in this portion of the footing are due to the weight of the backfill soil immediately above the footing plus the weight of the footing heel extension. We ignore the bearing pressure acting along the bottom of the heel extension, which is conservative. Thus, the heel extension is designed as a cantilever beam, as shown in Figure 24.17.

There is no need to develop shear and moment diagrams for the footing. Simply compute the factored shear and moment along a vertical plane immediately below the back of the wall and use this information to check the shear capacity of the footing and to select the required flexural reinforcement. Since the weight of the backfill is a dead load, use a load factor of 1.4.

The flexural analysis may be performed using Equation 9.13 with the effective depth obtained from the anchorage analysis. The steel area must satisfy the minimum and maximum steel requirements described earlier in this section. The reinforcing steel should be placed 70 mm (3 in) from the top of the footing, and should extend to 70 mm (3 in) from the end. The shear analysis should be based on Equations 9.2 and 9.9.

The toe extension also is designed as a cantilever beam. However, it bends in the opposite direction (i.e., concave upward), so the flexural steel must be near the bottom. In this case, the load is due the bearing pressure acting on the bottom of the footing and the weight of the footing toe extension. For conservatism and simplicity, we ignore any soil that may be present above this part of the footing. Since this bearing pressure is primarily due to the lateral earth pressure acting on the stem, use a load factor of 1.7.

The footing also should have longitudinal steel. A steel ratio of 0.0015 to 0.0020 is normally appropriate. There is no need to do a longitudinal flexure analysis.

Example 24.2—Part C

Anchorage of vertical steel from stem

This analysis must be based on the footing concrete, which is cast separately from that in the stem. Use $f_c' = 3000$ lb/in^2 and $f_y = 60,000$ lb/in^2

---

[1]This list only includes modification factors that are applicable to anchorage of vertical steel in retaining wall footings. ACI 12.5.3 includes additional modification factors that apply to other situations.

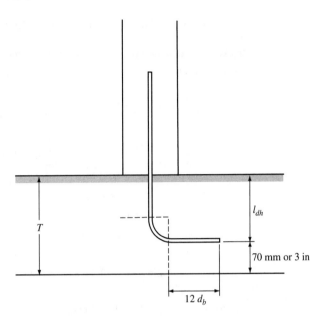

Figure 24.17   Loading for design of heel and toe extensions.

$$l_{dh} = \frac{1200 d_b}{\sqrt{f'_c}} \times \text{factors}$$

$$= \frac{1200(0.875)}{\sqrt{3000}} \times 0.7$$

$$= 13.4 \text{ in}$$

$$T \geq l_{dh} + 3 \text{ in} = 13.4 + 3 = 16.4 \text{ in}$$

Heel extension—shear analysis with $T = 18$ in

$$W_{soil}/b = 12{,}293 \text{ lb/ft (from Example 24.1—Part D)}$$

$$W_{footing}/b = (7.5 \text{ ft})(1.5 \text{ ft})(150 \text{ lb/ft}^3) = 1687 \text{ lb/ft}$$

Use a load factor of 1.4 because this load is due to the weight of the overlying soil and the footing.

$$V_u/b = 1.4(12{,}293 \text{ lb/ft} + 1{,}687 \text{ lb/ft}) = 19{,}572 \text{ lb/ft}$$

$$V_n/b = 2b_w d\sqrt{f'_c} = 2(12 \text{ in})d\sqrt{3000 \text{ lb/in}^2} = 1315 \, d \text{ lb/ft}$$

$$\phi V_n/b = (0.85)(1315 d) = 1117 \, d \text{ lb/ft}$$

Using these equations with $V_u/b \leq \phi V_n/b$ gives $d \geq 17.5$ in. Assuming #8 bars ($d_b = 1.00$)

$$T \geq d = d_b/2 + 3 = 17.5 + 1/2 + 3 = 20.5 \text{ in}$$

The shear analysis controls the required footing thickness, so use $T = 21$ in (a multiple of 3 in), $d = T - 3$ in $- d_b/2 = 21 - 3 - 0.5 = 17.5$ in. Although this is larger than the 15 inch thickness used in the external stability analysis, there is no need to redo that analysis.

Heel extension—flexure

$$M_u/b = (1.4)(12,293 \text{ lb/ft} + 1,687 \text{ lb/ft})\left(\frac{7.5}{2}\text{ ft}\right)(12 \text{ in/ft}) = 881,000 \text{ in-lb/ft}$$

$$A_s/b = \left(\frac{f_c'b}{1.176 f_y}\right)\left(d - \sqrt{d^2 - \frac{2.353\,M_u/b}{\phi f_c'b}}\right)$$

$$= \left(\frac{(300)(12)}{(1.176)(60,000)}\right)\left(17.5 - \sqrt{17.5^2 - \frac{(2.353)(881,000)}{(0.9)(3000)(12)}}\right)$$

$$= 0.987 \text{ in}^2/\text{ft}$$

$$\min A_s/b = 0.0018A_g = (0.0018)(18)(12) = 0.389 \text{ in}^2/\text{ft}$$

Use #8 bars at 9 in OC ($A_s/b = 1.05$ in^2/ft)

Note: This is the same spacing as the vertical stem steel, which will avoid interference problems and facilitate tying the steel together

Longitudinal steel (per discussion in Chapter 9):

$$A_s = 0.0018A_g = (0.0018)(21 \text{ in})(8.5 \text{ ft})(12 \text{ in/ft}) = 3.86 \text{ in}^2$$

Use 14 #5 bars ($A_s = 4.34$ in^2).

Final design:

The final design is shown in Figure 24.18.

QUESTIONS AND PRACTICE PROBLEMS

24.7 The first part of Section 24.2 discusses the advantages and disadvantages of concrete vs. masonry stems. Based on this discussion, explain in more detail why masonry stems are generally more cost effective for short walls, while concrete stems are usually preferred for taller walls.

24.8 The cantilever retaining wall in Figure 24.19 satisfies all external stability requirements. Select appropriate values of f_c' and f_y, then develop a structural design that satisfies the internal stability requirements and show your final design in a sketch.

24.9 The cantilever retaining wall in Figure 24.20 satisfies all external stability requirements. Select appropriate values of f_c' and f_y, then develop a structural design that satisfies the internal stability requirements and show your final design in a sketch.

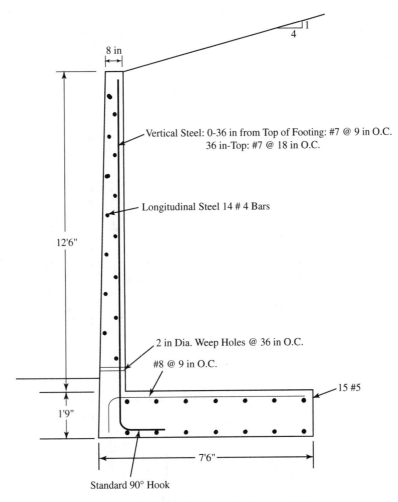

Figure 24.18   Loading design for Example 24.2.

24.4   DRAINAGE AND WATERPROOFING

Example 23.4 in Chapter 23 demonstrated the impact of groundwater on lateral earth pressures. Because groundwater increases earth pressures so dramatically, engineers provide a means of drainage whenever possible to prevent the groundwater table from building up behind the wall. Two types of drains are commonly used: *weep holes* and *perforated pipe drains*. Both are shown in Figure 24.21. It is also helpful to include a means of intercepting water and bringing it to the drain, possibly using products such as that shown in Figure 24.22.

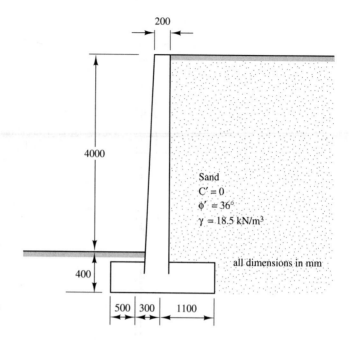

Figure 24.19   Retaining wall for Problem
24.8.

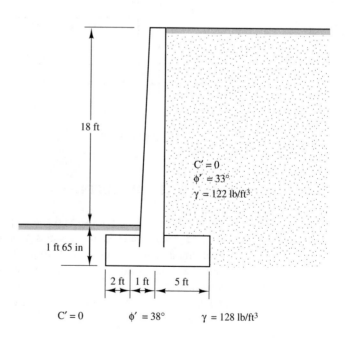

Figure 24.20   Retaining wall for Problem
24.9.

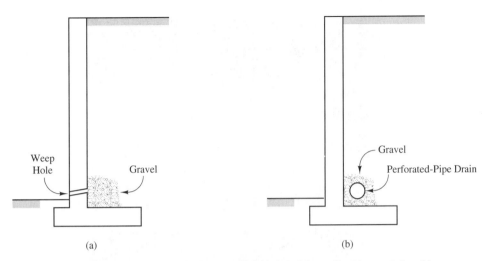

Figure 24.21 Methods of draining the soil behind retaining walls: (a) weep holes; (b) perforated pipe drains.

In addition to draining groundwater, it also is important to control the migration of moisture through the wall. This is especially important on basement walls because excessive moisture inside basements can be both a nuisance and a health hazard. Even in exterior walls, excessive moisture migration can be an aesthetic problem, as shown in Figure 24.23. These materials may be categorized as follows (Meyers, 1996):

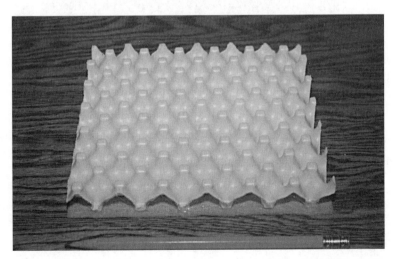

Figure 24.22 A filter fabric is attached to this eggcrate-shaped plastic material to form a prefabricated drain which can be installed behind a retaining wall.

(a)

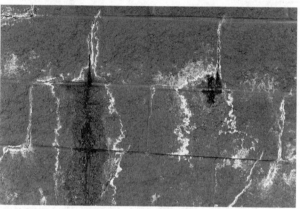

(b)

Figure 24.23   Moisture migration through this retaining wall has caused unsightly water
stains. a) overall view of wall; b) close-up view. This problem could have been avoided
by use of proper waterproofing or dampproofing behind the wall.

- *Dampproofing* is the treatment of the wall surface to retard dampness or water penetration under nonhydrostatic conditions. These methods are used when the wall is
always above the groundwater table, but the adjacent soil may be subject to moisture from rainfall, irrigation, capillary rise, or other sources.

- *Waterproofing* is the treatment of the wall surface to prevent passage of water under
intermittent or full hydrostatic pressure conditions. These methods are used when
the wall is occasionally or permanently below the groundwater table.

Dampproofing methods include:

- Thin bituminous coatings
- Thin cementitious coatings

Figure 24.24 Installation of bentonite waterproofing panels on a concrete retaining wall. These panels consist of dry bentonite clay encased in a cardboard-like panel, and are placed on the side of the wall that will receive the backfill. After the backfill has been placed, the bentonite absorbs water and swells to form a nearly impervious barrier (Colloid Environmental Techonologies Company).

- Acrylic latex coatings
- 6-mil polyethylene sheets

Waterproofing methods are more expensive, and more effective. They include:

- Bituminous membranes, which consist of multiple layers of mopped-on bituminous material (i.e., asphalt or coal tar) with alternating layers of reinforcing fabric or felt. The final product is similar to built-up roofing.
- Liquid-applied elastometric waterproofing, which consists of special chemicals applied to the wall.
- Sheet-applied elastometric waterproofing, which are prefabricated sheets applied to the wall.
- Cementitious waterproofing that consists of several coats of a special mortar.
- Bentonite clay panels (see Figure 24.24)
- Troweled on mixtures of bentonite clay and a binding agent

Alternatively, special admixtures may be used with the concrete in the wall to make it more impervious. This method is sometimes called *integral waterproofing*. However, its success depends on the absence of significant cracks, which is difficult to achieve. In addition, this method requires special *waterstops*, which are plastic or rubber strips, at construction joints, such as that between the stem and the footing.

24.5 AVOIDANCE OF FROST HEAVE PROBLEMS

Concrete is a relatively poor insulator of heat, so retaining walls located in areas with frost heave problems[2] may be damaged as a result of the formation of ice lenses behind the wall. In some cases, ice lenses have caused retaining walls to move so far out of position that they became unusable.

Reduce frost heave problems by using all of the following preventive measures:

- Incorporate good drainage details in the design to avoid the buildup of free water behind the wall. Deep perforated drainage pipes are usually a better choice than weep holes because they are less likely to become blocked by frozen water.
- Use non frost-susceptible soil for the portion of the backfill immediately behind the wall. This zone should extend horizontally behind the wall for a distance equal to the depth of frost penetration in that locality.

Even with these precautions, engineers working in cold climates often design retaining walls with an additional 5 to 10 kPa (100–200 lb/ft^2) surcharge pressure acting on the ground surface. The resulting lateral earth pressure from this surcharge results in a stronger wall design that is better able to resist frost heave.

SUMMARY

Major Points

1. Cantilever retaining walls are the most common type of earth retaining structure.
2. Cantilever retaining walls must satisfy two major requirements: *external stability* (to avoid failure in the soil) and *internal stability* (to avoid failure in the structure).
3. External stability requirements include the following:
   - Sliding
   - Overturning
   - Resultant in middle third
   - Bearing capacity
   - Deep-seated shear
   - Settlement
4. Internal stability requirements are satisfied by providing a structural design that satisfies the following requirements:
   - Sufficient flexural strength in the stem
   - Sufficient shear strength in the stem
   - Adequate connection between the stem and the footing
   - Sufficient anchorage of the stem steel into the footing
   - Sufficient flexural strength in the footing

---

[2]See discussion of frost heave in Chapter 8.

5. Walls also must have proper drainage and waterproofing details.

6. In regions with cold climates, walls must be protected against frost heave problems.

Vocabulary

Bentonite waterproofing panel	Key	Structural integrity
Cantilever retaining wall	Limit equilibrium	Tieback anchor
Dampproofing	Minimum steel ratio	Tiedown anchor
Deep-seated shear failure	Overturning	Toe
Drainage	Perforated pipe drain	Wall-soil unit
External stability	Resultant in middle third	Waterproofing
Heel	Sliding	Weep hole
Internal stability	Steel ratio	
	Stem	

COMPREHENSIVE QUESTIONS AND PRACTICE PROBLEMS

24.10 Explain the fundamental difference between external stability requirements and internal stability requirements.

24.11 As discussed in Chapter 23, field measurements indicate the resultant force from the active earth pressures acting on the back of a cantilever retaining wall is located at about 0.4 H from the bottom of the wall, not 0.33 H as indicated by theory (Duncan et al., 1990). Nevertheless, it is common practice to design walls using the theoretical location (as demonstrated in this chapter). Discuss the impact of using 0.4 H in the design process, and describe what portions of the design would change as a result. Assume the magnitude of the resultant force remains unchanged.

24.12 The retaining wall in Figure 24.25 is made of unreinforced cobbles and mortar. A vertical crack has developed in this wall near the point where it makes a 90-degree bend (it is visible directly above the middle of the stairs). This crack is 20 mm wide at the top, and gradually tapers to 1 mm at the bottom. Explain the cause of this crack.

24.13 A cantilever retaining wall is needed at the location shown in Figure 24.26. Using $f_c' = 30$ MPa and $f_y = 420$ MPa, develop a complete design for this wall, considering both external and internal stability, and show your final design in a sketch.

24.14 A cantilever retaining wall is needed at the location shown in Figure 24.27. Using $f_c' = 4000$ lb/in^2 and $f_y = 60$ k/in^2, develop a complete design for this wall, considering both external and internal stability, and show your final design in a sketch.

    Note: You will need to consider the surcharge load when computing the lateral earth pressures.

24.15 A cantilever retaining wall is to be built at a site with the following characteristics:

    Wall height = 8 ft

    Ground surface above wall: level

Figure 24.25 Retaining wall for Problem 24.12.

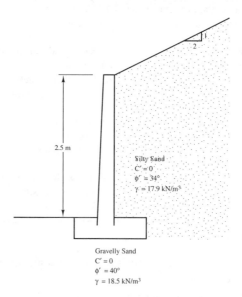

Figure 24.26 Retaining wall for Problem 24.13.

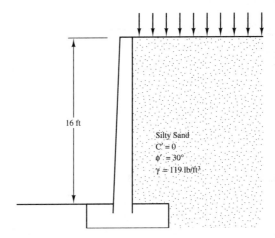

Figure 24.27   Retaining wall for Problem 24.14.

Backfill soils: Well graded sand, $c' = 0$, $\phi' = 37°$, and $\gamma = 128$ lb/ft^3

Soils below foundation: Same as backfill soils

Material properties: $f_c' = 3000$ lb/in^2 in footing and 5000 lb/in^2 in stem, $f_y = 60$ k/in^2

Using the retwall spreadsheet, develop a design that satisfies external stability requirements, then use the techniques described in this chapter to develop a structural design that satisfies internal stability requirements. Show your final design in a sketch, and include a printout from your spreadsheet.

24.16 A cantilever retaining wall is to be built at a site with the following characteristics:

Wall height = 2 m

Ground surface above wall: Sloped upward at 3H:1V

Backfill soils: Silty sand, $c' = 0$, $\phi' = 32°$, and $\gamma = 19.0$ kN/m^3

Soils below foundation: Same as backfill soils

Material properties: $f_c' = 30$ MPa, $f_y = 420$ MPa

Using the retwall spreadsheet, develop a design that satisfies external stability requirements, then use the techniques described in this chapter to develop a structural design that satisfies internal stability requirements. Show your final design in a sketch, and include a printout from your spreadsheet.

24.17 A cantilever retaining wall is to be built at a site with the following characteristics:

Wall height = 16 ft

Ground surface above wall: Level

Backfill soils: Compacted gravelly sand, $c' = 0$, $\phi' = 39°$, and $\gamma = 132$ lb/ft^3

Rock below foundation: Sandstone bedrock, $c' = 6000$ lb/ft^2, $\phi' = 46°$, and $\gamma = 145$ lb/ft^3

Material properties: $f_c' = 5000$ lb/in^2, $f_y = 60$ k/in^2

Using retwall spreadsheet, develop a design that satisfies external stability requirements, then use the techniques described in this chapter to develop a structural design that satisfies internal stability requirements. Show your final design in a sketch, and include a printout from your spreadsheet.

25

Sheet Pile Walls

The final solution to a problem is the one in hand
when time and money run out.

Sheet piles are rolled plate-steel structural members that are driven into the ground to form a *sheet pile wall*, as shown in Figure 25.1 These walls are used in many applications, including:

- Earth-retaining structures, including waterfront applications, bridge abutments, and temporary support for construction excavations.
- Cellular cofferdams
- Cut-off walls in levees, dams, or other applications. These cutoffs are intended to restrict the underground flow of water.

This chapter discusses the use, design, and construction of sheet pile walls.

25.1 MATERIALS

Steel

The vast majority of sheet piles are made of steel, primarily because of its strength, ease of handling, and ease of construction.

Figure 25.1   Sheet pile wall in a bridge abutment (Photo courtesy of Skyline Steel Corporation).

Standard Sections

Steel sheet piles are available in various cross-section shapes, as shown in Figure 25.2, and each shape is available in a range of sizes. Unfortunately, unlike steel H-piles, there are no widely-accepted standards for these various cross-sections. Each manufacturer sets its own standards for the dimensions and identification of the sections it produces. DFI (1998) contains a compilation of sheet piles produced by various manufacturers, along with relevant engineering properties. Table 25.1 presents this data for Z-section sheet piles available from one manufacturer.

Corrosion

Steel sheet piles can have problems with excessive corrosion, especially when used in waterfront applications. Salt water is especially corrosive, and the presence of contaminants in the water also can be a contributing factor. The greatest corrosion rates occur in the splash zone (the area between the lowest water surface and the upper limit of wave action) because of the cyclic wetting and drying that occurs there.

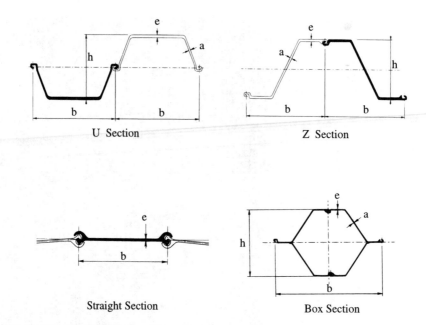

U Section

Z Section

Straight Section

Box Section

Figure 25.2   Typical cross sections for steel sheet piles. The exact shapes, dimensions, and section identifications vary between manufacturers.

TABLE 25.1   *Z*-SECTION STEEL SHEET PILES SOLD BY SKYLINE STEEL CORPORATION
(Data used with permission)

Section	Width b (mm)	Height h (mm)	Thickness e (mm)	Thickness a (mm)	Mass (kg/m² of wall)	Section Modulus S/b (cm³/m)	Moment of Inertia I/b (cm⁴/m)	Allowable Flexural Capacity[a] M_b/b (kN-m per m)	Allowable Flexural Capacity[a] M_b/b (ft-k per ft)
AZ 13	670	303	9.5	9.5	107.0	1300	19,700	234	52.4
AZ 18	630	380	9.5	9.5	118.0	1800	34,200	324	72.6
AZ 26	630	427	13.0	12.2	155.0	2600	55,510	468	104.8
AZ 36	630	460	18.0	14.0	194.0	3600	82,800	648	145.1
AZ 48	580	482	19.0	15.0	240.6	4800	115,670	864	193.5

[a]Flexural capacity is based on ASTM A328 steel, which has a yield strength, F_y, of 270 MPa (39,000 lb/in²) and an allowable stress of 0.667 F_y. Other grades of steel also are available, and they have correspondingly different flexural capacities.

Corrosion problems can be reduced by incorporating various preventive measures in the design. These might include one or more of the following (ASCE, 1996b):

- Coating the sheet piles with coal tar epoxy, paint, or some other protective material
- Using sheet piles made from special corrosion-resistant steel alloys
- Installing a cathodic protection system
- Using a thicker steel section than required, thus allowing for some of the thickness to be lost to corrosion

Usually these measures are sufficient. Alternatively, the wall could be made of some other material.

Other Materials

Sheet piles also may be made of other materials, including the following:

- **Reinforced concrete** sheet piles consist of precast concrete panels, typically about 200-mm (8 in) thick, which are driven into the ground. These panels typically have tongue-and-groove joints to maintain alignment with the adjacent panels. Reinforced concrete walls are more watertight and more resistant to corrosion than steel walls, but the panels are very heavy, bulky, and difficult to handle. Therefore, these walls are usually very expensive. However, various methods of constructing cast-in-place concrete walls can be very cost-effective. These cast-in-place walls are beyond the scope of this chapter.
- **Wood** sheet piles are made of flat interlocking tongue-and-groove wood sheets. This material is less expensive than steel, but its flat cross section has a small moment of inertia, and wood has a small modulus of elasticity and flexural strength. In addition, wood is subject to rotting. Therefore, wood sheet piles are limited to short wall heights and are used only as temporary structures.
- **Aluminum** sheet piles have cross sections similar to those used with steel. They are more resistant to corrosion, and thus are sometimes used on waterfront projects, especially those exposed to salt water. However, aluminum is more expensive than steel, and has a smaller modulus of elasticity and flexural strength.
- **Fiberglass, vinyl, and polyvinyl chloride (PVC)** sheet piles are another alternative where corrosion resistance is important. As with aluminum, these materials have cross sections similar to those used with steel, but have a smaller modulus of elasticity and flexural strength, and a higher cost.

Although these materials are useful in special circumstances, the vast majority of sheet pile walls are made of steel, so the remainder of this chapter will only consider steel sheet piles.

Figure 25.3   Steel sheet piles being installed (Photo courtesy of Skyline Steel Corporation).

25.2   CONSTRUCTION METHODS AND EQUIPMENT

Steel sheet pile walls are normally driven into the ground in pairs using a pile-driving hammer similar to those used to drive foundation piles. Various techniques have been used, but the most common is to begin by installing a guide along the ground surface to keep the piles aligned properly. This guide also eliminates the need for pile leads, and allows the contractor to suspend the pile hammer from a crane, as shown in Figure 25.3.

Contractors often drive several pairs of sheet piles a short distance into the ground, thus establishing both the vertical and horizontal alignment. Then they return to the beginning and drive each pair to its final penetration. When used to support temporary excavations, sheet piles are eventually extracted using a vibratory hammer and reused on another project.

25.3   CANTILEVER SHEET PILE WALLS

A *cantilever sheet pile wall* is one that does not have any additional supports, and thus relies on its flexural strength and embedment to resist the lateral earth pressures. Figure 25.4 shows a typical cantilever sheet pile wall. These walls are simple to build because

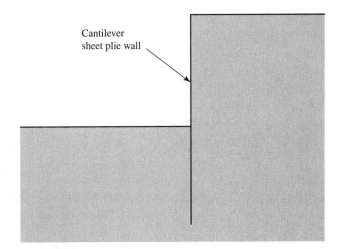

Figure 25.4   A cantilever steel sheet pile
wall.

they do not use bracing, anchors, or other structural elements. However, the lateral earth
pressures imposed on these walls create large flexural stresses in the steel, which means
they can generally be no more than 3 to 4 m (10–14 ft) tall. In addition, cantilever walls
experience greater lateral deflections, which may induce excessive settlement in the adjacent retained ground, and they are more susceptible to failure due to scour or erosion of
the supporting soils.

Cantilever Walls in Sand

Design Earth Pressures

Cantilever sheet pile walls in sand are normally designed using classical lateral earth pressure theories with drained strength parameters. This design method assumes active earth
pressures act along the back of the wall, and passive pressures along the front of the embedded portion. This is a simplification of the real earth pressures, but is sufficient for
these purposes. The design method also considers the effects of hydrostatic pressures and
surcharge loads.

Figure 25.5 shows the distribution of design earth pressures on a typical cantilever
sheet pile wall.

Stability

The wall must be embedded a sufficient depth into the ground to resist the applied loads.
This required depth of embedment may be determined as follows:

1. Using Rankine's or Coulomb's method, compute K_a and K_p for each soil stratum.
   Divide the K_p values by a factor of safety of 1.5 to 2.0.

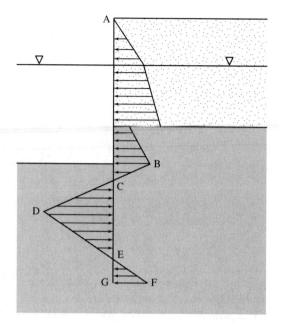

Figure 25.5   Design earth pressures on a cantilever steel sheet pile wall. In this case, the hydrostatic pressures on either side of the wall are equal and cancel out. However, if the water level on one side is lower, then the net hydrostatic pressure must be added to the earth pressures.

2. Using Equation 23.9, draw the pressure diagram from Point A to Point B. Notice how its slope changes at the groundwater table (because of effective stresses) and how it jogs when a new soil stratum is encountered (because of the change in K_a). The inclination of these lines with respect to the wall is γK_a above the groundwater table and $(\gamma - \gamma_w)K_a$ below the groundwater table.

3. If surcharge loads are present, use the techniques described in Section 23.7 to compute the resulting lateral earth pressures acting on the wall.

4. Below Point B, plot the net pressure acting on the wall, which is the difference between the passive and active pressures (or the difference between Equations 23.15 and 23.9). The inclination of these lines with respect to the wall is $\gamma(K_p - K_a)$ above the groundwater table and $(\gamma - \gamma_w)(K_p - K_a)$ below the groundwater table. Therefore, we can locate Point C by plotting a line at the appropriate slope from Point B.

5. Compute the total horizontal force between Points A and C (which is equal to the area under the pressure diagram) and the moment generated by this pressure diagram about Point C.

6. Lines CD and DF have slopes as defined in Step 4. Therefore, we can write equations for the shear and moment below Point C in terms of the dimensions CE and EG. For static equilibrium, this shear and moment must be equal to those computed in Step 5. We now have two equations and two unknowns, and thus can compute the required depth of embedment.

Structural Design

The next step in designing a cantilever sheet pile wall is to select the required sheet pile section. To do so, we compute the maximum moment, which occurs at the point of zero shear, than use it to select an appropriate sheet pile section from Table 25.1 or from some other source. If none of these sections provide sufficient flexural capacity, the wall is too tall for a cantilever design.

Example 25.1

Determine the required depth of embedment and select an appropriate sheet pile section for the cantilever wall shown in Figure 25.6.

Solution
$$K_a = \tan^2 (45 - 35/2) = 0.271$$

$$K_p/F = \tan^2 (45 + 35/2)/1.5 = 2.46$$

@ $z = 8$ ft
   $\sigma = (120)(0.271)(8) = 260$ lb/ft^2

@ $z = 20$ ft
   $\sigma = 260 + (123 - 62.4)(0.271)(12) = 457$ lb/ft^2

@ $z > 20$ ft
   slope of pressure diagram $= (123 - 62.4)(2.46 - 0.271) = 132.7$ lb/ft^2/ft
   $\sigma = 0$ @ $z = 20 + 457/132.7 = 23.44$ ft

@ $z = 23.44$ ft
   $V/b = (260)(8)/2 + (260)(12) + (457 - 260)(12)/2 + (457)(3.44)/2$
   $= 1040 + 3120 + 1182 + 786$
   $= 6{,}128$ lb/ft

   $M/b = (1040)(18.11) + (3120)(9.44) + (1182)(7.44) + (786)(2.29)$
   $= 58{,}880$ ft-lb/ft

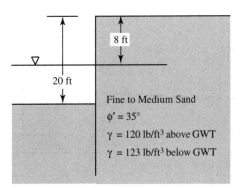

Figure 25.6   Sheet pile wall for Example 25.1.

@ $z > 23.44$ ft

    Let a = distance CE
    Let b = distance EG

$$V/b = 132.7\, a^2/4 - 132.7b^2/2$$
$$= 33.18\, a^2 - 66.35\, b^2$$

$$M/b = -(33.18\, a^2)(a/2) + (66.35\, b^2)(a + 0.667b)$$
$$= -16.59\, a^3 + 66.35\, ab^2 + 44.24b^3$$

Setting these two equations equal to 6128 and 58,880, respectively, and solving gives:

$a = 19.98$ ft
$b = 10.36$ ft

Therefore, the bottom of the sheet pile is at $z = 23.44 + 19.98 + 10.36 = 54$ ft

Use a 54 ft long sheet pile embedded 34 ft into the ground   ⇐ *Answer*

The maximum moment occurs where the shear equals zero, which is at $z = 33.43$ ft:

$$M/b = (1040)(28.10) + (3120)(19.43) + (1182)(17.43) + (786)(12.28)$$
$$- [(132.7)(9.99)(9.99)/2][9.99/3]$$
$$= 98,050 \text{ ft-lb/ft}$$

Per Table 25.1, select the AZ 26 section (allowable capacity = 104,800 ft-lb/ft).

Use an AZ 26 sheet pile section with ASTM A328 steel   ⇐ *Answer*

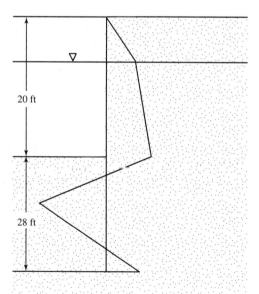

20 ft

28 ft

Figure 25.7 Pressure diagram for Example 25.1.

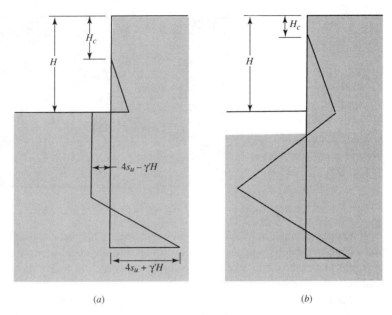

Figure 25.8 Design lateral earth pressures for cantilever walls in clay a) short-term analysis using undrained strengths; b) long-term analysis using undrained strengths.

Cantilever Walls in Clay

If some or all of the soils in contact with the wall are clay, then two stability analyses should be performed: one for short-term conditions and one for long-term conditions. The short-term analysis is based on the undrained strengths ($c = s_u$ and $\phi = 0$), while the long-term analysis is based on the drained strengths ($c = c'$ and $\phi = \phi'$).

The lateral earth pressures for these analyses, as shown in Figure 25.8, are computed using Equations 23.21 and 23.23. For the undrained condition, $K_a = K_p = 1$, so the first terms in these two equations cancel, and the net pressure diagram below the dredge line is uniform with depth. The critical height, H_c, is computed using Equation 23.20.

25.4   BRACED OR ANCHORED SHEET PILE WALLS

Most sheet pile walls include additional lateral support, using internal bracing or tieback anchors. These are known as *braced walls* and *anchored walls*, respectively. The additional support reduces the flexural stresses and lateral movements in the wall, which permits construction of walls much taller than is possible with cantilever designs. In addition, the depth of embedment, D, does not need to be as great. Finally, because the wall does not move as much, there is less settlement in backfill. Figures 22.9 and 25.9 show the configuration of internal bracing and tieback anchors.

Figure 25.9   A subway construction excavation with sheet piles and internal bracing (Photo courtesy of Skyline Steel Corporation).

Internal bracing may be horizontal or inclined. This method is often less expensive than tiebacks, especially in narrow excavations or when the adjacent soils are weak. However, the presence of these braces can hinder construction inside the excavation, so the total project cost may be greater.

Tieback anchors do not obstruct construction, and can be placed earlier in the excavation sequence, which reduces lateral movements in the wall. However, the placement of anchors is slower than that for braces, and must be done very carefully to maintain high reliability.

Design Methods

There are three principal design methods for braced or anchored walls:

- The *fixed-end method* requires a depth of embedment such that the bottom of the sheet pile is fixed against translation and rotation. This is similar to the criterion used to design cantilever walls.
- The *free-end method* permits the bottom of the wall to rotate or translate. This criterion permits shallower depths of embedment, and relies more heavily on the lateral support provided by the anchors or braces.
- The *beam on elastic foundation method* uses a technique similar to the *p-y* method described in Chapter 16.

All three methods are useful, but we will discuss only the free-end method.

Free-End Design Method

The free-end method is based on the assumption that the sheet pile is perfectly rigid and rotates about the tie rod level. The bottom of the sheet pile is embedded just far enough to provide the needed stability. Figure 25.10 shows the design earth pressures for walls in sand and clay.

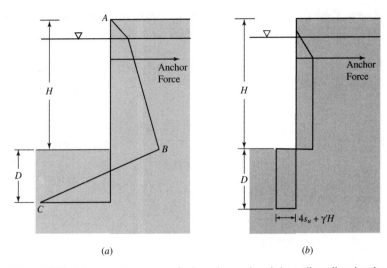

Figure 25.10   Design earth pressures for braced or anchored sheet pile walls using the free-end method a) walls in sand, b) walls in clay.

The design procedure is as follows:

1. Compute the coefficients of active and passive earth pressure, K_a and K_p using the methods described in Chapter 23, then divide K_p by a factor of safety of 1.5 to 2.0.
2. Using Equation 23.9, draw the pressure diagram from Point A to Point B. Notice how its slope changes at the groundwater table (because of effective stresses) and how it jogs when a new soil stratum is encountered (because of the change in K_a). The inclination of these lines with respect to the wall is γK_a above the groundwater table and $(\gamma - \gamma_w)K_a$ below the groundwater table.
3. If surcharge loads are present, use the techniques described in Section 23.7 to compute the resulting lateral earth pressures acting on the wall.
4. Below Point B, plot the net pressure acting on the wall, which is the difference between the passive and active pressures (or the difference between Equations 23.15 and 23.9). The inclination of these lines with respect to the wall is $\gamma(K_p - K_a)$ above the groundwater table and $(\gamma - \gamma_w)(K_p - K_a)$ below the groundwater table. Therefore, we can locate Point C by plotting a line at the appropriate slope from Point B.
5. Take moments about the tie rod and solve for the depth D.
6. Compute the tie rod force by taking Σ horizontal forces = 0.
7. Determine the maximum bending moment, which occurs at the point of zero shear, and use it to select an appropriate sheet pile section.

This method produces good values of the depth of embedment and tie rod force, but tends to overestimate the maximum bending moment. This is because the sheet pile is not truly rigid, and the actual earth pressures differ from the theoretical values. Therefore,

Rowe (1952, 1957, 1958) developed a procedure for adjusting the computed maximum moment using the following parameters:

$$\rho = \frac{(H + D)^4}{EI} \tag{25.1}$$

$$S = \frac{1.25c}{\gamma'(H + D)} \tag{25.2}$$

$$\alpha = \frac{H}{H + D} \tag{25.3}$$

Where:

ρ = flexibility number

H = height of sheet pile wall (ft)

D = depth of embedment (ft)

E = modulus of elasticity of sheet pile (lb/in^2) = 29,000,000 lb/in^2 for steel

I = moment of inertia per foot of wall (in^4)

S = stability number

c = soil cohesion

γ' = effective unit weight of soil (lb/ft^3) (above groundwater $\gamma' = \gamma$, below groundwater $\gamma' = \gamma - \gamma_w$)

Using these parameters, the ratio of the design moment to the computed moment may be determined from Figure 25.11.

Example 25.2

Determine the required depth of embedment and select an appropriate sheet pile section for the cantilever wall shown in Figure 25.12.

Solution

$$K_a = \tan^2 (45\text{-}33/2) = 0.295$$

$$K_p/F = \tan^2 (45\text{+}33/2) \,/\, 1.5 - 2.26$$

@ $z = 12$ ft

$\sigma = (121)(0.295)(12) = 428$ lb/ft^2

@ $z = 16$ ft

$\sigma = (121)(0.295)(16) = 571$ lb/ft^2

@ $z = 30$ ft

$\sigma = 571 + (124 - 62.4)(0.295)(14) = 826$ lb/ft^2

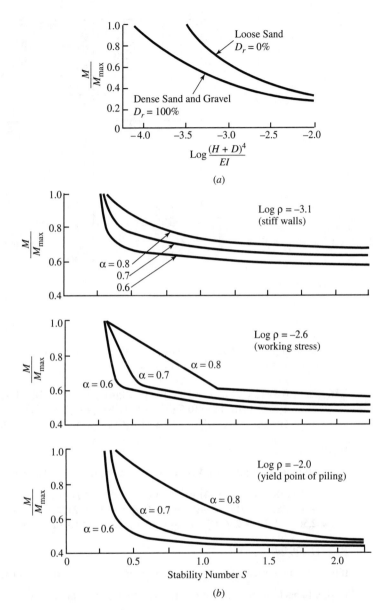

Figure 25.11   Rowe's moment reduction curves. (a) Sand, (b) clay.

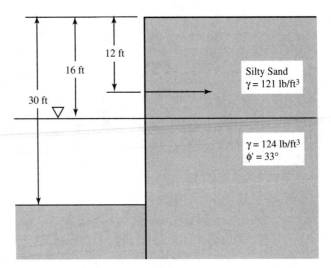

Figure 25.12   Sheet pile wall for Example 25.2.

@ $z > 30$ ft
 slope of pressure diagram = $(124 - 62.4)(2.26 - 0.295) = 121.0$ lb/ft^2/ft
 $\sigma = 0$ @ $z = 30 + 826/121.0 = 36.83$ ft

$\Sigma M_{tieback} = [(0.5)(12)(428)][(12)(0.333)] - [(428)(4)](2) - [(0.5)(4)(571\text{-}428)][(4)(2/3)]$
    $- [(571)(14)][11] - [(0.5)(826 - 571)(14)][4 + (14)(0.667)]$
    $- [(0.5)(826)(6.83)][18 + 6.83/3] + [(0.5)(121.0)D_1^2][24.83 + D_1/3] = 0$

$D_1 = 9.8$ ft

Use 40-ft long sheet pile with 10-ft embedment depth $\Leftarrow$ *Answer*

$\Sigma F_x = (0.5)(12)(428) + (428)(4) + (0.5)(4)(571\text{-}428) + (571)(14)$
   $+ (0.5)(826 - 571)(14) + (0.5)(826)(6.83) - (0.5)(121.0)(9.8)^2 - P_{tie} = 0$
   $= 2{,}568 + 1{,}712 + 286 + 7{,}994 + 1{,}785 + 2{,}821 - 5{,}810 - P_{tie} = 0$

$P_{tie} = 11{,}356$ lb/ft

If tiebacks are located 8 ft on center, load per anchor = $(11{,}356)(8) = 90{,}850$ lb

Design tieback anchors at 8 ft on center with 90,850 lb per anchor $\Leftarrow$ *Answer*

Note: Tieback anchors are often placed at a small downward angle. In that case, this value is the horizontal component of the tieback force.

The maximum moment occurs where the shear equals zero.

@ $z = 16$ ft $V/b = 2{,}568 + 1{,}712 + 286 - 11{,}356 = -6790$ lb/ft

$V/b = 0$ at depth z: $(z - 16)(571) + (0.5)(z - 16)^2(124 - 62.4)(0.295) = 6790$
$z_1 = 27.3$ ft

@ $z = 27.3$ ft

$M/b = (2568)(27.3 - 8) + (1,712)(27.3 - 14) + (286)(27.3 - 12 - 2.67)$
$\quad - (11,356)(27.3 - 12) + (571)(11.3)(11.3/2) + (0.5)(11.3^2)(121)(11.3/3)$
$\quad (124 - 62.4)(0.295)$
$\quad = 56,980$ ft-lb/ft

Per Table 25.1, an AZ 18 is required. However, with Rowe's reduction it might be possible to use an AZ 13.

$$I = (34{,}200 \text{ cm}^4/\text{m}) \left(\frac{1 \text{ in}}{2.54 \text{ cm}} \right)^4 \left(\frac{1 \text{ m}}{3.281 \text{ ft}} \right) = 250 \text{ in}^4/\text{ft}$$

$$\log \frac{(30 + 10)^4}{(29{,}000{,}000)(250)} = -3.5$$

Per Figure 25.11 = m/m_{max} = 0.8
M/b = 0.8(56,980) = 45,600 ft-lb/ft

Use AZ 13 sheet pile section with ASTM A328 steel   ⇐ *Answer*

SUMMARY

Major Points

1. Sheet pile walls are flat structural members that are driven into the ground. They may be made of a variety of materials, but most are made of steel.

2. The shapes and engineering properties (i.e., section modulus) of steel sheet pile sections vary from one manufacturer to another. Unlike H-piles, there are no widely accepted standards.

3. Although corrosion is a concern with steel sheet piles, it usually can be reduced to an acceptable level.

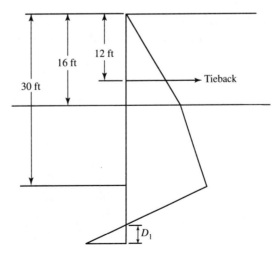

Figure 25.13   Pressure diagram for Example 25.2.

4. A cantilever sheet pile wall is one that does not have any additional supports. These walls rely entirely on their flexural strength and embedment into the ground to resist the applied loads. This design may be used only when the wall height is modest, and some lateral movement is acceptable.

5. The design of cantilever sheet pile walls must satisfy certain stability requirements. In addition, the designer must select a sheet pile section that has sufficient flexural strength.

6. A braced sheet pile wall is one that is partially supported by structural braces located inside the excavation. An anchored sheet pile wall is one that is partially supported by anchors that extend into the soil behind the wall. These designs are used for taller walls.

There are at least three methods of designing braced or anchored walls: the fixed-end method, the free-end method, and the beam on elastic foundation method. When using the free-end method, reduce the computed moment using Rowe's factors.

Vocabulary

Anchored sheet pile wall	Fixed-end method	Rowe's moment reduction
Braced sheet pile wall	Free-end method	Sheet pile wall
Cantilever sheet pile wall		

COMPREHENSIVE QUESTIONS AND PRACTICE PROBLEMS

25.1 Why are cantilever sheet piles practical only for short walls?

25.2 Redesign the cantilever sheet pile in Example 25.1, except increase the height of the wall from 20 ft to 25 ft. Comment on the impact this change has on the design.

25.3 Design a cantilever sheet pile wall for the cross-section shown in Figure 25.14. Your design should include the required depth of embedment and the required sheet pile section.

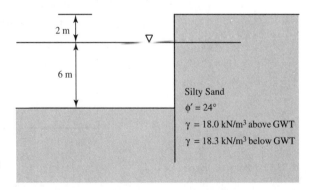

Figure 25.14 Proposed wall for Problems 25.3–25.5.

25.4 If the actual water level on the left side of the wall in Figure 25.14 was 1 m lower than shown, but the water level on the right side of the wall was as shown, would the stability of the wall be better or worse? Explain.

25.5 Redesign the wall described in Problem 25.3, but this time use a tieback anchor located 3 m from the top of the wall.

25.6 An 18-ft tall cantilever sheet-pile wall is to be built in a well-graded sand with $\phi' = 37°$ and $\gamma = 126$ lb/ft^3. There is no groundwater on either side of the wall. Determine the required depth of embedment and the required sheet pile section.

25.7 A 33-ft tall braced sheet pile wall is to be built in a poorly-graded silty sand with $\phi' = 32°$ and $\gamma = 121$ lb/ft^3. The brace will be located 8 ft from the top of the wall and will be 10 ft on center along the length of the wall. There is no groundwater on either side of the wall. Determine the required depth of embedment, the force per brace, and the required sheet pile section.

Appendix A

Unit Conversion Factors

ENGLISH UNITS

The following English units are commonly used in geotechnical engineering:

Table A1 COMMON ENGLISH UNITS

Unit	Measurement	Symbol
foot	distance	ft
inch	distance	in
pound	force or mass	lb
kip (kilopound)	force	k
ton	force or mass	t
second	time	s
pound per square foot	stress or pressure	lb/ft^2 or psf
pound per square inch	stress or pressure	lb/in^2 or psi
pound per cubic foot	unit weight	lb/ft^3 or pcf

SI AND METRIC UNITS

The following SI (Système International) units are commonly used in geotechnical engineering:

Table A2   COMMON SI UNITS

Unit	Measurement	Symbol
meter	distance	m
gram	mass	g
Newton	force	N
Pascal	stress or pressure	Pa
kilonewton per cubic meter	unit weight	kN/m^3
second	time	s

These units are often accompanied by the following prefixes:

Table A3   COMMON SI PREFIXES

Prefix	Symbol	Multiplier
milli	m	10^{-3}
centi	c	10^{-2}
kilo	k	10^3
mega	M	10^6

Some non-SI metric units also are used, especially in Europe. The most common example is the use of kg/cm^2 as a unit of stress.

CONVERSION FACTORS

The conversion factors in Tables B4 to B8 are useful for converting measurements between English, metric and SI units. Most of these factors are rounded to four significant figures. Those in **bold type** are absolute conversion factors (for example 12 inches = 1 ft). When units of force are equated to units of mass, the acceleration (F = ma) is presumed to be 9.807 m/s^2 (32.17 ft/s^2), which is the acceleration due to gravity on the earth's surface.

There are at least three definitions for the word "ton": the 2000 lb short ton (commonly used in the United States and Canada), the 2240 lb long ton (used in Great Britain), and the 1000 kg (2205 lb) metric ton (also known as a tonne).

A useful approximate conversion factor: 1 short ton/ft^2 ≈ 1 kg/cm^2 ≈ 100 kPa ≈ 1 atmosphere. These are true to within 2 to 4 percent.

Table A4   UNITS OF DISTANCE

To Convert	To	Multiply by
ft	in	**12**
ft	m	0.3048
in	ft	**1/12**
in	mm	25.40
m	ft	3.281
mm	in	0.03937

Table A5   UNITS OF FORCE

To Convert	To	Multiply by
k	kN	4.448
k	lb	**1000**
kg_f	lb	2.205
kg_f	N	9.807
kg_f	ton (metric)	**0.001**
kN	k	0.2248
lb	k	**0.001**
lb	kg_f	0.4536
lb	N	4.448
lb	ton (short)	**1/2000**
lb	ton (long)	**1/2240**
N	kg_f	0.1020
N	lb	0.2248
ton (short)	lb	**2000**
ton (long)	lb	**2240**
ton (metric)	kg_f	**1000**

Table A6   UNITS OF VOLUME

To Convert	To	Multiply by
ft^3	gal	7.481
gal	ft^3	0.1337

Table A7   UNITS OF STRESS
AND PRESSURE

To Convert	To	Multiply by
atmosphere	lb/ft^2	2117
atmosphere	kPa	101.3
bar	kPa	**100**
kg$_f$/cm^2	kPa	98.07
kg$_f$/cm^2	lb/ft^2	2048
kPa	atmosphere	0.009869
kPa	bar	**0.01**
kPa	kg$_f$/cm^2	0.01020
kPa	lb/ft^2	20.89
kPa	lb/in^2	0.1450
kPa	metric ton/m^2	0.1020
lb/ft^2	atmosphere	4.725×10^{-4}
lb/ft^2	kPa	0.04787
lb/ft^2	lb/in^2	**1/144**
lb/in^2	kPa	6.895
lb/in^2	lb/ft^2	**144**
lb/in^2	MPa	6.895×10^{-3}
metric ton/m^2	kPa	9.807
MPa	lb/in^2	145.0

Table A8   UNITS OF UNIT WEIGHT

To Convert	To	Multiply by
kN/m^3	lb/ft^3	6.366
kN/m^3	metric ton/m^3	0.1020
kN/m^3	Mg$_f$/m^3	0.1020
lb/ft^3	kN/m^3	0.1571
metric ton/m^3	kN/m^3	9.807
Mg$_f$/m3	kN/m^3	9.807

Appendix B

Computer Software

The author has developed the following four Microsoft Excel spreadsheets specifically for this book:

BEARING.XLS	Bearing capacity analysis of shallow foundations
SETTLEMENT.XLS	Settlement analysis of shallow foundations based on laboratory test results
SCHMERTMANN.XLS	Settlement analysis of shallow foundations using Schmertmann's method
RETWALL.XLS	External stability analysis of cantilever retaining walls

DOWNLOADING AND INSTALLATION INSTRUCTIONS

To download the software, use any web browser and the internet to log onto the following address:

http://www.prenhall.com/coduto

Follow the links to the *Foundation Design: Principles and Practices* software download page, then follow the on-screen instructions.

USING THE SOFTWARE

The software requires Microsoft Excel or another spreadsheet program that reads Microsoft Excel spreadsheet files.

References

Abbreviations for societies and organizations:

ACI   American Concrete Institute

ADSC   Association of Drilled Shaft Contractors

AISC   American Institute of Steel Construction

API   American Petroleum Institute

ASCE   American Society of Civil Engineers

ASTM   American Society for Testing and Materials

FHWA   Federal Highway Administration

PCA   Portland Cement Association

PCI   Precast/Prestressed Concrete Institute

PTI   Post-Tensioning Institute

AASHTO (1996), *Standard Specifications for Highway Bridges*, 16th ed., American Association of State Highway and Transportation Officials, Washington, D.C.

ACI (1974), "Recommendations for Design, Manufacture, and Installation of Concrete Piles," ACI 543R-74 (Reapproved 1980), ACI Journal *Proceedings*. Vol. 71, No. 10, p. 477–492; also printed in ACI *Manual of Concrete Practice*, American Concrete Institute

ACI (1980), *Recommendations for Design, Manufacture and Installation of Concrete Piles*, ACI 543R-74, American Concrete Institute, Detroit

ACI (1993), *Suggested Analysis and Design Procedures for Combined Footings and Mats*, ACI 336.2R-88 (Reapproved 1993), American Concrete Institute

ACI (1999), *Building Code Requirements for Structural Concrete (ACI 318-99)*, American Concrete Institute

ACI-ASCE (1962), "Report of Committee on Shear and Diagonal Tension," *Proceedings, American Concrete Institute*, Vol. 59, No. 1

AIRHART, T.P., COYLE, H.M., HIRSH, T.J. AND BUCHANAN, S.J. (1969), "Pile-Soil System Response in a Cohesive Soil," *Performance of Deep Foundations*, STP 444, p. 264–294, ASTM, Philadelphia

AISC (1989), *Manual of Steel Construction, Allowable Stress Design*, 9th ed., American Institute of Steel Construction, Chicago

AISC (1995), *Manual of Steel Construction, Load & Resistance Factor Design*, 2nd ed., American Institute of Steel Construction

ANAGNOSTOPOULOS, A.G. (1990), "Die Zusammendrückbarkeit nichtbindiger Böden (The Compressibility of Cohesionless Soils)," *Geotechnik*, Vol. 13, p. 181–187 (In German)

ANDERSON, J.N. AND LADE, P.V. (1981), "The Expansion Index Test," *ASTM Geotechnical Testing Journal*, Vol. 4, No. 2, June 1981, p. 58–67

ANDERSON, L.R.—see Moss, R.E.S.

API (1996), *Planning, Designing and Constructing Fixed Offshore Platforms-Working Stress Design*, 20th ed., American Petroleum Institute

API (1997), *Planning, Designing and Constructing Fixed Offshore-Platforms Load and Resistance Factor Design*, American Petroleum Institute

ARMSTRONG, R.M. (1978), "Structural Properties of Timber Piles," *Behavior of Deep Foundations*, ASTM STP 670, Raymond Lundgren, Ed., p. 118–152, ASTM, Philadelphia

ASCE (1972), "Subsurface Investigation for Design and Construction of Foundations of Buildings," ASCE *Journal of the Soil Mechanics and Foundations Division*, Vol. 98 No. SM5, SM6, SM7 and SM8

ASCE (1984), *Practical Guidelines for the Selection, Design and Installation of Piles*, ASCE

ASCE (1996a), *Minimum Design Loads for Buildings and Other Structures*, ANSI/ASCE 7-95, ASCE

ASCE (1996b), *Design of Sheet Pile Walls*, American Society of Civil Engineers

AVALLE, D.L.—see Mitchell, P.W.

AZAM, IR. TARIQUE—see Baker, Clyde N.

BACHUS, ROBERT C.—see Clough, G. Wayne

BAGUELIN, F., JÉZÉQUEL, J.F. AND SHIELDS, D.H. (1978), *The Pressuremeter and Foundation Engineering*, Trans Tech, Clausthal, Germany

BAKER, CLYDE N.—see Christopher, Barry R.

BAKER, CLYDE N., AZAM, IR. TARIQUE AND JOSEPH, LEN S. (1994), "Settlement Analysis for 450 Meter Tall KLCC Towers," *Vertical and Horizontal Deformations of Foundations and Embankments*, A.T. Yeung and G.Y. Felio, Eds., Vol. 2, p. 1650–1671, ASCE

BARDEN, L., McGOWN, A. AND COLLINS, K. (1973), "The Collapse Mechanism in Partly Saturated Soil," *Engineering Geology*, Vol. 7, p. 49–60

BARKER, R.M., DUNCAN, J.M., ROJIANI, K.B., OOI, P.S.K., TAN, C.K. AND KIM, S.G. (1991), *Manuals for the Design of Bridge Foundations*, National Cooperative Highway Research Program Report 343, Transportation Research Board, Washington, DC

BARTLETT, S.F. AND YOUD, T.L. (1992) *Empirical Analysis of Horizontal Ground Displacement Generated by Liquefaction–Induced Lateral Spreads*, Technical Report No. NCEER–92–0021, National Center for Earthquake Engineering Research, Buffalo, NY

BASMA, ADNAN A. AND TUNCER, ERDIL R. (1992), "Evaluation and Control of Collapsible Soils," ASCE *Journal of Geotechnical Engineering*, Vol. 118, No. 10, p. 1491–1504

BAUMANN, FREDERICK (1873), *The Art of Preparing Foundations, with Particular Illustration of the "Method of Isolated Piers" as Followed in Chicago*, Reprinted in Powell (1884)

BEECH, J.F.—see Kulhawy, F.H.

BELL, A.L. (1915), "The Lateral Pressure and Resistance of Clay, and the Supporting Power of Clay Foundations," p. 93–134 in *A Century of Soil Mechanics*, ICE, London

BERINGEN, F.L.—see DeRuiter, J.

BETHLEHEM STEEL CORP. (1979), *Bethlehem Steel H-Piles*, Bethlehem Steel Corp., Bethlehem, PA

BHUSHAN, KUL (1982), Discussion of "New Design Correlations for Piles in Sand," *Journal of the Geotechnical Engineering Division*, Vol. 108, No. GT11, p. 1508–1501, ASCE

BISHOP, ALAN W. AND BJERRUM, LAURITS (1960), "The Relevance of the Triaxial Test to the Solution of Stability Problems," *Research Conference on Shear Strength of Cohesive Soils*, ASCE

BJERRUM, L.—see Janbu, M.

BJERRUM, L.—see Skempton, A.W.

BJERRUM, LAURITS—see Bishop, Alan W.

BJERRUM, LAURITS (1963), "Allowable Settlement of Structures," *Proceedings, 3rd European Conference on Soil Mechanics and Foundation Engineering*, Vol. 2, p. 135–137, Weisbaden

BJERRUM, L., JOHANNESSEN, I.J. AND EIDE, O. (1969), "Reduction of Negative Skin Friction on Steel Piles to Rock," *Proceedings, Seventh International Conference on Soil Mechanics and Foundation Engineering*, Mexico City, Vol. 2, p. 27–34

BLACK, WILLIAM T.—see Luscher, Ulrich

BLASER, HAROLD D. AND SCHERER, OSCAR J. (1969), "Expansion of Soils Containing Sodium Sulfate Caused by Drop in Ambient Temperatures," *Effects of Temperature and Heat on Engineering Behavior of Soils*, Special Report 103, Highway Research Board, Washington, D.C.

BOCA (1996), *National Building Code*, Building Officials & Code Administrators International, Inc., Country Club Hills, IL

BOLENSKI, M. (1973), "Osiadania nowo wznoszonych budowli w zaleznosci of podloza gruntowego: Wyniki 20-letnich Badan w Instytucie Techniki Budowlanej" (Settlement of Constructions Newly Erected and Type of Subsoil: The Results of 20 Years Studies Carried Out in the Building Research Institute), Prace Instytutu Techniki Budowlanej, Warszawa (in Polish; partial results quoted in Burland and Burbidge, 1985)

BOLIN, H.W. (1941), "The Pile Efficiency Formula of the Uniform Building Code," *Building Standards Monthly*, Vol. 10, No. 1, p. 4–5

BOLLMAN, H.T.—see Brown, D.A.

BOUSSINESQ, M.J. (1885), *Application Des Potentiels, à l'Étude de l'Équilibre et du Movvement Des Solides Elastiques*, Gauthier-Villars, Paris (in French)

BOZOZUK, M. (1972a), "Downdrag Measurement on a 160-ft Floating Pipe Test Pile in Marine Clay," *Canadian Geotechnical Journal*, Vol. 9, No. 2, p. 127–136

BOZOZUK, M. (1972b), "Foundation Failure of the VanKleek Hill Tower Silo," *Performance of Earth and Earth-Supported Structures*, p. 885–902, ASCE

BOZOZUK, M. (1978), "Bridge Foundations Move," *Transportation Research Record 678*, p. 17–21, Transportation Research Board, Washington, D.C.

BOZOZUK, M. (1981), "Bearing Capacity of Pile Preloaded by Downdrag," *Proceedings, Tenth International Conference on Soil Mechanics and Foundation Engineering*, Stockholm, Vol. 2, p. 631–636

BRANDON, THOMAS L., DUNCAN, J. MICHAEL AND GARDNER, WILLIAM S. (1990), "Hydrocompression Settlement of Deep Fills," ASCE *Journal of Geotechnical Engineering*, Vol. 116, No. 10, p. 1536–1548

BRAUN, J.S.—see Thorson, Bruce M.

BRIAUD, J.L., TUCKER, L., LYTTON, R.L. AND COYLE, H.M. (1985), *Behavior of Piles and Pile Groups in Cohesionless Soils*, Report No. FHWA/RD-83/038, Federal Highway Administration, Washington, D.C.

BRIAUD, JEAN-LOUIS AND MIRAN, JEROME (1991), *The Cone Penetration Test*, Report No. FHWA-TA-91-004, Federal Highway Administration, McLean, VA

BRIAUD, JEAN-LOUIS (1991), "Dynamic and Static Testing of Nine Drilled Shafts at Texas A&M University Geotechnical Research Sites," *Foundation Drilling*, Vol. 30, No. 7, p. 12–14

BRIAUD, JEAN-LOUIS (1992), *The Pressuremeter*, A.A. Balkema, Rotterdam

BRIAUD, JEAN-LOUIS AND GIBBENS, ROBERT M. (1994), *Predicted and Measured Behavior of Five Spread Footings on Sand*, Geotechnical Special Publication 41, ASCE

BRIAUD, J.L.—see DiMillio, A.F.

BRIEKE, WERNER—see Massarsch, K.

BRINCH HANSEN, J. (1961a), "The Ultimate Resistance of Rigid Piles Against Transversal Forces," *Bulletin No. 12*, Danish Geotechnical Institute, Copenhagen

BRINCH HANSEN, J. (1961b), "A General Formula for Bearing Capacity," *Bulletin No. 11*, Danish Geotechnical Institute, Copenhagen

BRINCH HANSEN, J. (1970), "A Revised and Extended Formula for Bearing Capacity," *Bulletin No. 28*, Danish Geotechnical Institute, Copenhagen

BRINK, A.B.A. AND KANTEY, B.A. (1961), "Collapsible Grain Structure in Residual Granite Soils in South Africa," *Proceedings, Fifth International Conference on Soil Mechanics and Foundation Engineering*, p. 611–614

BRISSETTE, R.F.—see D'Appolonia, D.J.

BROMS, BENGT B. (1964a), "Lateral Resistance of Piles in Cohesive Soils," ASCE *Journal of the Soil Mechanics and Foundations Division*, Vol. 90, No. SM2, p. 27–63

BROMS, BENGT B. (1964b), "Lateral Resistance of Piles in Cohesionless Soils," ASCE *Journal of the Soil Mechanics and Foundations Division*, Vol. 90, No. SM3, p. 123–156

BROMS, BENGT B. (1965), "Design of Laterally Loaded Piles," ASCE *Journal of the Soil Mechanics and Foundations Division*, Vol. 91, No. SM3, p. 79–99

BROMS, BENGT B. (1972), "Stability of Flexible Structures (Piles and Pile Groups)," *Proceedings, 5th European Conference on Soil Mechanics and Foundation Engineering*, Madrid, Vol. 2, p. 239–269

BROMS, BENGT B. (1981), *Precast Piling Practice*, Thomas Telford, London

BRONS, K.F. AND KOOL, A.F. (1988), "Methods to Improve the Quality of Auger Piles," *Deep Foundations on Bored and Auger Piles*, p. 269–272, W.F. Van Impe, Ed., Balkema, Rotterdam

BROWN, D.A., MORRISON, C. AND REESE, L.C. (1988), "Lateral Load Behavior of Pile Group in Sand, *Journal of Geotechnical Engineering*, Vol. 114, No. 11, p. 1261–1276

BROWN, D.A. AND BOLLMAN, H.T. (1993), "Pile-Supported Bridge Foundations Designed for Impact Loading," Appended document to *Proceedings of Design of Highway Bridges for Extreme Events*, Crystal City, VA, p. 265–281

BROWN, PHILLIP R.—see Schmertmann, John H.

BROWN, P.T.—see Robertson, P.K.

BROWN, RALPH E. AND GLENN, ANDREW J. (1976), "Vibroflotation and Terra-Probe Comparison," ASCE *Journal of the Geotechnical Engineering Division*, Vol. 102, No. GT10, p. 1059–1072

BRUMUND, WILLIAM F., JONAS, ERNEST AND LADD, CHARLES C. (1976), "Estimating In-Situ Maximum Past (Preconsolidation) Pressure of Saturated Clays from Results of Laboratory Consolidometer Tests," *Estimation of Consolidation Settlement*, Special Report 163, p. 4–12, Transportation Research Board, Washington, DC

BRYANT, F.G.—see Peck, R.B.

BUCHANAN, S.J.—see Airhart, T.P.

BUCHIGNANI, A.L.—see Duncan, J.M.

BUDGE, W.D., ET AL. (1964), *A Review of Literature on Swelling Soils*, Colorado Department of Highways, Denver, CO

BULLET, P. (1691), *L'Architecture Pratique*, Paris (in French)

BURBIDGE, M.C.—see Burland, J.B.

BURDETTE, E.G.—see Marsh, M.L.

BURLAND, JOHN (1973), "Shaft Friction of Piles in Clay—A Simple Fundamental Approach," *Ground Engineering*, Vol. 6, No. 3, p. 30–42

BURLAND, J.B. AND WROTH, C.P. (1974), "Allowable and Differential Settlement of Structures, Including Damage and Soil-Structure Interaction," *Conference on Settlement of Structures*, p. 611–654, Pentech Press, Cambridge

BURLAND, J.B. AND BURBIDGE, M.C. (1985), "Settlement of Foundations on Sand and Gravel," *Proceedings, Institution of Civil Engineers*, Part 1, Vol. 78, p. 1325–1381

BURMISTER, D.M. (1962), "Physical, Stress-Strain, and Strength Response of Granular Soils," *Symposium on Field Testing of Soils*, STP 322, p. 67–97, ASTM

BURN, KENNETH N.—see Crawford, Carl B.

BUSTAMANTE, M. AND GIANESELLI, L. (1982), "Pile Bearing Capacity Prediction by Means of Static Penetrometer CPT," *Second European Symposium on Penetration Testing (ESOPT II)*, A. Verruijt et al., Ed., Vol. 2, p. 493–500

BUTLER, F.G. (1975), "Heavily Over-Consolidated Clays," p. 531ff in *Settlement of Structures*, Halstead Press

BYRN, JAMES E. (1991), "Expansive Soils: The Effect of Changing Soil Moisture Content on Residential and Light Commercial Structures," *Journal of the National Academy of Forensic Engineers*, Vol. 8, No. 2, p. 67–84

CALIENDO, J.A.—see Moss, R.E.S.

CALIENDO, JOSEPH A.—see Tawfiq, Kamal S.

CAMPANELLA, R.G.—see Robertson, P.K.

CAPANO, C.—see Kulhawy, F.H.

CAPPER, R.—see McVay, M.

CARRIER, W. DAVID—see Christian, John T.

CCC (1995), The *National Building Code of Canada*, Canadian Codes Center, Ottawa

CGS (1992), *Canadian Foundation Engineering Manual*, 3rd ed., Canadian Geotechnical Society, BiTech, Vancouver

CHAN, C.K.—see Seed, H.B.

CHANDLER, NEIL—see Shields, Donald

CHELLIS, ROBERT D. (1961), *Pile Foundations*, 2nd ed., McGraw Hill, New York

CHELLIS, R.D. (1962), "Pile Foundations," p. 633–768 in *Foundation Engineering*, G.A. Leonards, editor, McGraw-Hill, New York

CHEN, FU HUA (1988), *Foundations on Expansive Soils*, 2nd ed., Developments in Geotechnical Engineering Vol. 54, Elsevier, Amsterdam

CHRISTIAN, JOHN T. AND CARRIER, W. DAVID III (1978), "Janbu, Bjerrum and Kjaernsli's Chart Reinterpreted," *Canadian Geotechnical Journal*, Vol. 15, p. 123–128

CHRISTOPHER, BARRY R., BAKER, CLYDE N. AND WELLINGTON, DENNIS L. (1989), "Geophysical and Nuclear Methods for Non-Destructive Evaluation of Caissons," *Foundation Engineering: Current Principles and Practices*, ASCE

CHUNG, RILEY M.—see Seed, H. Bolton

CIBOR, JOSEPH M. (1983), "Geotechnical Considerations of Las Vegas Valley," *Special Publication on Geological Environment and Soil Properties*, ASCE Annual Convention, Houston, TX, p. 351–373

CLAYTON, C.R.I. (1990), "SPT Energy Transmission: Theory, Measurement and Significance," *Ground Engineering*, Vol. 23, No. 10, p. 35–43

CLAYTON, R.J.—see Rollins, K.M.

CLEMENCE, SAMUEL P. AND FINBARR, ALBERT O. (1981), "Design Considerations for Collapsible Soils," ASCE *Journal of the Geotechnical Engineering Division*, Vol. 107, No. GT3, p. 305–317

CLEMENCE, SAMUEL P., ED. (1985), *Uplift Behavior of Anchor Foundations in Soil*, ASCE

CLOUGH, G. WAYNE, SITAR, NICHOLAS , BACHUS, ROBERT C., AND RAD, NADER SHAFII (1981), "Cemented Sands Under Static Loading," *Journal of the Geotechnical Engineering Division*, Vol. 107, No. GT3, pp. 305–317, ASCE

CLOUGH, G. WAYNE—see Duncan, J. Michael

COBURN, SEYMOUR K.—see Dismuke, Thomas D.

CODUTO, DONALD P. (1994), *Foundation Design: Principles and Practices*, 1st Ed, Prentice Hall, Upper Saddle River, NJ

CODUTO, DONALD P. (1999), *Geotechnical Engineering: Principles and Practices*, Prentice Hall, Upper Saddle River, NJ

COLLINS, K.—see Barden, L.

CORNELL, C.A. (1969), "A Probability-Based Structural Code," *Journal of the American Concrete Institute*, Vol. 66, No. 12

COSTANZO, D., JAMIOLKOWSKI, M., LANCELLOTTA, R. AND PEPE, M.C. (1994), *Leaning Tower of Pisa: Description and Behaviour*, Settlement 94 Banquet Lecture, Texas A&M University

COULOMB, C.A. (1776) "Essai sur une application des règles de maximis et minimis à quelques problèmes de statique relatifs à l'architecture," *Mémoires de mathématique et de physique présentés à l'Académie Royale des Sciences*, Paris, Vol. 7, p. 343–382 (in French)

COYLE, H.M—see Airhart, T.P.

COYLE, HARRY M.—see Bruner, Robert F.

CRAVEN, ROBERT F.—see Lankford, William T.

CRAWFORD, CARL B. AND BURN, KENNETH N. (1968), "Building Damage from Expansive Steel Slag Backfill," *Placement and Improvement of Soil to Support Structures*, p. 235–261, ASCE

CRAWFORD, C.B. AND JOHNSON, G.H. (1971), "Construction on Permafrost," *Canadian Geotechnical Journal*, Vol. 8, No. 2, p. 236–251

CROWTHER, CARROLL L. (1988), *Load Testing of Deep Foundations*, John Wiley and Sons, New York

CRSI (1992), *CRSI Handbook*, Concrete Reinforcing Steel Institute, Schaumburg, IL

CUMMINGS, A.E. (1940), "Dynamic Pile Driving Formulas," *Journal of the Boston Society of Civil Engineers*, Vol. 27, No. 1, p. 6–27

CURTIN, GEORGE (1973), "Collapsing Soil and Subsidence" in *Geology, Seismicity and Environmental Impact*, Special Publication, Association of Engineering Geologists, Douglas E. Moran et al., ed., University Publishers, Los Angeles

D'APPOLONIA, D.J., D'APPOLONIA, E.D. AND BRISSETTE, R.F. (1968), "Settlement of Spread Footings on Sand," ASCE *Journal of the Soil Mechanics and Foundations Division*, Vol. 94, No. SM3, p. 735–759

D'APPOLONIA, DAVID J., POULOS, HARRY G. AND LADD, CHARLES C. (1971), "Initial Settlement of Structures on Clay," ASCE *Journal of the Soil Mechanics and Foundations Division*, Vol. 97, No. SM10, p. 1359–1377

D'APPOLONIA, E.D.—see D'Appolonia, D.J.

DAVIES, MICHAEL P.—see Robertson, Peter K.

DAVIS, E.H. AND POULOS, H.G. (1968), "The Use of Elastic Theory for Settlement Prediction Under Three-Dimensional Conditions," *Géotechnique*, Vol. 18, p. 67–91

DAVIS, E.H.—see Poulos, H.G.

DAVIS, STANLEY R.—see Richardson, E.V.

DAVISSON, M.T. (1970), "Lateral Load Capacity of Piles," *Highway Research Record No. 333*, p. 104–112, Highway Research Board, Washington, D.C.

DAVISSON, M.T. (1973), "High Capacity Piles" in *Innovations in Foundation Construction*, proceedings of a lecture series, Illinois Section ASCE, Chicago

DAVISSON, M.T. (1979), "Stresses in Piles," *Behavior of Deep Foundations*, ASTM STP 670, Raymond Lundgren, Ed., p. 64–83, ASTM, Philadelphia

DAVISSON, M.T. (1989), "Driven Piles," *Foundations in Difficult Soils: State of Practice*, Foundations and Soil Mechanics Group, Metropolitan Section, ASCE, New York

DEBEER, E.E. AND LADANYI, B. (1961), "Experimental Study of the Bearing Capacity of Sand Under Circular Foundations Resting on the Surface," *Proceedings, 5th International Conference on Soil Mechanics and Foundation Engineering*, Vol. 1, p. 577–585, Paris

DEERE, DON U.—see Nordlund, Reymond L.

DELOS-SANTOS, E.O.—see Shepherd, R.

DEMELLO, V. (1971), "The Standard Penetration Test—A State-of-the-Art Report," *Fourth PanAmerican Conference on Soil Mechanics and Foundation Engineering*, Vol. 1, p. 1–86

DEMPSEY, J.P. AND LI, H. (1989), "A Rigid Rectangular Footing on an Elastic Layer," *Géotechnique*, Vol. 39, No. 1, p. 147–152

DERUITER, J. AND BERINGEN, F.L. (1979), "Pile Foundations for Large North Sea Structures," *Marine Geotechnology*, Vol. 3, No. 3, p. 267–314

DERUITER, J. (1981), "Current Penetrometer Practice," *Cone Penetration Testing and Experience*, p. 1–48, ASCE

DEWOLF, JOHN T. AND RICKER, DAVID T. (1990), *Column Base Plates*, American Institute of Steel Construction, Chicago, IL

DFI (1990), *Auger Cast-in-Place Piles Manual*, Deep Foundations Institute, Sparta, NJ

DFI (1998), *Driven Sheet Piling*, Deep Foundations Institute, Englewood Cliffs, NJ

DICKEY, WALTER L.—see Schneider, Robert R.

DICKIN, E.A. AND LEUNG, C.F. (1990), "Performance of Piles with Enlarged Bases Subject to Uplift Forces," *Canadian Geotechnical Journal*, Vol. 27, p. 546–556

DIMILLIO, A.F., NG, E.S., BRIAUD, J.L., O'NEILL, M.W., ET AL. (1987a), *Pile Group Prediction Symposium: Summary Volume I: Sandy Soil*, Report No. FHWA-TS-87-221, Federal Highway Administration, McLean, VA

DIMILLIO, A.F., O'NEILL, M.W., HAWKINS, R.A., VGAZ, O.G. AND VESIĆ, A.S. (1987b), *Pile Group Prediction Symposium: Summary Volume II: Clay Soil*, Report No. FHWA-TS-87-222, Federal Highway Administration, McLean, VA

DISMUKE, THOMAS D., COBURN, SEYMOUR K. AND HIRSCH, CARL M. (1981), *Handbook of Corrosion Protection for Steel Pile Structures in Marine Environments*, American Iron and Steel Institute, Washington, D.C.

DONALDSON, G.W. (1973), "The Prediction of Differential Movement on Expansive Soils," *Proceedings of the Third International Conference on Expansive Soils*, Vol. 1, p. 289–293, Jerusalem Academic Press

DOWRICK, DAVID J. (1987), *Earthquake Resistant Design: A Manual for Engineers and Architects*, 2nd ed., John Wiley, New York

DUDLEY, JOHN H. (1970), "Review of Collapsing Soils," ASCE *Journal of the Soil Mechanics and Foundations Division*, Vol. 96, No. SM3, p. 925–947

DUNCAN, J.M. AND BUCHIGNANI, A.L. (1976), *An Engineering Manual for Settlement Studies*, Department of Civil Engineering, University of California, Berkeley

DUNCAN, J. MICHAEL, CLOUGH, G. WAYNE AND EBELING, ROBERT M. (1990), "Behavior and Design of Gravity Earth Retaining Structures," *Design and Performance of Earth Retaining Structures*, Philip C. Lambe and Lawrence A. Hansen, Ed., p. 251–277, ASCE

DUNCAN, J.M., WILLIAMS, G.W., SEHN, A.L. AND SEED, R.B. (1991), "Estimation Earth Pressures due to Compaction," ASCE *Journal of Geotechnical Engineering*, Vol. 117, No. 12, p. 1833–1847 (errata in Vol. 118, No. 3, p. 519)

DUNCAN, J.M.—see Barker, R.M.

DUNCAN, J. MICHAEL—see Brandon, Thomas L.

DUNCAN, J.M.—see Evans, L.T.

EBELING, ROBERT M.—see Duncan, J. Michael

EDINGER, PETER H.—see Krinitzsky, Ellis L.

EDWARDS, T.C. (1967), *Piling Analysis Wave Equation Computer Program Utilization Manual*, Texas Transportation Institute Research Report 33–11, Texas A&M University, College Station, TX

EIDE, O.—see Bjerrum, L.

EISENSTEIN, Z.—see Morgenstern, N.R.

EL-EHWANY, MOSTAFA—see Houston, Sandra L.

ELSON, W.K.—see Fleming, W.G.K.

ESLAMI, ABOLFAZI AND FELLENIUS, BENGT H. (1997), "Pile Capacity by Direct CPT and CPTu Methods Applied to 102 Case Histories," *Canadian Geotechnical Journal*, Vol. 34, p. 886–904

EVANS, L.T. AND DUNCAN, J.M. (1982), *Simplified Analysis of Laterally Loaded Piles*, Report No. UCB/GT/82-04, Department of Civil Engineering, University of California, Berkeley

FAKHROO, M.—see Tumay, M.T.

FARLEY, J. (1827), *A Treatise on the Steam Engine*, London (quoted in Golder, 1975)

FELLENIUS, BENGT H. (1980), "The Analysis of Results from Routine Pile Load Tests," *Ground Engineering*, Vol. 13, No. 6, p. 19–31

FELLENIUS, BENGT H., RIKER, RICHARD E., O'BRIEN, ARTHUR J. AND TRACY, GERALD R. (1989), "Dynamic and Static Testing in Soil Exhibiting Set-Up," ASCE *Journal of Geotechnical Engineering*, Vol. 115, No. 7, p. 984–1001

FELLENIUS, BENGT H. (1990), *Guidelines for the Interpretation and Analysis of the Static Loading Test*, Deep Foundations Institute, Sparta, NJ

FELLENIUS, BENGT H. (1996), *Basics of Foundation Design,* BiTech, Richmond,

FELLENIUS, BENGT H. (1998), "Recent Advances in the Design of Piles for Axial Loads, Dragloads, Downdrag, and Settlement," *ASCE and Port of NY and NJ Seminar* (also available from www.unisoftltd.com)

FELLENIUS, BENGT H. (1999), *Basics of Foundation Design*, 2nd ed., BiTech, Richmond, BC

FELLENIUS, BENGT H.—see Eslami, Abolfazi

FENG, TAO-WEI—see Mesri, Gholamreza

FEWELL, RICHARD—see Spanovich, Milan

FINBARR, ALBERT O.—see Clemence, Samuel P.

FINNO, R.J.—see Holloway, D.M.

FLAATE, KARRE S.—see Olson, Roy E.

FLEMING, W.G.K., WELTMAN, A.J., RANDOLPH, M.F. AND ELSON, W.K. (1985), *Piling Engineering*, John Wiley & Sons, New York

FOCHT, JOHN A.—see McClelland, Bramlette

FOOTT, ROGER AND LADD, CHARLES C. (1981), "Undrained Settlement of Plastic and Organic Clays," ASCE *Journal of the Geotechnical Engineering Division*, Vol. 107, No. GT8, p. 1079–1094

FRAGASZY, RICHARD J.—see Lawton, Evert C.

FRANK, R. (1994), "Some Recent Developments on the Behavior of Shallow Foundations," *Deformations of Soils and Displacements of Structures*, Proceedings, Tenth European Conference on Soil Mechanics and Foundation Engineering, p. 1115–1140, A.A. Balkema, Rotterdam

FREDLUND, D.G. AND RAHARDJO, H. (1993), *Soil Mechanics for Unsaturated Soils*, John Wiley, New York

FROST J.D.—see Leonards, G.A.

GARDNER, WILLIAM S.—see Brandon, Thomas L.

GARDNER, WILLIAM S.—see Greer, David M.

GARNIER, JACQUES – see Shields, Donald

GHAZZALY, OSMAN L.—see Viyayvergiya, V.N.

GIANESELLI, L.—see Bustamante, M.

GIBBENS, ROBERT M.—see Briaud, Jean-Louis

GIBBS, HAROLD J. (1969), Discussion of Holtz, W. G. "The Engineering Problems of Expansive Clay Subsoils," *Proceedings of the Second International Research and Engineering Conference on Expansive Clay Soils*, p. 478–479, Texas A&M Press, College Station, TX

GLENN, ANDREW J.—see Brown, Ralph E.

GOBLE, G.G. (1986), "Modern Procedures for the Design of Driven Pile Foundations," Short Course on Behavior of Deep Foundations Under Axial and Lateral Loading, University of Texas, Austin

GOBLE, G.G.—see Hannigan, P.J.

GOBLE, G.G.—see Rausche, F.

GOLDER, H.Q. (1975), "Floating Foundations," Ch. 18 in *Foundation Engineering Handbook*, H.F. Winterkorn and H. Fang, Ed., Van Nostrand Reinhold, New York

GOODMAN, RICHARD E.—see Kulhawy, Fred H.

GOULD, JAMES P. (1970), "Lateral Pressures on Rigid Permanent Structures" in *Lateral Stresses in the Ground and the Design of Earth-Retaining Structures*, p. 219–270, ASCE

GOULD, JAMES P.—see Krinitzsky, Ellis L.

GRAHAM, JAMES (1985), "Allowable Compressive Design Stresses for Pressure-Treated Round Timber Foundation Piling," American Wood Preservers Association Proceedings, Vol. 81

GRANT, R. ET AL. (1974), "Differential Settlement of Buildings," ASCE *Journal of the Geotechnical Engineering Division*, Vol. 100, No. GT9, p. 973–991

GREEN, R.K.—see Holloway, D.M.

GREENFIELD, STEVEN J. AND SHEN, C.K. (1992), *Foundations in Problem Soils*, Prentice Hall, Englewood Cliffs, NJ

GREER, DAVID M. AND GARDNER, WILLIAM S. (1986), *Construction of Drilled Pier Foundations*, John Wiley, New York

GRIGORIU, M.D.—see Phoon, K.K.

GROMKO, GERALD J. (1974), "Review of Expansive Soils," ASCE *Journal of the Geotechnical Engineering Division*, Vol. 100, No. GT6, p. 667–687

HAIN, STEPHEN J. AND LEE, IAN K. (1974), "Rational Analysis of Raft Foundation," ASCE *Journal of the Geotechnical Engineering Division*, Vol. 100, No. GT7, p. 843–860

HANDFORD, S.A. (1982), *The Conquest of Gaul*, revised by Jane F. Gardner, Penguin Classics

HANDY, RICHARD L. (1980), "Realism in Site Exploration: Past, Present, Future, and Then Some—All Inclusive," *Site Exploration on Soft Ground Using In-Situ Techniques*, p. 239–248, Report No. FHWA-TS-80-202, Federal Highway Administration, Washington, D.C.

HANINGER, E.R.—see Shipp, J.G.

HANNIGAN, P.J., GOBLE, G.G., THENDEAN, G., LIKINS, G.E. AND RAUSCHE, F. (1997), *Design and Construction of Driven Pile Foundations*, Report No. FHWA-HI-97-013, Federal Highway Administration, Washington, D.C.

HANNIGAN, PATRICK J. (1990), *Dynamic Monitoring and Analysis of Pile Foundation Installations*, Deep Foundations Institute, Sparta, NJ

HANSON, WALTER E.—see Peck, Ralph B.

HARDCASTLE, JAMES H.—see Lawton, Evert C.

HARDER, LESLIE F. AND SEED, H. BOLTON (1986), *Determination of Penetration Resistance for Coarse-Grained Soils Using the Becker Hammer Drill*, Report No. UCB/EERC-86/06, Earthquake Engineering Research Center, Richmond, CA

HARDER, L.F.—see Seed, H. Bolton

HARR, MILTON E. (1996), *Reliability-Based Design in Civil Engineering*, Dover Publishers

HARRISON, LAWRENCE J.—see Richardson, E.V.

HART, STEPHEN S. (1974), *Potentially Swelling Soil and Rock in the Front Range Urban Corridor, Colorado*, Colorado Geological Survey publication EG-7, Denver

HART, STEVEN S.—see Holtz, Wesley G.

HARTMAN, JOHN PAUL—see Schmertmann, John H.

HAWKINS, R.A.—see DiMillio, A.F.

HEAD, K.H. (1982), *Manual of Soil Laboratory Testing*, Vol. 2, Pentech, London

HEARNE, THOMAS M., STOKOE, KENNETH H. AND REESE, LYMON C. (1981), "Drilled Shaft Integrity By Wave Propagation Method," ASCE *Journal of the Geotechnical Engineering Division*, Vol. 107, No. GT10, p. 1327–1344

HEIJNEN, W.J. (1974), "Penetration Testing in Netherlands," *European Symposium on Penetration Testing (ESOPT)*, Stockholm, Vol. 1, p. 79–83

HEIJNEN, W.J. AND JANSE, E. (1985), "Case Studies of the Second European Symposium on Penetration Testing (ESOPT II)," *LGM-Mededelingen*, Part XXII, No. 91, Delft Soil Mechanics Laboratory, Delft, The Netherlands

HEINZ, RONEY (1993), "Plastic Piling," *Civil Engineering*, Vol. 63, No. 4, p. 63–65, ASCE

HERTLEIN, BERNARD H. (1992), "Selecting an Effective Low-Strain Foundation Test," *Foundation Drilling*, Vol. 31, No. 2, p. 14–22, Association of Drilled Shaft Contractors, Dallas, TX

HETÉNYI, M. (1974), *Beams on Elastic Foundation*, University of Michigan Press, Ann Arbor

HETHERINGTON, MARK D.—see Lawton, Evert C.

HEYMAN, JACQUES (1972), *Coulomb's Memoir on Statics: An Essay in the History of Civil Engineering*, Cambridge University Press, London

HIRSCH, CARL M.—see Dismuke, Thomas D.

HIRSH, T.J.—see Airhart, T.P.

HODGKINSON, ALLAN (1986), *Foundation Design*, The Architectural Press, London

HOLDEN, J.C. (1988), "Integrity Control of Bored Piles Using SID," *Deep Foundations on Bored and Auger Piles* (Proceedings of the 1st International Geotechnical Seminar on Deep Foundations on Bored and Auger Piles), p. 91–95, W.F. VanImpe, Ed., Balkema, Rotterdam

HOLLOWAY, D.M., MORIWAKI, Y., FINNO, R.J. AND GREEN, R.K. (1982), "Lateral Load Response of a Pile Group in Sand," *Second International Conference on Numerical Methods in Offshore Piling*, p. 441–456, ICE

HOLTZ, W.G. (1969), "Volume Change in Expansive Clay Soils and Control by Lime Treatment," *Proceedings of the Second International Research and Engineering Conference on Expansive Clay Soils*, p. 157–173, Texas A&M Press, College Station, TX

HOLTZ, WESLEY G. AND HART, STEPHEN S. (1978), *Home Construction on Shrinking and Swelling Soils*, Colorado Geological Survey publication SP-11, Denver

HOLTZ, WESLEY G.—see Jones, D. Earl

HORVATH, JOHN S. (1983), "New Subgrade Model Applied to Mat Foundations," ASCE *Journal of Geotechnical Engineering*, Vol. 109, No. 12, p. 1567–1587

HORVATH, JOHN S. (1992), "Lite Products Come of Age," ASTM *Standardization News*, Vol. 20, No. 9, p. 50–53

HORVATH, JOHN S. (1993), *Subgrade Modeling for Soil-Structure Interaction Analysis of Horizontal Foundation Elements*, Manhattan College Research Report No. CE/GE-93-1, Manhattan College, New York

HOUGH, B.K. (1959), "Compressibility as the Basis for Soil Bearing Value," ASCE *Journal of the Soil Mechanics and Foundations Division*, Vol. 85, No. SM4, p. 11–39

HOUSTON, SANDRA L., HOUSTON, WILLIAM N. AND SPADOLA, DONALD J. (1988), "Prediction of Field Collapse of Soils Due To Wetting," ASCE *Journal of Geotechnical Engineering*, Vol. 114, No. 1, p. 40–58

HOUSTON, SANDRA L. AND EL-EHWANY, MOSTAFA (1991), "Sample Disturbance of Cemented Collapsible Soils," ASCE *Journal of Geotechnical Engineering*, Vol. 117, No. 5, p. 731–752. Also see discussions, Vol. 118, No. 11, p. 1854–1862

HOUSTON, SANDRA L. (1992), "Partial Wetting Collapse Predictions," *Proceedings, 7th International Conference on Expansive Soils*, Vol. 1, p. 302–306

HOUSTON, SANDRA L.—see Houston, William N.

HOUSTON, WILLIAM N. AND HOUSTON, SANDRA L. (1989), "State-of-the-Practice Mitigation Measures for Collapsible Soil Sites," *Foundation Engineering: Current Principles and Practices*, Vol. 1, p. 161–175, F.H. Kulhawy, Ed., ASCE

HOUSTON, WILLIAM N. (1991), Personal communication

HOUSTON, WILLIAM N.—see Houston, Sandra L.

HUBER, FRANK (1991), "Update: Bridge Scour," ASCE *Civil Engineering*, Vol. 61, No. 9, p. 62–63

HUBER, TIMOTHY R.—see Mitchell, James K.

HUMMEL, CHARLES E. (1971), *Tyranny of the Urgent*, InterVarsity Press, Downers Grove, IL

HUNT, HAL W. (1987), "American Practice in the Design and Installation of Driven Piles for Structure Support," *Second International Deep Foundations Conference*, p. 13–28, Deep Foundations Institute, Sparta, NJ

HUNT, ROY E. (1984), *Geotechnical Engineering Investigation Manual*, McGraw Hill, New York

HUSSEIN, M.H., LIKINS, G.E. AND HANNIGAN, P.J. (1993), "Pile Evaluation by Dynamic Testing During Restrike," *Eleventh Southeast Asian Geotechnical Conference*, Singapore, p. 535–539

HUSSEIN, M., LIKINS, G. AND RAUSCHE, F. (1996), "Selection of a Hammer for High-Strain Dynamic Testing of Cast-in-Place Shafts," *5th International Conference on the Application of Stress Wave Theory to Piles*, p. 759–772

ICBO (1997), *Uniform Building Code*, International Conference of Building Officials, Whittier, CA

ICC (2000), *International Building Code*, International Code Council

INGELSON, I.—see Broms, B.B.

INGOLD, TERRY (1979), "Retaining Wall Performance During Backfilling," ASCE *Journal of Geotechnical Engineering*, Vol. 105, No. GT5, p. 613–626

ISAACS, D.V. (1931), "Reinforced Concrete Pile Formula," *Transactions of the Institution of Australian Engineers*, Vol. 12, p. 312–323

ISMAEL, NABIL F. AND VESIĆ, ALEKSANDAR S. (1981), "Compressibility and Bearing Capacity," ASCE *Journal of the Geotechnical Engineering Division*, Vol. 107, No. GT12, p. 1677–1691

JACKSON, CHRISTINA STAS—see Kulhawy, Fred H.

JAMIOLKOWSKI, M.—see Costanzo, D.

JANBU, N., BJERRUM, L AND KJAERNSLI, B. (1956), *Veiledning ved losning av fundamenteringsoppgaver*, Norwegian Geotechnical Institute Publication 16, p. 30–32, Oslo (in Norwegian)

JANSE, E.—see Heijnen, W.J.

JENNINGS, J.E. AND KNIGHT, K. (1956), "Recent Experience with the Consolidation Test as a Means of Identifying Conditions of Heaving or Collapse of Foundations on Partially Saturated Soils," *Transactions, South African Institution of Civil Engineers*, Vol. 6, No. 8, p. 255–256

JENNINGS, J.E. AND KNIGHT, K. (1957), "The Additional Settlement of Foundations Due to a Collapse Structure of Sandy Subsoils on Wetting," *Proceedings, Fourth International Conference on Soil Mechanics and Foundation Engineering*, Section 3a/12, p. 316–319

JENNINGS, J.E. (1969), "The Engineering Problems of Expansive Soils," *Proceedings of the Second International Research and Engineering Conference on Expansive Clay Soils*, p. 11–17, Texas A&M Press, College Station, TX

JENNINGS, J.E. (1973) "The Engineering Significance of Constructions on Dry Subsoils," *Proceedings of the Third International Conference on Expansive Soils*, Vol. 1, p. 27–32, Jerusalem Academic Press, Israel

JENNINGS, J.E. AND KNIGHT, K. (1975), "A Guide to Construction on or with Materials Exhibiting Additional Settlement Due to 'Collapse' of Grain Structure," *Sixth Regional Conference for Africa on Soil Mechanics and Foundation Engineering*, p. 99–105

JÉZÉQUEL, J.F.—see Baguelin, F.

JIANG, DA HUA (1983), "Flexural Strength of Square Spread Footing," ASCE *Journal of Structural Engineering*, Vol. 109, No. 8, p. 1812–1819

JOHANNESSEN, I.J.—see Bjerrum, L.

JOHNSON, LAWRENCE D. AND STROMAN, WILLIAM R. (1976), *Analysis of Behavior of Expansive Soil Foundations*, Technical report S-76-8, U.S. Army Waterways Experiment Station, Vicksburg, Miss.

JOHNSTON, G.H., ED. (1981), *Permafrost Engineering Design and Construction*, John Wiley, New York

JOHNSON, G.H.—see Crawford, C.B.

JONAS, ERNEST—see Brumund, William F.

JONES, C.J.F.P.—see O'Rourke, T.D.

JONES, D. EARL AND HOLTZ, WESLEY G. (1973), "Expansive Soils—The Hidden Disaster," *Civil Engineering*, Vol. 43, No. 8, ASCE

JONES, D. EARL AND JONES, KAREN A. (1987), "Treating Expansive Soils," *Civil Engineering*, Vol. 57, No. 8, August 1987, ASCE

JONES, KAREN A.—see Jones, D. Earl

JOSEPH, LEN S.—see Baker, Clyde N.

KAGAWA, TAKAAKI—see Kraft, Leland M.

KANTEY, BASIL A. (1980), "Some Secrets to Building Structures on Expansive Soils," *Civil Engineering*, Vol. 50, No. 12, December 1980, ASCE

KANTEY, B.A.—see Brink, A.B.A.

KATTI, R.K.—see Mitchell, James K.

KENNEDY, C.M.—see Sowers, G.F.

KERISEL, JEAN (1987), *Down to Earth, Foundations Past and Present: The Invisible Art of the Builder*, A.A. Balkema, Rotterdam

KIM, S.G.—see Barker, R.M.

KJAERNSLI, B.—see Janbu, M.

KNIGHT, K.—see Jennings, J.E.

KNODEL, PAUL C. (1981), "Construction of Large Canal on Collapsing Soils," ASCE *Journal of the Geotechnical Engineering Division*, Vol. 107, No. GT1, p. 79–94

KOERNER, ROBERT M. (1990), *Designing with Geosynthetics*, 2nd ed., Prentice Hall, Englewood Cliffs, NJ

KOOL, A.F.—see Brons, K.F.

KOSMATKA, STEVEN H. AND PANARESE, WILLIAM C. (1988), *Design and Control of Concrete Mixtures*, 13th ed., Portland Cement Association, Skokie, Illinois

KOTZIAS, PANAGHIOTIS C.—see Stamatopoulos, Aris C.

KOVACS, W.D.; SALOMONE, L.A. AND YOKEL, F.Y. (1981), *Energy Measurements in the Standard Penetration Test*, Building Science Series 135, National Bureau of Standards, Washington, D.C.

KRAFT, LELAND M., RAY, RICHARD P. AND KAGAWA, TAKAAKI (1981), "Theoretical *t-z* Curves," ASCE *Journal of the Geotechnical Engineering Division*, Vol. 107, No. GT11, p. 1543–1561

KRAMER, STEVEN L. (1991), *Behavior of Piles in Full-Scale, Field Lateral Loading Tests*, Report No. WA-RD 215.1, Federal Highway Administration, Washington, D.C.

KRINITZSKY, ELLIS L., GOULD, JAMES P. AND EDINGER, PETER H. (1993), *Fundamentals of Earthquake Resistant Construction*, John Wiley, New York

KULHAWY, FRED H. AND GOODMAN, RICHARD E. (1980), "Design of Foundations on Discontinuous Rock," *Structural Foundations on Rock*, P.J.N. Pells, Ed., Vol. 1, p. 209–220, A.A. Balkema, Rotterdam

KULHAWY, F.H., TRAUTMANN, C.H., BEECH, J.F., O'ROURKE, T.D., McGUIRE, W., WOOD, W.A. AND CAPANO, C. (1983), *Transmission Line Structure Foundations for Uplift-Compression Loading*, Report No. EL-2870, Electric Power Research Institute, Palo Alto, CA

KULHAWY, FRED H. (1984), "Limiting Tip and Side Resistance: Fact or Fallacy?" *Analysis and Design of Pile Foundations*, p. 80–98, ASCE

KULHAWY, FRED H. (1985), "Uplift Behavior of Shallow Soil Anchors—An Overview," *Uplift Behavior of Anchor Foundations in Soil*, Samuel P. Clemence, Ed., ASCE

KULHAWY, FRED H. AND JACKSON, CHRISTINA STAS (1989), "Some Observations on Undrained Side Resistance of Drilled Shafts," *Foundation Engineering: Current Principles and Practices*, p. 1011–1025, ASCE

KULHAWY, F.H. AND MAYNE, P.W. (1990), *Manual on Estimating Soil Properties for Foundation Design*, Report No. EL-6800, Electric Power Research Institute, Palo Alto, CA

KULHAWY, FRED H. (1991), "Drilled Shaft Foundations," Chapter 14 in *Foundation Engineering Handbook*, 2nd ed., Hsai-Yang Fang, Ed., Van Nostrand Reinhold, New York

KULHAWY, FRED H. AND PHOON, KOK KWANG (1996), "Engineering Judgement in the Evolution From Deterministic to Reliability-Based Foundation Design," *Uncertainty in the Geologic Environment*, Charles D. Shakelford, Priscilla P. Nelson, and Mary J.S. Roth, Eds., American Society of Civil Engineers

KULHAWY, FRED H.—see Mayne, Paul W.

KULHAWY, F.H.—see Phoon, K.K.

KULHAWY, F.H.—see Stas, C.V.

KULHAWY, FRED H.—see Turner, John P.

KUMBHOJKAR, A.S. (1993), "Numerical Evaluation of Terzaghi's N_g," ASCE *Journal of Geotechnical Engineering*, Vol. 119, No. 3, p. 598–607

KUWABARA, F. AND POULOS, H.G. (1989), "Downdrag Forces in a Group of Piles," ASCE *Journal of Geotechnical Engineering*, Vol. 115, No. 6, p. 806–818

LADANYI, B.—see DeBeer, E.E.

LADD, C. C., FOOTE, R., ISHIHARA, K., SCHLOSSER, F. AND POULOS, H. G. (1977), "Stress-Deformation and Strength Characteristics," State-of-the-art report, *Proceedings, Ninth International Conference on Soil Mechanics and Foundation Engineering*, Vol. 2, p. 421–494, Tokyo

LADD, CHARLES C.—see Brumund, William F.

LADD, CHARLES C.—see D'Appolonia, David J.

LADD, CHARLES C.—see Foott, Roger

LADE, P.V.—see Anderson, J.N.

LAM, IGNATIUS PO—see Martin, Geoffrey R.

LAMBE, T.W. AND WHITMAN, R.V. (1959), "The Role of Effective Stress in the Behavior of Expansive Soils," *Quarterly of the Colorado School of Mines*, Vol. 54, No. 4, p. 44–61

LAMBE, T. WILLIAM AND WHITMAN, ROBERT V. (1969), *Soil Mechanics*, John Wiley, New York

LAMBE, T. WILLIAM—see Mitchell, James K.

LANCELLOTTA, R.—see Costanzo, D.

LANKFORD, WILLIAM T., SAMWAYS, NORMAN L., CRAVEN, ROBERT F. AND McGANNON, HAROLD E. (1985), *The Making, Shaping and Treating of Steel*, 10th ed., United States Steel Corp.

LAWTON, EVERT C., FRAGASZY, RICHARD J. AND HARDCASTLE, JAMES H. (1989), "Collapse of Compacted Clayey Sand," ASCE *Journal of Geotechnical Engineering*, Vol. 115, No. 9, p. 1252–1267

LAWTON, EVERT C., FRAGASZY, RICHARD J. AND HARDCASTLE, JAMES H. (1991), "Stress Ratio Effects on Collapse of Compacted Clayey Sand," ASCE *Journal of Geotechnical Engineering*, Vol. 117, No. 5, p. 714–730. Also see discussion, Vol. 118, No. 9, p. 1472–1474

LAWTON, EVERT C., FRAGASZY, RICHARD J. AND HETHERINGTON, MARK D. (1992), Review of Wetting-Induced Collapse in Compacted Soil," ASCE *Journal of Geotechnical Engineering*, Vol. 118, No. 9, p. 1376–1394

LEE, A.R. (1974), *Blastfurnace and Steel Slag: Production, Properties and Uses*, John Wiley, New York

LEE, IAN K.—see Hain, Steven J.

LEONARDS, G.A. (1976), "Estimating Consolidation Settlements of Shallow Foundations on Overconsolidated Clay," *Estimation of Consolidation Settlement*, Special Report 163, Transportation Research Board, Washington, D.C.

LEONARDS, GERALD A. (1982), "Investigation of Failures," The 16th Terzaghi Lecture, ASCE *Journal of the Geotechnical Engineering Division*, Vol. 108, No. GT2, p. 185–246

LEONARDS, G.A. AND FROST, J.D., (1988), "Settlement of Shallow Foundations on Granular Soils," ASCE *Journal of Geotechnical Engineering*, Vol. 114, No. GT7, p. 791–809, (also see discussions, Vol. 117, No. 1, p. 172–188)

LEUNG, C.F.—see Dicken, E.A.

LI, H.—see Dempsey, J.P.

LIAO, S.S.C. AND WHITMAN, R.V. (1985), "Overburden Correction Factors for SPT in Sand," *Journal of Geotechnical Engineering*, Vol. 112, No. 3, p. 373–377, ASCE

LIAO, S.S.C. (1991), *Estimating the Coefficient of Subgrade Reaction for Tunnel Design*, Internal Research Report, Parsons Brinkerhoff, Inc.

LIKINS, GARLAND—see Goble, G.G.; Hannigan, P.J.; Hussein, M.; and Rausche, F.

LINDNER, ERNEST (1976), "Swelling Rock: A Review," *Rock Engineering for Foundations and Slopes*, Vol. 1, p. 141–181, ASCE

LITKE, SCOT (1986), "New Nondestructive Integrity Testing Yields More Bang for the Buck," *Foundation Drilling*, Sept/Oct, p. 8–11

LO, DOMINIC O. KWAN—see Mesri, Gholamreza

LUSCHER, ULRICH, BLACK, WILLIAM T. AND NAIR, KESHAVAN (1975), "Geotechnical Aspects of Trans-Alaska Pipeline," ASCE *Transportation Engineering Journal*, Vol. 101, No. TE4, p. 669–680

LUTENEGGER, ALAN J. AND SABER, ROBERT T. (1988), "Determination of Collapse Potential of Soils," *Geotechnical Testing Journal*, Vol. 11, No. 3, p. 173–178

LUTENEGGER, A.J.—see Miller, G.A.

LYTTON, R.L. AND WATT, W.G. (1970), *Prediction of Swelling in Expansive Clays*, Research Report 118-4, Center for Highway Research, University of Texas, Austin

MACGREGOR, JAMES G. (1992), *Reinforced Concrete: Mechanics and Design*, 2nd ed., Prentice Hall, Englewood Cliffs, NJ

MACGREGOR, JAMES G. (1996), *Reinforced Concrete: Mechanics and Design*, 3rd ed., Prentice Hall

MAHMOUD, HISHAM (1991), *The Development of an In-Situ Collapse Testing System for Collapsible Soils*, PhD Dissertation, Civil Engineering Department, Arizona State University

MARCHETTI, SILVANO (1980), "In Situ Tests by Flat Dilatometer," ASCE *Journal of the Geotechnical Engineering Division*, Vol. 106, No. GT3, p. 299–321 (also see discussions, Vol. 107, No. GT8, p. 831–837)

MARSH, M.L. AND BURDETTE, E.G. (1985), "Anchorage of Steel Building Components to Concete," *Engineering Journal*, Vol. 22, No. 1, p. 33–39, American Institute of Steel Construction, Chicago, IL

MARTIN, GEOFFREY R. AND LAM, IGNATIUS PO (1995), "Seismic Design of Pile Foundations: Structural and Geotechnical Issues," *Proceedings: Third International Conference on Recent Advances in Geotechnical Earthquake Engineering and Soil Dynamics*, Vol. III, p. 1491–1515

MASSARSCH, K. RAINER, TANCRÉ, ERIC AND BRIEKE, WERNER (1988), "Displacement Auger Piles with Compacted Base," *Deep Foundations on Bored and Auger Piles*, p. 333–342, W.F. Van Impe, Ed., Balkema, Rotterdam

MATLOCK, H. AND REESE, L.C. (1960), "Generalized Solution for Laterally Loaded Piles," ASCE *Journal of the Soil Mechanics and Foundations Division*, Vol. 86, No. SM5, p. 63–91

MATYAS, ELMER L.—see Taylor, Brian B.

MAYNE, PAUL W. AND KULHAWY, FRED H. (1982), "K_0-OCR Relationships in Soil," ASCE *Journal of the Geotechnical Engineering Division*, Vol. 108, No. GT6, p. 851–872

MAYNE, P.W.—see Kulhawy, F.H.

McCLELLAND, BRAMLETTE AND FOCHT, JOHN A. JR. (1958), "Soil Modulus for Laterally Loaded Piles," *Transactions of ASCE*, Vol. 123, p. 1049–1086, Paper No. 2954

McCORMAC, JACK C. (1998), *Design of Reinforced Concrete*, 4th ed., Addison Wesley, Menlo Park, CA

McDONALD, D.H.—see Skempton, A.W.

McDOWELL, C. (1956), "Interrelationships of Load, Volume Change, and Layer Thicknesses of Soils to the Behavior of Engineering Structures," *Proceedings, Highway Research Board*, Vol. 35, p. 754–772

McGOWN, A.—see Barden, L.

McGUIRE, W.—see Kulhawy, F.H.

McVAY, M., CAPPER, R. AND SHANT, T-I (1995), "Lateral Response of Three-Row Groups in Loose to Dense Sands in 3D and 5D Pile Spacing," *Journal of Geotechnical Engineering*, Vol. 121, No. 5, p. 436–441

MEHTA, P.K. (1983), "Mechanism of Sulfate Attack on Portland Cement Concrete–Another Look," *Cement and Concrete Research*, Vol. 13, p. 401–406

MEIGH, A.C. (1987), *Cone Penetration Testing: Methods and Interpretation*, Butterworths, London

MESRI, GHOLAMREZA, LO, DOMINIC O. KWAN AND FENG, TAO-WEI (1994), "Settlement of Embankments on Soft Clays," *Vertical and Horizontal Deformations of Foundations and Embankments*, Vol. 1, p. 8–56, ASCE

MESRI, GHOLAMREZA—see Terzaghi, Karl

MEYERHOF, G.G. (1953), "The Bearing Capacity of Foundations Under Eccentric and Inclined Loads," *Proceedings, 3rd International Conference on Soil Mechanics and Foundation Engineering*, Vol. 1, p. 440–445, Zurich (Reprinted in Meyerhof, 1982)

MEYERHOF, G.G. (1955), "Influence of Roughness of Base and Ground-Water Conditions on the Ultimate Bearing Capacity of Foundations," *Géotechnique*, Vol. 5, p. 227–242 (Reprinted in Meyerhof, 1982)

MEYERHOF, G.G. (1956), "Penetration Tests and Bearing Capacity of Cohesionless Soils," ASCE *Journal of the Soil Mechanics and Foundations Division*, Vol. 82, No. SM1, p. 1–19 (Reprinted in Meyerhof, 1982)

MEYERHOF, GEORGE GEOFFREY (1963), "Some Recent Research on the Bearing Capacity of Foundations," *Canadian Geotechnical Journal*, Vol. 1, No. 1, p. 16–26 (Reprinted in Meyerhof, 1982)

MEYERHOF, GEORGE G. (1965), "Shallow Foundations," ASCE *Journal of the Soil Mechanics and Foundations Division*, Vol. 91 No. SM2, p. 21–31 (Reprinted in Meyerhof, 1982)

MEYERHOF, GEORGE GEOFFREY (1976), "Bearing Capacity and Settlement of Pile Foundations," ASCE *Journal of the Geotechnical Engineering Division*, Vol. 102, No. GT3, p. 197–228

MEYERHOF, G.G. (1982), *The Bearing Capacity and Settlement of Foundations*, Tech-Press, Technical University of Nova Scotia, Halifax

MEYERHOF, G.G. (1983), "Scale Effects of Pile Capacity," *Journal of the Geotechnical Engineering Division*, ASCE, Vol. 102, No. GT3, p. 195–228

MEYERHOF, G.G. (1995), "Development of Geotechnical Limit State Design," *Canadian Geotechnical Journal*, Vol. 32, No. 1, p. 128–136

MEYERS, LARRY R. (1996), "Waterproofing," Chapter 9 in course notes *Effective Below Grade and Plaza Deck Waterproofing*, University of Wisconsin, Madison

MIKESELL, R.C.—see Rollins, K.M.

MILLER, DEBORA J.—see Nelson, John D.

MILLER, G.A. AND LUTENEGGER, A.J. (1997), "Influence of Pile Plugging on Skin Friction in Overconsolidated Clay," *Journal of Geotechnical and Geoenvironmental Engineering*, Vol. 123, p. 525–533

MIRAN, JEROME—see Briaud, Jean-Louis

MITCHELL, JAMES K., VIVATRAT, VITOON AND LAMBE, T. WILLIAM (1977), "Foundation Performance of Tower of Pisa," ASCE *Journal of the Geotechnical Engineering Division*, Vol. 103, No. GT3, p. 227–249 (discussions in Vol. 104 No. GT1 and GT2, Vol. 105 No. GT1 and GT11)

MITCHELL, JAMES K., ET AL. (1978), *Soil Improvement: History, Capabilities and Outlook*, ASCE

MITCHELL, JAMES K. (1978), "In-Situ Techniques for Site Characterization," *Site Characterization and Exploration*, p. 107–129, C.H. Dowding, Ed., ASCE

MITCHELL, JAMES K. AND KATTI, R.K. (1981), "Soil Improvement: State-of-the-Art," *Proceedings, 10th International Conference on Soil Mechanics and Foundation Engineering*, Stockholm

MITCHELL, JAMES K. AND HUBER, TIMOTHY R. (1985), "Performance of a Stone Column Foundation," ASCE *Journal of Geotechnical Engineering*, Vol. 111, No. 2, p. 205–223

MITCHELL, JAMES K. (1993), *Fundamentals of Soil Behavior*, 2nd ed., John Wiley, New York

MITCHELL, J.K.—see Seed, H.B.

MITCHELL, P.W. AND AVALLE, D.L. (1984), "A Technique to Predict Expansive Soil Movements," *Proceedings, Fifth International Conference on Expansive Soils*, Institution of Engineers, Australia

MOE, JOHANNES (1961), *Shearing Strength of Reinforced Concrete Slabs and Footings Under Concentrated Loads*, Portland Cement Association Bulletin D47, Skokie, Illinois

MORGAN, M.H. (1914), Translation of Marcus Vitruvius in *The Ten Books on Architecture*, Harvard University Press, Cambridge, MA

MORGENSTERN, N.R. AND EISENSTEIN, Z. (1970), "Methods of Estimating Lateral Loads and Deformations" in *Lateral Stresses in the Ground and the Design of Earth-Retaining Structures*, p. 51–102, ASCE

MORIWAKI, Y.—see Holloway, D.M.

MORRISON, C.—see Brown, D.A.

MOSS, R.E.S., RAWLINGS, M.A., CALIENDO, J.A. AND ANDERSON, L.R. (1998), "Cyclic Lateral Loading of Model Pile Groups in Clay Soil," *Geotechnical Earthquake Engineering and Soil Dynamics III*, Geotechnical Special Publication 75, p. 494–505, ASCE

MTO (1991), *Ontario Highway Bridge Design Code*, 3rd ed., Ontario Ministry of Transportation

MÜLLER-BRESLAU, H. (1906), "Erddruk auf Stützmauern," Alfred Kröner Verlag, Stuttgart (in German)

MURILLO, JUAN A. (1987), "The Scourge of Scour," *Civil Engineering*, Vol. 57, No. 7, p. 66–69, ASCE

MURPHY, E.C. (1908), "Changes in Bed and Discharge Capacity of the Colorado River at Yuma, Ariz.," *Engineering News*, Vol. 60, p. 344–345

NAGAO, YOSHIKAZU, ET AL. (1989), *Development of New Pavement Base Course Material Using High Proportion of Steelmaking Slag Properly Combined with Air-Cooled and Granulated Blast Furnace Slags*, Nippon Steel Technical Report No. 43, Oct 1989

NAHB (1988), *Frost-Protected Shallow Foundations for Houses and Other Heated Structures*, National Association of Home Builders, Upper Marlboro, MD

NAHB (1990, draft) *Frost-Protected Shallow Foundations for Unheated Structures*, National Association of Home Builders, Upper Marlboro, MD

NAIR, KESHAVAN—see Luscher, Ulrich

NEATE, JAMES J. (1989), "Augered Cast-in-Place Piles," *Foundation Engineering: Current Principles and Practices*, Vol. 2, p. 970–978, F.H. Kulhawy, Ed., ASCE

NEELY, WILLIAM J. (1989), "Bearing Pressure–SPT Correlations for Expanded Base Piles in Sand," *Foundation Engineering: Current Principles and Practices*, Vol. 2, p. 979–990, F.H. Kulhawy, Ed., ASCE

NEELY, WILLIAM J. (1990a), "Bearing Capacity of Expanded-Base Piles in Sand," ASCE *Journal of Geotechnical Engineering*, Vol. 116, No. 1, p. 73–87

NEELY, WILLIAM J. (1990b), "Bearing Capacity of Expanded-Base Piles with Compacted Concrete Shafts," ASCE *Journal of Geotechnical Engineering*, Vol. 116, No. 9, p. 1309–1324

NEELY, WILLIAM J. (1991), "Bearing Capacity of Auger-Cast Piles in Sand," ASCE *Journal of Geotechnical Engineering*, Vol. 117, No. 2, p. 331–345

NELSON, JOHN D. AND MILLER, DEBORA J. (1992), *Expansive Soils: Problems and Practice in Foundation and Pavement Engineering*, John Wiley, New York

NEWMARK, NATHAN M. (1935), *Simplified Computation of Vertical Pressures in Elastic Foundations*, Engineering Experiment Station Circular No. 24, University of Illinois, Urbana

NG, E.S.—see DiMillio, A.F.

NIXON, IVAN K. (1982), "Standard Penetration Test State-of-the-Art Report," *Second European Symposium on Penetration Testing* (ESOPT II), Amsterdam, Vol. 1, p. 3–24, A. Verruijt, et. al., eds.

NORDLUND, REYMOND L. AND DEERE, DON U. (1970), "Collapse of Fargo Grain Elevator," ASCE *Journal of the Soil Mechanics and Foundations Division*, Vol. 96, No. SM2, p. 585–607

NOTTINGHAM, L.C. (1975), *Use of Quasi-Static Friction Cone Penetrometer Data to Predict Load Capacity of Displacement Piles*, PhD Thesis, Department of Civil Engineering, University of Florida

NOTTINGHAM, LARRY C. AND SCHMERTMANN, JOHN H. (1975), *An Investigation of Pile Capacity Design Procedures*, Research Report D629, Dept. of Civil Engr., Univ. of Florida, Gainesville

NOWAK, ANDRZEJ S. (1995), "Calibration of LRFD Bridge Code," ASCE *Journal of Structural Engineering*, Vol. 121, No. 8, p. 1245–1251

NRCC (1990), *Canadian National Building Code*, National Research Council of Canada, Ottawa, Ontario

O'BRIEN, ARTHUR J.—see Fellenius, Bengt H.

O'NEILL, MICHAEL W. AND POORMOAYED, NADER (1980), "Methodology for Foundations on Expansive Clays," ASCE *Journal of the Geotechnical Engineering Division*, Vol. 106, No. GT12, p. 1345–1368

O'NEILL, MICHAEL W. (1983), "Group Action in Offshore Piles," *Geotechnical Practice in Offshore Engineering*, p. 25–64, Steven G. Wright, Ed., ASCE

O'NEILL, MICHAEL (1987), "Use of Underreams in Drilled Shafts," *Drilled Foundation Design and Construction Short Course*, Association of Drilled Shaft Contractors, Dallas, TX

O'Neill, Michael W. and Reese, Lymon C. (1999), *Drilled Shafts: Construction Procedures and Design Methods*, Federal Highway Administration

O'NEILL, M.W.—see DiMillio, A.F.

O'NEILL, MICHAEL W.—see Reese, Lymon C.

O'ROURKE, T.D. AND JONES, C.J.F.P. (1990), "Overview of Earth Retention Systems: 1970–1990," *Design and Performance of Earth Retaining Structures*, Geotechnical Special Publication No. 25, p. 22–51, P.C. Lambe and L.A. Hansen, eds., ASCE

O'ROURKE, T.D.—see Kulhawy, F.H.

OLSON, ROY E. AND FLAATE, KARRE S. (1967), "Pile-Driving Formulas for Friction Piles in Sand," ASCE *Journal of the Soil Mechanics and Foundations Division*, Vol. 93, No. SM6, p. 279–296

OLSON, LARRY D. AND WRIGHT, CLIFFORD C. (1989), "Nondestructive Testing of Deep Foundations With Sonic Methods," *Foundation Engineering: Current Principles and Practices*, ASCE

OOI, P.S.K.—see Barker, R.M.

OSTERBERG, JORJ O. (1984), "A New Simplified Method For Load Testing Drilled Shafts," *Foundation Drilling*, August, 1984, Association of Drilled Shaft Contractors, Dallas

PAIKOWSKY, SAMUEL G. AND WHITMAN, ROBERT V. (1990), "The Effects of Plugging on Pile Performance and Design," *Canadian Geotechnical Journal*, Vol. 27, p. 429–440

PANARESE, WILLIAM C.—see Kosmatka, Steven H.

PCA (1991), "Durability of Concrete in Sulfate-Rich Soils," *Concrete Technology Today*, Vol. 12, No. 3, p. 6–8, Portland Cement Association

PCI (1984), "Standard Prestressed Concrete Square and Octagonal Piles," Drawing No. STD-112-84, Prestressed Concrete Institute, Chicago

PCI (1993a), "Recommended Practice for Design, Manufacture and Installation of Prestressed Concrete Piling," *PCI Journal*, Vol. 38, No. 2, p. 14–41, Precast/Prestressed Concrete Institute, Chicago, IL

PCI (1993b), *Prestressed Concrete Piling Interaction Diagrams*, Precast/Prestressed Concrete Institute, Chicago

PDCA (1998), *Design Specifications for Driven Bearing Piles, Load and Resistance Factor Design*, Pile Driving Contractors Association, St. Louis, MO

PECK, RALPH B. (1942), Discussion of "Pile Driving Formulas: Progress Report of the Committee on the Bearing Value of Pile Foundations," *ASCE Proceedings*, Vol. 68. No. 2, p. 323–324, Feb.

PECK, RALPH B. (1948), *History of Building Foundations in Chicago*, Bulletin Series No. 373, University of Illinois Engineering Experiment Station, Urbana, IL

PECK, R.B. AND BRYANT, F.G. (1953), "The Bearing-Capacity Failure of the Transcona Elevator," *Géotechnique*, Vol. 3, p. 201–208

PECK, RALPH B. (1958), *A Study of the Comparative Behavior of Friction Piles*, Highway Research Board Special Report 36, Washington, D.C.

PECK, RALPH B., HANSON, WALTER E. AND THORNBURN, THOMAS H. (1974), *Foundation Engineering*, 2nd ed., John Wiley, New York

PECK, RALPH B. (1975), *The Selection of Soil Parameters for the Design of Foundations*, Second Nabor Carrillo Lecture, Mexican Society for Soil Mechanics, Mexico City (reprinted in *Judgment in Geotechnical Engineering*, Dunnicliff, John and Deere, Don U., eds, John Wiley, 1984)

PECK, RALPH B. (1976), "Rock Foundations for Structures," *Rock Engineering for Foundations and Slopes*, Vol. II, p. 1–21, ASCE

PECK, RALPH B.—see Terzaghi, Karl

PEPE, M.C.—see Costanzo, D.

PHOON, K.K., KULHAWY, F.H. AND GRIGORIU, M.D. (1995), *Reliability-Based Design of Foundations for Transmission Line Structures*, Report TR-105000, Electric Power Research Institute, Palo Alto, CA

PHOON, KOK KWANG—see Kulhawy, Fred H.

POLSHIN, D.E. AND TOKAR, R.A. (1957), "Maximum Allowable Non-Uniform Settlement of Structures," *Proceedings, 4th International Conference on Soil Mechanics and Foundation Engineering*, Vol. 1, p. 402–405

POORMOAYED, NADER—see O'Neill, Michael W.

POULOS, H.G. AND DAVIS, E.H. (1974), *Elastic Solutions for Soil and Rock Mechanics*, John Wiley, New York

POULOS, H.G. AND DAVIS, E.H. (1980), *Pile Foundation Analysis and Design*, John Wiley, New York

POULOS, H.G.—see Davis, E.H.

POULOS, H.G.—see Kuwabara, F.

POWELL, GEORGE T. (1884), *Foundations and Foundation Walls*, William T. Comstock, New York

PRANDTL, L. (1920), "Über die Härte plastischer Körper (On the Hardness of Plastic Bodies)," *Nachr. Kgl. Ges Wiss Go̅ttingen, Math-Phys. Kl.*, p. 74 (in German)

PRAWIT, ALFRED AND VOLMERDING, BRITTA (1995), "Great Belt Bridge: A Project of Superlatives," *Quality Concrete*, Vol. 1, No. 2, p. 41–44

PTI (1980), *Design and Construction of Post-Tensioned Slabs-on-Ground*, Post-Tensioning Institute, Phoenix, AZ

PYE, KENNETH (1987), *Aeolian Dust and Dust Deposits*, Academic Press, London

RAD, NADER SHAFII—see Clough, G. Wayne

RANDOLPH, M.F. AND WROTH, C.P. (1982), "Recent Developments in Understanding the Axial Capacity of Piles in Clay," *Ground Engineering*, Vol. 15, No. 7, p. 17–32

RANDOLPH, M.F.—see Fleming, W.G.K.

RANKINE, W.J.M. (1857), "On the Stability of Loose Earth," *Philosophical Transactions of the Royal Society*, Vol. 147, London

RAO, KANAKAPURA S. SUBBA AND SINGH, SHASHIKANT (1987), "Lower-Bound Collapse Load of Square Footings," ASCE *Journal of Structural Engineering*, Vol. 113, No. 8, p. 1875–1879

RAHARDJO, H.—see Fredlund, D.G.

RAUSCHE, F., MOSES, F. AND GOBLE, G. (1972), "Soil Resistance Predictions from Pile Dynamics," ASCE *Journal of the Soil Mechanics and Foundations Division*, Vol. 98, No. SM9

RAUSCHE, F. AND GOBLE, G.G. (1978), "Determination of Pile Damage by Top Measurements," *Behavior of Deep Foundations*, ASTM STP 670, p. 500–506, Raymond Lundgren, Ed., ASTM

RAUSCHE, FRANK, GOBLE, GEORGE G. AND LIKINS, GARLAND E. (1985), "Dynamic Determination of Pile Capacity," ASCE *Journal of Geotechnical Engineering*, Vol. 111, No. 3, p. 367–383

RAUSCHE, F.—see Hannigan, P.J. and Hussein, M.

RAUSCHE, FRANK—see Seitz, Jörn M.

RAWLINGS, M.A.—see Moss, R.E.S.

RAY, RICHARD P.—see Kraft, Leland M.

REESE, LYMON C. (1984), *Handbook on Design of Piles and Drilled Shafts Under Lateral Load*, Report No. FHWA-IP-84-11, Federal Highway Administration

REESE, LYMON C. AND WANG, SHIN TOWER (1986), "Method of Analysis of Piles Under Lateral Loading," *Marine Geotechnology and Nearshore/Offshore Structures*, ASTM STP 923, R.C. Chaney and H.Y. Fang, ed., p. 199–211, ASTM

REESE, LYMON C. AND O'NEILL, MICHAEL W. (1988), *Drilled Shafts: Construction Procedures and Design Methods*, Report No. FHWA-HI-88-042, Federal Highway Administration

REESE, L.C. AND O'NEILL, M.W. (1989), "New Design Method for Drilled Shafts From Common Soil and Rock Tests," *Foundation Engineering: Current Principles and Practices*, p. 1026–1039, Fred H. Kulhawy, Ed., ASCE

REESE, LYMON C. AND WANG, S.T. (1997), *Technical Manual of Documentation of Computer Program LPILE PLUS 3.0 for Windows*, Ensoft, Inc., Austin, TX

REESE, L.C.—see Brown, D.A.

REESE, LYMON C.—see Hearne, Thomas M.

REESE, L.C.—see Matlock H.

REEESE, L.C.—see O'Neill, Michael

REMPE, D.M. (1979), "Building Code Requirements for Maximum Design Stresses in Piles," *Behavior of Deep Foundations*, p. 507–519, ASTM STP 670, Raymond Lundgren, Ed., ASTM, Philadelphia

RICHARDS, B.G. (1967), "Moisture Flow and Equilibria in Unsaturated Soils for Shallow Foundations," *Permeability and Capillarity of Soils*, p. 4–33, STP 417, ASTM

RICHARDSON, E.V., HARRISON, LAWRENCE J. AND DAVIS, STANLEY R. (1991), *Evaluating Scour at Bridges*, Report No. FHWA-IP-90-017, Federal Highway Administration, McLean, VA

RICHART, FRANK E. (1948), "Reinforced Concrete Wall and Column Footings," *Journal of the American Concrete Institute*, Vol. 20, No. 2, p. 97–127, and Vol. 20, No. 3, p. 237–260

RICKER, DAVID T.—see DeWolf, John T.

RIKER, RICHARD E.—see Fellenius, Bengt H.

RIX, GLENN J.—see Sutterer, Kevin G.

ROBERTSON, P.K. AND CAMPANELLA, R.G. (1983), "Interpretation of Cone Penetration Tests: Parts 1 and 2," *Canadian Geotechnical Journal* Vol. 20, p. 718–745

ROBERTSON, P.K., CAMPANELLA, R.G. AND BROWN, P.T. (1984), *The Application of CPT Data to Geotechnical Design: Worked Examples*, Department of Civil Engineering, University of British Columbia

ROBERTSON, P.K. AND CAMPANELLA, R.G. (1988), *Guidelines for Using the CPT, CPTU and Marchetti DMT for Geotechnical Design*, Volume II, Report No. FHWA-PA-87-023+84-24, Federal Highway Administration

ROBERTSON, P.K. AND CAMPANELLA, R.G. (1989), *Guidelines for Geotechnical Design Using the Cone Penetrometer Test and CPT With Pore Pressure Measurement*, 4th ed., Hogentogler & Co., Columbia, MD

ROBERTSON, PETER K., DAVIES, MICHAEL P. AND CAMPANELLA, RICHARD G. (1989), "Design of Laterally Loaded Driven Piles Using the Flat Dilatometer," ASTM *Geotechnical Testing Journal*, Vol. 12, No. 1, p. 30–38

ROJIANI, K.B.—see Barker, R.M.

ROLLINS, K.M., CLAYTON, R.J. AND MIKESELL, R.C. (1997), "Ultimate Side Friction of Drilled Shafts in Gravels," *Foundation Drilling*, Vol. 36, No. 5, Association of Drilled Shaft Contractors, Dallas, TX

ROMANOFF, MELVIN (1962), "Corrosion of Steel Pilings in Soils," *Journal of Research of the National Bureau of Standards*, Vol. 66C, No. 3; also in National Bureau of Standards Monograph 58, Washington, DC

ROMANOFF, MELVIN (1970), "Performance of Steel Pilings in Soils," *Proceedings, 25th Conference, National Association of Corrosion Engineers*, March 1969

ROWE, PETER W. (1952), "Anchored Sheet Pile Walls," *Proceedings, Institution of Civil Engineers*, London

ROWE, PETER W. (1957), "Sheet Pile Walls in Clays," *Proceedings, Institution of Civil Engineers*, London

ROWE, PETER W. (1958), "Measurements on Sheet Pile Walls Driven into Clay," *Proceedings, Brussels Conference 58 on Earth Presure Problems*, Vol. II, p. 127–330, Belgian Group of the International Society of Soil Mechanics and Foundation Engineering

SANTAMARINA, J. CARLOS (1997), "Cohesive Soil: A Dangerous Oxymoron," *Electronic Journal of Geotechnical Engineering*, published at web site geotech.civen.okstate.edu/magazine/oxymoron/dangeoxi.htm

SAXENA, S.K.—see Vesić, A.S.

SBCCI (1997), *Standard Building Code*, Southern Building Code Congress International, Inc., Birmingham, AL

SCHEIL, THOMAS J. (1979), "Long Term Monitoring of a Building Over the Deep Hackensack Meadowlands Varved Clays," presented at 1979 Converse Ward Davis Dixon, Inc. Technical Seminar, Pasadena, CA

SCHIFF, M.J. (1982), "What is Corrosive Soil?" *Proceedings, Western States Corrosion Seminar*, Western Region, National Association of Corrosion Engineers

SCHMERTMANN, J.H. (1970), "Static Cone to Compute Settlement over Sand," ASCE *Journal of the Soil Mechanics and Foundations Division*, Vol. 96, No. SM3, p. 1011–1043

SCHMERTMANN, JOHN H., HARTMAN, JOHN PAUL AND BROWN, PHILLIP R. (1978), "Improved Strain Influence Factor Diagrams," ASCE *Journal of the Geotechnical Engineering Division*, Vol. 104, No. GT8, p. 1131–1135

SCHMERTMANN, JOHN H. (1978), *Guidelines for Cone Penetration Test: Performance and Design*, Report FHWA-TS-78-209, Federal Highway Administration, Washington, D.C.

SCHMERTMANN, J.H. (1986a), "Suggested Method for Performing the Flat Dilatometer Test," *Geotechnical Testing Journal*, Vol. 9, No. 2, p. 93–101

SCHMERTMANN, JOHN H. (1986b), "Dilatometer to Compute Foundation Settlement," *Use of In-Situ Tests in Geotechnical Engineering*, Samuel P. Clemence, Ed., ASCE, p. 303–319

SCHMERTMANN, JOHN H. (1988a), "Dilatometers Settle In," *Civil Engineering*, Vol. 58, No. 3, p. 68–70, March 1988

SCHMERTMANN, JOHN H. (1988b), *Guidelines for Using the CPT, CPTU and Marchetti DMT for Geotechnical Design*, Vol. I-IV, Federal Highway Administration

SCHMERTMANN, JOHN H.—see Nottingham, Larry C.

SCHNEIDER, ROBERT R. AND DICKEY, WALTER L. (1987), *Reinforced Masonry Design*, 2nd ed., Prentice Hall, Englewood Cliffs, NJ

SCHULTZE, EDGAR (1961), "Distribution of Stress Beneath a Rigid Foundation," *Proceedings, 5th International Conference on Soil Mechanics and Foundation Engineering*, p. 807–813, Paris

SCOTT, RONALD F. (1981), *Foundation Analysis*, Prentice-Hall, Englewood Cliffs, NJ

SEED, H.B. AND CHAN, C.K. (1959), "Structure and Strength Characteristics of Compacted Clays," ASCE *Journal of the Soil Mechanics and Foundations Division*, Vol. 85, No. SM5

SEED, H.B., MITCHELL, J.K. AND CHAN, C.K. (1962), "Studies of Swell and Swell Pressure Characteristics of Compacted Clays," *Bulletin No. 313*, Highway Research Board, p. 12–39

SEED, H. BOLTON (1970), "Soil Problems and Soil Behavior," Chapter 10 in *Earthquake Engineering*, Robert L. Wiegel, Ed., Prentice Hall, Englewood Cliffs, NJ

SEED, H. BOLTON, TOKIMATSU, K., HARDER, L.F. AND CHUNG, RILEY M. (1985), "Influence of SPT Procedures in Soil Liquefaction Resistance Evaluations," ASCE *Journal of Geotechnical Engineering*, Vol. 111, No. 12, p. 1425–1445

SEED, H. BOLTON—see Harder, Leslie F.

SEED, R.B.—see Duncan, J.M.

SHEPHERD, R. AND DELOS-SANTOS, E.O. (1991), "An Experimental Investigation of Retrofitted Cripple Walls," *Bulletin of the Seismological Society of America*, Vol. 81, No. 5, p. 2111–2126

SHIELDS, DONALD, CHANDLER, NEIL AND GARNIER, JACQUES (1990), "Bearing Capacity of Foundations on Slopes," ASCE *Journal of Geotechnical Engineering*, Vol. 116, No. 3, p. 528–537

SHIPP, J.G. AND HANINGER, E.R. (1983), "Design of Headed Anchor Bolts," *Engineering Journal*, Vol. 20, No. 2, p. 58–69, American Institute of Steel Construction, Chicago, IL

SIMONS, KENNETH B. (1991), "Limitations of Residential Structures on Expansive Soils," ASCE *Journal of Performance of Constructed Facilities*, Vol. 5, No. 4, p. 258–270

SIMONS, N.E. (1987), "Settlement," Chapter 14 in *Ground Engineer's Reference Book*, F.G. Bell, ed. Butterworths, London

SINGH, SHASHIKANT—see Rao, Kanakapura

SITAR, NICHOLAS—see Clough, G. Wayne

SKAFTFELD, KEN (1998), "The Failure and Righting of the Transcona Grain Elevator," *Geotechnical News*, Vol. 16, No. 2, p. 61–63, BiTech Publishers, Vancouver

SKEMPTON, A.W. (1951), "The Bearing Capacity of Clays," *Proceedings, Building Research Congress*, Vol. 1, p. 180–189, London

SKEMPTON, A.W. AND MCDONALD, D.H. (1956), "Allowable Settlement of Buildings," *Proceedings, Institute of Civil Engineers*, Vol. 3, No. 5, p. 727–768

SKEMPTON, A.W. AND BJERRUM, L. (1957), "A Contribution to the Settlement Analysis of Foundations on Clay," *Géotechnique*, Vol. 7, p. 168–178

SKEMPTON, A.W. (1986), "Standard Penetration Test Procedures and the Effects in Sands of Overburden Pressure, Relative Density, Particle Size, Aging and Overconsolidation," *Géotechnique*, Vol. 36, No. 3, p. 425–447

SMITH, E.A.L. (1951), "Pile Driving Impact," *Proceedings, Industrial Computation Seminar*, Sept. 1950, International Business Machines Corp., New York

SMITH, E.A.L. (1960), "Pile Driving Analysis by the Wave Equation," ASCE *Journal of Soil Mechanics and Foundations*, Vol. 86, No. SM4, p. 35–61

SMITH, E.A.L. (1962), "Pile Driving Analysis by the Wave Equation," ASCE *Transactions*, Vol. 127, Part 1, p. 1145–1193

SMITH, TREVOR D. (1989), "Fact or Fiction: A Review of Soil Response to a Laterally Moving Pile," *Foundation Engineering: Current Principles and Practices*, Vol. 1, p. 588–598, Fred Kulhawy, Ed., ASCE

SNETHEN, D.R. (1980), "Characterization of Expansive Soils Using Soil Suction Data," *Proceedings of the Fourth International Conference on Expansive Soils*, ASCE

SNETHEN, D.R. (1984), "Evaluation of Expedient Methods for Identification and Classification of Potentially Expansive Soils," *Proceedings of the Fifth International Conference on Expansive Soils*, p. 22–26, National Conference Publication 84/3, The Institution of Engineers, Australia

SODERBERG, L. (1962), "Consolidation Theory Applied to Foundation Pile Time Effects," *Géotechnique*, Vol. 12, p. 217–225

SOWERS, G.F. AND KENNEDY, C.M. (1967), "High Volume Change Clays of the Southeastern Coastal Plain," *Proceedings of the Third Panamerican Conference on Soil Mechanics and Foundation Engineering*, Caracas

SOWERS, GEORGE F. (1979), *Introductory Soil Mechanics and Foundations: Geotechnical Engineering*, 4th ed., MacMillan, New York

SPADOLA, DONALD J.—see Houston, Sandra L.

SPANOVICH, MILAN AND FEWELL, RICHARD (1968), Discussion of Crawford and Burn (1968), *Placement and Improvement of Soil to Support Structures*, p. 247–249, ASCE

STAMATOPOULOS, ARIS C. AND KOTZIAS, PANAGHIOTIS C. (1985), *Soil Improvement by Preloading*, John Wiley, New York

STEPHENSON, RICHARD (1995), "Shallow Foundations," Section 2B in *Practical Foundation Engineering Handbook*, Robert Wade Brown, ed., McGraw Hill, New York

STOKOE, KENNETH H.—see Hearne, Thomas M.

STROMAN, WILLIAM R.—see Johnson, Lawrence D.

SUROS, OSCAR—see York, Donald L.

TALBOT, ARTHUR N. (1913), *Reinforced Concrete Wall Footings and Column Footings*, Bulletin No. 67, University of Illinois Engineering Experiment Station, Urbana

TAN, C.K.—see Barker, R.M.

TANCRÉ, ERIC—see Massarsch, K.

TAWFIQ, KAMAL S. AND CALIENDO, JOSEPH A. (1994), "Bitumen Coating for Downdrag Mitigation in Cohesionless Soils," *Transportation Research Record No. 1447*, p. 116–124, National Research Council

TAYLOR, BRIAN B. AND MATYAS, ELMER L. (1983), "Influence Factors for Settlement Estimates of Footings on Finite Layers," *Canadian Geotechnical Journal*, Vol. 20, p. 832–835

TAYLOR, DONALD W. (1948), *Fundamentals of Soil Mechanics*, John Wiley, New York

TENG, WAYNE C. (1962), *Foundation Design*, Prentice Hall, Englewood Cliffs, NJ

TERZAGHI, K. (1934a), "Die Ursachen der Schiefstellung des Turmes von Pisa," *Der Bauingenieur*, Vol. 15, No. 1/2, p. 1–4 (Reprinted in *From Theory to Practice in Soil Mechanics*, Bjerrum, L. et. al., eds., p. 198–201, John Wiley, New York, 1960) (in German)

TERZAGHI, KARL (1934b), "Large Retaining Wall Tests," a series of articles in *Engineering News-Record*, Vol. 112; 2/1/34, 2/22/34, 3/8/34, 3/29/34, 4/19/34, and 5/17/34

TERZAGHI, KARL (1936), Discussion of "Settlement of Structures," *Proceedings, First International Conference on Soil Mechanics and Foundation Engineering*, Cambridge, MA, Vol. 3, p. 79–87

TERZAGHI, KARL (1939), "Soil Mechanics—A New Chapter in Engineering Science," *Proceedings of the Institution of Civil Engineers*, Vol. 12, p. 106–141

TERZAGHI, KARL AND PECK, RALPH B. (1967), *Soil Mechanics in Engineering Practice*, 2nd ed., John Wiley, New York

TERZAGHI, KARL (1942), Discussion of "Pile Driving Formulas: Progress Report of the Committee on the Bearing Value of Pile Foundations," *ASCE Proceedings*, Vol. 68. No. 2, p. 311–323, Feb.

TERZAGHI, KARL (1943), *Theoretical Soil Mechanics*, John Wiley, New York

TERZAGHI, KARL (1955), "Evaluation of Coefficients of Subgrade Reaction," *Géotechnique*, Vol. 5, No. 4, p. 297–326

TERZAGHI, KARL, PECK, RALPH B. AND MESRI, GHOLAMREZA (1996), *Soil Mechanics in Engineering Practice*, 3rd ed., John Wiley

THENDEAN, G.—see Hannigan, P.J.

THORNBURN, THOMAS H.—see Peck, Ralph B.

THORNTHWAITE, C.W. (1948), "An Approach Toward a Rational Classification of Climate," *The Geographical Review*, Vol. 38

THORSON, BRUCE M. AND BRAUN, J.S. (1975), "Frost Heaves—A Major Dilemma for Ice Arenas," ASCE *Civil Engineering*, Vol. 45, No. 3

TOKAR, R.A.—see Polshin, D.E.

TOKIMATSU, K.—see Seed, H. Bolton

TOMLINSON, M.J. (1957), "The Adhesion of Piles Driven into Clay Soils," *Proceedings, 4th International Conference on Soil Mechanics and Foundation Engineering*, Vol. 2, p. 66–71

TOMLINSON, M.J. (1987), *Pile Design and Construction Practice*, 3rd ed., Palladian Publications, London

TOURTELOT, H.A. (1973), "Geologic Origin and Distribution of Swelling Clays," *Proceedings of Workshop on Expansive Clay and Shale in Highway Design and Construction*, Vol. 1

TRACY, GERALD R.—see Fellenius, Bengt H.

TRAUTMANN, C.H.—see Kulhawy, F.H.

TRB (1984), *Second Bridge Engineering Conference*, Transportation Research Record 950, Vol. 2, Transportation Research Board, Washington, DC

TSCHEBOTARIOFF, GREGORY P. (1951), *Soil Mechanics, Foundations, and Earth Structures*, McGraw Hill, New York.

TUMAY, M.T. AND FAKHROO, M. (1981), "Pile Capacity in Soft Clays Using Electric QCPT Data," *Proceedings,* Conference on Cone Penetration Testing and Experience, St. Louis, MO, October 26–30, 1981, p. 434–455

TUNCER, ERDIL R.—see Basma, Adnan A.

TURNER, JOHN P. AND KULHAWY, FRED H. (1990), "Drained Uplift Capacity of Drilled Shafts Under Repeated Axial Loading," ASCE *Journal of Geotechnical Engineering*, Vol. 116, No. 3, p. 470–491

U.S. NAVY (1982a), *Soil Mechanics*, NAVFAC Design Manual 7.1, Naval Facilities Engineering Command, Arlington, VA

U.S. NAVY (1982b), *Foundations and Earth Structures*, NAVFAC Design Manual 7.2, Naval Facilities Engineering Command, Arlington, VA

ULRICH, EDWARD J. (1995), *Design and Performance of Mat Foundations, State-of-the-Art Review*, SP-152, American Concrete Institute

UNITED STATES STEEL (1974), *Steel Sheet Piling Design Manual*, United States Steel Corporation, Pittsburgh, PA

UPPOT, JANARDANAN O. (1980), "Damage to a Building Founded on Expansive Slag Fill," *Proceedings, 7th African Regional Conference on Soil Mechanics and Foundation Engineering*, Vol. 1, p. 311–314

VESIĆ, ALEKSANDAR (1963), "Bearing Capacity of Deep Foundations in Sand," *Highway Research Record*, No. 39, p. 112–153, Highway Research Board, National Academy of Sciences, Washington, DC

VESIĆ, A.S. AND SAXENA, S.K. (1970), "Analysis of Structural Behavior of AASHTO Road Test Rigid Pavements," *National Cooperative Highway Research Program Report 97*, Highway Research Board, Washington

VESIĆ, ALEKSANDAR S. (1973), "Analysis of Ultimate Loads of Shallow Foundations," ASCE *Journal of the Soil Mechanics and Foundations Division*, Vol. 99, No. SM1, p. 45–73

VESIĆ, ALEKSANDAR S. (1975), "Bearing Capacity of Shallow Foundations," *Foundation Engineering Handbook*, 1st ed., p. 121–147, Winterkorn, Hans F. and Fang, Hsai-Yang, eds., Van Nostrand Reinhold, New York

VESIĆ, ALEKSANDAR S. (1977), *Design of Pile Foundations*, National Cooperative Highway Research Program, Synthesis of Highway Practice #42, Transportation Research Board, National Research Council, Washington

VESIĆ, ALEKSANDAR S. (1980), "Pile Group Prediction Symposium, Summary of Prediction Results," *Proceedings, Pile Group Prediction Symposium*, Federal Highway Administration, Washington

VESIĆ, A.S.—see DiMillio, A.F.

VESIĆ, ALEKSANDAR S.—see Ismael, Nabil F.

VGAZ, O.G.—see DiMillio, A.F.

VIJAYVERGIYA, V.N. AND GHAZZALY, OSMAN I. (1973), "Prediction of Swelling Potential for Natural Clays," *Proceedings of the Third International Conference on Expansive Soils*, Vol. 1, p. 227–236, Jerusalem Academic Press, Israel

VIVATRAT, VITOON—see Mitchell, James K.

VOLMERDING, BRITTA—see Prawit, Alfred

WAHLS, HARVEY E. (1981), "Tolerable Settlement of Buildings," ASCE *Journal of the Geotechnical Engineering Division*, Vol. 107, No. GT11, p. 1489–1504

WAHLS, HARVEY E. (1994), "Tolerable Deformations," *Vertical and Horizontal Deformations of Foundations and Embankments*, A.T. Yeung and G.Y. Felio, Eds., Vol. 2, p. 1611–1628, ASCE

WAHLS, H.E. (1985), "Comparisons of Predicted and Measured Settlements," in *The Practice of Foundation Engineering: A Volume Honoring Jorj O. Osterberg*, p. 309–318, Krizek, Raymond J. et. al., eds., Dept. of Civil Engr., Northwestern Univ., Evanston, IL

WANG, SHIN TOWER—see Reese, Lymon C.

WARRINGTON, DON C. (1992), "Vibratory and Impact—Vibration Pile Driving Equipment," *Pile Buck* newspaper, Second October Issue, Pile Buck, Inc., Jupiter, FL

WARRINGTON, DON C. (1997), *Closed Form Solution of the Wave Equation for Piles*, MS Thesis, University of Tennessee at Chattanooga

WATT, W.G.—see Lytton, R.L.

WELLINGTON, ARTHUR M. (1887), *The Economic Theory of the Location of Railways*, 6th ed., John Wiley, New York

WELLINGTON, A.M. (1888), "Formulas for Safe Loads of Bearing Piles," *Engineering News*, Dec. 29, 1888; Reprinted in Wellington (1893), p. 22–33

WELLINGTON, A.M. (1892), "Mr. Foster Crowell on Pile Driving Formulas," *Engineering News*, Oct. 27, 1892; Reprinted in Wellington (1893), p. 52–73

WELLINGTON, A.M. (1893) *Piles and Pile-Driving*, Engineering News Publishing Co., New York

WELLINGTON, DENNIS L.—see Christopher, Barry R.

WELTMAN, A.J.—see Fleming, W.G.K.

WESTERGAARD, H.M. (1938), "A Problem of Elasticity Suggested by a Problem in Soil Mechanics: Soft Material Reinforced by Numerous Strong Horizontal Sheets," *Contributions to the Mechanics of Solids, Dedicated to Stephen Timoshenko*, p. 268–277, MacMillan, New York

WHITAKER, THOMAS (1976), *The Design of Piled Foundations*, 2nd ed., Pergamon Press, Oxford

WHITE, LAZARUS (1936), Discussion No. H-10, *International Conference on Soil Mechanics and Foundation Engineering*, Volume III, p. H-9

WHITE, L. SCOTT (1953), "Transcona Elevator Failure: Eyewitness Account," *Géotechnique*, Vol. 3, p. 209–214

WHITE, ROBERT E. (1962), "Caissons and Cofferdams," Chapter 10 in *Foundation Engineering*, G.A. Leonards, Editor, McGraw Hill

WHITMAN, R.V.—see Lambe, T.W.

WHITMAN, R.V.—see Liao, S.S.C.

WHITMAN, ROBERT V.—See Paikowsky, Samuel G.

WHITNEY, CHARLES S. (1957), "Ultimate Shear Strength of Reinforced Concrete Flat Slabs, Footings, Beams, and Frame Members Without Shear Reinforcement," *Journal of the American Concrete Institute*, Vol. 29, No. 4, p. 265–298

WILLIAMS, G.W.—see Duncan, J.M.

WINKLER, E. (1867), *Die Lehre von Elastizität und Festigkeit* (On Elasticity and Fixity), H. Dominicus, Prague (in German)

WOOD, W.A.—see Kulhawy, F.H.

WRIGHT, CLIFFORD C.—see Olson, Larry D.

WROTH, C.P.—see Burland, J.B.

WROTH, C.P.—see Randolph, M.F.

YOKEL, F.Y.—see Kovacs, W.D.

YORK, DONALD L. AND SUROS, OSCAR (1989), "Performance of a Building Foundation Designed to Accommodate Large Settlements," *Foundation Engineering: Current Principles and Practices*, Vol. 2, p. 1406–1419, F.H. Kulhawy, Ed., ASCE

YOUD, T.L.—see Bartlett, S.F.

ZEEVAERT, LEONARDO (1957), "Foundation Design and Behavior of Tower Latino Americana in Mexico City," *Géotechnique*, Vol. 7, p. 115–133

Index